Global Perspectives on Shifting Cultivation

NIPA® GENX ELECTRONIC RESOURCES & SOLUTIONS P. LTD.
New Delhi-110 034

About the Authors

Dr U. C. Sharma obtained his Ph.D degree in Soil Science from Punjab Agricultural University, graduation course from International Rice Research Institute, the Philippines and Advance course in Research Management from National Academy of Agricultural Sciences, Hyderabad, India. He is specialized in Soil Science and Agricultural Chemistry, watershed management, integrated water resources and catchment management, agro-ecosystem and environment management etc. He has held top positions in Indian Council of Agricultural Research system including Principal Scientist, Joint Director and Director of a Research Institute and National Coordinator (All India) of National Agricultural Technology Project (World Bank Project).

Dr U. C. Sharma has more than 400 research papers and 7 books to his credit, published in National and International Journals of repute. He has written more than 100 book chapters in books published at International level. He is Fallow of National Academy of Agricultural Sciences, Indian Society of Agronomy, Indian Potato Association and Indian Association of Hill Farming. He has won many National Awards including, Fakhruddin Ali Ahmed award for Natural Resources Conservation and Management, Phosphate and Potash Institute of Canada & Fertilizer Association of India award for generating technologies for maximization of crop yields.

Dr Sharma was Vice President, International Commission on Water Quality (IAHS) from 2003 to 2011, Chairman, Research Advisory Council of a project of International Fund for Agriculture Development (IFAD) project in India. He has been member and chairman of many committees of ICAR, Ex-office Advisor to North Eastern Council (Ministry of Home Affairs) for Northeastern States, Panelist United Nations Programme on 'Climate Change and Adaptation', Bangkok and Consultant, North Eastern Development Finance Corporation (NEDFi).

Dr Sharma is a member of, International Association of Hydrological Sciences Asia-Oceania Geo-science Society, Singapore. The Asia Pacific Association of Hydrology and Water Resources (APHW), Japan and member of many National and International scientific societies. Dr Sharma is reviewer for many Indian and International Journals.

Dr M. Datta had passed M.Sc (Ag) and Ph.D (Ag) in Soil Science and Agricultural Chemistry from University of Calcutta, West Bengal, India. After getting selected through Agricultural Research Service Examination, Dr Datta had started working in Indian Council of Agricultural Research (ICAR) as Scientist and thereafter in different capacities as Senior Scientist, Principal Scientist and Joint Director in a total duration of 38 years of Service.Thereafter, Dr Datta had also worked as OSD/ Principal of College of Agriculture, Tripura Central University, Tripura, India for a period of 2.6 years. During his service in North East India, Dr Datta has 41 years of Research Experience in Soil Management and Allied Activities, experience in Agri. Education and Research Extension as well as administrative experience. Dr. Datta is a Fellow of Indian Association of Hill Farming (FIAHF), has obtained prestigious D.N. Borthakur Award for outstanding Contribution in Farming System Research and Bharat Excellence Award by Frienship Forum of India. Dr Datta was also associated with various Committee in Agri and allied activities either as Chairman or Member in ICAR and Govt. of Tripura , India. In addition to his Teaching experience in Soil Science, Farming Systems and Forestry to Undergraduate/Post Graduate Students at the University, Dr Datta had served as a Guide for M.Sc (Ag) students at College of Agriculture and Technology, Kanpur, Co- Guide in Ph.D in Fisheries in ICAR-CIFE (Mumbai), Co-guide in Agronomy in Biswabharati, Santiniketan.

In addition to his special attainments in Technology Development in Soil Amelioration and fertility enhancement, Farming System Research and Agroforestry Models, Transfer of Technology in Soil Health Card and Agro-Meteorology Services, Dr. Datta had extensively worked in Shifting Cultivation areas in North East Himalayan region in order to find out any erudite solution in this cultivation being practiced in Neolithic period. Dr Datta had published 90 research papers in National and International Journals, 17 Books /Research Bulletins, 23 Book Chapters, 45 Abstract/ Conference Paper, 14 Popular Article, 15 Leaflet/Folders with Research Interest Score of 1439, Citations of 1640, Reads of 80, 428, Recommendations of 318 and H-Index of 20. Dr Datta had also visited Morocco in connection with Pulse Production Technology of ICARDA and visited Ivory Coast (Cote d'lvoire) in West Africa to assess the possibility and scope of lowland rice cultication. Dr Datta was also involved in Research Projects of the concerned SMD of Crop Improvement, Horticulture, NRM and Livestock Production, sanctioned by DBT, NAIP, NATP, NICRA, Tribal Sub Plan, Bill Gates Foundation, ICARDA, Department of Earth Sciences, Govt. of India etc. Dr Datta was Editor of Clay Research published by Clay Minerals Society of India and is a erudite reviewer in many International Journals.

Prof. Vikas Sharma is Head, Division of Soil Science and Agricultural Chemistry at the Sher-e-Kashmir University of Agricultural Sciences and Technology of Jammu, India. He obtained his doctorate from the Punjab Agricultural University, Ludhiana, India and did Post doc under the Raman Fellowship from West Virginia University, USA. His Research interests include soil carbon, Digital Soil Mapping, Pedometric, soil conservation and land degradation. Most of his work pertains to soils in Indian Himalayas and he has worked extensively on soil carbon, including carbon pool assessment in different land uses, carbon sequestration potential of different systems, soil carbon losses from different management systems, carbon storage in reclaimed degraded soil over chrono-sequence and so on. He has worked out the Carbon emissions from the first proposed carbon neutral village 'Palli' in J&K and has been involved in framing the carbon neutral plan. He has worked on soil erosion risk mitigation from the Himalayan soils and nutrient dynamics in the erosion prone foothill Himalayas. He has utilized his expertise in remote sensing and GIS in assessing and mapping soil resources in the region. He is presently focusing on utilizing machine learning models in soil mapping especially soil organic carbon. He is also involved in teaching and advising students' Research. He has executed multiple projects and presently is the PI of the prestigious NAHEP and HADP projects at SKUAST Jammu. He has over 90 Research articles and 30 book chapters to his credit.

Global Perspectives on Shifting Cultivation

U.C. Sharma
M. Datta
Vikas Sharma

NIPA® GENX ELECTRONIC RESOURCES & SOLUTIONS P. LTD.
New Delhi-110 034

NIPA® GENX ELECTRONIC RESOURCES & SOLUTIONS P. LTD.

101,103, Vikas Surya Plaza, CU Block
L.S.C.Market, Pitam Pura, New Delhi-110 034
Ph : +91 11 4386 0225, 9717133558, 9540816132
E-mail: newindiapublishingagency@gmail.com
Website: www.niparesources.com

Print ISBN: 978-93-58871-75-3

ebook ISBN: 978-93-58878-67-7

Composed and Designed by NIPA®.

Foreword

I am extremely happy to have been requested by Dr U.C. Sharma, a close friend and colleague for last four and half decades, to write the Foreword of his book "Global Perspectives on Shifting Cultivation". The Institute, ICAR Research Complex for North eastern Hills Region, Meghalaya, where we worked together, is known for its pioneering research on Shifting Cultivation (SC) and Dr U.C. Sharma, being a soil scientist, was very much involved in studying the different facets of SC which is also called Slash and Burn method of agriculture. A policy document on SC has also been written by him which was published by the National Academy of Agricultural sciences, New Delhi. This book is an outcome of his decade long study on SC particularly in North eastern region of India.

In the practice of SC, the mid to high land farmers select a tract of land, clears and burn the vegetation with the objective of using the ash as a nutrient supplier and cultivate on this tract until the fertility status supports the crops and then abandon it and migrate to another such plot and then to yet another. Meanwhile the first plot is naturally rejuvenated with seeds and vegetations from surrounding areas over a period of 15-20 years and that is when the farmers come back to that site and the cycle repeats. As such, the system is eco-friendly and fits very well in the emerging concept of Natural Farming.

Shifting cultivation system of food production has had its appreciators and critics alike. While some understood its contribution to both nature and the human living there, some others misunderstood it as destroyer of environment as could be known from different papers written on the system. However, the fact that the system is still in practice right from Roman Civilization time speaks of its hidden and to be explored benefits in agri-food production system - specially under harsh and challenging environment. As the modernization has crept in, the systems of shifting cultivation is becoming vulnerable to unexpected spikes in population expansion, socioeconomic changes, industrialization, digitization and other technologies altering lifestyles around the world. The authors have very nicely dealt with the issue that the system does not degrade forest resources but helps rejuvenate its health through a process of natural regeneration of fertility and vegetation built up.

As the readers will flip through the pages of this book, they will find that the authors have tried to bring up several issues of concern related to SC right from its origin in 7000 BC to its present conflict with policy makers of different countries where it is practiced. On every aspect of SC written, there is scope for debate and discussion, proper planning and strategy framing taking the woos, agonies and aspirations of the people engaged in SC. Though SC is blamed for maximum soil erosion impacting downstream eco-system negatively, an analysis of natural landslides under the influence of heavy precipitation will give a different picture. To me, the often repeated question as to whether SC should be abandoned or improved must now find an answer specially based on its different dimensions covered in this book which is not merely a copy paste book but a book supported by several research findings on SC.

Further, quantifying the past, present, and future contributions of land use to climate change has been a major focus of recent international efforts to compare and synthesize various earth system models. However, a better knowledge of the megatrends affecting the change in land use systems is necessary for accurate prediction of emissions based on land use. The retention and improvement of these regionally tailored production and processing methods while incorporating cutting-edge modern technologies is essential for effective soil management in hilly locations without disrupting the preexisting ecological balance and eco-system. I am particularly pleased that the writers considered these factors when writing this book and included extra fundamental knowledge about soils needed to design effective strategies to make the most of the region's scarce natural resources for food and nutritional security.

Overall the book goes very well with its title and I am sure it will benefit the scientists, researchers, teachers, students, the officers from the development departments, entrepreneurs and policy makers responsible for deciding the future of Shifting Cultivation specially in India.

Lastly I would like to complement and congratulate Dr U C Sharma and his co-authors for writing this important book which I believe will be a reference book on SC for strategic planning.

Dated 24th July, 2022
Guwahati, Assam, India

(Prof. K.M. Bujarbaruah)
Former Deputy Director General
Indian Council of Agricultural Research
&
Vice Chancellor
Assam Agricultural University

Preface

Shifting cultivation is known by many various names around the world, including "Slash-and-Burn Agriculture", "Swidden", and "Rotational Bush Fallow agriculture." Nearly 30% of the world's uplands from Africa, Latin America, Oceania, and South and Southeast Asia are covered by it and practiced in the wet tropics. It is the most prevalent kind of prehistoric land use on the planet and is found in hilly tropical regions, mainly in South and Southeast Asia, Africa, and the Americas. Since the beginning of the Neolithic period, which began around 7000 BC, this has been happening. Large swaths of various regions of the world continue to practice the practice. Scholars' perspectives on changing cultivation might vary greatly. While some see it as a wasteful and unprofitable practice, others believe it to be good for ecology and the ecosystem. Given the current situation, where the population is rapidly growing, the amount of land available per person is decreasing, and the climate change is a reality, it is high time we took a closer look to grasp the relevance of shifting agriculture.

Global agricultural systems are expected to shift as a result of climate change. Although climate has a significant impact on where and how farmers produce, centuries of agricultural ingenuity have greatly enlarged the areas where crops can be grown. Understanding the effects of climate change on agriculture requires evaluating how climate interacts with human activity to influence the kind of crops that farmers can grow. Despite the negative effects, shifting farming will continue to be the primary source of tribal livelihood in the vulnerable hill agroecosystems. Thus, both indigenous farmers and policymakers will find it effective to strengthen the current agriculture practice while making wise use of resources that might increase livelihood security. Both indigenous farmers and policymakers will find the existing agricultural approach and resource-wise use, which can boost livelihood security, to be a realistic strategy. Many academics throughout the world have produced some excellent research on this Neolithian practice. They have largely described their own interpretations of shifting cultivation based on their work experience in a small portion of the world, and they hardly ever discuss the practice as a whole. Few scientists, though, have examined the practice on a more worldwide level and as a result, they have conflicting opinions. Many theories and views have been developed as a result of this. Prior to the 1950s, the communication network in a global context and restricted mobility were the main obstacles to having a wider view. We made an effort to write about the current status of shifting cultivation in the world by connecting the multi-directional strings together. The current endeavour might not be the last.

Given the wealth of literature in the area, a broad consensus has now emerged, aiding in the comprehension of the depths of diverse operations, methodology, socio-economic elements, backgrounds, attitudes, and the effects of shifting cultivation. The practice of shifting cultivation is, however, terribly misunderstood and occasionally misrepresented. It's probable that casual observers are unable to understand the purpose of celebrations, the observance of rites, the full range of the system, and the clearing and burning of forest. They may also be unable to identify often environmentally durable cycles of cropping and fallowing. It has been noted that shifting cultivation systems are essentially vulnerable to rapid spikes in population growth as well as to socioeconomic changes, industrialisation, and increases in demand as a result of changing lifestyles, all over the world. We have seen that shifting cultivators have typically been blamed for the degradation of forest resources. But if we carefully examine the various shifting cultivation/ swidden operations, we can see that the practices that contributed to the rapid deforestation in the tropics and elsewhere at the end of the 20th century were the same practices that led to deforestation in Europe as a result of urbanisation, industrialization, population growth, and geographic expansion. Investigations into small landowners, dispersed shifting cultivators, and their interactions with their surroundings show that the root of the current issues lies deeply ingrained in human behaviour because, even in these uncomplicated communities, rivalry, competition, and conflict can be seen as the primary forces driving people to engage in conflict with their surroundings and natural resources.

Shifting cultivation has been the subject of some excellent research by scientists all around the world, who have also offered worthwhile alternatives and ideas for resolving the practice's long-standing problems. We gratefully recognise their excellent work and acknowledge the contribution of this data to our publication.

Since we have worked for about three decades in the areas where shifting cultivation is the major and primary land use and understood the various controversies and concurrences associated with the practice, it was felt to come out with a book to highlight the global perspectives on shifting cultivation and suggest measures to either leave the practice with sound alternatives or improve the existing practice to make it profitable and sustainable to raise the living standard of the population and ensure enduring food security and livelihood. We have made an effort to express what, in the context of today, we believe to be suitable; however, this may not be our final effort to eliminate obstacles to improvement or the quest for new methods to replace this 7000 years old practice.

Very interesting information regarding shifting cultivation, including its definition, scope, ecological and nutrient diversity, deforestation and soil degradation, dabbling, decline, and demise, as well as development with suitable alternatives, has been provided in various chapters of the book in a clear and concise manner. It would be interesting and very helpful to the reader. The book will act as a helpful manual for scientists, extension specialists, government officials, students, and farmers who want to manage natural resources and shifting cultivation areas sustainably. Readers suggestions are always welcome, dear readers.

We sincerely thank and are appreciative of all the writers whose works are cited in the book. We genuinely appreciate their assistance and fully acknowledge it.

Jammu
28 February, 2024

Dr. U.C. Sharma
Dr. M. Datta
Dr. Vikas Sharma

Contents

List of Colour Plates

Abbreviations

ADB	:	Asian Development Bank
AF	:	Agroforestry
AIPP	:	Asia Indigenous Peoples' Pact
AMY	:	Amylase
AQI	:	Air Quality Index
ASA	:	Aryl:sulphatase
B/C ratio	:	Benefit/Cost ratio
BCE	:	Before Common Era
BD	:	Bulk Density
CAR	:	Central Africa Republic
CBD	:	Convention on Biological Diversity
CBFM	:	Community Based Forest Management System
CEC	:	Cation Exchange Capacity
CHT	:	Chitagong Hills Tract
CIDA	:	Canadian International Development Agency
CIFOR	:	Center for International Forestry Research
CNRS	:	National Center for Scientific Research
COP	:	Conference of Parties
DGVM	:	Dynamic Global Vegetation Model
DHA	:	Docosahexaenoic Acid
DLDD	:	Drought, Land Degradation and Desertification
DRC	:	Democratic Republic of Congo
DRC	:	Democratic Republic of Congo
EPA	:	Environmental Protection Agency
FAO	:	Food and Agricultural Organization
FCPF	:	Forest Carbon Partnership Facility
FFL	:	Forest Floor Litter
		French Agricultural Research Centre for International Development (CIRAD)
FSI	:	Forest Survey of India
GCF	:	Green Climate Fund
GDAY	:	General Decomposition Yield
GEF	:	Global Environment Facility

GHG	:	Greenhouse Gases
GSA	:	Glucosidase
HFLD	:	High Forest Cover and Low Deforestation
HKH	:	Hindu: Kush Himalayas
IAEA	:	International Atomic Energy Agency
IAEG	:	Inter:Agency and Expert Group
IBSRAM	:	International Board for Soil Research and Management
ICAR	:	Indian Council of Agricultural Research
ICEF	:	Indo:Canada Environment Facility
ICFR	:	Indian Council of Forestry Research
ICIMOD	:	International Centre for Integrated Mountain Development
ICRAF	:	International Council for Research in Agroforestry
IEA	:	International Energy Agency
IEK	:	Indigenous Ecological Knowledge
IFFN	:	International Forest Fire News
INPE	:	Instituto Nacional de Pesquisas Espaciais
IPCC	:	Intergovernmental Panel on Climate Change
IRD	:	National Research Institute for Sustainable Development
ITK	:	Indigenous Technical Knowledge
IWGIA	:	International Work Group for Indigenous Affairs
LAO	:	Laos
Lao PDR	:	Lao People's Democratic Republic
LDN	:	Land Degradation Neutrality
LEG	:	Least Developed Countries Expert Group
LETK	:	Lack of Experience and Technical Knowledge
LFSC	:	Long Fallow Shifting Cultivation
LMB	:	Lower Mekong Basin
LTK	:	Local Technical Knowledge
LUCF	:	Land-use Change and Forestry
LULUCF	:	Land-use, Land-use Change and Forestry
MARD	:	Ministry of Agriculture and Rural Development
MBC	:	Microbial Biomass Carbon
MBN	:	Microbial Biomass Nitrogen
MBP	:	Microbial Biomass Phosphorus
MEPED	:	Nagaland Environment Protection for Economic Development
MMSEA	:	Mountain Mainland of South East Asia
MOAF	:	Ministry of Agriculture and Forestry
MoEF	:	Ministry of Environment and Forest

MoEFCC	:	Ministry of Environment, Forest and Climate Change
MONRE	:	Ministry of Natural Resources and Environment
MoSPI	:	Ministry of Statistics and Programme Implementation
MRV	:	Monitoring, Reporting and Verification
MSEA	:	Mountain areas of South East Asia, Mainland of South East Asia
NAEB	:	National Afforestation and Eco development Board
NAEB	:	National Afforestation and Eco:Development Board
NAFRI	:	National Advanced Fire & Resource Institute
NAP	:	National Adaptation Plam
NDC	:	Nationally Determined Contributions
NERCORMP	:	North Eastern Region Community Resource Management Project
NHB	:	National Horticulture Board
NLCB	:	National Land use Conservation Board
NRSC	:	Natural Regeneration in Shifting Cultivation
NTFP	:	Non:Timber Forest Product
NWDB	:	National Wasteland Development Board
OM	:	Organic Matter
PEN	:	Poverty and Environment Network
PHA	:	Acid:phosphatase
PRO	:	Protease
PSF	:	Principal Starchy Food
FCPF	:	Forest Carbon Partnership Facility
REDD	:	Reducing Emissions from Deforestation and Forest Degradation
R:PP	:	Readiness Preparation Plan
SALT	:	Sloping Agricultural Land Technology
SC	:	Shifting Cultivation
SDG	:	Sustainable Development Goals
SDI	:	Sustainable Development Indicators
SHGs	:	Self Help Groups
SLUB	:	State Land Use Board
SOC	:	Soil Organic Carbon
SOM	:	Soil Organic Matter
SQI	:	Soil Quality Index
TEKP	:	Technical Ecological Knowledge and Practice
TFSP	:	Tropical Forest Timber Plantation
UNCCD	:	United Nations Convention to Combat Desertification
UNCTAD	:	United Nations Conference on Trade and Development
UNDP	:	United Nations Development Programme

UNEP	:	United Nations Environment Programme
UNFCCC	:	United Nations Framework Convention on Climate Change
UNFF	:	United Nations Forum on Forests
VTGs	:	Vulnerable Tribal Groups
WB	:	World Bank
WDPSCA	:	Watershed Development Project in Shifting Cultivation Areas
WHC	:	Water Holding Capacity
WOCAT	:	World Overview of Conservation Approaches and Technologies
WRI	:	World Resources Institute

1

Introduction

1.1 GENERAL DESCRIPTION

There is a folk tale from India which is mentioned under relevant situations. This tale teaches how different perspectives lead to different points of view. The tale of the blind men and an elephant originated in the ancient India, from where it has been widely spread to different parts of the world. This famous ancient tale mentions about six blind men who had never come in contact with an elephant during their lifetime and who know and visualize what the elephant appear like by touching it and feeling. In the tale it is highlighted that every blind man touches a different part of the elephant's huge body, but only one part, such as the leg, ear, tail, body or the tusk. These blind men then go on to narrate about the elephant on the basis of their touch and feeling and their recount of the elephant varied widely from each other. They were told that elephants could trample forests, carry huge loads and is very big in size. The six blind men were taken to the elephant and asked how it look like (Fig.1.1). The first blind man reached out and touched the side of the huge animal. "An elephant is smooth and solid like a wall!" he declared. "It must be very strong and powerful." The second blind man put his hand on the elephant's trunk. "An elephant is like a giant snake," he said. The third blind man felt the elephant's pointed tusk. "I was right," he decided. "This creature is as sharp and deadly as a spear." The fourth blind man touched one of the elephant's big ear and announced "I believe an elephant is like a huge fan or may be a magic carpet that can fly over mountains and treetops". The fifth blind man touched one of the four legs. "What we have here," he said, "is pillar of a house." The sixth blind man gave a tug on the elephant's coarse tail. "Why, this is nothing more than a piece of old rope and it appeared to him dangerous, indeed".

Fig. 1.1. Different blind men touching different parts of the body of the elephant.

The allegory of blind men and elephant has been mentioned in different ways in many ancient India scriptures belonging to Buddhist, Hindu and Jain communities. The lessons which can be learnt from this legend and perception and the importance of the tale can be a great inspiration for the present and next generations. The apologue has been used to encompass an array of truths and myths. If we think calmly, the saga implies that subjective experiences can be reliable, but such thinking is inherently constrained by its failure many a time and we not be able to know the totality of actual truth. The legend has given an insight into the belief that there's no absolute truth, ambiguity or impassive condition of truth, the approach of specialists in fields of disputable theories, the necessity for deeper understanding, and reliability for different attitudes on the same gadget of scrutiny.

Shifting cultivation on the global scale could be described on similar lines. A number of workers have done very good work on this Neolithian practice in different parts of the world. By and large, they have described their own versions of shifting cultivation based on their work experience in limited area of the globe and hardly talk about the practice in totality. However, there are few scientists who have viewed the practice on a larger global scale and thus, hold different views. This has given birth to so many theories and beliefs. Before 1950s, the main bottleneck in taking a broader look, was the communication network in global context and low mobility. By joining the strings together, we have attempted to write about the state of shifting cultivation in the world. This may not be the final effort.

Different eminent scientists, scholars etc. have worked in different parts in the world on shifting cultivation or swidden agriculture or slash and burn cultivation, and came out with excellent results. These systems of cultivation are widely distributed across the world, confining mostly to humid tropics. However, the methodology followed is as varied as the systems themselves in various parts of the world. Most of the work produced, though of very high standard, but pertains to a particular area, under a particular situation and as performed by a particular group of people. So, various scholars have given the account of their work as they experienced in the areas they worked. With lot of literature in the field, now a broad consensus has developed, which has helped in understanding the fathoms of various operations, methodologies, socio-economic aspects, backgrounds, attitudes, compulsions and consequences of these practices. However, shifting cultivation is a grossly misunderstood, and sometimes, misinterpreted practice. It is possible that informal viewers are not able to appreciate the clearing and burning of forest and cannot distinguish often ecologically durable cycles of cropping and fallowing, the intent behind celebrations, observance of rituals and complete gamut of the system. It has been observed that the systems of shifting cultivation are essentially prone to the sudden spurts in population growth and to socio-economic changes, industrialization, increase in demands due to changing life styles, across the globe. We have witnessed that in most of the cases of forest resources destruction, the shifting cultivators have been held responsible. But if we carefully scan the various operation in shifting cultivation/swidden, the practices responsible the fast loss of forests in tropics and other places at the end of the 20th century are the same practices that resulted in the deforestation in Europe due to urbanization, industrialization, increase in population and geographical expansion. Investigations about small land-holders, remotely located shifting cultivators and their relationships with their surroundings indicate that The basis of the current problems lie deep in human attitude, because even in these straight-forward communities, rivalry, competitiveness and conflict can be identified as the main force pushing them into dispute with their surroundings and natural resources.

1.2 WHAT IS SHIFTING CULTIVATION

Shifting cultivation, commonly termed as "slash-and-burn agriculture," "swidden," and "rotational bush fallow agriculture," is called by different names the world over. It is practiced in the wet tropics and covers almost 30% of the world's uplands from Africa, Latin America, Oceania, and South and Southeast Asia (FAO 2010). Shifting cultivation is the most dominant primitive land use of the globe, practiced in the hilly areas of the tropics particularly, in South and Southeast Asia, Africa and Americas. It involves clearing of mountainous land by cutting the vegetation (tree or bushes), leaving the biomass in situ for drying, and finally burning it for production of charred material for soil fertility enrichment. Subsequently, multiple crops (annual and perennial) are grown as mixture on the same land until soil fertility is depleted and farmers move to a new forest patch and repeat the cycle (Ramakrishnan 1992). This has been going on since the beginning of Neolithic period from about 7000 BC. In shifting cultivation, the fields are cropped, usually for a couple of years, and then left without cultivation to let the land revert to their natural regeneration, while the farmer selects another field for farming to produce food for his family. The spell of farming is usually cut short when the soil is unable to supply enough nutrients to the growing plants for optimum growth and depicts signals of exhaustion, yields are poor or, more commonly, due to various other reasons such as; when the field becomes infested with perpetual weeds. The shifting cultivation areas are mostly dominated by mountainous and undulating and inhabit divergent cultural groups and clans. These mountain groups of cultivators practiced shifting cultivation in large forests areas, having low population density, and this permitted these communities to synchronize with surrounding environment and live in harmony with their ecosystem (Sharma and Bujarbaruah 2014). It is also considered by some scholars that shifting cultivation, involving cutting and burning of vegetation by the cultivators in their run to land preparation in the system, is a major cause of deforestation in the tropics and has drawn extreme denunciation from many quarters. Mutual antagonism, deep suspicion, and open conflict have characterized the relationship between The relationship between foresters and shifting cultivators is full of antipathy, deep scepticism and open dissension. The recent policies of the governments towards shifting cultivators in some countries are discouraging. The authorities want complete eradication of shifting cultivation and facilitate adoption of more permanent forms of agriculture. This remains on high priority of the national governments across Southeast Asia and other regions, including India. Along with resource depleting agricultural practices, the requirement for different commodities like; fuel, fodder and timber has further depleted the forest cover and caused gradual impoverishment of this valuable treasure. The delicate areas of tropics are under continuous pressure from different sides and there is likelihood of its being disintegrating if suitable interventions, based on a mix of traditional wisdom and state of the art new technologies in the field, are not introduced as early as possible (Sharma and Bujarbaruah 2014). There is need to introduce sound technologies, which are adoptable by the shifting cultivators, that is which suit to their socio-economic fabric, affordable, workable under local situations and sustainable in the long run. Deforestation has to be stopped with feasible, viable, realistic and reasonable measures, lest it can cause more harm than good. Gourou (1953) rightly pointed out, "The disasters brought on by agricultural methods which have taken no account of the treasures of wisdom and experience accumulated in the old tropical system are a sufficient proof of latter's value, it can be improved only if the reasons for its processes are fully understood". Unwavering misuse of forests in the world has caused natural disparity and disproportion. Mainly due to high energy efficiency, the shifting cultivation has survived in Northeast India. The energy efficiency (output of energy/ input of energy) ranged from 34 in a 30-year cycle to around 47 in 10-year and 5-year cycles (Toky and Ramakrishnan, 1982). Sharma (1993) pointed towards considerable agricultural prospects in Northeast India provided appropriate, environmentally safe and socio-culturally justifiable farming systems are adopted. The shifting cultivators, though not against to newer technologies and interventions, are not ready to embrace them due to socio-economic reasons.

Shifting cultivation has long been appreciated as a real “issue”. This practice was started as early as 6000 BC. With increase in demographic pressure, the practice has become uneconomical and causes natural resources degradation. It is high time that a debate is initiated to decide whether we should continue with it in the future or find an alternate use of land (Sharma and Bujarbaruah 2016). Shifting cultivation is the major farming system in the mountainous areas of Southeast Asia, as mentioned by Li *et al.* (2014). Because it makes available required subsistence produce to indigenous farmers and take care of their food security. The local farmers are mostly the poor ethnic minority communities. In this regard, a powerful method to identify and appraise the sites and area of shifting cultivation would need accessibility to Landsat historical imagery along with remote sensing based techniques, which can offer an effective way to tackle many inquisitiveness. Li *et al.* (2014) reported that many procedures for studying fire phenomenon and burnt area can be offered for shifting cultivation mapping due to the common attribute of fire relatedness. Broadly four techniques involving spectral signatures, phenological features, statistical theory and landscape environment were mentioned for characterization of shifting cultivation areas. Five research priorities about shifting cultivation involving remote sensing techniques, spatial pattern, transpose, catalysts and footprints were suggested consequently.

Shifting cultivation or slash-and-burn farming or swidden agriculture, is a primitive land use system of subsistence nature in the tropical landscapes of the world (Brady 1996, Comte *et al.* 2012). There was a change in the 1990s, when many authors started to describe that the disadvantages of shifting cultivation are more than the advantages. This was around the period when demographic pressure in humid tropics increased tremendously and the practice of shifting cultivation became unable to sustain this pressure. Many scholars then put forward many alternatives for this unsustainable practice across the tropical regions (Brady 1996). However, according to van Vliet (2012) mentioned that the practice was still a dominant agricultural system in the tropics.

Here, we try to mention our own experience with three terms, commonly used for the age old practice' known by different names of such as; 'shifting cultivation', 'swidden agriculture' and 'slash and burn' cultivation'. We have not only studied shifting cultivation but have lived with different communities practicing this type of farming since time immemorial, thus having deep insight into their attitudes, habits, rituals, festivities, approach towards practicing shifting cultivation and improvements done so far in isolated pockets for intensification, resource conservation and enhancing crop productivity to feed ever growing population. We can call the traditional agriculture system followed in Northeastern states of India as the 'shifting cultivation' as it is a complete agricultural system in which all the relevant operations/parameters, viz.; clearing, burning, cropping, fallowing, no-till, rotational, land tenure and observance of rituals exist. Based on our own experience and study of existing literature, out of eight basic parameters (mentioned above; broadly speaking, four mandatory and four optional, (Fig. 1.2) to describe this Neolithic system of cultivation, broadly shifting cultivation encompasses all the eight, 'slash and burn' has the first four and 'swidden agriculture' has the first six parameters. So, though we use these terms interchangeably, one has to be careful in their use when describing a particular method of cultivation for a specific region.

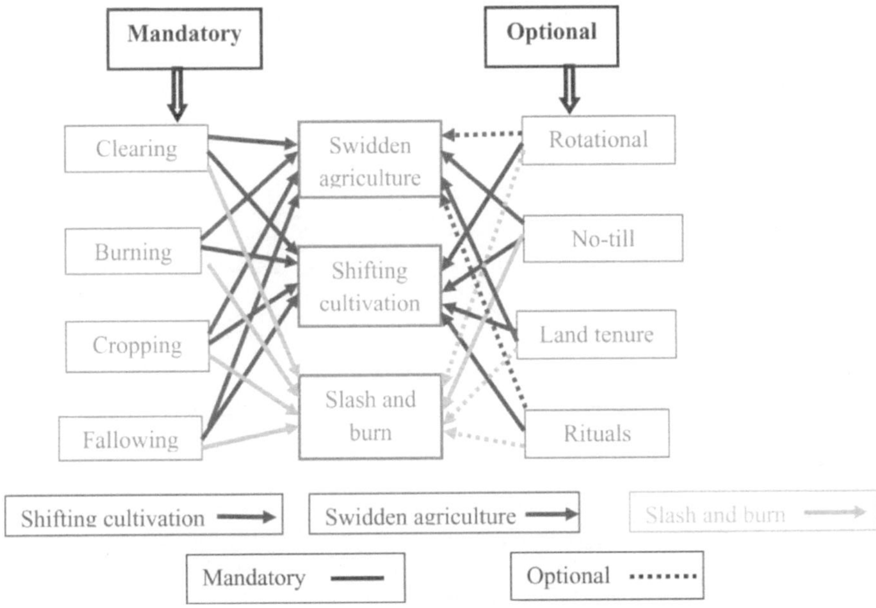

Fig. 1.2. Major facets of shifting cultivation, swidden agriculture and, slash and burn agriculture. (See colour version on page 651)

We may mention here that slash and burn is one of the component of the practice of shifting cultivation. Farming systems in which a cultivation time and fallow period are cyclic in nature, that is, wherein the cropping period of 1 to 3 years is followed by a longer fallow period of about 5 to 20 years for regeneration of vegetation consisting of grasses, bushes and woody trees and the vegetation is cut, slashed and burnt before next cropping, for the addition of ashes containing lot of nutrients. This practice has been mentioned as shifting cultivation by Nye and Greenland (1960), Aweto (2013), swidden agriculture by Mertz *et al.* (2009) and slash and burn agriculture by Palm *et al.* (2005). This is a system of crop cultivation using small amounts of labour and capital in relation to area of land being cultivated. The shifting cultivation has been kept alive over long period of time by augmentation over uncultivated land as a result of demographic pressure, in comparison to more developed farming systems, where the vegetation is mixed with the soil through ploughing or other incorporating equipment. When minutely studied, it is found that the practice encompasses an array of systems (Schlippe, 1956; Conklin, 1957; Dove, 1985; Schmidt-Vogt, 1999; Mertz *et al.*, 2009; Cairns 2014) which differ in describing the various operations attached to the shifting cultivation like; clearance of the ecosystem and use of fire.

Over a period of time, fields are cultivated for a relatively short time and allowed to recover for a longer time. This ratio of cultivation and recovery period is shortening of late due to increase in population and demand as well as, decrease in availability of area for the practice. Fields in established and stable shifting cultivation systems are cultivated and fallowed in rotation. The fallow fields are not unproductive if soil and water conservation measures are adopted during the cultivation phase. In many studies it is mentioned that shifting cultivation causes huge soil erosion, so much so that many a times even bedrock surface is exposed. In such situations, the fallow phase is not productive as re-growth of vegetation is impossible. However, if proper soil conservation measures are followed during the cultivation phase, the fallow phase can be highly productive. The successive species of trees are often used by shifting farmers during the fallow phase for fencing, timber and construction of hutments as well as for, firewood and other common uses. Some of these species of trees like; alder (Alnus nepalensis) fix nitrogen, thus, enriching the soil for good crop production. It has also been established that the resurgence of bushy and trees vegetation during fallow period protect soil from erosion and extract crop nutrients from deep layers of soil profile and bring to the soil surface for crop use.

Shifting farmers consider the forest plantation as an agricultural system of plots at different stages in a systematic cycle of events. It should be borne in mind that stable shifting agriculture systems are highly variable and not uniform, strictly adapted to local micro-environment and are properly managed by farmers during both the cropping and fallow stages. It has been observed that shifting farmers may acquire and possess a highly improved technique and knowledge of their local environments as well as of the crops and local plant species they make use of. The secondary forests, generated as a result of shifting agricultural systems are generally richer in plant and animal assets, mostly useful to humans, than primary forests, even though they are much less bio-diverse.

The shifting cultivation, swidden agriculture and slash and burn cultivation generally, by and large, cover a particular distance in the well recognized eight parameters describing the age old Neolithic period practice of land use system, such as; (i) clearing, (ii) burning, (iii) cropping, (iv) fallowing, (v) no-till, (vi) land tenure, (vii) rotation of fields and (viii) observance of rituals and associated festivals. Mostly, the slash and burn covers up to four, swidden agriculture cover up to six, while shifting cultivation runs up to the end (Fig. 1.3). There could be exceptions here and there. But, scheme of presentation, given in Fig. 1.3, need International recognition for clarity and better understanding of the scientists, scholars, practitioners and even laymen not associated with the practice.

In this book, we will use three terms, such as; shifting cultivation, swidden agriculture and slash and burn agriculture.

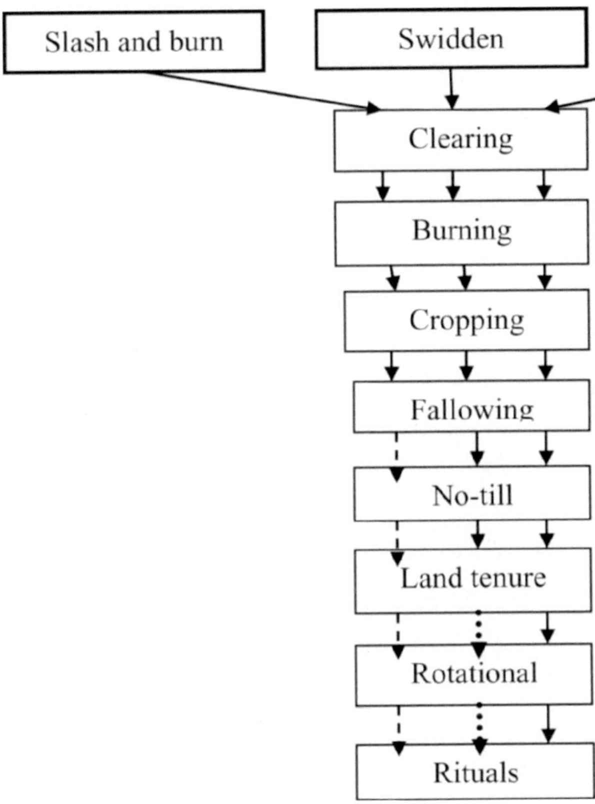

Fig. 1.3. Various parameters mostly found under 'slash and burn', 'swidden agriculture' and 'shifting cultivation'. (Yes, ——▶ ; May or may not be, ▪▪▪▪▶))

1.3 GOVERNMENT PERCEPTIONS

The government perceptions of shifting agriculture as a farming system are that, it is being performed by those who are considered outsiders, not representing the mainstream culture of the country. This preconception has demoralized the coming up of a more realistic scenario of shifting cultivation in many countries across the globe. Therefore, a farming system is considered based on who is practicing it, rather than on its own advantages and disadantages. In Asia and Latin America The perception of shifting agriculture in Asian and Latin American countries is compounded by the fact that it is, presently being performed not just by the tribals or indigenous population, but also by the farmers who are landless and the migrant. There is almost an uniform pattern in Southeast Asia of keeping permanent fields in the valley areas and shifting cultivation plots in the hilly locations. When the farmer migrates into the forest, one other type of partial shifting cultivation system is found. The cultivator spends all his time and agricultural endeavour in creating a plot with little previous understanding of shifting cultivation techniques. In many cases, these partial shifting cultivators make an embryonic shifting cultivation without proper knowledge to generate a shifting cultivation system that can be sustained for long (Conklin 1957). Regrettably, many governments do not or can not make a difference between the cultivators as to which are pioneer and which are integral. This differentiation is required to be made because the two systems have variable impacts on the environment and local ecosystems. By and large, when demolition of the tropical forest occurs, usually it is due to the pioneer and not the integral cultivator. It has been observed that in some countries, the migrants, without an adequate and proper knowledge of integral shifting cultivation that would provide them the techniques to manage the forest ecosystem, are joining, cultivating and deteriorating the forest areas (Olafson 1981, Moran 1983).

1.4 THE PRACTICE

1.4.1 Extent

Heinimann *et al.* (2017) estimated that the current area under shifting cultivation in the world is about 280 million hectares, which includes both farmed and fallow land. There are many variable estimates of shifting cultivation area in the world. As high as one thousand million ha area under shifting cultivation has been estimated by some scholars. Dixon *et al.* (2001) gave estimates of about 1000 million ha under shifting cultivation at the global level. Similar estimates were given by Sanchez *et al.* (2005) and Davidson *et al.* (2008). A possible strong decrease in shifting cultivation over the next decades has been predicted, based on expert survey and historical trends estimates.

1.4.2 The evolution of the systems

Shifting cultivation is one of the oldest agricultural land uses. Practiced worldwide, from the tropics to temperate areas, it has been abandoned in many regions, and alternative land uses developed (Brookfield, 2001; Mazoyer and Roudart, 2006; Cairns, 2007). The farmers adopt two major plans when they face the limits of slash-and-burn systems. Either they follow other land uses or professions along with shifting cultivation or they slowly transform the shifting cultivation system itself. Shifting cultivation supplies one or two mast preferred items of food, such as, rice, maize, beans, vegetables and a host of other products. Fruit trees are generally seen near the hutments, while most fertile soils are used for preferred crops. Livestock substitutes to wild animals when hunting is difficult, that is, when population pressure increases. When the fallow period shortens and the practice is unable to provide sufficient food items, all activities are curtailed except hunting, fishing and gathering which become the most important sources of food and income. The environmental impact of shifting cultivation is an arguably debatable issue, biased by political positions, government policies, site of study and, on the position taken on both sides of the debate. Shifting cultivation involves clearing of vegetation of primary forests or secondary forests before clearing it for repeating the cycle. This is bound to affect the environment and related issues would arise to remain at the center of the debate and challenged by the strong opposition and critique from organizations advocating for biodiversity conservation.

1.5 PREAMBLE

The length of time that a field is cultivated has become shorter than earlier, because of increase in population pressure and consequently need for more food. Shifting cultivation is a primitive practice of cultivation and people involved in such cultivation are called Shifting cultivators. Shifting cultivation has long been appreciated as a real "issue". This practice was started as early as 6000 BC. With increase in demographic pressure, this practice has become uneconomical and cause degradation of natural resources.

There has been always a discussion about the encroachment of forest lands for farming, which has become a possible cause of tropical deforestation. Some scientists view shifting cultivation as an agent of forest degradation, though the practice is one of the indigenous forms of farming in the forest-agriculture continuum in the tropics. Shifting cultivation still followed on a significant area of the forestlands in tropical nations. It has been advocate by some scientists that the share of the shifting cultivation in deforestation in the world is 45% (UNEP cited in Angelson. 1995) to 60% (Myers 1995). Many believe that since shifting cultivation is the cause of deforestation, it results in the loss of biodiversity and enhanced soil erosion. Factors like poverty, demographic pressure and migration into the forest land by the shifting cultivators are quoted as the most important causes of deforestation. Shift in the natural balance has direct and indirect effects on human life, and therefore, study on the ecological aspects of shifting cultivation is very important.

1.5.1 Is shifting cultivation a function of population

Boserup (1981) and Geertz (1974) debated that the transformation from shifting cultivation to the intensification of farming is a result of the fast growing demographic pressure, technology advancement and robotic environment. The cultivators have been practicing shifting cultivation for a long time now to provide food to ever-growing population as shifting cultivation alone may not be able to fulfill the needs of the cultivators. Farmers are inclined towards cash crops, such as; cardamom farming, rearing of livestock and other subsidiary professions to earn livelihood. So, the shifting cultivators do not put more pressure on land as they might change subsistence type o farming to intensified agriculture.

1.5.2 Contribution of shifting cultivation to livelihood

Shifting cultivation is a heterogeneous socially connected and varied crops production system, and therefore, contributing to food security of the shifting cultivators who do not have to worry for rice and vegetables for most of the year. Many crops can be grown in the same piece of land without any extra investment on inputs in shifting cultivation which is not the case with other lowland agriculture practices. To cope with the challenges of food security and livelihood, land tenure and resource management, the shifting cultivators in the CHT, Bangladesh have adopted many good practices.

1.5.3 Economics of shifting cultivation

Economics of shifting cultivation has been studied by many scientists, who have different views about the economic suitability of the practise. It includes economics of land use, production methods and productivity both in energy and monetary terms. Mostly, the results have been influenced by the method used, population depending on a particular piece of land, soil quality, original nutritional status of soil, crops grown, whether livestock farming is also practiced along with crop husbandry and many other factors but, by and large, the arguments advanced against the economics of shifting cultivation always state that it is a sub-optimal method of cultivation, output is low and labour intensity is very high. Sharma and Prasad (1995) reported a benefit/cost ratio of 0.6 in case of shifting cultivation as compared to 2.1, 1.9 and 1.8 in case of livestock rearing land use, Agri-horti-silvi-pastoral land use and agriculture land use, respectively.

Although researchers have arrived at different estimates to determine self-sufficiency among shifting cultivators, the general conclusion is that they are able to feed themselves adequately. Dove *et al.* (1989) have detailed out the requirements of food grains for different groups of family sizes and there annual requirement. Similarly Chin (1984) and Ramakrishnan (1992) have given their own estimated under different situations and consumption needs.

1.6 TERMINOLOGIES

It is a well known fact that there exists no commonly accepted term to describe, what is called, 'shifting cultivation', 'swidden farming' or 'slash and burn agriculture'. Besides a number of local names (Table 1.1), these three terms are generally used for an ancient agriculture systems in which a sizeable amount of vegetation is burnt in the field to supply nutrients to the soil where sowing of crops is to be done. The farmers may or may not shift to alternate area after the previously cultivated land is exhausted of its nutritive status and crop yields show a downward trend. This also depends on the availability of land to shift. On a broader basis, the term 'shifting cultivation' is generally used in Southeast Asia, 'swidden farming' in Europe and 'slash and burn agriculture' in Africa and Latin America. However, though these three terms were mostly used on spatial scale, it stands mixed-up now. If we take a universal view, these three terms have been, sometimes, loosely used by the scholars for this Neolithian agriculture practice. Somehow, these three terms have been mixed-up in scientific literature, but beauty of the matter is that almost all can understand about what one is going to talk about. So, we can say that these terms are now used interchangeably without caring for what type of practice is going to be discussed. There is a broad understanding among the scholars. However, when minute details are to be discussed, we must take cognizance of the practices involved in a system and then identify it accordingly.

Table 1.1. List of local names of shifting cultivation

Country	Local name	Country	Local name
Bhutan	Tseri, Pangzhing	Madagaskar	Tavi
Brazil	Roka		
Congo	Masole	Mexico	Comile
Equitorial African countries	Fang	Mexico and Central America	Milya
India (NE)	Jhum	Myanmar Nepal	Taungya Khoriya, Basme
India (Bundelkhand)	Vevar and Dahiyaar, MP	Philippines	Caingin
India (Bastar, MP)	Deepak	Sri Lanka	Chena
India (Southern staes)	Zara, Erka	Thailand	Tamrai
India (SE Rajasthan)	Batra	Uganda, Zambia and Zimbabwe	Chetemini
India (Andhra)	Podu	Venzuela	Konuko
India (Western Ghats of Kerala)	Kumari	Vietnam	Ray
India (Odisha)	Kuman, Vinga, Dhavi	Western Africa	Logan
Guadeloupe	Echalin	Yucatan and Guatemala	Milpa
Indonesia	Huma, Ladang	-	-

1.7 HISTORICAL

For over a million of years during the most of the period of the human's existence on this planet, their food source was by hunting and gathering. They did not know how to culture the plants and animals. This system is still prevalent in Africa and Southeast Asian countries as subsistence agriculture, where it is known as Chena, Conuco, Jhum, Kaingin, Lading, and Milpa (Chapman, 1975; Andriesse, 1989). In India, shifting cultivation or slash and burn agriculture is widely practised in the North-East Himalayan Region, Orissa, Andhra Pradesh,and in some pockets of Chattisgarh. It is known as 'Jhum' in the NE region, 'Podu' in Andhra Pradesh, 'Bewar or Daihya' in Chattisgarh and 'Podu or Dunger Chasa' in Orissa. Indeed, it may be regarded as a successful form of agro-ecology in the hills because the system is also one of low input practices in terms of energy and technology (Rappaport, 1971). It may be pertinent to mention that the hilly area in the NE region comprises 72.3% (2,02,212 km^2) of the total geographical area of 2,62,040 km^2 of the region. In shifting cultivation, the vegetation is cut and put on fire after drying (Fig. 1.4)

Fig. 1.4. Burning of forest vegetation to add ash to soil for improving soil fertility (See colour version on page 651)

Governmental approach on relocation of migrated population, conversion of reserve forest, and encouragement to generate secondary leadership and governance methodology, is continually eroding the common resources of the world. Research conducted in the last few decades has shown that the continuance of the shifting cultivation system despite the governmental prohibiting policies and the related issues, are far more complex than the simple regulatory effort to suggest otherwise. The major cause of environmental deterioration can be attributed to deforestation. While environmental degradation goes unbriddled with higher rate with such policies, the primitive communities are challenged with their livelihood practices.

Is shifting cultivation - a historical obsession

The practice of shifting farming is followed in entire humid tropics, commencing from Southeast Asia to Latin American countries and African nations. It was practiced in some European countries also at one point of time. Shifting cultivation with its different names, is considered as one of the earliest forms of agricultural production system and often originated in the transition from food gathering to food production. It is done by slashing the vegetation and burning it on the land surface to add ashes to enhance and maintain soil fertility. That is why it is also known as slash and burn agriculture also (Sadeque 2000). The name itself signifies that it is not folloed on a single field on a perpetual basis, and that highlights the shifting nature of the practice. Now, if we sit back and calmly imagine about the spread of shifting cultivation throughout the globe under the situations, when there was no communication system, no frequent contact among cultivators, no thought-over government policies and no interventions by the extension set-up, how this practice emerged with almost similar set of operations in the entire world. There must be some common features and thinking on how to get good food production under the prevailing set of conditions like, terrain, precipitation and other climatic parameters, soil fertility, slope etc. We reach the conclusion that this common set of operations came out of compulsions imposed by prevailing set of geographical and historical situations. So, the practice of shifting cultivation or slash and burn or swidden emerged out of historical compulsions and was the best food production system of the time until its sustainability or stability eroded with population increase and change of needs and requirements over time.

Fig. 1.5. A typical shifting cultivation field in Northeast India. (See colour version on page 651)

1.8 SHIFTING CULTIVATION- A ROTATIONAL AGROFORESTRY

A wide range of agricultural practices in the world are covered under the terms of slash and burn, swidden agriculture and shifting cultivation (Karkhoff and Sharma 2006). Mostly, these practices have a short 'cropping phase' of a 1-3 years followed by a longer phase called 'regeneration phase' or generally called the 'fallow'. Through cyclic operations in shifting cultivation, that is, agricultural crops followed by 'regeneration' or fallow phase during which woody vegetation takes birth, amounts to a complete rotation. That is why, we call the system a 'rotational agroforestry'. The rotational agroforestry system is a direct innate product from shifting cultivation. It amalgamates production of food and forest products, and provides ecosystem services like, soil reconstruction, soil erosion control and carbon sequestration in forests.

The prominent achievements that have been gained in agriculture and forestry on global scale during the past few decades were primarily due to some selected plant species that were cultivated in monoculture to produce desired commodities. The indigenous practice of growing crops and trees together was not taken into account and ignored in this single-commodity paradigm. Though agriculture and forestry are often interrelated in the ecosystem and have many common objectives, these have often been treated separately. The shifting cultivation encourages these two sectors to come together in a cyclic manner, giving birth to rotational agroforestry. Agroforestry offers a distinctive set of chances for arresting soil degradation, and provisioning of ecosystem services in different countries. The major thrust of agroforestry is on removing poverty, insuring nutritional security and arresting land degradation, particularly limited resources and input situations.

1.9 STAGES OF SHIFTING CULTIVATION

Broadly, there are six stages in the shifting cycle at which the shifting cultivator has to be careful in making choice and taking appropriate decisions for the ultimate success of the system. These are; (i) selection of site and clearing of vegetation etc., (ii) burning, (iii) planting, (iv) weeding and protecting, (v) harvesting, and (vi) succession. An inappropriate decision can spoil the whole cycle of shifting cultivation which may eventually lead to poor and smaller returns in the shape of harvests and sometimes no harvest at all.

1.9.1 Land selection and vegetation cutting

How do shifting cultivators select their fields? If the land is to be allotted by the community or head of a tribal clan, the cultivator has almost no choice. But a wise distributor will allot the fields and area according to the size of the family of the cultivator and his needs, whether a widow or infirm and old or young, distance from his settlement and other parameters set out by the head or clan. An integral cultivator usually has the right to make the field anywhere in the forest. In Southeast Asia and the Amazon, it has been observed that the sharing of food occurs within the community and it is liked also, but the produce goes to those who prapare and manage the plot. Okigbo (1982) mentioned that a cultivator in the humid areas of Southeast Asia and the Amazon basin can generally make a choice between primary forest and secondary forest, however, in Africa, the availability of primary forest is rare for making the plots for shifting cultivation. Because in many shifting communities a field may be planted more than once, it should be kept in mind that the choice will have to fulfill present and future requirements. The site selection depends not only on soil fertility needs, but also on accessibility of the site, potential crops, ease of labour availability, as well as supernatural constraints (sacred groves, presence of spirits, etc.) (Dove 1983; Warner 1981).

1.9.2 Burning

Burning of slashed or cut vegetation becomes necessary for a good crop production with minimum of labour. Rambo (1981) gave six important benefits of burning:

i) ***Clearance and removal of unwanted vegetation from the field***

The fields become clear of unwanted vegetation and planting and other field operations becomes easy. The weeds do not compete with the growing crop for a long period of time, thus, ensuring good productivity of various crops.

ii) ***Change in soil structure***

The heat generated by the fire changes the soil texture more friable, making it easy for seeding/planting with a dibble stick (Conklin 1957; Tivy 1987).

iii) ***Enhancement of soil fertility***

Due to burning of vegetation, large amounts of nutrient rich ash is added to the soil along with crop nutrients (Sanchez 1976, Dove 1983).

iv) ***Decrease in soil acidity***

There is an increase in soil pH due to burning as the plant ashes are generally alkaline in nature. The increase in soil pH will make even the native nutrients to the crop plants. Aluminum toxicity is also reduced due to increase in soil pH (Moran 1981).

v) ***Increase in availability of crop nutrients***

The heating of the soil makes the soil nutrients available to plants (Nye and Greenland 1960).

vi) ***Sterilization of soil***

There is sterilization of soil with heating due to burning which do away with weeds, insects and pathogens (Glass and Thurston 1978).

1.9.3 Cropping phase

The cropping phase is one of the most important and challenging phase in the whole system of shifting cultivation. Selection of right crops and right time of seeding/planting are the keys to a good harvest. By and large, the shifting cultivators know the techniques gained through the experience or sometime they take the advice of the elders of the village or clan. The choice of crops and how much area to be planted under a particular crop depends on the needs tastes of the family. If the physical condition of the soil is good after clearing and burning, the seeds are usually scattered on the land surface or dibbled into holes with a digging stick. Conklin (1957) categorised the cropping segment into grain and non-grain cropping (root or tree crop farming). The cropping phase is the most convoluted, because many crops are cultivated concurrently or in an overlapping design, giving a mosaic of inter-cropped cultivable plants in the system. Cropping in the shifting cultivation has a great significance of time and space for obtaining a continuous flow of crops during the whole year. Therefore, mixed and intercropping are common practices in the shifting cultivation system.

1.9.4 Fallowing

When the land is abandoned after cropping for 1-3 years in shifting cultivation, and allowed to regenerate naturally or through well managed reforestation, it is known as the 'fallow period' or fallow phase. This period is generally longer than the cropping phase. The enrichment of the soil would depend on the amount of vegetative matter produced, cut and burnt to add ashes. So, management of fallow land is as important as that of cropped land during cropping phase, if not more. Different types of vegetation in the fallow phase (grasses, bushes, trees etc.) take up the nutrients from various depths of the soil, depending on the root system, and bring them to the soil surface and the vegetation growing over it. This period is important for restoring soil fertility and health as the vegetation will be cut and burnt on the soil surface and nutrients are added to the soil before the cropping phase starts in the cycle of shifting cultivation. The sustainability of a shifting cultivation system would depends upon how well the fallow period re-establishes soil fertility.

The loss of weed seeds produces less weeds, requiring less labour for removing them during crop growth as well as less competition of with growing crops or sunlight, water and nutrients. With burning, there is also loss of beneficial microbes in the soil. Burning of vegetation improves the soil health and yields of crops as well as reduces the time used in weeding. Though the decision on the time to burn is generally taken by the individual, in some of the hill tribes of Thailand the decision is made by the elderly people and the entire village burns its fields on the same day (Kunstadter 1987). A poor burn can spoil the field and addition of ashes is reduced.

1.10 SUCCESSION

For the swidden agroecosytem continuance, it is necessary that the old fields have to be allowed to become part of the forest again (Moran 1981). The abandoned land has to be looked after well for regeneration. This would require good fallow management, afforestation with destroyed species of trees and safety from wild animals etc. Manner (1981) was of the view that natural succession is witnessed during regeneration of forest after the shifting cultivation fields are abandoned past cropping phase in humid tropics, provided the field has not been cropped for longer period as well as, the field is not too big. There are many questions, such as; how long will a forest succession take and what are its mechanisms, which need to be answered. In an Amazonian area study, it was observed that initially the abandoned plot was colonized by herbaceous annuals, but after one year pioneer woody species began to come up. The seed for initial woody plants was transported from the adjoining forests through the help of wind, water movement and insect/pests. The regeneration of the woody species was assisted by microhabitats, such as, fruit trees and timber logs, that provide a congenial environment for the seeds (Uhl 1983; Wilken 1972). The shade of the trees take care of the seedlings and protect them from the direct sun, and in course of time these "islands of woody vegetation" grow till they cover the whole field. Uhl (1983) estimated that it would take 100 years for the field to return to primary forest. The early views indicated that the role of a shifting cultivator was minimal in the regeneration of secondary forests as he believed "just let nature take its course" and "abandoned" the shifting fields to let them regenerate naturally. However, this view stands challenged now as some studies have shown that the shifting cultivators are actively involved in the management of fallow succession. The mature forests of Southeast Asia may not be of primary origin (Spencer 1966), as does Richards (1973) for Africa. Denevan *et al.* (1984) revealed that "in the past large areas of the Amazon forest may actually have been at different stages of productive swidden fallow." Evidences show that tropical forests appear to have been modified by inhabitants for diversity of species needed by them, as cluster of trees have been found, which may not normally occur in a natural succession (Denevan, 1984; Getahun *et al.,* 1982; Okigbo and Lal 1979).

1.11 SOCIAL SECURITY

There are as many as 65 or more ethnic groups in India practicing shifting cultivation. The shifting cultivation areas are mostly colonized by various tribes and they have sequentially practiced hunting, nomadism, pastoralism, shifting cultivation and now, settled cultivation up to some extent. These operations have taken care of their food and livelihood security. The socio-cultural relationships and tribal beliefs in these areas were able to keep stability between resource potential and their utilization for quite a longer period but due to the indiscriminate increase in population and misuse of natural resources, the stability has declined and almost vanished from some areas. The spurt in population has compelled the food production base to crumble under its load. To satisfy their needs, the inhabitants have mismanaged and made a mess of natural resources. To sustain the fast growing population, there is strong need for integrated approach for natural resources management and livelihood security. The major issue is to advance the protection and long-term use of natural resources which may encourage sustainable economic growth and increase in entrepreneurial capabilities, along with being unprejudiced and environmentally adaptable (Sharma and Bujarbaruah 2016).

The shifting cultivation also provides full social security to the members of a clan. In earlier days of hunting and food gathering, the entire expedition used to share whatever was collected, though the brave and the leaders could get a major share, but those who were left behind such as, old and infirm or widows, had also a well-defined share in the collection (Sharma 1981, Sharma and Bujarbaruah 2016). The individual's welfare was the social responsibility of the society and it continue to be honoured in shifting cultivation phase also. The clearing of forest is a community operation, where everybody contributes according to his ability. The old may only accompany the group and tender advice out of their past experience, while the youth may provide bulk of the physical labour. The old, infirm and widows are given fields in the middle and the young on the periphery. Many operations in the shifting cultivation are community based, where everybody works according to his capacity but, gets according to his needs. Thus, they are not dependent on their own children for their needs.

1.12 SUSTAINABILITY OF SHIFTING CULTIVATION

In shifting cultivation, the land is farmed for 2-3 years and the kept fallow for regeneration of vegetation. The period for which the land is kept fallow is of utmost importance for the sustainability of the system. The land gets exhausted of nutrients during the cultivation period and replenishment of nutrients will depend on the amount of biomass burnt, sufficient to add the required nutrients through addition of ash. The amount of biomass produced during the fallow stage is directly related to the number of years for which the land is kept fallow, nature of vegetation and level of management of fallow land. The loss of organic matter, decrease in CEC and other nutrients during cropping phase will depend on the period for which the field is cropped and the level of total production of various crops, that is, removal of nutrients by the crops. For the sustainability of shifting cultivation it has to be ensured that the practice synchronize with the local environment, judiciously managed by the cultivators during both the cropping and fallow stages and a proper ratio of fallow period/cropping period is maintained. It is likely that the cultivators engaged in shifting cultivation may have a state-of-art knowledge and appreciation of their regional environments and of the array of crops and local plant species they use. It has been observed that many complex and variable land tenure systems exist across the globe. Other than self ownership, the land desired for shifting cultivation belongs either to the village head, whole community together or head of the tribal clan. The shifting cultivators have skillfully integrated food and cash into some shifting cultivation systems for food and livelihood security.

Other important factor involved in sustainability of shifting cultivation is carrying capacity of the system. The number of persons depending on the system (shifting cultivation) for their livelihood per unit of land, is important feature to ensure sustainability of shifting cultivation under specific situations. Though this is related to the fertility status of the land in use, a load of 15 persons per km^2 is the safest number, which could be stretched to 30 persons km^2, depending on soil health and efficient management of the system. Beyond this load, the practice of shifting cultivation is always vulnerable to collapse.

A method of calculation of sustainability index of shifting cultivation system has been devised by us after scanning lot of literature and our own experience with regard to shifting cultivation. Four major parameters factors were considered for sustainability of of the system, such as, total fallow period in years, population density level, soil fertility level and ownership rights. Each level of magnitude of a parameter was allotted points as given in Table 1.2. The scored points are multiplied to get sustainability index (SI). It is evident from Table that minimum multiplication value or Sustainability Index (SI) will be 1 (1x1x1x1, when fallow period is only 1 year, population density is high, Soil fertility level is low and the shifting cultivator has usufructuary right over land he is cultivating) and maximum 90 (5x4x3x1.5, when fallow period is only 24 years, population density is very low, Soil fertility level is high and the shifting cultivator has ownership right over land he is cultivating). Maximum sustainability index has been capped at 90% as it is impossible to achieve 100% because there are many other factors (other than those described above) such as; social, economical, institutional, marginality, resource heterogeneity etc., which may affect sustainability. We have categorized the type of sustainability according to sustainability index (SI) or multiplication value as; Highly sustainable (>65), medium sustainable (50-65), marginally sustainable (40-50), vulnerable (30-40) and highly vulnerable (<30). Say, for example, a shifting cultivation plot had 12 years of fallow (3 points), population density is low with 15 to 30 persons km^{-2} carrying capacity (3 points), soil fertility level is high (3 points) and cultivators has ownership rights (1.5 points. The SI will be, 3 x 3 x 3 x 1.5 = 40.5. The system can be categorized as marginally sustainable.

Table 1.2. Parameter and their assigned value for calculating multiplication value

Fal-low	Period	Population	Density	Soil	Fertility	Land	Tenure
Years	**Points**	**Level**	**Points**	**Level**	**Points**	**Right**	**Points**
1	1	Very low (<15 persons km^{-2})	4	Low	1	Usufructuary	1.0
6	2	Low (15-30 persons km-^{2})	3	Medium	2	Self-ownership	1.5
12	3	Medium (30-50 persons km^{-2})	2	High	3	-	-
18	4	High (>50 persons km^{-2})	1	-	-	-	-
24	5	-	-	-	-	-	-

The sustainability index and percent sustainability percent of shifting cultivation has been shown in Fig. 1.6. There is a linear relationship between SI and sustainability percent. The sustainability line gets bolder as the magnitude of sustainability increases or the sustainability of shifting cultivation increases.

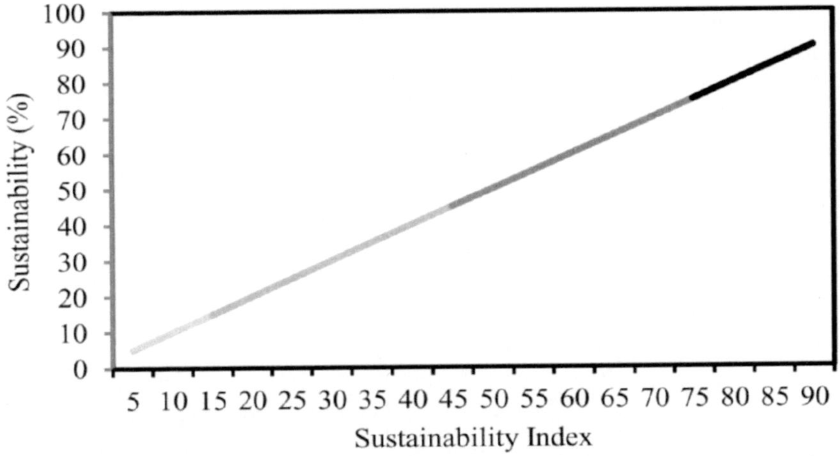

Fig. 1.6. Sustainability Index and sustainability percent of shifting cultivation.

1.13 FUTURE OF SHIFTING CULTIVATION

For Southeast Asia, Schmidt-Vogt *et al.* (2009) compiled available published shifting cultivation land assessment for some countries, such as, Laos, Cambodia, Thailand, Myanmar,Vietnam, Malaysia, and Indonesia. Broadly, it has been found that there has been a fast decrease in shifting cultivation area in sizeable parts of Southeast Asia and it is continuing. It was also predicted that a major share of the area under shifting cultivation will vanish by 2030, and the remaining areas are likely to be almost entirely vanished by 2060. Heinimann *et al.* (2017) reported on the basis of a survey results about the trends in area under shifting cultivation in Vietnam and Laos, that shifting cultivation is likely to be largely reduced by 2030 and totally disappeared by 2060. In Myanmar, shifting cultivation is likely to disappear between 2060 and 2090 if squabbles between government and ethnic armed groups are solved; the practice is thought to disappear from Borneo and Sulawesi sometime between 2030 and 2060; from India and Bangladesh by 2030 and; from Papua New Guinea by second half of this century, perhaps even until 2090.

We have constructed a schematic diagram and tried to predict the demise of shifting cultivation in the world, after broadly studying the trends in the practice in different areas of the world (Fig.1.7). Considering the present pace of decline in the area and population involved in the world, it would take another 40 to 60 years for complete eradication of shifting cultivation, that is, between 1960 and 1980. This period will be reduced to 30 to 40 years or eradication by 1950 to 1960, due to socio-economic reasons and lack of interest of the new generation in shifting cultivation practice. This period of completely weaning away the shifting cultivators from this age old practice will be further reduced to up to a maximum of 30 years or by 1950, because of strong government policies in this regard, intensification of agriculture due to increased demographic pressure and provision of incentives to the shifting cultivators to leave the practice. Intensification of agriculture itself means transition to normal agriculture, where shifting cultivation automatically stands eradicated.

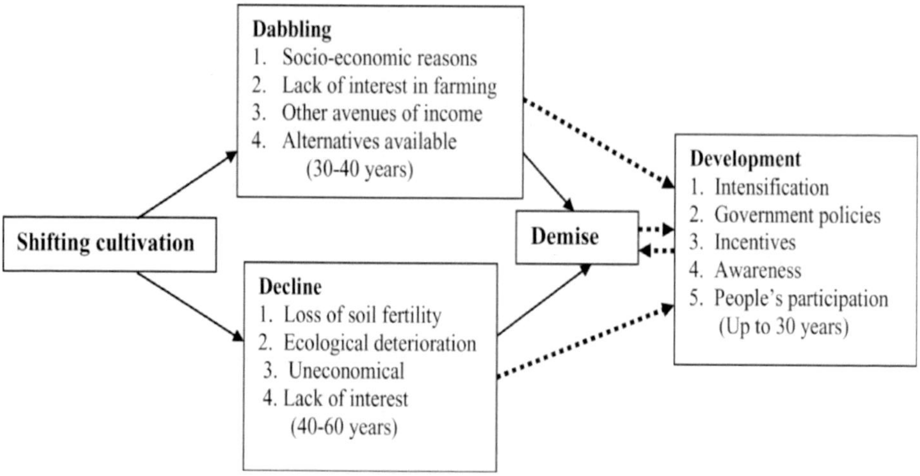

Fig. 1.7. Schematic diagram of the fate of shifting cultivation on global scale.

1.14. THE ELEVEN D's

We have identified 11 D's which matter to deal with shifting cultivation (Fig. 1.8), such as; definition, data base, dynamics of soils, diversity (bio/socio-cultural practices), deforestation, degradation of resources, decriminalization, dabbling, decline, demise and development. We have tried to tailor all major aspects of shifting cultivation around these titles. The shifting cultivation at global level is a vast field of study and it is not easy to cover it in one volume. However, we have tried to fill and ocean in a small pot with our own views and comments.

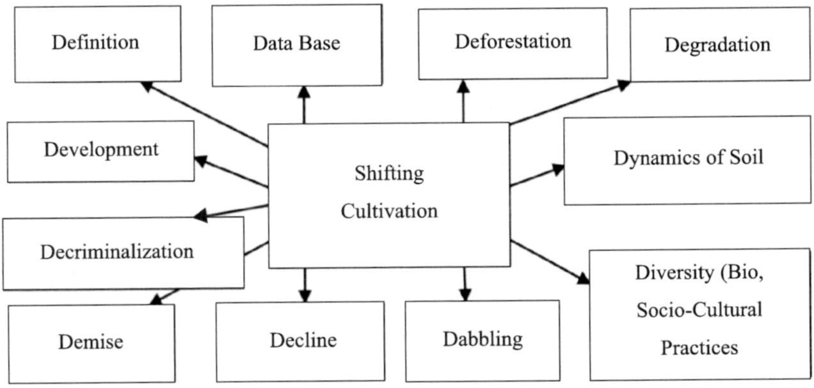

Fig. 1.8. The eleven D's of shifting cultivation

References

Allan, W. 1965. The African Husbandman. London: Oliver and Boyd.

Allan, W. 1972a. Ecology, techniques and settlement patterns. In Ucko, P.J., Tringham, R. and Dimbleby, G. W. Man Settlement and Urbanism. London: Duckworth & Co. Ltd. pp. 211 - 226.

Andriesse JP. Nutrient management through shifting cultivation. In: van der Heide, J. (Ed.), Nutrient Management for Food Crop Production in Tropical Farming Systems. Institute for Soil Fertility, Haren, The Netherlands, 1989, 29-62.

Angelson. Ariled (1995). 'Shifting cultivation and deforestation: A study from Indonesia', World Development. Vol. 23: 1713 -1729.

Aweto, A. O. 2013. Shifting cultivation and secondary succession in the tropics. Wallingford, CABI.

Benneh, G. (1972). Systems of agriculture in tropical Africa. Economic Geography 48(3): 244 - 257.

Boserup, Ester (1981). Population and technological change: a study of long-term trends. Chicago: University Of Chicago Press. ISBN 9780226066745.

Boserup, Ester (1981). Population and technology. Wiley-Blackwell. ISBN 9780631133711.

Brady NC (1996). Alternatives to slash-and-burn: a global imperative. Agric. Ecosyst. Environ., 58: 3–11.

Brookfield, H. 2001. Exploring agrodiversity. New York, Columbia University Press.

Cairns, M. (Ed). (2007). Voices from the forest: Integrating indigenous knowledge into sustainable farming. Washington, DC: Resources for the Future.

Cairns, M. F. (Ed.). (2014). Shifting cultivation and environmental change: Indigenous people, agriculture and forest conservation. London: Routledge.

Chapman, E.C. (1975) Shifting agriculture in tropical forest of South East Asia. Proceedings of regional meeting on use of ecological guidelines for development in tropical forest areas of South East Asia. IUCN Publication, New Series 32:120-135.

Chin SC (1984) 'Do shifting cultivators deforest', in V. Shiva (ed.) Forest Resources: Crisis and Management (1992), Natraj Publisher: Dehradun.

Comte, I.; Davidson, R.; Lucotte, M.; de Carvalho, C.J.R.; de Assis Oliveira, F.; Da Silva, B.P.; Rousseau, G.X. (2012) Physicochemical properties of soils in the Brazilian Amazon following fire-free land preparation and slash-and-burn practices. Agric. Ecosyst. Environ., 156, 108–115.

Conklin, H. C. 1957. Hanunoo agriculture: A report on an integral system of shifting cultivation in the Philippines (FAO Forestry Development Paper No 12). Rome, Food and Agriculture Organization of the United Nations.

Davidson EA, de Abreu Sa TD, Reis Carvalho CJ, de Oliveria Figueiredo R, Kato M do SA, Kato OR, et al. (2008). An integrated greenhouse gas assessment of an alternative to slash-and-burn agriculture in eastern Amazonia. Glob Change Biol.;14: 998–1007.

Denevan, W. M. (Ed). 2001. Cultivated landscapes of native Amazonia and the Andes. Oxford, Oxford University Press.

Denevan, W. M., Treacy, J. M., Alcorn, J. B., Padoch, C., Denslow, J., and Paitan, S. F. 1984. Indigenous agroforestry in the Peruvian Amazon: Bora Indian management of swidden fallows. Interciencia 9(6): 346 - 357.

Dixon J, Gulliver A, Gibbon D. (2001) Farming Systems and Poverty—Improving Farmers' Livelihoods in a Changing World. Rome: Food & Agriculture Organization of the UN (FAO), ftp://ftp.fao.org/docrep/fao/003/y1860e/y1860e00.pdf

Dove, M. R. 1983a. Forest preference in swidden agriculture. Tropical Ecology: 24(1): 122 - 142.

Dove, M. R. 1983b. Theories of swidden agriculture, and the political economy of ignorance. Agroforestry Systems 1: 85 -99.

Dove, M. 1985. Swidden agriculture in Indonesia: The subsistence strategies of the Kalimantan kantu'. Berlin and New York, Mouton.

Dumond, D.E. (1961) Swidden agriculture and the rise of Maya civilization. Southwest. J. Anthropol., 17, 301–316.

Ellen, R. (2012) Studies of swidden agriculture in Southeast Asia since 1960: An overview and commentary on recent research and syntheses. Asia Pac. World 2012, 3, 18–38.

Fairhead, J. & Leach, M. 1998. reframing deforestation. global analyses and local realities: studies in West Africa. Routledge, London. pp. xxv + 238. ..

FAO. Global Forest Resources Assessment 2010, FAO Forestry Paper No. 163; FAO: Rome, 2010.

FAO/UNEP. 1982. Tropical Forest Resources. Forestry Paper no. 30. Rome: FAO. Cited by OTA 1984. Farnworth, E. and Golley, F. (eds.). 1974. Fragile Ecosystems: Evaluation of Research and Applications in the Neotropics. New York: Springer-Verlag.

Geertz. C. 1974 Agricultural involution, The process of ecological change in Indonesia. University of California Press. Berkley.

Getahun, A., Wilson, G. F. and Kang, B. T. 1982. The role of trees in farming systems in the humid tropics. In MacDonald, L.H. (ed.) Agroforestry in the African humid tropics. Tokyo: UNU. pp. 28 - 35.

Glass, E. H. and Thurston, H. D. 1978. Traditional and modern crop protection in perspective. Bioscience 28 (2), 109 - 115.

Gourou, P. (1953). The Tropical World (Trans. By E. D. Labords), London. 156pp.

Grandstaff, T. The development of swidden agriculture (shifting cultivation). Dev. Chang. 1978, 9, 547–579.

Greenland, D. J. 1974. Evolution and development of different types of shifting cultivation. In Shifting Cultivation and Soil Conservation in Africa. Soil Bulletin 24. Rome: FAO.

Harris, D. R. 1973. Swidden systems and settlement. In Tringham, R. (ed.), Ecology and Agricultural settlements: an ethnographic and archaeological perspective. Andover, Mass.: Warner Modular Publications.

Harris, D.R. (1971) The ecology of swidden cultivation in the upper Orinoco Rain Forest, Venezuela. Geogr. Rev., 61, 475–495.

Heinimann A, Mertz O, Frolking S, Christensen AE, Humi K, Sedano F, Chini LP, Sahajpal R, Hansen M and Hurtt G, 2017. A global view of shifting cultivation: Recent, current, and future extent. PLoS ONE, 12 (9), https://doi.org/10.1371/journal.pone.0184479

IBSRAM (International Board for Soil Research and Management). (1987). Report of fhe First Regional Seminar on Management of Vertisols under Semi-Arid Conditions in Africa and Southwest Asia, Nairobi, Kenya, 1-6 December 1986. Bangkok,Thailand:

Kerkhoff E and Sharma E, 2006. Debating Shifting Cultivation in the Eastern Himalayas Farmers' Innovations as Lessons for Policy, International Centre for Integrated Mountain Development (ICIMOD), Kathmandu, Nepal

Kleinman, P.; Bryant, R.B.; Pimentel, D. 1996 Assessing ecological sustainability of slash-and-burn agriculture through soil fertility indicators. Agron. J., 88, 122–127.

Kunstadter, P. 1987. Swiddeners in transition: Lua' farmers in Northern Thailand. In Turner, B. L. II and Brush, S. B. (eds.). Comparative Farming Systems. New York: Guilford Press. pp. 130 - 155.

Lanly JP. 1985. Defining and measuring shifting cultivation. Unasylva.;37: 17–21.

Li P, Feng Z, Jiang L, Liao C and Zhang J, 2014. A Review of Swidden Agriculture in Southeast Asia. Remote Sensing, 6: 1654-1683; doi:10.3390/rs6021654

Manner, H. I. 1981. Ecological succession in new and old swiddens of montane Papua New Guinea. Human Ecology 9, (3): 359 - 377.

Mazoyer M and Roudart L, (2006); A History of World Agriculture: From the Neolithic Age to the Current Crisis.. London: Earthscan,. Pp. 528. .

Meine, V.N.; Elok, M.; Niken, S.; Fahmuddin, A. Swiddens in Transition: Shifted Perceptions on Shifting Cultivators in Indonesia; World Agroforestry Centre: Bogor, Indonesia, 2008; p. 64.

Mertz, O. (2009) Trends in shifting cultivation and the REDD mechanism. Curr. Opin. Environ. Sustain. 2009, 1, 156–160.

Mertz, O.; Padoch, C.; Fox, J.; Cramb, R.; Leisz, S.; Lam, N.; Vien, T. (2009) Swidden change in southeast Asia: Understanding causes and consequences. Hum. Ecol., 37, 259–264.

Moran, E. F. 1981. Developing the Amazon. Indiana University Press: Bloomington.

Moran, E. F. (ed.) 1983. Dilemma of Amazonian Development. Boulder, Colorado: Westview.

Myers N (1995). 'Tropical deforestation: rates and patterns' in Katrina Brown & David W Pearce (ed.) The Causes of Tropical Deforestation, The Economic and Statistical Anajysis of Factors giving Rise to the Loss of Tropical Forests. UCL Press London & Research Press New Delhi.

Nakano, K. An ecological study of swidden agriculture at a village in Northern Thailand. Southeast Asian Stud. 1978, 16, 411–446.

Newton, K. Shifting cultivation and crop rotations in the tropics. Papua and New Guinea Agric. J. 1960, 13, 81–118.

Nye P.H. and Greenland D.J. 1960. The soil under shifting cultivation. Commonwealth Bureau of Soils Technical Communication 51, Harpenden, U.K.

Nye, P. H. and Greenland, D. J. 1964. Changes in the soil after clearing tropical rainforest. Plant and Soil 21: 101 - 112.

Okigbo, B. N. 1982. Impact of agricultural systems and rural development on Nigerian forests. In MacDonald, L. H. in the African Humid Tropics. Tokyo: United Nations University. pp. 41 - 45.

Okigbo, B.N., and R. Lal. 1979. Soil fertility maintenance and conservation for improved agroforestry systems in the lowland humid tropics. In Soil Research in Agroforestry eds. H.O.

Olofson, H. (1981) Swidden and kaingin among the southern Tagalog: A problem in Philippine upland ethno-agriculture. Philipp. Q. Cult. Soc., 8, 168–180.

Palm, C. A., Vosti, S. A., Sanchez, P. A., and Ericksen, P. J (Eds). 2005. Slash-and-burn agriculture: The search for alternatives. New York, Columbia University Press.

Pollini, J. 2007. Slash-and-burn cultivation and deforestation in the Malagasy rain forests: Representations and realities. PhD dissertation. Ithaca, Cornell University. http://www.scribd.com/doc/26790856/Deforestation-in-Madagascar

Posey, D. A. 1983. Indigenous ecological knowledge and development of the Amazon. In Moran, E. F. (ed.), Dilemma of Amazonian Development. Boulder, Colorado: Westview, pp. 225 - 257.

Ramakrishnan, PS (1992) Shifting agriculture and sustainable development: An interdisciplinary study from North Eastern India. MAB Series, Volume 10. Paris, France: United Nations Educational, Scientific and Cultural Organization (UNESCO)

Rambo, A. T. 1981. Fire and the energy efficiency of Swidden agriculture. EAPI Preprint. Honolulu: East-west Center.

Rambo, A. T. 1983. Why shifting cultivators keep shifting: understanding farmer decision-making in traditionalagroforestry systems. Paper prepared for the Joint East-West Environment and Policy Institute/ United Nations Food and Agriculture Organization Regional Worksop on Community Forestry. Nakorn Rachasima, Thailand, August 22-29, 1983.

Rappaport, R.A. (1971) Scientific American ,225:116-132.

Richards, P. W. 1973a. The Tropical Rain Forest. Scientific American 229 (6): 58 - 67.

Russell, W. M. S. 1988. Population, swidden farming and the tropical environment. Population and Environment 10(2): 77 - 94.

Sadeque SZ, 2000. Shifting Cultivation in Eastern Himalayas: Regulatory Regime and Erosion of Common Pool Resources. Paper presented at the 8th Conference of the International Association for the study of Common Property, Bloomington, Indiana, May 31 to 4 June.

Sanchez, P.A. 1976. Properties and Management of Soils in the Tropics. New York, N.Y.: John Willey and Sons.

Sánchez, P.A., C.A. Palm, S.A. Vosti, T.P. Tomich, and J. Kasyoki. 2005. Alternatives to slash and burn: Challenge and approaches of an international consortium. In Slash-and-burn agriculture: The search for alternatives, edited by C.A. Palm, S.A. Vosti, P.A. Sanchez, and P.J. Ericksen. New York: Columbia University Press, 3-37.

Schlippe, P. de. (1956). Shifting cultivation in Africa : The zande system of agriculture. London, Routledge.

Schmidt-Vogt, D. (1999). Swidden farming and fallow vegetation in Northern Thailand. Wiesbaden, Germany, Franz Steiner Verlag.

Schmidt-Vogt, D.; Leisz, S.J.; Mertz, O.; Heinimann, A.; Thiha, T.; Messerli, P.; Epprecht, M.; Cu, P.V.; Chi, V.K.; Hardiono, M.; et al. (2009) An assessment of trends in the extent of swidden in southeast Asia. Hum. Ecol. 2009, 37, 269–280.

Sharma, U. C. (1993). Agricultural production potential in North Eastern hill states and role of fertilizer nutrients in its realization. Indian Fertilizer Scene Annual, 19-28.

Sharma, U. C. and Prasad, R.N. (1995). Socio-economic aspects of acid soil management and alternate land use systems for north eastern states of India. In: Plant-soil Internations at Low pH, Principles and Management (ed. by R. A. Date, N. J. Grundon, G. E. Rayment & M.E. Probert) Kluwer Academic Press, The Netherlands: 689-696.

Sharma U C and Bujarbaruah KM (2014). Issues and Challenges in Shifting Cultivation and its Relevence in the Present Context, National Academy of Agricultural Sciences, Policy Paper No. 83.

Sharma,T.C.(1976)The pre-historic background of shifting cultivation. In : Proceedings of seminar on shifting cultivation in NE India. NE India Council for Social Sciences Rsearch, Shillong, Meghalaya.

Silva JMN, Carreiras JMB, Rosa I, Pereira JMC, (2011). Greenhouse gas emissions from shifting cultivation in the tropics, including uncertainty and sensitivity analysis. Journal of Geophysical Resarch; Atmos.116: 1-21.

Sopher, D.E. (1964) The swidden/wet-rice transition zone in the Chittagong Hills. Ann. Assoc. Am. Geogr., 54, 107–126.

Spencer JE (1966). Shifting Cultivation in Southeast Asia. University of California Publication in Geography, voL 19. Berkeley: University of California Press.

Thrupp L. A., Hecht S. B. and Browder J. O. (1997), The diversity and dynamics of shifting cultivation: Myths, realities, and policy implications. World Resources Institute Washington, DC.

Tivy, J. 1987. Nutrient cycling in agro-ecosystems. Applied Geography 7: 93 - 113.

Toky, O.P. and Ramakrishnan, P.S. (1982). A comparative study of the energy budget of hill agro-ecosystems with emphasis on the slash and burn system (Jhum) at lower elevations of north-eastern India. Agricultural Systems, 9: 143-154.

Uhl, C. 1983. You can keep a good forest down. Natural History 92(4): 69 - 79.

Van Vliet, N.; Mertz, O.; Heinimann, A.; Langanke, T.; Pascual, U.; Schmook, B.; Adams, C.; Schmidt-Vogt, D.; Messerli, P.; Leisz, S.; et al. Trends, drivers and impacts of changes in swidden cultivation in tropical forest-agriculture frontiers: A global assessment. Glob. Environ. Chang. 2012, 22, 418–429.

Warner K, 1991. Shifting cultivators Local technical knowledge and natural resource management in the humid tropics. Food and Agriculture Organization of the United Nations, Rome.

Whitmore, T. M. and B. L. Turner II (Eds). 2001. Cultivated landscapes of Middle America on the eve of conquest. Oxford, Oxford University Press.

Wilken, G. C. 1972. Microclimate management by traditional farmers. Geographical Review. 62: 544-560.

2

Defining Shifting Cultivation

2.1. GENERAL DESCRIPTION

Forests covered only 31 per cent of the land in the world in 2011, while land used for agriculture, (cultivable land, perennial crops and pastures), covered 38 per cent (FAO 2011). In the history of humankind, agriculture is a recent invention (Mazoyer and Roudart, 2005). With the transition from the hunter-gatherer lifestyle to one of agriculture during the Neolithic Revolution, farmers quickly expanded their activity into forest areas. They developed shifting cultivation, an agricultural practice with temporal rotations in forest ecosystems. Shifting cultivation still exists today in tropical regions and is the livelihood basis for hundreds of millions of people (Mertz *et al.*, 2009). The range in the population estimate is extremely wide, because of difficulties in counting smallholders spread out under the tree canopy at the margins of tropical forests, as also due to confusion that exists in differentiating shifting cultivators from other farmers involved in small-scale agriculture (Thrupp *et al.*, 1997).

There is no particular and specific definition of shifting cultivation and that is why so many definitions have been offered to denote a farming practice. Having many operations in common throughout the world, the practice of shifting cultivation is quite variable in its causes, conditions, techniques and purposes. Shifting cultivation encompasses many operations, varying from region to region and continent to continent. Every scholar has tried to define the practice in his own way. Perhaps no definition fits well to define a complex and dynamic practice like shifting cultivation. Even then, some definitions are very close to perfection. Before something is said about definition, it may be worthwhile to understand the practice of shifting cultivation.

The characteristic of shifting cultivation are; rotation of field, crops grown on hill slopes receiving high rainfall, all operations are generally done manually use of family labour or community operations, the fallow period is longer than cropping period, use of simple implements like dao (big unserrated sickle) dibbling stick, hoe, sickle, basket etc., growing of a mixture of crops, growing of creeper vegetable crops like pumpkin, bottle gourd, cucumber, bitter gourd etc; which can climb the trees and start giving produce during the cropping period and above all the shifting cultivation is a way of life.

Slash-and-burn agriculture (Peters and Neuenschwander, 1988; Palm *et al.*, 2005), swidden agriculture (Mertz *et al.*, 2009), and shifting cultivation (Nye and Greenland, 1960, Aweto 2013) are all Neolithic agricultural systems. In this system, a cropping phase of 1-3 years is followed by a 'regeneration phase' the time length of which needs to sufficient enough to enable the fallow land to be covered by dense forest vegetation, and from where the cultivator can get enough biomass to be cut and burnt, before the next cultivation cycle. With increase in population, more and more uncultivated forest land was brought under cultivation for livelihood security. This was possible only in those areas where availability of land was not a limitation. In areas where the expansion was not possible due to land limitations, the practice of shifting cultivation proved to be disastrous. The shifting cultivation has a variable of sub-systems (Schlippe, 1956; Conklin, 1957; Pollini, 2007; Mertz *et al.*, 2009; Cairns 2014) that differ by different operations involved in the practice, rituals being followed and ratio of cropping and fallow period.

The information when passed on to descendants turns to "belief" with wisdom. If we continue with this practice and live with this information, it is identified as our culture which is our routine conviction and doctrine about certain events, sharing same place or ecological situations. The shifting cultivation acquired this status among the cultivators since its existence fron Neolithic period, about 7000 BC.

Shifting agriculture is also known as "swidden" or "slash-and-burn" farming. However, the words shifting cultivation and slash-and-burn agriculture can be distinguished. Slash-and-burn agriculture refers to farming practises that involve cutting and burning the forest. Traditional shifting cultivation refers to systems with long fallow periods to allow for forest regeneration, whereas slash-and-burn agriculture more broadly refers to farming practises that involve cutting and burning the forest. Slash-and-burn agriculture can be classified as large-scale or small-scale, and it may or may not involve extensive fallow periods. Traditional shifting farming is practised by indigenous peoples who have lived in isolated forest areas for a long time in many parts of the tropics, whereas migrant farmers living near the forest border may undertake small-scale slash-and-burn agriculture without including long fallow periods (Sanchez *et al.* 2005). Small-scale agricultural activities, on the other hand, contribute greater tropical forest loss in Africa than large-scale commercial agriculture, according to DeFries *et al.* (2010) and Fisher (2010). Deforestation rates in Africa are relatively low (Hansen *et al.* 2008). There isn't as much deforestation as in many Southeast Asian and Amazonian countries, and Central African countries with substantial tropical forests are known as "high forest, low deforestation" countries (da Fonseca *et al.,* 2007). However, in Central Africa, large-scale economic development, such as industrial logging has been related to deforestation (Laporte *et al.,* 2007).

2.2 RENEWAL OF FERTILITY AND BIOMASS ACCUMULATION

For normal agriculture systems, the level of fertility needed by the crop plants is generally added to the soil through chemical fertilizers, animal manure, alluvial deposits, etc. However, the renewal of fertility is based on lateral transfer and concentration of nutrients towards the cultivated field (Mazoyer and Roudart, 2005). Shifting cultivation is different from this general rule, wherein, the soil fertility comes from the biomass accumulated on-site during the fallow period. In the fallow period, the organic matter is stored in the above ground vegetation, as well as also in the soil. The accumulated biomass is mobilized for the cultivation phase by its mineralization. The burning of the slashed biomass following the clearing of the field, catalyses and speeds up the mineralization. The burning affects the soil by: (i) increasing the amount of available nutrients like potassium and phosphorus; (ii) the alkalinization, which increases the bioavailability of phosphates; and (iii) the thermal disinfection which cleans the top ground of pests (insects, nematodes, fungus, etc.) and of dormant weed seeds (Gerold *et al.*, 2004; Nye and Greenland, 1960; Roder, 2001).

When a new fallow phase begins, with biomass slowly accumulating above and below ground, fertility renewal depends on rotation in shifting cultivation. Fertility, by and larhe, is proportional to the fallow duration (Ramakrishnan 1992). The more the land is kept fallow, more the addition of soil fertility elements. However, there is a limit to this increase and keeping land fallow beyond a certain period may not prove useful. Also, the cultivators may not be in a position to keep their land fallow for longer periods. The progressive build-up of biomass has been proven, but recent research works challenge the oversimplified interpretation of linear progression. The staple crop yield is not strictly proportional to the fallow length (Mertz *et al.*, 2008; Roder, 2001) but overall people have often been confused about the use of terms shifting cultivation, swidden agriculture and slash and burn cultivation. Some times it is thought that these terms have originated from different areas of the World, say for example -, in India we never say swidden agriculture or slash and burn agriculture. In India this practice is termed only as 'shifting cultivation', of course local names are used in different states of the country. Actually the whole practice in totality, comprise of various operations namely; rotation of fields, no use of drought animals, observance of some rituals, land tenure in general, use of human labour, use of simple farm implements, burning of vegetation and relatively long fallow periods as compared to cropping. Even if they are not as ecologically diverse as virgin forests, it has been discovered that'regenerated forest vegetation' in abandoned land after shifting cultivation for a few years is usually more rich in floral and faunal wealth, which is important for the community. The shifting cultivators show little regard for the agricultural landscape and regard the forest plots at various stages of growth as natural in a predictable cycle.

2.3 EMERGENCE AND EXPANSION OF AGRICULTURE

Farming systems appeared in most continents by early to mid-Holocene (Holocene age began 11700 years ago) by a number of different passages in as many as 24 convincing independent centers of early habituation, with later diversion to adjoining areas (Fuller 2010, Fuller *et al.* 2011). In China, slow but regular history of extensification at early stages and later intensification of rice production, is supported by historical evidences, even though advanced technologies were available (Ellis and Wang 1997). Although current estimates of the Earth's potential to support further population and consumption increase are positive, there are still a variety of societal difficulties to overcome in convincing farmers to adopt appropriate land systems to satisfy rising demands (Tilman *et al.* 2011). The most evident concern is that some of the world's most productive regions are being converted to urban areas on a huge scale (Lambin *et al.* 2001), which, according to periodic assessments, may impede crop production. The connectivity between world regions has increased due to global market integration, with areas with maximum requirement for land resources generally leaving agriculture and reforestation and relegating their production needs to under-developed areas (Meyfroidt *et al.* 2010). Future land transformation and our capacity to meet community needs for various requirements, such as food, feed, housing, and energy, will most likely be determined by the rate of agricultural intensification (Tilman *et al.* 2011).

2.3.1 The three land use systems

In the tropical areas, shifting cultivation in its many different forms, continues to be a pervading practice. The fact that the shifting cultivation has survived for a long time, that is, Neolithic period to modern age, is enough proof of its being flexible and capable of embracing many modifications from time to time and maintaining crop production. It was one of the first forms of agriculture, however, it has remained a totally misjudged practice. Unfortunately, many academics fail to comprehend the ecologically stable cycles of cropping and fallowing that exist beyond the clearing and burning of forest vegetation. All the same, shifting cultivation is particularly susceptible to fast demographic pressure and to socio-economic change in the societies at global level. Though almost everyone points the finger at shifting cultivation for the depletion of forest resources, the drivers behind the rapid loss of tropical forests at the end of the twentieth century are the same as those that triggered the loss of forests in Europe, namely, urbanisation, industrialization, population explosion, geographical expansion, and the application of modern techniques to produce more from the ecosystem in the hounding for money and political power by contesting for land. We must recognise that those who work in agriculture are at the mercy of the social system. The roots of the recent problems lie down below in human attitude, as is evident from the studies of isolated groups and their relationship with the ecosystem. The main drive propelling people into mismatch with their environs can be defined as competition and collision.

2.3.2 Types of shifting cultivation

Conklin (1957) differentiated between "integral" and "partial" shifting cultivation. "Integral" shows the method of shifting cultivation as a fundamental part of a simple way of living of the cultivators, whereas; "Partial" shifting cultivation, as the name indicates, shows the interventions of the experts through modern techniques for cash crops or in simple words, a mix of subsistence and agribusiness farming. The classification of Conklin can be mentioned as under:

1. Integral shifting cultivation: (a) Innovator agriculture where substantial part of fully grown vegetation is cleared and cut annually; and (b) Traditional agriculture, where little or no fully grown vegetation is cleared annually.
2. Partial shifting cultivation: (a) Supporting agriculture, in which a cultivator who practises continuous agriculture devotes a portion of his efforts to shifting cultivation; and (b) Embryonic agriculture, in which a cultivator clears forest land for continuous farming but may be forced to shift after a few years due to low production.

Watters (1971) also gave a somewhat matching classification, that is, (i) traditional shifting cultivation, the farming system which has been practiced traditionally by indigenous tribal people whose lives and religious leanings have been closely related to the natural surroundings in which they live, and, (ii) imposed shifting cultivation which became a necessity, where the farmer is not related to a tribal community and usually owns no or not enough land to support permanent agriculture. The farmer in this situation usually clears forests for farming and uses shifting cultivation routes. Watters (1971) used the term 'land-hunger' shifting agriculture to describe this sort of shifting farming. The main goal of a grouping, according to Spencer (1966), should be to explain the generative evolution of integral farming. He divided integral shifting cultivation into three categories based on whether it was grown vegetatively, through direct sowing, or through a combination of both. He further subdivided these groups based on how the cultivators shift, whether they also alter their hutments, and whether they plant with a dribbling stick or a hoe.

2.4 STAGES OF SHIFTING CULTIVATION

The practice of shifting cultivation has basically two stages; such as, (i) removing and burning of old vegetation and, (ii) management of new vegetation. Removing and burning of old vegetation, includes, (a) clearing, (b) drying and, (c) burning of vegetation, while management of new vegetation encompasses, (a) cropping, (b) harvesting and, (c) fallow management (Fig. 2.1).

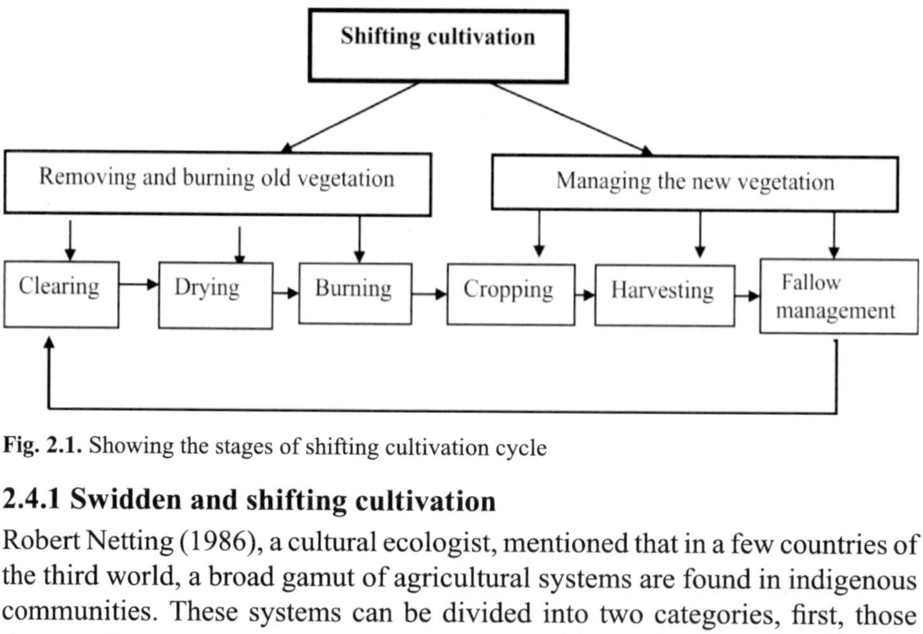

Fig. 2.1. Showing the stages of shifting cultivation cycle

2.4.1 Swidden and shifting cultivation

Robert Netting (1986), a cultural ecologist, mentioned that in a few countries of the third world, a broad gamut of agricultural systems are found in indigenous communities. These systems can be divided into two categories, first, those that entails continuous movement from one plot to another within primary forests and, second category of those systems which are assisted by irrigation for production all the year-round, also termed as intensive agriculture. Agricultural intensification may necessitate an increase in the number of fields allocated to short-fallow cropping in locations where population growth and houses obstruct the availability of sustainable forest plots. Boserup (1965) defined the forest fallow, bush fallow, and short fallow on the basis of land-use categories and the overall time of agriculture in relation to periods of fallow. A single shifting cultivation plot may be cultivated for one to two years in the forest fallow, after which the same field may lay fallow for twenty to twenty-five years. Unlike forest fallow, bush fallow can have one to eight years of cultivation followed by six to 10 years of fallow. Short fallows differ from forest and bush fallows in that they have a shorter and more variable period of cultivation (months or crop seasons) followed by one to two years of fallow. The amount of yield maintained in these systems is determined by the system's characteristics as well as the regeneration of tropical vegetation and its subsequent cutting and burning.

2.4.2 Indigenous knowledge

There have been several cases of setbacks to environment due to certain activities performed at large scale. In support, we may cite the examples of, the environmental setback due to Industrial Revolution in Europe and agricultural setback due to Green Revolution in India (Singh 2000), which were prominent. Anthropogenic climate change and its impact is a fact (IPCC 2007), though it varies intensity-wise in space and time. There has been a number of evidences from snow and ice melt, increase in average ocean temperatures, and rising of sea levels. Indigenous Traditional Knowledge (ITK) is an important part of indigenous farming that has been overlooked for a long time and labelled as primitive, but is now receiving more attention. The indigenous knowledge system's long-term viability is determined by how thoroughly they were tested through trial and error and how forcibly they were passed down orally or through shared practical experiences to future generations (Ohmagari and Berkes 1997). Though shifting agriculture has been viewed as a possible climate threat (FAO 1957, Dove 1983, Fox *et al.* 2009), it is actually a reliable mechanism for climate change prevention and adaptation.

2.4.3 Characterization

Shifting cultivation is practiced on a large scale, and many a times is an unavoidable methodology to practice farming in forest ecosystem. Many crops need full sunlight for optimum growth; therefore the forested land has to be cleared of shady vegetation to form new plots which could get direct sunlight without obstruction. When the cleared vegetation is dried and burnt, the ash, full of plant nutrients, is added to the soil. After two to three years of cultivation, many weeds, pests and diseases again proliferate and nutrients are either utilized by the crops, or lost with runoff or due to leaching. The plots are then abandoned and kept fallow, only to be cultivated again after the secondary forest vegetation has come up. The fallow period vary according to need and availability of land and usually varies 3 to 30 years, relying on meteorological and soil and drainage situation and the methodology being used. Its aim is to regenerate a amount of vegetation that can be propelled or used again by setting it on fire, and to do away with unwanted plants like weeds and pathogens. A shorter period is usually amounts to low biomass and absence of woody vegetation, which triggers a fertility crisis within a shorter period of time. To minimize the crop nutrients and carbon losses, the use of fire is generally avoided or slow burning is done when the biomass predominantly consists of grasses.

There are three methods followed to have woody vegetation; (i) the big woody trees from primary forest are not cut and allowed to grow, (ii) big woody trees are also planted and they remain growing for longer time during the fallow period and, (iii) whole vegetation of primary forest is cut and burnt but during the fallow phase some woody species grow naturally. Some times perpetual crops like cocoa or coffee can also be raised with annual crops after clearing of the forest vegetation. These perennial crops continue to grow after the harvest of annuals, in the same way as fallow vegetation otherwise, which generally takes the shape of a forest with a closed canopy, remains as a permanent agricultural system known as agro-forestry.

2.4.4 Economic rationality

Extensification of shifting cultivation is the most sensible land use in the humid tropics, and it can be sustained despite population expansion as long as forest land is available. When land is accessible and there is a low population, the availability of labour becomes a strong limiting factor in putting more land into production. Clearing the forest necessitates a significant amount of labour. Weed growth is reduced in shifting agriculture with long fallows because weeds are often found in low numbers in primary or ancient secondary forests. Under a short fallow time, this may not be practicable. Many times, weed proliferation is the reason for quitting the plot after a few years of gardening. It may or may not be linked to a loss of soil fertility.

2.5 REGENERATION OF BIOMASS

On global scale integral shifting cultivation practices have been discouraged by the authorities in the government. This job is accomplished by ignoring traditional land & resource rights, granting concessions for logging & export cash crop plantations on ancestral lands of shifting cultivators. The time required for regeneration of full biomass as reported by Uhl (1990) is shown in Fig. 2.2.

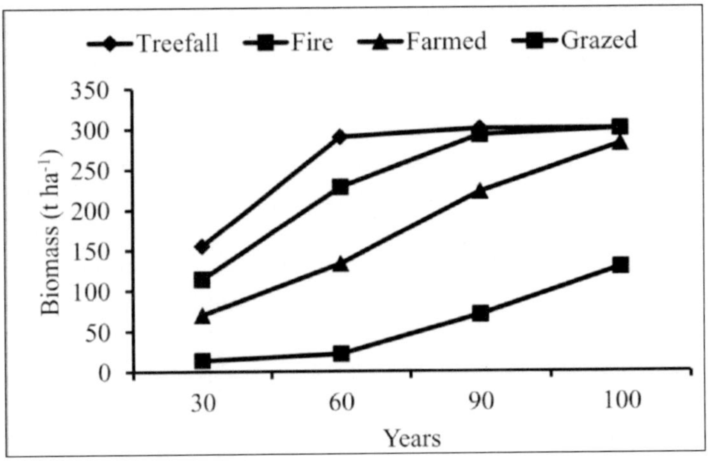

Fig. 2.2. Regeneration of biomass following disturbance in integral swidden
Source: Uhl (1990)

2.6 DEFINITIONS OF SHIFTING CULTIVATIONS

Shifting cultivation is a common name for agricultural operations where primary or fallow vegetation is cut, dried and burnt before planting/seeding. "Swidden" is an English word for a burning and clearing; and therefore, swidden farming is a synonym for shifting cultivation. This definition encompasses only those methods which have both, cropping and fallow phases. Such cases where crops are continuously grown, are ignored. Also such systems where light or heavy machinery is used for clearance and irrigated systems are expunged from the 'swidden'.

2.6.1 Defining shifting cultivation

Problems in defining

In the literal sense, the word 'shifting cultivation, is self-explainatory in understanding the characteristics of the practice. It may sound easy to understand the meaning, however, the complexity of operations involved in the system make it difficult to define the practice in a sentence or two. Slash-and-burn, swidden, and shifting cultivation all seem to describe the same thing, but the processes are generally different. In a slash-and-burn system, the biomass is slashed after clearing and burned, as opposed to a slash-and-mulch system, which is used in locations where there is no dry season and the biomass is left to degrade (Thurston 1997). The act of burning biomass in order to grow it is referred to as "swidden." It looks a lot like slash-and-burn, but it doesn't refer to the behaviours of "slashing" and "burning." However, it is rarely utilised in slash-and-burn systems with short fallow periods. Slash-and-burn, swidden, and shifting cultivation all seem to describe the same thing, but the processes are generally different. In a slash-and-burn system, the biomass is slashed after clearing and burned, as opposed to a slash-and-mulch system, which is used in locations where there is no dry season and the biomass is left to degrade (Thurston 1997). The act of burning biomass in order to grow it is referred to as "swidden." It looks a lot like slash-and-burn, but it doesn't refer to the behaviours of "slashing" and "burning." However, it is rarely utilised in slash-and-burn systems with short fallow periods. Shifting cultivation, on the other hand, is a practise in which fields are rotated based on the soil fertility and land availability. It includes both slash-and-burn and slash-and-mulch systems, but is rarely used in systems with short fallow, when cultivation is near to permanent, as it is for swidden cultivation. Many studies claim that after 2 to 3 years of cultivation, the land is abandoned and left idle in shifting cultivation, which is often not the case. In this case the abandoned land is allowed for re-growth of secondary vegetation, which eventually takes the shape of a forest, better than the primary forests. Not only in shifting cultivation areas, but in other areas also, the farmers burn the range grasses and bushes as their growth in subsequent years is better than the original.

For many shifting cultivators, Food production has always been and remained a top priority and major concern for many shifting cultivators. They see shifting cultivation as the only appropriate option on the steep hill slopes. In the changed scenario, increasing demands and needs are apparent, yet little attention has been paid to suitable research and development issues in the past. Additionally, in mountainous areas, the employment opportunities outside agriculture are very rare, and unless one moves out from home to plain and urban areas, leaving families behind, there is no mean to earn a livelihood. So, the practice of shifting cultivation emerged, spread, remained alive and sustained also. Though there has been increase in the population density and availability of land has reduced, the practice has remained alive over the years with little or no modifications and with different names and connotations, waiting for the appropriate alternatives. If followed with appropriate mix of traditional knowledge and modern technologies, the shifting cultivation does offer benefits to the shifting cultivator and society at large. The realization of this fact is important because many of the government efforts to wean away shifting cultivators from shifting cultivation and replace it with permanent cultivation have often caused increased poverty and degradation. The strong and determined bias against shifting cultivation is not deserving, and the current problems related to shifting cultivation are as a result of ineffectual policies and improper land use practices (Kerkhoff and Sharma 2006).

The term swidden makes it difficult to come up with a universally relevant and acceptable description due to the diversity of land and resource management. Although the terms swidden cultivation, shifting cultivation, and slash-and-burn agriculture are frequently used to describe the same activity, both orally and in writing, there appears to be little agreement on the term. The meanings are not similar. We also know that the "Swidden' word has been coined by Scandinavian people, meaning "land cleared by burning". At the same time there is one system in the Pacific Islands, which is without fire but, mulch-based systems suitably called shifting cultivation. On the other hand, Slash-and-burn is generally applied for a wide array of farming practices where no shifting of plots is done. Many definitions have been offered, published and debated during the last few decades. The satisfying aspect of the whole web is that we, at least, understand and can make-out what the term or definition want to convey. We may, however, draw our own conclusions as per our understanding or even want to understand according to our suitability. The FAO's call to eliminate shifting cultivation gave an appropriately negative definition: "Shifting cultivation is the custom of cultivating clearings scattered in the reservoir of natural vegetation (forest or grass-woodland) and of abandoning them as soon as the soil is exhausted" (FAO Staff 1957). Pelzer's (1945) definition that "shifting cultivation can be defined by the rotation of fields rather than crops, a short cropping period (1–3 years) succeeded by a long fallow period (5–20 years), and clearing by means of slash and burn" still imprison well the indigenous subsistence based systems, but ignore the reality that parts of the fallow areas are generally planted with useful tree crops for subsistence as well as cash needs. This area often put under the swidden cycle when trees are cut and upland rice planted again. Ruthenberg (1980), gave the most strict definition which mentions that if less than 33% of the land is cultivated and more than 67% of land is kept fallow, then the system is defined as 'shifting cultivation'.

A very elaborative division, including the difference between linear and cyclic-shift systems that largely match with Conklin's established categories was suggested by Spencer (1966). These distinctions, on the other hand, do not obscure the dynamic nature of swidden systems and the boundaries within which they are used. This shift can be seen in newer definitions. Rambo (1998) defines "composite swiddening" as a farming strategy that combines permanent wet rice plots with rotating swidden plots into a single household resource system. Swidden farming, according to some scientists, encompasses a wide range of agricultural practises in which the fallow phase is the primary means of maintaining soil fertility and crop productivity (Andriesse 1989; Brookfield *et al.* 1995), regardless of whether the fallow is fortified with productive species or not. In many regions, growers are 'assisting' these fallow features by employing fertilisers and pesticides, further complicating the situation, although such practices are not precluded from this broad description.

According to Fox *et al.* (2009), there are six major factors that have forced the change in swidden systems in Southeast Asia, including:

i) Categorising swidden cultivators as ethnically different and ancient;
ii) The swidden agroforestry system's absence from maps depicting woods and crops;
iii) In wooded areas where swidden is practised, government regulation and conservation activities;
iv) Relocation;
v) The transfer of ownership rights from government to private sector. and,
vi) The promotion for commercial agriculture.

According to Ziegler *et al.* (2009), swidden agricultural practises in Montane Southeast Asia are often relatively benign in their effects on the physical environment in their original form.

2.6.2 Defining small-scale farming systems

Small-scale farming entails cultivating crops according to the family's needs and requirements. Almost all of the food is consumed by the members of each family, and agriculture is a key source of income. Subsistence farming, on the other hand, means that farm products are produced purely to meet the needs of the family and are not sold in a market. However, some small farmers have been discovered selling their crops in local markets. Shifting agriculture is a method of small-scale farming that entails clearing ground, burning biomass, cultivating and harvesting crops, and then abandoning the field before relocating to a new location. Depending on the slope, fertility, and environment, small-scale farmers plant a variety of crops. In Latin America, women are increasingly participating in small-scale farming (World Bank 2007).

Because of the various variable determinants involved in shifting cultivation; such as, topography, agro-climate and institutional factors, it is not easy to undertake a systematic advance to define the practice. The practice has come up in forested hilly areas where plough agriculture is difficult to attempt. The plots are cleared of vegetation, burnt and cultivated for few years and left fallow for a longer period to allow the vegetation and soil to rejuvenate and revive the lost soil fertility. Shifting cultivation is not at anyway comparable to'slash-and-burn agriculture.' Slash-and-burn, according to Kerkhoff and Sharma (2006) is a land clearance strategy that is commonly employed for other agricultural and non-agricultural land use systems. The shifting cultivation practise has been referred to by a variety of names. According to Kelly (1975), around 60% of scholars use the term "shifting cultivation," whereas only about 15% use the term "swidden." About 48% of English names that have been used to describe shifting cultivation or swidden have been recorded (Spencer 1966). Cut-and-burn, fire-field, land-rotation, nomadic, slash-and-burn, and temporary farming are a few examples. The word "swidden" comes from a mediaeval English word that means "a clearing on a moor that has been cleaned by burning or still exhibits indications of burning" (Ekwall 1955). Despite the fact that an earlier word, "swithen," meant "to bum superficially," most swidden phrases indicate to ground that has been cleaned by burning.

2.6.3 Sequence of definitions

Conklin (1961) defined shifting cultivation as a perpetual agricultural system in which a plot is cropped for a relatively shorter period than it is fallowed. Such definitions are over simple and do not generally differentiate between the fallow and shifting systems. “Shifting cultivation is our cultural tradition, and it is sustainable. Shifting cultivation is the mother of agriculture. Swidden agriculture, is conventionally defined as “an agricultural system in which temporary clearings are cropped for fewer years than they are allowed to remain fallow” (Sanchez, 1976). Because the fallow time is long, well-managed, and often farmed with perennial plants rather than annuals, it is also referred to as forest farming. Swidden is a cyclic kind of agriculture that relies on natural vegetative processes to restore soil fertility while also limiting weed growth.

FAO (1982) defines shifting cultivation as an agricultural method characterised by relatively short periods of cultivation followed by relatively long periods of fallow. Shifting cultivation is described as "an economy" characterised by field rotation rather than crop rotation, fire clearing, the absence of draught animals and manuring, the sole use of human labour, the use of a dibbling stick or hoe, and short periods of soil occupancy interspersed with long fallow periods (Satapathyetal 2003). It includes horticulture and annual cropping, perennial tree crops, animal husbandry, and forest and fallows management in sequential or rotational cycles (Thruppetal, 1997). Shifting agriculture is type of farming in which a plot of land is cleared and cultivated for a brief period of time before being abandoned and left to return to its natural state while the farmer moves on to another plot. Slash-and-burn agriculture is a type of farming that involves cutting and burning plants in a forest or woodland to produce a swidden field. Cutting down trees and woody plants in an area is the first step in the procedure.

Integral long fallow swidden systems have gone down the memory lane. Integral swidden systems may still exist in a few remote parts of Amazonia and Southeast Asia, but they have perished elsewhere owing to social change, land scarcity, new market opportunities, deforestation, the emergence of huge plantations, and government laws. Emerging slash and burn is is not sustainable, spread weeds and often creates a wave of forest conversion. The difference between Integral swidden and slash and burn is:

Integral swidden	**Slash and burn**
Permanent or long-term settlers	Recent migrants
It has become a part of culture	Not percolated down to culture
Full of Indigenous Technical knowledge	Little or no ITK
Well regulated and properly managed	Mostly unregulated and poorly managed
Cultivation followed by long fallow	The fallow period is short or no fallow
Sustainable	The system is not sustainable
Land is not a constraint	Land is a constraint

The determinants and driving forces responsible for tropical deforestation and degradation have been debated and discussed for decades and, literally several books and articles have been published on the topic. Geist and Lambin (2002) provided one of the most considerate and valuable explanation. On the basis of a critical analysis of 152 empirical studies, they suggest that it is important to differentiate between imminent causes and basic driving forces, such as; the socioeconomic and/or political. Geist and Lambin (2002) concluded that there is no universal cause:effect relationships which could explain deforestation and/or degradation and that causes and driving forces are highly spatial and variable.

Mertz *et al.* (2009) were of the view that there are no commonly recognized and useful definitions of swidden cultivation, shifting cultivation and slash-and-burn agriculture in the scientific arena and that all these terminologies are generally mixed up. On the other hand, Rambo (2010) questioned the motive behind using these terms interchangeably. Between swidden agriculture and shifting cultivation, there appears to be a lot of disagreement (Inoue 2000). The word swidden was coined by Swedish anthropologist Izikovitz in 1951 (Russel 1988) to describe burned woody vegetation areas (Therik 1999). "Swidden was not invented by Karl Gustav Izikowitz (1903-1985). The Swedish anthropologist concentrated on economic issues during his fieldwork in northern Laos. Izikowitz recalled a similar tactic used in his home country called svedja, which means "to burn a field," during his return to Sweden in 1938. Shifting agriculture encompasses the complete system of swidden plots and fallow vegetation, as well as agricultural operations where fields are planted for rice and other crops before being left fallow (Therik 1999, Vongvisouk 2014). (Vongvisouk 2014). Even with relatively established towns, shifting farming was often performed by shifting or semi-nomadic people for impermanent fields, according to Inoue (Inoue 2000). Swidden agriculture, on the other hand, is defined by Conklin (1957) as any farming technique in which fields are burned and farmed for a shorter amount of time or for a longer period of time than the subsequent fallow.

2.7 SUBSISTENCE FARMING DEFINITION

Subsistence agriculture is a type of farming in which practically all of the field crops grown or domestic animals kept are needed for the farmer's and his family's survival, leaving little or no surplus material for sale or barter.

2.7.1 Characteristics of subsistence farming

The main characteristics of this traditional farming are; (i) generally small traditional farms, the size of which depends on the availability of land and/or need or producing enough food for user and his family. The produce of these small farming units is used mainly for consumption of the family., (ii) The level of methodologies mostly used are simple and hardly productive, (iii) In general, family labour is the source of power and energy. The members of the family may add to their income by working outside the farm part, (iv) the subsistence farming is characterized by low use of inputs which are arranged from the farm itself, like farm wastes, growing of leguminous crops, addition of compost etc., though this results in poor yields. (v) The livestock play a dominant role in subsistence farming system. (vi) the income and standard of living of the subsistence farmers is mostly below poverty line, (vii) decision making power of subsistence farmers is limited, (viii) animal manure is important source of nutrition, (ix) There is wastage of resources and the component of uncertainty is very high.

There are two types of subsistence farming:

1. Primitive Subsistence Farming
2. Intensive Subsistence Farming

2.7.2 Primitive or ancient subsistence farming

Primitive farming is the ancient way of farming and continue to be practiced in some isolated pockets of the globe. From ancient gathering from forests and other areas, some farmers have learnt the ingenuity of habituation of plants and their economy appears to be looking up. This style of agriculture is self-sufficient, and growers exclusively grow food for themselves and their family. However, it has been seen that some minor surplus produce is either bartered or sold for cash to other poor families at the market. Despite the fact that there is little hope for significant development, there is a high level of rural self-reliance and independence because farmers are not governed by landlords. Under humid tropical circumstances, primitive farming is the most ideal technique of agricultural cultivation, and many tribes, particularly in Africa, South and Central America, and Southeast Asia, rely on it. This agricultural method is appropriately referred to as "shifting agriculture."

2.7.3 Intensive subsistence agriculture

The term, ‘intensive subsistence farming’ is employed to highlight a system of farming which generates more produce per unit of land but comparatively or analogously less produce for each worker. This system of farming has undergone a complete change at many locations as it is no more subsistence at present. Despite these changes, the phrase "intense subsistence" remains popular to describe agricultural systems that are more polished than traditional farming. It is also known as'monsoon style farming' in some areas and on occasion. This style of agriculture is nearly only found in Asia's monsoonal regions. China, Japan, Korea, India, Pakistan, Sri Lanka, the majority of continental Southeast Asia, and sections of insular Southeast Asia are all home to this species. The main goal of this type of agriculture is to make farming as intensive as possible in both the wet lowlands and the terraced uplands in order to maintain a large population. The major features of the intensive subsistence farming are; relatively small holdings, intensive farming, manual labour, use of animal and plant manure and rice as the dominance crop for food production. The land plots have been fragmented many times through many generations so it have become small and uneconomic.

Characteristics

Small holdings, intensive farming, physical labour, the use of animal and plant manure, and rice as the dominant crop for food production are the key characteristics of intensive subsistence agriculture. The land plots have been fragmented numerous times over generations, resulting in incredibly small areas that are sometimes uneconomic to maintain. In Japan, an average farm is about 0.6 hectares, and farms in India and other parts of Asia may be even smaller. Individual peasants raise crops mostly to feed their own families, however some surplus is available for sale in some locations. In China, however, major agricultural developments occurred during the agrarian revolution of 1949, when small farms were merged into vast plots under communist administration.

Cultivators in Monsoon Asia are so limited on land that they use even little plots of ground to grow their crops. Only thin, handcrafted ridges and walkways separate the fields, allowing the farmers to move about their farms. Draining marshy regions and terracing hill slopes to create flat areas suited for paddy cultivation provide additional land for cultivation. Only the most barren places and the highest slopes are left uncultivated. Farming is so intensive that it is common to practise double- or triple-cropping, which means growing multiple crops on the same area over the course of a year. When just one crop of paddy can be grown, the fields are usually used to grow other food or cash crops during the dry season. Wet rice cultivation, known as 'Panikheti' in Northeast India, has traditionally required a lot of manual labour. The paddy is planted in precise rows by the ladies, the fields are raked by hand, harvesting is done with sickles, and threshing is done by hand. The tools utilised are extremely basic. Of course, machinery has evolved in recent years that can work in flooded fields, and distinct machines can plough, plant, and harvest paddy. These machines are widely utilised in more affluent Japan and are gradually spreading across Monsoon Asia.

2.8 SIMPLE DEFINITION OF SHIFTING CULTIVATION

There are innumerable forms of shifting cultivation at global level. The practice is associated with number of operations varying from place to place, region to region, country to country and continent to continent. It is difficult to offer any perfect definition of shifting cultivation encompassing all its associated operations. In some cases, in India, the farmers keep the fields fallow for 1-2 years to enhance the soil fertility through natural means and then, reverting back to these fields after above mentioned period to have a good harvest. So, on the basis of our experience with the practice and making it simpler, we define shifting cultivation as "a traditional land use in which a cultivator shifts from one field to another, when the first field is unable to support optimum crop productivity due to exhaustion of plant nutrients and escalation of weeds, and the soil fertility is restored in the next field through burning of regenerated in-situ vegetation or that of a primary forest". A simple form of shifting cultivation and its sustainability is shown in Fig. 2.3

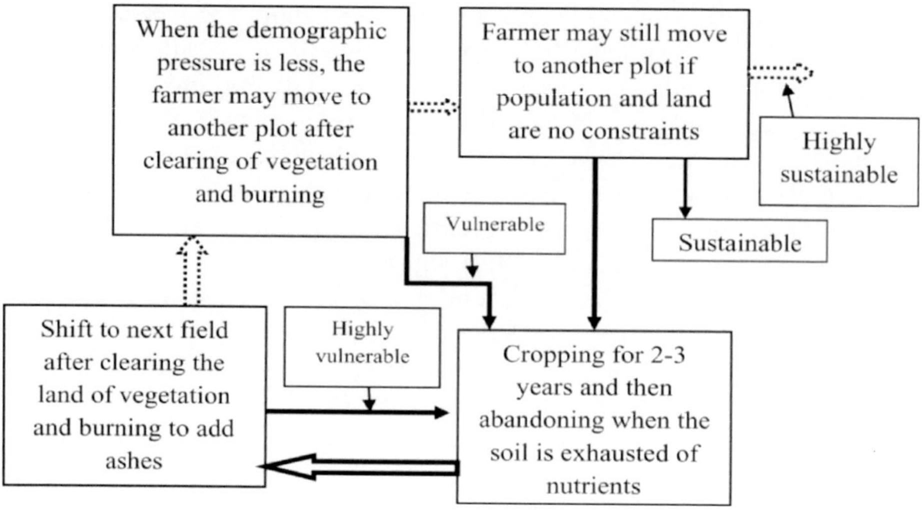

Fig. 2.3. Showing the period of shifting from one plot to another and sustainability of the system

References

Andriesse, J. P. (1989). Nutrient management through shifting cultivation. In van der Heide, J. (ed.), Nutrient Management for Food Crop Production in Tropical Farming Systems. Institute for Soil Fertility, Haren, pp. 29–62.

Aweto, A. O. 2013. Shifting cultivation and secondary succession in the tropics. Wallingford, CABI.

Boserup E. 1965. The Conditions of Agricultural Growth: The Economics of Agrarian Change under Population Pressure. Earthscan: London.

Brookfield H, Potter L, Byron Y. 1995. Place of the Forest: Environmental and Socio-Economic Transformation in Borneo and the Eastern Malay Peninsula. United Nations University Press: Tokyo.

Cairns, M. F. (Ed.). (2014). Shifting cultivation and environmental change: Indigenous people, agriculture and forest conservation. London: Routledge.

Conklin, H.C. 1957. Hanunoo Agriculture: A report on an integral system of shifting cultivation in the Philippines. FAO, Rome.

Conklin. H C (1961). 'The Study of Shifting Cultivation'. Curren/Anthropology. Vol. 2. p. 27-61.

da Fonseca, G.A.B., C.M. Rodriguez, G. Midgley, J. Busch, L. Hannah, and R.A. Mittermeier. 2007. No forest left behind. PLoS Biology 5: 1645-1646.

DeFries, R., Rosenzweig, C., 2010. Toward a whole-landscape approach for sustainable land use in the tropics. PNAS 107, 19627–19632.

Dove, M. (1983). Theories of Swidden Agriculture, and the Political Economy of Ignorance. Agroforestry Systems, 1(2): 85-99.

Ekwall, E. 1955. 'Slash-and-burn' Cultivation: A Contribution to Anthropological Terminology. Man 55: 135-136.

Ellis EC and Wang SM (1997). Sustainable traditional agriculture in the Tai Lake Region of China. Agriculture Ecosystems and Environment 61:177–193.

FAO, 1982. Forest Resources of Tropical Asia. Rome, Italy.

FAO Staff (1957). Shifting cultivation. Unasylva 11: 9–11.

Fisher, B. 2010. African exception to drivers of deforestation. Nature Geoscience 3: 375-376.

Fox J, Fujita Y, Ngidang D, Peluso N, Potter L, Sakuntaladewi N, Sturgeon J and Thomas D, 2009. Policies, Political-Economy, and Swidden in Southeast Asia. Human Ecology. 37:305–322, DOI 10.1007/s10745-009-9240-7

Fuller J (2010) Refining Virtual Co-Creation from a Consumer Perspective. California Management Review, 52: 98-122, DOI:10.1525/cmr.2010.52.2.98

Fuller DQ, van Etten J, Manning K and Castilo CC (2011) The contribution of rice agriculture and livestock pastoralism to prehistoric methane levels: An archaeological assessment. The Holocene 21(5):743-759 DOI:10.1177/0959683611398052.

Geist, H, and E. Lambin. 2002. Proximate causes and underlying driving forces of tropical deforestation. BioScience 52: 143-150.

Gerold G., Fremerey M. and Guhardja E. (eds, 2004). Land use, nature conservation and the stability of rainforest margins in Southeast Asia. Springer, Berlin.

Hansen, M.C., S.V. Stehman, P.V. Potapov, T.R. Loveland, J.R.G. Townshend, R.S. DeFries, K.W. Pittman, B. Arunarwati, F. Stolle, M.K. Steininger, M. Carroll, and C. DiMiceli. 2008. Humid tropical forest clearing from 2000 to 2005 quantified by using multitemporal and multiresolution remotely sensed data. Proceedings of the National Academy of Sciences 105: 9439-9444.

Inoue, M. 2000 Mechanism of Changes in the Kenyah's Swidden System: Explanation in Terms of Zensification Theory. In Rainforest Ecosystems of East Kalimantan—El Niño, Drought, Fire and Human Impacts; Guhardja, E., Fatawi, M., Sutisna, M., Mori, T., Ohta, S., Eds.; Springer-Verlag: Tokyo, Japan pp 167–184

IPCC, Summary for Policymakers, Fourth Assessment Report (AR4). (2007). New York, Cambridge University Press

Karl Gustav I. (1903-1985). Swidden definition and meaning. National Library of Sweden, National Library of the Czech Republic.

Kelly L. (1975). The role of crop yields in shifting cultivation. Ph.D. dissertation, University of Kansas, Lawrence, Kansas.

Kerkhoff E and Sharma E, 2006. Debating Shifting Cultivation in the Eastern Himalayas Farmers' Innovations as Lessons for Policy, International Centre for Integrated Mountain Development (ICIMOD), Kathmandu, Nepal

Lambin, E. F., Turner, B. L., Geist, H. J., Agbola, S. B., Angelsen, A., Bruce, J.W., (…), and Xu, J. (2001). The Causes of Land-Use and Land-Cover Change: Moving Beyond the Myths. Global Environmental Change 11: 261–269.

Laporte, N.T., J.A. Stabach, R. Grosch, T.S. Lin, and S.J. Goetz. 2007. Expansion of industrial logging in Central Africa. Science 316: 1451.

Mazoyer M. and Roudart L. (2005), A History of world agriculture: from the Neolithic age to current crisis. Monthly Review Press, New York City, NY.

Mertz, O., Padoch, C., Fox, J., Cramb, R.A., Leisz, S.J., Nguyen, T.L., & Vien, T.D. (2009). Swidden change in Southeast Asia: Understanding causes and consequences. Human Ecology, 37, 259–264.

Mertz, O., Wadley, R. L., Nielsen, U., Bruun, T. B., Colfer, C. J. P., de Neergaard, A., Jepsen,M. R., Martinussen, T., Zhao, Q., Noweg, G. T., and Magid, J. (2008). A Fresh Look at Shifting Cultivation: Fallow Length an Uncertain Indicator of Productivity. Agricultural Systems 96: 75–84. doi:10.1016/j.agsy.2007.06.002.

Meyfroidt, P.; Vu, T.P.; Hoang, V.A. (2013) Trajectories of deforestation, co_ee expansion and displacement of shifting cultivation in the Central Highlands of Vietnam. Glob. Environ. Chang, 23, 1187–1198.

Nye. PH and Greenland DJ (1960). The Soil under Shifting Cultivation. Technical communication no. 51. Farnham Royal. England: Commonwealth Agricultural -Bureaux.

Ohmagari, K. & Berkes, F. (1997). Transmission of indigenous knowledge and bush skills among the Western James Bay Cree women of subarctic Canada. Human Ecology 25: 197-222

Palm C. A., Vosti S. A., Sanchez P. A. and Ericksen P. J. (eds, 2005), Slash-and-burn agriculture: The search for alternatives. Columbia University Press, New York City, NY.

Pelzer, K. J. (1945). Pioneer Settlement in the Asiatic Tropics. American Geographic Society, New York.

Peters, J.W and Neuenschwander, L.F. (1988). Slash and Burn: Farming in the Third World Forest. University of Idaho Press. USA.

Pollini, J. 2007. Slash-and-burn cultivation and deforestation in the Malagasy rain forests: Representations and realities. PhD dissertation. Ithaca, Cornell University. http://www.scribd.com/doc/26790856/Deforestation-in-Madagascar

Ramakrishnan, P. S. 1992. Shifting agriculture and sustainable development: An interdisciplinary study from north-eastern India. Carnforth, Parthenon Publishing Group and UNESCO.

Rambo, A. T. (1998). The composite swiddening agroecosystem of the Tay Ethnic Minority of the Northwestern Mountains of Vietnam. In Patanothai, A. (ed.), Land Degradation and Agricultural Sustainability: Case Studies from Southeast and East Asia, Regional Secretariat, The Southeast Asian Universities Agroecosystem Network (SUAN). Khon Kaen University, Khon Kaen, pp. 43–64.

Rambo, A.T. (2010). Shifting agriculture in Asia: Implications for environmental conservation and sustainable livelihood. Mt. Res. Dev., 30, 56–57.

Robert M. Netting, (1986) Cultural Ecology (2nd Ed.). Waveland Press, Prospect Heights, IL.

Roder, W. (2001). Slash-and-Burn Rice Systems in the Hills of Northern Lao PDR: Description, Challenges, and Opportunities. International Rice Research Institute, Los Banos.

Russell, W. M. S. 1988. Population, swidden farming and the tropical environment. Population and Environment 10(2): 77- 94.

Ruthenberg, H. (1980). Farming Systems in the Tropics, Clarendon Press, Oxford.

Sanchez, P. A. 1976. Properties and management of soils in the tropics. John Wiley and Sons, New York, New York, USA.

Sanchez, P. A., Palm, C. A., Vosti, S. A., Tomich, T., and Kasyoki, J. (2005). Alternatives to slash and burn. Challenges and approaches of an international consortium. In Palm, C. A., Vosti, S. A., Sanchez, P. A., and Ericksen, P. J. (Eds.), Slash-and-Burn Agriculture. The Search for Alternatives. Columbia University Press, New York, pp. 3–37.

Satapathy KK, (2005). "Runoff production on hill slopes under different land use systems," in Agroforestry in North East India: Opportunities and Challenges, B. P. Bhatt and K. M. Bujarbaruah, Eds., pp. 451–459, ICAR Research Complex for NEH Region, Umiam, Meghalaya,

Schlippe PD, 1956. Shifting cultivation in Africa, Routledge and Kegan Paul, London.

Singh KB, Savant PV (2000) Social forestry for rural development in Mizoram. Linkmen Publications, Aizawl, p 57

Spencer, J.E. (1966). Shifting cultivation in South-eastern Asia. Publication in Geography, No 199, University of California Press, USA.

Therik, T. (1999). The Role of Fire in Swidden Cultivation: A Timor Case Study. In Fire and Sustainable Agricultural and Forestry Development in Indonesia and northern Australia; Russell-Smith, J., Hill, G.J.E., Djoeroemana, S., Myers, B.A., Eds.; Australian Centre for International Agricultural Research (ACIAR): Canberra, Australia, 1999; pp. 77–79.

Thrupp, L.A., Hecht, S.B. & Browder, J.O. (1997): The Diversity and Dynamics of Shifting Cultivation: Myths, Realities, and Policy Implications. World Resources Institute, Washington.

Thurston, H. D. 1997. Slash/Mulch systems: Sustainable methods for tropical agriculture. Boulder, Westview Press.

Tilman, D., Balzer, C., Hill, J., and Befort, B. (2011). Global food demand and the sustainable intensification of agriculture. Proceedings of the National Academy of Sciences.

Uhl, C., *et al.* 1990. Studies of ecosystem response to natural and anthropogenic disturbances provide guidelines for designing sustainable land-use systems in Amazonia. In: A. Anderson (ed). Alternatives to Deforestation. Columbia Univ. Press, NY. pp. 24-42.

Vongvisouk, T., Mertz, O., Thongmanivong, S., Heinimann, A., Phanvilay, K., 2014. Shifting cultivation stability and change: contrasting pathways of land use and livelihood change in Laos. Appl. Geogr. 46, 1–10.

Watters RF (1971). Shifting Cultivation in Latin America. FAO Forestry Report Development Paper No. 17, Rome.

World Bank (2007) Sustaining Indonesia's Forests: Strategy for the World Bank, 2006–2009, World Bank Office, Jakarta.

Ziegler AD, Bruun TB, Guardiola-Claramonte M, Giambelluca TW, Lawrence D, Lam NT (2009). Environmental Consequences of the Demise in Swidden Cultivation in Montane Mainland Southeast Asia: Hydrology and Geomorphology. Hum. Ecol., 37: 361–373, DOI 10.1007/s10745-009-9258-x.

3

Dimensions of Shifting Cultivation

3.1 GENERAL DESCRIPTION

Indigenous communities depending mainly on hunting, collection of food from the forests and managing natural resources to meet their requirements, have lived intimately with nature. Shifting cultivation has helped these communities to earn their livelihood for centuries to become to be food secure. Besides being a source of subsistence, the shifting cultivation has also been a repository of agro-biodiversity. Not considering this, the authorities have tried to impose the practice of intensive agriculture on indigenous people for conversion to mainstream agriculture (Singh *et al.* 2016). The way of life of indigenous people has been grossly impacted by the stress of market oriented development, government schemes, and interventions by some NGOs.

The 'dimensions of shifting cultivation' not only includes the area under shifting cultivation and the number of shifting cultivators in the world but, various forms, shapes, methods and other aspects of the practice on spatial and temporal basis as well (Fig. 3.1).

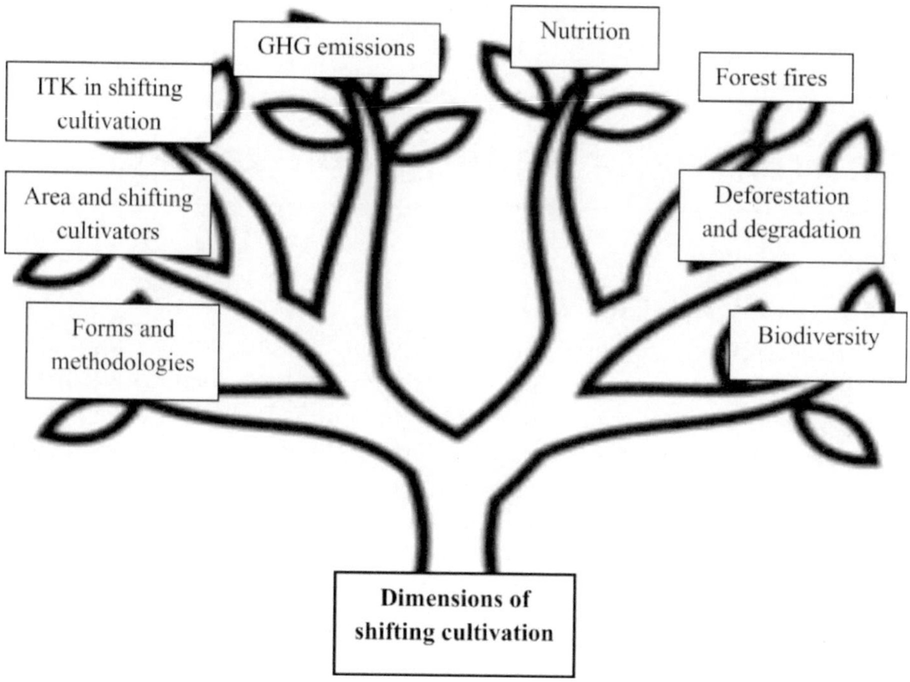

Fig. 3.1. Showing range of dimensions of shifting cultivation.

Particularly from the last quarter of the 20th century, the tribal people are being constantly advised to observe globalized market economy, leaving aside their traditional subsistence livelihoods practices. This may negatively affect their cultural ethos and thus, compromising their traditional way of life, so close to their hearts. Particularly Vulnerable Tribal Groups (PVTGs) or tribal communities are facing misfortune and revamping due to ecosystem unsteadiness, transitional and socio-economic oppression, and incongruous government policies. For a long time, these tribal tribes relied on shifting cultivation, forest product collection, hunting and gathering, animal husbandry, and, more recently, wage labour to support their way of life. These activities have become entwined with their social and cultural practises, rituals, and festivities over such a long period of time. Shifting cultivation has been a key source of income for many hill people around the world, but it has also been a focus of environmentalists' attention due to its'slash and burn' component, which is thought to be contributing to global warming (Biswal & Kumar 2013). To counter the objections raised by the environmentalists, indigenous knowledge researchers like Dove (2006) suggest that shifting cultivators have classic techniques of knowledge that offers more successful development interventions. This traditional knowledge of the hill tribes has been well recognized by the United Nations as is evident from the principle 22 of the 1992 Rio Declaration.

Shifting cultivation has been noticed as a road block in the development of indigenous hill people in the tropics for a long period. Continuance of this practice, its abandoning or improvement has remained as one of the unresolved debatable issues, and it has hampered the progress of tribal societies and other people in the tropics where this practice is the main stay of life. Rosamann (2010) mentioned that the historical evidences indicate that human population multiplied constantly due to livestock rearing and cultivating land to produce food, fiber, and shelter. In indigenous tribes, there was a noticeable shift from food collection to food production. Shifting cultivation was supposed to have begun between the years 1300 and 3000 B.C. (Ninan, 1992), however Sharma (1976) suggested that it began approximately 7000 B.C., during the Neolithic Period.

Shifting cultivation is done in various areas of the world across different cultures across around 36 million km2 of land, or about 30% of the world's exploitable land, and is considered an early stage of agricultural evolution. More than 250 million people throughout the world rely on shifting cultivation for a significant portion of their sustenance. Shifting cultivation is not only practiced in south Asian and Southeastern countries, but it is widely persistent among the indigenous communities, particularly in African, Latin American and North European countries. In northeast India, this practice is the major form of agriculture which has sustained for many centuries and still the common method of cultivation (FAO, 2014). Shifting cultivation is a rotational land use which includes agricultural crops, horticulture, livestock rearing, and forest and fallow management. The Land preparation is done by clearing of plots by cutting trees and bushes and burning the vegetation. This farming system is called "swidden" agriculture in many regions of Europe (Pelzer 1978). There is a great diversity in practices in different parts of the globe with which this land use system is operated. Shifting cultivation is practiced in areas where high-tech systems of agriculture have not been possible due to socio-economic and land configuration. The steadiness and dependability of the practice lies in keeping a minimum length of the fallow period before the cultivator comes back to the same plot to recultivate it (Darlong and Bhattacharyya 2014). Constituting highly diverse field operations that vary from highlands to lowland, and from tropical forests to grasslands ecosystems (Spencer 1966; Sharma 1976).

It is difficult to stop shifting cultivation completely as the; cultural ethos, belief, attitude and self-image of the tribal people are linked to different operations of the practice. Stopping the practise of shifting cultivation is critical to safeguarding the ecosystem and forest resources. It is necessary to ensure that new laws and regulations are socially acceptable, commercially feasible, and environmentally sustainable when they are enacted. Shifting Cultivation is a means of subsistence, thus it can't be readily abandoned. There have been endless frequent debates and the issue remains as it used to be half a century ago. The practice is on the decline and, perhaps, we may have to keep patience till it reaches its natural death or till some appropriate solutions are worked-out. The latter is an uphill complex task due to variability of methods followed across the globe in doing this practice and socio-economic and ecological implications involved. Shifting cultivation directly affects the socio-cultural life of the cultivators and many rituals and festivals are directly linked to the practice. Leaving the shifting cultivation for indigenous societies is just like snatching away their moments of happiness, celebrations and festivitie, and adding dullness and solitude to their lives. Bunding, terraces, agroforestry, residue cycling, integrated soil and water conservation measures and high yielding varieties, etc. are alternatives for improving shifting cultivation. This would would help in natural resources conservation, enhance soil fertility and will ensure sustainable production. Good sites for shifting cultivation in wooded environments are usually picked from secondary and, on occasion, primary forest (Peters and Neuenschwander, 1988).

Shifting cultivators make use of the land throughout the fallow time. During the fallow time, shifting cultivators employ various vegetation species for fencing, construction, firewood, thatching, ropes, clothes, tools, carrying devices, and medicines. Interestingly, several of these species fix nitrogen from the atmosphere. The trees during the fallow period help in arrestin soil erosion as well as draw nutrients to the soil surface from depths of soil profile. The high ratio of the period land is fallowed and cultivated is important for stability of shifting cultivation systems. During the cropping period, there is decrease in SOM, cation-exchange-capacity, N and P and, soil porosity and infiltration capacity is reduced. Fallow period is helpful in recovering these losses to make the land as productive as it was before the cropping or even better than that stage.

Friedman (1979, 1982) mentioned that if social mechanisms have the aptitude to increase in complication, they will come into "contradiction" with their environments. What comes out of this "contradiction" will adjudicate the the nature and limit of the environmental degradation. Ability of the society to change is important to innovate technologically and sociologically, to overcome the "contradiction" without causing environmental degradation or social disintegration. Transitions in ecological and social dispositions on the other hand, do not occur at the same rate. Natural selection and, to a lesser extent, human interventions and adaptability determine the rate of phylogenetic evolution. Humans can learn and convey their knowledge to one another, as well as pass it down from generation to generation. The difficulties in mapping shifting agriculture habitats and the lack of approaches to discriminate between shifting cultivation, settled cultivation, and forests severely limit attempts to comprehend shifting cultivation ecosystems. There are some fundamental concerns to consider while defining shifting cultivation, establishing an acceptable mapping framework, and selecting the methodical planning that follows (Kurien *et al.* 2019).

3.1.1 Shifting cultivation system

Shifting agriculture is thought to have begun around 7000 BC. Shifting agriculture is more common in India's North Eastern states than in any other region a region known locally as Jhum. This form of cultivation is the significant source of the staple food for the local communities living in the remote areas of the tropics. The GEF-UNDP-Government of Nagaland Project "Sustainable Land and Ecosystem Management in Shifting Cultivation Areas of Nagaland for Ecological and Livelihood Security" was launched in 2009 to address issues in the state. The initiative is focused on the state's three districts of Mon, Mokokchung, and Wokha (Jayahari and Sen 2015). The project's goal is to create and demonstrate sustainable land management strategies using an ecosystem approach to solve the issue of shifting cropping. The project activities can be broadly classified as a) measures to improve the economic benefit from Jhum cultivation by increasing productivity as well as improving market linkages, b) improve Jhum productivity through soil and water conservation measures, c) improve soil fertility during the fallow period so that the land will be more fertile during the next cultivation season and d) provide the community alternate livelihood through Integrated Farm Development and Self Help Group based activities.

Originating during 7000 BC and surviving for more than 9000 years, shifting cultivation has blended with the tradition and culture of the people traditionally practicing it (Maithani 2005). An enormous amount of work has been carried out by national and regional governments and other conservation agencies for halting conventional shifting cultivation across the world but the efforts have instead contributed to deforestation and decrease in carbon storage (Geertz 1963, Fox 2000). Most of these policies were counterproductive and therefore new, more effective and socially acceptable policy options that help to improve shifting cultivation are needed, instead of replacing the practice of shifting cultivation (Kerkhoff and Sharma 2006). The Shillong Declaration (2004) recommends that The policy makers should review the policies in place and make suitable amendments to these policies so as to implement only the beneficial policies for overall improvement of shifting cultivation (Shillong declaration 2004).

Shifting agriculture is in transition across the world. Primitive land use aggravated by destitution and lack of optimum technical knowledge is the main reason for the continuation of unsustainable practice. Increase in population with meager availability of land, poor education, people's non-participation in policy planning and implementation influence farmers' decision to continue shifting cultivation. Shifting cultivators have the right to practise shifting cultivation and it is for the government to support to make it generative and profitable (Panda *et al.* 2016).

Shifting cultivativation, despite constraints, has been practiced effectively for centuries, fully adapted to the specific meteorological situations across the regions of the world. Population increase, increase in area under cultivation, partly due to introduction of cash crops, restricted access to land and globalization have brought social change in Africa. Despite of human impact on Africa's tropical forests, the significant question of whether or not the practice is actually unsustainable today still remains (Ickowitz, 2004).

3.1.2 Labour input

The household people generally comprise the working force in shifting cultivation fields. Weed control in uplands is the major problem faced by the shiting cultivators as the rainfall is usually high, giving rise to weed proliferation. Weeding consume about for 50-60% of total crop labour input, followed distantly by slashing (11%), harvesting/thrashing (11%) and planting (10%). Weeding is usually required 3-4 times, with an average input of 45-55 labour days ha^{-1}. The majority of the heavy lifting is done by men, while weeding and harvesting is usually done by women and children.

3.2 HISTORICAL

In grasslands and forests, the shifting cultivation has been practiced throughout the planet Earth. It is known that during the Neolithic insurrection, there were agricultural advancements and some hunter-gatherers tamed different plants and animals, allowing them to settle and practice agriculture, which ensured more food security to indigenous communities than hunting and gathering, as experienced in the river valleys of Egypt and Mesopotamia. Some indigenous communities were able to plant their crops in open plots near the river valley. Since there were vast areas under forest cover, communities used the method of clearing the land of vegetation and burning it to get rid of this huge vegetative mass easily to make the land ready for planting crops. So, the practice of slash and burn started for converting forests into plots for growing crops. Fire is an important component, used as a tool for clearance by hunter-gatherers up to present times. Similat method was used by old American civilizations, like the Inca, Maya and Aztecs. Many-a-times, the flames from the burning vegetation spread and become a potential cause of forest fires, leading to human and animal loss. In Sweden the shifting cultivation is called svedjebruk. With shifting cultivation, the evergreen and deciduous forests supported by Southern European Mediterranean climates, were able to regenerate slowly than those north of the Alps. The difference in northern Europe and southern Europe was that in former case only one crop was usually taken and then grass was permitted to grow, whereas; in latter case, the land was usually cultivated for many years until soil was exhausted. The forests remained intact for a long time, however, there was drastic reduction in forest cover with new settlements, which moved-in regularly. The causes behind for this arrangement of settlements and transition to permanent agriculture from the late Viking period on, or from shifting cultivation to sedentary agriculture are not well established. During earlier period, the farming communities favoured good forests on hillocks with good relief. Shifting cultivation was a thing of the past in Italy by the common era. After the Roman Empire and before the Viking Age, during the migration period in Europe, the peoples of Central Europe moved to new forests after the soils exhausted of fertility. Sometimes, the deforestation was done to burn vegetation to create pastures. Stewart (1956) mentioned that less availability of timber led to its cost escalation, resulting in more stone construction in the Roman Empire. The forests have survived in the Nordic countries though, these slowly but steadily decreased in northern Europe. The tribal population, before Roman Empire in Italy, ostensibly lived in temporary settlements. They used to cultivate small plots of land and rear sheep and cattle and sometime, fought battles and developed their identities as settlers and warriors around 900 BCE (Before Comon Era).

The shifting cultivation and agriculture itself are almost at par in their evolution. The origin of shifting cultivation can be traced back to about 8000 BC in the Neolithic period as evidenced by archeology and radio-carbon dating. The practice viewed the noticeable and drastic change in human's way of food production from hunting and gathering to systematic food production. While in the prehistoric period shifting cultivators used fire-stone, axes and hoes, the present shifting cultivators use iron dribbling stick, hand hoe, dao and knives in place of the stone tools.

3.2.1 What are tropics

Shifting cultivation is the most extensive traditional land use system in the tropics. The Tropics are a region of the Earth on northern and southern side of the Equator, and lying between 23.43663°N and 23.43663° S. It accounts for 38 per cent of the Earth's total surface area and are host to approximately 80 per cent of the world's biodiversity. The sunlight is usually intense in the tropics but the variations in climate are not significant. The main features of the tropical environment have been appropriately described by Gourou (1958), Langsberg, *et al.* (1963), Hare (1973), Sanchez (1976) and Walters (1977). With urbanization and industrialization as a result of population growth, the tropical region is confronted with many challenges such as climate change, deforestation, migration to urban areas, and population pressure. Tropical areas are facing more some of these challenges than the rest of the world. In a period of 38 years (between 1980 and 2018), the percentage of people living in tropical cities increased from 30 to 47, as compared to percent increase in world's urbanization from 39 to 55. More people live in rural than in urban areas in the tropics. About 85 per cent of the extreme poor live in tropical countries. During the year 2018, about 670 million people lived in extreme poverty in the tropics. Though having 54% of the planet's renewable water resources, yet about half of population in the tropics is vulnerable to water stress. As per the 'State of the Tropics Reports (2014, 2020) the emissions of CO_2 have been increasing at a greater rate in the tropics compared to rest of the world, although the tropics continues to be a much lower emitter of CO_2 (20%) both in total and per capita emission.

3.2.2 Land area and population in the tropics

The spread of probable cultivable lands in the tropics is presented in Table. 3.1 Tropics are highly remote areas having equitably high temperature all the year round. Annual rainfall varies from under 100 mm in the inland desert areas at the higher latitudes to over 4000 mm in the most humid parts of the tropics.

Table. 3.1. Land use (million ha) and population (million) patterns in tropical zone of different continents of the world

Continent	Total area	Cultivated area	Pastures and meadows	Population	Per capita cultivated area (ha)
Americas	1683	81	272	492	0.164
Asia and Pacific	931	286	18	2265	0.126
Africa	1992	187	523	983	0.196
Total	4606	554	813	3740	0.148

3.3 STAGES OF SHIFTING CULTIVATION

Shifting cultivation typically has three basic stages: (i) removing the old vegetation or clearing; (ii) burning of the vegetation; (iii) and cropping. The concept of fallow management is, generally, lacking in India.

3.3.1 Clearing

Clearing of the plot intended to be used for cultivation starts usually with the beginning of the dry season after monsoon, since many plants dry slowly and have to be cut long before they can be burnt. Clearing comprise two main stages, such as; (i) slashing of tree brnches and cutting of grasses and bushes completely; and (ii) felling of large trees which takes time. Sometimes important trees like medicinal, fodder trees are left-out. After trimming and pollarding, the material is spread uniformly in the plot and left to dry before burning. The size of the land selected for clearing varies, depending on the number of people to be supported, soil fertility, and the size and energy of the labour force.

3.3.2 Burning

The quickness of burning depends on the cut-size and kind of vegetation debris, and the weather. If the material is dry, it may be burnt immediately after cutting. The best time for burning depends upon the dryness of the debris. Generally four to five weeks time is sufficient for drying. If the field remains unburnt for a longer period, weeds and shoots come-up, which lessen the chances of a complete burn.

3.3.3 Cropping

If the physical, chemical and biological properties of the soil are favourable after burning, the seeds are broadcasted on the top soil or put into holes made with a digging stick or hoe. Raised beds are constructed if roots are planted as a first crop. The cropping phase is the most demanding and shifting cultivators are very careful in sowing or planting their crops as a successful crop only would ensure their livelihood and food security. Many crops are cultivated simultaneously or in an overlapping method. Mixed cropping is the most common practice, highlighting the spatial and temporal relations for getting the produce throughout the year.

3.3.4 Fallowing

Fallowing is a period in which the land is left unattended for natural regeneration. Such lands are known as abandoned fallow land. The regeration of fallow vegetation is important for storing nutrients drawn from various soil depths depending on the root system. A substantial portion of these nutrients is accumulated in the vegetation and a part returned to the soil surface in the shape of leaf trash and rain-wash. The nutrients stored in the vegetation during the fallow phase, return to the soil after the vegetation is burnt. The sustainability of a shifting cultivation system depends upon how perfectly the fallow phase restores and maintains soil fertility. However, if the fallow time is not sufficiently long enough, there is low addition of organic material causing decadence in soil fertility. The unique feature of shifting cultivation is that before the start of the field preparation, the family pay obeisance to evil spirits and sacrifices are made for excellent cropping season ahead and well being of the family.

3.3.5 Issues in shifting cultivation

There are variable views about the hostile effects of shifting cultivation on the ecology and environment. Some of the scholars hold the view that it is primordial and adversely affect the natural resources like land, water and vegetation and that, since shifting cultivation harm the ecosystems, it should be discontinued immediately. However, there are some scholars who believe that the practice does not encourage soil erosion and inflict no damage on the ecosystem and that, it can be continued with necessary modifications. As no tillage of soil is done, the soil remains intact during agricultural operations. Moreover, shifting cultivation lands are commonly steep slopes on which sedentary farming is difficult to develop. We can say that shifting cultivation is a way of life, evolved in response to the physiographical character of land under peculiar environment. The practice is in vogue for livelihood security and not without the information of its harmful effects. It may be stressed here that the shifting cultivation cannot be eradicated altogether, rather it will be important to modify it for making it more productive to face the challenges posed by demographic pressure. The main path to eliminate the harmful effects of shifting cultivation is to go for sedentary form of cultivation. Construction of terraces in hilly areas under shifting cultivation could be the immediate step to arrest soil loss through erosion. However, this will depend on the nature of land tenure prevalent in a particular region. If the land is not in the individual names of shifting cultivators, they will not like to invest huge amounts on land modification. Other than terraces, other water and soil conservation measures like bunding, trenching, half-moon terracing, grassed water ways and gully plugging, can be undertaken according to the local land configuration.

3.3.6 Management of livestock

Two livestock management systems can be distinguished in the study areas, which are strongly linked to the ethnicity of the people (Aryal *et al.* 2010). The Limbus and Rais people in Yasang, Lawajin and Langluwa areas hold at least one or two cows, goats, sheep and pigs etc. for subsidiary source of income, and the livestock is an important component of the farming. The Sherpas of Sherpagaun village, however, have chauris (crossbreed of yak and cattle) in significant numbers, and their milk, churpi or hard dry cheese and ghee are a major source of income for the farmers. Livestock and crop production are intimately linked in shifting cultivation. The animal dung is used as a manure in the field foor enhancing the soil fertility for higher crop production.

3.4 SOURCE OF NUTRITION IN SHIFTING CULTIVATION

DeAngelo *et al.* (2004) were of the view that the shifting cultivation is practiced by small-scale and marginal farmers in many tropical countries. This practice usually has five major stages, such as; (i) clearing, in which large trees and other bushy vegetation is cut and uniformally distributed all over the field and put on fire, thus clearing the field for sowing/planting as well as releasing the nutrients for optimum crop growth, (ii) sowing the seeds in small holes made by a dibbler, (iii) inter-culturing and removal of weeds by the shifting cultivators so that these weeds do not compete with the regular crops for nutrition, space and water, (iv) during the cropping season, the crops are protected from pests and diseases. The farmers keep the best seed for next year planting, and (v) after cultivating the land for 2 to 3 years, the field are kept fallow for regeneration off secondary forests.

During decomposition, the organic materials in the soil increase and this boosts the soil fertility level. This proves quite beneficial for the growing crops. Because in most of the tropical areas, the rainfall is heavy, there are chances of leaching of nutrients as well as their loss due to erosion. So, sowing has to be done in time for giving maximum benefit to the crops of the added and released nutrients. The cleared land, after a couple of years cultivation, for regrowth of secondary vegetation and within a few years, a closed canopy of secondary forest develops. If sufficient time (at least 10 years) time is given to fallow land, then soil will regain the nutrients, allowing the shifting cycle to be repeated. Despite the relevance of shifting agriculture in tropical nations, there remains debate concerning its merits. Some experts, such as Dove (1983), Southgate (1990), and Pearce and Warford (1993), have criticised shifting cultivation, claiming that it is environmentally detrimental because the practice's land clearance activities result in tropical forest degradation. However, a second group of experts, such as Peters and Neuenschwander (1988) and Dufour (1990), have suggested that shifting cultivation with long fallow intervals in tropical forests can be environmentally and economically sustainable in some scenarios. Batabyal and Lee (2003) and Batabyal and Beladi (2004) have mentioned that the viability of shifting cultivation in the long run depends on the length of the fallow period until the natural fertility has been restored and fallowing at appropriate points in time. The natural fertility of the soil is reinstated in the course of the rest of the fallow time between crops for utilization during the succeeding period of crop growth. The shifting cultivation remains in ecological balance with the ecosystem and will not degrade the soil resource if adequate fallow period is permitted for restoration. Though population explosion may necessitate continuous, or nearly continuous, use of land, thereby making the fallow period shorter and shorter. This is particularly true humid tropics of Asia and in the savanna and forest zones in Africa (Okigbo 1984). The shifting cultivators in the tropics are, therefore, getting lower yields, resulting in shortage of food grains, livelihood insecurity and reduction in living standard. Other than this, shifting cultivation, as it is practised, is a wasteful method of cultivation of scarce land resources and frequently leads to intolerable erosion, particularly of hillsides and sloping lands.

It has been observed that during the short cultivation phase, a mixture of crops and species are grown for food and other requirement of daily use. The cultivation phase alternates with a longer fallow period, when the cropped area is left for regeneration. There appears significant variations in the lengths of cropping period as compared to fallow periods, which have been highlighted by different scholars (Nye and Greenland, 1960; Ruthenberg, 1971; Spencer, 1966; Watters, 1971; and Barrau, 1971). Shifting cultivation is considered environmentally sound and stable when ratios of fallow periods to cultivation phases are 10 or more, but when this ratio falls below 10, may be due to demographic pressure or other reasons the fallow period is radically reduced, there is every likelihood that the system will breaks down. There will be more soil erosion and loss of soil fertility and productivity. The methods of doing the practice of shifting cultivation in different areas of the tropics are under continuous change due to socio-economic and other pressures for intensification.

3.4.1 Historical and socio-economic setting for shifting cultivation

Shifting cultivators might grow a single commercial crop in a single field or combine it with other crops in different locations. As a result of increased production due to population pressure, soil fertility is lost due to loss of plant nutrients, soil erosion, and soil deterioration.

3.4.2 Economic drawbacks of shifting cultivation

There are a variety of agronomic and cost disadvantages to shifting agriculture. Some are inherent in the system, while others are the result of socioeconomic forces affecting places where agriculture is shifting. Some of the inherent disadvantages are as follows:

i) Shifting cultivators do not get a good price for their produce for lack of proper markets.
ii) Systematic destruction of forests and the degradation of forest soils where shifting cultivation is practiced. Some valuable resources are lost in the process.
iii) Low investment capacity of shifting cultivation.

The major impact of socio-economic aspects is because of two reasons such as, population growth and need for cash income. For example, during the 1970s, the African population increased by 2.7 percent but per capita food production declined by about 10 percent. Under such a situation, the farmers are unable to give optimum period for fallow phase, which further diminishes the output from shifting cultivation. Traditional farming practises, due to their poor productivity, are unable to generate enough to elevate farmer consumption above subsistence levels or to meet new cash-based needs.

3.4.3 Shifting cultivation and spontaneous change

During the last few decades, Africa is witnessing a social change encompassing; population increase, increase in area under cultivation, introduction of cash crops, decrease in per capita availability of land and, increase in access to markets. Changes in adoption of agricultural techniques are rather slow compared to decreasing availability of land. The practice of shifting cultivation is stable when cropping phase is followed by long periods of fallow, which necessitates large areas for cultivation to sustain the population. How high a population shifting cultivation can sustain is a matter of debate and many views have been expressed in this regard. Ruthenberg (1980) found that no more than 56 persons can be supported per km^2, however, Lassailly-Jacob (1983) estimated that up to 123 inhabitants can be sustained per km^2 in Beumi, Ivory Coast, without destroying the environmental balance. Our experience from Northeast India show that 20 families with average size of 5 members (100 persons), can be supported on one km^2 (100 ha) of land without affecting the ecosystem. This will give at least a minimum period of 20 years period for fallow phase after cropping for 2 years, which can be considered optimum for rejuvenation of forest vegetation. This would require best management of crops and soil during cropping phase and forest vegetation and soil during fallow phase. Simply abandoning the land after cropping phase at the mercy of God and not caring for it, as is happening in most of the cases in Northeast India, will be disastrous and may lead to irreversible soil degradation as well as degradation of environment.

Because of the scarcity of land, the shifting cultivator may switch to crop rotation, as in Tiv, Nigeria (Vermeer, 1970), and Baoulé, Ivory Coast (Lassailly-Jacob, 1983), or lengthen the cultivation phase and reduce the fallow season, as in northeastern Zambia (Lawton, 1982). This latter strategy, cannot be stretched far as it would result in decrease in yield and soil degradation. There has been long debates on why the shifting cultivators resist change until there are low yield of crops and deterioration in soil fertility, often irreversibly. Some of the reasons for not changing before soil degradation are:

i) Increased labour requirement in going for intensive cultivation,
ii) No security of land tenure,
iii) Limited options are available to the shifting cultivators because of rugged and undulating terrain.
iv) Adoption of new agricultural technique creates a risk which the farmers cannot afford,
v) Due to socio-cultural reasons, many rituals and festivities are attached with the practice of shifting cultivation. Leaving of the practice means leaving of all related rituals and festivals.
vi) Lack of capital investment which is necessary for new seed, fertilizers, pesticides etc.,
vii) Limited or no access to markets,
viii) Limited site-specific technological options to replace shifting cultivation,
ix) With new generation, the interest in farming itself has taken a dip. They want to go for white collar jobs and do not want to be a part of the rigorous routine and,
x) Lack of will

Richards (1977), when detailing the deteriorating conditions of the farmers in the Ikale region of southwestern Nigeria, where the population growth rate was around 7% between 1952 and 1963, mentioned a few of these issues. The environment of Northeast India is being threatened by changing farming. In 2005–06, the area under shifting agriculture in Northeast India was estimated to be at 8771.62 km^2, accounting for around 85.7 percent of the country's total shifting cultivation area. The key elements that influence this type of land use are personal, social, economic, institutional, and physical. It ia an economical practice, having some advantages due to which farmers prefer such type of farming (Ketterings *et al.* 1999). It is connected with ecological and socio-cultural life of indigenous people (Tangjang, 2009). However, the practice reduces biodiversity and increases the rate of soil erosion substantially

(Lianzela, 1997). Toky and Ramakrishnan (1983) reported the loss of original vegetation, soil fertility and biodiversity with fire when the land is prepared employing slash and burn technique. As most of such losses are ecological and social, they may not immediately change farmers' thinking towards shifting cultivation and adopt shifting cultivation practice (Schuck *et al.*, 2002).

The family income is generally considered to test the impact of level of living on the call for adoption of shifting cultivatin. Economic variables also include the nature of the principal occupation, the current farmed area, and access to land resources. The slope of the ground is one of the physical variables. The conceptual framework of the determinants of shifting cultivation is presented in Fig. 3.2.

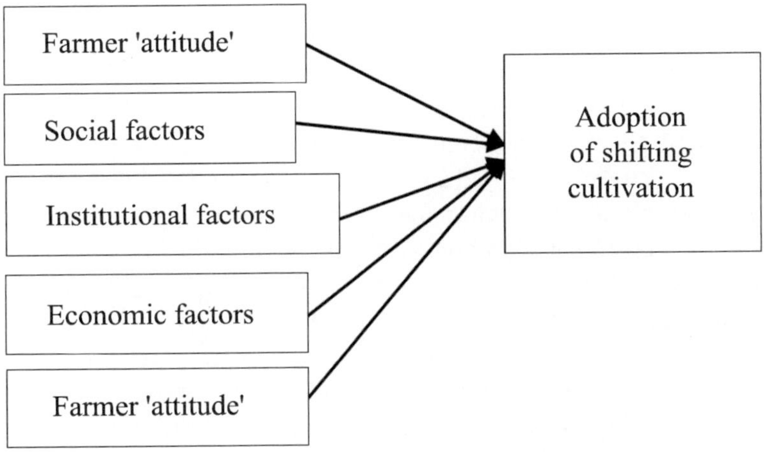

Fig. 3.2. Conceptual model of adoption of shifting cultivation.

3.5 DELINEATING SHIFTING CULTIVATION

A case study from Laos's Vien gkham district illustrates the methodology of Hett *et al.* (2012). The shifting cultivation areas have grown over the course of 30 years, and as a result of the move from lengthy to short fallow cycles, these regions have deteriorated. Hett *et al.* (2012) have demonstrated that the landscape mosaics methodology can be effectively utilised inside a future REDD+ framework's changing cultivation compatible monitoring, reporting, and verification (MRV) system.

Shifting cultivation is practiced with minor variations in methodology, market orientation, and policies in different countries of the world or within the same country as well (Nair, 1993; Sanchez, 1976). Padoch *et al.* (2007) indicated that although the importance of shifting cultivation in the tropics is beyond doubt, there is a general lack of spatially clear knowledge about the practice, causing failure of land cover/land use maps to entrap this practice from the world to the regional scales. Two explanations were given: I it is still considered an archaic and fledgling form of land use (Mertz *et al.*, 2009) and (ii) it is condemned due to environmental deterioration (Mertz *et al.*, 2009). (Fox, 2000; Fox *et al.*, 2000; Padoch *et al.*, 2007). It is regarded to be a key contributor to tropical deforestation, flora degradation, and soil fertility exhaustion (Lawrence *et al.*, 2010). (Arnason *et al.*, 1982; Sanchez, 1976). On a landscape, shifting agriculture and land use changes are difficult to distinguish (Padoch *et al.*, 2007). Remote sensing is a critical tool for determining land use change over time and providing a precise assessment of human impact on the environment (Bakr *et al.*, 2010). Shifting cultivation as a different land use does not appear in land use change analyses due to its complexity, which pose a challenge to the normally used remote sensing techniques (Schmidt-Vogt *et al.*, 2009; Sirén and Brondizio, 2009).

A method was developd by Yamamoto *et al.* (2009) to delineate shifting cultivation areas in Northern Laos. Using landscape mosaics, during the same period, Messerli *et al.* (2009) developed a methodology by the use of landscape mosaics, to identify shiting cultivation from other land use types in Laos. Deforestation in the tropics caused about 20% of world anthropogenic CO2 emissions in the 1990s (IPCC, 2007) and about 12% in 2008 (Le Quere *et al.*, 2009). Because of negative perception that shifting cultivation contributes to deforestation, it is considered as a main cause of the emissions by many tropical countries (Griffiths, 2008). The widespread perception that shifting cultivation is to blame for the swidden's downfall in the REDD+ arena provides a solid foundation for furthering the swidden's doom (Angelsen, 2008). Shifting cultivation, according to another school of thought, is a traditional style of agriculture that promotes biodiversity (Rerkasem *et al.*, 2009) and nourishes local people (Colfer, 2004).

3.5.1 Beneficial aspects of burning

The shifting cultivators consider ash as a source of nutrients for the crop and would never agree that burning of vegetation is a wasteful exercise. Burning of forest vegetation considerably increases the availability of nutrients to crop plants from the soil (Schmidt 1973). Burning vegetation is an important method of removing hazardous insects and fungus foci. It is desired in Liberia to have succulent vegetation that is 6 to 10 years old, especially if young males are not available to perform the clearing task (Carter 1982). Large trees in northern Zambia are frequently impossible to cut, thus fields that have been fallowed briefly may be better for quick clearing (Lawton 1982). Okigbo (1981) asserted that bush fallow systems might be productive, ecologically sound, and stable.

Scholars have not been able to come up with a better method for producing food in a tropical climate, according to Nye and Greenland (1960).

3.5.2 Carrying capacity of shifting cultivation

People generally want to know that how many persons can be supported by km-2 of land under shifting cultivation. According to Ruthenberg (1980) calculated that only up to 56 persons per square kilometre can be supported by shifting cultivation, with an assumption that land use is even, 15 yeears of fallow preceded by 3 years of cropping, 6 moves per cycle and 0.3 ha of crops per person. In the 1930s, the Bemba people of northern Zambia (Richards, 1939) relied on a 30-year fallow period, which they practised in an area with a population density of 3-4 people per square kilometre. Many Zambian farmers have left their fields to work as copper miners in the copper belt region, either temporarily or permanently.

Jean's (1975) classified cultivation techniques into three types:

1. Simple non-permanent farming with fallow;
2. Semi-permanent cultivation with fallow and use of fertilizer;
3. Sedentary farming with fertilizer application.

The availability of fertilizers will ensure change from 1 to 2 and 3 techniques. In the absence of fertilizers, animal manure, green manure etc., there is a possibility that increasing land pressure will cause soil degradation. At low yields from cultivation, the farmers may search for alternate jobs to earn livelihood or he may move to other place where the pressure on land is low, alongwith family members.

3.6 FORMS OF SHIFTING AGRICULTURE

Shifting cultivation is an important shape of agriculture necessary for livelihood, nutrition and as a survival for millions of people in the tropics (Cramb *et al.* 2009, Van Vliet *et al.* 2012). At the same time, the practice has also been criticized for being environmentally destructive both by scholars and policy makers alike, and is generally mentioned under wasteland category in government records (Myers 1992, FAO 1957, Schmidt-Vogt *et al.* 2009).

Brookfield (2001) identified a number of different types of shifting agriculture and discovered that the fallow intervals in most of them were typically a few decades long, allowing for large biomass regeneration and secondary forest formation before being removed and cultivated again. Various concerns have driven most of the literature on shifting cultivation. Some have expressed concern that shifting cultivation is a major cause of deforestation (Eckholm 1975), and development policymakers have viewed shifting cultivation as a primitive, unproductive kind of agriculture that required development (FAO 1957, 1974). (Greenland 1975). These positions have been moderated somewhat (Brady 1996, Myers *et al.* 2000), the debate has no end. One of the major concerns is that shifting cultivation may be intensified to ensure livelihood security for increasing population, which would result in shorter fallow periods, thereby affecting the biomass growth as well as changing its species composition, making cultivation ineffective and unmanageable (Dalle *et al.* 2011, Lawrence *et al.* 2010, Van Vliet 2012).

Using object-based image analysis, Riahtam *et al.* (2018), distinguished two types of shifting cultivation, that is, present and rejuvenating. They found a decreasing trend in relocating agricultural lands in Northeast India. It was discovered that most of the area used for shifting agriculture was situated on moderate to steep slopes, mostly on sloping plains with an easterly orientation. Slopes that faced west and were steeper showed less effect from the strategy. Their research indicates that shifting agriculture is becoming less common in areas with higher rainfall.

A precise evaluation of extent and vitality of shifting cultivation on global and regional levels is of utmost importance for ascertaining the procedures that account for the dimentions and change, as well as assessing their repercussions (Padoch 2014). Before the development of remote sensing technologies reports on shifting cultivation were primarily descriptive or qualitative in nature (Inoue 2000). Furni *et al.* (2013) state that the use of remote sensing to make accurate appraisals is particularly difficult in mountainous terrain, as Schmidt-Vogt (2009) points out.

Various forms of shifting cultivation are practised in Africa, Central and South America, Southeast Asia, and Oceania (Nye and Greeland, 1960; Barrau, 1971; Spencer, 1966; De Schlippe, 1956; FAO, 1974; Ruthenberq, 1980), and these systems are followed by people of various cultural backgrounds in these countries (Nye and Greeland, 1960; Barrau, 1971; Spencer, 1966; De Schlippe, 1956; FAO, 1974; Ruthen There are many similarities in the agricultural production systems of shifting cultivators and marginal farmers. All shitting cultivators can be found in tropical latitudes in impoverished countries throughout the world. Shifting agriculture, in its purest form, is carried out in remote places with or without animals (Okigbo 1984). The indigenous communities lack education and are isolated in clearings, hamlets and villages.

We are of the view that, broadly, the shifting cultivation now exists in eight different forms (Fig.3.3), such as; (i) in Spoiled or hotch potch/distorted form due to sudden increase in population and enhanced demand for food and other needs (For example in Mizoram at some places at low altitudes, the shifting agriculture in spoiled form exists and farmers have converted much of the subtropical evergreen and semi evergreen forests into scrub and bamboo bushes), (ii) Historical, as it started in the Neolithic period and did not experience any change because of inaccessibility, remoteness, low or no population increase as a result of warfare, head hunting etc., (iii) Innovative or imbibing best methods of cultivation at different period of its life due to receptive cultivators and good extension services from line departments. The farmers have switched to modern methods of cultivation which are more adapted to the current availability of resources and present day societal requirements like money. In some cases they have also introduced new crops, which help maintain fertility and also have good market, (iv) Fallow, where cultivators continue to derive maximum benefits from the fallow phase only like grazing of livestock, use of roots and tubers of some plant species as food supplement; (v) Traditional, as is found in some villages or areas which have not experienced much population pressure increase; (vi) Improved, for example in India the focus is on improving the shifting cultivation rather than to do away with the practice with which the farmers and their families are socio-culturally attached, as also many rituals and festivals are associated with the practice. The respective governments have introduces new projects and schemes for improvement of the practice rather than to wean away the shifting cultivators, (vii) Natural form, where the cultivators use a part of land for growth of natural vegetation as it comes on slopes and use it for themselves and the livestock for 2 to 3 years or until fertility of land stop supporting vegetative growth, soil erosion takes place leading to land degradation and the hill slopes suffer irreversible damages and, (viii) Growing or transitional form of shifting cultivation, where this practice is on the move and shapes itself in a form as per need of the farmer or according to the wishes of head of the community/clan who allots land to them for cultivation.

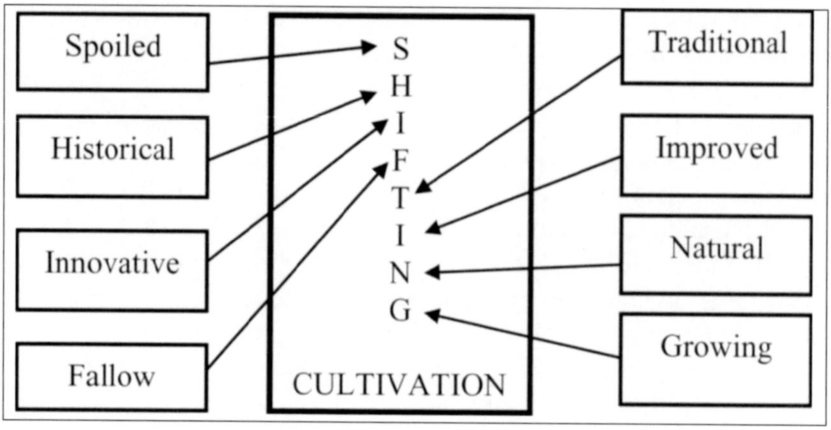

Fig. 3.3. Different forms of shifting cultivation encountered in various countries.

3.7 LAND USE CLASSIFICATION

Padoch *et al.* (2007) indicated that shifting cultivation often does not appear on land use maps. This is mostly due to (i) swidden is a complex and dynamic land use system that is difficult to map, and (ii), the process of dividing up land uses associated with swidden systems into different groups (Schmidt-Vogt *et al.* 2009). Another factor could be that swidden systems are being replaced by alternative land use practises in various parts of Southeast Asia. There is no solid information on the extent of shifting cultivation in Southeast Asia. To address this, solutions are being considered that could provide more credible data for future assessments (Schmidt-Vogt *et al.* 2009). Transitional land covers, changing over time in the course of succession impede the identification of these land use covers. This is exacerbated even more by the fact that shifting agriculture is frequently not the only land use in a region, but rather happens within a complex of other land uses.

According to Fox (2000), between 33 percent and 50 percent of "non-forested" land in Southeast Asia may be demoted to "unclassified" or "other" categories in several land-use classifications. It has also been attempted to estimate land use change and map shifting cultivation in Southeast Asia. According to a map showing shifting cultivation distribution in the Asia-Pacific region (Collins *et al.* 1991), shifting agriculture was the primary land use in roughly half of Southeast Asia's land area, based on Spencer's sectional examination of shifting cultivation in Southeast Asia in 1966. According to Uhlig *et al.* (1994), around 5.5 to 6 million acres of land in Sarawak, Sabah, and Thailand was under shifting cultivation in the early 1980s.

3.8 EXTENT OF SHIFTING CULTIVATION

The area under shifting cultivation and the soil loss through water erosion in the Northeastern states of India is shown in Fig. 3.4.

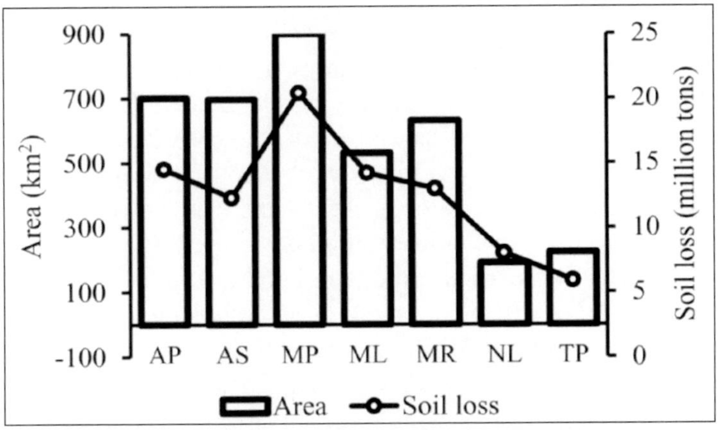

Fig. 3.4. Area and soil loss from different states of Northeast India, AP, Arunachal Pradesh; AS, Assam; MP, Manipur; ML, Meghalaya; MR, Mizoram; NL, Nagaland; TP, Tripura (Northeastern states of India)

3.8.1 What is shifting cultivation

Shifting cultivation was generally used historically in temperate regions for clearing of forest land for arable land to be used for agriculture. Now, slash and burn forms an important component of shifting cultivation in tropics too. This is usually known as environmentally catastrophic, but it could be suitable when undertaken by low populations in wide forest areas, where plots have enough time to recover before again being cut, burnt, and farmed. Spencer (1966) mentioned that slash-and-burn based shifting cultivation is a widely practiced historical practice in southeast Asia, however, some farmers may go for slashing without burning and sow the crop, waiting for the slashed vegetation to decompose.

Myers (1986) put the number of shifting cultivators in world at 250 million compared to 300 million shifting cultivators estimated by Russell (1988). In a world population of 7.9 billion, it might be insignificant. But what cannot be overlooked is the spread of shifting cultivators throughout the world and the vast area under the practice and its prevalence for such a long time in the history of mankind. This practice is the most widely distributed type of tropical soil management methd. During the mid-seventies (Hauck 1974, Sanchez 1976), different types of shifting cultivation used to be practiced on 30% exploitable soils. Many past investigations called the shifting cultivation as innately stable.

Generally considered as a waste of land and human resources, shifting cultivation is also a major cause of soil erosion and decadence. Use of the swidden field for a year or two after clearing the forest, and then move on to another forest area appears to be wasteful if the forest is anticipated as of timber values alone (Grinnell 1977). It is not the cutting of the forest alone which foresters consider, but the burning of the trees. The issue isn't so much who should benefit from deforestation as who should benefit from forest preservation. Governments see burning as a misallocation of resources from the national level to the most local, i.e. farmer level. Though the practice of shifting cultivation is practiced by cultivators in Africa, the long fallow practice has been slowly changed to intensive cultivation close to the settlements and long-term cyclic cultivation away from home (Chidumayo 1987; Getahun *et al.* 1982). The intensification of shifting cultivation is being done in the entire area, although with some variability in management methods and crops grown.

In Nepal, shifting cultivation is known as Khoriya and Bhasme. The communities remove and cultivate secondary forests in tiny plots before allowing them to regrow naturally through medium to long-term fallow periods (Fujisaka *et al.* 1996). The main characteristics of shifting farming in the tropics, according to Watters (1971), are the changing of plots rather than crops on the same field. He described that cultivation is done only for 1 to 3 years followed by longer fallow periods varying between 4 to 60 years. He further stated that the cutting and burning of the fallow vegetation is done at the beginning of each cropping period and human energy is commonly used in almost all land management operations. In case of Nepal, shifting cultivation practice is operated by 2 to 4 years of cultivation and 4 to 9 years of fallow period. Bajracharya *et al.*, (1993) observed that there is a marked tendency of cultivation of fields for longer periods and reduction in the period of keeping the land fallow, because of population increase.

3.8.2 Current extents of shifting cultivation

Only scant investigations mention about the exact area under shiting cultivation at regional or global level as mentioned by Heinimann *et al.* (2017) after review of literature on the subject. FAO estimated the global area under shifting cultivation in 1980s at 400 million hectares (Lanly 1985), while in 2011, with the goal of monitoring GHG emissions, estimated at 260 million hectares for the 2000s (Silva *et al.* 2011). Davidson *et al.* (2001) reported an area of 1,000 Mha, quoting Sanchez *et al.* (2005), who in turn had quoted Dixon *et al.* (2001). The estimate of 280 Mha (Heinimann *et al.* 2017) for the total area under shifting agriculture is extremely similar to the 259 Mha estimate provided by Silva *et al.* (2011). At the regional level, the value of 110

Mha for Asia estimated by Spencer (1966), which seems to be considerably bigger, is the only estimate to which the assessment of Heinimann *et al.* (2017) of 70 Mha for Asia can be matched. Given that Spencer's estimate is based on a hand-drawn map and that the area under shifting cultivation has significantly decreased since the map was created, the discrepancy seems to be fairly legitimate and logical.

3.8.3 Future trends in the extent of shifting cultivation

Predicting future trends in any sort of land use requires a great deal of caution (Mueller *et al.* 2014, Rounsevell *et al.* 2012). Heinimann *et al.* (2017) evaluated predicted shifts in shifting cultivation in various parts of the world by integrating observable trends from the Butler map and their own map with expert survey responses on future shifts in shifting cultivation areas. They expected that shifting cultivation would drop considerably in all locations during the following 20 years, and that it would be phased out by 2090 in all places (Table 3.2). They expected that due to economic development and structure and, farming practices in Asia, may result in complete elimination of shifting cultivation more rapidly than in Africa and the Americas. Further, shifting cultivation will persist for more time in Africa, especially in Central Africa, depending on the demographic pressure and alternative livelihoods opportunities (Mertz *et al.* 2002, 2009, Heinimann *et al.* 2013). Both situations apply to major areas of Central Africa.

Table 3.2. Between 2010 and the 2030s, 2060s, and 2090s (percent) ranges of losses relative to 2010, trends in percent area left under shifting cultivation (Based on Heinimann *et al.* 2010).

Region	2030	2060	2090
Central and South America	60-90	20-40	0-20
Africa	80-100	20-40	0-20
Asia	50-70	0-20	0

The survey outcomes on shifting cultivation trends for some specific Asian countries (Heinimann *et al.* 2017):

- Vietnam and Laos: The practice will rapidly reduce by 2030 and completely eradicated by 2060.
- Myanmar: the practice assessed to be mostly depart between 2060 and 2090 if conflicts are resolved.
- Borneo and Sulawesi: Shifting cultivation is likely to vanish between 2030 and 2060.
- India and Bangladesh: The practice is assessed to depart by 2030.
- Papua New Guinea: The practise could last long into the second half of this century, possibly until 2090.

- According to the findings, shifting agriculture is more likely to persist in Africa. It is expected to rise in Central Africa until 2025, after which it will begin to decline.
- Sierra Leone, Liberia, Guinea, Côte d'Ivoire, Ghana: The practice will decline rapidly by 2030 and largely vanish by 2060. If conflicts appear, the shifting cultivation would continue for some more time.
- Central Africa: Due to having large areas under forests, the practice may continue well into the 2060s or longer.
- Madagascar: The practice will persist beyond 2030, especially along the eastern cliff.
- Central and South America: The survey results show that in some areas expansion of area under shifting cultivation is predicted up to next few years, while in countries like Maxico and Brazil swift decrease in area is anticipated. Overall, a decline after 2030 and disappearance after 2060 is forecasted.

The elementary assessment of Heinimann *et al.* (2017) for area uner shifting cultivation indicate that it will decrease noteceably over the next decades. The immediate implications predicted are; reduced provision of ecosystem services and limited availability of land due to intensification, tree plantations, urbanization, and forest protection or restoration (Fox *et al.* 2009). However, if the development efforts are in accordance with the needs of shifting cultivators, this future may also make available better opportunities for crop production and income generation (van Vliet *et al.* 2013).

3.9 GLOBAL EXTENT

According to world estimates, about 200 and 500 million farmers practise shifting cultivation (Fig. 3.5). This form of agriculture, generally carried out in the tropical forest regions, has been criticized by the United Nations Food and Agriculture Organization (FAO) since the 1992 Earth Summit in Rio. The FAO has considered that this cultivation model contributes to deforestation and global warming by encouraging carbon release into the atmosphere (IRD 2008). Slash and burn farming, also known as swidden cultivation, is an agrarian system in which plots of land are cleared of vegetation cover by cutting and burning before temporary cultivation for about a couple of years. The period of fallow are longer than the period of cropping. Because the system is mainly established in forest areas, it is considered contributing actively to deforestation as it is thought to inflate the anthropogenic greenhouse effect. Since the Rio de Janeiro Earth Summit of 1992, the FAO has been promoting the replacement of this practice by permanent cultivation, considered to be less harmful for the environment.

Shifting cultivation has been practiced throughout the tropics since about 7000 BC. A "traditional, year-round, community-wide, largely self-contained and ritually sanctioned way of life" is the foundation of this essential practise, a land use system that is still popular among tribal communities in Southeast Asia and South America as well as a small, declining portion of African farmers (Conklin 1957). In addition to agriculture, this system involves forest-based pursuits like fishing, hunting, and gathering wood for burning. It also involves edible vegetation. Planting and protecting trees aids in the forest's restoration to its natural state, even though shifting agriculture requires destruction, clearing, and burning. Such shifting cultivation methods seem to be flexible in responding to changes in a variety of climatic conditions or to moving from one plot to another, rather than inflexible in their adaption. Previously, this land use was thought to be exploitation as opposed to management of the humid tropical region's natural resources. This perception has changed now. Integral shifting farmers' natural resource management focuses on preserving the forest ecosystem's variety (Warner 1991). Shifting cultivation can succeed where other land use strategies have failed by combining natural processes with in-depth knowledge of forest and plot microenvironments. The survival of the practice for about 9000 years, is enough proof of its sustainability, though some conditions may apply. Although successful in the past, shifting cultivation cannot serve as the model for the future of the tropics (Warner 1991).

Fig. 3.5. Shifting cultivation in the World

Adopted from Warner (1991)

The magnitude of large-scale deforestation of the tropics show that this may lead to flooding, drought, high scale erosion and global warming that would transform vast areas unsuitable for future use. The protection of forests is not only a national concern but, this ecosystem has to be managed internationally. As a result, the major question is what has been done to protect the forests? The shifting cultivators were the first recipients of getting the blame of destroying the forest. Attempts have been made by Different governments have made concerted efforts to forbid shifting cultivators to stop the practice of shifting cultivation by enacting laws, giving incentives, subsidies and, in some cases, jobs too. The new viewpoint of shifting cultivation has been gaining momentum due to the failure of agricultural development projects started by respective governments in the tropics. Ruthenberg (1976) defined shifting cultivation as an "agricultural system that involves an alternation between cropping for a few years on selected and cleared plots and a lengthy period when the soil is rested."

3.9.1 How many shifting cultivators are there?

No specific and sincere efforts have been made to estimate the number the total number shifting cultivators, number of families involved and area under shifting cultivation. For example in northeastern region of India, the available data on these aspects of shifting cultivators has remained unchanged for about six decades. Wherever new figures are available, these have been approximated not by visiting the sites but based on broad experience and hearsays only. There are many constraints also in actually measuring the number of farmers doing this practice. Due to increase in population, the number and sites are changing so frequently that it may be difficult to do this job. It may not be impossible but, really an uphill task if not taken up sincerely and purposefully. Three explanations were given by Padoch *et al.* (2007) for the difficulty in measuring the scope of shifting cultivation in Southeast Asia.

1. Shifting cultivation is "a diverse, complex, and dynamic practice and the collection of information is difficult to measure.
2. It is a "marginal farmers group, and government officials find adding up what is happening in all those spatially divergent and dynamic meager holdings far more difficult than summing up some monoculture farming".
3. Classifying and measuring the extent of shifting cultivation amounts to not only an admission of its survival but also forecasting that lands will continue to be used for shifting cultivation, generally showing

great variation in 'landscape visions' between government agents and cultivators".

Mertz *et al.* (2009), in their estimation of the numbers of shifting cultivators at the global level and in Southeast Asia mention about similar problems faced by them. The extent of the problem can be gauged from the fact that the estimated number of shifting cultivators by different agencies vary from about 40 million to 1000 million, but most lack the empirical data needed to substantiate the claim. The figure of existence of 500 million shifting cultivators by FAO (1982) is a very broad approach, and the actual numbers must have changed considerably since then. The number of shifting cultivators in Southeast Asia vary from 50 million, who make a living from shifting cultivation to 400 million forest-dependent people in Asia, the majority of whom practise some sort of shifting agriculture.

3.9.2 The natural resources of the humid tropics: forest and soils

Humid tropics have a natural vegetation which is closed and open forests (Richards 1977; Hadly and Lanly 1983). The closed forest is found where average annual rainfall is above 1600 millimeters and open forest where rainfall varies from 900-1200 mm. However, in areas having a rainfall varying from 1200-1600 mm, the vegetative cover may be of open or closed forest regulated by the period of the dry season, soils, etc. (OTA 1984). Tropical areas of Africa have the closed tropical forests of about 18%, in comparison to 66% of the world's open forest (Fig. 3.6). Congo basin and East Africa, having drier climate, have more areas under open forests. However, tropical America has about 57% of the planets' closed tropical forests. Most of these closed forest are found in the Amazon basin. Closed tropical forest area is only 25% in Asia, but almost half of it is in Indonesia (Hadly and Lanly 1983: OTA 1984).

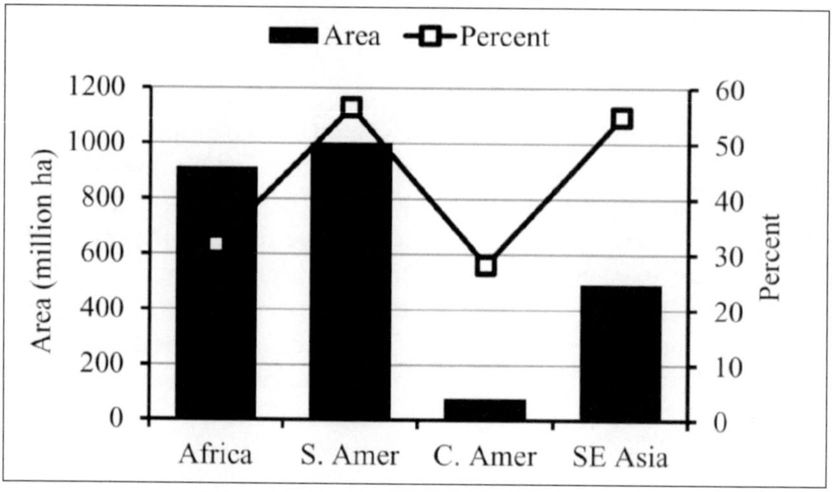

Fig. 3.6. Area under humid tropics in the World.
Source: FAO (1981; 1982)

Because the tropical forest ecosystem relies on a nearly closed and self-contained nutrient cycle, tropical soil fertility and health are often poor. With around 65 to 85 percent of the vegetation's root system lying within the topsoil layer, the forest operates like a giant squeegee in terms of nutrient recovery and recycling (Hadly and Lanly 1983; Uhl 1983; Moran 1981). The vegetative biomass, which serves as a storage for the nutrients, contains the nutrients that are cycled in the system.

The destruction of the tropical forest is, commonly, never caused by integral shifting cultivator or who has a long field experience of working as swiddner, but the forest destroyer are usually the "Land hungry" migrants, without a past experience of shifting cultivation practice (Olafson 1981, Moran 1987; Watters 1971). A community that has lived in a place for a long time will have a better understanding of the local ecology than migrants who apply agricultural techniques from their home country rather than those appropriate to the new settlement (Moran 1987). Cox and Atkins (1979) stated in response to the argument that "shifting cultivation is a wasteful practice that causes environmental degradation" that shifting cultivation seems to be the most effectual method for managing with the environmental realities of the tropical forest ecosystem.

Rambo (1981) mentioned six major beneficial effects of shifting cultivation and that burning is important for a good crop with a minimum of labour.

1. Burning removes the unwanted vegetation from the plot to be prepared for cropping.

2. It makes planting of crops easier dut to change of soil structure as the heat of the fire changes the texture of the soil making it more friable. The friable soil is easy to cultivate and plant with a dibble stick (Conklin 1957; Tivy 1987).
3. Addition of the nutrient rich ash makes the soil more ferile (Sanchez 1976: Tivy 1987).
4. There is increase in soil pH with the addition of ash as the ash is alkaline in nature. Most of the soil nutrients become available with increase in soil pH and aluminium toxicity is either reduced ar eliminated due to reduction in exchangeable aluminium (Moran 1981, Popenoe 1960).
5. Soil heating due to burning of vegetation makes the soil nutrients available to crops (Nye and Greenland 1960).
6. The weeds, insect, nematodes, and various pathogen populations are reduced due to heating (Glass and Thurston 1978).

3.9.3 Shifting cultivation in Europe

Shifting cultivation was practiced in many parts of Europe and eastern Siberia up to the end of the 19th century and in some areas of Siberia even up to first quarter of the 20th century (Steensberg 1993). The main swidden crops were; oats, wheat, turnips, barley, flax, millets, rye and radishes. Though cropping periods was generally one year, however, it was extended to 2 to 3 years on fertile soils. Fallow periods were very long and in some areas even up to 40 years. A tree similar to alder (Alnus nepalensis) was used to improve soil fertility. From the Neolithic to the middle of the twentieth century, shifting farming was practised. Fallow periods in Europe throughout the Middle Ages ranged from 20 to 40 years (Linnard 1970). In the twentieth century, shifting cultivation was common in Estonia, Poland, the Caucasus, Serbia, Bosnia and Herzegovina, Hungary, Switzerland, Austria, and Germany, according to Steensberg (1993). This indicated that the farmers did not want to leave shifting cultivation as it was able to provide food and livelihood to the population around well up to 1950s, in certain cases.

Around 1000 BC, the first recorded descriptions of forest devastation in Southern Europe appeared. Forests were cleared for cultivation and used for shipbuilding, urban construction, the manufacture of barrels, pitch, and charcoal. Many people claimed that goats were to blame for environmental degradation, but a more major source of forest destruction was the practise of awarding ownership rights to those who began permanent farming after clearing forests in some regions. Following the collapse of the Roman economy, tree cover recovered in many sections of the Roman empire between 400 BC and roughly 500 AD, indicating that deforestation was driven by numerous factors other than agriculture. Darby (1956) observed that by 400 AD "land that had once been tilled became degraded, and at many places the area was covered with secondary forests. Wild fires were also a major cause of forest destruction in the Mediterranean environment with its hot dry summers. Human interference in the forests supported this factor. It is evident from the palaeoecological and prehistory record from the Neolithic, that the use of stone tools was common in Central and Northern Europe. Warfare and the non-availability of agricultural land are generally mentioned as the causes of reduction, though not well documented (Meggers 1954; Dumond 1961; Turner 1974). The latest work indicate that the Maya may have, in appropriate areas, developed irrigation infrastructure and intensive agricultural practices (Humphries 1993). In mountainous areas, the Maya people made terraces and retaining walls on the steep hill slopes in mountain areas. These terraces helped in reduction of runoff and erosion and could be planted with maize or other crops. These people used canals system as a source of irrigation.

3.9.4 Situation in tropical Africa

The estimates of forest fallows areas for African countries are given in Table 3.3, according to the FAO/UNEP Tropical Forest Resources Assessment Project report for Africa. In many situations, the forest fallow areas were approximated using aerial pictures and satellite imagery, on which the R (C x 100)/(C + F) value is not measurable (Laseail1y-Jacob, 1983).

Table 3.3. Estimated areas of forest fallows as at end of 1980 in km2 (after FAO/UNEP, 1981)

Country	GA (km2)	Fallow area (km2)	% of GA	Country	GA (km2)	Fallow area (km2)	% of GA
Chad	1284000	8000	0.6	Angola	1246700	122500	9.82
Gambia	10400	2000	19.2	Cameroon	475442	61000	12.8
Mali	1204000	25000	2.1	C. A. Republic	622984	41000	6.6
Niger	126700	30000	2.4	Congo	342000	11000	3.2
Senegal	196722	17500	6.9	Equi. Guinea	26000	11650	44.8
Upper Volta	274200	45000	16.4	Gabon	267670	15000	5.6
N. Savanna Region	4236322	127500	3.00	Zaire	2344885	184000	7.8
Benin	112620	37570	33.4	Central Africa	5325681	446150	8.4
Ghana	238538	91800	38.5	Burundi	27834	240	0.8
Guinea	245857	29000	11.8	Ethopia	1221895	103000	8.4
Guinea. BiaBau	36125	5600	15.5	Kenya	580367	6050	1.0
Ivory Coast	322463	153300	47.5	Madagascar	590992	35000	5.9
Liberia	96320	55400	57.5	Malawi	118580	-	-
Nigeria	923768	126500	13.6	Mozambique	783030	132000	1.7
Sierra Leone	73326	42750	58.3	Rawanda	26338	650	2.5
Togo	56800	14500	25.5	Somalia	637539	500	0.001
W. Africa	2015817	556420	27.6	Sudan	2505813	116000	4.6
Botswana	574992	-	-	Tanzania	942345	41000	4.4
Namibia	824293	3300	0.4	Uganda	196840	16000	8.1
TS Africa	1399285	3300	0.2	Zambia	752612	76000	10.1
East Africa	7905400	491440	6.2	Zimbabwe	389367	-	-

CA Republic : Central African Republic

TS Africa: Tropical South Africa

In all the highland areas of Ethiopia, Kenya, Uganda, Rwanda, Burundi, and many parts of Tanzania, and Malawi, area under sedentary agriculture is on the rise. In Tanzania, expatriation has reduced fallow period and encouraged a transformation to quasi-permanent cultivation. Major part of the population concentrated along the rail-line in Zambia, changed to permanent oultivation during the last 30 years. Shifting cultivation is now found only in isolated areas in Tanzania and Zambia. In the Zaire basin, people have been concentrating round the towns, where sedentary farming is being practised. Permanently cultivated farms are usually found in south eastern Nigeria. Farming itself is often subsidiary to nonfarm occupations (FAO 1984). Middle belt of West Africa is the main hulid area where shifting cultivation is still prevalent.

3.10 SHIFTING CULTIVATION IN SOUTH AND SOUTHEAST ASIA

In many areas of South and Southeast Asia people depend on shifting cultivation, fully or partially, for their livelihood security. The number of shifting cultivators in Southeast Asia, has been assessed to vary between 14 and 34 million. Though for South Asia, no such assessment about number of shifting cultivators has been made made, but this figure may be several millions. These ethnic people in Asia comprise two thirds of the world's estimated 370 million of aboriginals (AIPP and IWGIA 2014). Shifting cultivation done by these indigenous people is as variable as the people who practice it. Common in these forms of shifting cultivation is:

1. Land clearance by cutting and burning of forest vegetation.
2. Cyclic nature of the system with cropping and fallow phases.
3. Shifting of plots in most cases.

Slash-and-burn based shifting cultivation is a widespread historical practice in southeast Asian countries. The period of cultivation depends on the productive capacity of the soil, its fertility status, availability of land and choice of the farmer to move. Generally the cultivation is stopped when the soil gets exhausted or, more frequently, when the weeds start creating problems. In this system of farming, the farmers may use the practice of slash-and-burn as one element of their farming cycle, employ land clearing without any burning, and in some cases, do not use any cyclical method when the cultivators are purely migratory. The common practice we observed in Northeast India is that, many a times, the shifting cultivators leave the fallow land on its own fate to regenerate and do not get any benefit from this land. Very often, this becomes the cause of land degradation. In some areas, the farmers use the successive vegetation for firewood, timber for fencing and construction, ropes, clothing, carrying devices and medicinal plants. Many of the species have been shown to fix nitrogen, such as alder (Alnus nepalensis), making the soils fertile and productive. The kinship between the time the field is used for crop production and the time it is left for regeneration are crucial to the steadiness of the shifting cultivation. Whether or not the shifting cultivation suffers a net loss of nutrients over a time frame depends on these parameters. It is considered that a system having a net loss of nutrients will consequently lead to degradation of resources if corrective measures are not in place to restrict the causes of degradation. There is loss of soil organic matter, nitrogen and soil P as well as increase in soil acidity and reduction in soil porosity and infiltration capacity, when the when the fields are cultivated for longer periods. The secondary forests in fallow lands are generally richer in floral and faunal resources and more useful to forest communities than primary forests, though they are less bio-diverse. Aplomb shifting cultivation systems are highly variable and closely adjusted to micro-environments. During cropping and fallow phases, these systems are judiciously managed by the farmers.

The land system in most of the Eastern India Himalayas (EIH) states is governed by conventions and traditions. The land is either owned by the community or by the clan, and everyone has the right to own it. The tenancy system is set up in such a way that most farmers have no rights to land, and the chief is in charge of allocating land for agriculture. Because tenants have a short-term relationship with the land, this has resulted in a loss of soil productivity. Shifting cultivators find it difficult to invest in soil and water conservation techniques, as well as other fields of modern technology, because they lack ownership or security of tenure. In many instances, the tenants seemed to have lost the capacity to meet their basic consumption requirements as a result of land degradation and reduction in agricultural production. They are much more interested either to get employment in cities or search for other jobs/ opportunities to get cash money to meet their requirements. Here, we can say, is not the case of 'shifting cultivation' but 'shifting cultivators' to a particular piece of land.

The Mizoram State Government has taken a positive step by stating that "all land belongs to the government," without harming the society's socioeconomic structure. The amount of land allotted to a specific farmer is determined by the duration of cultivation, which might range from temporary to permanent. The shifting cultivators have a sense of belonging and are concerned about soil degradation restoration. Following are the perspectives of shifting cultivation to draw the attention of the concerned.

Shifting cultivation, deforestation, faulty method of cultivation on hill slopes, construction of roads, etc. are the major operations responsible for soil degradation in northeast India are; shifting cultivation, deforestation, methods of cultivation on hill slopes and some developmental activities. About 90% of the soils of the region are acidic in nature and this acidity cause restricted availability of certain nutrients, particularly phosphorus, limiting the crop productivity. The existing best methods in some areas having sustainable productivity without land degradation also need to be researched to popularize them at other sites. Improved/enriched short fallows and combination of trees, shrubs with crop and livestock farming may be helpful in preserving the soil health as a promising improvement/alternative to shifting cultivation. Soil degradation involves a number of interrelated processes, leading to deterioration of the chemical, physical and biological condition of soils and affecting the socio-economic pattern of the farming community (Theng, 1991). Only those soil conservation practices as developed with the consent and cooperation of the local community, whose socio-economic aspirations receive due consideration, can be acceptable. Shifting cultivation with a fallow period of more than 10 years is sustainable from ecological, economical and energy efficiency points of views; and it should continue. The germplasm of native crops should be preserved for a long time preferably by cryopreservation. Species diversity of the region which at present is threatened due to heavy pressure on the natural forests should be preserved for future generations. To take the pressure off the lands under short cycles of 5 years or so, the valley cultivation should be strengthened for cereal production. Fig. 3.7 broadly show the areas under shifting cultivation in Southeast Asia.

Fig. 3.7. Shifting cultivation in Southeast Asia.

Plantation of bamboos *(Dendrocalamus hamiltonii)* is very useful as it effectively conserve potassium during early stage of secondary succession. Superior vegetative produced bamboo saplings should be provided to the farmers at a subsidized rate. Cairns and Garrity (1999) mentioned that the mountainous ecosystem of Southeast Asia, full of forests with sparsely populated diverse cultural communities, permitted these people to practice shifting cultivation. These communities remained in complete harmony with the environment they lived in. The annual cycle of cutting of forest vegetation and its burning has often been criticized as a major cause of deforestation in the tropics. They elaborated that the kinship between foresters and shifting cultivators is that of mutual animosity, deep skepticism, and conflictual interests. Discouraging shifting cultivation and adoption of more permanent forms of agriculture continues to be the policy of national governments across Southeast Asia. Conklin (1954, 1957), gave a much more suitable assessment of shifting cultivation through their work in the Philippines. He explained that shifting cultivation exhibits (i) a storehouse of traditional technical knowledge secured through centuries of experiences, (ii) a complex balance between product harvest and ecological resilience, and (iii) and lot of agrodiversity. The part played by shifting cultivators in protecting forest ecosystems and natural species has been suitably described by De Foresta and Michon (1997).

Not talking much of advantages and disadvantages of shifting cultivation, the premise that used to be the basis of the sustainability of the traditional practice, that is, low densities of population and vast availability of forest land, are non-existent now. There has been large scale increase in population from both endogenous increase and in-migration by large numbers of marginal farmers. These cultivators now have no access to enormous swaths of land classified as protected forests, as well as other competing land uses. Due to increased population pressure on land, shifting cultivators can no longer afford extensive fallows that allow secondary vegetation to regenerate and depleted soils to be restored. Shortening fallow seasons, lengthening cropping periods, and quickly deteriorating surroundings are all too common in Southeast Asia, putting farmers in poverty owing to severely low yields. Many a times, highly degraded areas are abandoned to make it prone to soil erosion and infestation with wild grasses and weeds.

3.10.1 System in practice

Shifting cultivation is known as suitable system of agriculture for the exploitation of certain areas and will have to persist but in improved form. In the early period of agriculture, the tribal population was limited and forest was abundant for shifting cultivation. Nineties Lewis, the Deputy Commissioner of Chittagaon hills reported in 1892 that the production of jhum crop was not only sufficient to meet the whole food requirement of a family but there was adequate surplus (Ghosh, 1992). Food from natural forests in the form of tubers, fruits, herbs, shrubs, wild animals and fish sustained the tribal families during natural calamities. A single crop of paddy is usually preferred in second year of jhuming (Borthakur, 1992). Some 8-35 crop species are grown together. Though shifting cultivation emit carbon when fallows are cleared and burnt, however, during regeneration of forest during fallow phase carbon is later sequestered.

3.10.2 Country estimates

The approximate area under shifting cultivation in HKH countries is shown in Table 3.4 (Silva *et al.*, 2011).

Table 3.4. Area under shifting cultivation

Country	Area under shifting cultivation (Mha)	Fallow period (years)
Afghanistan	na	na
Bangladesh*	1.508	12
Bhutan*	0.472	3, 5, 7, 8
China*	6.766	3, 5, 15, 30
India*	7.582	3, 4, 5, 12
Myanmar*	18.621	13, 15, 18
Nepal	0.540	na
Pakistan	0.007	na
Total	35.496	3 to30

*Silva *et al.* (2011)
na = Not available

Bhutan

Wangchuck and Tashi (2004) mentioned that the practice of shifting cultivation exist in about all areas of Bhutan. The practioners in Bhutan are primarily indigenous communities like as the Doyaps and Monpas, who value their environment and place a high value on natural resources. Even before formal rules were enacted in Bhutan in 1957, the government had authority over the tenure of shifting agriculture areas. Perhaps the state's presence was always present. The government granted individual families and groups permission to use land and other natural resources. By institutionalising private ownership of shifting cultivation land, formal regulations such as the Land Act of 1979 have rendered community-based common land tenure systems obsolete. The Bhutan Forest Act of 1969, according to Wangchuk (2001), eliminated all traditional rights and transferred traditional institutions such as'reesup' (a community-based system of forest protection) and'meesup' (a community-based system of forest fire protection) to government authorities. In 1985, the government, on the other hand, resurrected the traditional institution of reesup. The Forest and Nature Conservation Act of 1995 was responsible for dismantling the existing institutions for forest resource management. Despite this, customary customs and beliefs continue to have an impact on Bhutan's shifting agriculture lands.

Chittagaon Hills Tracts (CHT), Bangladesh

CHT, which includes the hill districts of Rangamati, Khagrachari, and Bandarban, is located in the southeast of Bangladesh between 21o25′ and 23o45′ N latitude and 91o45′ and 92o50′ E longitude (Ishaque 1971, Brammer 1986). The terrain is mostly hilly in this area (ADB, 2001; Brammer 1986). Shifting agriculture is still used by several ethnic groups in the mid-hills of Bangladesh's Chittagong Hills Tracts (CHT) as a primary source of food security. The CHT spans 13,184 square kilometres, with 12129 square kilometres of highlands, 264 square kilometres of medium highlands, 131 square kilometres of medium lowlands, and 660 square kilometres of homesteads and water bodies. Historically, the shifting cultivators had certain customary rights and a unique kinship with the area they cultivate for raising crops. Even though shifting agriculture has declined in popularity, it remains an important source of income for indigenous peoples and a dominating farming system in the Chittagong Hill Tracts' steep terrain (Roy 2000). A majority (87.6%) of the CHT's total population of 124,762 people were assessed to be living on the practise in 1901. (Roy 2000, cited in Hutchinson, 1906). Every year, roughly 41,485 ha (4.3 percent of the CHT's area) of land is cleared for shifting cultivation, according to a district gazetteer issued in 1971. (Ishaq 1971). According to Mohabbatullah *et al.* (2012), some 40,000 households are now engaged in shifting cultivation.

Nepal

Indigenous communities practice shifting cultivation in Nepal. Among the ethanic people, like as the Magars of Nawalparasi District, collective ownership of 'khoriya' (shifting cultivation) lands is still popular. Individual customary tenure of khoriya lands is practised by the Chepangs of Gorkha and Chitwan districts. The Limbus of Taplejung, who are assumed to be descendants of the ethnic Saawaayethaangs, employ 'taangsing-khoksing' (clearing and cultivating) land that is regulated by common tenure and institutions. The taangsing-khoksing land could not be sold or purchased, but it was jointly owned by Limbus and his clan.

The land which is kept fallow for a very long period, is likely to be more fertile and less weedy, is preffered the by the cultivators. The local farmers cut all the bushes and trees during the slashing phase. In some places, however, it is a common practice to leave some selected trees uncut to serve as mother trees and to reduce landslides occurrence and soil erosion. The slashed material is left to dry in the field, depending upon the intensity of sunlight (Aryal *et al.* 2010). After the biomass is fully dried, the burning commonly done in March. The biomass is, generally, never kept as such for decomposition as it would take much longer time. The biomass is put on fire only after properly ascertaining the weather condition, wind direction and ensuring that it has completely dried. The major crops of this area are maize, barley, wheat and vegetables. Earlier Potatoes were also grown, but in recent years this crop has moved to the rainfed areas (Aryal *et al.* 2010). Generally after one yar of cropping, the fields are left fallow. Rather than leaving the area fallow, farmers have recently begun to grow chiraito (medicinal plants) from the second to third year forward for three to five years. People who are poor choose to produce food crops, whereas those who are wealthy want to grow chiraito. The latter have far more land that they can share produce for this purpose. The forest fallows serve as the soil and water conservation measure in the areas studied (Aryal 2010). It was observed that poor families usually have 8 years of fallow period, in comparison to richer families who have fallow forests of 12-15 years. The farmers here use their fallows for grazing their livestock, and collect forest food items.

Myanmar

In the upland areas of Myanmar, shifting cultivation is a major practice. It is known as "shwe pyaung taung ya," in the country. Fallow lands in Myanmar are judicioulу managed with cover crops, planting of trees, and pollarding, or sometimes they may be allowed to regenerate naturally. Usually, the fallow periods vary fro 5 to 20 years and when shifting cultivation areas are managed according to traditional customary system, not much deforestation takes place as forests are protected from encroachment. A variety of subsistence and commercial crops are produced, maintaining soil fertility and generating biological diversity at the same time (Cairns 2007).

Combodia

Fox and Vogler (2005) mentioned that the shifting cultivation is mainly confined to the northeast of Combodia and that a fast decrease in the area was observed between 1965 and 1992, which was linked to a sharp increase in population of the country. FAO assessed that 5 to 6 million people were depending on forest in Cambodia (Ma 1999).

Laos PDR

Throughout the uplands of Laos, shifting cultivation is practiced by the indigenous communities. About 1.0 to 1.2 million people, were engaged in shifting cultivation during the early 1990s (Fujisaka 1991; Pravongviengkham 1998). But, FAO claimed during the same period that about 3.0 million people were forest dependent (Ma 1999). However, Chazee (1994) mentioned that about 0.3 million families were forests dependent in Laos and we consider six persons per family, then total number of forest dependent people amounts to 1.8 million people.

A study conducted in northern Laos, by a team of IRD and NAFRI researchers on upland rice cultivated using the swidden method, has revealed that slash and burn farming practised on the tropical forest slopes encourages soil carbon storage. If these 'shifting' cultivation strategies were to be replaced by uninterrupted crop growing, as FAO advocates, the soil's organic carbon storage capacity would diminish significantly. There have been many hitches in the process because of the vast remote areas to be covered and under-representation of the area by the farmers under the practice due to restrictions imposed by the government on shifting cultivation. Because of these issues with data acquisition, the data must be thoroughly scrutinised. Furthermore the Lao government has a political motivation to under-represent the area under shifting cultivation in order to show that the practise is being reduced and eventually eliminated. There is a great variation in the data collected by scholars and that given by the government.

Upland communities have been practicing shifting cultivation, which is one of the main reasons of deforestation in Lao PDR (DoF, 1996). The forest area declined from 17 million ha (about 71.7% of the land area) in the 1940 to about 11.4 million ha (48.1% of land area) in 1989 (Fig.3.8). According to the government of Lao PDR estimates 0.1 million ha of mature forest and 0.3 million ha of secondary forest areas are degraded by shifting cultivation each year (DoF, 1996). For millennia, Lao PDR has practised traditional shifting cultivation with a protracted fallow time (Watters, 1971). According to Roder *et al.* (1997), the average fallow periods in Northern Lao PDR during the 1950s, 1970s, and 1992 were 38, 20, and 5 years, respectively.

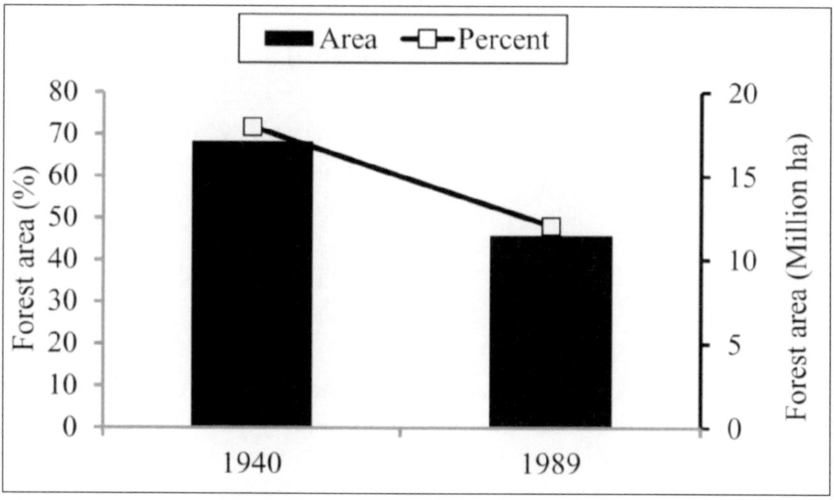

Fig. 3.8. Decrease in forest area in Lao PDR over a period of 49 years.

Malaysia

Sabah and Sarawak are the places in Malaysia where shifting cultivators are mainly concentrated and only a small number of Orang Asli farmers still do shifting cultivation. More than 50 ethnic groups of Sabah and Sarawak that predominantly have been living in the country side, practice shifting cultivation. For Sabah and Sarawak, no particular data about the number of shifting cultivators is available. However, according to Chan (1987), Sarawak's rural population in 1985 was 1.2 million, with a great majority of these being shifting cultivators. In Malaysia, the FAO estimated that 4 to 5 million people get their livelihood from forests (Ma 1999), however this number may have fallen since then.

Indonesia

In many parts of Indonesia, shifting cultivation is the major land use (Sulistyawati *et al.* 2005). A ccording to Sunderlin *et al.* (2000), about six million people are practising shifting cultivation in the five main outer islands of the country, including Kalimantan. In Kalimantan, shifting cultivation has been practised by indigenous people residing in remote areas whose main economic activity is growing rice through shifting cultivation. In many areas, community members are also doing work for wages and growing cash crops like, rubber and rattan (Weinstock, 1983; Dove, 1985). The question is would leaving shifting cultivation and drift to cash crops be more remunerative for shifting cultivators?

The development of Mayan society and economy began around 250 AD as per archeological scriptures. Major changes were brought to their environments by the communities before the industrialization and, for this agriculture was the driving force in the economy and shifting cultivation was the major form of land use. Mayas may have developed irrigation systems and intensive agricultural systems at appropriate locations (Humphries 1993). Kirch and Hunt (1997) mentioned that virtually similar methods were used by Polynesian settlers in New Zealand and the Pacific Islands. They arrived here around 1100 AD and within 500 years of their arrival, changed vast areas from forest to scrub lands and made many species of birds and animals extinct due to change in their habitats. The societies of shifting cultivators in the tropical forest of Yucatán and the highlands of New Guinea, increased in numbers and to develop layered and sometimes complex social order. Transformations in ecological and social systems are driven mainly by theory of evolution and partly by human interventions and alterations. The humans, however, have the capacity to learn and to communicate and transfer their knowledge to next generations.

Boserup (1965) argued that low intensity farming has lower labor costs than more intensive farming, though it remains debatable. She also argued that if asked, a human would always select a method which has the lowest labour requirement rather than the highest yield. FAO (1990) put the annual rate of deforestation in Southeast Asia at 34,000 km^2 and Indonesia alone accounted for 38.5% of this, the forest loss being 13,100 km^2 per year in the country.

The data from the Ministry of Forestry studies in 1983 is presented in Table 3.5 (Weinstock 1990). The total number of shifting cultivators in Indonesia, thus, stands at about 5.792 million during 1980, out of a population of 55.8 million of the country at that time.

Table 3.5. Showing the population of shifting cultivators in Indonesia

Region of Indonesia	Shifting cultivators ('000)
Sumatra (without Lampung)	1310.65
Kalimantan	1142.35
Sulawesi	1217.85
Moluccas	93.52
Papua (Irian Jaya)	598.20
Nusa Tenggara	64.80
East Nusa Tenggara	1189.85
East Timor	175.31
Total	5792.53

India

The North Eastern Region has rich natural resources but their disorganized use has caused ecosystem degradation because of large scale deforestation in the region (Mishra and Sharma 2001). The unsuitable agricultural practices to satisfy the needs for fuel, forage and timber have influenced the forest cover and caused destruction of this valuable resource. There is need to interject suitable methodologies to discourage deforestation. The region has high agricultural potential which is required to be exploited through suitable approaches (Sharma 1993, Sharma and Sharma 2009).

3.10.3 Present status, including past discussion and existing policy

In shifting cultivation, after cultivating the plot for 1-3 years, land is left fallow and process of slashing and burning is repeated on a new piece of land designated for the practice. With increase in requirements for increasing population, the shifting cycle has come down to 3-7 years fron 25 to 30 years, which is presumably bringing down the positive facets of shifting cultivation leading to land degradation and ecological problems. As per North Eastern Council, Shillong, Meghalaya, about 14.66 lakh ha area is affected by shifting cultivation in the North Eastern Region. No comprehensive and focused study has been made to ascertain the correct number of shifting cultivators and area they cultivate. Because shifting farming in the tropics is primarily done on nutrient-depleted soils, forest vegetation restoration is critical for supplying adequate agricultural nutrients for optimum crop growth. According to one FAO estimate, shifting agriculture covers over 360 million hectares, or 30% of the world's farmed land, and employs over 250 million people. Table 1 shows the total area of shifting cultivation mapped using land cover maps in selected tropical Asian nations (Silva *et al.* 2011). In India, the area under shifting agriculture in states other than the north-eastern region is estimated to be 2.32 million acres, with 2.53 million families involved (Table 3.6). Odisha leads with 5298 km^2 area, annually. Here it is called 'podu' cultivation. Annual area under shifting cultivation in north-eastern states of India is 3690 km^2, while total affected area is 14660 km^2 with shifting cycle from 3 to 7 years. Number of families involved in the practice in the region are 443.3 thousands.

Table 3.6. Shifting cultivation area and families involved in some states of India

State	Families Involved (million)	Area (million ha)
Andhra Pradesh	0.11	0.15
Bihar	0.23	0.19
Madhya Pradesh	0.19	0.38
Odisha	2.00	1.60
North-eastern states	0.44	1.46
Total	2.97	3.38

Source: Shifting cultivation in India, ICAR

Abandoned jhums are not given any attention. Normally in a village or clan the piece of land is selected, slashing and burning is done and cultivated for 2-4 years. The next cycle of slashing and burning is done when new piece of land is selected. Finding ways to discourage shifting cultivation and facilitate adoption of more permanent forms of agriculture continues to be a high priority policy of the central government.

The Government of India"s report of 1995 also accepted shifting cultivation as a major cause of land degradation because it degrades natural landscape through soil erosion and destroying forests and in the mean time agricultural productivity and income of the shifting cultivators has constantly declined. On the other hand the, Socio Economic implication' also believed shifting cultivation as a major source of livelihood and a way of life to the people. In the mean time consistent scientific studies have been suggesting an optimal and careful utilization of natural resources which will stabilise and sustain agriculture (Ramakrishna, 1992; Sharma, 1992). It is very important to control or improve shifting cultivation to safeguard ecology and raise crop productivity. Shifting cultivation is dominant in the highlands of Northeast India (Sati and Rinaawma 2014). It is practiced from mountains to lowland ecosystems and from tropical forests to grasslands (Spencer, 1966a).

3.10.4 Shifting cultivation in Indian states

A globally applicable definition of shifting cultivation is difficult to determine (Thrupp, Hecht, & Browder, 1997). The number of families involved and area under shifting cultivation as given by Indian Council of Agricultural Research (ICAR) for different states of India is presented in Table. A primitive system of agriculture, the shifting cultivation is the first step in transition from food gathering and hunting to food production, is nearly 9000 years old agricultural practice and believed to have originated during Neolithic period. This is extensively practised by the tribals throughout the tropical and subtropical regions and in India, about 5.0 million tribal families are practising this system on 4.37 million ha of land covering 11 states (Table 3.7).

Table 3.7. Shifting cultivation in different states of India

State	Tribal families (million)	Total area (million ha)	State	Tribal families (million)	Total area (million ha)
Andhra Pradesh	0.11	0.15	Meghalaya	0.61	0.47
Arunachal Pradesh	0.43	0.21	Mizoram	0.40	0.19
Assam	0.31	0.31	Nagaland	0.19	0.12
Bihar	0.23	0.19	Odisha	2.00	1.60
Madhya Pradesh	0.19	0.38	Tripura	0.19	0.49
Manipur	0.36	0.26	**Total**	**5.02**	**4.37**

Source : Shifting cultivation in India, ICAR

3.10.5 Shifting cultivation: extent and impact

A huge quantity of forest vegetation is burnt in the shifting cultivation. Nair and Fenandes (1985) was of the view that the forests are in danger of disappearing if no corrective measures were taken to arrest its degradation and destruction. Watters (1960) reported the loss of wild life, flora and fauna, wild plants of diverse gene pool due to burning of forest vegetation. The shifting cultivation practice prevalent in the region causes tremendous loss of soil and nutrients (Shahlace *et al.*, 1991) but, these losses can be minimized by managing the watersheds (Blackburn *et al.*, 1986, 1990; Singh and Singh 1981, Satapathi, 1996). Table 3.8 shows the losses of soil and nutrients due to shifting cultivation from the northeastern states of India. Annual soil loss has been estimated to be 88346 thousand tones and combined loss of N, P and K 17.092 thousand tones due to shifting cultivation (Sharma and Prasad, 1995).

Table 3.8. Area ('000 ha) and, soil and available nutrients ('000 t) loss due to shifting cultivation.

State	Area under SC	% of net sown area	Soil loss	Nutrient loss
Arunachal Pradesh	70.0	4.7	14490	3.111
Assam	69.6	2.6	12318	1.685
Manipur	90.0	64.2	20430	4.289
Meghalaya	53.0	27.0	14151	2.999
Mizoram	63.0	46.0	13041	2.738
Nagaland	19.0	34.5	7962	1.432
Tripura	22.3	8.2	5954	0.839
Total	386.9	-	88346	17.092

SC: Shifting cultivation

It is generally accepted that a large proportion (60-95%) of the nutrients in a tropical forest ecosystem are stored in the above ground biomass (Folster, 1986). These nutrients (in the ash) are released into the soil by burning the vegetation, making them available to the crop. Besides as a nutrient source, the alkaline ash causes soil pH to increase and as a result both soil acidity and Al/Mn toxicity are controlled (Sanches *et al.*, 1982). These positive effects are optimized by burning the biomass when it is not completely dry and at temperature below 3000C (Fehlberg, 1989). In this system ,growers only use the seeds as input and not any other inputs. The duration of fallow relative to the number of harvests is therefore an important factor affecting stability of the systems. Aweto (1981) opined a fallow-to-cropping ration of 5:1 to be minimum requirement for sustainability.

In planting cultivated crops, the shifting farmers create a remarkably stable sub system based upon knowledge of the growth requirements of every component so that full use is made of sunlight and soil nutrients, and of the complementarities among crops (Jackson, 2001). The large number of crops over space and time are effectively managed taking advantages of micro-environmental variations and for sequential harvesting during the year. Thus field is used for a year for mixed cropping and thereafter it is either abandoned or cultivated for one or two selected crops. It ensures self-sufficiency in food in terms of calories, and a great variety of foodstuffs. The shifting cultivators' diets are mainly from the foods gathered and hunted from the surrounding forest.

3.10.6 Farmers engaged in the practice of shifting cultivation in Nagaland

It may be noted that this diet is far superior to the diet of stale, contaminated (by pesticides and permitted food additives) processed foods on which present-day urban dwellers are surviving (Jackson. 2001). Most jhumias also traditionally produced small amounts of saleable commodities. The careful selection of crop species/varieties to adapt the local ecosystem get small damages by insect-pests and diseases. Cultivation lasts only from two to three years, and once the soil fertility of a particular land declines , the area is left for certain years to allow the vegetation to regenerate. Another patch of land is cleared and cultivated and so on. The mode of land allotment to individuals, varies from tribe to tribe, such as by village council in Mizo hills, lottery system in Manipur and Mizo hills, by village headman (Nokma) in Garo hills etc. The area for cultivation is decided on the basis of family size. When population pressure was low and abundant land was available, the forest (bush) fallow period might vary from 25 to 30 or more years. During which time farmer was cultivating in a series of sites elsewhere in the forest. Now with the increase in population and declared reserve forest areas, the fallow period has reduced between 2-5 years. Ramakrishnan and Toky (1981) recorded total biomass (root and shoot) yield from grain and seed yielding crops was 2.5 to 20.5 times more under a 30 year jhum cycle than that of under 10-and 5-year cycles. The overall productivity of the system was maximum under 10 years jhum cycle. In terms of maintenance of biodiversity, cultivation clearly scores higher marks than any system of sedentary agriculture in terms of reliance on adopted species / varieties. In a given area, perhaps in some cases, several varieties of the same crop are grown, having been selected over time for their adaptation to varying soil conditions and farmers' needs. Sarma and Sharma (2000) reported that 13 varieties of rice are grown by Lakhers of Mizoram.

3.10.7 Impact of shifting cultivation

The practice of shifting cultivation on hill slopes has led the elimination of water sources and increased the run-off water, resulting in soil erosion (Fig. 3.9). The destabilization of mountain slopes through shifting agriculture is one of the factors to increase the floods in Brahmaputra plains, extension of delta and formation of islands in the Bay of Bengal, massive siltation and reduction of the available water resources. In Manipur, Loktak lake is silting with sediment materials eroded from the vegetation less hills around it. The various adverse effect of shifting cultivation is summarised below (Borthakur, 1992).

Fig. 3.9. A view of shifting cultivation in hills of Tripura. (*Source*: Datta *et al.* 2004) (See colour version on page 652)

The major effects of shifting cultivation are presented in Table 3.9.

Table 3.9. Major effects of shifting cultivation

Major effects	Results
Soil erosion	1.Silting of reservoirs and stream causing floods 2. Loss of soil productivity 3 Low productivity, 4. Pressure on land
Denudation of forest	1.Ecological imbalance 2. Elimination of water resources
Low technology	1.Labour intensive 2. No subsidiary income
Social aspects	1.Social customs 2. Large family size

3.10.8 Prevalence of shifting cultivation in different states of India

Though shifting cultivation is considered environmentally destructive and economically unfeasible, it is still practiced by the tribal people of Northeast India. The government and pro-environment agencies are trying to replace shifting cultivation. However, for the shiting cultivators, it is more than just a method of cultivation as they are connected with it from generations and owe shifting cultivation to their traditional beliefs and the ceremonies and rituals are held at the end of a successful harvest. None-the-less several cultivators have adopted alternative farming methods in the form of integrated farming, poultry, piggery and terrace cultivation. They have also adopted other agricultural practices in the form of fishery, bee keeping, tea, rubber and floriculture. Yet the practice is still very much there. Perhaps, the economic condition and lesser labour required in shifting cultivation makes it a preferable farming method.

Eastern Himalayas of India

Eight Northeastern Himalayan states states of India constitute Eastern India Himalayas (EIH), with a total area of 26.2 million ha and elevation varying from 73 m to more than 8000 m amsl. The basic information about eight states namely; Arunachal Pradesh, Assam, Manipur, Meghalaya, Mizoram, Nagaland, Sikkim and Tripura is provided in Table 3.10,. The rate of growth of population is higher in these states compared to the country as a whole (Fig. 3.10). The density of population was 340 persons km-2 in Assam as compared to 13 persons km-2 in Arunachal Pradesh. Lowest percentage of area affected by shifting cultivation is in Arunachal Pradesh (3.7% of GA) while highest percentage of area is in Manipur (538% of GA) (Fig. 3.11). Families involved in shifting cultivation and the area under the practice is given in Fig. 3.12. The area per family is constantly decreasing, which has affected the food security in the region and the people involved are finding other avenues of income for sustenance.

Table 3.10. Features of North Eastern States

State	Area (km2)	Population ('000) (2011)	Density (persons/km2)	Growth rate 1981-91	(%) 1991-01	Sex Ratio
Arunachal	83,743	1,382	16	36.8	26.21	901
Assam	78,438	31,169	398	24.2	18.85	932
Manipur	22,327	2,721	122	29.3	30.02	978
Meghalaya	22,429	2,964	132	32.9	29.94	975
Mizoram	22,081	1,091	51	39.7	29.18	938
Nagaland	16,579	1,980	120	56.1	64.41	909
Tripura	10,486	3,191	350	34.3	15.74	950
Total	255,083	44,498	174	34.1	30.62	940

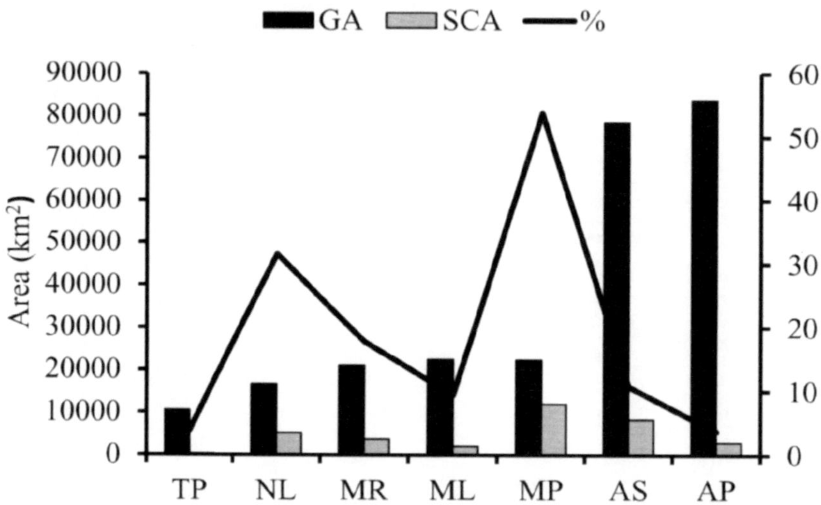

Fig. 3.10. Shifting cultivation area (SCA) as percent of geographical area (GA) in IEH states*. *AP, Arunachal Pradesh; AS, Assam; MP, Manipur; ML, Meghalaya; MR, Mizoram; NL Nagaland; SK, Sikkim; TP, Tripura.

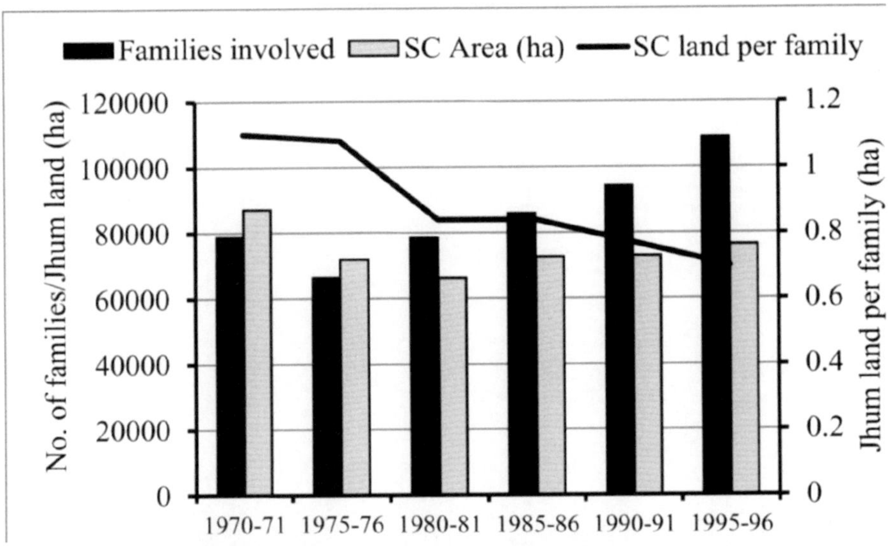

Fig. 3.11. Change in shifting cultivation area, families and cultivated area per family in EIH.

3.10.9 Area and extent

There is significant variability in area under shifting cultivation and families involved as different scholars have divergent views regarding this aspect. It is also expressed by some scholars that area under this indigenous practice and number of number of cultivators have not reduced in the last few decades. No comprehensive and focused study has been made in this direction. Tropical forests are important component of the ecosystem to deliver ecosystem services and livelihoods security of many indigenous peoples (Sharma and Bujarbaruah 2016). Shifting cultivation is a small scale farming and known to be a major causes of deforestation. During the fallow period, the forest vegetation re-generates and can be re-burned during the cycle, adding nutrients to the soil for future cropping. Since shifting cultivation in the tropics is mainly practiced on nutrient-poor soils, forest vegetation regrowth is important for addition of sufficient crop nutrients for optimum crop growth. In the practice of burning; weed, pest, microbe population decline and thus the crop diseases. Fallow periods in a shifting cultivation system vary and can be long enough for forests in abandoned plots to regenerate (Sharma and Bujarbaruah 2016). One FAO estimate reveals that shifting cultivation extends over approximately 360 million hectares or 30% of the cultivated area in the world, and supports over 250 million people. The longer a field is cropped, the greater the loss of soil organic matter, cation-exchange-capacity, nitrogen, phosphorus etc. and infiltration capacity is reduced. Table 3.11 shows the extent of shifting cultivation and soil loss in northeastern states of India.

Table 3.11. Shifting cultivation and, soil and nutrient loss in Northeast Indian states.

State	Annual area ('000 ha)	Total area involved ('000 ha)*	Fallow period (years)	No. of families Involved ('000)	Soil loss (Million tones)	Soil loss (t ha-1)	Land available per family (ha)
Arunachal Pradesh	70	210.0	3-10	54.0	14.5	69.0	1.30
Assam	90	139.2	2-10	58.0	12.3	88.3	1.20
Manipur	69	360.0	4-7	70.0	20.4	66.8	1.28
Meghalaya	53	265.0	5-7	50.0	14.2	53.6	1.01
Mizoram	63	189.0	3-4	54.3	13.0	68.8	1.26
Nagaland	19	191.3	5-8	111.1	11.0	57.5	0.17
Tripura	22	111.5	4-9	43.0	5.9	52.9	0.46
Total	386	1466.0	2-10	440.4	88.3	60.2	0.87

The rural economy of northeast India is mainly linked to shifting cultivation. With increase in population, the rotational cycle has come down to 3-10 years and the land does not get enough time for rejuvenation. As high as 130.2 t ha^{-1} of soil erosion has been reported in north-eastern hill region by shifting cultivation on a hills having a slope of 70% (Singh and Singh 1989). Shifting cultivation implies the whole nexus of people's religious belief, attitude, self-image and tribal identity. This kind of inter-connections between different elements and domains of social life restricts the cultivators to leave shifting cultivation unless suitable acceptable alternate options are available. Annual soil loss from the NE region is 88.3 million tonnes from shifting cultivation land, 90.7 million tonnes from other agricultural land and 422.7 million tonnes from non-agricultural land. It was estimated that annual loss of N, P, K, Mn, Zn, Ca and Mg is 686.0, 100.2, 511.0, 22.7, 14.0, 57.3 and 43.4 thousand tonnes, respectively (Sharma, 2001).

In north eastern region, the mountain with its restricting influence on settlement and movement, did not provide cultural advancement and, therefore, it has the considerable effect on the economic life of the inhabitants. They accepted the nature as it is and fought it valiantly, working on the mountain slopes for centuries in the way as their ancestor did. The prevalent primitive form of shifting agriculture (jhum) is the result of their urge for survival in the most difficult hilly terrains of the region. In 1983, the Task Force on shifting cultivation (Ministry of Agriculture) estimated that the total area under this system was 14,66,000 hectares (386,900 hectares, annually) and 440,000 tribal families were engaged. The area (1.77 % of TGA) under shifting cultivation is minimum in Assam.

On the other hand, Forest Survey of India (1999) estimated the cumulative area of 1.73 million ha affected by shifting cultivation during the 1987-1997. Fig. shows the total area and area per family in Northeastern states of India.

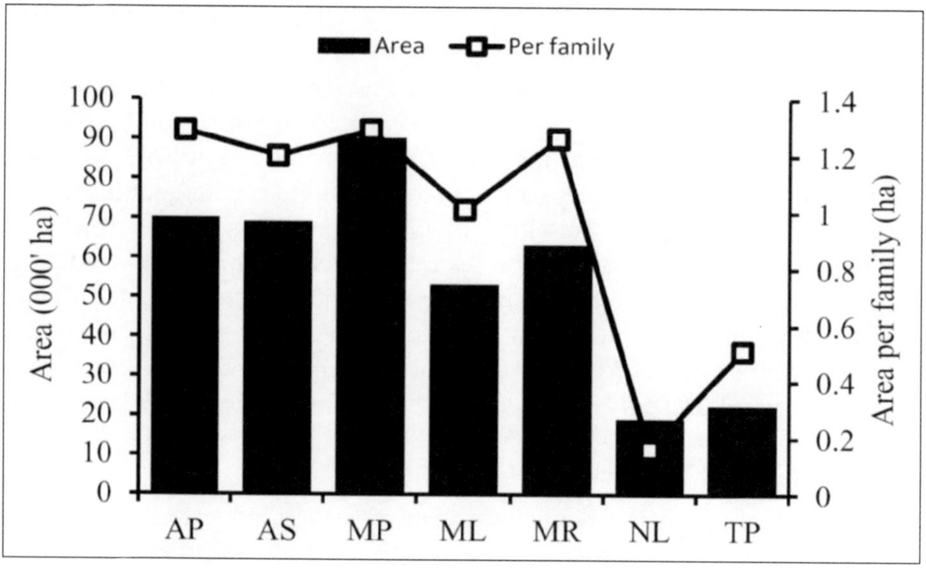

Fig. 3.12. Total area under shifting cultivation in NE states of India and area per family.

EIH is situated at the confluence of Indo-Malayan, Indo-Chinese, and Indian bio-geographical realms. Because of its geographical location, EIH represents diverse ecosystem types dominated by mountains and is endowed with rich forest resources. There are over 100 different tribes in EIH, differing linguistically and culturally (Ramakrishnan, 1992) practicing shifting agriculture for millennia. This agricultural practice has evolved as a part of the culture of the hill people of the region, and the agricultural procedures are closely linked with the sociocultural practices and religious beliefs (Ramakrishnan 1992, Sharma, 1999). However, increasing population pressure has drastically reduced the shifting cycle and has been the cause of poverty and food insecurity for the hill farmers and environmental degradation in the landscape. Land area measuring 1.5 M ha is under shifting cultivation in NEI (Roy *et al.*, 2012). Endangering the soil resilience are the soil erosion, nutrient loss and ecosystem disservices provoked by shifting cultivation, resulting in poverty and food insecurity among the farmers. Change in the area under shifting cultivation in Northeastern Indian states is presented through Fig. 3.13.

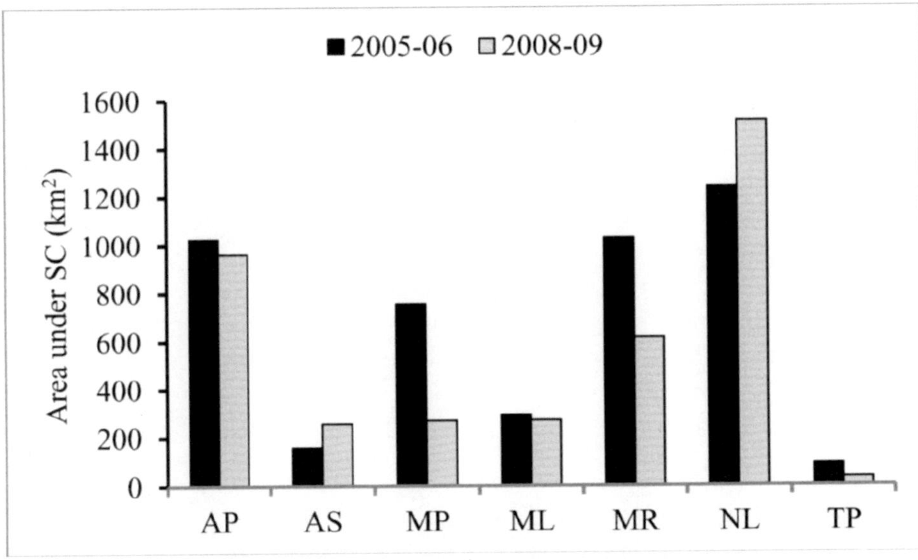

Fig. 3.13. Current area under shifting cultivation (SC) in Northeastern states of India AP-Arunachal Pradesh, AS-Assam, MP- Manipur, ML- Meghalaya, MR-Mizoram, NL-Nagaland, TP-Tripura

In northeast India, though government enacted many policies and programmes to stop shifting cultivation but the results were not encouraging (Deka and Sarmah, 2010). Rapid increase in population was one of the the main reason for unsustainability of the practice of shiting cultivation in the region. Table 3.12 gives a clear picture about the fast increase in population in northeastern states of India.

Table 3.12. Area and population of Northeastern states of India

State	**Area (km2)**	**Population**		**ACGR of population (%)**
		1991	**2011**	
Arunachal Pradesh	83743	864558	1382611	2.37
Assam	78438	22414322	31169272	1.66
Manipur	22327	1837149	2721756	1.98
Meghalaya	22429	1774778	2964007	2.59
Mizoram	22081	689756	1091014	2.32
Nagaland	16579	1209546	1980602	2.50
Sikkim	7096	406457	607688	2.03
Tripura	10486	2757205	3671032	1.44
NE Region	263179	31953771	45587982	1.79

3.10.10 State-wise shifting cultivation

Tripura

The Tripura state of India lies between 22056' N to 24032' N latitudes and 91009' E to 92020' E longitudes, with an area of 10492 km2. Topographically the state has a number of hill ranges, scattered with wide fields inbetween (Das and Das 2014). The state has humid climate and tropical evergreen deciduous forest. Tripura has 19 tribal groups with a tribal population of about 31 % (census 2011). The shifting cultivation is a clean operation and keeps the tribal in the open, enjoying, singing and dancing. No wonder the tribal refuses to go down to the valleys to lead a more 'settled' life on the paddy field! Work and leisure for him are not two distinct and mutually opposing entities; they are two sides of the same coin". Over 10,039 ha of land is under shifting cultivation in the state.

Pragmatic side of swidden

1. Shifting cultivation helps the indigenous communities in order to protect their cultural legacy.
2. Shifting cultivation is a labour intensive an assured source of food security.
3. The practice in its traditional form, conserve agro-biodiversity, native food, vegetables and fruit crops.
4. It is an effective form of land use satisfying almost all needs of shifting cultivators and rreduce their reliance on other activities and external inputs.
5. With shifting cultivation the cultivators also practice their religious beliefs and perform rituals so important for their ethos and mental satisfaction.
6. The shifting cultivation prevent soil erosion partially because of minimum tillage operations (Das and Das 2014).

Negative aspects of shifting cultivation

With positive effects of shifting cultivation, some negative effects are also associated.

1. In 1876, W.W. Hunter mentioned that "regression of forests had already started in hills because of shifting cultivation practiced by almost the whole population numbering less than 50000 who were all tribals". It causes deforestation and land degradation.
2. Deforestation due to shifting cultivation encourage GHG emissions.

3. Deforestation and burning of vegetation disturb the flora and fauna of the forest.
4. Shifting cultivation causes soil erosion on hill slopes and damage to soil system, if not managed properly.
5. Reduction in shiting cycle eventually converts the shifting cultivation plots into degraded wastelands (Das and Das 2014).
6. Excessive shifting cultivation affects the ecological balance of the ecosystem.

3.10.11 Suggestions and remedial measures

Considering the adverse effects of shifting cultivation, it is essential to adopt a broad based strategy for making the people aware of the baneful effects of the practice in order to change the mindset of the shifting cultivators of the region and ti divert them to a settled mode of cultivation. The following measures will be helpful to control shifting cultivation:

- To settle the shifting cultivators on an irrigated terraces by channeling water from mountain streams.
- Introduction of diversified farming comprising agriculture, horticulture along with animal husbandry, fishery, poultry farming etc.
- To take up intensive cultivation of plantation crops such as; tea, coffee, rubber, pineapple etc. as well as the development of non-forest wasteland in the region.
- To reduce the population pressure on cultivable land by adequate legal provisions to stop outside migration
- The concept of 'Sustainable Agricultural Development' can be an effective method for ensuring required supply of food, fibre, fuel and other amenities to the growing tribal population of the region.
- Introduction of soil and water conservation methods.

Manipur

Rapid population growth is responsible for bringing more land under shifting cultivation in Manipur state of India (Reimeingam 2017). There is need to strengthen land use and forest laws to protect forest land. According to Forest Survey of India (FSI) (2011), the loss of forest cover was 190 km2 during 2006-09 (satellite data) in Manipur. FSI (2013) attributed the loss in forest cover to the prevalence of shifting cultivation. Northeastern Council (NEC 2008) corroborated this by mentioning that shifting cultivation contributed to soil erosion, deforestation and destabilisation of the ecology.

Mizoram

Between the year 1997-98 and 2010-11, the area under shifting cultivation has decreased from 63,124 ha to 28,562 ha, a decrease of 54.6%, whereas; the area under wet rice cultivation increased from 9,446 ha to 12,130 ha, an increase of 28.4% (Fig. 3.14).

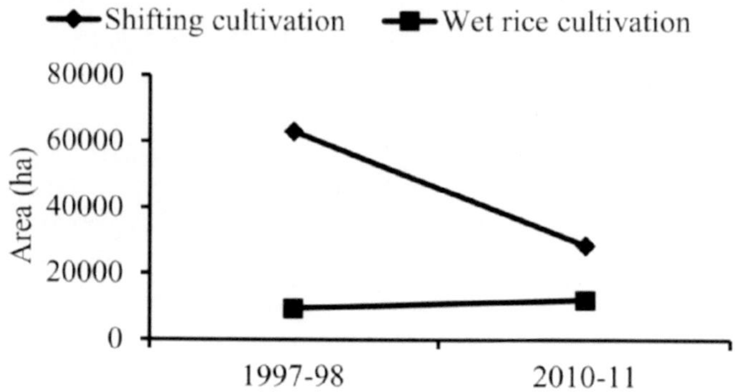

Fig. 3.14. Change in area under shifting cultivation and wet rice cultivation in Mizoram state of India.

3.10.12 Source of livelihoods

Singh and Ramakrishnan (1982) reported that shifting cultivators form about 82 per cent of the rural main workers in Mizoram. Maithani (2005a, b) observed that shifting cultivation is a major source of economy of the state and about 58 thousand families of the state are involved in the practice. The trend of decrease in the area under shifting cultivation is presented in Fig. 3.15 and number of families involve, area and decline in available area per family in Fig. 3.16.

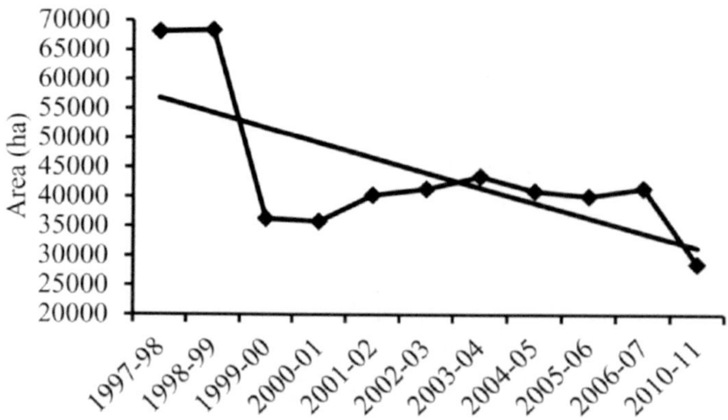

Fig. 3.15. Reduction of area under shifting cultivation in Mizoram state of India.

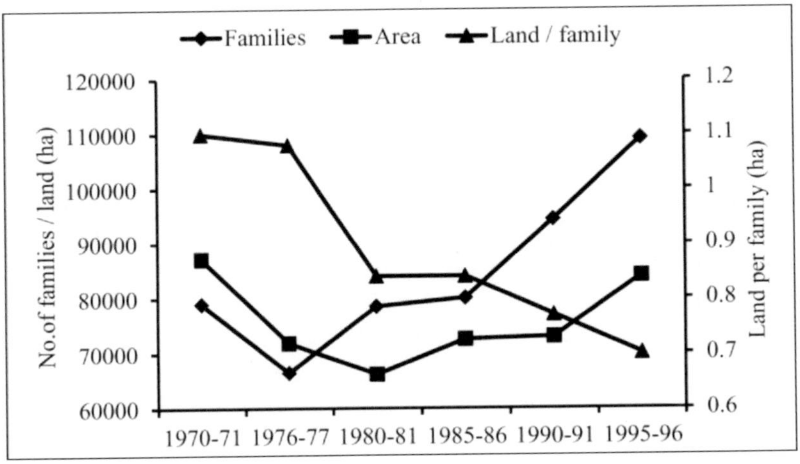

Fig. 3.16. Number of shifting cultivation families, land per family and area involved in practice in Northeast India

3.10.13 Recommendations

1. Multipurpose (usually nitrogen fixing) tree species should be left-out during the clearance.
2. Policies that should address poverty alleviation and biodiversity conservation.
3. Fallow management strategies should be improved to be more responsive to the needs of the increasing population.
4. To improve shifting cultivation land to enhance productivity so that cultivators do not shift to other areas immediately after one year of cultivation.
5. Alternatives to shifting cultivation that provide farmers with a livelihood to save forests.
6. Market development.

Meghalaya (Garo hills)

In 1980, the total area under shifting cultivation in the Garo hills (current + regenerated) was 8833 ha, but by 1990, it had dropped to 6230 ha. The area increased in size from 7487 hectares in 1995 to 7838 hectares in 2000 and 8916 hectares in 2005. From 4792 ha in 2010 to 3953 ha in 2016, the area has been shrinking (Riahtam *et al.* 2018). In 1980, the area under contemporary shifting cultivation in the East Garo Hills was around 6475 hectares, but by 1990, it had fallen to 4624 ha, and by 1995, it had decreased to 3497 ha. Current shifting cultivation area increased from 4687 ha to 5408 ha between 2000 and

2005, but then declined to 2480 ha in 2010 and 1742 ha in 2016. (Fig. 3.17). Regenerating shifting cultivation was observed to be diminishing in the East Garo Hills, from 2358 ha in 1980 to 1606 ha in 1990. The area grew to 3990 ha in 1995, then shrank to 3151 ha in 2000. The area increased to 3508 acres in 2005. Following that, the area shrank in 2010 (2312 hectares) and 2016 (2312 ha) (2211 ha). Non-shifting cultivation fields were converted to shifting agriculture zones in several regions. In some of the studied locations, there has been a movement from regenerating shifting cultivation to contemporary shifting cultivation areas since 1995. Both the area and percent change in area trend lines showed a decrease. Fig. 3.18 shows the area under shifting cultivation in East Garo Hills and per cent change.

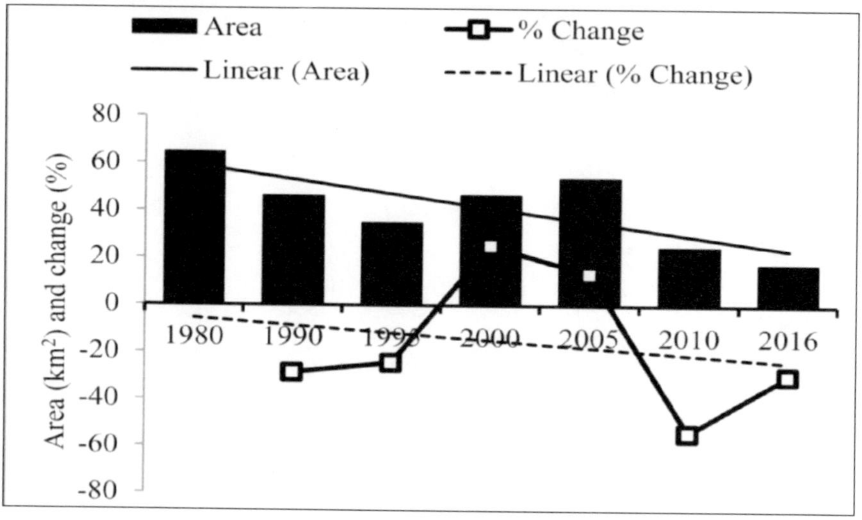

Fig. 3.17. Showing the land under the practice in Garo Hills of Meghalaya state of India and percent change over the previous estimate between 1980 and 2016.

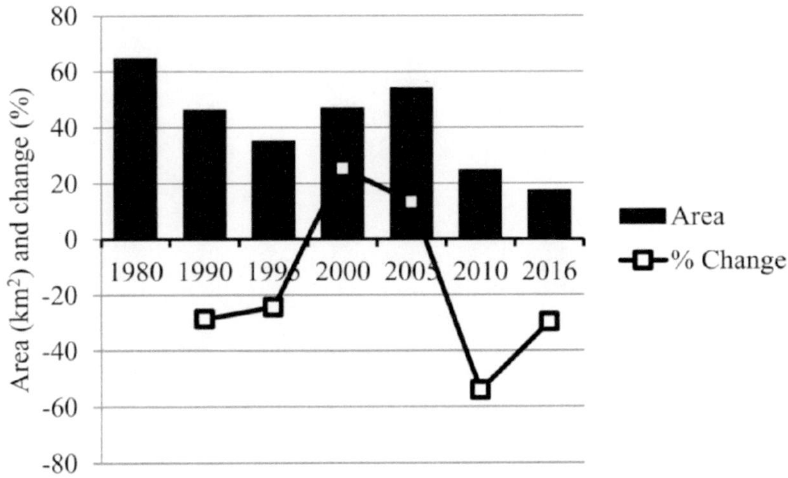

Fig. 3.18. Area under shifting cultivation in East Garo Hills and per cent change

3.11 MODELLING SWIDDEN AGRICULTURE SYSTEM

Depending on the requirement of the model and method to be followed, a number of models have been developed for shifting cultivation practice. In order to comprehend the relationship between the labour market and the cropping and fallow cycles in shifting cultivation, Dvorak (1992) developed an analytical model based on the attitudes of shifting cultivators and households in terms of resource allocation and decision-making. The findings demonstrated that the environmental and economic variables controlling the ways in which weeding, burning, and vegetation removal contribute to crop productivity also influence variations in cropping intensity. Similar models were constructed by Angelsen (1999) to assess the effects of several factors that lead to the spread of agriculture in forest land. Gilruth *et al.* (1995) created a dynamic spatial model of changing farming in the highlands of Guinea, West Africa. To investigate the long-term effects of shifting farming practises on the topography and forest vegetation, Wilkie and Finn (1988) developed a simulation model in the Ituri forest in northern Zaire. The forest composition varied mostly with population dynamics, land tenure, and fallow duration, according to their findings. There have also been some models built that consider shifting cropping in the Indonesian environment (Iskandar, 1994, Angelsen, 1999). The Kantu are a shifting cultivation group in West Kalimantan. Sulistyawati (2001) modelled their social structure and agricultural production method. The four modules of the model, which comprise productivity evaluation, vegetation dynamics, population dynamics, and land use decision-making, are shown in Fig. 3.19.

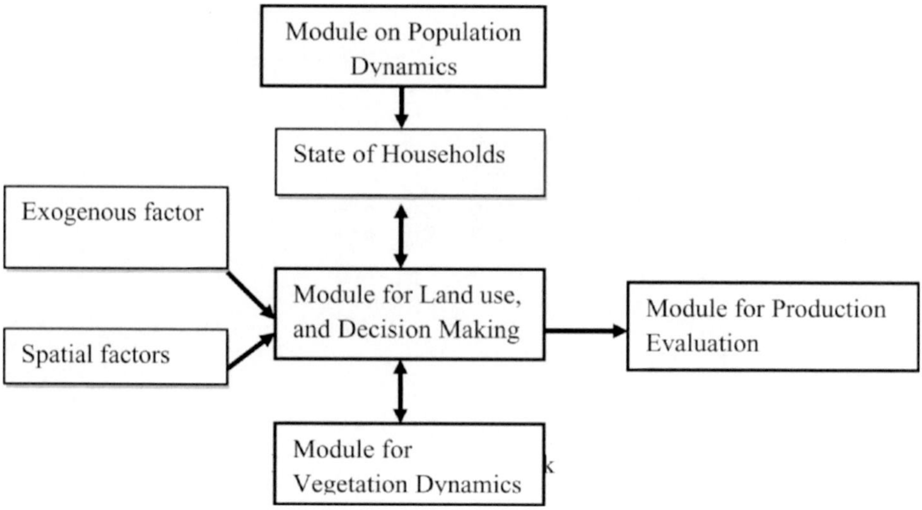

Fig. 3.19. The model showing four modules of social structure and agricultural production.

3.11.1 Important assumptions of the model

Module on population dynamics

Here, the population size is totally ascertained through births and deaths data and, simulated community is considered as a closed community.

Module for land use and decision-making

In this model, the quantity of agricultural production is determined from the needs of the household of a family living on a subsistence economy (Sahlins, 1972), and not on profit maximization. The demand for food grains depends on total requirement for home consumption, for seed and for paying debts.

Module for production evaluation

The module considers that how much rice is available to a household to meet its requirement. Here the total quantity available rice includes, the rice saved from previous year, rice produced in the current year and rice earned by working in other household practicing shifting cultivation.

Module for vegetation dynamics

In this model, the vegetation succession is simulated as an apparent change from one state into another.

3.11.2 Mathematical models

The average sediment yield per hectare was 20.58, 25.52 and 60.23 tha-1 from un-cultivated, cultivated and shifting cultivation areas, respectively (Sharma and Sharma 2003). A mathematical model was evolved to monitor the sediment yield from shifting cultivation areas. The sediment yield (Sy) prediction model involves a partial regression equation of the form: $S_y = a + b_1x_1 + b_2x_2 + b_3x_3 + b_4x_4$. Where, Sy is the sediment yield (t ha-1) and b1, b2, b3 and b4 are the partial regression coefficients for slope (x1) in %, annual rainfall (x2) in cm, vegetation cover (x3) in 1 to 5 scale and soil clay content (x4) in %, respectively. The observed and predicted values of sediment yield agreed relatively well. An extensive, multidisciplinary study has been conducted to develop sustainable, economically viable, and environmentally acceptable alternative land use methods to replace shifting cultivation and lessen runoff's sediment load. Comparing the sediment yield in the new land use systems to the shifting cultivation, which was 3621 t km^{-2}, the differences were 11.2 to 97.2 t km-2. Compared to shifting cultivation, where the in-situ retention of rainwater was only 60.2%, new land use systems retained 90.1% to 98.3% of rainfall, significantly lowering runoff and sediment production (Sharma and Sharma 2003).

The north-eastern region is rich in natural resources, but exploitation and mismanagement have resulted in soil erosion and land degradation, as well as silting and floods in river beds (Sharma 2001). The region's rapidly growing population has put strain on the food production base, and people have mismanaged and exploited water resources to suit their needs (Sharma, 2003). There is an annual loss of about 601 million tonnes of soil with runoff from 26.2 million ha of land area o the region at an average rate of 22.9 t ha-1 compred to about a loss of 88.3 million tones of soil from 1.466 million ha of shifting cultivation area at the rate of 60.2 t ha-1 and a loss of 512.7 Mt of soil from 24.73 million ha of uncultivated and cultivated area at an average rate of 20.7 t ha-1 (Sharma & Prasad, 1995) (Table 3.13). To sustain the fast increasing population in the region, there is strong need to integrated approach for natural resources management for livelihood security. The most significant matter is to promote the preservation and viable use of natural resources which may help in economic growth and productive capacity, as well as being equitable and environmentally acceptable (Bassam, 1997). The annual area under shifting cultivation, soil and nutrient loss is presented in Table (Sharma 1998).

Table 3.13. Loss of soil and nutrient from northeastern Indian states

State	Shifting cultivation			Total soil loss (million tonnes)	Nutrient loss (000' t)		
	Annual area (km2)	Fallow period (years)	Soil loss (million tonnes)		N	P	K
Arunachal pradesh	700	3-10	14.5	178.1	217	36.6	153
Assam	696	2-10	12.3	178.4	201	33.4	155
Manipur	900	4-7	20.4	64.0	76	7.4	63
Meghalaya	530	5-7	14.2	57.7	62	7.0	48
Mizoram	630	3-4	13.0	39.4	60	6.9	40
Nagaland	190	5-8	8.0	41.7	44	5.2	34
Tripura	223	4-9	5.9	15.4	26	2.7	18
Total	3869	2-10	88.3	601.2	686	99.2	511

There are roughly 443.3 thousand farm families impacted by shifting agriculture across a total area of 14660 km2 (Anonymous, 2000). The farmers are unable to abandon their practise unless appropriate alternative possibilities become available due to the relationships between many aspects of social life. In shifting cultivation, the maximum sediment yield was 45.8 t ha^{-2} when annual rainfall was 2705 mm, and the lowest sediment yield was 3.1 t km-2 when annual rainfall was 1992 mm (Sharma and Sharma 2004) (Table 3.14). The sediment production fluctuated depending on the amount of rain that fell in a given year and the type of plants present.

Table 3.14. Sediment yield as affected by land use and rainfall (t km^{-2}).

Land use	**Annual rainfall (mm)**						
	2195	**2705**	**2770**	**2599**	**2388**	**1992**	**Mean**
(Livestock) Fodders/ grasses	14.2	16.3	28.8	18.6	10.6	9.0	16.2
Forestry (trees)	60.1	115.4	141.1	131.7	69.9	65.3	97.2
Agro-forestry	35.4	70.2	75.6	74.3	37.6	27.8	53.4
Agriculture	3.9	9.8	24.3	22.7	3.7	3.1	11.2
Agri-horti-silvi-pastoral	20.1	37.4	36.0	36.5	18.2	11.6	26.6
Horticulture	65.0	101.4	124.8	80.2	70.5	51.7	82.2
Shifting cultivation	2950.0	4580.0	4499.7	3610.0	3419.1	2669.4	3621.3
Mean	447.2	704.2	704.3	567.7	518.5	405.4	
C.D. (p=0.05)	Land use = 77.6; Precipitation = 66.2; Land use x Precipitation = 162.5						

Water yield

The water yield from various watersheds with different land use systems varied considerably (Table 3.15). Insitu retention of rain water was significantly low in watersheds with shifting cultivation and runoff was relatively very high (Singh 1988).

Table 3.15. Effect of land use on water yield and soil loss from different watersheds.

Parameters	Land use					
	Livestock based	**Forestry**	**Agro-forestry**	**Agriculture**	**Agri-horti-silvi-pastoral**	**Shifting cultivation**
Total water yield (mm)	14.0	433.3	237.6	20.9	77.6	835.3
Per cent of rainfall	0.6	17.7	9.6	0.9	3.1	34.1
In situ rainwater retention (%)	99.4	82.3	90.4	99.1	96.9	65.9
Base flow (mm)	2.9	365.3	202.2	0.5	8.5	275.3
Surface flow (mm)	11.1	68.0	35.4	20.4	69.1	560.0
Soil loss (t ha-1)	0.2	2.0	1.9	0.1	1.8	42.4
Benefit/cost ratio	2.1	1.2	1.5	1.8	1.9	0.6

3.11.3 Modelling sediment yield

A mathematical model was evolved to monitor the sediment yield from shifting cultivation areas. The sediment yield (Sy) prediction model involves a partial regression equation of the form:

$$Sy = a + b1x1 + b2x2 + b3x3 + b4x4$$

The parameters that represent the partial regression coefficients for slope (x1) in %, yearly rainfall (x2) in cm, plant cover (x3) in 1 to 5 scale, and soil clay content (x4) in % are Sy, the sediment yield (t ha-1) and b1, b2, b3, and b4. The Dolittle approach, which produced a fairly high degree of accuracy in the forecast or calculation of sediment yield, was used to build the equation (Sharma and Sharma 2003). The equation can be solved by direct application of elementary methods involving successive elimination of the unknowns. The vegetation is the only independent variable for which the values (1 to 5) have to be based on visual estimates, 1 being the bare soil surface and 5, the most dense vegetation. More available data for calculating the values of the coefficients may result in an increase in the accuracy of predictions. In principle, the model can be used for predicting sediment yield under a wide range of climatic and physical conditions of the catchments involving shifting cultivation. In the northeastern region; slope, amount and intensity of rainfall, vegetation cover and soil texture are the most important factors determining the sediment yield from the shifting cultivation areas. In the regression equation, the values of a, b1, b2, b3, and b4 were, +37.65, +0.499, +0.125, -12.083, and -1.065, respectively (Sharma and Sharma 2003). The model was validated by comparing the predicted values with the observed values of the data obtained from different locations (Table 3.16). The sediment yield values are for the first year of shifting cultivation. There was difference in sediment yield for the second and third year of shifting cultivation and multiplying factors, 1.254 and 0.816, respectively, were evolved for these years. The sediment yield derived from the model is multiplied by these factors for the second and third years, respectively, of shifting cultivation. The sediment yield is higher during second year of cultivation which subsequently reduces. The shifting cultivators, generally, select and shift to a new site after third year. The observed and predicted values of sediment yield agreed relatively well.

Table 3.16. Observed and Predicted values of sediment yield

Location	Slope (%)	Rainfall (cm)	Vegetaion (1-5 scale)	Clay (%)	Sediment (t ha^{-2})	Yield
					Observed	Predicted
1	30	241	5	15	2.16	6.35
2	26	235	5	13	1.09	5.73
3	9	208	4	12	3.47	7.02
4	6	264	5	11	0.90	1.52
5	17	194	4	15	4.15	6.07
6*	70	254	2	14	130.20	65.32
7*	42	245	2	12	36.32	52.28

*Shifting cultivation

3.12 FOOD SECURITY AND SHIFTING CULTIVATION

3.12.1 Maintenance of the agroecosystem

The biotic community supplies the necessary nutrients for the forest vegetation. The same strategy is used by the shifting cultivators. The shifting cultivation creates a system of "fast decomposition" that forge the scheduling of nutrients in a tropical forest. Not depending on the natural decomposition of the tropical forest to provide nutrients, the shifting cultivators "accelerates natural decay by the burning of the slashed and felled vegetation" (Warner 1991). For wild animals, secondary forests have a larger carrying capacity than primary forests (Peterson 1981). Shifting cultivators believe that soil fertility is linked to forest growth, and a fully established forest is supposed to have areas that are better for agricultural cultivation (Dove 1983; Warner 1981).

In South and Southeast Asia, shifting cultivation met the subsistence needs of a sizable population for an extended period of time although population densities were low (Rasul and Thapa 2003). This has, however, been deliberated to be hostile to environment and uneconomic practice. The practice has been, by and large, replaced by permanent farming in Nepal, a gaint shift has taken place in Thailand, and restrained shifts have been noticed in Indonesia and Malaysia. Shifting agriculture, on the other hand, is still popular in Bangladesh's and Laos' hills, as well as northeastern India. Rasul and Thapa (2003) came to the conclusion that the shift from shifting to permanent agriculture, as stated by Boserup, cannot be explained solely by population pressure (1965). Favourable environment is important for a change and this condition is created by population growth, government will, ownership rights and provision of necessary support services.

Shiting agriculture occurs in a variety of contexts and forms (Concklin, 1957; Spencer, 1966; Ruthenberg, 1976), and involves simple tools and techniques, little investment, and subsistence-level production and consumption (Concklin, 1957; Spencer, 1966; Ruthenberg, 1976). (Geertz, 1963; Ruthenberg, 1976). Though intensive agriculture is gradually replacing shifting cultivation (Grigg, 1974), it is still practised in the highlands of South and Southeast Asia, as well as the tropics of Africa and the Americas. According to Sanchez (1996), shifting cultivators removed around 10 million acres of forest and scrubland every year across tropical Asia, Africa, and Latin America in the mid-1990s. Shifting agriculture was responsible for over half of the deforestation in South and Southeast Asia (FAO 1985).

3.12.2 Livelihood changes: drivers, constraints and opportunities

Due to necessity and choice, livelihoods in ethnic tribes have become more varied. Scarcity of land and market availability are the main outside drivers of change in current livelihood changes. Throughout Asia, government policies, in the name of resources and environment protection, are targeted towards elimination of shifting cultivation. Due to non-recognition of land rights and tenure insecurity, are the major constraints for many indigenous communitie to retain livelihood and food security. Land tenure, food security, and livelihood policies at all levels (national, regional, and global) must be strengthened based on the principle of equity. Awareness campaigns are required to dispel beliefs about shifting cultivation as a driver and cause of deforestation. There is also need to recognize and promote traditional knowledge of shifting cultivators.

3.13 FOREST FIRES

Fire is used for burning forest vegetation in the tropics (Eva and Lambin, 2000; Aragãoand *et al.*, 2008). Fire incidences during clearing, agricultural intensification and drought period is a matter of concern in the tropics (Cochrane, 2003). Fire is a predominant man induced disturbance in in shifting cultivation. In India, nearly 0.45 million families practice shiting cultivation who depend on fire for clearing (Lele and Joshi, 2009). With decrease in fallow period, the pattern of shifting cultivation has changed and fire frequency has increased (Ramakrishnan, 2001). The forest fire have increased due to increase in shifting cultivation in the moist deciduous forests.

According to Crutzen and Lelieveld (2001) and Badrinath *et al.* (2004), higher fire activity resulted in increased greenhouse gas emissions, reactive chemicals, and aerosol loading. As a result, global climate change has occurred (Harden, 2000). Frequent fires are likely to suffocate tree development and the regrowth of successional vegetation into secondary forests. It could also be harmful to soil fertility and carbon sequestration (Muller *et al.*, 2013). Knowledge of forest fire distribution trends is critical for optimal land development, evaluating REDD+ operations (Muller *et al.*, 2013), and determining the amount to which forest fire emissions contribute to global warming. Major increases in the frequency of forest fires have been seen in tropical places during extreme drought years (Alencar, 2006; Langner and Siegert, 2009), with the fire-affected area in Borneo being three times bigger on average than during normal weather circumstances (Langner and Siegert, 2009). In Amazon forest areas, decrease in deforestation rates and increase in fire frequency was observed between 2000-2007 (Luiz and Shimabukuro, 2010). It was revealed during the study that outflow of fires from land near to degraded forests was the main cause for increase in number of fire incidences.

Increased incidences of forest fires and its frequency have deterimental environmental and economic consequences for shifting cultivation aress. Increased risk of forest fire can make northeast India, a biodiversity hot spot, more vulnerable to climate change; as linkages between biomass burning in shifting cultivation areas and atmospheric aerosol loading have been reported by Badarinath *et al.* (2009). According to Rosenfeld (1999), increased fire activity releases more aerosols into the sky, which can have major implications for regional weather systems. Mishra and Ramakrishnan (1983) found that shorter fallow cycles had lower nitrogen concentrations than longer ones. False fire pixels may form as a result of increased canopy surface temperature variations caused by new cutting and the appearance of uniform dark char patches from recent fires (Schroeder *et al.*, 2008). Because of the presence of moisture, vegetation burning might take longer, and this continual burning leads to an increase in the number of fires in the area.

3.13.1 Intra-annual fire occurrence

According to Padalia and Mondalb (2014), tracking fire incidence changes is crucial for estimating the amount of carbon emissions caused by shifting cultivation. These fire incidences peaked during March, the time during which vegetation is burnt in shifting cultivation plots. In the Indian state of Nagaland, where shifting agriculture is a common land use, In the Indian state of Nagaland, where shifting agriculture is a common land use, Fig. 3.20 depicts the temporal distribution of fire pixels between 2001 and 2013. Fig. 3.21 shows the occurrence of fire pixels in different ecosystems. Highest fire pixels occurred in moist mixed deciduous forests (46%) and lowest (10%) in combined subtropical pine and wet temperate forests. On an average, out of total annual fire incidences of 8809, about 6896 or 78% incidences occurred during the monthe of February and Marcch, coinciding with the time during which burning is done in shifting cultivation plots.

These areas have inaccessible, rugged and undulating terrain. There are some rain-shadow areas in Nagaland, which are more prone to forest fire incidents. The rains during monsoon reduces the incidences of forest fires, but more fire incidents occurred during delayed rainfall years.

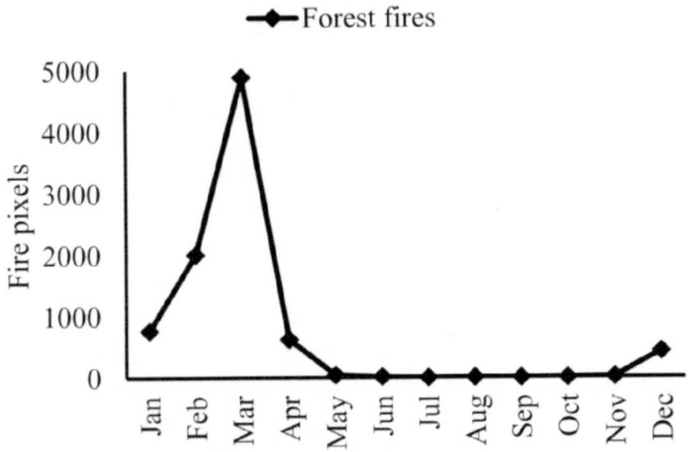

Fig. 3.20. Trends of forest fires in Nagland (2001-2013). The majority of fire pixels are dispersed in the months of February and March, as seen by the number of fires counts plotted against months covering the entire year.

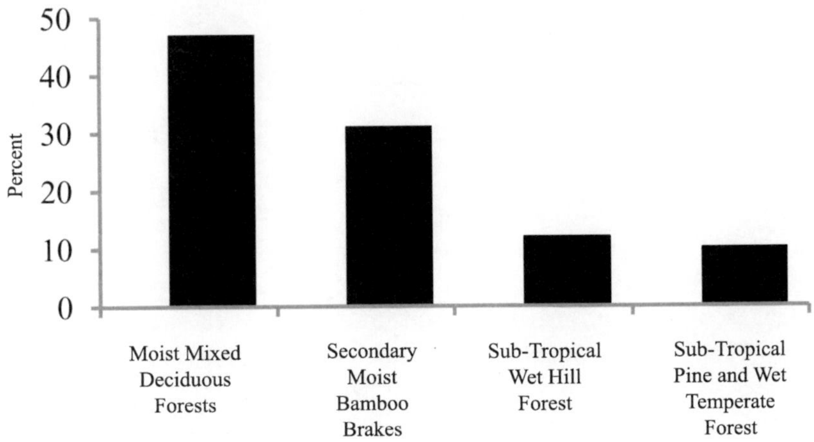

Fig. 3.21. Incidences of forest fires in different ecological zones (Percent)

3.14 ITK IN SHIFTING CULTIVATION

Farmers use their indigenous technical knowledge (ITK), which is accumulated inherited knowledge of the land, water, vegetation, animals, etc. in their farming environment, to make decisions and find solutions to issues that arise. This information assisted farmers in developing numerous techniques that have shown to be advantageous over an extended length of time. For the system to be sustainable, they are entirely dependent on locally accessible resources. IEK has the potential to have an impact on forest resources and their future conservation throughout Indonesia (Saihaya *et al.* 2016).

Understanding, farming techniques, and cultural heritage are frequently passed on from one generation to the next (Arkanudin 2009). Such knowledge and practises, according to Parrotta *et al.* (2016), are often intertwined around land use, biodiversity, and ecosystem functioning. Human cultures, in general, have the ability to both harmonise with and adapt their environments in order to survive. Based on their long standing experience of shifting cultivators with regard to their environment, crop growth, pest elimination, crop nutrition etc., they have standardized certain self-created practices due to their ingenuity and skill, which came to be known as 'Indigenous Technical Knowledge' (ITK). Also the shifting cultivators were fully aware of their environment and ecology, and set certain principles, usually called the 'Indigenous Ecological Knowledge' (IEK), to deal with different ecological situations. Based on this ITK, the shifting cultivators were able to successfully raise good crops, in absence of external inputs, particularly inorganic ones. However, whenever required, they apply animal dung and urine.

3.14.1 Natural resource management and local technical knowledge

Tribal minority in the Amazon and Southeast Asia have been extremely successful in utilising and conserving the tropical forest. Though it was always assumed that tropical forests should be maintained as a national natural resource, there is now a rising sentiment that they are a global resource whose management is of international relevance. Shifting cultivators were the first to bear the brunt of the tropics' destruction. Governments and international agencies tried to stop them because they were considered as recklessly destroyer of the forests. However, to blame them and enacting laws to stop shifting cultivators from cutting and burning forests was easy, discontinuing shifting cultivation was not. Based on available literature, much of the blame appears to be miscalculated and misdirected. Many shifting cultivators have been found actively reestablishing the forest in abandoned land after cropping was over. Deep knowledge of the tropical environment and good supervisory skills are needed for successful completion of the shifting cycle. The failure of agricultural development projects in the tropics is testimony to this.

Local technical knowledge (LTK) is a deep understanding of the ecosystem and its planning based on years of emotional experience (Bodley 1976). Knight (1980) advocated that baseline data on natural resources, including crops, be collected as shifting farmers' local technical expertise was studied. Environmental knowledge is used extensively by shifting cultivators in making judgments about what to do and when to do it. If the decions made based on the knowledge gained are successful, it goes to knowledge pool as 'local technical or indigenous technical knowledge', otherwise ignored. However, there could be some social, cultural or environmental constraints (Ellen 1982). The passion, with its complicated feedback gadget, give better idea of how the shifting cultivator blend the environment and the agricultural system to manage for better output (Gladwin 1983, Olafson 1983, Warner 1981, Benneh 1972). The use of ITKs is very common with small and marginal farmers, especially. Some of the major ITKs related to plant protection and used by shifting cultivators are mentioned below:

1. Bamboo perches and Eupatorium audoratum are used to control rice pests in rice fields.
2. Farmers frequently use wood ash on vegetable plants to suppress significant pests.
3. Controlling rice gundhi bugs by hanging dead frogs/crabs in rice fields.
4. Rice pests are controlled by night shooting in rice fields. The fire serves as a magnet for gundhi bugs (Leptocorisa acuta). When the insects go too close to the fire, they die.
5. Pumelo (Citrus maxima) application (Burm.) To suppress rice pests, use Merr. peel in the rice field during the vegetative stage. The stem borer is repelled by pumelo.
6. Application of mustard oil cake to control red ant and mole cricket in potato crop.
7. Neem leaf extract is used to control rice pests in rice fields.
8. Use of fish water to suppress the trunk borer in citrus trees. Red ants are attracted to fish water.
9. Fruit fly is suppressed in cucurbits by smoking.
10. Cow urine acts as a repellentin and is applied citrus to control citrus pests.

11. To control stored grain pests in godowns, dry neem leaves are used. It acts as an antifeedant. 12. Beating empty drum in field to control bird pests. Due to sound produced, birds are scared.
13. To manage rice case worm, a rope drenched in kerosene oil is run over the standing rice crop in the field. Kerosene oil is poisonous.
14. Tobacco leaf extract is used to control rice pests. Tobacco's alkaline composition functions as an antifeedant and pest repellant.
15. Pests in vegetable crops are controlled by sprinkling a detergent-tobacco powder combination.
16. Rotten lemons are kept in rice field to control rice pests.
17. Apply layer of cow dung in storage bin to control rice moths.
18. To manage stored grain pests, mix pulse seeds with black pepper powder and kerosene in a godown.
19. To manage mouse pest, place cut pieces of papaya stem and delicate fruits in rodent holes. Rodent pests are irritated by papaya latex.

Sharma and Prasad (1994) have documented some indigenous farming systems evolved by the indigenous tribes due to their experiences, ingenuity and skill; in northeast India where shifting cultivation is a major land use. The major ones are being discussed here.

3.14.2 Zabo system

Historical

The tribal population in northeastern India remained more or less static for a very long period until inter-tribal warfare, head hunting, and pestilence and epidemics got effectively controlled. Gowswami (1981) mentioned that one Mr. Peal, who was sent on exploratory expedition in early nineteenth century by the British Government observed in his notes "from internal evidence, the tribal population seems to have been stable for a very long period, perhaps centuries. The checks are all positive, such as constant warfare, want of food, inducing diseases etc. As a consequence of head hunting and its isolating influence, we thus see a community of some hundred people perched on a hill and depending exclusively on their own resources, constantly fighting others similarly isolated on all sides, yet thoroughly able to maintain themselves. Perhaps in no other part of the world can so complete a tribal isolation seen. The every 40 or 50 square miles, there are about 4 villages of, perhaps, one hundred families each, yet no more than an eighth or tenth of the land available can be cultivated at a time, and the population would seems to have reached it maximum". The equilibrium, however, got upset when the population started increasing paradoxically because of stopping of warfare amongst the tribes, health care and general awareness.

With decrease in shifting cultivation cycle as a result of increase in population, the crop productivity decreased considerably, thereby necessitating the tribal people to migrate from one place to another for an easy access to food and water. The tribal people of northeastern region of India have some social taboos and superstitions, which are to be followed religiously. One such tribe, known as Chakhesang, also migrated from the areas bordering Burma (now Myanmar) and settled in an area, now called Kikruma village in Phek district of Nagaland State of India. There is a legendary story attached with the migration of the tribe. According to the village elders, the inhabitants of kikruma village have possibly migrated during nomadic period. During migration, the people were accompanying their leader, the chief of the tribe, for settlement at some suitable place, which should have fertile land for crop production and pastures for grazing of the domestic animals. The leader was carrying a big stone on his head and a cock in his hands. The belief of the tribe was that they would settle at a place where the stone must fall from the leader's head and the cock must crow. Only then that place will be free from diseases, good from crop production and overall welfare of the tribe. During the movement, the tribe moved through many fertile lands, but neither stone fell down from leader's head and not the cock crowed. This was not a good omen, the leader thought. The tribe kept on moving and reached a place where the stone fell down from the leaders head and the cock also crowed. This was thought to be God's will for settling down of the tribe (Sharma, 1997). The place was, however, was not good for crop production and devoid of water source nearby. This is the place where now kikruma, village is situated (Sonowal *et al.*, 1991; Sharma, 1998) the village is sorrounded by two streams, 'Seidzu' and Khuzha, a few hundred meters down the settlement of the tribe on the hill. The tribe grew crops but the yields were poor in absence of irrigation. These was no drinking water around except to bring it from the streams after walking a couple of miles. The situation might have compelled the people to develop their own water resources due to their ingenuity and skill. This system of water harvesting and management is called 'Zabo' system.

The system

The system is a unique combination of forest, agriculture, livestock and fisheries with a water and soil conservation base (Sharma and Prasad, 1994, Sonowal *et al.*, 1989, 1994).

Forest land

The area on the hill top serves as catchment for rain water and is kept under natural vegetation. This area is normally not disturbed by cutting forest trees. The slope of this area is up to 100% or even more. Being permanent under vegetation, there is hardly any loss of soil through erosion.

Water harvesting system

Water harvesting ponds are dug-out with formation of earthen embankment below the catchment area. First two small ponds serve as silt retention tanks followed by a big tank down the slope. The size of these tanks depends on the catchment area but usual capacity is 300 to 600 m3 with a depth varying from 1.5 to 2.5m. Silt retention tanks are constructed in between forest land and the main water storage tank to store run-off water from the catchment area for few days in these tanks so that soil, humus and organic matter is retained here before passing on the water to the main tank. Soil retention tanks are desilted annually and the soil containing organic matter is put in the rice fields for enhancing fertility. The ponds are rammed and compacted from inner sides so as to avoid water loss through seepage. The main water-harvesting tank is plastered on inner surface with mud mixed with chopped paddy straw to minimize water losses due to seepage. The bottom of the tank is rammed and animals are let loose in the beginning to make the bottom surface hard to reduce water losses.

Livestock Enclosure

Enclosures are made with wood and bamboo for cattle and are managed by a group of farmers by stocking cattle on rotation basis. The enclosure is constructed on lower side of the pond. Buffaloes, cows, and pigs are the common animals with the farmers and about 30 to 50 of them are kept in one enclosure for a period of 10-15 days. Buffaloes and pigs are the common animals. During irrigation the water from the main tank is taken to the paddy fields by passing through the cattle enclosure and it carries dung and urine of the animals making the fields highly fertile.

Agricultural Land

Paddy and other crop fields are located at lower elevation. Application of cow dung and run-off through open cattle yard are the common methods of manuring. Besides, leaves and succulent branches of Alnus nepalensis and Albizia lebbeck are also added to the fields and decomposed for soil fertility. No inorganic source of nutrients is added, making the whole system as organic farming. There is no pollution of water and environment due to fertilizer use. Some farmers practise paddy-cum-fish culture and about 50 to 60 kg of fish is obtained per ha which is an additional source of income to the farmers. Thus, the Zabo system is clearly conservation-based and might be a viable alternative to shifting agriculture, however implementing terrace farming in a steep terrain like Nagaland could be problematic.

Construction of terraces

The terraces are constructed and supported with strong risers with width varying from 0.5 to 1.2m and height from 0.2 to 1.4m depending on the slope of the land and size of the slope (Fig. 3). Contour lines are drawn by separating the top soil upward so that sub-soil is exposed for firm foundation.. After removing the top soil, the sub-soil below the proposed riser is dug and heaped up for compaction. The leveling of the terrace is done by cut and fill method. When the terrace is leveled, the removed top soil is spread uniformly over the surface of the terrace. Perfect leveling of the terraces and well managed irrigation-cum-drainage system prevent soil erosion and loss of fertility. To maintain desired level of water in the field, an appropriate-size stone is kept in the opening at the riser top. The openings of water from one field to another are not kept straight but alternately on opposite points to reduce the velocity of gushing water and prevent soil erosion. The land is selected and cleared of vegetation. After removing the top soil (kept aside) the sub-soil below the proposed riser is dug and heaped-up for compaction with wooden clubs, vertically and horizontally to form the riser and the side bund (Fig. 3). The space left above the riser is filled-up with stones, removed during digging. The levelling of terrace is done by cut and fill method by compacting the filled soil with feet and sticks. The top soil removed earlier and kept aside is, then, spread uniformly over the surface of the terrace. Shales and soil clods are also collected and spread over the terrace which disintegrate into fine soil after coming in contact with water (Sonowal *et al.*, 1989). The land preparation in terraces is done manually with a small spade, in general, and by bullocks or buffaloes, in some cases. The tractors or power tillers are rarely used since the land holdings are very small. The crops are sown/planted in the terraces (low land rice, mostly) and inter-culturing and weeding is done manually with a local khurpa, iron blade fitted to a wooden handle. The rain water moves from one slope to another down the slope through openings made in the riser bunds at extreme ends, alternately. The extra runoff water moves down the slope through grassed water-ways on either sides of the slope.

The Zabo system, which conserves water and soil, has reduced the risk of floods as soil erosion is negligible when the method is properly followed. The Zabo is, therefore, best method for reducing risks due to natural and human induced disasters. About 25% rainwater is retained in situ in shifting cultivation compared to 65% in zabo system. Runoff generation is more in shifting cultivation leading to floods. About 40% of rainwater is collected in zabo farming system in water harvesting ponds, to be used for irrigation at critical stages of crop growth; whereas, whole runoff, along with sediment load, from shifting cultivation goes to river channels, inducing floods. Soil loss through erosion extremely high from shifting cultivation and very low from Zabo system (Table 3.17).

Table 3.17. Soil loss and approximate budgeting of rainwater.

Land use	**Soil loss (t ha-1)**	**Rainwater (%)**		**Runoff collected in ponds (%)**
		In situ	**Runoff**	
Shifting cultivation	30-130	20-25	75-80	Nil
Zabo	1-3	55-65	35-45	28-45

3.14.3 Agriculture with Alder' System

In Nagaland state, India, alder (*Alnus nepalensis*) is planted to increase soil fertility for crops such as millet, potatoes (*Solanum tubrosum* L.*)*, pumpkin (*Cucurbita* sp.), job's tears (*Coix lacryma-jobi*), maize (*Zea mays L.*), and barley (*Hordeum vulgare*). Non-leguminous, alder trees can thrive on terrain ranging in elevation from 800-3000 metres. The trees fix atmospheric nitrogen through nodules that form on the roots. This tree is known as a multipurpose tree (MPTS) because it serves as a fuel source and source of lumber in addition to increasing soil fertility.

Sharma and Singh (1994) investigated alder's ability to fix nitrogen and discovered that total litter yield is proportional to the number of plants, with N fixed ranging from 48.3 kg ha^{-1} (60 trees ha-1) to 184.8 kg ha^{-1} (625 plants ha^{-1}) (Fig.3.22). The amount of N fixed per tree dropped as the tree population per ha increased. In addition to fixing atmospheric nitrogen, alder tree litter delivered P, K, Ca, and other nutrients to the soil through the addition of biomass (Sharma and Singh, 1994).

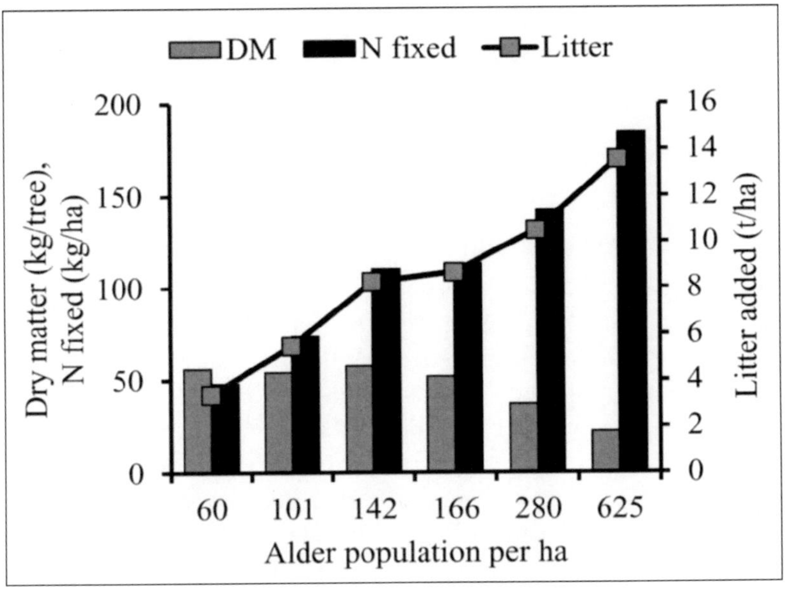

Fig. 3.22. Litter yield and nitrogen added through Alder
Source: Sharma and Singh, 1994.

3.14.4 Water management by Apatani tribe

The Apatani tribe of India's Arunachal Pradesh state has devised a unique water management method. Small streams that emerge from the surrounding hills are tapped shortly after they emerge from the forest (Chaudhary *et al.*, 1994, Prasad and Sharma 1994). While the stream continues its path, the water is allowed to flow in the first feeder channel. The feeder canal branches off at angles to send water via a succession of terraces, allowing any field to be watered or drained by closing or opening the connecting ducts (Huburs).

Paddy-cum-fish culture

Fish channels of about 30 cm x 150 cm with 30 cm depth are dug at various sites in the rice fields. Fish fingerlings are put in these channels at the start of the cropping season and the fields are flooded with water. When the rice fields are drained before the harvest of the crop, the water remains in the fish channels. Average weight attained by the fish in four months varies between 130 and 400g and in general, 180-185 kg fish is harvested per ha, per season.

Environmental preservation

The entire hills surrounding the valley and uplands surrounding the villages are kept green by forest vegetation. Apart from the utilization of rain water for irrigation and recycling of household wastes and by-products for fertility management, most of the important tools used in the *Apartani* farming are made of wood. The plot to plot outlets in rice fields are made with wood or bamboos. Fishing nets and fish catching baskets are also made of bamboos. Local baskets and containers called *yagli, Yopho* and *Rajhu* are also made of bamboos.

3.14.5 Drip irrigation through bamboo

Drip irrigation through bamboo is a unique example of efficient management of soil and water and reflects the ingenuity and skill of the tribal communities. This is practiced by the Khasi and Jaintia tribes of the Meghalaya state of India. The systems is based on the gravity flow and this aspect has been fully used by the farmers. The water is brought from upper reaches of hills to lower level through gravitational flow. The water is carried from the source to the place of use with the help of bamboo pipes cut into different shapes and forms (Singh, 1989) (Fig. 3.23). The flow of water can be controlled as per requirement by blocking and opening of bamboo slits at diversion points.

Fig. 3.23. A view of the well laid Bamboo Drip Irrigation system
Source: ICAR Research Complex for NEH Region. (See colour version on page 652)

The water is allowed to trickle down in drops at the plant site by having small holes at appropriate points. Once laid out, the system works round the clock, if desired. Farmers have the requisite skills to expertly build out the bamboo network, and the entire system runs smoothly and efficiently (Singh, 1989,

Singh and Singh 1978). Depending on the plant positions where the water is to be applied, the network of bamboo channels is made with diversion and water control devices. Bamboos of large diameter are used at the start and the size is subsequently reduced as per requirement. Irrigation by this system is generally provided to plantation crops. Since the topography of the region is undulating, the cost of constructing of concrete channels for carrying the water to the desired sites is very high and the area for irrigation is relatively low. Locally available bamboo is used for fabricating and laying-out the system. The fabricating cost and maintenance costs of the whole system are nominal. The tribal community takes care of the system together. Use of local resources has made these systems more cost effective.

3.14.6 Social life

Earlier called the eastern Angamis (Hutton, 1969), the Chakhesang is one of the most progressive tribes of the Nagaland. The word 'Chakhesang' comprise of three syllable; Cha-Khe-Sang, and each of them represent a tribal group. The 'Cha' is for Chokri tribe, 'Khe' for Kheza tribe and 'Sang' for Sangtam tribe. The Chakhesang people speak many dilects due to diversity in their composition. Apart from English and Nagamese, the Chakhesang people could easily speak in other tribal dialects. Chakhesang people are making progress in social, political and religious life. Being neighbours to many tribes, multilingual society and variable in their social composition, they are readily accepted by others. Most of the Chakhesang people are hard-working and economically self-reliant.

3.15 GHG EMISSIONS FROM SHIFTING CULTIVATION

3.15.1 GHG emission from shifting cultivation in Eastern Himalayas Indian states

When vegetation is burned in shifting agriculture, carbon dioxide (CO_2) and other trace gases are released into the sky. Table shows estimates for carbon dioxide (CO_2), methane (CH_4), and nitrous oxide (N_2O). Because rice crop residue burning is uncommon in the area, GHG emissions were not calculated. Annually, 1101 Gg of CO_2, 4.482 Gg of methane, and 0.1353 Gg of nitrous oxide were expected to be emitted (Table 3.18). This translates to 1.237 Tg CO_2e. Approximately 6.32 percent of GHG emissions in the region are attributed to shifting cultivation (Sharma and Sharma 2018).

Table 3.18. GHG emission from differet Northeastern states of India.

State	Annual area ('000 ha)	CO2	NH4 (Gg)	N2O	CO2e
Arunachal Pradesh	70	119.7	0.811	0.0245	224.326
Assam	69	196.8	0.799	0.0242	221.081
Manipur	90	256.7	1.042	0.0316	288.378
Meghalaya	53	151.2	0.613	0.0185	169.808
Mizoram	63	179.7	0.743	0.0221	202.154
Nagaland	19	54.2	0.220	0.0067	60.897
Tripura	22	62.2	0.254	0.0077	70.421
Total	386	1101.0	4.482	0.1353	1237.065

References

Achard, F., Eva, H. D., Stibig, H.-J., Mayaux, P., Gallego, J., Richards, T., and Malingreau, J. P. (2002). Determination of Deforestation Rates of the World's Humid Tropical Forests. Science 297: 999–1002. doi:10.1126/science.1070656.

ADB (Asian Development Bank) 2001. Chittagong Hill Tracts Region Development Plan. Final Report No. 3: Agriculture and Marketing in the CHT. ADB TA No. 3328. Rangamati.

AIPP and IWGIA, 2014 Shifting Cultivation, Livelihood and Food Security - New and Old Challenges for Indigenous Peoples in Asia. Published by Asia Indigenous Peoples' Pact (AIPP) and the International Work Group for Indigenous Affairs (IWGIA), Chiang Mai 2014

Alencar, Ane, Daniel Nepstad, Mariadel Carmen Vera Diaz, 2006. Forest Understory Fire in the Brazilian Amazon in ENSO and Non- ENSO Years: Area Burned and Committed Carbon Emissions. Earth Interactions, 10, pp. 1–17.

Altieri, M. A., Letourneau, D. K., and Davis, J. R. 1973. Developing sustainable agroecosystems. Bioscience 33(l): 45 - 49.

Andriesse, J. P. 1989.. Nutrient management through shifting cultivation. A comparative study on cycling of nutrients in traditional farming systems of Malaysia and Sri Lanka. In van der Heide, J. (ed.), Nutrient Management for Food Crop Production in Tropical Farming Systems. Institute for Soil Fertility and Universitas Brawijaya, Haren, pp. 29–61. 384 Hum Ecol (2009) 37:375–388

Andriesse, J.P. and Schelhaas, R.M., 1987. A monitoring study of nutrient cycles in soils used for shifting cultivation under various climatic conditions in tropical Asia. II. Nutrient stores in biomass and soil--results of baseline studies. Agric. Ecosyst. Environ., 19:285-310.

Angelsen, A. (2008). Moving ahead with REDD: Issues, options and implications. Center for International Forestry Research, Bogor, Indonesia.

Angelsen, A., 1999. Agricultural expansion and deforestation: modelling the impact of population, market forces and property rights. Journal of Development Economics 58, 185–218.

Angelsen A. 1995. Shifting cultivation and "deforestation": a study from Indonesia. World Dev. 23:1713–1729.

Anonymous (2000) Basic Statistics of the North-eastern Region. North Eastern Council, Shillong, Meghalaya, India.

Aragão LEOC, Malhi, Y., Barbier, N., Lima, A., Shimabukuro, Y., Anderson, L., and Saatchi, S., 2008. Interactions between rainfall, deforestation and fires during recent years in Brazilian Amazonia. Philosophical Transection of Royal Society of Britain, 363,pp.1779-1985

Arkanudin. 2009. Sistem Perladangan dan Kearifan Tradisional Orang Dayak Dalam Mengelola Sumber Daya Hutan [Shifting cultivation system and traditional wisdom of the Dayak people in managing forest resources]. Available from: http://arkandien.blogspot.com/2009/03/sistem-perladangan-dan-kearifan.html

Arnason, T., Lambert, J. D. H., Gale, J., Cal, J., & Vernon, H. (1982). Decline of soil fertility due to intensification of land use by shifting agriculturists in Belize, Central America. Agro-Ecosystems, 8 (1), 27-37.

Aryal KP, Kerkhoff EE, Maskey N and Sherchan R, 2010. Shifting cultivation in sacred Himalayan landscape: A case study in the Kangchenjunga Conservation Area @ Kathmandu, Nepal, First Edition.

Ascoli FD. (1918). Report on the Administration of the Chittagong Hill Tracts. The Bengal Secretariat Press: Calcutta.

Aweto, A.O. 1981.. Fallowing and soil fertility restoration in South west Nigeria. Malaysian Journal of Tropical Geography,3:1-6.

Badarinath, K.V.S., Latha, K. M., Chand, T.R.K., Gupta, P.K., Ghosh, A.B., Jain, S.L., Gera, B.S., Singh, R., Sarkar, A.K., Singh, N., Parmar, R.S., Koul, S., Kohli, R., Nath Shambhu,Ojha V.K., and Singh, Gurvir, 2004. Characterization of aerosols from biomass burning—a case study from Mizoram (Northeast), India. Chemosphere, 54,pp.167–175.

Badarinath, K.V.S., Latha, M.K., Kiran Chand, T.R.K., and Gupta, P.K., 2009. Impact of biomass burning on aerosol properties over tropical wet evergreen forests of Arunachal Pradesh, India. Atmospheric Research, 91,pp. 87–93.

Baird, I.G., and Shoemaker, B. 2007. Unsettling Experiences: Internal Resettlement and International Aid Agencies in Laos. Development and Change 38(5): 865-888.

Bajracharya KM, Malla KB and Thapa PB 1993. Shifting cultivation in Southern Gorkha. IIMG/GTZ Gorkha Dcvdopment Project Bajracharya et al., (1993)

Baker R. 1984. Protecting the environment against the poor. Ecologist 14: 53–60.

Bakr Ahmad Alserhan, Ingo Forstenlechner, Ahmad Al-Nakeeb, 2010. "Employees' attitudes towards diversity in a non-western context", Employee Relations, Vol.32 Iss:1, pp.42-55

Barrau, J. 1971. Subsistence agriculture in Polynesia and Micronesia. Hawaii, Bernip, Bishop Museum/New York, Kraus Reprint Co.

Bassam El, N (1997). Fundamentals of sustainability in agricultural production systems and global food security. In: Sustainable Agriculture for Food, Energy and Industry, (ed. by N.El Bassam, R.K. Behl & B.Prochnow), 1-12, James and James (Publishers, London, U.K.

Batabyal, A.A., and Beladi, H. 2004. Swidden Agriculture in Developing Countries, Review of Development Economics, 8, 255-265.

Batabyal, A.A., and Lee, D.M. 2003. Aspects of Land Use in Slash and Burn Agriculture, Applied Economics Letters, 10, 821-824.

Becker, B. K. 1995. Undoing myth: the Amazon, an urbanized forest. In Brazilian Perspectives on Sustainable Development of the Amazon Region, Vol. 15 (Eds, Clüsener-Godt, M. and Sachs, I.) UNESCO, Paris 53-89.

Beckerman, S. 1987. Swidden in Amazonia and the Amazon Rim. In Turner, B. L. and Brush, S. B. (eds.. Comparative Farming Systems. New York: Guilford Press.

Behera, M.D.; Tripathi, P.; Das, P.; Srivastava, S.K.; Roy, P.S.; Joshi, C.; Behera, P.R.; Deka, J.; Kumar, P.; Khan, M.L.; et al. 2018. Remote sensing based deforestation analysis in Mahanadi and Brahmaputra river basin in India since 1985. J. Environ. Manag. 206, 1192–1203.

Benneh, G. 1972. Systems of agriculture in tropical Africa. Economic Geography 48(3): 244 - 257.

Biswal D.K. and Kumar, S. 2013. Shifting cultivation and policies of sustainable development: a meaningless obsession. Journal of Economic & Social Development, 9: 132-138.

Blackburn, W. H.; Knight, R. W.; Wood, J. C. and Pearson, H. A. 1990. Storm flow and sediment losses from intensively managed forest watersheds in East Texas. Water Resources Bulletin. 26(3): 465-477.

Blackburn, W. H.; Wood, J. C. and Dehaven, M. G. 1986. Storm flow and sediment losses from site prepared forest land in East Texas. Water Resources Research. 22(5): 776-784.

Bodley, J. H. 1976. Anthropology and Contemporary Human Problems. Menlo Park, California: Benjamin/Cummings.

Borthakur, D.N. 1992. Agriculture of the North Eastern region, with Special Reference to Hill Agriculture. BEE CEE Prakashan, Guwahati.

Boserup E. 1965. The Conditions of Agricultural Growth: The Economics of Agrarian Change under Population Pressure. Earthscan: London.

Brady NC. 1996. Alternatives to slash-and-burn: a global imperative. Agriculture, Ecosystem and Environment 58: 3–11.

Brammer, H. 1986. Reconnaissance Soil and Land Use Survey: Chittagong Hill Tracts, Soil Resources Development Institute, Dhaka, 206 pp. (9 volume reports of Forestall).

Brookfield, H. Exploring Agrodiversity; Columbia University Press: New York, NY, USA, 2001; Volume 42, pp. 181–192.

Brookfield H, Potter L, Byron Y. 1995. Place of the Forest: Environmental and Socio-Economic Transformation in Borneo and the Eastern Malay Peninsula. United Nations University Press: Tokyo.

Bruun, T.B., A. de Neergaard, D. Lawrence, A.D. Ziegler. 2009. Environmental Consequences of the Demise in Swidden Cultivation in Southeast Asia: Carbon Storage and Soil Quality. Human Ecology 37:375-388.

Byron, N., and Arnold, M. 1999. What Futures for the People of the Tropical Forests? World Development 27: 789–805.

Cairns, M. ed. 2007. Voices from the forest: integrating indigenous knowledge into sustainable upland farming. Washington, DC: Resources for the Future.

Cairns, M., & Garrity, D. P. 1999. Improving shifting cultivation in Southeast Asia by building on indigenous fallow management strategies. Agroforestry Systems, 47(1), 37–48. Doi: 10.1023/A:1006248104991

Carter, J.E. and Mends-Cole, J. 1982. Liberian women: Their role in food production and their educational and le al status. Profile of Liberian women in development project. University of Liberia.

Castella J.C., Lestrelin G., Hett C., et al. 2013. Effects of landscape segregation on livelihood vulnerability. Human Ecology, http://dx.doi.org/10.1007/s10745-012-9538-8.

Cavendish, W. (2003). How Do Forests Support, Insure and Improve the Livelihoods of the Rural Poor? A Research Note. Center for International Forestry Research (CIFOR), Bogor.

Census of India 2011. Population census in India, Government of India, New Delhi.

Chan, K. E. 1987. Current and prospective urbanization in Malaysia. Malaysian Journal of Tropical Geography 15: 1–12.

Charlton CA. 1987. Problems and prospects for sustainable agricultural systems in the humid tropics. Applied Geography 7: 153–174.

Chazee, L. 1994. Shifting cultivation practices in Laos: present systems and their future. In Van Gansberghe, D., and Pals, R. (Eds.), Shifting Cultivation Systems and Rural

Development in the Lao P.D.R. Report of the Nabong Technical Meeting. Nabong Agricultural College, Lao People's Democratic Republic, July 14–16, 1993, Nabong Agriculture College Project. Nabong Agriculture College, Vientiane, pp. 66–97.

Chazee, L. 1998. Evolution des Systèmes de Production Ruraux en République Démocratique Populaire du Laos 1975-1995. L'Harmattan, Paris.

Chidumayo, E. N. 1987. A shifting cultivation land use system under population pressure in Zambia. Agroforestry Systems, 5: 15 - 25.

Choudhary, R.G.; Kochhar, S. and Datta, K.K. 1994. "Rice based farming system of Apatani Plateau In : Potential Indigenous Farming systems of northeastern region. (ed. By R. N. Prasad & U. C. Sharma), Indian Council of Agricultural Research complex, Northeastern region Meghalaya, India.

Christanty, L. 1986 Shifting cultivation and tropical soils patterns, problems and possible improvements. In: Marten, G. G., Traditional Agriculture in Southeast Asia: A human Ecology Perspective, West View Press (Boulder, Colorado) pp 229-240

CIFOR 2007. CIFOR Poverty and Environment Network (PEN). Center for International Forestry Research (CIFOR), Bogor.

Clarke, W. C. 1976. Maintenance of agriculture and human habitats within the tropical forest ecosystem. Human Ecology, 4(3): 247 - 259.

Cochrane, M. A., 2003. Fire science for rainforests. Nature, 421,pp. 913-19.

Colfer, C. J. P. 2004. The Equitable Forest: Diversity, Community, And Resource Management. In. RFF Press, Washington DC.

Collins, N. M, Sayer, J. A., Whitmore, T. C. (eds.) (1991). The Conservation Atlas of Tropical Forests: Asia and the Pacific. MacMillan, London.

Conklin H 1954. An ethnoecological approach to shifting agriculture. Transactions New York Academy of Sciences, Series II 17: 133–142

Conklin HC 1957. Hanunoo agriculture- An example of shifting cultivation in the Philippines. FAO,11: 172-173.

Cox, G. W. and Atkins, M. D. 1979. Agricultural Ecology. San Francisco: W. H. Freeman and Co.

Cramb RA, Colfer CJP, Dressler W, Laungaramsri P, Le QT, Mulyoutami E, Peluso NL, Wadley RL 2009. Swidden Transformations and Rural Livelihoods in Southeast Asia. Hum Ecol.; 37: 323-346.

Craswell, E.T., Sajjapongse, A., Howlett, D. J. B. and Dowling, A., J. 1997. Agroforestry in the management of sloping lands in Asia and the Pacific. Agrofor Syst 38 pp 121–137

Crutzen, P.J., Lelieveld, J., 2001. Human impacts on atmospheric chemistry. Annual Review Earth and Planetary Sciences, 29,pp. 17-45.

Cruz, C. J., Zosa-Feranil, I., and Goce, C. L. 1988. Population Pressure and Migration: Implications for Upland Development in the Philippines. Journal of Philippine Development 26: 15–26.

Cuc, L. T. 2002. Sustainable development in the mountain of Vietnam over the last 10 years: Statement and Problem (in Vietnamese). Vietnam National University, Hanoi.

Dalle, S.P.; Pulido, M.T.; De Blois, S. Balancing shifting cultivation and forest conservation: Lessons from a "sustainable landscape" in southeastern Mexico. Ecol. Appl. 2011, 21, 1557–1572.

Darby, H. C. 1956. The clearing of the woodland of Europe. Man's Role in Changing the Face of the Earth (ed. Thomas, W. L.) The University of Chicago Press, Chicago and London, 183-216.

Darlong L, Bhattacharyya D 2014. Phanera jampuiensis (Leguminosae: Caesalpinioideae: Cercideae), a new species from Tripura, India. Kew Bulletin 69: 9534–9539

Das GN. 2001. Swidden Cultivation and Development Programmes in North-East India. Akasha Publishing House: New Delhi.

Das S. and Das M. 2014. Shifting Cultivation in Tripura – A Critical Analysis. Journal of Agriculture and Life Sciences, 1: 48-54

Datta, M. Dhiman, K.R. and Singh , N. P. 2004. Shifting Cultivation – A case Study to Evaluate Soil Fertility, Pub. No 6, ICAR Research Complex for NEH Region, Tripura, India

Davidson, E.A., M.M.C. Bustamente, and A. de Siqueira Pinto. 2001. Emissions of nitrous oxide from soils of native and exotic ecosystems of the Amazon and Cerrado regions of Brazil. In Optimizing nitrogen management in food and energy production and environmental protection: Proceedings of the 2nd International Nitrogen Conference on Science and Policy. The Scientific World 1(S2):312–319.

DeAngelo, GJ. and Batabyal, AA., 2004. A Dynamic and Stochastic Analysis of Fertilizer Use in Swidden Agriculture. Available at SSRN: http://ssrn.com/abstract=557003

De Foresta H and Michon G 1997. The agroforest alternative to Imperata grasslands: when smallholder agriculture and forestry reach sustainability. In: Garrity DP (ed) Agroforestry Innovations for Imperata Grassland Rehabilitation, pp 105–120. Kluwer Academic Publishers, Dordrecht, the Netherlands and the International Centre for Research in Agroforestry, Nairobi, Kenya

DeFries, R.S., R.A. Houghton, M. C. Hansen, C.B. Field, D. Skole, and J. Townshend 2002. Carbon emissions from tropical deforestation and regrowth based on satellite observations for the 1980s and 1990s, Proc. Natl. Acad. Sci. U. S. A., 99, 14,256–14,261, doi:10.1073/pnas.182560099.

De Jong W, van Noordwijk M, Sirait M, Liswanti N, Suyanto, S. 2001. Farming secondary forest in Indonesia. Journal of Tropical Forest Science 13: 705–726.

Deka, P.R. and Sarmah, D. 2010. Shifting cultivation and its effects in regarding of perspective in Northern India. International Journal of Commerce and Business Management, 2 : 157-165.

den Biggelaar C, Lal R, Wiebe K, Breneman V, Reich P 2004. The global impact of soil erosion on productivity II: effects on crop yields and production over time. Adv Agron 81: 49-95.

Department of Social Development and Services 2002. Annual Report, Department of Social Development (Western Cape Government).

de Rouw A (1995). The fallow period as a weed-break in shifting cultivation (tropical wet forests). Agric, Ecosyst. Environ., 54: 31–43.

de Schlippe P. (1956) Shifting Cultivation in Africa: the Zande System of Agriculture. London: Routledge and Kegan Paul.

Dick H, Forbes D. 1992. Transport and communications: a quiet revolution. In The Oil Boom and After: Indonesian Economic Policy and Performance in the Soeharto Era, Booth A (ed.). Oxford University Press, Asia: Singapore; 258–282.

Dingkuhn M, Johnson DE, Sow A, Audebert AY 1999. Relationships between upland rice canopy characteristics and weed competitiveness. Field Crops Res., 61:79–95.

Dixon, J., Gulliver, A., and Gibbon, D. 2001. Farming Systems and Poverty. Improving Farmers' Livelihoods in a Changing World. FAO and World Bank, Rome.

DoF. (1996). Annual Report 1996. Department of Forestry (DoF), Ministry of Agriculture and Forestry, Vientiane, Lao PDR.

Dove, M. R. 2006. Indigenous people and environmental politics. Annual Review of Anthropology, 35, 191-208. Doi:10.1146/annurev.anthro.35.081705.123235

Dove, M.R., 1985. Swidden Agriculture in Indonesia: The Subsistence Strategies of the Kalimantan Kantu. Mouton Publishers, Berlin.

Dove MR. 1983. Theories of Swidden agriculture and the political economy of ignorance. Agroforestry Syst. 1:85–99.

Dufour, D.L. 1990. Use of Tropical Rainforests by Native Amazonians, BioScience, 40, 652-659.

Dumond, D. E. 1961. Swidden agriculture and the rise of Maya civilization. Southwestern Journal of Anthropology, 17301-316.

Dupin, B., de Rouw, A. and Phantahvong, K., B. and Valentin, C., 2009 Assessment of tillage erosion rates on steep slopes in northern Laos. Soil Tillage Res 103 pp 119–126

Dupriez, H, De Leener. 2002. Land use and life: Ways of water, Run off, irrigation and drainage; Tropical Handbook. Publisher: CTA and TERRES ET VTE, Netherlands; ISBN: 2-87105-011-2.

Dvorak, K.A., 1992. Resource management by west african farmers and the economics of shifting cultivation. American Journal of Agricultural Economics 74, 809–815.

Eastmond, A. and Faust, B. 2006. Farmers, fires, and forests: a green alternative to shifting cultivation for the conservation of the Maya forest? Land sc Urban Plan 74 pp 267–28

Eaton JM, Lawrence D 2009. Loss of carbon sequestration potential after several decades of shifting cultivation in the Southern Yucata. For. Ecol. Manage., 258: 949–958.

Eckholm, E.P. The Deterioration of Mountain Environments. Science 1975, 189, 764–770.

EIU 1991. Warbler, yearbook of Eastern Illinois University

Ellen, R. 1982. Environment, Subsistence, and Systems: The Ecology of Small-scale Social Formations. Cambridge and New York: Cambridge University Press.

Ellen, R. 2012. Studies of Swidden Agriculture in Southeast Asia since 1960: An Overview and Commentary on Recent Research and Synthesis. Asia Pac. World 2012, 3, 18–38.

Eva, H. and Lambin, E. F., 2000. Fires and land-cover change in the tropics: a remote sensing analysis at the landscape scale. Journal of Biogeography, 27(3), pp.765–776.

FAO 'Report of the sixth session of the committee on 1982. forestry' (Rome, 3-7 May 1982), FAO, Rome, Italy.

FAO 1980. World census of agriculture. Rome, Italy.

FAO 1984. Changes in shifting cultivation in Africa. FAO forestry paper 50, Food and Agriculture Organization of the United Nations, Rome, Italy.

FAO (Food and Agriculture Organisation). 1990. Report of the Independent Review of the Tropical Forestry Action Plan. FAO, Rome, Italy.

FAO, 1974. Shifting cultivation and soil conservation in Africa. FAO/SIDA/ARCN Regional Seminar. FAO Soils Bulletin 24. 248 p. Rome.

FAO. 1984c. Improved Production Systems as an Alternative to Shifting Cultivation. FAO. Soils Bulletin No 53, Rome, FAO.

FAO. 2014. Building a common vision for sustainable food and agriculture. Principles and approaches. Rome.

FAO. 1985. Tropical Forestry Action Plan. UN Food and Agricultural Organization: Rome.

FAO. Forest Resource Assessment 1990. Rome: FAO; 1996.

FAO/UNDP. Tropical forest resources. FAO; 1982.

FAO/UNEP. 1981. Tropical Forest Resources Assessment Project(GEMS): Tropical Africa, Tropical Asia, Tropical America (4 volumes). Rome: FAO. Cited by OTA 1984.

FAO/UNEP. 1982. Tropical Forest Resources. Forestry Paper no. 30. Rome: FAO. Cited by OTA 1984.

FAO/University of Ibadan. 1982. Recommendations arising from the Workshop on Shifting Cultivation and Extension. Rome.

FAO/World Bank 1982. 'Sierra Leone. Forestry Development Project Preparation Report', FAO, Rome, Italy,

FAO 1957. Shifting Cultivation. Unasylva 11:1

Fearnside, P. M. 2000. Global warming and tropical land-use change: Greenhouse gas emissions from biomass burning, decomposition and soils in forest conversion, shifting cultivation and secondary vegetation, Clim. Change, 46, 115–158, doi:10.1023/A:1005569915357.

Fehlberg HG 1989. Shifting cultivation: Some management aspects for its management. In : vander Heider J. (edn.) Nutrient Management for Food Crop Production in Tropical Farming Systems, Institute for Soil Fertility, Haren, The Natherlands.

Folester, H. 1986. Nutrient loss during forest clearing. In: Land Clearing and Development in the Tropics (eds. Lal, R., Sanchez, P.A. and Cummings, Jr. R.W.), pp. 241-246. Balkema, Rotterdam.

Foley J, DeFries R, Asner G, Barford C, Bonan G, Carpenter S, Chapin FS, Coe M, Daily G, Gibbs H, Helkowski J, Holloway T, Howard E, Kucharik C, Monfreda C, Patz J, Prentice I, Ramankutty N, Snyder P. 2005. Global consequences of land use. Science 309:570–4.

Forest Department, 2008. Ministry of Natural Resources and Environmental Conservation, Government of Myanmar

Forest Survey of India 2000. State of Forest Report, 2000. Forest Survey of India, Dehradun, Uttarakhand, India.

Forest Survey of India (FSI). 1999. India State of Forest Report; FSI: Dehradun.

Fox, J. (2000). How Blaming 'Slash and Burn' Farmers Is Deforesting Mainland Southeast Asia. Asia Pacific Issues 47: 1–8.

Fox, J., and Atok, K. 1997. Forest-Dweller Demographics in West Kalimantan, Indonesia. Environmental Conservation 24: 31–37. doi:10.1017/S0376892997000076.

Fox, J., and Vogler, J. 2005. Land-Use and Land-Cover Change in Montane Mainland Southeast Asia. Environmental Management 36: 394–403. doi:10.1007/s00267-003-0288-7.

Fox, J., D. M. Truong, A. T. Rambo, N. P. Tuyen, L. T. Cuc, and S. Leisz 2000. Shifting cultivation: A new old paradigm for managing tropical forests, BioScience, 50, 521–528, doi:10.1641/0006-3568(2000)050 [0521:SCANOP]2.0.CO;2.

Fox, J., Fujita, Y., Ngidang, D., Peluso, N. L., Potter, L., Sakuntaladewi, N., Sturgeon, J., and Thomas, D. 2009. Policies, Political- Economy, and Swidden, in Southeast Asia. Human Ecology. 37: 305-322, doi:10.1007/s10745-009-9240-7.

Friedman M 1979-82. "Lessons from the 1979-1982 Monetary Policy Experiment," The American Economic Review Papers and Proceedings, Vol. 74, No. 2, 1984, pp 397- 400.

FSI 2011. "India State of Forest Report", Dehra Dun: Ministry of Environment and Forest.

FSI 2013. "India State of Forest Report", Dehra Dun: Ministry of Environment and Forest.

Fujisaka, S. 1991. A Diagnostic Survey of Shifting Cultivation in Northern Laos: Targeting Research to Improve Sustainability and Productivity. Agroforestry Systems 13: 95–109.

Fujisaka S. 1986. Pioneer shifting cultivation, farmers' knowledge, and an upland ecosystem: co-evaluation and systems sustainability in Calminie, Philippines. Philippines Quarterly of Culture and Society 14: 137–174.

Fujita, Y., and Phanvilay, K. 2008. Land and Forest Allocation in Lao People's Democratic Republic: Comparison of Case Studies from Community-Based Natural Resource Management Research. Society and Natural Resources 21(2): 120-133.

Geertz C. 1963. Agricultural Involution: The Processes of Ecological Change in Indonesia. University of California Press: Berkeley, CA.

Getahun, A., Wilson, G. F. and Kang, B. T. 1982. The role of trees in farming systems in the humid tropics. In MacDonald, L.H. (ed.) Agroforestry in the African humid tropics. Tokyo: UNU. pp. 28-35.

Ghosh, A.K. 1992. Changing scenario of forests in Tripura. Indian Forester., 118:754-762.

Gilruth PT, Marsh S, Itami R 1995. A dynamic spatial model of shifting cultivation in the highlands of Guinea, West Africa. Ecol. Mod., 79: 179-197.

Gladwin, C. H. 1983. Contributions of decision-tree methodology to a farming systems program. Human Organization , 42(2): 146 - 157.

Glass, E. H. and Thurston, H. D. 1978. Traditional and modern crop protection in perspective. Bioscience 28 (2), 109 - 115.

Gliessman, S. R. 1985. Economic and ecological factors in designing and managing sustainable agroecosystems. In Edens, T. C., Fridgen, C. and Battenfield, S. L. (eds.) Sustainable agriculture and integrated farming systems. East Lansing: Michigan State University Press. pp. 56 - 63.

GOL. 2005. Forest Strategy to the Year 2020 of the Lao PDR. No. 229/PM. Vientiane, Laos: Prime Minister‘s Office.

Goldammer, J.G. Rural land-use and wildland fires in the tropics. Agrofor. Syst. 1988, 6, 235–252.

Goswami K, Choudhury HK and Saikia J, 2012. Factors influencing farmers' adoption of slash and burn agriculture in North East India. Forest Policy and Economics 15: 146–151

Gourou, P. 1953. The Tropical World (Trans. By E. D. Labords), London. 156pp.

Gowswami, P.C. 1981. Socio-economic aspects of soil conservation activities with special reference to northeast India. In: Socio-economic aspects of soil conservation, soil conservation Department, Government of Meghalaya, Shillong, India, pp 7-12.

Greenland, D.J. 1975. Bringing the Green Revolution to the Shifting Cultivator. Science 1975, 190, 841–844.

Griffiths, T. 2008. Seeing ‘REDD’? Forests, climate change mitigation and the rights of indigenous peoples and local communities. In. Forest Peoples Programme, Wales, England.

Grigg DB. 1974. The Agricultural Systems of the World: An Evolutionary Approach. Cambridge University Press: Cambridge.

Grinnell, H.R. 1975. A study of agrisilviculture potential in West Africa. Ottawa, International Development Research Centre.

Grogan, P., Lalnunmawia, F. and Tripathi, S. K. 2012 Shifting cultivation in steeply sloped regions: a review of management options and research priorities for Mizoram state, Northeast India, Agroforest Syst (2012) 84 pp 163–177

Guo, H., Padoch, C., Coffey, K., Aiguo, C., and Yongneng, F. 2002. Economic Development, Land Use and Biodiversity Change in the Tropical Mountains of Xishuangbanna, Yunnan, Southwest China. Environmental Science & Policy 5: 471–479. doi:10.1016/S1462- 9011(02)00093-X.

Hadi S. 1984. A comparative study on swidden cultivation in Asia. Report of a meeting of researchers, 19–23 September 1983, Thailand. UNESCO, ROEAP: Bangkok.

Hadley, M. and Lanly, J-P 1983. Tropical forest ecosystems: identifying differences seeking similarities. Nature and Resources 19: 1 (Jan-March): 2 - 19.

Hansen, P. K. 1998. Shifting cultivation development in Northern Laos. In: Upland farming system in the Lao PDR-Problem and opportunities for livestock.. Proceedings No 87, Canberra, Australia.

Harden, R. M. 2000. The integration ladder: A tool for curriculum planning and evaluation. Medical Education, 34, 551-557.

Hare, F.K. Climatic classification. pp. 97-109. In: 1973 G. McBoyle. Climate in Review. Boston, Houghton Mifflin Company.

Hartmans EH, (1981). 'Land development and management in tropical Africa', International Institute of Tropical Agriculture, Ibadan, Nigeria.

Hauck, F. W. (1974). Introduction. Shifting Cultivation and Soil Conservation in Africa. FAO, Rome, Soils Bulletin 24: 1–4.

Heinimann, A. (2006). Patterns of Land Cover Change in the Lower Mekong Basin: The Relevance of Mesoscale Approaches. Doctoral dissertation. Department of Geography. Berne University, Berne, Switzerland.

Heinimann A, Hett C, Hurni K, Messerli P, Epprecht M, Jørgensen L, et al. Socio-Economic Perspectives on Shifting Cultivation Landscapes in Northern Laos. Hum Ecol. 2013; 41: 51-62.

Heinimann A, Mertz O, Frolking S, Christensen AE, Hurni K, Sedano F, Chini LP, Sahajpal R, Hansen M and Hurtt G, 2017. A global view of shifting cultivation: Recent, current, and future extent. PLoS One. 12(9): e0184479. Published online . doi: 10.1371/journal.pone.0184479

Hett C., Castella J.C., Heinimann A., Messerli P., Pfund J.L., 2012. A landscape mosaic approach for characterizing swidden systems from a REDD+ perspective. Applied Geography 32(2):608-618.

Hlaing, M. (2004). Analysis of impacts of Taung-ya program on livelihoods of local people: a case study in Bago Yoma, Myanmar. Master's thesis, Agricultural Science, Tokyo University of Agriculture and Technology, Tokyo.

Hoa, D. V. (2002). Ten years of implementing fixed field, fixed residence, migration and establishing new economical zones in the Mountains (in Vietnamese). Department of Sedentarisation, Ministry of Agricultural and Rural Development, Hanoi.

Houghton, R. A. (2003), Revised estimates of the annual net flux of carbon to the atmosphere from changes in land use and land management 1850–2000, Tellus, Ser. B, 55, 378–390 doi:10.1034/j.1600-0889.2003.01450.x.

Houghton, R. A., and J. L. Hackler (2006), Emissions of carbon from land use change in sub Saharan Africa, J. Geophys. Res., 111, G02003, doi:10.1029/2005JG000076.

Htun, A. A. (2007). Local Communities and Environmental Conservation in Panglaung Watershed Area. PhD Dissertation. Department of Botany, University of Yangon Yangon, Myanmar.

Humphries S (1993) The Intensification of Traditional Agriculture among Yucatec Maya Farmers: Facing up to the Dilemma of Livelihood Sustainability. Human Ecology, 21: 87-102.

Huq MM. 2000. Government Institutions and Under Development: A Study of the Tribal Peoples of Chittagong Hill Tracts, Bangladesh. Centre for Social Studies: Dhaka.

Hurni K, Hett C, Heinimann A, Messerli P, Wiesmann U. (2013) Dynamics of Shifting Cultivation Landscapes in Northern Lao PDR between 2000 and 2009 Based on an Analysis of MODIS Time Series and Landsat Images. Hum Ecol.,41: 21–36. doi: 10.1007/s10745-012-9551-y

Ickowitz, A. 2004. Shifting Cultivation and Deforestation: questioning the dominant narrative in tropical Africa, University of California, Riverside.

IFAD, IDRC, CIIFAD, ICRAF, IIRR (2001). Shifting Cultivation: Towards Sustainability and Resource Conservation in Asia. International Institute of Rural Reconstruction, Cavite The Philippines.

Inoue, M. 2000 Mechanism of Changes in the Kenyah's Swidden System: Explanation in Terms of Zensification Theory. In Rainforest Ecosystems of East Kalimantan—El Niño, Drought, Fire and Human Impacts; Guhardja, E., Fatawi, M., Sutisna, M., Mori, T., Ohta, S., Eds.; Springer-Verlag: Tokyo, Japan pp 167–184

IPCC. (2007). Climate Change 2007: Impacts, Adaptation and Vulnerability. In. Cambridge University Press, Cambridge, UK.

IRD (Institut de Recherche Pour le Developpement) 2008. Role for slash and burn farming in greenhouse effect control. Actualite Scientifique, sheet No. 307.

Ishaq, Muhammad (ed.) 1971. Bangladesh District Gazetteers, Chittagong Hill Tracts, 1971. Ministry of Cabinet Affairs, Government of Bangladesh.

Iskandar, J., 1994. System dynamic modelling for prediction and evaluation of the future of natural resources use in Siberida, Sumatra, Indonesia. In: Sandbukt, O., Wiriadinata, H. (Eds.), Rain Forest and Resource Management, Proceedings of the NORINDRA Seminar, Jakarta, 25–26 May, 1993. Indonesian Institute of Sciences, Jakarta.

Jackson, M.G. (2001) Plants and Soils: Liebig's Legacy and Beyond. Asian-Agri-History 5:197-223.

Jayahari KM, Sen, M, 2015. 'Sustainable Land and Ecosystem Management in Shifting Cultivation Areas of Nagaland for Ecological and Livelihood Security', Socio-Economic and Ecological Impact Study of GEF-UNDP-Government of Nagaland Project, Final Report. Nagaland Human Development Report 2005.

Jean S, 1975. Les jachbres en Afrique tropicale, Interprétation technique et foncibre. memoires de l'Institut d'Ethnologie XIV, Musée de l'Homme, Paris, France.

Johnson, E.D. (1996). Weed management in small holder rice production in tropics. Natural Resources Institute, University of Greenwich Chatham, Kent, UK.

Jordan, C. F.(1982). Amazon Rain Forests. American Scientist 70: 394 - 401.

Juo ASR, Manu A (1996). Chemical dynamics in slash-and-burn agriculture. Agricul. Ecosyst. Envion. 58(1): 49-60.

Kafle G, 2011. An overview of shifting cultivation with reference to Nepal. International Journal of Biodiversity and Conservation, 3(5): 147-154,

Kar PC. 1982. Garos in Transition. Cosmo Publications: New Delhi.

Kato, M. S. A., Kato, O. R., Denich, M., and Vlek, P. L. G. 1999 Fire-free alternatives to slash-and-burn for shifting cultivation in the eastern Amazon region: the role of fertilizers. Field Crops Research 62 pp 225–237

Keck A, Sharma NP and Feder G, (1994). Population Growth, Shifting Cultivation, and Unsustainable Agricultural Development: A Case Study in Madagascar. The World Bank Washington, D.C.

Kerkhoff, Elisabeth and Eklabya Sharma (2006), "Debating Shifting Cultivation in the Eastern Himalayas: Farmers' Innovations as Lessons for Policy", International Centre for Integrated Mountain Development, Kathmandu, Nepal.

Ketterings, Q.M., Wibowo, T.T., Noordwijk, M.V., Penot, E., 1999. Farmers' perspectives on slash-and-burn as a land clearing method for small-scale rubber producers in Sepunggur, Jambi Province, Sumatra, Indonesia. Forest Ecology and Management 120, 157–169.

Kirch, P. V. and Hunt, T. L. (eds.) (1997) Historical Ecology in the Pacific Islands: Prehistoric Environmental Change and Landscape Change, Yale University Press, New Haven and London.

Kleinman, P.; Bryant, R.B.; Pimentel, D. (1996) Assessing ecological sustainability of slash-and-burn agriculture through soil fertility indicators. Agron. J. 88 pp 122–127

Knight, C. G. 1980. Ethoscience and the African farmer: Rationale and strategy (Tanzania). In Brokensha, D., Warren, D. M. and Werner, O. (eds.) Indigenous Knowledge Systems and Development. Lanham: University Press of America. pp. 203 - 230.

Kurien AJ, Lele SC and Nagendra H. 2019. Farms or Forests? Understanding and Mapping Shifting Cultivation Using the Case Study of West. Garo Hills, India. Land, 8: 133-158; doi:10.3390/land8090133

Lafond GP, May WE, Stevenson FC, Derksen DA (2005) Effects of tillage systems and rotations on crop production for a thin Black Chernozem in the Canadian Prairies. Soil Tillage Res 89: 232-245.

Landsberg, _et _al. _In: E. Rodenwalt and H. J. Justaz eds. Die 1963 Jahreszeitenklimate der Erde. Heidelberger Akademie der WW/issenshaften, Springler Verlag, Heidelberg. Lee, D.H.K. Climate and economic development in the tropics. 1957 Harper, New York.

Langner, A., Siegert, F., 2009. Spatio-temporal fire occurrence in Borneo over a period of 10 years. Global Change Biology, 15,pp.48-62.

Lanly, J. P. (1985). Defining and Measuring Shifting Cultivation. Unasylva 37: 17–21.

Lassailly-Jacob, V. (1982). 'Taungya in Sierra Leone'. In: MadDonald, L.H. (ed.) Agro-forestry in the African Humid Tropics, United Nations University, Tokyo, Japan.

Lauk, C., and K. H. Erb (2009), Biomass consumed in anthropogenic vegetation fires: Global patterns and processes, Ecol. Econ., 69, 301–309, doi:10.1016/j.ecolecon.2009.07.003.

Lawrence, D., Radel, C., Tully, K., Schmook, B., & Schneider, L. (2010). Untangling a Decline in Tropical Forest Resilience: Constraints on the Sustainability of Shifting Cultivation Across the Globe. Biotropica, 42 (1), 21-30.

Lawton, R.M. (1982). 'National resources of miombo \~oodland and recent changes in agricultural and land-use practices", Forest Ecology and Management, 4: 287-297.

Lehsten, V., K. Tansey, H. Balzter, K. Thonicke, A. Spessa, U. Weber, B. Smith, and A. Arneth (2009), Estimating carbon emissions from African wildfires, Biogeosciences, 6, 349–360, doi:10.5194/bg-6-349-2009.

Leisz, S. J. (2007) Reinterpreting the Uplands of Vietnam: Towards Understanding and Correcting aMisreading of Its Land Cover, Land Use, and Resource Management System. Ph.D. Dissertation. Geography Department, University of Copenhagen, Copenhagen, Denmark.

Lele, N., Joshi, P.K., 2009. Analyzing deforestation rates, spatial forest cover changes and identifying critical areas of forest cover changes in north-east India during 1972-1999. EnvironmentMonitoring and Assessment, 156, pp.159-170.

Le Quere, C., Raupach, M. R., Canadell, J. G., & Marland, G. (2009). Trends in the sources and sinks of carbon dioxide. Nature Geosci, 2 (12), 831-836.

Lestrelin, G., Castella, J.C., and Bourgoin, J. (2012). Territorializing Sustainable Development: The Politics of Land-use Planning in the Lao PDR. Journal of Contemporary Asia 42(4), 581-602.

Liang L, Shen L, Yang W, Yang X, Zhang Y (2009). Building on traditional shifting cultivation for rotational agroforestry: Experiences from Yunnan. China For. Ecol. Manage., 257: 1989–1994.

Lianzela, 1997. Effects of shifting cultivation on the environment: with special reference to Mizoram. International Journal of Social Economics 24, 785–790.

Liao C, Feng Z, Li P, Zhang J. (2015) Monitoring the spatio-temporal dynamics of swidden agriculture and fallow vegetation recovery using Landsat imagery in northern Laos. J Geogr Sci. ;25: 1218–1234. doi: 10.1007/s11442-015-1229-0.

Linnard (1970). Terms and techniques in shifting cultivation in Russia. FAO, AGRIS, 197p.

Luiz, E.O., Aragão, C., Shimabukuro, Y.E., 2010. The Incidence of Fire in Amazonian Forests with Implications for REDD, Science, 328 (5983), pp.1275-1278.

Lynch, O. J., and Talbott, K. (1988). Legal Responses to the Philippine Deforestation Crisis. Journal of International Law and Politics 20: 679–713.

Ma, Q. (1999). Asia-Pacific Forestry Sector Outlook Study: Volume I— Socio-economic, Resources and Non-wood Products Statistics. FAO, Rome.

MAF (1999). The Government's Strategic Vision for the Agricultural Sector Vientiane (Lao PDR). Ministry of Agriculture and Forestry, Vientiane.

MAF. (2010a). Ministerial Instruction to Prepare for the Complete Stopping of Slash and Burn and Shifting Cultivation in 2010. No. 0022/MAF. Vientiane, Laos.

Maithani B. P. (2005) Shifting cultivation in North East India, Policy issues and options, Mittal Publications, P 165.

Majid Z. 1983. Malaysia. In Swidden Cultivation in Asia, Vol. 2, Country Profiles: India, Indonesia, Malaysia, Philippines, Thailand, Atal Y,

Majumdar DN. 1990. An assessment of shifting cultivation in north-east India. In Shifting Cultivation in North-East India,Majumdar DN (ed.). Omson Publications: New Delhi; 22–32.

McGrath, D. G. 1987. The role of biomass in shifting cultivation. Human Ecology 15(2): 221 - 242.

Meggers, B. J. (1954) Environmental limitations on the development of culture. American Anthropologist, 56, 5, 801-824.

Mertz, O. (2009). Trends in shifting cultivation and the REDD mechanism. Current Opinion in Environmental Sustainability, 1 (2), 156–160.

Mertz, O., 2002. The relationship between fallow length and crop yields in shifting cultivation: a rethinking. Agroforestry Systems 55, 149–159.

Mertz, O., Wadley, R. L., Nielsen, U., Bruun, T. B., Colfer, C. J. P., de Neergaard, A., Jepsen,M. R., Martinussen, T., Zhao, Q., Noweg, G. T., and Magid, J. (2008). A Fresh Look at Shifting Cultivation: Fallow Length an Uncertain Indicator of Productivity. Agricultural Systems 96: 75–84. doi:10.1016/j.agsy.2007.06.002.

Mertz O, Birch-Thomsen T, Elberling B, Rothausen S, Bruun TB, Reenberg A, et al. (2012). Changes in shifting cultivation systems on small Pacific islands. Geogr J. 2012; 178: 175-187.

Mertz O, Leisz SJ, Heinimann A, Rerkasem K, Thiha, Dressler W, et al. (2009) Who Counts? Demography of Swidden Cultivators in Southeast Asia. Hum Ecol. 2009; 37: 281-289.

Messerli, P., Heinimann, A., & Epprecht, M. (2009). Finding homogeneity in heterogeneity - a new approach to quantifying landscape mosaics developed for Lao PDR. Human Ecology, 37 (3), 291-304., doi:10.1007/ s10745-009-9238-1.

Messerli, P., Heinimann, A., Epprecht, M., and Minot, N. (2008). Socio-economic Atlas of Lao PDR—an Analysis Based on the 2005 Census. Geographica Bernsis, Bern.

Millennium Ecosystem Assessment (2005). Ecosystem and human well-being. Washington DC: Island Press. p 137p.

Ministry of Forestry (MOF) (1995). Myanmar Forest Policy 1995. Yangon, Myanmar.

Mishra, A.K. and Sharma, U.C., 2001. Traditional wisdom in range management for resource and environment conservation in North eastern region of India, ENVIS Bulletin: Himalayan Ecological Development, 9 (1), p 32

Mishra, B.K., Ramakrishnan, P.S.,1983. Slash and burn agriculture at higher elevation of Northeastern India, II Soil fertility changes, Agri Ecosystem and Environ, 9-83-96. Cited by R.E. de la Cruz and N.J. Vergara Productive and ameliorative roles of agroforestry. An overview of agroforestry in Humid tropics, pp.18-19.

Mittelman, A. (2001). Secondary Forests in the Lower Mekong Subregion: An Overview of Their Extent, Roles and Importance. Journal of Tropical Forest Science 13: 4671–690.

MOAF (Ministry of Agriculture and Forestry) (1997), Government o India, New Delhi.

MOAF (Ministry of Agriculture and Forestry) (2001), Government o India, New Delhi.

Mohabbatullah Md. et al. 2012. A Report on Jhum Research in CHT. Hill Agricultural Research Station, Bangladesh Agricultural Research Institute, Khagrachari.

Molinario G, Hansen M, Potapov P. (2915) Forest cover dynamics of shifting cultivation in the Democratic Republic of Congo: a remote sensing-based assessment for 2000–2010. Environ Res Lett.;10: 94009.

Moran, E. F. 1987. Socioeconomic considerations in acid tropical soils research. In Management of Acid Tropical soils for Sustainable Agriculture: Proc. No.2. IBSRAM, Bangkok, Thailand, pp. 227 - 244.

Moran, E. F. 1981. Developing the Amazon. Indiana University Press: Bloomington.

Muller D, Sun Z, Vongvisouk T, Pflugmacher D, Xu J, Mertz O. (2014). Regime shifts limit the predictability of land-system change. Glob Environ Change.; 28: 75-83.

Myers, N. 1980. Role of Forest Farmers in Conversion of Tropical Moist Forest. A report prepared for the committee on research priorities in tropical Biology of the National Research Council. Washington, D.C: National Academy of Sciences.

Myers, N. 1986. Forestland farming in western Amazonia: stable and sustainable. Forest Ecology and and Management 15: 81 - 93.

Myers, N. 1992. Tropical Forests: The Policy Challenge. Environmentalist 12: 15–27. doi:10.1007/BF01267592.

Myers, N., Mittermeier, R.A., Mittermeier, C.G., da Fonseca, G.A.B., Kent, J., 2000. Biodiveristy hotspots for conservation priorities. Nature 40, 853–858.

Myers N. 1994. Tropical deforestation: Rates and patterns. In K. Brown and D.W. Pearce (eds.). The causes of tropical deforestation. UCL Press, London.

Müller, D., Suess, S., Hoffmann, A.A., and Buchholz, G., 2013. The Value of Satellite-Based Active Fire Data for Monitoring, Reporting and Verification of REDD+ in the Lao PDR. Human Ecology, 41,pp.7-20.

Nair, P. K. R. (1993). An Introduction to Agroforestry. Kluwer Academic Publishers, Dordrecht, The Netherlands.

Nair, P. K. R. and Fernandes, E. 1985. Agroforestry as an alternative to shifting cultivation. ICARF No. 17 (Reprinted from FAO Soils Bull. 53, FAO Roam, 1984) : 1-57.

NEC (North Eastern Council) (2008). Prioritised Schemes of Meghalaya for funding under the NEC's plan during 2008-09 & the 11th Five Year Plan, Shillong, Meghalaya.

Nietschmann, B. 1973. Between Land and Water. New York and London: Seminar Press.

Nikolic, N., Schultze-kraft, R., Nicolic, M., Böcker, R., and Holz, I. (2008). Land Degradation on Barren Hills: A Case Study in Northeast Vietnam. Environmental Management 42: 19–36. doi:10.1007/s00267-008-9099-1.

Ninan, K. N (1992), "Economics of Shifting Cultivation in India", Economic and Political Weekly, 27(13), A2-A6.

Nye, P.H. and Greenland, D.J. (1960) The soil under shifting cultivation. Technical communication No. 51, Commonwealth bureau of soils, Harpenden. Commonwealth Agricultural Bureau, Bucks, England.

Okigbo, B.N. 1981. 'Alternatives to shifting cultivation' Ceres Nov-Dec. 1981:41-45, FAO, Rome.

Okigbo B. N. 1984. Improved permanent production systems as an alternative to shifting. International Institute of Tropical Aqriculture, Ibadan, Nigeria. In: Okigbo B. N. Improved permanent production systems as an alternative to shifting. FAO Soils Bulletin No. 53, Food and Agricultural Organization of the United Nations.

Olafson, H. 1981. Introduction. In Olafson, H. (ed.) Adaptive Strategies and Change in Philippine Swidden-based Societies. Laguna, Philippines: Forest Research Institute. pp. 1 - 12. Olafson, H. (ed.) Adaptive Strategies and Change in Philippine Swidden-based Societies. Laguna, Philippines: Forest Research Institute. pp. 43 - 54.

Olafson, H. 1983. Indigenous agroforestry systems. Philippine Quarterly of Culture and Society 11: 149 - 174.

Olofson, H. (1982). An anthropological approach to social forestry: the study of indigenous agroforestry systems'. Paper presented to the Monthly Forum in Social Forestry of the Bureau of Forest Development, Asian Institute of Tourism, Quezon City, Philippines.

OTA (Office of Technical Assessment). 1984. Technologies to Sustain Tropical Forest Resources. Washington, D. C.: U. S. Congress.

Padaliaa H and Mondalb PP, 2014. Spatio-temporal trends of fire in slash and burn agriculture landscape: A case study from Nagaland India. ISPRS Annals of the Photogrammetry, Remote Sensing and Spatial Information Sciences, Volume II-8, ISPRS Technical Commission VIII Symposium, Hyderabad, India.

Padoch, C., Coffey, K., Mertz, O., J. Leisz, S., Fox, J., & L. Wadley, R. (2007). The demise of swidden in southeast Asia?: local realities and regional ambiguities. Geografisk Tidsskrift, 107 (1), 29-41.

Panda BK, Alam MA and Sarkar S, 2016. Shifting Cultivation in North-East India: Trend, Benefits and Challenges. 2: 85-115, ISSN 2455–7846.

Parrotta J, Youn YC, Camacho LD. 2016. Traditional knowledge for sustainable forest management and provision of ecosystem services. Int J Biodiversity Science, Ecosystem Serv Manag. 12:1–2.

Pearce, D.W., and Warford, J.J. 1993. World Without End: Economics, Environment, and Sustainable Development. Oxford University Press, Oxford, UK.

Pelzer KJ, 1978. Planter and Peasant: Colonial Policy and the Agrarian Struggle in East Sumatra, JSTOR 84:1863-1947, www.jstor.org/stable/10.1163/j.ctvbnm4v6. Accessed 18 Aug. 2020.

Peters, J.W and Neuenschwander, L.F. (1988). Slash and Burn: Farming in the Third World Forest. University of Idaho Press. USA.

Peterson, J. T. 1981. Game, Farming, and Interethnic relations in Northeastern Luzon, Philippines. Human Ecology 9(1): 1- 22.

Popenoe, H. l960. Effects of shifting cultivation on natural soil constituents in Central America. Ph.D. dissertation, University of Florida, IFAS. Cited in Moran 1981.

Potter WJ (1993). Cultivation Theory and Research-A Conceptual Critique. Human Communication Research, 19 (4): 564-601, https://doi.org/10.1111/j.1468-2958.1993.tb00313

Poulsen, G. 1978. Man and tree in tropical Africa: three essays on the role of trees in the African environment. Ottawa, Ont.: IDRC.

Prasad, R. N. and Sharma, U. C. (1994) "Potential Indigenous Farming System of Northeastern Region; Indian council of Agricultural Research complex for northeastern region, Meghalaya, India.

Pravongviengkham, P. P. (1998). The Role of Animal Husbandry and Aquaculture in Improvements of Swidden-Based Livelihood Systems in the Lao PDR, Thesis, School of Environment, Resources and Development, Asian Institute of Technology, Bangkok.

Rahman, H., Karuppaiyan, R., Kishore, K., and Denzongpa, R. 2008 Traditional Practices of Ginger Cultivation in Northeast India, Indian Jour. of Traditional Knowledge 8 pp 1, 23-28.

Raintree, J., & Warner, K. (1986). Agroforestry pathways for the intensification of shifting cultivation. Agroforestry Systems, 4 (1), 39-54.

Raintree, J.B. 1980. 'Conservation farming with multipurpose tree legumes: an underdeveloped branch of tropical agroforestry research'. In: Buck, L. (ed.) Proceedin s of the Kenya National Seminar on ALroforestry (12-22 Nov. 1980 ICRAF, Nairobi, Kenya.

Ramakrishanan, P.S., Saxena, K.G., Rao, K.S. 2006. Shifting agriculture and sustainable development of North-East India: Tradition in Transition, UNESCO and Oxford& IBH, New Delhi; p 495

Ramakrishnan, P.S., 2001. Ecology and Sustainable Development, National Book Trust, New Delhi, India.

Ramakrishnan, P.S.; Toky, O.P. 1981. Soil nutrient status of hill agro-ecosystems and recovery pattern after slash and burn agriculture (jhum) in north-eastern India. Plant Soil, 60, 41–64.

Ramakrishnan PS (1992). Shifting Agriculture and Sustainable Development: an interdisciplinary study from north-eastern India. MAB Series, Volume 10, UNESCO, Paris.

Ramakrishnan PS. 1984. The science behind rotational bush fallow agriculture system (jhum). Proceedings of the Indian Academy of Science (Plant Science) 93: 379–400.

Ramankutty, N., H. K. Gibbs, F. Achard, R. Defriess, J. A. Foley, and R. A. Houghton (2007), Challenges to estimating carbon emissions from tropical deforestation, Global Change Biol., 13, 51–66, doi:10.1111/j.1365-2486.2006.01272.x.

Rambo, A. T. 1981. Orang Asli adaptive strategies: Implications for Malaysia natural resource development planning. EAPI reprint. Honolulu: East West Center.

Rambo, A. T. 1983. Why shifting cultivators keep shifting: understanding farmer decision-making in traditionalagroforestry systems. Paper prepared for the Joint East-West Environment and Policy Institute/ United Nations Food and Agriculture Organization Regional Worksop on Community Forestry. Nakorn Rachasima, Thailand, August 22-29, 1983.

Ranjan, R., and Upadhyay, V.P. 1999. Ecological problems due to shifting cultivation. Current Science 77:1246–50.

Rarijaona, Rene. 1967. Le Concept de Propriete en Droit Foncier de Madagascar Paris: Cujas

Rasul, G., and Thapa, G. B. 2003 Shifting Cultivation in the mountains of South and South East Asia: Regional Patterns and Factors Influencing the Change, Land Degrad. Develop. 14: 495-508

Regmi BR, Subedi A, Aryal KP, Tamang BB (2005). Shifting cultivation systems and innovations in Nepal. Unpublished report. LIBIRD, Pokhara.

Reimeingam M, 2017. Shifting cultivation in manipur: land, labour and environment. Journal of Rural Development, NIRD&PR, Hyderabad. 36: 97-119

Rerkasem, K. 2003. Uplands land use. In Kaosa-ard, M., and Dore, J. (Eds.), Social Challenges for the Mekong Region. White Lotus, Bangkok, pp. 323–346.

Rerkasem, K., D. Lawrence, C. Padoch, D. Schimdt-Vogt, A.D. Ziegler, T.B. Bruun. 2009. Consequences of Swidden Transitions for Crop and Fallow Biodiversity in Southeast Asia. Human Ecology 37:347-360., doi:10.1007/s10745-009-9250-5.

Rerkasem K, Rerkasem B. 1994. Shifting cultivation in Thailand: its current situation and dynamics in the context of highland development. IIED forestry and land use series No. 4. International Institute for Environment and Development: London.

Riahtam N.B., Nongkynrih, J.M., Sarma, K.K., Raju, P.LN., Mishra, A.R., Lal, D., Kharsahnoh and Sahkha, D.J. 2018. Assessment of shifting cultivation dynamics in East Garo Hills District, Meghalaya, India. IOP Conf. Series: Earth and Environmental Science 169, 012104 doi :10.1088/1755-1315/169/1/012104.

Richards, A.I. 1939. Land labour and diet in Northern Rhodesia. An economic study of the Bemba tribe, Oxford University Press for the International Institute of African Languages and Cultures, London, New York, Toronto.

Richards, P. W. 1977. Tropical forests and woodlands: an overview. Agro-ecosystems 3: 225 -238.

Richards PW. 1952 The tropical rainforest. London: Cambridge University Press;

Roder, W. 1997. Slash-and-burn rice system in transition. In: Challenges for agricultural development in the hills of northern Laos. Mountain Research and Development, Vol.17. Nol, International Mountain Society and United Nation University pp.1-10.

Roder, W. 2001. Slash-and Burn Rice Systems in the Hills of Northern Lao PDR: Description, Challenges and Opportunities. International Rice Research Institute, Los Baños.

Rosenfeld, D.,1999. TRMM observed first directevidence of smoke from forest fires inhibiting rainfall. Geophysical Research Letters, 26 (20),pp. 3105-3108.

Rosmalah S, Widayati W, La rianda and Sidu D. 2019. Existence of Swidden Agriculture in a Small Island. The International Journal of Engineering and Science, 8: 68-75.

Rosmann MR, 2010. The Agrarian Imperative. Journal of Agromedicine, 15: 71-75. DOI:10.1080/10599241003630585.

Rounsevell MDA, Pedroli B, Erb K-H, Gramberger M, Busck AG, Haberl H, et al. (2012) Challenges for land system science. Land Use Policy; 29: 899-910.

Roy, D. (2000). Occupations and Economy in Transition: A Case Study of the Chittagong Hill Tracts. In International Labour Organization, Traditional Occupations of Indigenous and Tribal Peoples: Emerging Trends, (pp. 73-122). Geneva: ILO.

Roy, P.S.; Kushwaha, S.P.S.; Mirthy, M.S.R.; Roy, A.; Kushwaha, D.; Reddy, C.S.; Behera, M.D.; Mathur, V.B.; Padalia, H.; Saran, S.; Singh, S.; Jha, C.S.; Porwal, M.C. (2012). Biodiversity Characterization at Landscape Level: National Assessment; Indian Institute of Remote Sensing: Dehradun; 140 pp.

Roy RD. 1995. Land rights, land use and indigenous people in the Chittagong Hill Tracts. In Bangladesh Land Forest and Forest People, Gain P (ed.). Society for Environment and Human Development SEHD: Dhaka; 53–118.

Ruddle, K. and Manshard, W. 1981. Renewable natural resources and the environment. Dublin: Tycooly International (for UNU).

Russell, W. M. S. 1988. Population, swidden farming and the tropical environment. Population and Environment 10(2): 77- 94.

Ruthenberg, H. (1971). Farming Systems in the Tropics. Oxford: Clarendon Press.

Ruthenberg, H. (1980). Farming systems in the tropics, third edition (also second edition, 1976), Clarendon Press, Oxford, U.K. (see especially Chapter 3 'Shifting cultivation systems').

Sachchidananda, Pathak KN. 1983. India. In Swidden Cultivation in Asia. Vol. 2, Country Profiles: India, Indonesia, Malaysia, Philippines, Thailand. UNESCO Regional Office for Education in Asia and the Pacific: Bangkok; 1–107.

Sachchidananda 1989. Shifting Cultivation in India. Concept Publishing Company: New Delhi.

Sahlins, M.D., 1972. Stone Age Economics. Aldine-Atherton, Chicago.

Sam DD, 1994. Shifting Cultivation in Vietnam: its social, economic and environmental values relative to alternative land use. Land Use Working Group, Ministry of Forestry 123 Lo Duc Street - Hanoi – Vietnam and Forest Science Institute of Vietnam, Chem Tu Liem Hanoi Vietnam

Sanchez, H. P. 1976. Properties and Management of Soils in the Tropics. John Willey and Sons, New York, USA.

Sanchez, P. A., Palm, C. A., Vosti, S. A., Tomich, T., and Kasyoki, J. 2005. Alternatives to slash and burn. Challenges and approaches of an international consortium. In Palm, C. A., Vosti, S. A., Sanchez, P. A., and Ericksen, P. J. (Eds.), Slash-and-Burn Agriculture. The Search for Alternatives. Columbia University Press, New York, pp. 3–37.

Sanchez, P.A. 1996 Alternatives to Slash-and-Burn Agriculture Introduction (Special Issue). Agric. Ecosyst. Environ. 58: 1–2

Sanchez P.A., Couto W. & Buol S.W., 1982, The fertility applicability and modification. Geoderma, 27, 283-309.

Sargent, C., Palmer, J., and Morrison, E. 1991. Setting Priorities for Research in the Land Use Continuum in Vietnam. In National Seminar Proceedings, International Institute for Environment and Development, London, pp. 324–346.

Sarma, , B.K. and Sharma, U.C. 2000 Agriculture of the Lakhars of Mizoram. Asian Agri-History 4:21-30.

Satapathy, K. K. 1996. Hill slope runoff under conservation practices. Res. Bull. No. 14 ICAR Research Complex for NEH Region, Shillong 42pp.

Satapathy, K.K. and Sarma, B.K. 2002 Shifting cultivation in India: An overview. Asian Agri-History 6:121-139.

Sati VP and Rinawma P, 2014. Practices of Shifting Cultivation and its Implications in Mizoram, North-East India: A Review of Existing Research. Nature & Environment,19 (2): 179-187 (www.natureandenvironment.com).

Sawathvong, S and Vongleck, P. (1999). Sustainable use of natural resources of Lao PDR "How to balance conservation and development?" The papers presentations in Development of national forest policies and strategics international training courses (23 Aug - 24 Sep), Sida, Stockholm, Sweden.

Schmidt, D.R. 1973. Anthropological and ecological considerations regarding the transition of shifting cultivation in the tropics', African Soils/Sols Africains, XVIII, 2:59 68.

Schmidt-Vogt, D., 1999. Swidden Farming and Fallow Vegetation in Northern Thailand. Franz Steiner Verlag, Stuttgart.

Schmidt-Vogt D, Leisz SJ, Mertz O, Heinimann A, Thiha T, Messerli P, Epprecht M, Pham Van Cu PV, Chi VK, Hardiono M and Dao TM, 2009. An Assessment of Trends in the Extent of Swidden in Southeast Asia. Human Ecology, 37:269–280, DOI 10.1007/s10745-009-9239-0

Schroeder, R.G., Linderman, K., Liedtke, C. and Choo, A.S. (2008), "Six sigma: definition and underlying theory", Journal of Operations Management, Vol. 26 No. 4, pp. 536-554.

Schuck, E.C., Nganje, W., Yantio, D., 2002. The role of land tenure and extension education in the adoption of slash and burn agriculture. Ecological Economics 43, 61–70.

Seiler, W. and Crutzen, P.J. (1980) Estimates of Gross and Net Fluxes of Carbon between the Biosphere and the Atmosphere from Biomass Burning. Climatic Change, 2, 207-247.

Shahlace, A. K.; Nutter, W. L.; Burroughs, E. R. and Morris, L. A. 1991. Runoff and sediment production from burned forest sites Achard, F., H. D. Eva, P. Mayaux, H. J. Stibig, and A. Belward (2004), Improved estimates of net carbon emissions from land cover change in the tropics for the 1990s, Global Biogeochem. Cycles, 18, GB2008, doi:10.1029/2003GB002142.

Sharma, U. C. & Sharma, V. (2003), Mathematical model for predicting soil erosion by flowing water in ungauged watersheds, in Erosion Predictions in Ungauged Basins: Integrating Methods and Techniques, edited by D. Boer, F. Wojciech, T. Mizuyama & A. Pietroni, pp 79–83. IAHS Publ. 279. IAHS Press, Wallingford, UK.

Sharma, U. C. (1997b) Some indigenous farming systems of north-eastern India. Asian Agri-History 1, 267–273.

Sharma, U. C. (1999) Food Security in the northeast : new paradigms. In : Persrpective for Planning and Development in North-eastern India (ed. by R. C. Sundariyal, Uma Shankar & T. C. Upreti), 197-212, G.B. Pant Institute of Himalayan Environmental and Development, Almorah, India.

Sharma, U. C. (2001) Effect of farming system type on in-situ ground water recharge and quality in northeast India. In: Impact of human Activity on Groundwater Dynamics (ed.

by H. Gehrels, N.E. Peters, E. Hoehn, K. Jenson, C. Leibundgut, J. Griffioen, B. Webb & W. J. Zaadnoordijk), IAHS Publ. No.269, 167-169, IAHS Press, Wallingford, U.K.

Sharma, U. C. (2003) Impact of population growth and climate change on the quantity and quality of water resources in the northeast region of India. In: Water Resources System - Hydrological Risks, Management and Development. (ed. by G. Bloschl, S. Franks, M. Kumagia, K. Musiake & D. Rosbjerg), IAHS Publ. No. 281, 349-357, IAHS Press, Wallingford, U.K.

Sharma, U. C. 1993. Agricultural production potential in North Eastern hill states and role of fertilizer nutrients in its realization. Indian Fertilizer Scene Annual, 19-28.

Sharma, U. C. and Prasad, R. N. 1994. Potential indigenous farming systems of North-Eastern Hill region. ICAR Research Complex for NEH Region, Shillong 48pp

Sharma, U. C. and Vikas Sharma (2009). Plant-animal-soil-hydrology interactions in the northeastern region of India. Biologia, 64 (3): 460-464.

Sharma, U. C., (2001). Effect of farming system type on in-situ ground water recharge and quality in northeast India. In: Impact of human Activity on Groundwater Dynamics (eds.), IAHS Publ. No.269, 167-169, IAHS Press, Wallingford, U.K. (2001).

Sharma, U.C. & Prasad, R.N. (1995). Socio-economic aspects of acid soil management and alternate land use systems for northeastern states of India. In : Plant - Soil Interactions at Low pH (ed. by R.A. Date, N.J. Grundon, G.E. Rayment & M.E. Probert), 689-696, kluwer Academic Press, The Netherlands.

Sharma, U.C. & Sonowal, D. K. (1991) Zabo farming system of Nagaland, Proc. National Seminar on Shifting Cultivation, pp.110-126, Dimapur, Nagaland, India

Sharma, U.C. (1997a) Water Management by the tribes of northeastern region - the traditional wisdom and future strategies. Fert. News 43: 43-49

Sharma, U.C. (1997b). Some indigenous farming systems of northeastern India. Asian Agri-Histroy, 1:267-273.

Sharma, U.C. (1998) Water Management by the tribes of north-eastern region: The traditional wisdom and future strategies. Fertilizer News: 43:43-49.

Sharma, U.C. (2003) Impact of population growth and climate change on the quantity and quality of water resources in the northeast of India. In: Water Resources Systems - Hydrological Risks, Management and Development (ed. by G. Bloschl, S. Franks, M. Kumagai, K. Musiake and D. Rosbjerg) ,Proc. Sappore Symp. , 240-247. IAHS Publ. No. 281.

Sharma, U.C. and Singh, K.A. 1994. Agriculture with Alder- A potential indigenous farming system of Nagaland. In Proc. Of the Nat. Seminar on Developments in Soil Sci., Nov. 28 - Dec. 1, 1994 , New Delhi, India : 570-571.

Sharma, U.C., Prasad, R. N. and Sonowal (1994). An indigenous technique of soil and water conservation in north -eastern region- The Zabo system of farming. Soil and Water Conservation Challenges and Opportunities (Proceeding of 8th, ISCO conference. Ed. L. S. Bhushan, I. P. Abrol and M. S. Rama Mohan Rao). Oxford and IBH, publication Co. Pvt. Ltd., New Delhi (India). p. 969-975.

Sharma, U.C., Prasad, R. N. and Sonowal (1994). An indigenous technique of soil and water conservation in north -eastern region- The Zabo system of farming. Soil and Water Conservation Challenges and Opportunities (Proceeding of 8th, ISCO conference. Ed. L. S. Bhushan, I. P. Abrol and M. S. Rama Mohan Rao). Oxford and IBH, publication Co. Pvt. Ltd., New Delhi (India). p. 969-975.

Sharma, U.C., Prasad, R.N., 1994. Potential indigenous farming systems of North-Eastern Hill region. ICAR Research Complex for NEH Region, Shillong, India.

Sharma S K, Punitha P, Sailo B and Singh I M. 2017. Jhum Improvement for Sustaining Farm Livelihood and Natural Resource Conservation in North Eastern Hill Region : Vistas And Frontiers. 110 Pages, ICAR Research Complex for NEH Region, Meghalaya, India.

Sharma TC (1976). Shifting cultivation in north eastern India. NEI Council for Scientific Research, Shillong, 1-3.

Sharma U. C. and Sharma, V. (2004b). Implications of nutrient and soil transfer with runoff in the northeastern reghion of India. In: Sediment Transfer through the Fluvial System (ed. by V. Golosov, V. Belyaev and D. E. Walling). IAHS Publ. No. 288, 488-493, IAHS Press, Wallingford, U.K.

Sharma U. C. and Vikas Sharma (2004a). The 'Zabo' water management and conservation system: tribal beliefs in the development of water resources. In: In: Basis of Civilization-Water Science? (ed. by John C. Rodda and Lucio Ubertini). Pp 184-191, Rome, Italy, IAHS Publ. 286. IAHS Press, Wallingford, U.K.

Sharma UC and Bujarbaruah KM 2016. Issues and Challenges in Shifting Cultivation and its Relevence in the Present Context, National Academy of Agricultural Sciences, Policy Paper No. 83.

Sharma UC and Sharma Vikas (2018). "Greenhouse Gases Emissions from Agriculture Sector in Northeastern Region of India". Acta Scientific Microbiology 1.12: 36-43.

Sharon, M. F and Kenneth A. F. (1988). Deforestation in South-east Asia, The Asian Forum of environmental Journalist and ESC,4P, Bangkok, Thailand. Shifting Cultivation Practices By Hmong, Khamu and Lao ethnic categories In The Nam Nane Watershed, Nane District, Luang Prabang Province, Lao PDR 20040530 68

Shillong Declaration (2004) Regional Shifting Cultivation Policy Dialogue Workshop for the Eastern Himalayas held in Shillong, India, from 6-8th October, 2004.

Siahaya ME, Hutauruk TR, Aponno HSES, Hatulesila JW and Mardhanie AB, (2016). Traditional ecological knowledge on shifting cultivation and forest management in East Borneo, Indonesia, International Journal of Biodiversity Science, Ecosystem Services & Management, 12:1-2, 14-23, DOI: 10.1080/21513732.2016.1169559.

Silva JMN, Carreiras JMB, Rosa I and Pereira JMC, 2011. Greenhouse gas emissions from shifting cultivation in the tropics, including uncertainty and sensitivity analysis. Journal of Geophysical Research, 116: D20304, doi:10.1029/2011JD016056, 21p.

Singh, A. 1989. Bamboo drip irrigation system. ICAR Research Complex for NEH Region, Barapani, Meghalaya, India 20pp.

Singh, A. and Singh, M. D. 1981. Soil erosion hazards in North- Eastern Hill region. Res. Bull. No. 10, ICAR Research Complex for NEH Region, Shillong 30pp.

Singh, J. and Ramakrishnan, P. S., Proc. Indian Acad. Sci., 1982, B91, 269–280.

Singh A (1988) Annual Report, ICAR Research Complex for North Eastern Hill Region of India, Umiam, Meghalaya, India.

Singh A and Singh, M. D. (1978). Effect of various stages of shifting cultivation on soil erosion from steep hill slopes. Indian Forester, Vol.106:pp 115-121.

Singh S, Purohit JK and Bhaduri A (2016). Shifting cultivation in Odisha and Chhattisgarh: rich agro-biodiverse systems under risk. Jharkhand Journal of Development and Management Studies, 14(2): 7023-7036.

Sirén, A. H., & Brondizio, E. S. (2009). Detecting subtle land use change in tropical forests. Applied Geography, 29 (2), 201-211.

Sodarak, H. 1999. Shifting Cultivation Practices By Hmong, Khamu and Lao ethnic categories In The Nam Nane Watershed, Nane District, Luang Prabang Province, Lao PDR, Thesis submitted to Swedish University of Agricultural Sciences in for the requirement of degree of M.Sc., Uppsala, Sweden.

Soewardi B. 1983. Indonesia. In Swidden Cultivation in Asia. Vol. 2, Country Profiles: India, Indonesia, Malaysia, Philippines, Thailand.

Sonowal, D.K., Sharma, U.C. & Tripathi, A.K. (1994) Zabo Farming system. In: Potential Indigenous Farming Systems of Northeastern Hills Region (ed. by R.N. Prasad & U.C. Sharma), 5-12, Indian Council of Agriculture Research Complex for Northeastern Hills Region, Shillong, Meghalaya, India.

Sonowal, D.K., Tripathi, A.K.& V.Kirekha (1989). Zabo, an indigenous framing system of Nagaland. Indian J.Hill Res, 2:1-8.

Spencer, J. E. 1966 Shifting cultivation in south-eastern Asia. University of California Press, Berkeley 19, pp 1-179

Srisawas, N., and Suwan, M. (1985). Thailand-swidden cultivation in the north (White Meo). In Swidden Cultivation in Asia, UNESCO, Bangkok, pp. 269–349.

S Southgate, D. 1990. The Causes of Land Degradation along Spontaneously Expanding Agricultural Frontiers in the Third World, Land Economics, 66, 93-101.

State of the Tropics (2014). State of the Tropics 2014 Report. James Cook University, Townsville, Australia.

State of the Tropics (2020). State of the Tropics 2020 Report. James Cook University, Townsville, Australia.

Steensberg, A. (1993) Fire-clearance Husbandry: Traditional Techniques Throughout the World. Herning: Poul Kristensen.

Stewart O.C. 1956, Fire as the First Great Force Employed by Man, II. Thomas W.L. Man's Role in Changing the Face of the Earth, Chicago.

Sulistyawati, E., 2001. An agent-based simulation of land use in a swidden agricultural landscape of the Kantu in Kalimantan, Indonesia. Ph.D. Thesis, The Australian National University, Canberra.

Sulistyawati E, Noble I. R. and Michael L. Roderick M. L. 2001. A simulation model to study land use strategies in swidden agriculture systems. Agricultural Systems 85: 271–288 (Available online at www.sciencedirect.com)

Sunderlin, W. D., Resosiudarmo, I. A. P., Rianto, E., and Angelsen, A. (2000). The effect of Indonesia's economic crisis on small farmers and natural forest cover in the outer islands, CIFOR Occasional Paper No. 28(E). CIFOR, Bogor, Indonesia.

Suraswadi P, Thomas DE, Pragtong K, Preechapanya P, Weyerhauser H. 2000. Changing land use mosaics of (former) shifting cultivators in watersheds of North Thailand. In Alternatives to Slash- and -Burn Sanchez PA (ed.). American Society of Agronomy: Madison, WI.

Suryanata K. 1994. Fruit trees under contract: tenure and land use change in Upland Java, Indonesia. World Development 22: 1567–1578.

Suyanto S, Tomich TP, Otsuka K. 2001. Land tenure and farm management efficiency: the case of smallholder rubber production in customary land areas of Sumatra. Agroforestry Systems 52: 145–160.

Szott LT, Palm CA, Buresh RJ (1999). Ecosystem fertility and fallow function in the humid and subhumid tropics. Agrofor. Syst., 47: 163-196.

Talukdar, G.; Ghosh, S.; Roy, P.S. (2004) Landscape Dynamics in North East Region of India (Meghalaya State) Using Spatial Decision Tree Model. Geocarto Int., 19, 11–18.

Tangjang, S., 2009. Traditional slash and burn agriculture as a historic land use practice: a case study from the Ethnic Noctes in Arunachal Pradesh, India. World Journal of Agricultural Sciences 5 (1), 70–73.

Thapa GB, Paudel GS. 2002. Farmland degradation in the mountains of Nepal: a study of watershed 'with' and 'without' external intervention. Land Degradation & Development 13: 479–493.

The, B. D., Ha, D. T., and Chinh, N. Q. 2004. Rewarding upland farmers for environmental services. Experience, Constraints and Potential in Vietnam.World Agroforestry Centre (ICRAF), Bogor.

Thomas, D. 2004. Review of Policies and Practices in Upland Areas of the Lao PDR. Paper presented at the 2004 NAFRI Workshop on Shifting Cultivation and Poverty Eradication in the Uplands of the Lao PDR. Luang Prabang, Laos.

Thomas, D. Ziegler E. (2005). Review of policies and practices in upland areas of Lao PDR. In Bouahom, B., Glendinning, A., Nilsson, S., and Victor, M. (Eds.), Poverty Reduction and Shifting Cultivation Stabilization in the Uplands of Lao PDR: Technologies, Approaches and Methods for Improving Upland Livelihoods, National Agriculture and Forestry Research Institute, Vientiane, pp. 9–38.

Thomaz, E.L., 2013. Slash-and-burn agriculture: establishing scenarios of runoff and soil loss for a five-year cycle. Agric. Ecosyst. Environ. 168, 1–6.

Thrupp, L. A., Hecht, S. B., & Browder, J. O. (1997). The diversity and dynamics of shifting cultivation: Myths, realities, and policy implications. World Resources Institute. Retrieved from pdf.usaid.gov/pdf_docs/pnacd174.pdf

Tivy, J. 1987. Nutrient cycling in agro-ecosystems. Applied Geography 7: 93 - 113.

Toky, O.P. and Ramakrishna, P. S. 1981a. Runoff and infiltration losses related to shifting agriculture (Jhum) in North Eastern India. Environmental Conservation. 8: 313-321.

Toky OP, Ramakrishnan PS (1981) Cropping pattern and yields in agricultural systems of the northeastern hill region of India. Agroecosystem 7: 11-25.

Turkelboom, F., Poesen, J. and Trebuil, G., 2008. The multiple land degradation effects caused by land-use intensification in tropical steeplands: a catchment study from northern Thailand. Catena 75 pp102–116

Turner, B. L. (1974) Prehistoric intensive agriculture in the Mayan lowlands. Science, 185, 4146, 118-124.

Uhl, C. 1983. You can keep a good forest down. Natural History 92(4): 69 - 79.

Uhlig, J., Hall, C. A. S., and Nyo, T. (1994). Changing patterns of shifting cultivation in selected countries in Southeast Asia and their effect on the global carbon cycle. In Lange, O. L., Mooney, H. A., and Remmert, H. (eds.), Effects of Land-Use Change on Atmospheric CO2 Concentrations. Springer, New York, pp. 145–200.

Uhlig H. 1984. Spontaneous and Planned Settlement in Southeast Asia. Institute of Asian Affairs: Hamburg.

UNDP (2002). National Human Development Report Lao PDR 2001: Advancing Rural Development. United Nations Development Programme, Vientiane.

UNESCO/UNEP/FAO. 1978. Tropical forest ecosystems: A state-of-knowledge report. Paris: Unesco/UNEP.

Van Gansberghe, D. and Pals, R. (eds.) 1994. Shifting Cultivation Systems and Rural Development in the Lao PDR. Report of the Nabong technical meeting. Nabong Agriculture College, Lao PDR. July 14-16, 1993.

Van Keer K (2003). On-farm agronomic diagnosis of transitional upland rice swidden cropping systems in northern Thailand. Department of Land Management, Faculty of Agricultural and Applied Biological Sciences - Catholic University of Leuven.

Van Turkelboom FKK, Ongprasert S, Sutigoolabud P, Pelletier J. 1996. The Changing Landscape of the Northern Thai Hills: Adaptive Strategies to Increasing Land Pressure. Montane Mainland Southeast Asia in Transition. Chiang Mai University Agricultural Systems Programmes: Thailand; 436–461.

van Vliet N, Adams C, Vieira ICG, Mertz O. 2013. "Slash and Burn" and "Shifting" cultivation systems in forest agriculture frontiers from the Brazilian Amazon. Soc Nat Resour. 26:1454–1467.

van Vliet N, Mertz O, Heinimann A, Langanke T, Pascual U, Schmook B, et al. (2012) Trends, drivers and impacts of changes in swidden cultivation in tropical forest-agriculture frontiers: A global assessment. Glob Environ Change-Hum Policy Dimens; 22: 418-429.

Vergara, N.T. 1981. 'Integral agroforestry: a potential strategy for stabilizing shifting cultivation and sustaining productivity of the natural environment'. Environment and Policy Institute. The EastWest Centre, Honolulu, Hawaii

Vermeer, D.E. 1970. Population pressure and crop rotational changes among the Tiv of Nigeria. Ann. Ass. Amer. Geogr., 60 (2): 299-314.

Vongvisouk, T., Mertz, O., Thongmanivong, S., Heinimann, A., Phanvilay, K., 2014. Shifting cultivation stability and change: contrasting pathways of land use and livelihood change in Laos. Appl. Geogr. 46, 1–10.

Walters, H. Vegetation of the earth and ecoloqical systems of the 1977 geo-biosphere. Translated from the German by J. Wieser. New York, Springer-Verlag.

Wangchuk, P; Tashi, K (2004) Shifting cultivation in Bhutan. Thimphu, Bhutan: Policy and Planning Division, Ministry of Agriculture

Wangchuk, S (2001) 'Local resource management institutions: A case study on Sokshing management.' Journal of Bhutan Studies 3(1): 44 pp

Wangpakapattanawonga, P, Kavinchana, N, Vaidhayakarna, C Schmidt-Vogtb, D and Elliott S. 2010. Fallow to forest: Applying indigenous and scientific knowledge of swidden cultivation to tropical forest restoration. Forest Ecology and Management, 260: 1399–1406.

Warner, K. 1981. Swidden strategies for stability in a fluctuating environment: The Tagbanwa of Palawan. In Olafson, H. (eds.) Adaptive Strategies and Change in Philippine Swidden-based Societies. Laguna, Philippines: Forest Research Institute. pp. 13 - 28.

Warner K, 1991. Shifting cultivators, Local technical knowledge and natural resource management in the humid tropics. Food and Agriculture Organization of the United Nations Rome, Italy.

Wastelands Atlas, 2010. Categorywise and districtwise wasteland area. Ministry of Rural Development. National Remote Sensing Agency. Government of India.

Watson DJ. 1989. The evolution of appropriate resource management systems. In Common Property Resources: Ecology and Community-Based Sustainable Development, Beres FT (ed.). Belhaven: London; 55–69.

Watter, R.F. (1971). Shifting Cultivation in Latin America. FAO Forestry Development Paper No17, Rome, Italy.

Watters, R. F. 1960. The nature of shifting cultivation. Pacific View Point: 59-99.

Weinstock, J. A. (1990). Situation and Outlook of the Forestry Sector in Indonesia. Volume 4: Social, Environmental and Institutional Aspects. Utf/Ins/065/Ins:Forestry Studies Technical Report No. 1, Directorate General of Forest utilization, Ministry of Forestry, Government of Indonesian, Food and Agriculture Organization of the United Nations, Jakarta.

Weinstock, J.A., 1983. Rattan: ecological balance in a Borneo rainforest swidden. Economic Botany 37, 58–68.

Weyerhaeuser, H., Wilkes, A., and Kahrl, F. (2005). Local Impacts and Responses to Regional Forest Conservation and Rehabilitation Programs in China's Northwest Yunnan Province. Agricultural Systems 85: 234–253. doi:10.1016/j.agsy.2005.06.008.

Wilkie, D.S., Finn, J.T., 1988. A spatial model of land use and forest regeneration in the Ituri Forest of Northeastern Zaire. Ecological Modelling 41, 307–323.

World Bank (2001). Lao PDR: Production Forestry Policy - Status and Issues for Dialogue. World Bank, Vientiane.

World Bank. (1990). Indonesia: Forest, Land, and Water: Issues in sustainable development. Washington D.C: World Bank.

World Bank. 1988. "Forests Management and Protection Project: Democratic Republic of Madagascar." Staff Appraisal Report. Washington, D.C.: World Bank.

World Development Report 1992. New York: Oxford University Press. 1993. Strategy for Agricultural Growth, 1993. Unpublished World Bank Draft Document.

Xu, J. C. (2006). The Political, Social, and Ecological Transformation of a Landscape—the Case of Rubber in Xishuangbanna, China. Mountain Research and Development 26: 254–262. doi:10.1659/0276-4741(2006)26[254:TPSAET]2.0.CO;2.

Xu, J., Fox, J., Lu, X., Podger, N., Leisz, S., and Ai, X. (1999). Effects of Swidden Cultivation, State Policies, and Customary Institutions on Land Cover in a Hani Village, Yunnan, China. Mountain Research and Development 19: 123–132. doi:10.2307/3674253.

Yamamoto, Y., Oberthür, T., & Lefroy, R. (2009). Spatial identification by satellite imagery of the crop-fallow rotation cycle in northern Laos. Environment, Development and Sustainability, 11 (3), 639-654.

Yaso, T., 2000. Ethnobotany of the Htin and Lawa in Phuphaa subdistrict, Baugleua District, Nan province. M.Sc. Thesis, Department of Biology, Faculty of Science, Graduate School, Chiang Mai University, Thailand.

Younis AS and Dragun AK, (1993). Land and Soil Management. ISBN 9780429041631, 16p.

Zhang, Q., C. Justice, P. Desanker, and J. Townshend. 2002. "Impacts of Simulated Shifting Cultivation on Deforestation and the Carbon Stocks of the Forests of Central Africa." Agriculture, Ecosystems & Environment 90 (2): 203–9.

Ziegler, A. D., Bruun, T. B., Lawrence, D., and Nguyen, T. L. (2009). Environmental consequences of the demise in swidden agriculture in Montane Mainland SE Asia: hydrology and geomorphology. Human Ecology this issue.

4

Diversity in Shifting Cultivation

4.1 GENERAL DESCRIPTION

Most temperate forest habitats support a small number of animal and tree species with wide geographic distributions, whereas the montane forests support a large number of species with narrow geographic distributions. "For mankind's most prevalent disease, we have a multiplicity of diagnoses, each provided with maximum casualness," Prof Galbrith writes in one of his publications. Poverty is a difficult situation to be in. It would be beneficial if we knew what was causing the problem. Obviously, there isn't a single answer. Many explanations are presented because they include a small amount of truth. However, one reason of poverty is pervasive: the historical or present link between land and people. Food, clothing, and shelter are all provided by the land. Poverty exists if these are not available." This statement by Galbrith, which is relevant for shifting agriculture areas with excessive rains and turbulent rivers, summarizes the genesis and cause of every responsible individual's considerable worry for good land management and soil maintenance. To round out the aspects of this issue, there is need to add 'environment' and 'ecology' to this.

Shifting cultivation's environmental impact is a contentious topic that is skewed by social and political concepts, having both well wishers and adverseries. Shifting cultivation entails cutting primary forest and allowing a regeneration of vegetation to grow before cutting it again. These primary and secondary forests, used for cutting and burning, are appreciated by social scientists sympathetic to community or "traditional" ways of life, whereas; it is crititicised by organisations favouring biodiversity conservation. It may be advisable to use the techniques to produce more from less land, by the so-called "land-sparing" paradigm, if biodiversity is to be preserved.

Despite the fact that it is not legally recognized, it makes a significant contribution to smallholder rural farmers' livelihoods and food security (Mukul 2015). The bulk of research from the tropical Asia-Pacific area that was published had an emphasis on human ecology or anthropology. There was also a lot of heterogeneity in the findings on several forest/environmental characteristics. The oldest fallow sites had much higher tree species richness whereas the old-growth forest sites had significantly better variety and forest structure indices (Mukul 2015). In older fallow sites and old-growth forest, an uniform species composition was discovered.

Variation in species, wood products, aqua, and carbon capture and storage are just a few of the benefits that tropical forests provide for civilization and human well-being (Lewis *et al.*, 2015). Changes in biodiversity, land use socioeconomic environment, and climate will affect forests' capacity to deliver essential ecological services in the future (Trumbore *et al.*, 2015; Malhi *et al.* 2014; Metzger *et al.*, 2006). In remote, rural upland areas, most smallholder farmers engage in shifting cultivation, or kaingin (Lawrence, 1997). A study found that after shifting cultivation, secondary forests offer significant conservation benefits. The oldest fallow areas exhibited significantly higher species richness, whereas the old-growth forest had the highest indexes of species evenness, stem number, basal area, and leaf area, as well as the highest Shannon's diversity index.

4.2 ECOLOGICAL DIVERSITY

4.2.1 Shifting cultivation and biodiversity

The forest vegetation undergoes a change from the continuous shifting cycle into semi-evergreen, moist deciduous, deciduous forest, grasses, bamboos and a number of aquatic and semi-aquatic plant species (Ghosh, 1992). In the course of time, a dominant stand of bamboos with the exclusion of tree growth, gradually succeeds these. It is particularly due to the extensive growth of bamboo rhizomes particularly in muli (*Melocanna baccifera*). The successive short shifting cycle (2-5 years) and lengthening of cropping periods at the same site have even eliminated the growth of bamboos. In some tracts natural vegetation has been successively replaced by bamboo and thereafter by unproductive thatch grass (*Imperata cylindrica*), thus, ultimately leading to land abandonment.

Shifting agriculture has led to the greatest losses in biodiversity since it has deteriorated the habitat of species and forests. The recovery phase (abandoned times) sees the plants growing towards the original climax condition. If the area fully recovers before being planted again, the strategy might work. Toky and Ramakrishnan (1981, 1983) looked at how the dynamics of vegetation in an abandoned jhum field changed over time. Species diversity remained modest throughout the first five years, with herbaceous species dominating plots. As the vegetation transitioned from a bamboo (*Dendrocalamus hamiltonii*)-dominated forest to a bamboo (*Dendrocalamus hamiltonii*)-dominated forest between 5 and 15 years after abandonment, diversity grew significantly.

When compared to the species composition of sacred groves in the area, it eventually transitioned into a mixed broad-leaved forest nearing the climax type. However, they were unable to follow the process for more than 20 years.

The problem of shifting cultivation stems from the shortening of the fallow cycle and associated unsustainable practices in the management of shifting cultivation lands, rather than shifting cultivation practices themselves, which vary greatly between states, ethnic groups, and even within ethnic groups. The following are the major impediments to enhancing shifting cultivation systems in terms of fallow improvement: (a) remoteness or inaccessibility, (b) jurisdiction of several line departments, (c) prevalent land tenure, and (d) transitory land usage. Shifting cultivation disrupts habitats and threatens ecosystems, resulting in the extinction of valuable species of tree plants, shrubs, medicinal plants, and minor forest products (Sharma and Bujarbaruah 2016). Deforestation releases roughly 450 metric tons of carbon into the atmosphere per hectare of land. This carbon eventually makes its way to the Earth's ozone layer, where it contributes to global warming and the greenhouse effect. Forest fires, as well as the smoke they produce, pollute the air and water. However, shifting cultivation has been steadily intensified as a result of the rapid increase in human population densities and the high demand for cash crops. Cropping becomes more intensive and protracted, and fallow periods are reduced, causing the soil's physical state and nutritional status to deteriorate. During the period between burning and the installation of an effective cover by the first crops in the evergreen and semi-deciduous lowland forest zone, some runoff and erosion occurs, especially on steep soil. As a result, if more intense cropping is conducted on steep land, erosion may become a severe concern. Surface runoff can accelerate soil removal if the topsoil is exposed and disturbed during clearing, cropping, and weeding (Sharma and Bujarbaruah 2017). The studies conducted on the economics of shifting cultivation in Meghalaya, India, have indicated that the practice is not economical. Against output/input ratio of 2.1, 1.9 and 1.8 in livestock based, horticulture and agriculture farming system, respectively, the shifting cultivation has an output/input ratio of only 0.6 (Sharma 1995).

The issues involved in the problem of shifting cultivation are very complex and multifaceted. Over the years, various Working Groups and Committees have given considerable thought and recommendations. An open house seminar was held in Aizwal, Mizoram, India, in 1986; the task force on shifting cultivation in 1983, deliberation of National Board on Shifting Cultivation at Kohima in 1985, have suggested the extent of the problem and the means to tackle the problem. The approach to solve these problems need different pathways attitude and financial involvement (Sharma and Bujarbaruah 2016). This problem is deepening because of additional areas for SF not being available to increase the cyclicity due to increased pressure on land on one hand and non-remunerative return through SF on the other. The immediate challenge of shifting cultivation area rehabilitation is thus two-fold: scientific management of the fallow cycle to ensure the land's long-term viability, and the creation of competent local institutions to support such scientific interventions for long-term development. Furthermore, many farmers preserve the base for regeneration of the selected species while clearing the fallows for cropping, not only to allow wide dissemination of seeds/ propagules and rapid build-up of forest fallow after the cropping period is complete, but also to conserve the soil on the slopes (Sharma and Bujarbaruah 2016).

Specific conservation objectives are required to know whether the practice of shifting cultivation is advantageous or not (Kerkhoff and Erni, 2005). In Nepal, systematic studies on the biodiversity benefit of shifting farming are severely lacking. Participatory tools have been used to demonstrate the importance of shifting cultivation ecosystem for eco-diversity, and in many cases agrobiodiversity, by using available information (Aryal *et al.*, 2010; Aryal and Kerkhoff, 2008; Regmi *et al.*, 2005; Kerkhoff and Erni, 2005; Kerkhoff and Sharma 2006). However, these findings have not been backed up by empirical data or scientific research. The authors claim that shifting agriculture has better biodiversity than other agricultural systems, and that present techniques can be modified to fulfil specific conservation goals. To statistically verify these versions using data, scientific research are required. Rerkasem *et al.* evaluated the biodiversity in Southeast Asia's altering shifting agriculture ecosystems (2009). Traditional shifting cultivation in Southeast Asia is associated with high levels of both spontaneous and agricultural biodiversity. This variety is a result of agricultural households and communities constantly innovating to meet economic, ceremonial, and social demands in a highly diverse and dynamic environment. When shifting cultivation fields are converted to monocultures, they often lose their rich agro-biodiversity. As Southeast Asian landscapes change, our understanding of the dynamic processes of agrobiodiversity management, particularly diversity in fallows under shifting cropping practices, is plainly lacking.

Tropical rain forests have recently become a concern for ecological evolutionary biologists and conservationists. According to Ramakrishnan (1992), tropical rain forests encompass about 7% of the earth's land area yet contain more than half of the world's biota species. As a result, any harm to such forests would put a vast number of species in jeopardy, and in some cases, extinction many of which are not even listed or understood by biologists. However the impact on biodiversity varies depending on the sort of land use change. Species-area curves with an elasticity of the number of species with respect to area are frequently used to estimate biodiversity loss. There are considerable disparities in how different land use regimes define biodiversity (Angelson. 1995). Xu *et al.* (1993) in a Mengsong field investigation. When compared to mature and closed canopy secondary forests, plots classified as grass, bamboo, shrub, and open canop)' had a higher number of species and variety. Similar research in Kalimantan, Indonesia (Lawrence *et al.* 1998) found that plant and bird diversity in shifting agricultural areas was lower than in primary forests. According to Xu *et al.* (1999), there is a widespread assumption that shifting farming reduces not only species variety but also the quantity of desirable species. Shifting cultivators, on the other hand, often choose species they value the most and protect them during cutting, selective weeding, and afterwards replanting the useful species when the ground is left fallow. Conklin listed 77 different crops raised by the Hanunoo shifting cultivators in the Philippines (Hong. 1990). 84 plant varieties are cultivated in shifting agriculture among the Lua' hill farmers in northeastern Thailand, of which 70 are food species and at least 13 are grown for medicinal purposes. During the fallow period, these farmers recognize and use 482 species of plants for food, animal feed, medicine, construction, weaving, dyeing, firewood, fencing, and insect repellant, among other things (Kunstadter, 1978). According to Dove (1985), an average Kantu household in Kalimantan, Indonesia, grows roughly twenty-one different plants in shifting cultivation each year. In a research in northeastern India, Swidden. Ramakrishnan (1992) discovered that sequential harvesting of crops is an effective means of managing up to 35-40 crop species over time and space.

4.2.2 Ecological viability of shifting cultivation systems

If there is sufficient land available to have long enough fallow period (12-25 years) in shifting cultivation for restoration of soil fertility, without jeopardizing the livelihood security of the cultivators, the systems are ecologically viable. These systems are naturally and mutually sustainable with adverse and brittle environment and ecosystems of the tropics. It is for this reason that no viable alternatives to shifting cultivation have been discovered, and attempts in this direction have proven fruitless. There are a number of soil and climate constraints (Table 4.1) that must be overcome. As a result of these limits, soil degradation becomes a serious issue when changing from shifting cultivation to mechanised intensive agriculture. As a result, many large-scale mechanical farming plans that were implemented in the humid tropics have failed (Lal 2005).

Table 4.1. Ecological constraints to finding viable alternatives to shifting cultivation

Soil related constraints	Climate related constranits
Soil fertility is inherently low	Low radiation during the cropping period
Prone to soil erosion	Prevalence of high relative humidity
Severe degradation	Big losses during storage
Nutrient imbalance	Constaints in drying of farm produce (like grains)
	More pests and diseases
	Heavy rains

4.2.3 FORESTS OF SOUTH ASIA

Three of the world's 18 unique biodiversity 'hotspots,' which are home to almost 50,000 plant species, or around 20% of the world's flora, are found in the region (Himalaya; Western Ghats; and Sri Lanka). Over millions of years, migration from the Malayan, Chinese, and Australian regions, as well as the Caucasus, has enriched the floral compositions. Floristic compositions in the north-western regions of the region show significant Mediterranean components, whereas in the east there is a definite transition to Malayan flora. Almost half of the region's dicotyledons are endemic, with the majority occurring in the Himalayan region, which has been separated for a long time by orogenesis and glaciers. Invasive plants like as eupatoreum and lantana, on the other hand, have recently had a significant impact on the floral composition of many sections of the region, decreasing productivity and delaying natural succession. Other types of inward migration, on the other hand, have become naturalized and may even have attained a kind of transitional climax in the ecosystems they helped build. Prosopis juliflora is an excellent example of this type of naturalization (Brovkin *et al.* 2013).

4.2.4 Assessing and matching biodiversity

Ecological hypotheses concerning the effects of shifting cultivation and deforestation on Philippine biodiversity, as well as the impending disappearance of Philippine forests and the extinctions that follow, can be tested by analysing the number and types of species, as well as population sizes, when measuring biodiversity (Pimm *et al.* 1995). As a result of these findings, the Philippines has gained broad recognition as a biodiversity and endemism hotspot. Secondary forest, consisting of "27 tree species belonging to 22 genera and 15 families," was removed to make space for the cultivation of "53 species in 52 genera distributed among 30 families," according to Caringal and Panganiban (2008). Further breakdown reveals that these consisted of "at least 10 species and 6 families of vegetables, 7 species and 6 families of root crops, 13 fruit trees by 10 families, 5 species of legumes and pulses under 2 families, 7 spices under 4 genera and 3 families, 4 forage and pasture species belonging to 3 families and 7 species of valuable crops". Given the number and types of species, it is legitimate to question if kaingin is genuinely as benign as indicated in terms of biodiversity in this case. The biodiversity of primary and mid-montane forest, Imperata-Saccharum grasslands, and shifting upland farming was compared in the Mount Makiling area (Sajise *et al.* 2005).

4.2.5 Biomass

Shifting cultivation relies heavily on biomass production. The nutrients are stored and conserved in the biomass, which is then recycled to replace soil fertility. Secondary succession rehabilitates abandoned fields in slash and burn agriculture zones. At Burnihat in north-eastern India, the biomass, net primary productivity, and litterfall trends during a 20-year fallow following the change to agriculture were analyzed (Toky and Ramakrishnan, 1983a). In a 20-year-old fallow, succession was accompanied by increased plant variety, decreased dominance, and improved above-ground productivity, which reached 18 ton/ha/year. Litterfall increased with fallow age, reaching 10 tons per hectare per year in a 20-year fallow. A remarkable linear relationship exist not only with respect to community structure, but also with respect to functional aspects such as productivity and litterfall. This seems to be due to the great change in the structure from the initial herbaceous specie to the bamboos and other woody species. A variety of strategies is adopted by successional species among which stump, root and rhizome sprouts and invasion through seeds are common. With the shortening of shifting cultivation cycle to 5-year more important tree species such as *Bauhinia purpurea,Careya arborea Cedrela toona, Dillenia indica, Schima wallichii, Terminalia belerica, Vitex glabrata, Vitex peduncularis* and *Wrightia tomentosa*, and bamboos do not get opportunity to come up in the community.

The biomass produced by leafy and fruit vegetables was highest during a 5-year jhum cycle, whereas the biomass produced by tuberous and rhizomatous crops during a 10-year cycle was about double that of the other two cycles. Within a 30-year cycle, the maximum economic yield per hectare for rice, maize, and *Setaria italica* was attained, whereas yield reductions for these crops were 48 percent and 98 percent, respectively, under 10 and 5 year cycles. However, the economic yields of leaf and fruit vegetables were substantially higher in the 5-year cycle than in the longer periods. Jhum has thrived for thousands of years in the hills of north-eastern India, owing to its excellent energy efficiency (output of energy/ input of energy) that ranged from 34 in a 30-year cycle to around 47 in 10-year and 5-year cycles (Toky and Ramakrishnan, 1982).This high energy efficiency is due to natural subsidies in the form of biomass and soil fertility available free to the farmers, and also due to the traditional knowledge of mixed cropping gained by the farmers since ages (Ramakrishnan, 1992). From an economic and ecological standpoint, settled valley cultivation is viable in the north-eastern hill region. It is reliant on the natural inflow of nutrients from the surrounding hills into the valley. From an economic and ecological standpoint, settled valley cultivation is viable in the north-eastern hill region. It is reliant on the natural inflow of nutrients from the surrounding hills into the valley. Two crops of rice under valley cultivation are equivalent to one mixed cropping under 30-year cycle. In valley cultivation comparatively high efficiency of around 17 is explainable because nutrients washed from the hills are available to the good growth of two crops of rice.

4.2.6 Soil regeneration and species diversification

Soil elements vary dramatically over the fallow period, from removing forestland for agriculture to the establishment of a secondary forest. The effects of these adjustments, however, differ from one attribute to the next. According to the results of the fallow reduction study, soil attributes are roughly divided into two categories based on their behaviour in relation to fallow length. Short fallow systems improve pH, exchangeable bases, exchangeable acidity, and accessible phosphorus. Conversely, the morphological development of the A1 subhorizon, organic matter, soil texture, and bulk density are all negatively impacted by short fallow. At the beginning of the cropping season, the ashes from burned biomass of plants serve as a lime fertiliser and significantly influence the chemical properties of the soil. The liming effect of ashes is limited by crop harvesting, cation leaching, and sheet erosion caused by intense rainfalls on bare soil at the beginning of the cropping season. Many authors (Tulaphitak *et al.*, 1980; Prat, 1990) have stated that the greatest loss of ash nutrients during the cropping phase of shifting cultivation happens between burning and soil surface covered by crops. Throughout the whole farming season, this process of soil fatigue persists. The land is left fallow for two to three years after it has been cropped.

4.2.7 Secondary forests

Secondary forests – a prominent component of forest landscapes and small-scale farmers' slash-and-burn agriculture in the tropics – are not only expanding in size around the world, but also gaining attention for their potential ecological services and contributions to sustainable lifestyles. Improved regional ecological conditions are a result of secondary forests and forest fallows, including erosion management and watershed protection, in addition to global environmental benefits like carbon storage and biodiversity conservation (Smith and Scherr, 2003). Shifting cultivation's productivity is based on a transient increase in soil nutrient availability, the buffering capacity of ash, and the regenerative capability of fallow periods (Fujisaka *et al.*, 2000). Land is frequently turned to pasture after cropping, and agroforests are left fallow in expectation of future production. Shifting cultivation, which has been practiced in Panama since 5000 BC, is the main cause of current deforestation, which peaked at 51,000 ha per year in 1990. (Fischer and Vasseur, 2000). Swidden–fallow agriculture, once considered a singularly damaging and unsustainable technique, is now recognized a crucial component of peasant farming portfolios with the promise for sustainable development and improved incomes among the rural poor (Coomes *et al.*, 2000; Vosti and Witcover, 1996). As a result, enhanced swidden–fallow systems could not only help to alleviate rural poverty, but also contribute to the Kyoto Protocol's societal aims.

4.2.8 Cropping and fallow practices

Farmers' management attitudes, the availability of seeds and agricultural labor, fire incidence, and advantages from non-timber forest products are all elements that influence management decisions in crop–fallow systems (de Jong *et al.*, 2001). Most farmers in Ipet said they practice shifting cultivation on at least one area of their fields at least once a year. On cropped fields, the use of organic and mineral fertilizer was regarded to be the best management approach (38 percent and 31 percent, respectively). The following are the unanswered questions about enhanced crop–fallow systems: (1) What are the present C stocks in soils and aboveground biomass for various crop kinds and fallow areas of various ages? (2) What are the expected gains from changes in land use and management? and (3) Which farmers are most likely to make changes to their crop–fallow systems in order to encourage secondary forest fallow?

Shifting cultivation's most distinguishing positive attribute is that it is integrated into the pre-existing natural ecosystem. It has been suggested that fluctuating cultivation practices keep the natural environment in a state of dynamic equilibrium (Geertz 1974, Found 1987, Keesing & Strathern 1998). Shifting cultivation manifests itself in a variety of ways for various people. As a result, there is no single optimal way to categorise shifting agriculture. From sea level to 4,000 metres above sea level, II' is practised in sections of south.

Eastern Tibet and Southeast Asia's upper main land (Spencer 1966). However, there are two general points of view from an anthropological standpoint. First, there's the nonnative perspective, which focuses on the disadvantages of shifting cultivation, such as low productivity, a large land footprint, and unfavorable environmental consequences. Second, shifting cultivation is discussed as a rational response to current ecological, cultural, and economic challenges.

Three distinct methods to the study of shifting cultivation have been developed based on these ideas (Found 1987). The three approaches are ecological or environmental, cultural-historical, and economic study. The first approach, known as the common ecological approach, is based on the assumption that shifting farming exists in harmony with nature (Found 1987) with regard to the production in relation to labour input and land requirement and its impact on the ecology. Anthropologists frequently examine changing cultivation from a cultural-historical perspective, relating it to different types or phases of human civilisation. They remind out that the majority of shining cultivators employ primitive equipment and come from societies that are primitive in many aspects. Slash and bum agriculture is seen by some as more of an ancient method with a long history than a modern way of dealing with the requirement to generate food (Found 1987). A mixed strategy, on the other hand, might be the best way to deal with the problem at hand.

Agricultural encroachment by shifting cultivators, according to scholars, is at the heart of the debate over tropical deforestation. The impacts of shifting cultivation vary depending on the practices used. The inability to determine the amount of ecological problems produced by shifting cultivation has been hampered by a lack of understanding of the features of its various varieties. According to Padoch and Vayda (1983), "traditional resource use practices in the tropics were criticized in the past as wasteful and inefficient," but "in recent years, there have been perspectives that praise the stability and conservativeness of these technologies." Such a shift in perceptions of primordial man as a protector is not surprising, and it is at least partially justified.

Mountains are home to half of the world's 34 biodiversity hotspots, containing a wide range of species, habitat types, and ecological conditions. Mountains are ethno-culturally rich and home to various human societies, as they are at the heart of endemism, genetic variety, and agro-diversity (CBD 2011). Mountains have traditionally been a location of survival and give potential for climate change adaptation due to their topographic diversity and bio-cultural richness (Caprez et. al 2010). Conservation activities are becoming more participatory 2 as a result of a decentralised governance model that allows local populations to manage biodiversity. In several regions of the Hindu Kush Himalayan region, people-centered conservation and development techniques are now being used (Sharma *et al.* 2010). The COP-10 agreement on mountain biodiversity encourages national governments and relevant organisations to share best practises in biological resource conservation and management (CBD 2010).

4.3 AGROBIODIVERSITY DUE TO SHIFTING CULTIVATION

Attempts at conversion have instead resulted in a slew of'second-generation' issues, including a loss of agro-biodiversity and a significant erosion of ecosystem goods and services (Fox *et al.* 2000; Cramb *et al.* 2009; Rerkasem *et al.* 2009; Xu *et al.* 2009; van Vliet *et al.* 2012; Dressler *et al.* 2017). Changes in property regimes have accompanied agricultural revolution, depriving the poor of access to land and productive resources (Leduc and Choudhury 2012; Dressler *et al.* 2017). Furthermore, rising youth out-migration due to a lack of interest in agriculture has resulted in labor shortages in many farming areas, with worrisome rises in women's workload. Furthermore, even as the area historically under shifting cultivation shrinks, their incursion grows, resulting in "elite capture" of commons, a loss of customary land rights, and cultural deterioration as a result (Suyanto and Otsuka 2001; Leduc and Choudhury 2012).

Tropical rain forests have been increasingly important to ecological evolutionary biologists and environmentalists in recent years. Despite covering only 7% of the earth's land area, tropical rain forests are home to more than half of the world's biota. As a result, any damage to such woods would put a vast number of species in jeopardy, and in some cases, extinction. Many of which tropical biologists haven't even catalogued or comprehended (Ramakrishnan. 1992). Another widely held view is that shifting cultivation diminishes both species diversity and the number of desirable species. Shifting cultivators, on the other hand, tend to favor species they value the most, preserving them through cutting, burning, and selective weeding, and then replanting the useful species when the ground is left fallow (Xu *et al.* 1999). Aside from that, another notable element of shifting cultivation is the wide range of crops planted. Harold Conklin of the Philippines recorded 77 different crops that the Hanunoo shifting cultivators grew (Hong. 1990). In northeastern Thailand, among the Lua' hill farmers. The shifting cultivation field is home to 84 plant varieties, 70 of which are food species and at least 13 of which are grown for medicinal purposes. The Lua' recognize and employ 482 plant kinds in the fallow swiddens for a variety of purposes including food, animal feed, medicine, construction, weaving and dying, decoration, firewood, fencing, and insect repellent (Kunstadter. 1978 cited in Hong. 1990). Among the Kantu in Indonesia's Kalimantan. According to Michael Dove's research, the typical Kantu household grows roughly twenty-one different types of plants year in their swidden garden (Dove, 1985).

According to Punitha *et al.* (2018), shifting cultivators have a complex relationship with the land and have a great affection and respect for it because it is a way of life. Though there are certain patterns of change in shifting cultivation, the data is extremely variable. According to the research, several countries are transitioning from a subsistence-based system to a dual-economy one in diverse forms. The practice of shifting cultivation land use is controversial, as it is considered as an unscientific method of land use that results in land degradation and ecosystem fragility. As a result, erosion from steep slopes is causing massive losses of vegetation cover and top productive soils in the region. As a result, a comprehensive, multifaceted plan is needed to turn large-scale agriculturally degraded land from a financial and environmental liability into a profitable asset. However, in the North Eastern Region, a thorough and extensive study on the social dimension of shifting cultivation and methods for its improvement is required (Punitha *et al.* 2018). The Food and Agricultural Organization declared shifting cultivation to be the most important land use problem in the tropical globe in 1957, and many governments devised policies and regulations to combat it (IWGIA 2014). However, in spite of the best efforts, shifting cultivation is still continuing. Although, large volume of literature has been created since time immemorial, many aspects of the system are still poorly understood (Mertz *et al.* 2008). Hence, it has created interest to explore the various social dimensions of shifting cultivation. The social dimension of shifting cultivation and its cross cultural reflection in North East India is not well reported and documented systematically. However, we have analyzed the systematic and detailed study in the North Eastern Region on the social dimension of shifting cultivation as well as its improvement strategies. Fig. 4.1 shows the area under shifting cultivation in northeast India.

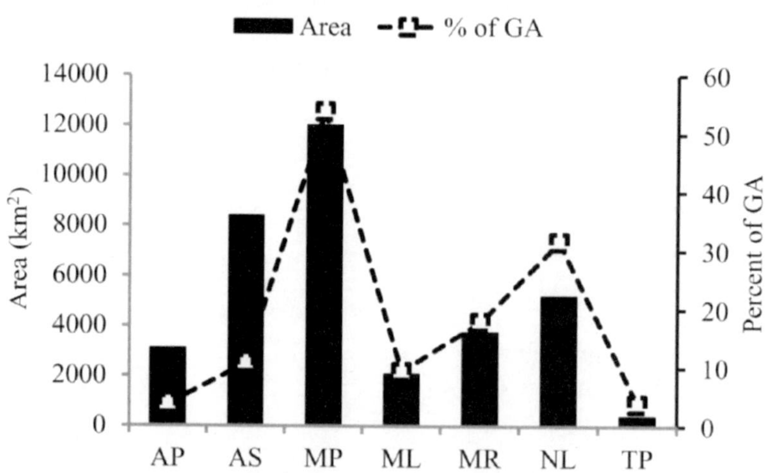

Fig.4.1. Area under shifting cultivation in Northeast India states and as percent of total geographical area.

4.3.1 Species

Over all it shows that floral species composition are unique in these three sampled habitats. As far as the total number of species and families are concerned, high number of species were observed in the unmanaged fallow areas and lowest were observed in the forest areas (Fig. 4.2). Rubiacea shows dominance in the forest areas followed by Fabacea, and the all other families are represented in a more or less equal manner. 57 percent of the species recorded in this habitat are trees.

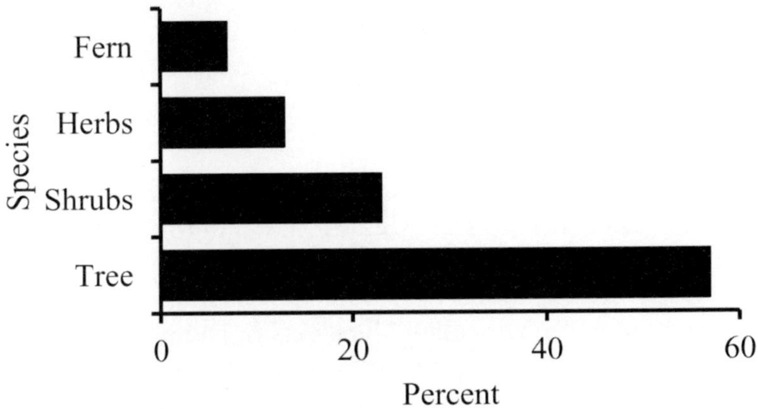

Fig. 4.2. Habit of species recorded from forest areas

Comparatively lower number of species and families in the managed fallow areas was as expected since more plantations have been carried out in the managed fallow areas and as a result many naturally occupying primary species are absent. The significant output of the floral analysis is that the project interventions have increased the biodiversity of the fallow areas as reflected by the higher Shannon index value for this habitat. It should also be noted that species composition of the managed and unmanaged fallow areas are completely unique, which means the interventions are also changing the floral composition of the growing forest in the area (Figure 2.10). Since the interventions are less than three years old comparison of present diversity with that of the climax vegetation (forest area) is not possible at the moment. However the higher diversity of the managed fallow areas along with the presence of more nitrogen fixing plants will improve the fertility of the fallow areas.

4.3.2 Biodiversity in northeast India

The Northeastern region of India, where shifting cultivation is practiced has an area of 262,189 km^2, which is 7.98% of the total area of the country.

However, out of 15,000 angiosperms species found in India, 8,000 or 53.3% are in the Northeastern region. The region has also 8,000 species of endangered flowering plants, which is about 55% of the country. Two of the 34 Global Biodiversity Hotspots, 39% of the Himalayan Hotspot, 8% of the Indo-Burma Hotspot, and 1/3 of all of India's biodiversity are found in this part of the country. There are 48 sanctuaries, 15 National Parks, and 2 Bio-Reserves (BRs) in the area. According to some, shifting cultivation is a primitive, unscientific, environmentally damaging, and commercially unviable mode of living that incorporates both material and non-material culture, a distinct pattern of resource ownership and utilisation, and the upholding of harmonious social standards. It therefore requires development that is centred on shifting cropping.

4.3.3 Biodiversity conservation

Shifting cultivation has more biodiversity than other agricultural systems, and present practises can be modified to meet specific conservation goals (Kerkhoff and Erni 2005; Kerkhoff and Sharma 2006, and Kerkhoff 2006). Compared to other methods of agriculture, shifting cultivation contributes to biodiversity by providing more forest cover, agricultural biodiversity, and wildlife habitat. Indigenous habits and knowledge contribute to resource conservation while also promoting cultural integrity and social security. The length of the fallow period is maximised to preserve forest cover. Forest fallows are the most significant aspect of the farming system, both as a source of soil productivity and as a source of a variety of forest products (in places where shifting cultivation is used). These fallows are rarely referred to as "abandoned land," as they are frequently actively managed by farmers. It begins during land clearing, when farmers decide which trees to cut down and which to leave standing and protect from burning in order to promote the regrowth of preferred species in the fallow. For the rest, food self-sufficiency lasts an average of 5.5 months every year (WWF 2002). People have used a variety of coping mechanisms to supplement their food supply during times of scarcity, ranging from wage and seasonal labor to other possibilities, with the major sources of revenue varying by village and ethnicity.

Human geographers and anthropologists, such as Pelzer (1945), Freeman (1970), and Conklin (1980), have undertaken the majority of studies on swidden agriculture in Southeast Asia (1957). On the other hand, compared to Africa or Latin America, intensive research by natural scientists into this topic is quite rare. Of course, human geographers and anthropologists interested in Southeast Asia can find a wealth of material regarding the natural histories of swidden places in their research, but they rarely characterise the ecological characteristics of swidden sites in quantitative terms (Nakano 1978).

Upland rice is the Karen people's principal food crop on their swiddens, with maize, sorghum, taro, and cucumber as secondary crops. They do not produce opium in most years. However, since the Meo people arrived in the area, Karen peasants have periodically planted opium crops, as depicted on the detailed map of Geddes (1976). They did, however, grow some opium poppies for their own enjoyment in a swidden in 1971. Robbins (1964) states, "In northern Thailand, lower montane forest begins at about 3,500 feet and thus, true montane forest does not occur in Thailand". Large trees are rarely felled by Karen swiddeners.

4.3.4 Primary vs. secondary forests

Shifting cropping models must address cleared primary forest and secondary forest fallow length. Although secondary forests provide fertile soil, main forests provide more fertile soil as well as new land. These significant differences must be taken into account while developing new farming models. In Latin America, shifting cultivation, or milpa, is defined as "an agricultural system in which temporary clearings are farmed for fewer years than they are allowed to lie fallow" (Sanchez, 1976). The practice has evolved in different ways around the world, and it changes according to local ecological, socioeconomic, and cultural constraints. Shifting cultivation systems have proven to be productive and long-lasting adaptations to difficult environmental conditions characterized by high worker productivity and low population density (Cairns, 2007). From colonial to modern times, shifting cultivation, has a bad reputation and has been deliberately repressed. Conklin (1957) distinguished between two types of shifting cultivation systems, viz.; (1) integral, which he defined as productive, sustainable, and carefully regulated and, (2) incipient, which are more intensive, destructive, and unsustainable practice. Plant succession after cultivation on hill soils is shown in Table 4.2 and chemical properties in Table 4.3.

Table 4.3. Plant succession following disturbance in hillside farms.

	Vegetation 7 weeks after disturbance					
	Forest	**Fallow**	**Grass**	**Fallow**	**Permanent**	**Farm**
	1	**2**	**3**	**4**	**5**	**6**
No. of species present	58	17	10	14	17	5
Tree species present	28	9	1	0	7	1
Exotic weed species	10	4	7	9	10	12
Exotic weed dominance (%)	22	50	98	94	95	99
Ground cover (%)	23	18	80	100	87	100

Based on Holm *et al.* (1979)
Siebert (1987)

Table 4.3. Soil response to cultivation in San Vicente swiddens

Hillside farms	**Ca**	**Mg**	**pH**	**Av. P**	**OM**
		100g			**%**
After burning	16.0	22.6	6.7	19.2	6.5
1 yeaar cultivation	14.2	20.7	6.2	3.2	5.8
2 years continuous cultivation	11.3	16.6	5.6	0	3.2
5 years continuous cultivation	5.2	9.9	5.0	0	3.2

Source: Siebert (1987)

4.3.5 Integral long fallow shifting cultivation

Long-term fallow shifting cultivation techniques are largely extinct now. Integral shifting cultivation may still be practised in a few remote areas of Amazonia and Southeast Asia, but it has vanished elsewhere due to cultural change, a lack of available land, emerging market opportunities, destructive logging practices, the establishment of large plantations, government policies, and competition with recent migrants who lack the necessary TEKP (Table). Integral shifting cultivation techniques have been supplanted by incipient shifting cultivators, or slash and burn, as Conklin described them. Slash and burn is only getting started. A lack of knowledge and experience (TEKP), continual and/or more intensive cultivation, extensive clearings with few/no trees retained, no or short fallows, and frequent burning are all characteristics of farming. Slash and burn in its early stages is unsustainable, leads to the spread of invasive weeds, and frequently results in a wave or front of forest conversion. The difference between Integral swidden and slash and burn is:

Integral swidden	**Slash and burn**
Generally permanent dwellers	Recent migrants
Rooted in routine life	Not found in routine life
More knowledge and experience	Less knowledge and experience
Self organized	Generally unorganized
Fallow time is long	Little or no fallow period
Sustainabble	Wasteful and uneconomic
Considerable land	Intensive land

4.3.6 Regeneration of Biomass following disturbance

In the majority of tropical hilly regions, shifting farming is the norm. Most studies show that, whereas pioneer tree species recover more quickly, mature forest tree woody biomass recovers over several decades after cultivation is stopped (Fig. 4.3).

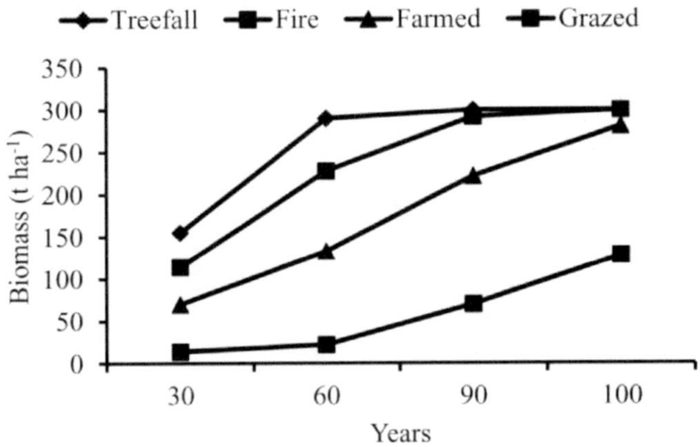

Fig. 4.3. Regeneration of biomass following disturbance in integral shifting cultivation
Source: Uhl (1990)

4.3.7 Myanmar forests and diversity

Myanmar is one of the most heavily wooded countries in Southeast Asia's mainland. These woods are home to a diverse range of valuable species and endemics, and they are vital to global biodiversity conservation efforts. To create a nationwide forest map and estimate deforestation, Landsat satellite imagery from the 1990s and 2000s was used. Although the country's forest cover has been relatively stable, it has decreased by 0.3 percent per year (Leimgruber *et al.* 2005). Deforestation rates differed greatly among administrative entities, with the biggest losses occurring in central and more populated states and divisions. Ten deforestation hotspots had annual deforestation rates that were significantly higher than the national average. Increased agricultural conversion, fuelwood use, charcoal manufacturing, commercial logging, and plantation growth were all major causes of forest loss in these regions. While Myanmar remains a bastion for closed canopy forests, a number of places have been severely deforested. This Indo-Burma region is home to a large number of rare and unique species, and it has been identified as having high biodiversity conservation value (Myers *et al.* 2000). The region's huge and largely undamaged forests are also said to be one of the remaining strongholds for large creatures like tigers and elephants (Leimgruber *et al.* 2003; Lynam 2003). Due to its political and physical isolation, Myanmar may have kept one of the highest levels of species richness and forest cover of any country in mainland Southeast Asia (UNEP 1995). There has been no systematic evaluation to identify Myanmar's remaining forest cover, fragmentation patterns, or rate of change in forest cover. Recent regional and coarse-scale forest surveys in Asia have yielded worrying estimates of present forest loss rates (FAO 2001a,

2003). Myanmar has been included as one of the ten tropical countries with the highest yearly deforestation rate in some of these assessments, despite once being known for its huge closed canopy forests (FAO 2001).

Biomass measurements were taken on trees, undergrowth, trash, and plant roots for each treatment. Physicochemical analyses were performed on soil samples. Although the average soil carbon stock in the MC and CF systems is 93 percent and 78 percent, respectively, it only accounts for 26 percent of total carbon in the VF system. There is a total carbon loss of 97 percent from aerial biomass and 17 percent from soil when a PF system is changed to MC. Following and converting to SF results in a 3.9 $TC/ha^{-1}.year^{1}$ carbon re-accumulation (Njomgang *et al.* 2011).

4.3.8 Shifting cultivation principles and fallow ecology

The practices of shifting cultivators vary widely depending on difference between the cultures, regions and ecosystems were it is practiced. Key authors have built up definitions and typologies for shifting cultivation (Cairns, 2015; Palm *et al.*, 2005; Ramakrishnan, 1992), which help us to reduce the diversity to a minimal common point: roughly, shifting cultivation comprises a short phase of cultivation followed by a long phase of fallow development.

As soon as the farmers stop weeding the field, the spontaneous vegetation tends to develop, beginning with pioneer herbaceous species, whose the seeds had arrived since the clearing and that farmers struggled to control by weeding during the cultivation phase. While the clearing and the subsequent cultivation phase are an allogenic succession (human disruption), the secondary succession after the plot is left to fallow corresponds to an autogenic one, with successive seral communities that vary widely according to regional biomes and local ecosystems (Gómez-Pompa *et al.*, 1991). In this chapter, the examples of species refer to a local ecological succession in Southeast Asia, around Phongsaly, Northern Laos. The first seral community comprises pioneer herbaceous plants (e.g. *Pennisetum* spp.*, Imperata cylindrica, Ageratum conyzoides, Miscanthus* spp.). Scrub species (e.g. *Chromolaena odorata*), quickly grow over the weeds and supplant them in a few years as the second seral community. After five years or more, the first trees appear (i.e. stand initiation), from heliophilous fast growing species (e.g. *Trema orientalis, Macaranga dendiculata, Mallotus paniculatus*). When the canopy closes, only shade-tolerant (sciaphilous) plants survive in the understory. The trees are engaged in a competition for light (stem exclusion), which modifies their habits and exposes them to rapid decay in windthrow, leaving room for the last seral community by stand-replacing of sciaphilous slow-growing trees (e.g. *Antidesma acidum, Alstonia scholaris, Spondias pinnata*).

4.3.9 Forest dynamics and impact of shifting cultivation

Agriculture emerged around 10,000 years ago in half a dozen of Neolithic centres, which were mostly in open ecosystems: steppes of the Fertile Crescent flood plains, peatlands, etc. (Mazoyer and Roudart, 2005). Quickly, expansion of agriculture faced forests, the dominant ecosystems, which had been in extension after the last Glacial Period (Williams, 2006). To conquer those forest frontiers, the farmers had to design new tools, new techniques and new farming systems lead to the appearance of shifting cultivation. For example based on artefacts archaeologists conclude that influence of shifting cultivation progressed at a pace of one kilometre per year in Neolithic Europe. However should we imagine a pioneer front, inexorably progressing from southeast to northwest, with forest ahead and open field behind? This is unlikely. Clearing a primary dense forest, either a temperate or a tropical one, requires ten times more work than a 15–20 year-old fallow. With a population limited to less than 100 million people at the dawn of our era, labour was the scarcest factor of production (Mazoyer and Roudart, 2005). Neolithic farmers were definitely more interested in living in sedentary settlements and in basing their livelihood on rotational shifting cultivation around their village, with long fallows, than in being nomadic along a pioneer deforestation front. Such a rationale still prevails nowadays, when population densities remain low and compatible with long fallows. The forest space included in the rotation tends to increase with the needs of a growing population. To keep the yield levels – i.e. maintain the duration of the fallow – the villagers had to expand their territory in the rotational system. However, that was not a permanent solution: eventually they faced impassable obstacles (e.g. large river, mountains) or borders of forests cultivated by other villages; they might be reluctant to increase to an unbearable level the access time to the fields; some might reject the pressure of community rules enforced by more and more distant village or clan authorities. Therefore, some people left the village to found a new community settlement further into the primary forest, while most of the households stayed. The inclusion of primary forests into shifting cultivation rotation combines these two processes of crown expansion and new settlement leap; it ends when the pristine forest resource is no longer available, or too far away. In a mature shifting cultivation system, the villages' agroforests are adjoining, leaving no room for expansion. To face continual demographic growth, farmers must then rely on speeding up the rotation to increase the crop area in a limited accessible space. The early stages are hardly noticeable, but, with shorter and shorter fallows, (i) the biomass accumulation is more and more limited – the yields drop due to inadequate fertility; (ii) the fallow time becomes insufficient for forest secondary successions. The forest is more and more degraded, and finally vanishes while people convert to permanent farming (Mazoyer and Roudart, 2005; Williams, 2006).

4.3.10 Shifting cultivation and biodiversity: a complex nexus

With deforestation, another facet of the environmental impact of shifting cultivation is biodiversity. At the plot level, the issue is clear: the biodiversity level of an old-growth forest exceeds those of a secondary seral community (Cairns, 2015; Gerold *et al.*, 2004; Ramakrishnan, 1992). The drop can be huge when the shifting cultivation faces the demographic crisis, with poor grass and scrub fallows that dominate the landscape. However, the comparison becomes more complicated when biodiversity is compared at wider scales, either village or regional ones. The biodiversity level of an old-growth forest is nearly isotropic; it marginally increases with the size of the sample plot. In a shifting cultivation landscape, the association of species differs from place to place according to the time since the last cultivation phase. The biodiversity level is the total for the different stages of the fallow mosaic plus that of the cultivated field (Rerkasem *et al.*, 2009). The number of species in the fallow varies with the number of seral communities comprising the mosaic of ecosystems (Gerold *et al.*, 2004; Schroth *et al.*, 2004). As those seral communities are successive, the species differ in the rotation; the total number of species in a shifting cultivation landscape (i.e. a fallow matrix) may top those of an old-growth forest (Babin, 2004). However, the biodiversity of an ecosystem cannot be reduced to the number of species it comprises. Specialist species, localized and often rare, may be superseded by generalist species when old-growth forests are converted into secondary forests. In the shifting cultivation field, the biodiversity combines inter- and intra-specific agrobiodiversity. Up to 30 or 50 different species can be planted in a swidden (Conklin, 1957) and, for each, farmers tend to manage a large set of varieties and cultivars to adapt the production to (i) the micro-specificities of the plot; (ii) pest and climate hazards, reducing sensitivity to the hazards by mixing cultivars; (iii) the current household needs; (iii) the level of the former harvest (Brookfield, 2001). The stocks for the different cultivars are managed at household level, but also at the community one with exchanges between households (Rerkasem *et al.*, 2009; Xu Jianchu and Mikesell, 2003). Kept away from the Green Revolution, shifting cultivation has suffered less from the erosion of genetic diversity for the main cereals (rice, maize, wheat). For example, more than the half of the 90,000 rice germplasms stored at the IRRI world gene-bank (Los Baños, Philippines) come from the uplands of two countries, India and Laos (Xu Jianchu and Mikesell, 2003).

4.3.11 Diversity of shifting cultivation and fallow secondary forests

The diversity of shifting cultivation

When it comes to agro-technical, socio-cultural, and politico-economic characteristics, swidden agriculture comprises a wide range of techniques. This variation makes identifying general trends more difficult, but a number of authors have attempted to grapple with it. Some of the first attempts to classify swidden agriculture, such as Ruthenberg's (1976) classification based on the swidden length/fallow length ratio, are related to the evolution stages of shifting cultivation. Several endogenous or exogenous variables are at work in this scenario, causing rapid changes and resulting in swidden agricultural differentiation. Alternatively, the sort of shifting cultivation practised is determined by relatively stable biophysical circumstances or slowly changing socio-cultural influences. Population pressures and logging activities conducted by city-based entrepreneurs, as well as the quality of the soils and the state of the forest, all contribute to variation in swidden agriculture in northeast India. In relatively remote locations of northeast India, Ramakrishan and Kushwaha (2001) report 60-year cycles and a complex mosaic of woods of various ages. With rising population pressure and land degradation closer to metropolitan areas, swidden agriculture has evolved into a bush fallow rotating system or weedy fallow vegetation. This is increasingly being replaced by various sedentary systems (Ramakrishnan & Kushwaha 2001). In tropical Asia, there are numerous examples of similar variation in shifting cultivation along multiple continuums linked to diverse sources of change (de Jong *et al.* 2001).

Thailand has experienced the impact of socio-cultural elements on shifting cultivation methods, maybe in conjunction with more dynamic forces of change. The three types of shifting cultivation described are; rotating shifting cultivation, pioneer shifting cultivation, and supplementary shifting cultivation, are associated with various ethnic groups, environments, and external incentives (Schmidt-Vogt 2001). The prevalent indigenous land and tree property rights systems in Indonesia may have a significant impact on the type of swidden fallow management that is carried out (Weinstock 1989). Dayak groups on Borneo's island employ swidden agriculture with varying degrees of sedentarism, owing to cultural preferences (de Jong *et al.* 2001). These variations in swidden agriculture, according to De Jong *et al.* (2001), are nonetheless "typical" shifting cultivation systems of long-established populations in the region, as opposed to the shifting cultivation practises of recent migrants. The shifting cultivation practises of recent migrants are sometimes blamed for the widespread prevalence of Imperata cylindrica grasslands. In Sri Lanka, the impact of biophysical conditions as a variable is established.

From India to Sri Lanka, there are differences in shifting cultivation fallow secondary forest conditions that are related to the current biophysical conditions. Secondary vegetation in the highlands of northerneastern India begins with herbaceous weedy formations for the first 8-10 years, then bamboo forest for the next 10-20 years. They then develop into early successional mixed broad-leaf forests at lower elevations, or pine and broad-leaf mixed forests at higher elevations, over the course of 10-30 years. Finally, depending on the elevation, they grow into broad-leaved climax forests, which can be sub-tropical or sub-temperate (Ramakrishnan and Kushwaha 2001). As a result, the secondary vegetation in the region's usual 12-15 year fallows is quite similar to that which existed before farming. It's likely that repeated cycles of cutting and burning over recent history have resulted in a selection for this form of secondary forest (Perera 2001). Shifting cultivation is performed in a variety of methods, which leads to variance in swidden fallow secondary forests. Secondary forests that emerge on the fallow land of pioneer swiddeners in Thailand, for example, are frequently stunted trees. Bamboo grows out of renewing woody vegetation and quickly takes over shrubs and trees. Bamboo may grow dominant in some areas, even suppressing tree reproduction (Schmidt-Vogt 2001).

4.3.12 Development pathways of shifting cultivation fallow secondary forests

Chokkalingam *et al.* (2001) suggest a three-stage land use intensification model for tropical Asia at the extensive use stage, intensive exploitation stage, and forest depleted stage. External pressures, according to this paradigm, are the primary drivers of change in shifting cultivation and related secondary forest management. The Asian model differs from the Latin American model. The main drivers of development in Latin America are colonists' relocation to forested areas and the following integration of agricultural frontiers into the larger economy. Exogenous stimuli do not always lead to degeneration whereas endogenous variables lead to good evolution.

In mainland Southeast Asia, shifting cultivation is the principal agent of primary forest to secondary forest conversion, but there is deterioration in practice due to land use intensification. Intensive shifting cultivation or permanent farming is becoming more popular among Highland farmers. The loss of shifting cultivation is resulting in net degradation and a reduction in fallow-related secondary forest cover, as well as losses in biodiversity and economic potential. In Thailand, there is no regulatory or legal framework in place that allows for the lawful and sustainable usage of these forests. The classification of secondary vegetation as degraded scrub and a highland development policy that prohibits communities from using forests are major roadblocks to the formation of such a framework (Schmidt-Vogt 2001).

4.3.13 Shifting cultivation and secondary forests in Southeast Asia

Many secondary forests in the mainland Southeast Asia have originated from shifting cultivation. These are "forests regenerating largely through natural processes after significant human disturbance of the original forest vegetation at a single point in time or over an extended period, and displaying a major difference in forest structure and/or canopy species composition with respect to nearby primary forests on similar sites" (Chokkalingam *et al.* 2000). Until the 1960s, swidden farming, which was practiced by minority groups known as hill tribes or hill people, was the primary mode of land use in mainland Southeast Asia. Except for northern Thailand, this is still the situation across the majority of this region. Lowlanders have just recently begun to intrude into the highlands to undertake supplementary shiting cultivation, which involves short cultivation and fallow times, resulting in badly damaged secondary growth dominated by thorny species.

There are two types of shifting cultivation practiced by highland minority. Short cultivation and extended fallow seasons are used in rotational shifting cultivation, allowing shifting cultivation fallow secondary forests to grow quickly through numerous successional stages. "Forests recovering mostly by natural processes in woody fallows of shifting cultivation for the goal of restoring the land for cultivation again" is how shifting cultivation fallow secondary forests are characterized (Chokkalingam *et al.* 2000). Shifting cultivation has been transformed or replaced entirely by other land uses as a result of changing conditions. This has an impact on secondary forest cover, condition, and ecological and economic functions.

4.3.14 Secondary forest in shifting cultivation

The ecology of secondary succession

Although they can also develop on frequently exploited regions by pioneer shifting cultivators, secondary forests are a component of rotational systems. The original vegetation mix, site conditions, and land use choices all influence the development, structure, and composition of these forests. The establishment and structure of shifting cultivation fallow secondary forests in rotational systems are determined by the duration of the cultivation and fallow seasons, together with the shifting cultivation practices. The most important component is the fallow period's duration, which should be at least ten years (Sabhasri 1978). It's also essential that there be stumps and roots remaining in the fields, as these can serve as a breeding ground for coppice shoots and root suckers. Rotational shifting cultivators also leave behind some massive trees known as "relict emergents" during clearing in shifting agriculture. There is variation in the number of relict emergents that ethnic groups abandon on new territory.

The Lawa practise selective felling, leaving intact any tree larger than 15 cm in diameter. On average, they leave 244 relict emergents per hectare. According to Schmidt-Vogt (1999), the Karen seem to lack a logical selection principle, destroying both huge and small trees at random.

Pioneer swiddening systems: secondary forests and secondary succession

Sprouting is the primary cause of woody plants in early successional phases in rotational shifting cultivation fallows. Because of the lower possibility of seed regeneration due to a protracted dry season in the seasonally dry tropics, trees are even more prone to reproduce by sprouting than in the wet tropics (Ewel 1980). The availability of stumps and rootstocks determines how quickly a woody fallow develops. The quantity of resprouting plants decreases in intensive shifting cultivation systems with extended and intensive cultivation. As a result, perennial herbs like Chromolaena odorata survive far longer in pioneer swidden fallows than in rotational swidden fallows. Grasslands and bamboo groves are distinctive features of pioneer shifting cultivators' fallow land. Intensive cultivation over several years and fire eliminate most other plants, favoring grasses that can survive fires thanks to their rhizomes. The formation of secondary forests is slowed but not completely stopped by grassland cover. Bamboo grows out of renewing woody vegetation and quickly takes over shrubs and trees. Bamboo may grow dominant in some areas, even suppressing tree reproduction.

Secondary woods' socioeconomic worth and indigenous knowledge

Anderson (1993) summarised the findings of various ethnobotany investigations conducted among shifting cultivators in northern Thailand. The ethnobotany of fallow vegetation and woods was investigated by Kunstadter (1978) and Kunstadter *et al.* (1978) among the Lawa of northwestern Thailand, while Schmidt-Vogt (1999) did so among the Lawa, Karen, and Akha. Plants are used extensively by the Lawa for food, construction, fuel, medicine, animal fodder, clothing and decoration, and rituals. 295 species were used for food, 119 for medicine, 44 for weaving and dyeing, and 27 for fuel, according to Kunstadter *et al.* (1978). Swidden fallows provided the majority of valuable species. According to Kunstadter (1978), the Lawa have improved the availability of beneficial plants by preserving a mosaic of microenvironments or successional stages that would not exist in the natural environment through protracted fallow shifting cultivation.

Cultural and ritual significance of secondary forests

In cultures that practice cyclic shifting cultivation, secondary woods have a religious importance. The Lawa's ceremonial life revolves around the forest which serves as the site of shifting cultivation. Before clearing land for shifting cultivation, before rice planting, during the growing season, during harvest, after harvest, and when the fields change into fallows, ceremonies including animal sacrifices are held to request permission from the forest spirits to expand into their territory. However, due to the impact of Christian missionaries, exposure to Thai culture, and the demise of the swiddening way of life, these rituals are dwindling (Kunstadter 1983b).

Crop diversity

The three tribes of Meghalaya, Northeast India, employ shifting farming, which is a mixed cropping method that maintains high crop diversity. Despite the fact that the majority of the crops raised by the farmers are for personal consumption, the surplus produce is sold on the market for cash. Potato and turmeric are the main cash crops in War-War and Khrang villages, while ginger is the main cash crop in Khulia and Kuswai villages, followed by potato, colocasia, pumpkin, and chilli to a lesser extent.

Fallow management

The Khasi of War-War and Khrang villages in India fallow their lands for 8 to 15 years, while the Garo fallow their lands for 7 to 9 years and the Karbi for only 3 to 5 years. The land is set aside at this time to allow for vegetation regeneration. Previously, the fallow was completely ignored, with no form of management in place. The land tenure structure, as well as farmers' ignorance of the relevance and advantages of fallow land management, are likely to be important factors in the fallow's neglect. Farmers were not only limited in terms of resources, but they also lacked access to education, training, outside contacts, and media coverage. Farmers in the Khasi community (Khrang village) are now planting bay leaf, jackfruit, and black pepper on fallow fields, while Karbi farmers (Kuswai village) are planting banana and broom grass.

4.3.15 Soil regeneration, forest growth and species diversification

Soil elements vary dramatically over the fallow period, from removing forestland for agriculture to the establishment of a secondary forest. The effects of these adjustments, however, differ from one attribute to the next. According to the results of the fallow reduction study, soil attributes are roughly divided into two categories based on their behavior in relation to fallow length. Short fallow systems improve pH, exchangeable bases, exchangeable acidity, and accessible phosphorus. Short fallow, on the other hand, harms the A1 subhorizon's morphological development, organic matter, soil texture, and bulk density. At the start of the cropping period, ashes from burned vegetation biomass operate as a lime fertiliser and have a significant impact on specific chemical soil properties. Crop harvesting, cation leaching, and sheet erosion caused by heavy and severe rains on bare soil at the start of the cropping season reduce the liming effect of ashes. The most significant loss of ash nutrients during the cropping phase in shifting cultivation, according to numerous writers (Tulaphitak *et al.*, 1980; Prat, 1990), occurs between burning and soil surface covered by crops. This soil exhaustion process continues throughout the agricultural season. The land is left fallow for two to three years after it has been cropped.

4.3.16 Existing shifting cultivation systems encourage innovation

Shifting cultivation is an extremely broad set of practices. Examples of its full diversity can be found elsewhere in Shifting Cultivation Policies: Balancing Environmental and Social Sustainability. However, in order to highlight the positive potential of swidden in contemporary agricultural innovation, Xu *et al.* (2017) provided a brief outline of some existing shifting cultivation practices.

The Taungya system

The taungya system originated in the mountainous borderlands between China and the rest of mainland Southeast Asia. As with many swidden systems, forest land is cleared for the cultivation of annual food crops. An important further feature of taungya is the planting of a preferred species of timber tree that develops to maturity in the fallow. In some systems, annual crops are planted for only one or two years, but there are also indigenous systems in which the number of years of food cropping in the understorey of the growing timber plantation is extended by using a rotation of increasingly shade-tolerant crops. This allows farmers to continue annual crop production while they wait for fallow timber to mature (Menzies, 1988). Subsequent colonial and post-colonial forestry sectors in Myanmar have continued to use various versions of taungya teak production (Win and Kumazaki, 1998), while the success of the system in Myanmar led to its adaptation and utilization elsewhere in the tropics (Kalame *et al.*, 2011). Michael Dove (1993) later described how rubber cultivation was integrated into shifting cultivation in Borneo, and how it was able to complement the labour demands of the pre-existing system. In the shifting cultivation cycle, there are three peaks in demand for labour; planting, weeding and harvesting, that account for about four months of work, leaving eight months of relative labour surplus every year. Rubber cultivation was well suited to exploiting this surplus of labour because it could be exploited intermittently, and latex flows could even benefit from pauses in tapping that occurred when labour was required for other swidden work.

4.3.17 Dimensions of ecological systems (Xu *et al.*, 2017)

Local climatic, hydrologic, edaphic, and geomorphologic elements, as well as biotic interactions, have a significant impact on ecological processes and species abundance and distribution. Appreciating such place-specific factors is important not only for the innovation process, but also for the process of scaling-up. In this respect, the spread of the taungya system across the world demonstrates the potential for scaling-up of shifting cultivation-based innovations.

1. Many shifting cultivation systems involve the deliberate management of biomass through practices such as notching, slashing, felling, trimming and pollarding to promote tree regeneration. As well as having the potential to increase or accelerate the production of exploitable resources, enhanced biomass production may also enhance the potential for carbon sequestration.
2. The use of controlled burning of selected areas is an important part of many, though not all, swidden systems. As well as controlled fire, a further vital dimension to swidden agriculture can be fire prevention, and fallow recovery in particular. Understanding and innovating intensified swidden systems that might maximize, among other things, productivity and carbon sequestration will require analyses of fire regimes across all stages of the shifting cultivation, from clearing to fallow recovery.
3. Most shifting cultivation systems incorporate a diverse mix of crops and trees in order to harness positive inter-organism, as well as plant-soil, interactions. Understanding and optimizing these interactions requires an appreciation of the broad-scale ecosystem-level effects that exist between particular perennial species and networks of interacting species.
4. Sub-dimensions to fallow succession include the type, intensity and duration of annual cropping, as well as the nature of management of the subsequent fallow. Each of these dimensions shapes the characteristics of the ecological communities fostered by shifting cultivation fields.
5. The dynamics of people, groups, and ecosystems are influenced by the size, shape, and spatial interactions of land-cover types. The ecology of shifting cultivation systems must therefore be approached at landscape level and include an analysis of surrounding land-cover types. This will involve assessment of non-shifting cultivation areas such as water-source forests, stream buffer forests and protected cemeteries or sacred forests.

4.3.18 Ecological implications

Agriculture scientists, philosophers, economics, and anthropologists have long been concerned about the impact of changing cultivation on the environment. Deforestation induced by shifting cultivation is frequently cited as one of Southeast Asia's most serious environmental issues (FAO, 1995). Tawnenga (1990), Tawnenga & Tripathi (1996), and Tawnenga *et al.*, (1997), all from Mizoram, conducted studies on the ecological impact of shifting cultivation. Climate change, biodiversity loss, decreased timber supply, flooding, siltation, soil degradation, and a shift in forest vegetation from primary to secondary, and finally grassland, can all occur from clearing forests for shifting cultivation (Holden, 2001). Deforestation is primarily caused by clearing forests and burning them for shifting agriculture (Monela and Abdallah, 2007; Zahabu, 2008). Soil erosion is exacerbated by the removal of vegetation cover. Soils are especially vulnerable to erosion in hilly places (Shoaib *et al.*, 1998; Sfeir-Younis and Dragun, 1993). The shifting farming cycle in Mizoram has been decreased from 20-30 years to 3-4 years. It has exacerbated soil erosion and nitrogen loss from the top soil. A total of 16.84 metric tons of soil per hectare is lost each year. As a result, soil fertility has been compromised. The shifting cultivation has resulted in low cereal crop production and per hectare yields. In contrast to the negative effects of shifting agriculture mentioned above several research have indicated that shifting cultivation can be used to restore ecological. According to a report from the Ministry of Environment and Forest (1997), shifting cultivation has gained 1700 square kilometres of scrubby vegetation. This expansion also aided in preventing soil erosion on high slopes. Forest clearing followed by protracted fallow periods replicates natural, small-scale, high-intensity disturbances in forests, and could be a sustainable land use option (Andrade & Rubio-Torgler, 1990). It has the potential to increase biodiversity in the landscape by establishing new habitats (Gadgil & Guha 1992; Kricher & David, 1992).

Shifting cultivation has devastating ecological consequences. It is a major contributor to deforestation and soil degradation. Because of the shorter fallow periods, natural vegetation regeneration is hampered, causing irreversible damage to the forest ecology. Fallow periods used to be around 20-30 years long, but they have since been cut to just 5 years. According to the Government of India Task Force report on shifting agriculture from 1983, the fallow period in Andhra Pradesh is three years, the shortest in the country. According to the Environmental Protection Agency (EPA 1990), forest fires are responsible for around 25% of global warming. The resulting ecological imbalance will very certainly have a negative influence on the tribals' socioeconomic situation.

4.3.19 Impact of shifting cultivation on non-timber forest products (NTFP)

There have been documented positive and negative effects on the availability of NTFPs as a result of land use change for a variety of species. Endopleura uchi, for example, becomes scarcer as land is converted from forest to other uses (Shanley *et al.*, 2002). Palm plants like Orbygnia phalerata (May *et al.*, 1985; Pinheiro, 2004) and Astrocaryum tucuma, on the other hand, benefit from the transition from forest to other land uses (Schroth *et al.*, 2004). Xa'an leaves have also been utilised to thatch tourist structures in the Mexican Caribbean for the past four decades (Caballero *et al.*, 2004). This species thrives in closed canopy forests, although it may also thrive in high-light situations (Zona, 1990). Pulido and Caballero (2006) investigated the availability of xa'an palm tree leaves in the Yucatec Maya's changing farming methods. Since prehispanic times, the adult leaves of this palm species have been widely used for roofing; in the year 2000, roughly 90 thousand people used S. yapa in Quintana Roo, and 360 thousand utilized Sabal spp. on the Yucatan peninsula. One of the implications of shifting cultivation, according to their findings, is a decline in the availability and quality of this NTFP. Variations in population composition and leaf production rates between farmland and woodland regions cause decreasing availability. As agriculture becomes more intense these negative consequences may become more pronounced. According to the findings, S. yapa harvesting in natural systems is viable for another nine decades and will be compatible with shifting cultivation as long as land tenure remains communal, long fallow times are observed, and other characteristics stay constant (Pulido and Caballero 2006). Improving our understanding of anthropogenic effects on tropical ecosystems is a top objective (Bawa *et al.* 2004).

One of the main inducers of land-use change is shifting cultivation, which involves alternating between a brief time of agriculture and a long period of fallow (Ruthenberg, 1976). In tropical places, anthropogenic disturbance is quite frequent; it is the major food production method, with at least 240 million people using it (Attiwill, 1994, Juo and Manu, 1996). Shifting cultivation is a common land use in tropical forested areas, and it is frequently blamed by policymakers for forest degradation. Secondary fallow woods that re-grow following shifting cultivation are not generally thought to be good targets for biodiversity conservation and carbon storage (Mukul *et al.* 2016).

Two of the most important threats to Southeast Asia's forests and biodiversity are deforestation and forest degradation (Sodhi *et al.*, 2010). Deforestation and forest degradation in this area have long been blamed on shifting cultivation

(Ziegler *et al.*, 2011). In this region, which supports an estimated 14-34 million people, shifting cultivation has been practised for decades (Dressler *et al.*, 2015). As a result, secondary forests regrowing after shifting agriculture are swiftly becoming a prominent forest type in Southeast Asia, accounting for around 63 percent of total forest cover following shifting cultivation, logging, and other disturbance (Kettle *et al.*, 2010, Koh, 2007; de Jong *et al.*, 2001). The Philippines is a biodiversity hotspot and a megadiverse country, making it one of the top conservation priority countries (Posa *et al.*, 2008). (Conservation International, 2013). The country has one of the worst rates of deforestation in Southeast Asia, and it was one of the first to embark on large-scale reforestation and forest restoration projects to combat the rapid loss of forest cover (Pulhin *et al.*, 2007).

4.4 DIVERSITY IN SHIFTING CULTIVATION RELATED ISSUES

Population expansion would necessitate extending the agricultural phase and reducing the forest fallow period based on the presumptions of perpetual land scarcity, food self-sufficiency, and land production. This expansion would eventually change shifting farming into continuous cropping with no fallow or inactive periods (Liang *et al.*, 2009). In two watersheds in Nepal's Chitwan and Tanahun districts, Balla *et al.* (2000) discovered that shifting cultivation is unsustainable and unable to support livelihood systems within agricultural communities. Governments across many regions have been deeply resentful of shorter fallow periods because they are believed to cause environmental harm through soil mining and faster erosion, which is compounded by national interests in preserving forest resources for other uses (Dove, 1996; Fujisaka 1991). On the other hand, long-term shifting agricultural systems with a forest fallow time are becoming rare in most regions, and traditional shifting farming is under pressure both internally and externally to adapt (Cairns and Garrity 1999).

4.4.1 Sustainability of shifting cultivation

Shifting cultivation is regarded as destructive and disadvantageous because it not only harms the ecosystem but also has a negative economic impact. Some of the negative consequences of such cultivation include: helping springs dry up, causing soil erosion, destroying precious timber, creating very heavy floods, silting tanks and fields, and agricultural damage (Biswal and Kumar 2013).

4.4.2 Socio-economic constraints

Schlippe (1955) in his monograph on shifting cultivation in Africa has commented that agriculture is that sector of human activity in which there is greatest interaction between the environment and the culture which has grown in and from it. All cultures are intimately linked with the mode of subsistence. The environment, local conditions and values are invariably connected with the agricultural practices that have evolved in the area. It is, therefore, not surprising that shifting cultivation; which had evolved in the Neolithic age, would be very closely linked up with the socio-religious or socio-cultural life of the people. It should also be appreciated that the people connected with the type of cultivation, have remained mostly cut-off from the main stream of civilization. The socio-economic conditions of the people and the areas are also responsible for their adherence to this system of cultivation.

Shifting cultivation represent an identifiable distinct stage in the history of mankind's economic development.The erlier stage of hunting and food gathering required a very large area for subsistence. As the population increased this method of food collection was not able to sustain the population and a new food growing method had to be established. This necessitated clering of forest landsand grow crops useful for the population in the cleared area. However they learnt through experience that the level of production from these areas was good only for a limited period.The fields became exhausted of nutrients useful for supporting crop production and only way left was to move to another forest area. The vegetation was cleared, burnt in-situ to enhance soil fertility for growing luxuriant crop for a year or two. This practice is called shifting cultivation, or slash and burn agriculture or swidden agriculture. This practice has continued for a long time as it supported the population for a long time in the history of mankind. The rate of growth of population was not very high and there was sufficient forest land to extend the area under this system as also for adjustment in the intensity of utilization of area under command by reducing the cucle of shifting cultivation. A challenging stage in shifting cultivation was reached when the shifting cycle became so small that this system of cultivation began to exhaust the soil resource. The growth of vegetation was not sufficient to add required nutrients to the soil after burning for optimum crop production. This stage generally occurs when the shifting cycle is less than 10 years or so. In the absence of any other source of livelihood, the process continues and the shifting cultivation gets converted to a form of 'extensive cultivation' wher a piece of land may be allowed to remain fallow for one or two years. It was at this stage that a beginning of intensive cultivation was perceived which signifies the end of the practice of shifting cultivation finally.

It is still the larger base of the forest around, which provides secondary, or even primary source of livelihood. In case of natural calamity, the forest resource-base could be more intensively utilized. The shifting cultivation is so organized sector that each field provides all requirements of the family, such as, cereals pulses, oilseeds, fodder for livestock etc. Before sowing, the crop seeds are carefully mixed in due proportions according to requirement of the family. Each member of the shifting cultivators community gets full sustenance from his field which may be supplementd by collection of forest produce, hunting etc. The shifting cultivation in the early stages represent optimal utilization of natural resources. The forests do not have a commercial value in a closed tribal economy. The trees in natural forests grow and die according to their natural cycle. Instead of this natural regeneration and decay cycle, the human hand intervenes and fell the forests once in a number of years and then allows the nature to work again.

4.4.3 A way of life

Shifting cultivation not only represent a complete economic system but is representative of a way of life. The very existence of shifting cultivation implies that the community practicing this system has a recognized right over an area which is well defined. A number of shifting cultivators communitie may occupy a large territory. One important characteristics of the stage of shifting cultivation is the right of the community over an area, that is, territorial delimitation. Another important characteristics of the shifting cultivation is that it is community as a whole which acts and participates in the total economic activity in the area. The social organization itself is built around the conept of community ownership and joint responsibility. The forests are the property of the community and each individual can get as much land as he requires according to the size of the family as well as his own capacity to work. The basic point is that there are no permanent claims and the indivisual family has a right to use the land assigned to it for the specified period. While the community ownership is the general rule amongst shifting cultivators, there are some extension as well. Among many tribes, the land formally belongs to the Chief, who has the right to distribute the land to cultivators as he may wish.

4.4.4 Social security

The system of shifting cultivation also provides full social security. In the early stages of hunting and food gathering, the entire expedition shares whatever is collected,though they brave and the leader, gets a bigger share but those who were left behind, like old and infirm, or women have also a well defined share in the collection. The individual's welfare is social responsibility of the society continue to be honoured in the shifting cultivation stage also. The clearing of the forest is a community operation where each one contributes according to his capability. The old persons may only accompany the group and give advice while young may provide bulk of the physical labour. Protection of crops from wild and other animals is a community operation and old, infirm and widows are generally allotted plots in the centre and young on the periphery. One interesting fact is that, generally sowing operation start from the centre and so the harvesting, thus giving full protection to old and infirms. The old people share in this community system where each is expected to contribute according to his capacity but gets according to his needs. The old people and widows are not dependent on their children for their upkeep and livelihood.

4.4.5 Customary laws governing shifting cultivation

Each ethnic group has its own set of customary laws governing shifting agriculture (and other natural resource management), which might even differ between neighboring villages. Although the following explanation is based on the Ao tribe's system in Nagaland, India's Mokokchung district, the essential ideas apply to practically all tribal tribes (Choudhury *et al.* 2003; Choudhury 2006; Erni 2021). Each village has a number of land parcels or blocks set aside for shifting cultivation - one less than the length of the fallow cycle; for example, if the fallow cycle is 15 years, the community that practices one year of cultivation in a plot will have 14 land chunks. The cycle is influenced by the length of cultivation in each plot and the amount of moving cultivating land. The more land parcels available for shifting cultivation, the more land chunks there are, and thus the longer the fallow cycles. Despite slight differences between tribes or even villages, the fundamental concepts are the same among groups in NE India, with similarities in Southeast Asia as well (Erni 2021). Upland shifting agriculture groups in Myanmar's northern Chin state and India's northeastern state of Mizoram, however, deviate from this common structure. All households in these communities are granted annual access to a plot using a lottery system that is customarily used for plot access (GRET 2019). While a family is guaranteed the right to a plot, they may not always be able to access the same plot each year because the family's plot is determined by a lottery. A family's chance to enter the lottery is granted to each family that moves into the village to become permanent residents; families that leave the community forfeit their chance to enter. This agreement acknowledges the rights of widows and women, including homes headed by women or when the male has moved abroad in search of employment. Figures. 4.4 to 4.7 Show the various operations in shifting cultivation farming.

Fig. 4.4. Forest vegetation

Fig. 4.5. Burning of vegetation

Fig. 4.6. Readying to prepare land
(See colour version on page 653)

Fig. 4.7. Land preparation

4.4.6 Social characteristics of shifting cultivation

Based upon ethnic values and culture, both private and common property land tenure regime is found in existence in the shifting cultivation system. In Odisha, there exists both private and common property regime (Dash 2006). In case of Meghalaya, while the private land ownership pattern does exist among the Khasi and Karbi tribes, the Garo tribes, on the other hand, resort to community land ownership pattern (Deb *et al.* 2013). Further, shifting cultivation based ethno-agrarian economy is self sufficient and habitually such shifting cultivators either depend minimally or not at all on outside for mitigation of their livelihood needs (Gupta 1994). The use of human labour is the key input in the shifting cultivation system and the practice of mutual exchange of labour amongst these shifting cultivator communities is also not uncommon. In practices like harvesting, the groups of families join hands together to harvest the crop of each plot one by one. This practice of community labouring reduces drudgery and increases work productivity. During this harvesting time, it is said that farmers exchange their innovations and look at each other's innovations as they work. Further, it is customary to arrange food and drinks by the family of the plot which is being harvested. The community recognizes these harvest periods as festive events which makes their work enjoyable. The case of village named Nawalparasi, Nepal, shows that shifting cultivators work in an innovative way in their community in which the part of the village land is set aside and is cultivated by the members of whole community jointly. The benefit they harvest from this area is used to fund for teachers' salaries in community school (Kherkhoff and Sharma 2006). Further, the close knit and kinship ties are not only necessary for labour sharing but also important for technology diffusion among the members (NAEB 2008).

The division of labour exists between men and women. It varies from place to place. The Manipur shifting cultivators characteristics portray that the women predominate in seed selection and planting, weeding and other operations, the activities like cutting of the trees in jungle, clearing, burning are generally done by men. Both men and women participate in harvesting of produce and so harvested produces are carried to the villages by head loading (Anonymous 2015). The research shows that the shifting cultivation economy once was considered to be an egalitarian one, nevertheless, there was social hierarchy noticed as folkways in many such societies (Gupta 1994). The village headman exercises power and authority over their fellow shifting cultivators. This is exhibited while distributing shifting cultivation lands among the fellow social members in his village. Even during the regime of maharajas in the past, he could partake a certain amount of money collected for those rulers. He used to allot larger area of shifting cultivation land for himself by virtue of having more labour. Further, by virtue of his position, he could act as an agent of traders and money lenders in the shifting cultivators society (Gupta 1994).

4.4.7 Institutions and shifting cultivation communities

Traditional institution plays an important role in the shifting cultivation society. In sociological terms, the institutions "comprise shared norms, values, traditions, beliefs, religion, rules, regulations, laws, civil society organizations, and government agencies (North 1990). The traditional institutions were nothing but the village councils which have been responsible for managing land resources at community level. Each and every clan was represented in this village council. However, the composition of the traditional institution varies between the tribes. In Nagaland state, the communities in Mokokchung district practice collective decision making on land use for different purposes like demarcation of area for burning, digging of fire lines and controlled burning which were responsible for continuation of both sustainable and equitable use of land and resources (IWGIA 2014). Similarly in Kachin state of Myanmar the community cooperation took place in fire management and fire control. In Chittagong hill tract of Bangladesh, the community decision making was involved in natural resources management.

Social security was also one of the main functions of local institutions of shifting cultivators. They provide farmers with the social capital which they require for their livelihood. The hangvu in Manipur, a customary village authority were like in other shifting cultivation communities. The hangvu is comprised of chieftain and a council of clan elders (Kherkhoff and Sharma 2006). The indigenous knowledge occupies an important place among the communities of shifting cultivators. This information is also inextricably linked to cultural norms, values, beliefs, and rituals. Land management, biodiversity conservation, water management, health care, and medicine are only a few of the activities covered by indigenous knowledge (Chin 2006). This indigenous knowledge has been acquired over centuries by community members through careful observation and experimentation, and it is passed down verbally or via experience learning.

The land allocation is an unique characteristics of this society. The land allocation meant that the land is distributed to every family member of the village to eke out the livelihood by cultivating the allocated shifting cultivation plot. The allocation of plot is based on the family size. The size of the cultivation plot possessed by the family is based on the food requirement and the available labour in the family. The larger the family, more hands will be there to extend their help in the shifting cultivation plot. The land allotment exhibits slight variation among the North Eastern Region. For example in a village named Phuldangshai in Tripura adjoining Mizoram, village council decides the hillock to be selected for shifting cultivation. The areas selected for jhuming in the village are classified into big, small and medium size ones. Similarly, families who are interested in shifting cultivation are also divided on the basis of adult members. Within each category, lotteries are put and the land is allotted accordingly (Gupta 1994).

Distribution of land was based on the traditional institutions existing in the village. In case of Tripura's tribals, when the shifting cultivation activities were about to start, the group of headman from close by villages decides a meeting for the Ker worship. The institution of Ker worship exists in the Tripura tribals in which the group of villagers exercise their joint rights over a specific area they would shifting cultivation in a particular year (Gupta 1994). After the Ker worship, the individuals were free to choose any hillock he likes. This was based on the number of family members he had. The labour support taken from other family for his work in his shifting cultivation plot has to be returned in terms of work days as much as was taken by him from others. In many villages of North Eastern India, the lands were divided into different shifting cultivation blocks (Kherkhoff and Sharma 2006). The shifting cultivation blocks were based on the shifting cultivation cycle. For example, if the shifting cultivation cycle were of 10 years, there will be ten shifting cultivation blocks and author further stressed that sometimes these shifting cultivation blocks have been named by the communities. The plot selection was slightly different in other countries. The selections of the sites in the phangzhing type of shifting cultivation in Bhutan were done by women as they manage the fields and crops (Kherkhoff and Sharma 2006). Further, there are certain norms and values followed in this shifting cultivation system. The shifting cultivators regard some patches of primary forest for religious and ecological purposes. Also the shifting cultivators' village territory were not encroached by the outsiders. They respect and the violation of the rules and regulations posed by the society would harm the others with punishment (Rahman *et al.* 2011).

4.4.8 Social causes of shifting cultivation

There were very few researchers who looked upon the causes of shifting cultivation system. Rahman *et al.* (2011) elaborated the underlying causes of the shifting cultivation. They discovered that older and less educated farmers have a predisposition to use shifting cultivation as a result of their heritage. Variables including high family size and a lack of off-farm income opportunities considerably impacted the farmers' ability to choose shifting cropping. The likelihood of permanent agriculture increases as more investment and finance becomes available. Shifting agriculture is less likely to be practiced by younger farmers with small households. Shifting cultivation was used by the farmers because it provided a generally consistent net cash flow year after year. In addition, his logistic regression study demonstrated that tradition and custom play a critical role in the continued cultivation of shifting crops.

4.4.9 Initiatives of the government to curb shifting cultivation in India

The Pilot project for control of shifting cultivation was launched by the Government of India during the fifth five year plan (1975-80). This was the first ever specific programme for controlling shifting cultivation in which each family was assisted with a ha of irrigated or unirrigated terraced land and one ha as developed plantation and targeted settlement of 1700 families practising shifting cultivation within a period of 3 years (Maithani 2005). Subsequently river basin scheme for control of shifting cultivation was initiated with eight pilot schemes by the North Eastern Council during the fifth five year plan period with a common objective of controlling shifting cultivation but the component varied.

The task force on shifting cultivation programmes implemented by North Eastern Council and the state government reported that in the first phase there was no concentrated effort and secondly in several cases the involvement of the target group was totally absent (Maithani 2005).

After this task force report, model schemes were introduced with 100 per cent financial assistance, taking the important elements from the earlier projects. It concentrated on both young and older generations. For older generations the technologies like terracing, cash crop plantations, supply of agricultural inputs, livestock were introduced. The vocational training for the development of non-farm skills was introduced for the youths to wean them away from jhum. Emphasis was given on infrastructure creation for community development. A group of 25 Jhumia families or so in a village or hamlet constituted the basic unit of development. The midterm review of the model scheme in 1989-90 suggested adopting holistic approach rather than group or family units promoted under the scheme (Maithani 2005). Subsequently the Watershed Development Project for Shifting Cultivation Area (WDPSCA) was introduced from 1995 to 2012 with 100 per cent financial assistance from the central government with a holistic development and emphasized the soil and water conservation measures in the shifting cultivation area. Further, the National Mission on Greening India 2010, implemented by Ministry of Environment and Forest has proposed different activities for rehabilitation of shifting cultivation areas. The mission lays emphasis on fast growing species used by the community (Deb *et al.* 2013). Nowadays in case of Manipur no separate project is being implemented in shifting cultivation areas. But WDPSCA has been subsumed in Rashtriya Krishi Vikas Yojana and is being implemented as a part of the Project. Further, through North Eastern Council, NERCORMP (North East Region Community Resource Management Project) is working with the shifting cultivation farmers through formation of natural resource management group and other livelihood improvements of the shifting cultivation farmers since 2000 in various states of North East India.

4.4.10 Improving shifting cultivation and livelihood of the shifting cultivators

The research portrayed that "shifting cultivation is to be accepted as a rational land use system" and that author calls for green revolution in shifting cultivation area to bring intensive and sustainable production (Mertz *et al.* 2008). The farmers should be educated to enrich the biodiversity of fallows and to increase the soil productivity to avoid the problems of shortened fallow periods and new policies to be designed to protect the land of shifting cultivators (Fox 2000), as in India shifting cultivation has been recognized under wasteland category (NRSC 2011).

The livelihood benefits demonstrated by the project by and large has been stated in the earlier sections of the impact assessment document. Even though the over-all increase in the livelihood will benefit the women of the community also, the project also implemented certain measures targeting the women of the community. In order to understand the role of women in the day to day life of traditional societies existing in the villages of the targeted districts detailed interactions were carried out. Women are more involved in agriculture and marketing of the produces than men even though men voluntarily takes up responsibilities and works which require more physical strength like slashing and burning of the fallow areas, land preparation etc. The specific activities targeted towards women or significant to women, identified during the impact study were – Market sheds and support to SHGs Since marketing of the agricultural produce is predominantly the responsibility of the women, marketing sheds provide a platform for the women in the village itself to shelf and display the agricultural produces and market them. This saves the time and efforts of the women in marketing and also thereby increase their income.

Marketing shed constructed at Lampongsheanghah village of Mon district provides support for women from 80 households of the village. The average income from agricultural marketing has been doubled. During the time the women spend in the marketing shed they also do basket making from the market sheds and earn an additional income of atleast Rs. 200 per month. One of the major means of supporting the women-folk of the targeted community is through support to self-help groups. The project has supported self-help groups through different means. The impacts of these measures are as follows:

4.4.11 Documentation of traditional customs and practices

18 villages have been surveyed for documenting the customs and practices prevailing among communities in detail. The main objective of this exercise was to understand the social systems prevailing in the targeted villages and also to identify the presence of any socially marginalized groups.

4.4.12 Village councils

The village councils in the state are legally empowered by Nagaland Village Council Act 1978. The strength of the village councils in the villages of Mon district varied from 9 – 45. The selection of the members is unique in each village. In 3 villages the selection of the members is based on representation from different clans and in another three villages the representation was found to be from different Morungs (families) and in one village the selection was based on the qualities of the individual beyond the clan or morung to which the person belongs. In two villages the tenure of councils were found to be unlimited where the members were replaced on voluntary retirement of the existing members or whenever the performance of a member was found to be poor. In all the villages in Mon District the tenure of the Village Council Chairman was 5 years. In one village the chairmanship was found to be rotated among the clans in every term of the council and two villages had minimum and maximum age limits for the members. The strength of the village councils in Mokokchung district were found to range between 8 to 40. Except for one of the sampled villages where the selection of the members was based on the merits of the candidates beyond the clan representations, in all other villages the members of the council were representatives of different clans. The tenure of the council also varied from 5 to 30 years whereas the chairman's tenure was 5 years. In one village the chairmanship was being rotated among the clans within a total tenure of 30 years of the council. No age limits were observed in these villages of Mokokchung District to qualify to be a member. In Wokha District the strength of the village councils varies from 20 to 40 members. The selection of members is mostly clan based. In one village it was reported to be based on representation from a fixed number of households. In all sampled villages the tenure of the council and chairman was 5 years.

4.4.13 Indigenous knowledge and traditional practices

Traditional knowledge of plant species in local communities has paved the way for a wide range of ethnobotanical research. Farmers are well-versed in the plants that are good for manure, meat, and milk. Shifting farming is common in the eastern Himalayan region, including North-East India and the Chinese-Myanmar borderlands. The shifting cultivators here rely on livestock farming for a living. Despite the fact that a variety of shifting cultivation alternatives (terrace cultivation and agroforestry intervention) have been proposed and tested, there has been little success in discovering viable alternatives. As a result, the goal of this article is to create a complete mechanism that protects forest ecosystems while also improving the livelihood security of hill farmers in Northeast India. As an alternative to shifting cultivation, this item posits the viability of adopting a "grains for forest management program" by providing food grains to farmers. The suggested program also aims to meet the goals such as; (i) soil restoration and ecosystem resilience; (ii) food security through grain supply under grain for forest management; and (iii) carbon offset under climate negotiations (Nath *et al.* 2016).

4.4.14 Sustainability of shifting cultivation

The sustainability of shifting cultivation at a particular place can be determined using the parameters like; fallow period, population density, land tenure, slope of the land and economic condition of the cultivators (Table 4.4, Fig. 4.8). We tried to evolve this model to achieve the objective. The sustainability Index (SI) can be easily calculated if we have the requisite data on these aspects.

Table 4.4. Contribution of different parameters to the sustainability Index (SI) of shifting cultivation

Parameter	**Scoring**					
Fallow period	Years	25	18	10	5	1
	score	37	29	21	13	5
Population density	Persons km^{-2}	<15	15-30	31-45	46-60	>60
	Score	22	17	12	7	2
Land tenure	Right over	land Self	Community	Village chief	Usufructuary	Annual basis
	Score	12	10	8	6	4
Slope	(%)	<10	10-30	30-50	50-75	>75
	Score	9	7	5	3	1
Poverty/ marginality	Level	Affluent	Poor	Slightly poor	Very poor	Extremely poor
	Score	10	8	6	4	3

Source: Sharma (2023)

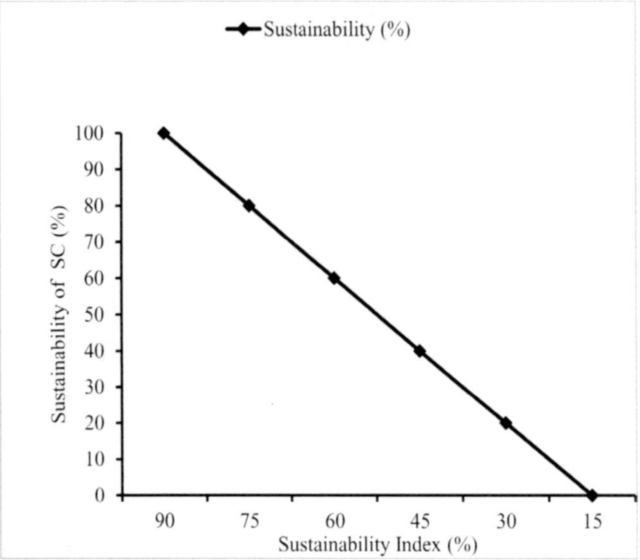

Fig. 4.8 Sustainability of shifting cultivation systems.

The sustainability can be categorized as; Highly Sustainable (75-90%), Sustainable (60-75%), Poorly Sustainable (45-60%), Unsustainable (30-45%) and Dangerously Unsustainble (<30%).

From the figure it is evident that an SF score of 60 is necessary for shifting cultivation to become sustainable and, higher the score, high will be the sustainability level.

Because of its compatibility with the socio-physical environment, which includes a sparse population, special types of community land tenure, undulating geography, poverty, and limited resources, the shifting agriculture system has survived. All of these civilizations are inextricably related to a certain style of existence. The agricultural practises that have evolved in the area are inextricably linked to the environment, local conditions, and value systems. Shifting cultivators are used to manage a wide range of cultivated and wild plants. According to Descola (1994), a single plot can include over a hundred species per hectare. Dove (1993) uses the Kantu of Kalimantan, Indonesia, as an example of a people who plant over 44 types of rice, with an average of 17 per household. Shifting growers also raise a variety of rice varietals in Asia and Africa. In Bangladesh's jhums, farmers cultivate about 40 different kinds. Within the species, there is also a lot of genetic variety (more than 20 rice varieties/cultivars). Farmers also use about 50 wild plant species as food plants, in addition to the agro biodiversity (Alam & Mohiuddin, 2009).

4.4.15 Social dimensions of forest dynamics under shifting cultivation

Based on manual work and low yields, shifting cultivation provides only limited labour productivity, as well as land productivity compared to input-intensive farming systems. Over the basic needs of the family, the surplus-labour is so limited that the accumulation of capital at the household level is minimal, as is the possibility of a social body to capture it. Shifting cultivation cannot support large and dense populations, in contrast with other agrarians systems (Mazoyer and Roudart, 2005). In History, societies that based their economy on shifting cultivation were not able to build strong states, contrary to hydraulic empires for example. It has caused dissymmetric relationships, balances of power and conflicts with neighbouring states based on openfield agriculture, which have progressively taken control of the forest people's resources: land, forest, water and workforce. Shifting cultivation populations have been marginalized (Scott, 2009), forced to submit to disrupting levies in metal (gold or iron tributes), in kind (grains, animal, wood tributes) or even in labour (slavery, forced labour), despised and heavily criticized by dominant powers, based in the openfield plains. The Western European colonialism globalized such an approach.

4.4.16 Variation in shifting cultivation

Shiting cultivation agroecosystems vary in their emphasis on different subsistence subsystems even within the same locations. Fishing is significant in some shifting cultivation systems, while collecting is vital in others; homegardens can range from highly productive to nearly non-existent. Despite the fact that shifting cultivation systems have different subsystems, they all have the same concept of having prospective subsystems that can be strengthened as needed. Only when other subsystems fail can these subsystems be used. Gathering from the forest is a common subsystem, however the amount of gathering depends on the situation. Food taken from the forest may be limited to specially preferred fruits, vegetables, or "snacks" if the yield of cultivated crops is good.

4.4.17 Causes of fallow length reduction

- Non-availability of sufficient land and people's rights over land.
- Demographic pressure
- Crop production type and reasons
- Soil Al-toxicity reduction
- Labour shortage to clear forest land

4.4.18 Importance of fallowing in shifting cultivation

The two main goals for fallowing are to reduce weeds and increase ecosystem fertility. The fallowing phase is critical for restoring soil fertility that was lost during the previous cropping season (Liang *et al.*, 2009). Many studies have shown that when the fallow period is short, weeds become more problematic (de Rouw, 1995; Dingkuhn *et al.*, 1999). The carbon stock in continuous yearly cultivation is just 1-4 t/ha. To improve traditional shifting farming practises crop and fallow management practises are widely understood and performed. The available studies on shifting cultivation in Bangladesh are insufficient in terms of determining the practice's status, distribution, management strategies and practical ramifications (Hossain eta. 2011). Systematic research on shifting cultivation are urgently needed in Bangladesh, and results-based suggestions will be presented to the government for inclusion in national plans, policies and priorities.

4.5 DIVERSITY IN FOOD AND LIVELIHOODS

4.5.1 Diversity in food habits

Three groups of shifting cultivatorss in central Africa's forest areas have common food traditions that include ball-shaped principle starchy food (PSF) and a sauce side dish with stewed components (Komatsu 1998). While all three cultures cultivate vegetatively reproducing crops in and around the tropical forest, the materials employed in their diet differ. The ratio for each food material of the main starchy food and the side dish is computed and compared from the food diaries kept for a complete year. The characteristics of availability, usefulness, and acceptability are used to evaluate food selection in each category. While all three groups express a strong desire to keep their diets varied, each group does so in a unique way.

There is a wide variety of plant and animal food resources in the African tropical forest. Ngandu people, shifting cultivators in the Congo jungle, considered 91 plant species out of 414 documented as food, and 287 animal species out of 437 recognized by the people themselves, according to Takeda and Sato (1993). Almost half of the species considered edible made it into the annual food diary. Mbuti people, hunter-gatherers living in the Ituri forest, utilised more than 100 plant and animal species as food, according to Ichikawa (1993), despite the fact that there were many more potential resources. As a result, the issue is the principles by which people choose their daily dietary items from such a diverse range.

4.6 DIVERSITY IN LAND USAGE IN SHIFTING CULTIVATION

4.6.1 Land, a cultural domain

People and the land have a deep bond since land is widely regarded as the primary source of life. It is regarded not merely as an object or property, but also as a gift from God, making it sacred to them. It's also sacred because their forefathers and mothers lived and toiled on the same land, which was passed down to them to carry on to future generations. Despite the fact that Christianity is the prevalent religion, individuals swear to settle disagreements by summoning the spirits of the land.

4.6.2 Land ownership and management of the Ao Nagas of India

Different communities – even within the Ao Nagas – have "their own unique customary laws regulating ownership and access to land and resources," making land tenure difficult (Darlong 2004). Despite the wide range of tenures within villages, there are no conflicts between the communities in terms of how they manage their land. Homestead land, clan forests, permanent farms, shifting cultivation land and fallow forests, conservation zones, and other land-use patterns are developed by each village in its own unique way. The guarantee of land ownership rights guaranteed by the Indian Constitution continues to offer a favourable and supportive environment for shifting cultivators, not just in Mokokchung District, but throughout Nagaland. Land in Ao society belongs to either the individual or the community, according to custom. While individual land is well recognized, the term 'community' in 'community land' can be confusing because it might refer to an extended family, a clan, a hamlet, or even a group of people. There is little or no cadastral survey of land in rural areas, except for private land holdings in metropolitan centers. As a result, land records are certified and maintained by the community, and land documents produced by village councils are legally recognized by the state government. Following is a brief summary of the Ao Naga community's land ownership classifications, as seen in the two case study villages.

4.6.3 Private or individual land

This is land acquired through the transfer of ownership rights for various reasons, such as land held against a mortgage, that an individual purchases from someone or receives from parents. While individual land ownership entails absolute rights, there is also a type of private 'ownership' of land in which a person is allowed to utilise or develop land as the lone occupier even if the land does not belong to them. When a family or clan allows a person to utilise or develop land, this type of arrangement – which is not the same as a lease – occurs frequently. In such circumstances, while the individual may enjoy full usufruct rights to the land's resources, the property cannot be sold, mortgaged, or leased. Regardless of the situation, the individual in question retains control and ownership of such property, with the exception of the selection of an area for shifting agriculture. For shifting cultivation, the community's decision – made by the Village Council – is ultimate, and one must obey the community's decision in terms of site selection, vegetation slashing, and fire lighting.

4.6.4 Community land

This category includes two distinct forms of land: I clan land, which is under the jurisdiction and ownership of a clan, and (ii) lineage land, which belongs to and is controlled by an extended family belonging to a single lineage (or relatives land). Forests, residential grounds on which clan or family members create homesteads, farmland, and so on are examples of such land. The elders of either the clan or the family are in charge of such territory. While some of this land is used for shifting farming, the most of it is preserved as woods or managed as bamboo groves. When a member of the clan or family makes a request, portions of such property are sometimes granted to them for development as farms and orchards. Individuals are given trees or bamboos from such lands and woods in extreme situations, either for personal use or to settle creditors. As a result, these shared property resources serve as a stable social safety net in both emergency and non-emergency situations. For example, self-help groups (SHGs) and other similar organizations lease such land (group land) to engage in income-generating activities such as cultivating vegetables and other cash crops. In all of these cases, the rights to such land are largely limited to usufruct rights.

4.6.5 Fallow patterns shifts

Until the late 1970s, nearly 90% of farmers in Nagaland's Sungratsü and Chuchuyimpang villages used shifting cultivation, with a 15-year or longer fallow period. Fallow land was used to develop orchards, tea plantations coffee farms, tree plantations, spices, and other cash crop plantations, such as floriculture. They recalled how converting land to permanent land-use regimes increased land competitiveness while putting pressure on fallow land. As a result, the typical shifting of cultivated lands away from cultivating food crops has been 'diverted' even further. However, in Mokokchung District, the usual fallow cycle is now extending to ten to twelve years. According to the farmers in the study communities, unless severe land-use changes are made, fallow land will continue to grow in most regions, including their villages. The increase in fallow land is mostly due to a combination of decreased rice cultivation area and a decrease in the number of farmers who practice traditional shifting cultivation. In such cases, only about half of the typical shifting cultivation blocks are utilised. The conventional blocks are further divided into two or more regions as a result of this practice, lengthening the fallow cycle.

4.6.6 Current policy processes

Provinces, towns, districts, villages or communes, and hamlets make up Vietnam's administrative and organisational structure. The province is in charge of all economic, cultural, and social activities in the communities it oversees. The government, like many other countries, has a large number of ministries. Each ministry is accountable to the state for its particular branch's actions. The legislation are considered and approved by the National Assembly. As a result, all State policies are developed as a result of the brainwork and real-world operations of ministries, branches, provinces, localities, and people's representatives.

4.6.7 Carrying capacity of land under shifting cultivation

The carrying capacity of land under shifting cultivation is also an important factor. The carrying capacity of land in Mizoram state of India was reported to be 19.5 persons km-2 (Bose 1967). Ganguly (1969) found that carrying capacity of Tripura state is barely 10-11 persons km-2. Saha (1970) showed that providing a minimum shifting cycle of 15 years, the maximum capacity that can be supported km-2 at a minimum standard of living comes to only 5 persons in Garo hills, 6 persons in Mizoram and 9 persons in Karbi Anglong of Assam. Shifting cultivation is an agroecosystem that is today found primarily in the humid tropics, but having been employed in temperate forest climates in the past. The humid tropics are defined as a region that includes the following features: 1) all months with monthly mean temperatures over 18°C, 2) 24 hour mean temperatures above 20°C during the growing season, 3) longer than a 180-day growing period.

4.6.8 Socio-economic implication in land management

The 'Socio Economic Implication,' on the other hand, claimed that land is communally owned and cultivated among tribal people. The number of persons supported by shifting agriculture can be used to estimate its productivity. Traditionally, shifting agriculture has aided in the protection of agro-biodiversity, served as an efficient type of land use, and maximized space utilization. The majority of shifting cultivation research in India has been done from an ecological standpoint, with only a few studies on the socioeconomic aspect of shifting cultivation. Recent research (Ramakrishnan 2003) and his group have looked into the significance of traditional ecological knowledge in the practice's variety. Shifting cultivation isn't only a historical agricultural method; it's also deeply entwined with the culture and identity of those who practice it (Schendel 1992). As long as shifting agriculture is intertwined with their cultural diversity, shifting cultivation helps to preserve the rich culture and heritage of more than 200 tribal races that live in the north-east region. As a result, the tribal people wisely utilised natural and forest resources, and they have the experience to deal with the instability and unpredictability that all ecosystems face (Kanagaraj 2014; Zaitinvawra & Kanagaraj 2008). Shifting cultivation has also been demonstrated to be quite sustainable in terms of ecological restoration in some studies. According to a report from the Ministry of Environment and Forest (1997), shifting cultivation has gained 1700 square kilometers of scrubby vegetation. This expansion also aided in preventing soil erosion on high slopes. It has the potential to increase biodiversity in the landscape by establishing new habitats (Gadgil & Guha, 1992; Kricher & David, 1992). Shifting cultivation supports 23 persons per square kilometer, which is more than double the amount of people supported by commercial cultivation in Kalimantan, Indonesia. However, the focus on modern methods labels indigenous people's shifting cultivation as a primitive method (Das 2006), as well as traditional knowledge systems and their ability to improve existing methods (Lalengzama & Kanagaraj 2012; Zaitinvawra & Kanagaraj 2008). Shifting cultivation has gained 1700 square kilometres of scrubby vegetation, according to a report from the Ministry of Environment and Forest (1997). This growth also helped to reduce soil erosion on steep slopes. By developing new habitats, it has the potential to boost biodiversity in the landscape (Gadgil & Guha, 1992; Kricher & David, 1992). In Kalimantan, Indonesia, shifting cultivation supports 23 individuals per square kilometre, which is more than double the number of people supported by commercial production. Traditional knowledge systems and their ability to improve existing methods, on the other hand, are labelled as primitive by the concentration on modern ways, as instance indigenous people's shifting cultivation (Das 2006). (Lalengzama & Kanagaraj 2012; Zaitinvawra & Kanagaraj 2008). Because the fallow time of shifting cultivation was extensive, shifting cultivation was able to last for decades (Cochet 2014; Luoga, 2000; Goswami 1980; Nye & Greenland 1960). However, the fallow period for shifting agriculture has been decreased from 20-25 years to 2-3 years, possibly as a result of growing population pressure leading in greater demand for grains and forest products (Mwampamba, 2009; Batterbury 2007). Fallow land reduction has been advocated as a management strategy to mitigate the detrimental effects of shifting agriculture in the tropics (Nye and Stephens, 1962). Introducing cash crops is not the greatest strategy for stabilizing shifting farming since it necessitates careful consideration of the socioeconomic situation and appropriate research (Ducourtieux *et al.*, 2005).

4.7 THE HUMID TROPICS AND RESOURCES

The total area under humid tropics is; 911.7, 1001.5, 76.3 and 491.8 million ha in Africa, South America, Central America and Southeast Asia, respectively. This amounts to 31.7, 56.5, 28.1 and 54.8% of the total area of the above regions, respectively (FAO 1980, 1982).

4.7.1 Stability of tropical forest ecosystem

The stability of the tropical forest ecosystem is the result of its capacity to "withstand climate and other hazards of the natural environment" (Richards 1977). Several characteristics of the tropical forest create this stability:

1. The humid tropical forest is home to a diverse range of plant and animal species. The forest ecosystem's stability is based on its high level of species diversity.
2. Tropical forests are the most complex terrestrial ecosystems on the planet (Connell 1978). Within the tropical forest ecosystem, plants and animals are inextricably intertwined. Animals in the tropical forest provide the same seed distribution and pollination functions as wind in the temperate forest (Hadly and Lanly 1983). The tropical forest reestablishes and sustains local variety since the species are significantly more diversified and the animals are not far range.
3. Due to the lack of nutrients in tropical soils, the tropical forest ecosystem relies on a self-contained, nearly closed nutrient cycle. The nutrients cycling through the system are stored in the biomass, which acts as a type of vegetative storage. With 65 to 85 percent of the vegetation's root system contained within the topsoil layer, the forest behaves like a big "sponge" in terms of nutrient recovery and recycling (Hadly and Lanly 1983; Uhl 1983; Moran 1981).

4.7.2 Soils

Despite the fact that the humid tropics have a wide range of soil types, the vast majority of the region's soils are nutrient poor (Jordan 1985). Phosphorus shortage, aluminum toxicity, drought stress, and low inherent fertility are all widespread and well-known concerns in the humid tropics of Africa, Southeast Asia, and the Amazon (Sanchez 1987; Lal 1989; Moorman and Kang 1978). The amount of rainfall appears to be the cause of the region's bad soils, since if an area's rainfall surpasses 1000 mm, the soils are frequently acidic and leached (Sanchez 1987). Tropical soil nutrient shortages are a major limiting factor in tropical productivity. Tropical rainforests are supported by these "old, severely weathered, and excessively leached" soils, yet the forests are not reliant on the soil for nutrition (Lal 1987).

4.7.3 Shifting cultivation and tropical soils

Shifting agriculture is increasingly being recognized as "ingenious adaptations to unfavorable conditions, based on a fairly thorough understanding of local ecology and soil potential" (Allan 1972a). Around the world, acid tropical soils cover one billion hectares of land. Almost majority of the one billion hectares are in the developing world, with 700 million hectares in the humid tropics and 300 million hectares in the savanna (IBSRAM 1987). The shifting cultivator lives in a humid tropical climate with acid soils. The use and preservation of the forest to restore soil fertility is "the pivot of every system of agriculture," and the swiddeners of the tropics have evolved a technique that works (Benneh 1972). The differences in soil fertility are used to create microsites that will be planted with various crops based on the swiddeners' understanding of which crops benefit from rich soils and which are unaffected by poor soils. The field is left fallow after the cropping cycle is completed (typically 1 - 4 years), while tree crops may be harvested for years. If the site is not utilized for a long time, it will recover its fertility; if it is used too soon, degeneration will occur. The degradation may not be quite visible, particularly if it is occurring gradually, perhaps over several generations. With swiddeners it is especially difficult since they "appear to be so self-sustaining, so well integrated with their environment" (Street 1969).

4.8 ENERGY INPUTS

Energy inputs for various shifting periods is given in Table 4.5.

Table 4.5. Energy inputs (k.eal./ha/yr) for agriculture under different shifting cycles

Agricultural operations	Energy input (k.cal. ha^{-1}yr^{-1})		
	30 years	**10 years**	**5 years**
Labour input (total)	193728	135872	66,560
Clearing under Storey vegetation	26112	31424	7488
Cutting trees and bamboos	55040	30080	-
Burning debris	6720	4608	640
Dibbling, broadcasting and Transplantation	6144	6720	6720
Hut construction	3584	3584	3584
Weeding	5448	10112	11520
Watching of the field from wild animals	11968	11968	10688
Rearing caterpillars	18880	-	-
Harvest	24640	26112	22720
Transportation	3584	2048	1600
Threshing	11712	3776	640
Shelling	16986	5440	960
Seed input	204132	175396	174617
Total energy input into the system	397860	311268	241177

Source: Om Prakash (1980)

4.9 SOCIO-CULTURAL DIVERSITY

The shifting cultivators observe a number of agricultural rituals. They begin their agricultural efforts according to the fortunate period indicated in Panchang (a religious document issued annually). They use Panchang to refer to a variety of agricultural issues. Land preparation is a key and important aspect of any crop production, and the folklore practise of magi-ulime (deep ploughing) that began during the Hindu calendar's Ugadi festival is still done in some parts of India today. Farmers begin land preparation by completing a land pooja ceremony on the auspicious festival of Ugadi, according to tradition. In order to determine the germination potentiality or viability of the seeds, the farmers employ traditional germination assays. One of the most prevalent tactics used by farmers in this area is to correlate seed germination with a significant event. One of the most essential and common customs during this festival is to sow pulses, cereals, and other seeds in a container that is watered for nine days on the first day. Throughout the nine days, this pot is revered. These pots are regarded as a symbol of growth and prosperity. During the celebrations, the seeds begin to sprout. The seeds mature into seedlings near the end of the 9-day celebrations, large enough to give farmers a good idea of the vigor of the seeds they would use for the following cropping season. The leaves of the Banni tree (Prosopis Cineraria) are distributed among each other during Vijayadashami celebrations. This serves as a symbol of mutual love and respect, as well as a desire to put the past behind us. This tree's blossoms, bark, and pod have therapeutic characteristics and are widely utilized in Ayurveda and traditional medicine. Depending on the location, the Banni tree is also referred to as Shami, Khijro, Jhand, Jat, Khar, Kanda, and Jammi. It is known as Khejri or Khejro in Rajasthan. In the Middle East, it is known as Ghaf.

Even now, traditional methods of animal branding with a hot iron rod are common legend. These animals were healthy and had full access to grain and grass. This folkloric practise allowed each village/region to keep a breeding animal. This method was developed to emulate the regularly used artificial insemination procedure. During Hori habba, these animals were scarified to prevent inbreeding and to introduce fresh animals to enhance genetic diversity. Hori Habba, also known as Hatti Habba or Kobbari Hori Competition, is a rural sport in which hundreds of highly trained and decorated draught animals and bulls are forced to race through large crowds (Shambulingappa and Mansur 2019). Catchers aim to subdue the animals and steal their treasures, which include copra, cash, and gift items. During the Deewali festival, the sports are mostly done in the Karnataka districts of Shivamogga, Haveri, and Uttara Kannada. Similarly, animal branding with a hot iron rod is still a frequent folkloric practise today. People think that branding the region and applying a mixture of turmeric powder, ghee, and cow urine to it helps to keep flies away from the wound. This heat branding procedure has been scientifically demonstrated to prevent skin infections and improve cell-mediated immunity in animals, allowing them to fight off a variety of viral, bacterial, and parasite diseases. Farmers in the area light Jyoti (oil lamps) in front of houses and temples for almost a week in order to attract beneficial insects from farmland to the hamlet. Because insects are significant pollinators in plants during this season, farmers think that lighting Jyoti during Deepawali helps to attract many sorts of insects for pollination.

4.9.1 Land preparation

Land preparation is an essential component of any crop cultivation. Farmers begin land preparation by executing a land pooja (worship) ceremony on the auspicious day of Ugadi, according to tradition. Farmers use this traditional tradition to keep soil moist, encourage termites to eat crop wastes, and promote soil health by reducing water logging and soil erosion during monsoon rains. Slowly, as people began to use modern agricultural practises like as tractors, tillers, and other machinery, the traditional practise fell out of favour and was reduced to a bare minimum to one or two lane ploughing in recent years. This, combined with longer rabi (winter season crops) farming seasons, allows for more crops per acre of land.

4.9.2 Soil health and pest management

The festival Nagapanchami, which takes place in August, places a specific emphasis on soil health. In the name of the snake god, raw milk and ghee were poured into termite-built dirt bunds during Nagapanchami. This draws a variety of ants to the field, which aid in the prevention of termite breeding throughout the harvesting season. Farmers create bunds in the field after harvesting the crop, which is a proactive action conducted in the traditional fashion to prevent termite breeding.

4.9.3 Implications of shifting cultivation

There are two schools of thought on the influence of shifting cultivation on the socio-economic and environmental systems. Shifting cultivation, according to the first school, is a wasteful approach that damages the natural landscape by causing soil erosion and forest depletion. 'Shifting cultivation is a key cause of land degradation that has consistently decreased agricultural output and hence income of the farmers,' according to a 1995 report by the Government of India (Sati and Rinawma 2014). Other researchers, on the other hand, believe it is a vital source of livelihood and a way of life for the people. Scientific investigations have found that the shifting cropping regime makes optimal use of natural resources, which is beneficial to the stability and sustainability of agriculture in the highlands (Ramakrishna, 1993; Sharma, 1992). Controlling shifting agriculture becomes necessary not only to promote environmental quality but also to increase crop output.

4.9.4 Socio-economic implications

Shifting cultivation's economic productivity has been investigated in many ways, with some finding it to be productive and others finding it to be ineffective. The economic productivity of shifting cultivation can be estimated by measuring the number of people employed. Shifting agriculture is beneficial, as it has proven to be incredibly successful in a number of regions with considerable populations (Thrupp *et al.*, 1997). According to the Agriculture Department Report (2009-2010), more than 20% of Mizoram's population is directly or indirectly involved in shifting farming. It is their primary source of revenue for them.

In Kalimantan, Indonesia, shifting cultivation sustains 23 people per square kilometer, more than twice the number supported by commercial farming. In Mesoamerica, Mayan shifting farming supports 100-200 people per square kilometer. Jhum is deeply ingrained in the history and legacy of the north-more east's than 200 tribal races, therefore changing cultivation helps to conserve the region's rich cultural diversity (Tripathi, *et al.*, 2003 a). Shifting agriculture in its traditional form aids agrobiodiversity conservation. It is a cost-effective method of land utilization. It makes the most of the space provided. Around 60 distinct crop varieties are cultivated in a given period and place. In addition, shifting agriculture has a detrimental impact on the economy. Several scholars have labeled shifting cultivation as inefficient and uneconomical from the aspect of production. The main causes were the high frequency and intensity of shifting agriculture. In recent decades, the shifting agriculture fallow period has been extensive. As a result, it wasn't as unsustainable as before (Luoga, 2000a). Because of growing population pressure, high grain demand, and the emergence of urban markets for forest goods, the fallow period for shifting agriculture has been reduced from 20-25 years to 2-3 years (Luoga, 2000b; Mwampamba, 2009).

Beliefs are inseparable elements of a culturally and traditionally bounded society. It refers to the socio-cultural norms of the society inferred through it. It gives a mark of distinctiveness to a community. In fact, it can be said that beliefs are unwritten laws, which keep the society intact. There are substantial numbers of beliefs prevalent among the Assamese and the Oriya people pertaining to the Ambuvaci and the Raja festival. It is the common belief of the people of Assam that Mother Earth undergoes through her menstrual period once in a year. During this time people forfeit themselves from regular activities like ploughing, plantation of seedlings, sowing of seeds, cutting trees, climbing in to it, digging land, even scratching on it also prohibited. Before the advent of the Ambuvaci all clothes and garments except those, which are absolutely necessary, are kept separately and nobody is allowed to touch these during the defilement period. If these are touched or come in contact with the Earth, they are treated to be impure.

4.9.5 Rice and its associated rituals, festivals and beliefs

Rice is conceived not only as an integral part of the Tangkhul (tribe) diet in the everyday life and rituals but it has also become intertwined with traditional cultural production, festivals that celebrate the bounty of the rice harvest and all the processes from early preparation of fields to harvest and post-harvest. Rice rituals are primarily performed to ensure bountiful rice production and food sustainability. The annual calendar of festivals and rituals has many celebrations based on the agriculture cycle of rice cultivation. There is a strong emic values placed on the consumption of rice for the people of Khambi and Nungbi village. A day begins with morning rice and ends with evening rice. There is a common greetings; "Maphazahaira?" which means, "Have you eaten rice?". Rice culture can be seen in people's ways of life, in food habit, medicines, rituals, festivals, myths and the likes. Its production and consumption are expressed extensively in festivals, beliefs, rituals, folksongs, myths and socio-economic relations of the people. For instance, folksongs were composed in praise of rice grains. The myths and rituals that surround planting and harvest attest to the co-evolution of cultivation, culture, and people's life itself. It has enjoyed a unique status in people's life, because of the special qualities it possessed. It is a fact that majority of their traditions and practices centred on paddy cultivations (Tangvah, 2017), which is deeply rooted in the every aspect of the people.

4.9.6 Life cycle rituals

The practice of rituals is central aspect of traditional religion or way of life for shifting cultivators. There is a purpose, function and a meaning behind every ritual action. It is often argued that rituals are either "supporting social structure by directly representing it, or legitimising social authority by concealing it. Thus, ritual's social role is either to bolster, or conceal, the prevailing social order" (Mitchel, 2008). Rice is more than a food to the people. It is integral to their life, be it in festivals, marriage, birth ceremonies and other rituals, it is incomplete without serving rice. The life cycle of the people from birth to death is surrounded by a number of rituals, sacrifices and ceremonies performed at different stages of life, where rice is one of the major sacrificial ingredients, which is served as the only food and rice beer as the only drink.

4.9.7 Annual paddy cultivation rituals and festivals

There are various rituals and rites performed by the people of the present villages, these are mostly offerings to the spirits to protect the paddy and invokes his/her blessing for good harvest of the year. It is agriculture around which most of their festivals and rituals revolve. According to Mills (1937), a Naga's very life depends on his crops. It is therefore not surprising that from childhood to old age he spends most of his day in his fields and that almost all the ceremonies of his religion are designed to protect and increase his crops. Likewise, every stage of jhum and terrace cultivations is marked by the village kaphani/genna (village-wide taboo). All the village festivals are celebrated at the behest of the chief who has an important role to play. He announces the date and reason of the festivals by observing the lunar calendar. To have a succinct knowledge of the people's life and culture, it is essential to know annual rituals and festivals connected with their principal economic activity which is paddy cultivation. Since they depended completely on paddy cultivation for their livelihood they observe series of rituals and festivals connected with it, beginning from the field site selection or soil tilling to harvest and post-harvest activities. Details of the annual rituals and festivals associated with shifting cultivation and terrace cultivations are shown in Table 4.6.

Table 4.6: Shiting cultivatin festivals with annual agricultural calendar

Month	**Tangkhul**	**Name of festival**	**Activity**	**Who performs**
January	Tharao	Ramhao	Search of field site	Men
February	Marun	-	Clearing of forest	Men
March	Mayo	Meikasa-shar	Burning of forest	Men
April	Khaying	Sangkha	For smooth sowing	Men
May	Kharam	Mangkhap	Completion of sowing	Men
June	Makha	Luiphung sakhamatha	Beginning of weeding	Men
July	Morang	Kashongkahao	Protecting crop from animals, pests, diseases	Men
August	Phei	Mawonzai	Completion of weeding	Men
September	Pi	Dharthei Tharshant	Eating of new rice	Men
October	Tharshant	Maakahat	Harvesting	Men
November	Ngaphei	Chumpha	Opening of grainary	Women
December	Khayon	Wonyaithing	Keeping implements used in harvesting at one place.	Men

4.9.8 Seed sowing festival (Ramhao/Luira)

Ramhao or Luira festival which is translated as the 'beginning of movement of the field' is the seed sowing festival and New Year of the Tangkhul Nagas. At present, it is the most popularly celebrated traditional festival. It is celebrated following the lunar calendar during the full moon of January. There are slight variations in date and time of celebration of this festival between jhum and terrace cultivators. The field site searching festival among the shifting cultivators is known as "Ramhao phanit" which literally means "field-beginning festival" or the "beginning of the movement of the field". Among the terraced paddy cultivators, the festival is called "Luira phanit" which means "field-wet festival". The slight difference in their terminology is the reflection of the two different types of cultivations. Ramhao refers to the "dry-land" and "Luira" refers to the "wet-land", but the general principle is the same.

4.9.9 Ramhao festival

To begin the shifting cultivators process 'Ramhao' festival is observed. It is celebrated during the full moon of the first or second week of January that lasted for four days. Village authority headed by Awunga (chief) has to search new field site for the whole village. After the field site is chosen, a date is fixed for ritual ceremony; the purpose of this ritual is to see omens and to invoke to the field god for blessings. On approaching the date, male folk especially married men gathered at chief's house at dawn in the morning and together they would go to the field. After reaching the field, they chose one beautiful tree of medium size somewhere in the middle of the field. After which, the bark of the chosen tree is peeled off by the chief. It is peeled in such a way that the peel-off bark does not cut short, but skilfully peeled to make it long enough. The peel-off bark comes out in a curve shape, like flowers.

Women and children are not allowed to participate in this ritual; they are even tabooed to go to the field on this day. The villagers are well informed not to step into the field during these ritual days. If the omens are considered positive, general announcement has to be made to the villagers by the Luplaka (village mantri). After the proper declaration, the site is publicly opened for cultivation. Traditionally, each household head or the eldest son select plot for his family or sometimes each head of the household clears a token patch. The chief has to select the plot in the first place and in the second place comes, the Khambu1 (next to the chief), the councillors and then the common people. Two days after the necessary rituals, they can start falling off trees. Despite mass conversions to Christianity, Ramhao festival remains the greatest agriculture festival celebrated by the people till today with certain modifications.

4.9.10 Luira festival (Terrace cultivation)

'Luira' festival is observed by the terrace cultivators. In the past 'luira' festival prolonged for a period of ten to fifteen days, which was fixed on a full moon day. Traditionally, on the first day of the festival, the villagers would kill pigs and cattle. Husband or the head of every household mend his hearth properly with doughmud for the purpose of cooking rice and meat for luira ritual, whereas, womenfolk would clean her house inside and outside and their utensils before the festive meal. In the early morning of 'sorei' day or second day, the husband would light up fresh fire to start the preparation of festival foods. On this day all the households would take their morning meal at their courtyard or outside their houses. On the day of the sacrifice, the village Awunga (chief) would take the pig meant for the sacrifice to the sacrificial place. He would dig out the buried grains from the ground and smears or sprinkled over them with blood of pig. Later these seeds were sown on the sacrificial field or house yards. Luira festival still remains the greatest agriculture festival celebrated by the people till today, devoid of any ritual sacrificial parts. Today, Nungbi Khullen village celebrate Luira festival in the month of January (27th-31st). On the first day of the festival, domestic pigs and buffaloes are killed by the villagers meant to consume during the celebration, almost every household would kill a pig or is shared by two to three households.

4.9.11 Field plot burning ritual (Lui-Meikasa)

As the use of fire is an integral part of jhumming cultivation that enables them to manage soil fertility, control weeds and pests in a labour efficient manner, the jhumias of Khambi observed series of genna (village-wide-taboo) known as "kaphani" by the Tangkhuls during the burning ritual. The village hangva (councils) headed by the chief fixed the date with the common consent of all the villagers, after which proper announcement was made by the 'village crier' to the entire village, a week or at least two days ahead, the announcement of the date has to be made. Proper information for firing the plot is required for two reasons: firstly, villagers are to get themselves ready for the rites; secondly, it is the duty of the menfolk to see that the fire does not get out of control and spread to the plot of other neighbouring villages, as firing is done during the months in which the atmosphere becomes very dry and strong winds blow, and thus the danger of the spreading to adjacent areas always remains high. Firing the cleared jungle is considered a man's work, it is like going to battle field so weak, and women are taboo to join them. Likewise, a warrior is to abstain from physical union with his wife in order to keep himself pure and well prepared for any eventualities. Failure to comply with this law would mean death for the man and defeat for the entire village. If discovered the guilty is punished with death (Horam, 1975). No man has the courage to break this law for fear of inviting the wrath of both God and man. Men folk who would go to the field are also tabooed from eating fresh meat, yam, bird, bee and crab. Since, fresh meat contains fresh blood and yam is slippery, it may affect the proper burning of the vegetation. Whereas, eating bee and crab may lead the fire spread beyond the boundary like they fly and crab. There is a particular clan who can light the fire; this ownership is passed down from generation to generation to the next.

4.9.12 Completion of transplantation festival (Mangkhap)

Mangkhap festival is celebrated after the arduous work of transplantation of paddy seedlings in the terrace field and sowing/planting seeds in jhum cultivation. Mankhap festival is also known as rasit festival. It is a festival on which every household finds time to rest and spent time singing and feasting. On the completion of the busiest season of the year everyone is exhausted and as such, they take rest and spend their days in feasting and resting mode. This festival usually falls in the month of July. In the past mangkhap festival is celebrated for five to six days or for a week. During this festival every family would kill their domestic animals like buffalo, cow, pig and chicken which ever they can afford. As a sign of happiness and joy, every family lights up resonated pipe-wood in front of their house on this day, which symbolically signifies that from the dark and gloomy busiest days, they have come to happiness and a bright future (Ruivah, 1993). It is also a festival to drive out the destructive insects and pest. For this, they symbolically kill a cock, observed the omens and the meat is used for ritual sacrifice.

4.9.13 Protection ritual (Kashong Kahao)

Kashong Kahao festival is observed in the month of Marang (July) to protect the standing crops from the danger of insects and pests. The day is fixed by the village priest in consultation with the village council. On the first day, no one is allowed to leave the village. The village priest performs the ceremony outside the village gate by sacrificing a chicken to the deity of fruitfulness, who grows and protects the crops from any danger. After the performance of the ceremony by the priest, a young man considered to be one of the best men physically, mentally and morally, is chosen to cut down the selected tree by a single stroke. If he could fell the tree by a single stroke, it is considered to be a good omen and thus the year is believed to have a bountiful harvest.

4.9.14 Eating new rice festival (Dharthei)

It is a festival that is observes before the harvest. It falls usually in the month of October. In Tangkhul "dhar" means "new" and "thei" means "grain". Thus dharthei is a festival of eating newly harvested rice of the year. During this festival, every household disinfect or cleanse the granary and takes out sickles and chonham (mat) for threshing. On the first day, the chief has to go to the field and bring a basketful of paddy. He performs rituals by appeasing the spirit saying, "heiyun ungmeilu, ewui chumli ungzanglu" meaning, "let the harvest be plentiful, come to my granary". On the second day, head of the family or husband goes to the field and brings a basket of paddy with necessary rituals having performed. After which is then dried on the mat-rack, grinded and cooked. This cooked rice is then eaten by all the family members after the completion of necessary rituals, the meal is considered as sacred so any outsiders other than the family members are not welcome for this meal. It is believed that outsiders may take away the good fortunes of the family. This belief, however, no longer exists today.

4.9.15 Socio-cultural aspects of shifting cultivation

Shifting cultivation is an integral component of the tribal people's socio-cultural life in North-East India, and all of its operations are inextricably related to religious rites and festivals such as Agalmaka, Miamua, Rongchuglu, and Ahiaom. Despite its negative effects on the environment and low productivity, it is still seen as a necessary evil due to the following inherent characteristics, as revealed in other analyses: (a) the majority of labour force management and capital comes from households; (b) production is either consumed on the farm or traded in local markets; (c) decision-making is hampered by limited access to marketing and political institutions; and (d) The majority of farmers live at or below the culturally specified subsistence level. The rationale for the system's persistence is its compatibility with the physico-social environment of sparse community land tenure, undulating and steep topography, short crop cycles, rainy season and subsequent acute moisture stress during the post-monsoon period, meager resources with the farmers, as well as minimal risk and income variability. Some festivals associated with shiftifting cultivation are shown in Table 4.7.

Table 4.7: List of festivals related to shifting cultivation in North Eastern India

State	Festival	Activity
Meghalaya	Achiroka/ Agalmika	It is observed to propitiate God for rains.
	Jamegapovahaua	It is observed at the time of sowing of Seeds. They offer boiled grains of paddy to the God of Harvest to protect the crops and for bumper yield.
	Mangela	It is celebrated after harvest of Jhum produce. Jhum products are offered to the God/ Goddess.
Manipur	Nungchunga	The worship of Nungchungha is permed at the beginning of the agricultural operation.
	Sabuhog	The ceremony is observed in the month of October-November. The worship of Saubuhog is carried out when the paddy stalks appear.
	Senamahi	In this festival the first fruits of the year are offered to God Senamahi by the priest.
Arunachal Pradesh	Mopun/papet	The ceremony is observed in the months of February and April. After 10 days of this festival sowing are performed.
	Etter Festival	The festival is observed in the months of April and June. Etter (Fencing) are completed and winter and autumn paddy and millets etc are sown. The ceremony is observed during mid June to mid August. It is performed for prosperity of cattles and general welfare.
	Solung Aran	The ceremony is observed during mid December to mid February and is popularly known as harvest festival. Arun is celebrated by sacrificing pigs, chickens to propitiate. Goddess "Kenedene (the Goddess of Agriculture and wealth).
	Kopkfmt/ challiwn	It is celebrated before sowing operation
Mizoram	Chapchar Kut	This the most important spring festival held in between the cutting of jungles and burning of Jhums usually in the last part of February and first part of March every year. The meaning of the chapchar kut is chapbamboo, trees are cut down and awaiting for burning of Jhum. Mizo believes that the festival would ensure a good harvest in their Jhum field.
	Powl Kut	It is celebrated when rice harvest is almost reaped. This festival is mainly for the children who are dressed in their best attire and are fed with rice, meat and boiled eggs.
	Min Kut	It is observed during August and September when maiie harvest is reaped. This festival is celebrated with feasting, singing and dancing.
Assam	Rongkher Bishu	Festival of Karbis Festival of Dimasa

Sources: Complied with

1. Shifting Cultivation in North East India, Majumdar, D.N.,ed;Omsons Publication, New Delhi
2. Tribes of Assam, B.N.Bordoloi, G.C. Sharma Thakur, M.C. Saikia published by Tribal Research Institute, Guwahati
3. The Dimasa Kacharis of Assam, B.N.Bordoloi, Tribal Research Institute, Guwahati, Assam

4.9.16 Rituals associated with cultivation

Doing away with the practice and finding other ways of food and livelihood security for the shifting cultivatiors may not be so easy as is said. Studying the practice of swidden agriculture in Andhra Pradesh state of India, Stephen Tyler (1974), observed that it is not just an alternate form of land use or a set of agricultural practices; it implies the whole nexus of people's religious belief, attitudes, self-image and tribal identity. This kind of interconnections between different elements and domains of social life is very much there in case of the shifting cultivators. Among the Naga tribes in general and Angami tribe in particular there are many great cycles of rituals distinctly associated with disease and general well-being of clans and with the cultivation (Das Gupta and Das 1981). In every Angami village there are some main authorities who play main role in matters of many rituals. They are Kemovo (a sort of earth priest), Zhevo, Tsakro (the first sower) and Lidepfu (the first reaper). Tsakro and Lidepfu are particulary connected with agricultural operations. The Tsakro is the old man whose duty is to begin the sowing and unless the Tsakro has inaugurated the sowing, it is a taboo for any other person to start sowing. There may be different Tsakro for different crops. The Lidepfu is the old women who starts reaping of the crop during harvesting.

The community ceremonies of Terhengi and Sekrengi are of great significance as the Terhengi marks the completion of agricultural year and the Sekrengi is responsible for the health of the clan. Another ceremony, Gnongi, marks the completion of sowing in shifting cultivation, The ceremony of Thekrengi marks the completion of transplanting of paddy, Likwengi marks the start of transplanting and Lideh, is observed to indicate opening of the paddy harvest. Among the Angamis there are certain persons who conduct some rituals to produce rain. The rain-making ritual among Angami tribe indicates a sort of 'Earth' and ;Sky' cult. A few drops of water are dropped twice from the leaf. The 'rain-maker' symbolically wears a dress which is a sort of a rain protecting dress made of dried grass.

4.9.17 Head hunting

The practice of head-hunting by the Naga tribes as reported by the Britsh administrators is a historical fact. The dead, whose life-substance was conceived of as forming a continuous cycle of reproduction, passing from men to cereal sown, and then back to man through grain eaten, or the flesh of the animals that had eaten it, was closely associated with crop fertility and village prosperity in all Naga tribes. It is this theory that formed the philosophic basisfor head hunting (Hutton 1928), and while one form of social status was acquired by taking heads and bringing additional life-substance to the community in that way, another was acquired by the distribution of acquired wealthand an attempt thus made by a successful man to inform his family, clan or fellow villagers with his superfluity of that life-substance which was the secret of his success (Hutton 1969). Hutton (1969) also described about a form of sacrifice in which the victim was cut down so that his blood fell from above to the ladder (a notched pole by which the house of the sacrifice was entered). This ceremony was performed by prosperous families for the benefit of the community at the time of harvest. Among the Aos and certain of the Trans-Dikhu tribes human sacrifices were done to ensure good crops. When the vegetation was ready to be put on fire, a slave was used to be tied-up in the middle of them. The vegetation was then lighted and slave was burnt to death. This ensured a good crop, according to the people's belief (Hutton 1969).

4.9.18 The complexity of the problem

Shifting cultivators are not aware of the long-term disadvantages of shifting cultivation, especially when the lands are not too far eroded. When the land becomes eroded over time, they move out to 'new pastures' or become daily wagers (Barthakur 1990). Even if they are told that there land would be eroded its top soil in a decade or even sooner, they might not be convinced. The shifting cultivators would not realize of the gravity of the situation. The isses involved in the problems of shifting cultivation are very complex and multifaceted. Over the years, various working groups and committees, organizations and individual authors have expressed their concern with regard to the problem.

4.9.19 Rain making ritual

Among some tribes, there are authorities who conduct certain rituals to produce rain. It indicates a sort of 'Earth' and 'Sky' cult. 'Rain maker' symbolically wears a rain protecting dress made of dried grass. This is done to make the rain God happy so that rains are in plenty.

4.9.20 Fertility cult

The dead were closely associated with the fertility of the soil and the prosperity of the village in Naga tribes, whose life substance was conceived as forming a continuous cycle of reproduction, passing from men to cereals sown and then back to man through grains eaten or the flesh of the animals that had eaten it. It is the theory on which head hunting is based (Hutton, 1928). Human sacrifices were performed among the Ao tribe to ensure bountiful crops (Hutton, 1969).

4.9.21 Soil and water conservation

Some tribal societies believe in soil and water conservation. Though the tribal societies had poor knowledge in scientific conservation of soil and water their natural wisdom made them conscious of their conservation and regeneration. They have village land having sacred forests, which are thought to be the abode of spirits and cutting of vegetation or timber is considered a taboo. Among Chakhesang and Angami tribes, terracing of fields is done to ensure best use of available water and control soil erosion through run-off. Care is taken to regulate water in and out the fields. There are three types of terraced fields to regulate water flow viz. (i) Dzutse, the water supply to these fields is regular throughout the year and water which is not needed in one terrace is conveyed to the next through channels; (ii) Khuso, where water supply depends from streams or from overflows of Dzutse and (iii) Vakhra, involving much labour for regulating water as approach for water is relatively difficult.

4.9.22 Ethnic minorities in Vietnam and shifting agriculture

The 'ethnic minorities' of Vietnam can be split into 54 groups, 50 of which use shifting farming. Ethnic minorities have a population of around 9 million people. According to data from 1989, there were 2,879,685 persons involved in shifting cultivation, living in 482,512 families. Approximately 3.5 million hectares are under shifting agriculture. Shifting agriculture is practiced by 7% of persons in the Tay ethnic group, 16% of the Nung ethnic group, 45% of the Thai ethnic group, and 100% of the Thai ethnic group (almost all the remaining ethnic groups). The ethnic groups who practice shifting farming in Vietnam are typically found in altitudinal belts. Shifting agriculture is primarily practiced in the Central Highlands of south Vietnam. The terrain remains somewhat fertile, with exceptionally fertile basalt soil strewn across the Central Highlands. Water rice cultivation is still foreign to the ethnic tribes of the Central Highlands, as is the use of dung as manure. Previously, the majority of ethnic minorities had a self-sufficient economy. Except for ethnic groups living at low altitudes, such as the Tay, Nung, Muong, and others, there was very little commodities exchange. Apart from shifting cultivation, the people's lives were intertwined with the forest through the collection of minor forest products and fuelwood, hunting, fishing, and a variety of handicrafts.

4.9.23 The traditional Dayak Jalai farming system

The Dayak Jalai's survival strategy has always included farming. Their farming expertise and skills have been passed down through the years. Farmers have traditionally had a mixed system in which their rice farms provided their main food while their rubber plantations offered revenue to cover their other expenses. The family's livelihood was ensured by this combination of farming and rubber tapping. In addition to farming, the Dayak Jalai could readily supplement their diet with various food sources such as edible ferns, mushrooms, bamboo shoots, and other forest plants that were common in their daily diet. Forests also provided food for wild creatures such as boars and deer. During one cycle of farming, the phases of physical farming activities are detailed below.

1. Menyapat tahun and menyandam, two traditional ceremonies, precede Menabas. Menabas refers to the clearance of bushes and small trees in a region to allow for more land clearing for farming. The land's border is usually delineated prior to clearance. This exercise can be done by the entire family with machetes.
2. Menyakat is chopping down smaller trees (less than 10 centimetres in diameter) and cleaning the plot of clumps of bamboo trees. Axes and huge machetes are used by both men and women in menyakat.
3. Menabang is chopping down the larger trees. Menabang has become easier and faster with the use of chain saws. Menabang is usually carried out by adult males.
4. Meradaq is chopping branches and stems from trees that have already been felled to allow for effective burning in the following phase. Meradaq is mainly carried out by men with axes and machetes.
5. Hamparan Raban is a period of time during which farmers cease working to allow the cut plants to dry and prepare for burning.
6. Meladang is constructing the farm's perimeter firebreaks. Firebreaks are usually two to five metres long and run along the edge of the land. When the drought is very severe and the forests are easily combustible, the width of the firebreak is increased by two metres. Everyone in the family participates in this activity.
7. Mencucul is the burning of plants on a farm plot that has dried out. The farm owner studies and anticipates the direction of the wind before lighting the fire. To guarantee that the fire does not grow too large, it is started from the side of the farm that faces the opposite direction of the wind.
8. Memanit is piling the remaining unburned foliage and cutting and burning it. Both men and women participate in this hobby.
9. Menugal is the simultaneous spreading of rice seeds, corn, vegetables, and other annual crops by men and women. Men use a dibble stick to

construct seed-sowing holes. The ladies, on the other hand, sow rice seeds in the holes.

10. Menggurun is removing weeds from the entire farm plot that are obstructing rice development. 11. Betanam hubiq-tabuq is the practice of growing crops other than rice, mainly cassava and sugar cane, which are not planted at the same time as rice in order to avoid cassava and sugar cane growth interfering with rice development.
12. Memansau is removing bamboo shoots from the farm plot.
13. Behanyiq is harvesting rice, which is a female-dominated activity. Men assist in transporting harvested rice baskets to paddy barns.

4.9.24 Dimensions of social systems

1. Shifting cultivation systems (SCS) are often intimately tied to broader indigenous values, knowledge systems and spiritual practices. Properly understanding indigenous SCS therefore requires an understanding of this broader ethnological context. We cannot take an exclusively economic approach in order to understand motivations for adopting particular practices, and must adopt ethnological and ethnobotanical methods that will allow us to appreciate the broader values and meanings of SCS to their practitioners.
2. In order to fully benefit from the lessons of existing swidden systems, we must understand the indigenous knowledge underlying their management. This will include eliciting indigenous taxonomic systems and the relation of these taxonomies to, among other things, the selection of domesticated species for cultivation and the management of naturally occurring species. In many cases, it will also be important to understand and learn from indigenous knowledge of ecological and climatic indicators, for example, naturally occurring species that indicate high- or low-quality soils.
3. Customary institutions dimension relates to the social organization of technical knowledge, and the collective management of shifting cultivation. Focusing on this dimension will allow us to understand indigenous decision-making patterns and processes, and to appreciate how local people arrive at individual and collective decisions about the management and use of natural resources.
4. Shifting cultivators constantly search for novel variations and manage a dynamic portfolio of practices and plant varieties. Shifting cultivation research should engage this dynamism by identifying not only existing

knowledge, but also the ways in which swidden farmers generate further knowledge and refine their existing management of natural and agricultural resources.

5. Shifting cultivation research should incorporate studies of how long-standing practices, as well as novel innovations, are communicated intra- and inter-generationally. This will allow us to develop an understanding of how people learn from each other and share information about natural and agricultural resources.
6. A vital part of any agricultural system is the exchange of goods and services. In the contemporary context, research into agricultural transactions will need to consider both global and local value chains and understand existing and potential access to world and local markets.

4.9.25 Socio-ecological systems approach to agricultural research for development

All too often in the past, otherwise nobly intended and meticulously designed development programmes have failed because the knowledge and experience of local populations has been ignored. Nowadays, scientists are studying and incorporating the experience-based knowledge and insights of upland swidden cultivators into their research designs. Two realizations have driven this new trend. First, the knowledge of specific micro-environments acquired by local populations over centuries of learning by discovery has intrinsic value and, in fact, it may be the only existing knowledge of some complex micro-environments. Second, knowing and understanding the concepts and language forms of indigenous knowledge enables scientists and technicians to communicate more effectively with local people.

Building on these positive developments, further efforts should be made to encourage the closing of knowledge gaps between scientists and indigenous peoples by encouraging more extensive and intensive interaction in the field. There is also a need to promote effective interaction between agents of government bureaucracies dealing with health, irrigation, natural resources and agriculture and local leaders who hold specialized indigenous knowledge in these areas and are willing to share it. To make this possible, the indigenous knowledge of local specialists must be documented and scientists and government officers trained not only in its use, but also in effective methods of acquiring more indigenous knowledge. The framework described above, which focuses on six ecological and six social dimensions of the swidden socio-ecological system, could help us to carry out this documentation of indigenous swidden knowledge, as well as to identify divergences in government, research and local perspectives. Our socio-ecological systems framework can therefore provide the basis for coordinating focused interactions between stakeholders with differing perspectives, and facilitating innovation processes that engage multiple stakeholders. Integrating indigenous knowledge with scientific knowledge can help us to better understand shifting cultivation and develop innovative solutions for the sustainable intensification of agriculture. Indeed, many of the core issues identified by the Humid tropics programme – especially the marginalization of upland peoples and environmental degradation that has followed the replacement of diverse traditional agricultural and forestry systems with modern monocrop plantations – are issues that can be addressed through greater engagement with indigenous technical knowledge with regard to shifting cultivation.

4.9.26 Culture and ritual in Iban shifting cultivation

Rice production is intertwined with the worldview, values, and social organization of Malaysia's indigenous Iban shifting growers (Majid, 1983). All components of shifting agriculture, from appeasing the "spirits of the soil forest" with the manggo rite before clearing to ceremonies linked with the storage of produced rice, are infused with ritual and religion. The rice is considered sacred in and of itself. Various rituals performed before and during harvesting ensure that the paddy's spirit is not scared away, that there will be enough rice for the following year, and that the crop will be plentiful and easy to harvest. Rice is also picked so that the reaper's paddy spirit does not become disoriented. "Rice is not just a staple food; it has a spirit, a soul, and the proper rituals must be followed in order to win the esteem and favor of the paddy spirits, for a plentiful supply of grains" (Majid, 1983). Because the Iban regard the world as one shared with various orders of beings, each of which plays a critical role in the success of their agricultural, land use practices and religious rites are tightly linked. Changes to fanning systems unavoidably influence features of native cultures because of this tight interconnectedness.

The economy of the shifting cultivators may conveniently classified into hunting, nomadism, pastoralism, shifting cultivation and, now, settled cultivation in some cases (Tyler 1974). The homo-sapiens had been under the awe of the natural objects, such as; fountains, rivers, forests etc. from the pre-historic age. Human being very small in number, primitive in technology and their needs were extremely small. Slowly and slowly their number increased, technology improved and their need multiplied (Sinha 1981). Such a development was fraught with depletion of natural resources, causing ecological balance, creating numerous hazards to the human settlements.

Quite a large number of studies have been conducted with regard to socio-economic aspects of shifting cultivation and almost all point out the socio-economic problems associated with change over from this age-old practice. However, many of theses studies do not suggest some tangible solutions to the problems. It is very easy to say that any solution to the problems of shifting cultivation should not disturb their social values and customs. What is important is how to achieve this. Moreover, all solutions offered may not be perfect and useful for different areas of the globe where shifting cultivation is practiced. Technologically also there is inadequacy of information. Knowledge about proper land use, reliable data on soil and nutrients loss on various slopes and climatic conditions, production technology on terraced condition under a particular or specific situation, suitable cropping patterns to replace shifting cultivation, desirable farming systems based on proper management practices, is below satisfactory level. Another aspect which we bsselieve is necessary to mention is that, by and large, all the areas or countries where shifting cultivation is followed are relatively poor and high level of research was not possible to identify the exact solutions to the socio-economic problems. Except a few good studies, most of the work is either repetition or below desired level and impracticable to extend beyond the site where it was conducted.

(i) Land tenure system

In many areas of Northeast India, bulk of the land is commonly owned by the people. The power of distribution of land to the villagers for cultivation is usually vested in village headman or the village council. The distribution of land for shifting cultivation is done mainly according to need and availability of family labour in a household. In Khasi hills of Meghalaya state, the areas are divided into 'Rajs' which are ruled by Syiems (King), Lyngdoh (Priest), Pator etc., who are administrators of each Raj. Each member of the Raj is entitled to cultivate as much land as he pleases. As long as he continues to cultivate, the land belongs to him. If a man does not cultivate land under his possession for three consecutive years, the land is reverted back to Raj and some other person from the Raj can cultivate it (Gohain 1967). In Garo hills of Meghalaya state, the hill mouzas are divided into several Akhings. The land is in the name of Maharis or clans. These lands are called 'Akhings' and is managed by the Nokma or head of the clan, who technically owns the land on behalf of his clansmen. For cultivation, plots are allotted among the clan members belonging to the Akhing. The Mizo tribe of Mizoram state had a system of hereditary Chieftainship in the villages. The 'Lal' (village chief) used to own the land during British period. But, now, the Lalship has been abolished and the management of the village land is vested in the village council which s an elected body. The land for shifting cultivation is now distributed by the village council to the cultivators (Borkotoky and Saha, 1968). The karbis (Karbi tribe) of Assam state practise community ownership of the land. A Karbi can cultivate a plot from this area according to his choice. This tribe usually shift their home and hearth when they move to a new place for shifting cultivation. If the village becomes very big or if there is any dispute between two groups of people of the same village, that cannot be settled, the aggrieved group move to a new place under a new leader of their choice. The new village is named after new headman (Saikia 1968). In Arunachal Pradesh, land is the roperty of village community, however, individual ownership is also recognized. An outsider can cultivate a plot of land with the permission of the headman of the village and payment of the rent (Dutta and Borkotoky 1972). Of late, private ownership of land has been there, mostly confined to urban land, land under horticultural crops and settled cultivation.But suitable areas for settled cultivation, particularly wet rice cultivation is very limited. In parts of Meghalaya state, the practice of leasing out of land is found where wet rice or other forms of settled cultivation is there. When the fallow period is cut to four years, the time-averaged aboveground carbon stock drops by 88-90 percent, while soil organic carbon drops by 0-27 percent. A switch to continuous yearly planting resulted in a 95-99 percent loss in aboveground carbon and a 13-40 percent reduction in soil carbon.

References

Abizaid, C., Coomes, O.T., 2004. Land use and forest fallowing dynamics in seasonally dry tropical forests of the southern Yucatán Pensinsula, Mexico. Land Use Policy 21, 71–84.

Adedipe NO. (1984). Environmental Considerations of Shifting Cultivation in Africa and the Task ofUniversities, A.H. Bunting and E. Bunting (Eds.). Proceedings of the International Workshop on Shifting Cultivation: Teaching and Research at the University Level, July 4-9, 1982. University of Ibadan, Nigeria. Food and Agricultural Organization, Rome, 192.

Ahrends, A., Hollingsworth, P.M., Ziegler, A.D., Fox, J.M., Chen, H., Su, Y., Xu, J., 2015. Current trends of rubber plantation expansion may threaten biodiversity and livelihoods. Glob. Environ. Change 34, 48–58.

Aide, T. M. and Cavelier, J. 1994. Barriers to lowland tropical forest restoration in the Sierra Nevada de Santa Marta, Columbia. Restoration Ecology 2: 219-229.

AIPP and IWGIA, 2014 Shifting Cultivation, Livelihood and Food Security - New and Old Challenges for Indigenous Peoples in Asia. Published by Asia Indigenous Peoples' Pact (AIPP) and the International Work Group for Indigenous Affairs (IWGIA), Chiang Mai 2014

AIPP/IWGIA/IKAP (2009). Shifting cultivation and climate change. Briefing paper for UNFCCC Intersessional Meeting, Bangkok. AIPP, IWGIA, IKAP.

Alam, M. K., & Mohiuddin, M. (2009). Shifting cultivation (jhum) agrobiodiversity at stake: Bangladesh situation. ISHA Acta Horticulturae, 806, 709-716. Doi: 10.17660/ActaHortic.2009.806.88

Albrecht, A., Kandji, S.T., 2003. Carbon sequestration in tropical agroforestry systems. Agriculture, Ecosystems and Environment 99, 15–27.

Alcorn JB. 1990. Indigenous agroforestry strategies meeting farmers' needs. In: Anderson A, editor. Alternatives to deforestation: steps towards sustainable use of the amazon rain forest. New York (NY): Columbia University Press.

Allan, W. 1972a. Ecology, techniques and settlement patterns. In Ucko, P.J., Tringham, R. and Dimbleby, G. W. Man Settlement and Urbanism . London: Duckworth & Co. Ltd. pp. 211 - 226.

Anderson, E. F. 1993. Plants and People of the Golden Triangle: Ethnobotany of the Hill Tribes of Northern Thailand. Dioscorides Press, Portland.

Andrade G.I. and Rubio-Torgler H. (1990): Sustainable use of the tropical rain forest: evidence from the avifauna in a shifting cultivation in habitat mosaic in the Colombian Amazon. Conservation Biology, 8: 545-554.

Andriesse, J.P., 1980. Nutrient level changes during a 20-year shifting cultivation cycle in Sarawak (Malaysia). In: K.T. Joseph (Editor), Proc. Conf. on Classification and Management of Tropical Soils, 1977. Malaysian Society of Soil Science, Kuala Lumpur, pp. 479-49 I.

Angelsen, A. 1996. "Shifting Cultivation and Deforestation: A Study from Indonesia." World Development. 23 (10): 1713-1729.

Angelsen. A. 1994 Shifting Cultivation and "Deforestation" II Study from Sumatra, Indonesia. Chr'. Michelsen Institute.

Anonymous. 2015. Land use and Jhum cultivation in Manipur. planningmanipur.gov.in/ pdf/ MSDR/Chapter%206_Land.pdf.

Aryal, K.P. and E.E. Kerkhoff (2008). The right to practice shifting cultivation as a traditional occupation in Nepal. A case study to apply ILO Conventions Nos. 111 (Employment and Occupation) and 169 (Indigenous and Tribal Peoples), Kathmandu, International Labour Organization.

Aryal KP, Kerkhoff EE, Maskey N, Sherchan R (2010). Shifting Cultivation in the Sacred Himalayan Landscape: A Case Study in the Kanchenjunga Conservation Area. WWF Nepal, Kathmandu.

Asia. Westview Press, Boulder, CO, pp. 226-240. Conklin, H.C., 1961. The study of shifting cultivation. Curr. Anthropol., 2(1): 27-61.

Attiwill, P.M., 1994. The disturbance of forest ecosystems: the ecological basis for conservative management. Forest Ecol. Manag. 63, 247–300.

Babin D. (ed., 2004). Beyond tropical deforestation: From tropical deforestation to forest cover dynamics and forest development. Unesco/Cirad, Paris.

Balick, M., A. Anderson, and P. May. 1991. The Subsidy from Nature. New York: University of Columbia.

Balla MK, Awasthi KD, Shrestha PK, Sherchan DP, Poudel D (2000). Degraded Lands in Mid-Hills of Central Nepal: A GIS Appraisal in Quantifying and Planning for Sustainable Rehabilitation. LIBIRD.

Balmford A, Bond W. Trends in the state of nature and the implications for human well-being. Ecol Letts 2005; 8: 1218-1234.

Barbier, E. B. (2004). Agricultural expansion, resource boom and growth in Latin America: implications for long-run economic development. World Development, Vol. 32, No. 1, 137-157.

Barthakur IK, 1990. A strategy to reduce and replace shifting cultivation. Paper presented at North-Eastern Regional seminar on 'Land use planning and watershed management for control of shifting cultivation' held at Dimapur, Nagaland, 8-9 March, 1990, Ministry of Agriculture, Department of Agriculture and Cooperation, Government of India.

Bawa, K.S., Kress, W.J., Nadkarni, N.M., 2004. Beyond Paradise-meeting the challenges in tropical biology in the 21th century. Biotropica 36, 276–284.

Beckerman, S. 1987. Swidden in Amazonia and the Amazon Rim. In Turner, B. L. and Brush, S. B. (eds.. Comparative Farming Systems. New York: Guilford Press.

Behera R. N., Nayak D. K., Andersen P. and Maren I. E. 2015 From jhum to broom: Agricultural land-use change and food security implications on the Meghalaya Plateau, India. Ambio. doi: 10.1007/s13280-015-0691-3.

Bene, J.G., Beall, H.W., and de J. Hart, R.A. 1977. Trees, Food, and People: Land Management in the Tropics. Ottawa: IDRC.

Benneh, G. 1972. Systems of agriculture in tropical Africa. Economic Geography 48(3): 244 – 257

Biswal, D. K., & Kumar, S. (2013). Shifting cultivation and policies of sustainable development: A meaningless obsession. Journal of Economic and Social Development, 9(1), 132-138.

Bodley, J. H. 1976. Anthropology and Contemporary Human Problems . Menlo Park, California: Benjamin/Cummings.

Borkotoky MD and Saha N, 1968. Hmunpui – Socio-economiv survey of a Mizo village. Report of Agro-Economic Research Centre, Jorhat.

Bose S, 1967. Carrying capacity of land under shifting cultivation. The Asiatic Society, Calcutta.

Brookfield H, Padoch C. 1994. Appreciating agrodiversity: a look at the dynamism and diversity of indigenous farming practices: Environment. 36:6–11.

Brookfield H. C. 2001, Exploring agrodiversity. Columbia University Press, New York City, NY.

Brooks TM, Pimm SL, Collar NJ. Deforestation predicts the number of threatened birds in insular Southeast Asia. Conservation Biol 1997; 11: 382-394.

Brovkin V, Boysen L, Arora VK, Boisier JP, Cadule P, Chini L, Claussen M, Friedlingstein P, Gayler V, Van Den Hurk BJJM, Hurtt GC, Jones CD, Kato E, Noblet-Ducoudre ND, Pacifico F, Pongratz J, and Weiss M, 2013. Effect of Anthropogenic Land-Use and Land-Cover Changes on Climate and Land Carbon Storage in CMIP5 Projections for the Twenty-First Century. American Meteorological Society, DOI: 10.1175/ JCLI-D-12-00623.1

Brown S., 1997, Estimating biomass and biomass change of tropical forests: a primer. FAO Forestry Paper 134. Rome: FAO.

Brown, S. & Lugo, A. E. (1990). Tropical secondary forests. Journal of Tropical Ecology, Vol. 6, No. 1, 1-32.

Brown, S., C.A.S. Hall, W. Knabe, J. Raich, M.C. Trexler, and P. Woomer. 1993. Tropical forests: Their past, present and potential future role in the terrestrial carbon budget. Water, Air, Soil Pollut. 70:71–94.

Bruun TB, de Neergaard A, Lawrence D and Ziegler AD (2009). Environmental Consequences of the Demise in SwiddenCultivation in Southeast Asia: Carbon Storage and Soil Quality.. Human Ecology 37: 375-388, DOI 10.1007/s10745-009-9257-y

Bruun, T.B., Egay, K., Mertz, O., Magid, J., 2013. Improved sampling methods document decline in soil organic carbon stocks and concentrations of permanganate oxidizable carbon after transition from swidden to oil palm cultivation. Agric. Ecosyst. Environ. 178, 127–134.

Bryant, R. L. 1997. Scientific Forestry: Control and Resistence, 1956-1881. InThe Political Ecology ofForestry in Burma 1824-1994, pp. 43-76. Honolulu: University of Hawai'i Press.

Caballero, J., Pulido, M.T., Martinez, A., 2004. El uso de la palma de guano (Sabal yapa) en la industria turı´stica de Quintana Roo, Mexico. In: Alexiades, M., Shanley, P. (Eds.),

Productos forestales, medios de subsistenciay conservacion. Estudios de caso sobre sistemas de manejo de productos forestales no maderables, Indonesia, 365–385.

Cai, Z., G. Xing, X. Yan, H. Xu, H. Tsuruta, K. Yagi, et al. 1997. Methane and nitrous oxide emissions from rice paddy as affected by nitrogen fertilizers and water management. Plant Soil 196:7–14.

Cairns M and Garrity D P. 1999. Improving shifting cultivation in South East Asia by building on indigenous fallow management strategies. Agrofor. Syst. 47: 37–48.

Cairns M. F. ed., 2007. Voices from the forest: integrating indigenous knowledge into sustainable upland farming. Earthscan, London.

Cairns M. F. ed., 2015. Shifting cultivation and environmental change: Indigenous people, agriculture and forest conservation. Earthscan, London.

Caringal, AM; Panganiban, MG. 2008 'Ecological features of swidden fragments in the biodiversity corridor of Southern Batangas, Luzon Island'. Journal of Nature of Studies 7(1): 193–202

CBD (Convention on Biological Diversity) (2010). Strategic plan for biodiversity 2010-11.

Chave, J. et al. Improved allometric models to estimate the aboveground biomass of tropical trees. Glob. Change Biol. 20, 3177–3190 (2014).

Chave, J. et al. Tree allometry and improved estimation of carbon stocks and balance in tropical forests. Oecologia 145, 87–99 (2005).

Chazdon R. 1998. Tropical f orests (log 'em or leave 'em). Science 281: 1295–1296.

Chazdon, R. L., Guariguata, M.R. 2016. Natural regeneration as a tool for large-scale forest restoration in the tropics: prospects and challenges. Biotropica, 48: 844–855.

Chazdon, R. L., Peres, C. A., Dent, D., Sheil, D., Lugo, A. E., Lamb, D., Stork, N. E. & Miller, S. E. (2009). The potential for species conservation in tropical secondary forests. Conservation Biology, Vol. 23, No. 6, 1406-1417

Chin Y W, Balunas M J, Chai H B and Kinghorn A D .2006. Drug discovery from natural sources. AAPS J 8: 239–53.

Chin, (1985) 'Do shifting cultivators deforest', in V. Shiva (ed.) Forest Resources: Crisis and Management (1992), Natraj Publisher: Dehradun.

Chokkalingam, U., De Jong, W., Smith, J. & Sabogal C. 2000. Tropical secondary forests in Asia: introduction and synthesis. Paper prepared for the "Tropical secondary forests in Asia: Reality and perspectives' workshop". 10–14 April 2000. Samarinda, Indonesia. Center for International Forestry Research, Bogor, Indonesia.

Chokkalingam, U. and de Jong, W. 2001. Secondary forest: a working definition and typology. International Forestry Review.

Chokkalingam, U. et al. One century of forest rehabilitation in the Philippines: Approaches, outcomes and lessons. Center for International Forestry Research (CIFOR), Bogor, Indonesia (2006).

Choudhury D. 2006 Equitable resource access under shifting cultivation systems in Northeast India. In: ANGOC and ILC. Enhancing Access of the Poor to Land and Common Property Resources. A Resource Book. Asian NGO Coalition for Agrarian Reform and Rural Development, Quezon City, Philippines, and International Land Coalition, Rome.

Choudhury D. 2012 Why do jhumiyas jhum? Managing change in shifting cultivation areas in the uplands of north-eastern India. In: Sumi Krishnan (ed.) Agriculture in a Changing Environment: Perspectives on North-eastern India. Routledge, New Delhi. pgs. 78-100.

Choudhury, D., Ingty, D. and Jamir, Sentikemba 2003 Managing marginalisation in shifting cultivation areas of Northeast India: Examples of community innovations and initiatives. In: Tang Ya and Pradeep M. Tulachan (eds.) Mountain Agriculture in the Hindu Kush Himalayan Region. ICIMOD, Kathmandu, pgs. 207-212.

Conklin HC 1957. "Hanunoo Agriculture: A Report on an Integral System of Shifting Cultivation in the Philippines." FAO Development Paper No. 12. Rome: FAO.

Conklin, H. 1975. Hanunóo agriculture: A report on an integral system of shifting cultivation in the Philippines. FAO Forestry Development Papers, No. 12. Rome: Food and Agriculture Organization

Conklin, H. C. 1980. Ethnographic Atlas of the Ifugao: A Study of Environment, Culture and Society in Northern Luzon. New Haven and London: Yale University Press.

Connell, J. H. 1978. Diversity in tropical rain forests and coral reefs. Science 199: 1302-1310.

Coomes, O. T., Takasaki, Y. & Rhemtulla, J. (2011). Land-use poverty traps identified in shifting cultivation systems shape long-term tropical forest cover. Proceedings of the National Academy of Sciences of the United States of America, Vol. 108, No. 34, 13925-13930.

Coomes, O.T., Grimard, F., Burt, G.J., 2000. Tropical forests and shifting cultivation: secondary forest fallow dynamics among traditional farmers of the Peruvian Amazon. Ecological Economics 32, 109–124.

Coomans M. 1987. Manusia Dayak, Dahulu, Sekarang, Masa Depan. Jakarta: Gramedia [Dayak people, past, present, future. Jakarta: Gramedia].

Costanza R, d'Arge R, de Groot R, Farber S, Grasso M, Hannon B, Limburg K, Naeem S, O'Neill RV, Paruelo J, Raskin RG, Sutton P, van den Belt M. The value of the world's ecosystem services and natural capital. Nature 1997; 387: 254-260.

Cowgill, UM 1960. Agriculture -and population density in the southern Maya lowlands. 7 pp., duplicated. New Haven. x

Cramb, R., 1998. Environment and development in the Philippine uplands: the problem of agricultural land degradation. Asian Studies Review, 22: 289-308.

Cramb R A, Colfer C J P, Dressler W, Laungaramsri P, Trung L Q, Mulyoutami E, Peluso N L, Wadley R L. 2009. Swidden transformations and livelihoods in South East Asia. Hum. Ecol. doi: 10.1007/s10745-009-9241-6.

Cuaron AD (2000). A global perspective on habitat disturbance and tropical rainforest mammals. Conserv. Biol., 14: 1574–1579.

Darlong, Vincent D. 2004. To Jhum or Not to Jhum. Guwahati, Assam: The Missing Link (TML-India).

Das Gupta PK and Das NK, 1981. Socio-cultural aspect of cultivation and soil conservation in Northeast India. Paper presented in a seminar on 'Socio-economic aspects of soil conservation activities with special reference to Northeastern India' held at Tura, Meghalaya, India, on 20-21 June, 1981.

Dash, B. (2006, July). Shifting cultivation among the tribes of Orissa. Orissa Review, 76-84.

Davidson, E.A., P.A. Matson, and P.D. Brooks. 1996. Nitrous oxide emission controls and inorganic nitrogen dynamics in fertilized tropical agricultural soils. Soil Sci. Soc. Am. J. 60:1145–1152.

Davidson, E.A., M. Keller, H.E. Erickson, L.V. Verchot, and E. Veldkamp. 2000. Testing a conceptual model of soil emissions of nitrous and nitric oxide. BioScience 50:667–680.

de Jong, W, van Noordwijk,M, Sirait, M., Liswanti, N, Suyanto. 2001. Farming secondary forests in Indonesia. Journal of Tropical Forest Science.

de Rouw A (1995). The fallow period as a weed-break in shifting cultivation (tropical wet forests). Agric, Ecosyst. Environ., 54: 31–43.

DeFries, R.S., C.B. Field, I. Fung, G.J. Collatz, and L. Bounoua. 1999. Combining satellite data and biogeochemical models to estimate global effects of human-induced land cover change on carbon emissions and primary productivity. Global Biogeochem. Cycles 13:803–815.

Dent, D. H. & Wright, S. J. (2009). The future of tropical species in secondary forests: a quantitative review. Biological Conservation, Vol. 142, No. 12, 2833-2843.

Descola, P. (1994). In the society of nature: A native ecology in Amazonia. Cambridge: Cambridge University Press.

Detwiler, R. P., and Hall, C. A. S. 1988. Tropical forests and the global carbon budget, Science 239: 42–47.PubMedCrossRefGoogle Scholar

Dewar, R.E. 1997. Were People Responsible for the Extinction of Madagascar's Subfossils, and How Will We Ever Know? In: Natural Change and Human Impact in Madagascar (eds. S.M. Goodman and B.D. Patterson), pp. 364–377. Smithsonian Institution, Washington D.C.

DeWasseige C, Flynn J, Louppe D, Hiol Hiol F and Mayaux P 2014 The Forests of the Congo Basin—State of the Forest 2013 (Belgium: Weyrich) (http://observatoire-comifac.net/?l=en)

DeWasseige C, Marcken P, de, Bayol N, Hiol Hiol F, Mayaux P, Desclée B, Nasi R, Billand A, Defourny P and Eba'a Atyi R 2012 The Forests of the Congo Basin—State of the Forest 2010 (Luxembourg: Publications Office of the European Union) (http://carpe.umd.edu/Documents/2010/SOF_2010_ EN_Intro.pdf)

Dingkuhn M, Johnson DE, Sow A, Audebert AY (1999). Relationships between upland rice canopy characteristics and weed competitiveness. Field Crops Res., 61:79–95.

Dlugokencky, E. J., B. P. Walter, K. A. Masarie, P. M. Lang, and E. S. Kasischke (2001), Measurements of an anomalous global methane increase during 1998, Geophys. Res. Lett., 28, 499– 502, doi:10.1029/ 2000GL012119.

Dove MR. 1983. Theories of Swidden agriculture and the political economy of ignorance. Agroforestry Syst. 1:85–99.

Dove, M. 1985. Swidden Agriculture in Indonesia: The Subsistence Strategies of the Kalimantan Kantu'. Mouton, Berlin.

Dove, M. R. (1993). A revisionist view of tropical deforestation and development. Environmental Conservation 20(1), 17-24, 56.

Dove MR (1996). So far from power, so near to the forest: a structural analysis of gain and blame in tropical forest development, In: Padoch C, Peluso NL (eds.), Borneo in Transition: People, Forests, Conservation, and Development, Oxford University Press, Kuala Lumpur, pp. 41-58.

Dressler W, Wilson D, Clendenning J, Cramb R, Mahanty S, Lasco R, Keenan R, To P and Gevana D. 2015. Examining how long fallow swidden systems impact upon livelihood and ecosystem services outcomes compared with alternative land-uses in the uplands of Southeast Asia. Working Paper 174. Bogor, Indonesia: CIFOR.

Dressler, W. et al. 2017. 'The impact of swidden decline on livelihoods and ecosystem services in Southeast Asia: A review of the evidence from 1990 to 2015' in Ambio, April 2017, 46(3): 291-310.

Driessen, P.M., Buurman, P. and Permadhy, 1976. The influence of shifting cultivation on a "podzolic" soil from Central Kalimantan. In: Peat and Podzolic Soils and their Potential for Agriculture in Indonesia. Soil Research Institute, Bogor, pp. 95-115.

Du Puy, D.J. and J. Moat. 1996. A Refined Classification of the Primary Vegetation of Madagascar Based on the Underlying Geology, Using GIS to Map its Distribution and to Access its Conservation Status. In: Biogrographie de Madagascar (ed. W.R. Lourenco), pp. 205–218. ORSTOM, Paris.

Ducourtieux, O., Laffort, J.-R., and Sacklokham, S. 2005. Land Policy and Farming Practices in Laos. Development and Change, 36(3): 499-526.

Dufiles, J.M. 2003. Remaining Forest Cover. In: The Natural History of Madagascar (eds. S.M. Goodman and J.P. Benstead), pp. 88–96. University of Chicago Press, Chicago.

Dutta SK, Singh D, Sood A (2011) Effect of soil chemical and physical properties on sorption and desorption behaviour of lead in different soils of India. Soil Sediment Contam 20:249–260. https://doi.org/10.1080/15320383.2011.560979

Dutta J, Gangadharappa N R and Biradar G S. 2014b. Livelihood status of tribal people practicing shifting (jhum) cultivation in Tripura state of North-East India. Trop. Agric. Res. 25(3): 316–26.

Eaton JM, Lawrence D (2009). Loss of carbon sequestration potential after several decades of shifting cultivation in the Southern Yucata. For. Ecol. Manage., 258: 949–958.

Ellen, R. F. 1982. Environment, subsistence and system. Cambridge: Cambridge University Press.

EPA (Environmental Protection Agency). 1990. Greenhouse gas emission from agricultural systems, vols. 1 and 2. Washington, DC: EPA.

Epprecht, M., Minot, N., Messerli, P., and Heinimann, A. (2008). Poverty and inequality in the Lao PDR: Spatial patterns and geographic determinants. Geographica Bernensis, Bern.

Erni C. 2021 Persistence and Change in Customary Tenure Systems in Myanmar. MRLG Thematic Study Series # 11. POINT MRLG, Yangon.

Erskine, P.D., Lamb, D., Bristow, M., 2006. Tree species diversity and ecosystem function: Can tropical multi-species plantations generate greater productivity? Forest Ecology and Management, 233, 205–210.

EWEL, J. 1980. Tropical succession: manifold routes to maturity. Pp. 2–7 in Ewel, J. (Ed.) Tropical succession. Biotropica, supplement to Volume 12(2).

Escamilla A, Sanvicente M, Sosa M, Galindo-Leal C (2000). Habitat mosaic, wildlife availability and hunting in the tropical forest of Calakmul, Mexico. Conserv. Biol., 14: 1592–1601.

FAO 'Report of the sixth session of the committee on (1982) forestry' (Rome, 3-7 May 1982), FAO, Rome, Italy.

FAO (1995): Alternatives to Shifting Cultivation.

FAO (2001) Woodfuel production and marketing in Myanmar. Regional Wood Energy Development Programme Report No. 56, FAO, Bangkok, Thailand.

FAO (Food and Agriculture Organization). 1997. Estimating biomass and biomass change of tropical forests: A primer. FAO Forestry Paper 134. FAO, Rome, Italy.

FAO, 1996. Forest Resource Assessment 1990. FAO Forestry Paper, vol. 130. FAO, Rome.

FAO (2001a). State of the World's Forests. FAO, Rome.

FAO, 2003. Trade reforms and Food Security. Conceptualizing the linkages. Rome:

Fearnside, P.M., and W.M. Guimaraes. 1996. Carbon uptake by secondary forests in Brazilian Amazonia. For. Ecol. Manage. 80:35–46.

Finegan, B. 1996. Pattern and process in neotropical secondary rain forests: the first 100 years of succession. TREE 11: 119-124.

Fischer, A., Vasseur, L., 2000. The crisis in shifting cultivation practices and the promise of agroforestry: a review of the Panamanian experience. Biodiversity and Conservation 9.

Folster, H., 1986. Nutrient loss during forest clearing. In: R. Lal, P.A. Sanchez and R.W. Cummings, Jr. (Editors), Land Clearing and Development in the Tropics. A.A. Balkema, Rotterdam, pp. 241-256.

Forman R.T.T., 1995, Land mosaics: the ecology of landscapes and regions. Cambridge, UK: Cambridge University Press.

Found WE 1987 Shifting cultivation in Jamaica: an invisible dilemma. Development Discussion Paper. No. 238, Harvard institute for International Development, Harvard University.

Fox J, 2002. Understanding a dynamic landscape: Land use, land cover, and resource tenure in Northeastern Cambodia, in S. Walsh and K. Crews-Meyer (Eds) Linking People, Place, and Policy: A GIScience Approach. Boston: Kluwer Academic Publishers, pp. 113-130

Fox J, Fujita Y, Ngidang D, Peluso N L, Potter L, Sakuntaladewi N, Sturgeon J and Thomas D. 2009. Policies, Political- economy, and swidden, in Southeast Asia. Hum. Ecol. doi: 10.1007/ s10745-009-9240-7.

Fox J, Truong DM, Rambo AT, Tuyen NP, Cuc LT, Leisz S. Shifting cultivation: a new old paradigm for managing tropical forests. BioScience. 2000; 50:521–528.

Fox J. 2000. How blaming slash and burn farmers is deforesting mainland South East Asia. https://www.researchgate. net/29738770_How_blaming_'slash_and_ burn'_farmers.

Fox, J., Castella, J.C., Ziegler, A.D., 2014. Swidden, rubber and carbon: Can REDD+ work for people and the environment in Montane Mainland Southeast Asia. Global Environmental Change, 29, 318-326.

Freeman, D. 1970. Report on the Iban. London School of Economics Monographs of Social Anthropology, No. 41, University of London, The Athlone Press, New York: Humanities Press Inc.

Fujisaka, S., C. Castilla, G. Escobar, V. Rodrigues, E.J. Veneklaas, R.J. Thomas, et al. 1998. The effects of forest conversion on annual crops and pastures: Estimates of carbon emissions and plant species loss in a Brazilian Amazon colony. Agric. Ecosyst. Environ. 69:17–26.

Fujisaka, S., Escobar, G., Veneklaas, E.J., 2000. Weedy fields and forests: interactions between land use and the composition of plant communities in the Peruvian Amazon. Agriculture, Ecosystems and Environment 78 (2), 175–186.

Fujisaka, S. 1991. A Diagnostic Survey of Shifting Cultivation in Northern Laos: Targeting Research to Improve Sustainability and Productivity. Agroforestry Systems 13: 95–109.

Gadgil M. and Guha R. (1992): This fissured land. Oxford University Press, Delhi.

Gafur A, Borggaard OK, Jensen JR, Peterson L (2003). Run off and loses of soil and nutrients from watershed under shifting cultivation (Jhum) in the Chittagong Hill Tracts of Bangladesh. J. Hydrol., 279:293-309.

Gafur, A., Koch, C.B., Borggaard, O.K., 2004. Weathering intensity controlling sustainability of ultisols under shifting cultivation in the Chittagong Hill Tracts of Bangladesh. Soil Sci. 169, 663–674.

Galbraith JK (1999). The Affluent Society. Penguin Publishers, USA, (ISBN13: 9780140285192), 276p.

Ganguly J. B. 1969. Economic problems of the jhumias of Tripura, Bookland Pvt. Ltd., Calcutta,

Geddes, W. R. 1976. Migrants of the Mountains: The Cultural Ecology of the Blue Miao (Hmong Njua) of Thatland. Oxford Univ. Press, Ely House, London.

Geertz. C. 1974 Agricultural involution: The process of ecological change in Indinesia. University of California Pres, Berkley.

Gerold G., Fremerey M. and Guhardja E. (eds, 2004). Land use, nature conservation and the stability of rainforest margins in Southeast Asia. Springer, Berlin.

Ghosh, A.K. (1992) Changing Scenario of Forests in Tripura. Indian Forester., 118:754-762.

Gladwin, C. H. 1983. Contributions of decision-tree methodology to a farming systems program. Human Organization , 42(2): 146 - 157.

Gliessman, S. R. 1985. Economic and ecological factors in designing and managing sustainable agroecosystems. In Edens, T. C., Fridgen, C. and Battenfield, S. L. (eds.) Sustainable agriculture and integrated farming systems. East Lansing: Michigan State University Press. pp. 56 - 63.

Godfrey, L.R., W.L. Jungers, K.E. Reed, E.L. Simons and P.S. Chatrath. 1997. Subfossil Lemurs: Inferences about Past and Present Primate Communities in Madagascar. In:

Natural Change and Human Impact in Madagascar (eds. S.M. Goodman and B.D. Patterson), pp. 218–256. Smithsonian Institution, Washington D. C.

Gohain D, 1967. Mawtnum – Socio-economic survey of a village in Khasi hills. Report of Agro-Economic Research Centre, Jorhat.

Gómez-Pompa A., Whitmore T.C. and Hadley M. (eds, 1991). Rain forest regeneration and management. Unesco, Paris.

Goswami. P. C. 1985. "Importance of Socioeconomic Factors and Role of Incentives in Controlling Shifting Cultivation in Northeast India." Indian Forester 111:1-11.

Goudie, A. 1984. The Human Impact: Man's role in Environmental Change . Oxford: Blackwell.

Grange, I., Kansuntisukmongkol, K., 2003. Effect of fallow length on soil structure, hydraulic properties, and soil organic C in a swidden cultivation system of western Thailand. Trop. Agric. 80, 246–251.

Gregorio, N., Herbohn, J., Harrison, S., Smith, C. 2015. A systems approach to improving the quality of tree seedlings for agroforestry, tree farming and reforestation in the Philippines. Land Use Policy, 47: 29–41.

GRET 2019 Farming Systems in Myanmar: Methodological Background and Synthesis of Field-based Studies across 5 States and Regions of Myanmar. LIFT, Yangon.

Gupta M D. 1994. Continuity and change in tribal societies: A study of the economic aspects of tribal jhumia society in Tripura. (In) Sangma M B (Ed) Essays on North East India, Indus publishing company, New Delhi, pp 147–64.

Hadley, M. and Lanly, J-P 1983. Tropical forest ecosystems: identifying differences seeking similarities. Nature and Resources 19: 1 (Jan-March): 2 - 19.

Hariyadi B. and Ticktin T. 2012 From shifting cultivation to cinnamon agroforestry: Changing agricultural practice among the Serampas in the Kerinci Seblat National Park, Indonesia. Human Ecology 40: 315-325.

Harrison, S., Emtage, N.F., Nasayao, E.E., 2004. Past and present forestry support programs in the Philippines, and lessons for the future. Small-scale Forest Economics, Management and Policy, 3, 303-317.

Hattori, D., Sabang, J., Tanaka, S., Kendawang, J.J., Ninomiya, I., Sakurai, K., 2005. Soil characteristics under three vegetation types associated with shifting cultivation in a mixed dipterocarp forest in Sarawak, Malaysia. Soil Sci. Plant Nutr. 51, 231–241.

Heinimann A, Hett C, Hurni K, Epprecht M, Messerli P, Joegesnsen L, Breu T 2013. Socio□ economic perspectives on shifting cultivation landscapes in Northern Laos. Human Ecology. 41 (1):51□62

Herbohn, J. L., Vanclay, J. K., Ngyuen, H., Le, H.D., Harrison, S.R., Cedamon, E., Smith, C., Firn, J., Gregorio, N.O., Mangaoang, E., Lamarre, E. 2014. Inventory procedures for smallholder and community woodlots in the Philippines: methods initial findings and insights. Small-scale Forestry, 13: 79–100.

Hett C, Castella J, Heinimann A, Messerli P and Pfund J. 2012. A Landscape Mosaics Approach for Characterizing Swidden Systems from a REDD+ Perspective. Applied Geography 32(2): 608–618.

Holden S. 2001: A century of Technological Change and Deforestation in the Miombo woodland of Northern Zambia. In: Angelsen, A and Kaimowitz, D (Editors). Agricultural Technologies and Tropical Deforestation. CABI and CIFOR publications, pp. 251-269

Hong E 1990. 'Swidden as a Rational Agricultural System' in Return to the Good Earth, Damaging Effects of Modern Agriculture and the Case for Ecological Farming, Third World Network. Malaysia.

Hossain, M.A. 2011. An overview on shifting cultivation with reference to Bangladesh. Scientific Research and Essays, 6: 6509-6514. DOI: 10.5897/SRE11.1282.

Houghton RA, Skole DL, Nobre CA, Hacker JL, Lawrence KT, Chometowski WH 2000 Annal fluxes of carbon from deforestation and regrowth in the Brazilian Amazon. Nature 403:301–304

Houghton, R., 2012. Carbon emissions and the drivers of deforestation and forest degradation in the tropics. Current Opinion in Environmental Sustainability, 4, 597-603.

Houghton, R.A. 1997. Terrestrial carbon storage: Global lessons from Amazonian research. Ciencia Cultura 49:58–72.

Houghton, R.A., R.D. Boone, J.R. Fruci, J.E. Hobbie, J.M. Melillo, C.A. Palm, et al. 1987. The flux of carbon from terrestrial ecosystems to the atmosphere in 1980 due to changes in land use: Geographic distribution of the global flux. Tellus 39B:122–139.

Hurni K, Hett C, Heinimann A, Messerli P, Wiesmann U. 2012 Dynamics of Shifting Cultivation Landscapes in Northern Lao PDR between 2000 and 2009 Based on an Analysis of MODIS Time Series and Landsat Images. Hum Ecol.,41: 21–36. doi: 10.1007/s10745-012-9551-y

Hutton, J.H. 1928 The significance of head – hunting in Assam. J. Royal Anthropological Inst., 58:12-32.

Hutton, J.H.(1969). The Angami Nagas. Oxford University Press, Bombay, India.

IBSRAM (International Board for Soil Research and Management). (1987). Report offhe First Regional Seminar on Management of Vertisols under Semi-Arid Conditions in Africa and Southwest Asia, Nairobi, Kenya, 1-6 December 1986. Bangkok, Thailand: IBSRAM.

Ichikawa, M. 1993. Diversity and selectivity in the food of the Mbuti hunter-gatherers in Zaire. In (C.M. Hladik, A. Hladik, O.F. Linares, H. Pagezy, A. Semple& M.Hadley, eds.) Tropical Forests, People and Food: Biocultural Interactions and Applications to Development, pp. 487-496, UNESCO, Paris and The Parthenon Publishing Group, Lanes and New York.

IPCC (Intergovernmental Panel on Climate Change). 2001. Climate change 2001: The scientific basis. J.T. Houghton, Y. Ding, .D.J. Griggs, M. Nogues, P.J. van der Linden, K. Dai, et al. (eds.). Cambridge Univ. Press, Cambridge.

IWGIA. 2014. Shifting cultivation, livelihood and food security new and old challenges for indigenous peoples in Asia. International Work Group for Indigenous Affairs. AIPP Shifting cultivation livelihood security. www.iwgia.org./0694.

Jahn, R. & Asio, V. B. Climate, geology, geomorphology and soils of the tropics with special reference to Leyte Islands (Philippines). In: 8th International Seminar and Worshop on Tropical Ecology. Visaya State College of Agroculture, Baybay, Leyte, 25–43 pp (2001).

Janzen D. 1998.Gardenification of wildland nature and the human footprint. Science 279: 1312–1313.

Janzen, D.H. 1973. Tropical agro ecosystems. Science, 182, 1212-1219.

Jarosz, L. 1993. Defining and explaining tropical deforestation: shifting cultivation and population growth in colonial Madagascar (1896–1940). Economic Geography 69(4):366–379.

Jordan, C. F. 1985. Nutrient cycling in Tropical Forest Ecosystems . New York: John Wiley and Sons.

Jordan, C. F.1982. Amazon Rain Forests. American Scientist 70: 394 - 401.

Jordan, C.F. (Editor), 1989. An Amazon Rain Forest: The Structure and Function of a Nutrient Stressed Ecosystem and the Impacts of Slash-and-Burn Agriculture. Man and Biosphere Series, Vol. 2. UNESCO, Paris, 176 pp.

Juo ASR, Manu A (1996). Chemical dynamics in slash-and-burn agriculture. Agricul. Ecosyst. Envion. 58(1): 49-60.

Justice C, Wilkie D, Zhang Q, Brunner J and DonoghueC2001 Central African forests, carbon and climate change Clim. Res. 17 229–46

Kalame, F. B., Aidoo, R., Nkem, J., Ajayie, O. C., Kanninen, M., Luukkanen, O. and Idinoba, M. (2011) 'Modified taungya system in Ghana: A win–win practice for forestry and adaptation to climate change?', Environmental Science and Policy 14(5), pp519-530

Kammesheidt, L. 1999. Forest recovery by root suckers and above-ground sprouts after slash-and-burn agriculture, fire and logging in Paraguay and Venezuela. Journal of Tropical Ecology 15: 143-157.

Kasischke, E. S., and L. P. Bruhwiler (2002), Emissions of carbon dioxide, carbon monoxide, and methane from boreal forest fires in 1998, J. Geophys. Res., 107, 8146, doi:10.1029/2001JD000461. [printed 108(D1), 2003]

Kasischke ES, Hyer EJ, Paul C. Novelli PC, et al. (2005). Influences of boreal fire emissions on Northern Hemisphere atmospheric carbon and carbon monoxide. Global Biogeochemical Cycles, 19: 1-16, GB1012, doi:10.1029/2004GB002300, 2005

Kauffman J.B., Cummings D.L., Ward D.E. & Babbit R., 1995, Fire in the Brazilian Amazon: 1. Biomass, nutrient pools and losses in slashed forests. Oecologia, 104: 397-408.

Keesing, R.M. & Strather 1998 (3rd. cd.) Cultural Anthropology: ../ Contemporary Prospective. Harcourt Brace College Publication.

Keller, M. 1986. Emissions of N2O, CH4 and CO2 from tropical forest soils. J. Geophys. Res. 91:11791–11802.

Kennard, D. K. 2002. Secondary forest succession in a tropical dry forest: patterns of development across a 50-year chronosequence in lowland Bolivia. Journal of Tropical Ecology 18: 53-66.

Kenzo, T., Ichie, T., Hattori, D., Kendawang, J.J., Sakurai, K., Ninomiya, I., 2010. Changes in above- and belowground biomass in early successional tropical secondary forests after shifting cultivation in Sarawak, Malaysia. Forest Ecology and Management, 260, 875–882.

Kerkhoff, E., Erni, C. (eds) (2005) Shifting cultivation and wildlife conservation: A debate in indigenous affairs, 2/05: 22-29. (International Work Group for Indigenous Affairs), Copenhagen

Kerkhoff E, Sharma E (2006). Debating Shifting Cultivation in the Eastern Himalayas. Farmers' Innovations as Lessons for Policy. International Centre for Integrated Mountain Development (ICIMOD), Kathmandu.

Kettle, C.J., 2010. Ecological considerations for using dipterocarps for restoration of lowland rainforest in Southeast Asia. Biodiversity and Conservation, 19, 1137-1151.

Kharas, H and Rogerson A. 2012. Horizon 2025: Creative destruction in the aid industry. London: Overseas Development Institute.

Kirby, K., 2005. Land-use in the Neotropics: regional-scale predictors of deforestation and local effects on carbon storage and tree-species diversity. Masters of Science Thesis. Department of Biology, Neotropical Environments Option. McGill University Montreal. 117 pp.

Knight, C. G. 1980. Ethoscience and the African farmer: Rationale and strategy (Tanzania). In Brokensha, D., Warren, D. M. and Werner, O. (eds.) Indigenous Knowledge Systems and Development. Lanham: University Press of America. pp. 203 - 230.

Koh, L.P., 2007. Impending disaster or sliver of hope for Southeast Asian forests? The devil may lie in the details. Biodiversity and Conservaton, 16, 3935–3938.

Komatsu K. 1998. The food cultures of the shifting cultivators in central africa: the diversity in selection of food materials. African Study Monographs, Suppl. 25: 149- 177.

Kotto-Same J., Woomer P.L., Moukam A. & Zapfack L., 1997, Carbon dynamics in slash-and-burn agriculture and land use alternative of the humid forest zone in Cameroon. Agriculture, Ecosystems & Environment, 65, 245-256.

Kramer R, van Schaik C, Johnson J (eds.) (1997). Last stand: protected areas and the defense of biodiversity. Oxford University Press, London.

Krautkraemer, J. A. (1994). Population growth, soil fertility, and agricultural intensification. Journal of Development Economics, Vol. 44, No. 2, 403-428.

Kricher J.C. and Davis W.E. (1992): Patterns of avian species richness in disturbed and undisturbed habitats of Belize, Pp. 240-246 in J. M. Hagan and D. W. Johnston, editors. Ecology and conservation of neo-tropical migrant land birds. Smith Sorian Institution Press, Washington, D. C.

Kumar, S. (2008). Report of the Inter-Ministerial National Task Force on Rehabilitation of Shifting Cultivation Areas.

Kummer, D.M. 1992. Upland agriculture, the land frontier and forest decline in the Philippines. Agroforestry Systems, 18: 31-46.

Kunstadter, P. 1978. Ecological modification and adaptation: an ethnobotanical view of Lua' swiddeners in north-western Thailand. Pp. 168–200 in Ford, R. I. (Ed.) The Nature and Status of Ethnobotany. Anthropological Papers 67, Ann Arbor, Michigan.

Kunstadter,, P. 1983b. Animism, Buddhism and Christianity: religion in the life of the Lua people of Pa Pae, north-western Thailand. Pp. 135–154 in McKinnon, J. & Bhruksasri, W. (Eds.) Highlanders of Thailand. Oxford University Press, Kuala Lumpur.

Lal, R. 1987. Needs for, approaches to, and consequences of land clearing and development in the tropics. In Tropical land clearing for sustainable agriculture, Proc. No. 3 IBSRAM, Bangkok, Thailand, pp. 15-28.

Lal, R. 1989. Agroforestry systems and soil surface managment of a tropical alfisol. Agroforestry systems 8: 1 – 6.

Lal, R, 2005. Encyclopedia of Soils in the Environment, Pages 488-497

Lalengzama C and Kanagaraj E (2012) Shifting Cultivation to Settled Agriculture : Agrarian Change and Tribal Development in Mizoram. Journal of Social Worrk and Social Development, 5: 14-43.

Langrand, O. and L. Wilmé. 1997. Effects of Forest Fragmentation on Extinction Patterns of the Endemic Avifauna on the Central High Plateau of Madagascar. In: Natural Change and Human Impact in Madagascar (eds. S.M. Goodman and B.D. Patterson), pp. 280–305.

Lanly, J. P. 1985. "Defining and Measuring Shifting Cultivation." Unasylva 37(147): 17-21.

Lasco, R. D. & Pulhin, F. B. Carbon budgets of forest ecosystems in the Philippines. J. of Environ. Sci. Manage. 12, 1–13 (2009).

Lasco, R. D., Visco, R. G. & Pulhin, J. M. Secondary forests in the Philippines: formation and transformation in the 20th century. J. Trop. For. Sci. 13, 652–670 (2001).

Lasco, R.D., Guillermo, I.Q., Cruz, R.V.O., Bantayan, N.C., Pulhin, F.B., 2004. Carbon stocks assessment of a secondary forest in Mount Makiling Forest Reserve, Philippines. Journal of Tropical Forest Science, 16, 35-45.

Lasco, R.D., MacDicken, K.G., Pulhin, F.B., Guillermo, I.Q., Sales, R.F., Cruz, R.V.O., 2006. Carbon stocks assessment of a selectively logged dipterocarp forest and wood processing mill in the Philippines. Journal of Tropical Forest Science, 18, 166–172.

Lawrence D, Radel C, Tully K, Schmook B, Schneider L. Untangling a decline in tropical forest resilience: Constraints on the sustainability of shifting cultivation across the globe. Biotropica. 2010; 42.1: 21–30.

Lawrence DL, Peart DR and Leighton M, 1998. The impact of shifting cultivation on a rainforest landscape in West Kalimantan: spatial and temporal dynamics. Landscape Ecology 13: 135–148.

Lawrence, A., 1997. Kaingin in the Philippines: is it the end of the forest? Rural Development Forestry Network Paper 21, London, UK.

Lawrence, D. 2004. Erosion of tree diversity during 200 years of shifting cultivation in Bornean rainforest. Ecological Applications 14: 1855-1869.

Lawrence, D., 2005. Biomass accumulation after 10-200 years of shifting cultivation in Bornean rain forest. Ecology, 86, 26-33.

Lawrence, D., P. D'Odorico, L. Diekmann, M. DeLonge, R. Das, and J. Eaton. 2007. Ecological feedbacks following deforestation create the potential for a catastrophic ecosystem shift in tropical dry forest. Proceedings of the National Academy of Sciences USA 104:20696–20701.

Lawrence, D., Radel, C., Tully, K., Schmook, B., Schneider, L., 2010b. Untangling a decline in tropical forest resilience: constraints on the sustainability of shifting cultivation across the globe. Biotropica 42, 21–30.

Le, H.D., Smith, C., Herbohn, J., 2014. What drives the success of reforestation projects in tropical developing countries? The case of the Philippines. Global Environmental Change, 24, 334–348.

Le, H.D., Smith, C., Herbohn, J., 2015. Identifying interactions among reforestation success drivers: A case study from the Philippines. Ecological Modelling, 316, 62-77.

Leduc, B. and Choudhury, D. 2012 Agricultural transformations in shifting cultivation areas of Northeast India: Implications for land management, gender and institutions. In: Dev Nathan and Virginius Xaxa (eds.) Social Exclusion and Adverse Inclusion: Development and Deprivation of Adivasis in India. Oxford University Press, New Delhi, pgs. 237-258.

Leimgruber P, Kelly DS, Steininger MK, Brunner J, Müller T and Melissa S, 2005. Forest cover change patterns in Myanmar (Burma) 1990•2000. Environmental Conservation 32: 356-364, doi:10.1017/S0376892905002493

Leimgruber, P., Gagnon, J.B., Wemmer, C, Kelly, D.S., Songer, M.A. & Selig, E.R. (2003) Fragmentation of Asia's remaining wildlands: implications for Asian elephant conservation. Animal Conservation 6: 347•359.

Lestrelin G., Vigiak O., Pelletreau, Keohavong B. and Valentin C. 2012 Challenging established narrative on soil erosion and shifting cultivation in Laos. Natural Resources Forum 36: 63-75.

Lewis, S.L., Edwards, D.P., Galbraith, D., 2015. Increasing human dominance of tropical forests. Science, 349: 827-832.

Liang L, Shen L, Yang W, Yang X, Zhang Y (2009). Building on traditional shifting cultivation for rotational agroforestry: Experiences from Yunnan. China For. Ecol. Manage., 257: 1989–1994.

Luizao, F., P. Matson, G. Livingston, R. Luizao, and P. Vitousek. 1989. Nitrous oxide flux following tropical land clearing. Global Biogeochem. Cycles 3(3):281–285.

Luoga E.J. (2000 a&b): Subsistence use of wood products and shifting cultivation within miombo woodland of eastern Tanzania, with special notes on commercial uses. South African Journal of Botany, 66: 72-85.

Lynam, A.J. (2003) A national tiger action plan for the Union of Myanmar. Wildlife Conservation Society, New York, USA: 57 pp.

Lynch Owen J, 1992b. "Colonial Legacies in a Fragile Republic: a History of Philippine Land Law and State Formation." New Haven: Yale University Law School (Doctoral Dissertation).

Maithani B P. 2005. Shifting cultivation in North-East India-policy issues and options. Mittal Publications, New Delhi.

Majid, Zuraina. 1983. "Malaysia." 157-238. InAtal Y. and P. L. Bennagen, eds. Swidden Cultivation in Asia, Vol. 2, Y.. Bangkok: UNESCO Regional Office for Education in Asia and the Pacific.

Maki F, Mamoru K, Thein HM and Minn Y, 2007. Recovery Process of Fallow Vegetation in the Traditional Karen Swidden Cultivation System in the Bago Mountain Range, Myanmar. Southeast Asian Studies, 45: 317-333.

Malhi, Y., Gardner, T.A., Goldsmith, G.R., Silman, M.R., Zelazowski, P., 2014. Tropical forests in the Anthropocene. Annual Review of Environment and Resources, 39, 125-159.

Malhi, Y., Wood, D., Baker, T.S., Wright, J., Phillips, O.L. et al., 2006. The regional variation of aboveground live biomass in old-growth Amazonian forests. Global Change Biology, 12, 1107-1138.

Matson, P.A., R. Naylor, and I. Ortiz-Monasterio. 1998. Integration of environmental, agronomic, and economic aspects of fertilizer management. Science (Washington, DC) 280:12–115.

May, P.H., Anderson, A.B., Fraza˜o, J.M.F., Balick, J.M., 1985. Babassu palm in the agroforestry systems in Brazil's Mid-North region. Agroforest. Syst. 3, 275–295.

Mayaux P et al. 2013 State and evolution of the African rainforests between 1990 and 2010 Phil. Trans. R. Soc. B 368 20120300

Mazoyer M. and Roudart L. (2005), A History of world agriculture: from the Neolithic age to current crisis. Monthly Review Press, New York City, NY.

Mehta PS, Sharma AK, Negi KS. (2010) Indigenous knowledge system and sustainable development with particular reference to folklore practices in Kumaon hills, Uttarakhand. Indian Journal for Traditional Knowledge.; 9(3):547-550.

Menzies, N. (1988) 'Three hundred years of taungya: A sustainable system of forestry in south China', Human Ecology 16(4), pp361-376

Mertens B. & Lambin E.F., 2000, Land-cover change trajectories in southern Cameroon. Annals of the Association of American Geographers, 90 (3), 467-494.

Mertz O, Christensen E A, Hojoskov P Thomsen B T. 1999. Subsistence or cash: strategies for change in shifting cultivation system. https://tidsskrift.dk/index.php/geografisktidsskrift/article/view/2566 /4569.

Mertz O, Padoch C, Fox J, Cramb RA, Leisz SJ, Lam NT, et al. (2009) Swidden change in Southeast Asia: understanding causes and consequences. Hum. Ecol; 37.3: 259–264.

Mertz O, Wadley R L, Nielsen U, Bruun T B, Colfer C J P, Neergaard D A, Jepsen R M, Martinussen T, Zhao Q, Noweg G T and Magid J. 2008. A fresh look at shifting cultivation: fallow length an uncertain indicator of productivity. Agricultural Systems. doi:10.1016/j.agsy.2007.06.002.

Mertz, O., Müller, D., Sikor, T., Hett, C., Heinimann, A., Castella, J.C., Lestrelin, G., Ryan, C.M., Reay, D.S., Schmidt-Vogt, D., Danielsen, F., Theilade, I., van Noordwijk, M., Verchot, L.V., Burgess, N.D., Berry, N.J., Pham, T.T., Messerli, P., Xu, J., Fensholt, R., Hostert, P., Pflugmacher, D., Bruun, T.B., de Neergaard, A., Dons, K., Dewi, S., Rutishauser, E., Sun, Z. 2012. The forgotten D: challenges of addressing forest degradation in complex mosaic landscapes under REDD+. Danish Journal of Geography, 112: 63-76.

Messerli, P., Heinimann, A., Epprecht, M., and Minot, N. (2008). Socio-economic Atlas of Lao PDR—an Analysis Based on the 2005 Census. Geographica Bernsis, Bern.

Metzger, M.J., Rounsvell, M.D.A., Acosta-Michlik, L., Leemans, R., Schroter, D., 2006. The vulnerability of ecosystem services to land-use change. Agriculture, Ecosystems & Environment, 114, 69-85.

Dove, M. R. (1993). A revisionist view of tropical deforestation and development. Environmental Conservation 20(1), 17-24, 56.

Mills, J. P. (1937). The Rengma Nagas. London: Macmillan.

Ministry of Environment and Forest (1997): The state of the forest report – 1997. Forest Survey of India, Government of India, Dehradun.

Molinario G, Hansen MC and PotapovPV, 2015. Forest cover dynamics of shifting cultivation in the Democratic Republic of Congo: a remote sensing-based assessment for 2000–2010. Environ. Res. Lett. 10 094009 doi:10.1088/1748-9326/10/9/094009

Monela G.C. and Abdallah J.M. (2007): External Policy Impacts on Miombo Forest Development in Tanzania. In: Dube, Y.C. and Schmithusen, F. (Editors). Cross-sectoral Policy Development in Forestry. FAO and CABI publications, pp.117-125.

Montagnini, F., Nair, P.K.R., 2004. Carbon sequestration: an underexploited environmental benefit of agroforestry systems. Agroforestry Systems 61–62, 281–295.

Moorman FR and Kang BT, 1978. Microvariability of soils in the tropics and its agronomic implications with special reference to West Africa: In diversity of soils in the tropics. ASA Special PubL. 34: 29-43.

Moran, E. F. 1981. Developing the Amazon. Indiana University Press: Bloomington.

Mukul SA, 2015. Shifting cultivation in the upland secondary forests of the Philippines: Biodiversity and carbon stock assessment, and ecosystem services trade-offs in land-use decisions. A thesis submitted for the degree of Doctor of Philosophy at The University of Queensland, School of Agriculture and Food Sciences.

Mukul SA and Herbohn J, 2016. The impacts of shifting cultivation on secondary forests dynamics in tropics: A synthesis of the key findings and spatio temporal distribution of research. Environmental Science & Policy 55: 167–177.

Mukul SA, Herbohn J and Firn J, 2016. Tropical secondary forests regenerating after shifting cultivation in the Philippines uplands are important carbon sinks. Sci. Rep. 6. 22483, https://doi.org/10.1038/srep22483.

Mwampamba H.T. (2009): Forest Recovery and Carbon Sequestration under Shifting Cultivation in the Eastern Arc Mountains, Tanzania: Landscape and Land use Effects. Un published dissertation submitted in partial satisfaction of the requirements for the degree of Doctor of philosophy in Ecology in the Office of graduate studies of the University of California.183 p.

Myers N. 1989. Deforestation rates in tropical forests and their climatic implications. London: Friends of the Earth.

Myers, N., 1992. The Primary Source: Tropical Forests and Our Future. W.W. Norton, New York, 416 pp.

Myers, N., Mittermeier, R.A., Mittermeier, C.G., da Fonseca, G.A.B. and Kent, J., 2000. Biodiversity hotspots for conservation priorities. Nature, 403, 853-858.

NAEB (National Afforestation and Eco development Board). 2008. Report of the inter-ministerial national task force on rehabilitation of shifting cultivation areas, naeb.nic.in/.../ Shifting%20Cultivation%20Task%20Force%20Repo.

Nakano K, 1978. An Ecological Study or Swidden Agriculture at a Village in Northern Thailand. South East Asian Studies, Vol. 16: 411-446.

Namgyel U, Siebert SF, Wang S (2008). Shifting Cultivation and Biodiversity Conservation in Bhutan. Conserv. Biol., 22:1349–1351, DOI: 10.1111/j.1523-1739.2008.01019.x.

Nath AJ, Brahma B, Lal R and Dass AK, 2016. Soils under jhum cultivation in Eastern Himalayas, Encyclopedia of Soil Science, Third Edition DOI: 10.1081/E-ESS3-120053748

Naughton-Treves L, Salafsky N (2003). Wildlife conservation in agroforestry buffer zones: opportunities and conflict. In G. Schroth, editor. Agroforestry and biodiversity conservation in tropical landscapes. Columbia University Press, New York.

Naughton-Treves, L., 2004. Deforestation and carbon emissions at tropical; frontiers: a case study from the Peruvian Amazon. World Development 32, 173–190.

Neergaard, A., Magid, J., Mertz, O., 2008. Soil erosion from shifting cultivation and other smallholder land use in Sarawak, Malaysia. Agric. Ecosyst. Environ. 125, 182–190.

Nelson, R. and N. Horning. 1993. AVHRR-LAC estimates of forest area in Madagascar, 1990. International Journal of Remote Sensing 14:1463–1475.

Ngo, K.M., Turner, B.L., Muller-Landau, H.C., Davies, S.J., Larjavaara, M., Nik Hassan, N.F.,Lum, S., 2013. Carbon stocks in primary and secondary tropical forests in Singapore. Forest Ecology and Management, 296, 81–89.

Nguyen, H., Lamb, D., Herbohn, J., Firn, J., 2014. Designing Mixed Species Tree Plantations for the Tropics: Balancing Ecological Attributes of Species with Landholder Preferences in the Philippines. PLoS ONE, 9, e95267.

Njomgang R, Yemefack M, Nounamo L, Moukam A and J. Kotto-Same J, 2011. Dynamics of Shifting Agricultural Systems and Organic Carbon Sequestration in Southern Cameroon. TROPICULTURA, 29: 176-182

North D. 1990. Institutions, institutional change and economic performance. Cambridge University Press, New York.

NRSC (National Remote Sensing Centre). 2011. Wasteland Atlas of India, Department of Land Resources. dolr.nic..in/dolr/ wasteland_atlas.asp.

Nye P. H. and Greenland D. J. (1960), The soil under shifting cultivation. CABI, Wallingford, UK.

Nye PH, Stephens D (1962). Soil fertility. In: Wills JB, ed, Agriculture and Land Use in Ghana, pp 127--143. Oxford University Press, London.

Olafson, H. l983. Indigenous agroforestry systems. Philippine Quarterly of Culture and Society 11: 149 – 174.

Oldeman L.R., Hakkeling R.T.A. & Sombroek W.G., 1990, World Map of the status of human-induced soil degradation: an exploratory note. Wageningen: ISRIC/UNEP Nairobi. 27.

Olofson, H. 1980. Swidden and kaingin among the southern Tagalog: a problem in Philippine upland ethno-agriculture. Philippine Quarterly of Culture and Society, 8: 168-180.

Osman, K.S., Jashimuddin, M., Haque, S.M.S., Miah, S., 2013. Effect of shifting cultivation on soil physical and chemical properties in Bandarban hill district, Bangladesh. J. For. Res. 24, 791–795.

OTA (Office of Technical Assessment). 1984. Technologies to Sustain Tropical Forest Resources . Washington, D. C.: U. S. Congress.

Padoch & Vaidya 1983 Pattern of Resource Use and I-Iuman Settlement ill Tropical Forest. In F.B. Golley (cd.) Tropical Rain Forest Ecosystems: Structure and Function. Elsevier. Amsterdam.

Padoch and Peters 1993. Managed Forest Gardens in West Kalimantan, Indonesia. Potter, C.S. et al. (eds.). Perspectives on biodiversity: Case studies of genetic resource conservation and development. AAAS Washington.

Padoch, C., Coffey, K., Mertz, O., Lesiz, S.J., Fox, J., Wadley, R.L., 2007. The demise of swidden in Southeast Asia? Local realities and regional ambiguities. Danish Journal of Geography, 107, 29-41.

Palm C. A., Vosti S. A., Sanchez P. A. and Ericksen P. J. (eds, 2005), Slash-and-burn agriculture: The search for alternatives. Columbia University Press, New York City, NY.

Parrotta, J.A., Wildburger, C., Mansourian, S. (eds.), 2012. Understanding relationships between biodiversity, carbon, forests and people: The key to achieving REDD+ objectives, A global assessment report. Global Expert Panel on Biodiversity, Forest Management, and REDD+, World Series Volume 31. International Union of Forest Research Organizations (IUFRO), Vienna, Austria.

Patnaik, N. 1982. "Shifting Cultivation in Orissa." In K. S. Singh, ed. Economies of the Tribes and Their Transformation. New Delhi: Concept Publishing Company.

Pelletier, J., Codjia, C., Potvin, C., 2012. Trditional shifting agriculture: tracking forest carbon stock and biodiversity through time in western Panama. Global Change Biology, 18, 3581-3595.

Pelzer, K.]. 1945. Pioneer Settlement in the Asiatic Tropics: Studies in Land Utilization and Agricultural Colonization in Southeastern Asia. American Geographical Society, Special Publication, No. 29, American Geogr. Society, New York.

Perera, G.A.D. 2001. Secondary forest situation in Sri Lanka: A review. Journal of Tropical Forest Science.

Peroni N. and Hanazaki N. 2002 Current and lost diversity of cultivated varieties especially cassava under swidden cultivation systems in Brazilian Atlantic forests. Agriculture, Ecosystems and Environment 92: 171-183.

Persoon GA, van Weerd M. Biodiversity and natural resource management in insular Southeast Asia. Island Studies J 2006; 1: 81-108.

Pimm SL, Russell GJ, Gittleman JL, Brooks TM. The future of biodiversity. Science 1995; 269: 347-350.

Pinheiro, C.U.B., 2004. A palmeira babac¸u (Orbignya phalerata Martius) e sua explorac¸a˜o na regia˜o dos cocais, Maranha˜o, nordeste do Brasil. In: Alexiades, M., Shanley, P. (Eds.), Productos Forestales, Medios de Subsistencia y Conservacio´n. Centro para la investigacio´n forestal internacional, Indonesia, pp. 163–180.

Popenoe, H., 1960. Effects of shifting cultivation on natural soil constituents in Central America. Ph.D. Thesis, University of Florida.

Posa MRC, Sodhi NS. (2006) Effects of anthropogenic land use on forest birds and butterflies in Subic Bay, Philippines. Biological Conservation; 129: 256-270.

Posa, M.R., Diesmos, A.C., Sodhi, N.S., Brooks, T.M., 2008. Hope for threatened tropical biodiversity: lessons from the Philippines. BioScience, 58, 231-240.

Potapov P V, Turubanova S A, HansenMC, Adusei B, Broich M, Altstatt A, Mane L and Justice CO2012 Quantifying forest cover loss in Democratic Republic of the Congo, 2000–2010, with landsatETM+data Remote Sens. Environ. 122 106–16

Prat, C. (1990). Relation entre érosion et systèmes de production dans le bassin-versant sud du lac Managua (Nicaragua). Cah. Orstom, sér. Pédol., Vol. XXV, no 1-2, 1989-90: 171-182.

Pringle, (1970) Rajahs and rebels. Cited in 'Do shifting cultivators deforest', by S.C. Chin, in V. Shiva (ed.) Forest Resources: Crisis and Management (1992), Natraj Publisher: Dehradun.

Pulhin, J.M., Inoue, M., Enters, T. 2007. Three decades of community-based forest management in the Philippines: Emerging lessons for sustainable and equitable forest management. International Forestry Review, 9: 865-883.

Pulido, M.T. and Caballero, J. 2006. The impact of shifting agriculture on the availability of non-timber forest products: the example of Sabal yapa in the Maya lowlands of Mexico. Forest Ecology and Management 222: 399–409.

Raharimalala, O., Buttler, A., Schlaepfer, R., Gobat, J.M., 2012. Quantifying biomass of secondary forest after slash-and-burn cultivation in central Menabe, Madagascar. Journal of Tropical Forest Science, 24, 474-489.

Rahman S A, Rahman F and Sunderland T. 2011. Causes and consequences of shifting cultivation and its alternatives in the hill tracts of eastern Bangladesh. Agrofor. Sys. doi 10.1007/10457-011-9422-3.

Ramakrishnan, P.S. and Kushwaha, S.P.S. 2001. Secondary forests of Himalaya with emphasis on north-eastern hill region of India. Journal of Tropical Forest Science.

Ramakrishnan PS (1992). Shifting Agriculture and Sustainable Development: an interdisciplinary study from north-eastern India. MAB Series, Volume 10, UNESCO, Paris.

Ramakrishnan, P.S., 1993. Shifting agriculture and sustainable development: an interdisciplinary study from Northeastern India. Unesco and Oxford University Press, New Delhi.

Ramakrishnan P S. 2003. Linking natural resource management with sustainable development of traditional mountain societies. Trop. Ecol. 44: 54.

Raman, T. R. S., Rawat, G. S. and Johnsingh, A. J. T. 1998. Recovery of tropical rainforest avifauna in relation to vegetation succession following shifting cultivation in Mizoram, North-East India. Journal of Applied Ecology 35: 214-231.

Rambo T. A. 2007 Observations on the role of improved fallow management in swidden agricultural systems. In Cairns M. F. (ed.) Voices from the Forest: Integrating Indigenous Knowledge into Sustainable Upland Farming. Resources for the Future, Washington, D C. pgs. 780-802.

Regmi, BR; Subedi, A; Aryal, KP; Tamang, BB (2005) Shifting cultivation systems and innovations in Nepal. Local initiatives for biodiversity, research and development. Report submitted to ICIMOD, Kathmandu, Nepal

Rerkasem K., Lawrence D., Padoch C., Schmidt-Vogt D., Ziegler A. and Bruun T. (2009), 'Consequences of swidden transitions for crop and fallow biodiversity in Southeast Asia', Human Ecology, vol 37, no 3, pp. 347–360.

Reynders, J.J., 1961. Some remarks about shifting cultivation in Netherland New Guinea. Neth. J. Agric. Sci., 9(1): 36-40.

RFD (Royal Forest Department). 1998. Forest Statistics of Thailand 1998. Royal Forest Department, Bangkok.

Richards, P. W. 1973. The Tropical Rain Forest. Scientific Americ an 229 (6): 58 - 67.

Richards, P. W. 1977. Tropical forests and woodlands: an overview. Agro-ecosystems 3: 225-238.

Rickart EA, Heaney LR, Goodman SM, Jansa S. (2005) Review of the Philippine genera Chrotomys and Celaenomys (Murinae) and description of a new species. J Mammalogy; 86: 415-428.

Robbins, R. G. 1964. "The montane habitat in the tropics," The Ecologv of Man in the Tropical Environment (Proc. of 9th Technical Meeting of IUCN, Nairobi, September 1963), IUCN Publications, new series, No.4, pp. 163-171.

Rodenburg, J., Stein, A., Noordwijk, M., Ketterings, Q.M., 2003. Spatial variability of soil pH and phosphorus in relation to soil run-off following slash-and-burn land clearing in Sumatra, Indonesia. Soil Till. Res. 71, 1–14.

Ruddle, K. and Manshard, W. 1981. Renewable natural resources and the environment. Dublin: Tycooly International (for UNU).

Russell, W. 1988. "Population, Swidden Farming, and the Tropical Environment." Population and Environment 10(2): 77-94.

Ruthenberg, 1976. Farming systems in the tropics. Clarendon press, Oxford.

Sabhasri, S., 1978. Effects of forest fallow cultivation on forest production and soil. In: P. Kunstadter, E.C. Chapman and S. Sabhasri (Editors), Farmers in the Forest: Economic Development and Marginal Agriculture in Northern Thailand. East-West Center, Honolulu, HI, pp. 160-184.

Saha N. 1970. Economics of shifting cultivation in Assam, Ph D Thesis (unpublished), Gauhati University, Gauhati, Assam.

Saikia PD, 1968. Changes in Mikir society. Report of Agro-Economic Research Centre, Jorhat.

Sajise PE, Ticsay MV, Gruezo S, Gonzales JCT, Dans AT, Francisco HA, Torres CS, Vergara DK, Velasco V. (2005) Biodiversity in the Mount Makiling forest reserve, Laguna, Luzon. In: Dove MR, Sajise PE, Doolittle AA, eds.Conserving Nature in Culture, Case Studies from Southeast Asia, New Haven: Yale University Press: 241-278.

Sanchez, P. 1976. Properties and Management of Tropical Soils. New York: John Wiley.

Sanchez P.A., 1977, Soil management under shifting cultivation., in A review of soils research in tropical Latin America, P.A. Sanchez, Editor. Rosaline Njomgang, Cameroonian, Doctorate in Soil Chemistry, Director of Research at the Institute of Agricultural Research for Development (IRAD), Yaounde, Cameroon.

Sanchez, P.A. (1981) Soils of the humid tropics. Studies in Third World Societies, 14, college of Williams and mary, Virginia, USA.

Sanchez, P. A. 1987. Management of acid soils in the humid tropics of Latin America. In Management of Acid Tropical soils for Sustainable Agriculture: Proc. No.2.IBSRAM, Bangkok, Thailand, pp. 63 - 107.

Saner, P., Loh, Y. Y., Ong, R. C. & Hector, A. Carbon Stocks and Fluxes in Tropical Lowland Dipterocarp Rainforests in Sabah, Malaysian Borneo. PLoS One 7, e29642 (2012).

Santisuk, T. 1988. An account of the vegetation of northern Thailand. Geoecological Research, Volume 5. Franz Steiner Verlag, Stuttgart.

Sati VP and Rinawma P, 2014. Practices of Shifting Cultivation and its Implications in Mizoram, North-East India: A Review of Existing Research. Nature & Environment,19 (2): 179-187 (www.natureandenvironment.com).

Saurez RK and Sajise PE, 2010). Deforestation, Swidden Agriculture and Philippine Biodiversity. Philippines Science Letters, 3: 91-99.

Saxena, K. G. and Ramakrishnan, P. S. 1984. Herbaceous vegetation development and weed potential in slash and burn agriculture (jhum) in N.E. India. Weed Research 24: 135-142.

Schendel van W.(1992). The Invention of the 'Jummas': State Formation and Ethnicity in Southeastern Bangladesh. Modern Asian Studies 26(01):95 – 128,

Schlippe, P.D. (1956). Shifting Cultivation in Africa: The Zande Sjystem of Agriculture. Routledge & Kegan Paul, London.

Schmidt-Vogt D. (1999), Swidden farming and fallow vegetation in northern Thailand. Franz Steiner, Stuttgart, Germany.

Schmidt-Vogt D, 2001. Secondary forests in swidden agriculture in the highlands of Thailand. Journal of Tropical Forest Science, 13: 748-767.

Schroth G., Da Fonseca G. A. B., Harvey C. A., Gascon C., Vasconcelos H. L. and Izac A.-M. N. (eds, 2004), Agroforestry and biodiversity conservation in tropical landscapes. Island Press, Washington, DC.

Scott JC. 2009. The art of not being governed: an anarchist history of upland Southeast Asia. New Haven (CT): Yale University Press.

Seavoy, Ronald E (1973), "The Shading Cycle in Shifting Cultivation", Annals of the Association of American Geographers, 63(4), 522-528.

Sfeir-Younis A, Dragun AK. 1993. Land and Soil Management: Technology, Economics and Institutions.Westview Press: Boulder, CO; 49–50.

Shambulingappa BG and Mansur CP (2019). Rituals followed in different festivals related to agriculture and animal husbandry by Haveri district farmers in Karnataka. International Journal of Chemical Studies, 7: 3939-3943.

Shanley, P., Luz, L., Swingland, R., 2002. The faint promise of a distant market: a survey of Bele´m's trade in non-timber forest products. Biodivers. Conserv. 11, 615–636.

Sharma, U.C. (1995). Yield sustainability and strategies to meet food needs of NE states of India. Productivity J. 30: 682-690.

Sharma UC and Bujarbaruah KM 2016. Issues and Challenges in Shifting Cultivation and its Relevence in the Present Context, National Academy of Agricultural Sciences, Policy Paper No. 83.

Sharma, U.C., Datta, M. and Sharma, V. (2010). Minimizing agricultural land use induced pollution in the northeast India: farming systems approach through integrated watershed management. Paper presented at World Water Week, held in Stockholm, Sweden, Sept. 05-11, 2010.

Sharma, N. P. (1992). Managing the world's Forests: Looking for Balance Between Conservation and Development. Kendall/Hunt Publishing company. Iowa, U.S.A.

Shoaib JU, Mostafa G, Rahman M (1998). A Case Study on Soil Erosion Hazard in Hilly Regions of Bangladesh, Annual Report, SRDI. Dhaka.

Shrcstha TB, 1989. Development ecology of Arun Basin in Nepal. International Centre lor Intigratcd Mountain Dcvelopment (ICIMOD). Kathmandu, Nepal.

Siahaya ME, Hutauruk TR, Aponno HSES, Jan W. Hatulesila JW and Mardhanie AB, (2016) Traditional ecological knowledge on shifting cultivation and forest management in East Borneo, Indonesia, International Journal of Biodiversity Science, Ecosystem Services & Management, 12:1-2, 14-23, DOI: 10.1080/21513732.2016.1169559

Siebert, S.F. 1987. Land use intensification in tropical uplands: Effects on vegetation, soil fertility and erosion. Forest Ecology and Management 21:37-56.

Silver, W.L., R. Ostertag, and A.E. Lugo. 2000. The potential for carbon sequestration through reforestation of abandoned tropical agricultural and pasture lands. Restoration Ecol. 8:394–407.

Sinha AC, 1981. A note on forest and forest dwellers: A policy approach. Paper presented in a seminar on 'Socio-economic aspects of soil conservation activities with special reference to Northeastern India' held at Tura, Meghalaya, India, on 20-21 June, 1981.

Skole D, Salas W, Silapathong C. 1998. Interannual variation in the terrestrial carbon cycle: Significance of Asian tropical forest conversion to imbalances in the global carbon budget. Pages 162–186 in Galloway J,Melillo J, eds.Asian Change in the Context of Global Change. New York: Cambridge University Press.

Smith, J., Scherr, S.J., 2003. Capturing the value of forest carbon for local livelihoods. World Development 31, 2143–2160.

Smith J, van de Kop P, Reategui K, Lombardi I, Sabogal C, Diaz A (1999) Dynamics of secondary forests in slash-and-burn farming: interactions among land use types in the Peruvian Amazon. Agric Ecosyst Environ 76:85–98

Sodhi, N.S.; Koh, L.P.; Brook, B.W.; Ng, P.K. Southeast Asian biodiversity: An impending disaster. Trends Ecol. Evol. 2004, 19, 654–660.

Sodhi, N.S., Koh, L.P., Clements, R., Wanger, T.C., Hill, J.K., Hamer, K.C., Clough, Y., Tscharntke, T., Posa, M.R.C., Lee, T.M., 2010. Conserving Southeast Asian forest biodiversity in human-modified landscapes. Biological Conservation, 143, 2375–2384.

Spencer, J.E. (1966). Shifting cultivation in South-eastern Asia. Publication in Geography, No 199, University of California Press, USA.

Steppan SJ, Zawadski C, Heaney LR. Molecular phylogeny of the endemic Philippine rodent Apomys (Muridae) and the dynamics of diversification in an oceanic archipelago. Biol J Linn Soc 2003; 80: 699-715.

Stiles, D. 1994. "Tribals and Trade: a Strategy for Cultural and Ecological Survival." Ambio. 23(2): 106-111.

Street, J. M. 1969. An evaluation of the concept of carrying capacity. The Professional Geographer 21(2): 104 – 107.

Struhsaker T (1998). A biologist's perspective on the role of sustainable harvest in conservation. Conserv. Biol., 12: 930–932.

Suarez, R.K., Sajise, P.E. 2010. Deforestation, swidden agriculture and Philippine biodiversity. Philippine Science Letters, 3: 91-99.

Sugam RK, Choudhury P, Hartl J. (2016) Promoting Neo-traditional Agriculture to Achieve Food and Livelihood Security, and Climate Change Adaptation. Council on Energy, Environment and water. New Delhi, India,.

Suratno, W., D. Murdiyarso, F.G. Suratmo, I. Anas, M.S. Saeni, and A. Rambe. 1998. Nitrous oxide flux from irrigated rice fields in West Java. Environ. Pollut. 102(S1):159–166.

Suyanto S. and Otsuka K. 2001 From deforestation to development of agroforests in customary land tenure areas of Sumatra. Asian Economic Journal 15: 1-17.

Takasaki, Y. (2007). Dynamic household models of forest clearing under distinct land and labor market institutions: can agricultural policies reduce tropical deforestation? Environment and Development Economics, Vol. 12, No. 3, 423- 443.

Takeda J. and H. Sato 1993. Multiple subsistence strategies and protein resources of horticulturalists in the Zaire Basin: the Ngandu and the Boyela. In (C.M. Hladik, A. Hladik, O.F. Linares, H. Pagezy, A. Semple & M.Hadley, eds.) Tropical Forests, People and Food: Biocultural Interactions and Applications to Development, pp.497-504, UNESCO, Paris and The Parthenon Publishing Group, Lanes and New York.

Tawnenga, U. Shankar & R. S. Tripathi. 1997. Evaluating second year cropping on jhum Fallows in Mizoram, Northeastern India: Energy and economic efficiencies. Journal of Biosciences 22: 605-613.

Tawnenga. 1990. Studies on Ecological Implications of Traditional and Innovative Approaches to Shifting Cultivation in Mizoram. Ph. D. Thesis, North Eastern Hill University, Shillong.

Taylor, P., Asner, G., Dahlin, K., Anderson, C., Knapp, D., Martin, R., Mascaro, J., Chazdon, R., Cole, R., Wanek, W., Hofhansl, F., Malavassi, E., Vilchez-Alvarado, B., Townsend, A., 2015. Landscape-scale controls on aboveground forest carbon stocks on the Osa Peninsula, Costa Rica. PLoS ONE, 10, e0126748.

Teegalapalli K., Gopi, G. V. and Prasanna K. Samal. 2009. Forest recovery following shifting cultivation: an overview of existing research. Tropical Conservation Science Vol.2(4):374-387.

Terborgh J (1999). Requiem for nature. Island Press, Washington, D.C.

Thomaz, E.L., 2013. Slash-and-burn agriculture: establishing scenarios of runoff and soil loss for a five-year cycle. Agric. Ecosyst. Environ. 168, 1–6.

Thrupp L. A., Hecht S. B. and Browder J. O. (1997), The diversity and dynamics of shifting cultivation: Myths, realities, and policy implications. World Resources Institute, Washington, DC.

Toky, O.P. and Ramakrishnan, P.S. (1981). Cropping and yields in agricultural systems of the northeastern hill region of India. Agro-Ecosystems 7, 11-25.

Toky, O.P. and Ramakrishnan, P.S. (1982). A comparative study of the energy budget of hill agro-ecosystems with emphasis on the slash and burn system (Jhum) at lower elevations of north-eastern India. Agricultural Systems 9: 143-154.

Toky, O.P. and Ramakrishnan, P.S. (1983). Secondary succession following slash and burn agriculture in north-eastern India.1. Biomass, litter fall and productivity. Journal of Ecology 71, 735-745.

Toledo, V.M., Ortiz-Espejel, B., Cortés, L., Moguel, P., Ordoñez, M.d.J., 2003. The multiple use of tropical forests by indigenous peoples in Mexico: a case of adaptive management. Conservation Ecology 7 (online).

Tripathi R.S. and Barik S.K. (2003): Shifting in North East India, Approaches for Increasing Agricultural Productivity in Hill and Mountain Ecosystem (Ed) Bhatt B.P., Bujarbaruah K.M., Sharma Y.P. and Patiram, Pp. 317-322.

Trumbore, S., Brando, P., Hartmann, H., 2015. Forest health and global change. Science, 349, 814- 818.

Tschakert P, Coomes OT and Potvin C, 2006. Indigenous livelihoods, slash-and-burn agriculture, and carbon stocks in Eastern Panama. Elsevier, pp 807-820, doi:10.1016/j. ecolecon.2006.02.001.

Tschakert P, Coomesb OT and Potvin C, 2007. Indigenous livelihoods, slash-and-burn agriculture, and carbon stocks in Eastern Panama. Ecological Economics, 807-820. doi:10.1016/j.ecolecon.2006.02.001

Tulaphitak, T., Pairintra, D. and Kyuma, K. (1985). Changes in soil fertility and tilth under shifting cultivation: changes in soil nutrient status. Soil Science and Plant Nutrition 31 (2), 239-249.

Tyler SA, 1974. Fields are for planting: Swidden agriculture among the Koyas of South India. Anthropological Society, 8:17-38.

Uhl, C. 1987. Factors controlling succession following slash-and-burn agriculture in Amazonia. Journal of Ecology 75: 377-407.

Uhl, C. 1983. You can keep a good forest down. Natural History 92(4): 69 - 79.

Uhl, C., et al. 1990. Studies of ecosystem response to natural and anthropogenic disturbances provide guidelines for designing sustainable land-use systems in Amazonia. In: A. Anderson (ed). Alternatives to Deforestation. Columbia Univ. Press, NY. pp. 24-42.

UNEP (1995) Land cover assessment and monitoring Myanmar, UNEP/EAP, TR/95-06. UNEP Environment Assessment Programme for Asia and the Pacific, Bangkok, Thailand.

United Nations Development Programme, 1992. "Global Alternatives to Slash and Burn Initiative." Project Brief, GEF Technical Assistance Projects: New York: UNDP.

Van Gemerden, B. S., Shu, G. N. and Olff, H. 2003. Recovery of conservation values in central African rainforest after logging and shifting cultivation. Biodiversity and Conservation 12: 1553-1570.

Van Keer K (2003). On-farm agronomic diagnosis of transitional upland rice swidden cropping systems in northern Thailand. Department of Land Management, Faculty of Agricultural and Applied Biological Sciences - Catholic University of Leuven.

Van Noordwijk Meine, Elok Mulyoutami, Niken Sakuntaladewi, Fahmuddin Agus (eds.) 2008. Swiddens in transition: shifted perceptions on shifting cultivators in Indonesia. Occasional Paper No. 9. Bogor, Indonesia: World Agroforestry Centre

Van Noordwijk Meine, Thomas P. Tomich, Renato Winahyu, Daniel Murdiyarso, Suyanto, Soetjipto Partoharjono and Ahmad M. Fagi (eds.) 1995. Alternatives to Slash-and-Burn in Indonesia. Summary of Report of Phase 1. ASB-Indonesia Report Number 4. Bogor.

van Vliet N, Adams C, Vieira ICG, Mertz O. 2013. "Slash and Burn" and "Shifting" cultivation systems in forest agriculture frontiers from the Brazilian Amazon. Soc Nat Resour. 26:1454–1467.

van Vliet N, Mertz O, Heinimann A, Langanke T, Pascual U, Schmook B, Adams C, Schmidt-Vogt D, Messerli P, Leisz S, et al. 2012. Trends, drivers and impacts of changes in Swidden cultivation in tropical forest-agriculture frontiers: a global assessment. Global Environ Chang. 22:418–429.

Veldkamp, E., and M. Keller. 1997. Nitrogen oxide emissions from a banana plantation in the humid tropics. J. Geophys. Res. 102:15889–15898.

Verchot, L.V., E.A. Davidson, J.H. Cattânio, I.L. Ackerman, H.E. Erickson, and M. Keller. 1999. Land use change and biogeochemical controls of nitrogen oxide emissions from soils in eastern Amazonia. Global Biogeochem. Cycles 13:31–46.

Vitousek, P.M., and W.A. Reiners. 1975. Ecosystem succession and nutrient retention: A hypothesis. BioScience 25:376–381.

Vitousek, P., P. Matson, and C. Volkmann. 1989. Nitrous oxide flux from dry tropical forests. J. Geophys. Res. 102:15889–15898.

Vosti, S.A., Witcover, J., 1996. Slash-and-burn agriculture—household perspectives. Agriculture, Ecosystems and Environment 58, 23–38.

Wangchuk S (2005). Indigenous natural resource management institutions in Bhutan. DSB Publication, Thimphu, Bhutan.Yin, 2001.

Warner, K. 1981. Swidden strategies for stability in a fluctuating environment: The Tagbanwa of Palawan. In Olafson, H. (eds.) Adaptive Strategies and Change in Philippine Swidden-based Societies. Laguna, Philippines: Forest Research Institute. pp. 13 - 28.

Weerd van MJ, van der Ploeg JD and Whitaker R (2006). Communitybased crocodilian conservation workshop. In: Crocodiles. Proceedings of the 18th working meeting of the crocodile specialist group. IUCN, Gland: 392-402.

Weinstock, J.A. and Sunito S. 1989. Review of shifting cultivation in Indonesia. Directorate General of Forest Utilization, Forestry Studies, Field Document II-1, Jakarta..

Went, G. W. and Stark, N. l968. The biological and mechanical role of soil fungi. Proceedings of the National Academy of Science 60: 497 - 504. Cited in Beckerman l987.

Wikramanayake, E., Dinerstein, E., Loucks, C, Olson, D., Morrison, J., Lamoreux, J., McKnight, M. & Hedao, P. (2001) Terrestrial Ecoregions of the Indo-Pacific: a Conservation Assessment. Washington, DC, USA: Island Press: 643 pp.

Williams M. (2006), Deforesting the Earth: From prehistory to global crisis, an abridgment. University Of Chicago, Chicago, IL, US.

Win, S. and Kumazaki, M.(1998) 'The history of taungya plantation forestry and its rise and fall in the Tharrawaddy Forest Division of Myanmar (1869-1994)', Journal of Forest Planning 4(1), pp17-26

Wood, S., K. Sebastian, and S.J. Scherr. 2000. Pilot analysis of global ecosystems: Agroecosystems. IFPRI and WRI, Washington, DC.

WWF (2002) Socio-economic survey of Kangchenjunga Conservation Area. Taplejung, Nepal. Kathmandu, WWF Nepal.

Xu, J.C., Li, Y.H., Pei, S.J., Chen, S.Y. & Wang, K.L. (1993): Swidden-fallow succession in the Mengsong area of Xishuangbanna, Yunnan Province, China. Pp 183-196 in: Pei, S. & Sajise, P.E. (eds.) Regional Study on Biodiversity: Concepts, Frameworks, and Methods. Yunnan University Press, Kunming, China.

Xu, H., G. Xing, Z. Cai, and H. Tsuruta. 1997. Nitrous oxide emissions from three rice paddy fields in China. Nutr. Cycling Agroecosyst. 49:23–28.

Xu J, Fox J, Xing L, Podger N, Leisz S and Xihui A (1999). 'Effects of Swidden Cultivation, State Policies and Customary Institutions on Land Cover in a t-lani Village. Yunan. China'. Mountain Research and Development. Vol. 19. No.2.

Xu J., Lebel L. and Sturgeon J. 2009 Functional links between biodiversity, livelihoods and culture in a Hani swidden landscape in Southwest China. Ecology and Society 14: 20. Retrieved on 21st January 2021 from http:// www.ecologyandsociety.org/vol14/iss2/art20/

Xu J, Mclelian T and Hiwasaki L, 2017. Integrating swidden agricultural knowledge systems into sustainable intensification in the Central Mekong region. World Agroforestry Centre, East and Central Asia

Xu Jianchu and Mikesell S. 2003, Landscapes of diversity: Indigenous knowledge, sustainable livelihoods and resource governance in montane mainland Southeast Asia. Yunnan Science and Technology Press, Kunming, China.

Zahabu E. 2008: Sinks and Sources: A strategy to involve forest communities in Tanzania in global climate policy. A dissertation submitted to obtain the degree of doctor at the University of Twente on 3rd December 2008. P. 241.

Zaitinvawra D and Kanagaraj E 2008. Shifting Cultivation to Settled Agriculture: Rural Livelihood Patterns in Mizoram. The Eastern anthropologist 6(2):201-226

Ziegler AD, Phelps J, Yuen JQ, Webb EL, Lawrence D, Fox JM, Bruun TB, Leisz SJ, Ryan CM, Dressler W et al. 2012. Carbon Outcomes of Major Land-Cover Transitions in SE Asia: Great Uncertainties and REDD plus Policy Implications. Global Change Biology 18(10):3087–3099. doi:10.1111/ j.1365-2486.2012.02747.x.

Ziegler A. D., Fox J. M. and Jianchu Xu 2009, 'The rubber juggernaut', Science, Vol 324, pp. 1024–1025.

Ziegler A. D., Fox J. M., Webb E. L., Padoch C., Leisz S. J. and others (2011), 'Recognizing contemporary roles of swidden agriculture in transforming landscapes of southeast Asia', Conservation Biology, vol 25, no 4, pp. 846–848.

Zinke, P., Sabhasri, S. & Kunstadter, P. 1978. Soil fertility aspects of the Lua' forest fallow system of shifting cultivation. Pp. 134–159 in Kunstadter, P., Chapman, E. C. & Sabhasri, S. (Eds.) Farmers in the Forest. University of Hawaii Press. Honolulu.

Zona, S., 1990. A Monograph of Sabal (Arecaceae: Coryphoideae). Aliso 12, 583–666.

5

Dynamics of Soils Under Shifting Cultivation

5.1 GENERAL DESCRIPTION

As more is discovered about tropical soils, shifting cultivation has gained popularity as a symbol of change and showing "ingenious adaptations to unfavourable environments, based on a remarkably complete knowledge of local ecology and soil potential" (Allan 1972). The acidic tropical soils are found in about one billion ha of land in the World. Out of this, about 700 million ha of land is in the high rainfall tropics and, 300 million ha in the savanna, and practically all of it is in developing countries (IBSRAM 1987). The shifting cultivator lives in a humid tropical area with acidic soil. The use and protection of the forest to restore soil fertility is "at the heart of every agricultural system," and the tropics' swiddeners have devised a way that works (Benneh 1972). The integrated swiddener prefers field locations with existing mature forest, either "primary" or "secondary," because it recognises that it is vegetation which supply nutrients to the crop (Dove 1983a; Allan 1965; Rambo 1983; Posey 1983). The availability of nutrients to food crops increases after a burn, but these nutrients fast deplete due to leaching and erosion (Nye and Greenland 1960 and 1964). Nye and Greenland (1964) discovered the soil within the shifting cultivation to be quite variable due to fallen timber, termite mounds, and irregular ash distribution following the burning. These differences will be used to create microsites that will be planted with various crops based on the swiddeners' understanding of which crops benefit from rich soils and which are unaffected by poor soils. The field is left fallow after the cropping cycle is completed (typically 1 - 4 years), while tree crops may be harvested for years. If the site is not utilised for a long time, it will recover its fertility; if it is used too soon, degeneration will occur. It may be difficult to detect degeneration, particularly if it occurs gradually over multiple generations. It's especially tough with Shifting cultivators since they "look to be so self-sufficient, so well integrated with their surroundings" (Street 1969). The assumption is that the longer the land is left fallow, the greater the soil recovery will be. The system should be sustainable if extended fallows can be maintained. Shifting cultivators' response to the requirement to produce food without using manures, fertilisers, or alluvial deposition is soil replenishment by fallowing (Greenland 1974). The technique works if a long fallow time is maintained; if the fallow period is shortened, soil fertility falls. Shifting agriculture is considered environmentally sound while population pressure is mild, but it is environmentally damaging when land scarcity precludes a long enough fallow interval to recover soil productivity. It's crucial to distinguish between traditional shifting cultivation and shifting agriculture in a state of disequilibrium (Watters 1971).

Shifting cultivation exists in ASEANs, Africa and Latin America and is known by various local names in different countries (Thakuria and Sharma 2014). It still supports the livelihood of an estimated 300-500 million people world wide. In fact, soil resources of shifting cultivation lands could be the potential natural production system of today's world in contributing to the food basket of the everincreasing population. In North East India, out of 4.0 million hectare (Mha) of net sown areas, about 1.6 Mha area is under the practice with very low average yields. It is thought that fallowing the land will regenerate the shiftting cultivation soils and restore their fertility (Ramakrishnan and Toky 1981) Furthermore, the chemical content of forest floor litters (FFLs) appears to differ significantly depending on the length of the fallow period (Cornwell *et al.* 2008). Some litters contain more labile carbon (C), while others contain more lignin or recalcitrant carbon (C) that is difficult to decompose (Gessner *et al.* 2010). Reintroduction of fire through slash and burn practices is also known for favouring certain plant communities with particular sets of functional traits that drive the soil communities (Pausas *et al.* 2004).

The relationship between below-ground food-web community and nutrient cycling potentials in soils of forest ecosystem was also previously observed in relation to fire induced shifts in functional plant community (Kardol and Wardle 2010). Therefore, it can be hypothesized that the slash-cum-burn induced modifications in the photosynthesis and consumerbased subsystem within terrestrial ecosystem can bring about drastic changes on the below-ground food-web community and thereby possible alterations in the nutrient cycling potentials of shifting cultivation soils. Understanding such alterations in nutrient cycling processes in relation to the length of fallow phase is pertinent for devising appropriate nutrient management systems for increasin shifting cultivation yields.

The fallow phase of the shifting cultivation cycle stands to exist as a living example in studying the effect of burning on nutrient cycling that drives by interrelationship of plant species succession and soil biota community shifts (Thakuria and Sharma 2014). Thus, different degree of relationship may exist in the pattern of FFLs (forest floor litter) accumulation, soil nutrient availability and microbiologically induced biochemical reactions in shifting cultivation soils depending upon the length of fallow phase. The amount of nutrients accumulated, according to Bruun *et al.* (2006), is principally determined by the nutrient concentration of the vegetation, the temperature of the relevant nutrient elements, as well as the nature of the burning. Not only does fire consume the cut biomass, but it also consumes the litter layer and the highest humus layer, both of which are important resources and habitats for soil organisms. Further, the changes in below-ground biota community are thought to be tightly linked with the processing of biological pools of soil C, nitrogen (N) and phosphorus (P) through their control over soil enzyme activities (Harper *et al.* 2005). But it is still unclear how forest floor litter accumulation, soil nutrient availability and enzyme activities vary according to the length of fallow phases especially in shifting agriculture agroecosystems. Soil fertility was reduced by a significant reduction in soil pH, organic carbon, and available and total major nutrients after a 3-5-year fallow period. The fertility level of all landuses was dramatically recovered with an increase in fallow duration of 8 or more years. Soil repair begins at the age of 8-10 years in an agroforestry system dominated by pulse crops and deep-rooted nitrogen-fixing trees which enrich soil, and is significantly improved over a 25-year-old pine-dominated undisturbed secondary forest setting. As a result, improved appropriate landuses, longer fallow cycles (5-7 years), and the proper combination of soil enriching crops and trees in northeastern India's degraded shifting cultivation land have a significant potential for recovering soil fertility and sustaining production (Devi and Choudhury 2013).

5.1.1 Distribution of soils in American Humid Tropics

The distribution of soils under different soil orders in American humid tropics is shown in Table 5.1 and a comparison of crop yields in tropics and temperate regions in Table 5.2.

Table 5.1. Area under different soils orders in the America's humid tropics.

Soils	Area (M ha)	Soils	Area (M ha)
Oxisols	332	Fluvents	6
Ultisols	213	Psamments	6
Inceptisols	61	Lithics	19
Aquepts	42	Alfisols	18
Andepts	2	Spodosols	10
Tropepts	17	Vertisols	1
Entisols	31	Total	666

Sanchez and Cochrane (1980)

Table 5.2. Average yield (t ha-1) in temperate and tropical areas

Crop	Temperate	Tropical	3 as % of 2
1	2	3	4
Rice	4.11	1.96	48
Wheat	2.98	1.36	46
Maize	4.00	1.35	34
Sorgham	2.27	1.25	55
Soybean	1.98	1.04	64
Beans (dry)	1.08	0.64	59
Groundnot	1.66	1.03	62
Potato	18.00	8.70	48
Sweet potato	13.60	6.88	51
Cassava	11.84	9.10	77
Sugarcane	61.19	53.32	87

Source: Haws et al. 1983

5.2 TROPICAL FOREST SOILS AS AFFECTED BY SHIFTING CULTIVATION

Bruun *et al.*, (2009) and Mertz *et al.*, (2009) have conducted several studies on the on the tropical soils as effected by shifting cultivation. The nature and magnitude of impact depend on the phases of the practice such as; clearing, cropping and fallow as well as on physical, chemical and biological properties. To evaluate the effect of shifting cultivation on tropical forest soils, Filho *et al.* (2013) hinted at some soil quality criteria parameters for estimating this farming operations in the soils of tropical forests. Some of the studies have focused on global climate change (Lal *et al.*, 1995; Kauffman *et al.*, 2009) and the cutting of tropical forests (Fearnside, 2005). Shifting cultivation, main subsistence activity of marginal communities in tropics, has been in practice for many years (Altieri, 1999; Bellwood, 2005). Variable assessments, varying from three hundred fifty million to one thousand million people making their living from shifting cultivation, are currently available (Kleinman *et al.*, 1996; IFAD *et al.*, 2001; Sanchez *et al.*, 2005).

5.2.1 Sustainability of shifting cultivation

The stability and resilience of shifting cultivation is dependent on the preservation of tropical forest condtion and portrayed by opposing situations (Pedroso-Junior *et al.*, 2004, 2009). Aside from deforestation, there was anxiety that the impact on soils could jeopardize forest biodiversity (FAO, 1985; Bandy *et al.*, 1993) and, tracts used for shifting cultivation may contribute to global warming (Fearnside, 2005). In many tropical countries the policies with regard to elimination of shifting cultivation were framed keeping in view the negative impacts of the practice (Ziegler *et al.*, 2009). Namgyel *et al.* (2008) mentioned that shifting cultivation was inharmonious with biodiversity preservation and the management of reserve forests (Namgyel *et al.*, 2008). Nonetheless, a number of academics endorsed shifting cultivation and listed its benefits for the environment (Mertz *et al.*, 2009). Various academics have posited that certain forest ecosystems might have coevolved with human activities like shifting cultivation (Brown and Lugo, 1990; Willis *et al.*, 2004). As a result, given the temporal scale of human/tropical forest interactions, these ecosystems might be classified as cultural forests (Balée, 1989).

As far as impact of shifting cultivation on soils is concerned, the result of studies in this regard show degenerative processes in the soil (Rasul *et al.*, 2004), while other scholars disagree with this plea because this lacked sound evidences (Mertz *et al.*, 2009; Bruun *et al.*, 2009). These lines of evidence include the maintenance of a range of other environmental processes, such as hydrology (Ziegler *et al.*, 2009), biodiversity protection (Rerkasem *et al.*, 2009), and potential carbon sequestration in certain agricultural areas (Bruun *et al.*, 2009). One of the positive outcomes described is an improvement in soil fertility after clearance, with increases in P, Ca, and Mg, showing that these macronutrient reserves are unaltered (Zarin *et al.*, 1998; Johnson *et al.*, 2001; Frizano *et al.*, 2003). Following the departure of the crop area, the nutrient collection mechanisms of the dead organic matter on the forest floor and in the fine roots reconstitute in about five years (Oliveira, 2008). Other studies have found detrimental effects on soil fertility and an increase in erosion (McDonald *et al.*, 2002). A negative impact has been observed as the disruption of soil/plant nutrient cycles as a result of increased slash-and-burn and cultivation cycles (Ewel *et al.*, 1981; Gafur *et al.*, 2003; Davidson *et al.*, 2007). Changes in the physical characteristics of soils have been discovered by some authors to exacerbate the leeching process, resulting in a decrease in the amount of organic matter in the soil (Roder *et al.*, 1994; Frizano *et al.*, 2003). However, Pedroso-Junior *et al.* (2004, 2009) suggested that depending on the spatial and temporal scale evaluated, the effects of shifting agriculture on the environment could be good or detrimental. The debate about the stability of shifting agriculture and its consequences on soils appears to be unsolved.

5.2.2 Impact of shifting cultivation on litter accummulation

Saplalrinliana *et al.* 92016) investigated whether the above-ground biological inputs like vegetative biomass and addition of litter in the shifting cultivation has any impact on the soil activity such as, soil enzyme processes. They considered shifting cycles of 5, 10, and 15 years from Mizoram and 5, 10, and 20 years from Nagaland states of India. Increase in fallow phase increased the amount of litter produced. Bulk density, pH, electrical conductivity, available P, and K in soils under burned sites were found to be higher than in unburned sites, but soil organic carbon (SOC) and avail-N showed the opposite trend. Except for BD, the longer fallow phase had substantially higher values of these soil parameters than the shorter fallow phase. Between the burnt and unburnt stages, the activity of soil enzymes such as amylase (AMY), -glucosidase (GSA), dehydrogenase (DHA), aryl-sulphatase (ASA), acid-phosphatase (PHA), and protease (PRO) altered dramatically (P0.05, Kruskal-Wallis test). With the exception of burnt circumstances, AMY and GSA activities were substantially lower in the protracted fallow phase than in the short fallow phase. Finally, burning biomass releases a bigger quantity of plant-available nutrients in jhum soils for a short period of time, allowing for improved crop development during prolonged fallow periods. Burning, on the other hand, had a negative impact on the biochemical processes of shifting cultivation soils, showing a steady deterioration in the native nutrient processes potentials of soils in shifting cultivation cycles (Saplalrinliana *et al.* 2016).

5.2.3 Fallow cycles under different landuses

The examined soils were significantly acidic in reactivity (Ph 5.0) regardless of land uses of various eras. However, as the age of post-burning periods increased, the amount and pattern of change in major soil fertility, metrics differed substantially across the three other land uses (recurrent agriculture, agro forestry, and abandoned shiting cultivation) employed in shifting farmed area. At the surface layer 0-15 cm, the influence of land uses and fallow periods on soil chemical characteristics was stronger. According to Yemefack *et al.* (2010), analysis of the farming system led to the construction of a conceptual model of shifting agriculture's spatio-temporal dynamics. The farming approaches have a 30-35 percent impact on the 0-20 cm top soil layer variance. pH, calcium, accessible phosphorus, bulk density, and organic carbon are five soil factors that are extremely responsive to agricultural methods.

Unsustainable land use practises in the humid tropics, such as shifting cultivation, have posed a threat to tropical forest preservation and global climate fitness. To optimise root growth and nutrient uptake, organic matter, as well as physical and chemical markers, must be controlled. Ploughing, as opposed to mulching, causes quick and long-term soil organic matter losses, as well as perpatual concretion, soil deterioration, and anthropic savannization; as a result, it should be avoided. An alternative must be created that favours the use of local resources to increase fertiliser supply and offset processes that diminish fertiliser efficiency (de Moura *et al.* 2016).

Sustainable agriculture can be defined as "the management and conservation of the natural resource base, and the orientation of technological change in such a manner as to ensure the attainment of continued satisfaction of human needs for present and future generations. Sustainable agriculture conserves land, water, and plant and animal genetic resources, and is environmentally non-degrading, technically appropriate, economically viable and socially acceptable" (FAO 1988). In Amazonian areas, improper land management is the biggest threats to conservation of the rainforest and sustaining a healthy ecosystem, because after declining of soil fertility in an area, the cultivators generally move to nutrient rich rainforest area. This operation is replicated again and again by marginal farmers for producing needed crops or by big farmers for the creation of pastures. Due to fast decline in soil fertility, these land uses are practiced for a short period. This has led to vanishing of the plant species which are sensitive to burning and leaving behind more resistant species to be in supreme position. In a worldwide context, it is estimated that uncontrolled fires account for 36 percent of all CO2 generated by Brazil, which is one of the world's top five emitters (Moura *et al.* 2013). Small landholder farms in the humid tropics, particularly the Amazon basin and its environs, are increasingly recognizing this continual process of moving cultivated area as a major driver of deforestation and worsening food security. As a result, for individuals concerned about the health of the rainforest ecosystem, sustainable agricultural soil management is a critical challenge (Agnair *et al.* 2011). The quality of soil is measured in terms of specific soil functions. In the humid tropics, some crop nutrients, particularly P and K, as well as cation base depletion, soil organic matter loss, and soil rootability deterioration, have an impact on system performance. These factors have an impact on biological variety, biomass productivity, and, as a result, environmental quality (Agnair *et al.* 2013). Due to high temperatures and rainfall in soils formed by clastic sedimentary rocks, effective soil management by retaining fertility is difficult in most tropical settings (Dove *et al.* 2011). It's worth emphasising that the huge agrosystems and monocultural agriculture developed for other tropical regions, such as southeastern Brazil, are insufficient for humid tropic circumstances because they don't meet the region's family farmers' food and income needs (Moura *et al.* 2013).

5.2.4 Physical attributes for tenable use of soil in humid tropics

The nutrients, water, air, and space for root growth are the major needs for the growth of plants and the ability of soil to make available these essentials in the root micro-environment is referred to as fertility (Aweke *et al.* 2015). Other than these requirements, the pore space size and distribution, impact many essential physical events, such as storage movement, the availability of plant food, and nature of root system, and this affects crop productivity (ordan *et al.* 2013, Moura *et al.* 2016). Despite the fact that the impacts of physical indicators on the growth of some plants have been examined, there are few examples of physical indicators that have a close relationship to plant output, unlike chemical fertility indicators (Mishra *et al.* 2015). Mulumba and Lal (2008) discovered that mulch boosted field moisture capacity by increasing overall porosity, soil aggregation, and moisture content. The effect of mulch rate on soil bulk density, however, was not linear. When compared to no mulch, mulch rates as low as 2 Mg/ha resulted in considerable increases in soil porosity. Similarly, no substantial increases in soil accessible water capacity were seen above 8 Mg/ha of mulch. Crop residue mulch may only be useful near the soil surface in terms of improving soil structural qualities (Johan and Hakansson 2014). Furthermore, according to Moura *et al.* (2013), the direct effect of residues on soil characteristics did not extend deeper than 15 cm.

5.2.5 Soil organic matter and its significance

The soil soil organic matter (SOM) is a key index of soil quality and its physical, chemical and biological dimensions. As far as soil physical dimension is concerned, SOM content is negatively linked with bulk density (Chaudari *et al.* 2013, Blanco-conqui *et al.* 2013) and positively linked with water infiltration (Blanco-conqui *et al.* 2013), water retention (Maans and Berg 2014, Yang *et al.* 2014), and volume and soil aeration (Zaffar and Lu 2015). Organic forms of nutrients predominate over inorganic forms in the soil chemical dimension (Blume *et al.* 2016), and SOM is linked to lower P-fixation (Flores *et al.* 2014) and reduced Al-toxicity (Hagvall *et al.* 2015) due to organic complexation of free Al. SOM content, on the other hand, is not clearly connected to soil pH. As a result of the physico-chemical reasons outlined above, soil biota is strongly reliant on and concentrated inside the SOM (Dungait *et al.* 2012), and this, in turn, has a significant impact on it. Organic nutrients support macro- and mesofauna, as well as soil bacteria and fungus, whereas labile and dissolved organic matter provides energy to decomposer organisms (Whalen 2014). (Cotrufo *et al.* 2013). Plant root colonisation is stimulated by increased food availability in topsoil organic matter, which in turn boosts microbial activity in the "rhizosphere" via root exudation, cell slaying, and dead roots (Hinsinger *et al.* 2009, Jones *et al.* 2009). In humid tropics soils, SOM is even more important than in semiarid tropics soils (Lal 2009); in contrast to topsoil organic matter, highly weathered mineral sub-soils have poor nutrient availability and nutrient-retention capacity, and suffer from severe P-fixation (Gama-Rodrigues *et al.* 2014), as well as Al- and Mn-toxicity issues (Cristancho *et al.* 2014).

5.2.6 SOM in soils

The content of organic matter due to input from vegetative material, and exudates drives organic matter breakdown, which can take anywhere from a few days to several years (Dias *et al.* 2013). Lignins and polyphenols with a high molecular weight degrade much more slowly (Talbot *et al.* 2012). SOM contents can be affected by species-specific changes in organic matter inputs, chemical composition, and decomposition kinetics (Russel and Raich 2012), albeit these impacts could be of the unstable portion of soil organic matter (Gentile *et al.* 2011). SOM concentration is positively and approximately linearly related to soil clay and silt content (Desjardins *et al.* 2004) and negatively correlated with (especially coarse) sand content (Feller and Beare 1997) in humid tropical soils, as in any soil. The higher OM concentration in clayey/silty soils compared to sandy soils is due to the formation of "organo-mineral complexes," which impede SOM decomposition (Feng *et al.* 2014).

5.2.7 Shifting cycle and yield of crops

After low to medium intensity burns, large levels of inorganic N have been discovered in the ash (Sanchez 1976; Andriesse 1989; van Reuler and Janssen, 1993; DeBano *et al.* 1998). Ramakrishnan and Toky (1981) (Table) reported an increase in accessible N and P in soil with increased fallow period, as well as an increase in crop production by Bruun *et al.* (2006). Swamy and Ramakrishnan (1988) reported a similar observation, finding that burning added more base cations in lengthy fallow (more than 20 years) than in 5 years. In the short term, burning increases the availability of specific nutrients like phosphorus and potassium, and it may boost crop productivity the next season (Erenstein 2002). (Haider, 2012). However, it can result in the loss of plant nutrients such as nitrogen, potassium, and sulphur, as well as a negative impact on the surrounding microbial population and organic carbon (Gupta *et al.*, 2004; Heard *et al.*, 2006). (Heard *et al.* 2006). Table 5.3 shows the data on the effect of fallow length on soil nutrient concentration.

Table 5.3. Impact of fallow period and conservation on soil properties before sowing of relay crops. (Average of 2 years)

Treatment	pH	OC (%)	Av. N (kg ha^{-1})	Av. P_2O_5 (kg ha^{-1})	Av. K_2O (kg ha^{-1})
			Fallow	**cycle**	
5 years	4.56	1.84	264.5	3.86	102.0
10 years	4.82	1.90	293.2	5.42	112.8
20 years	5.15	2.05	312.5	7.54	122.3
CD (P = 0.05)	0.02	0.01	1.35	0.19	2.86
			Resource	**conservation**	
Mulcuing	4..60	1.80	273.7	4.89	103.0
Burning	5.09	2.05	306.5	6.33	121.8
CD (p = 0.05	0.46	NS	3.73	0.49	2.39

Source: Ramakrishnan and Toky, 1981.

5.2.8 Significance of the Fallow Period

Despite fertility constraints, Tulaphitak *et al.* (1985) stated that even after two years of continuous cropping, soil fertility reduction may not be the primary motivator for farmers to fallow their fields. In certain experiments, it was shown that most nutrients were in more quantity after harvest than before planting, and that the supply of nutrients from ash was greater than the cropping loss. These findings indicate that soil fertility recovery was not the only, or even the most important, motivation for farmers leaving their fields fallow. If the fallow period is fewer than 6 years, local farmers consider it to be short. They believe that in short-fallow fields, weeds are more numerous. Short-fallow farmers must weed four times per crop, whereas long-fallow farmers only need to weed twice per crop. Roder *et al.* (1994) found that as the fallow period was reduced from 38 to 5 years, the weeding need increased from 1.9 to 3.9 times. Weed control is critical for long-term crop yields, and weed control is critical. The germination of seeds of annual herbs or grasses is suppressed by burning, while seed germination of woody plants is encouraged (Kamada *et al.* 1987).

5.2.9 Land use for miniature fallow plots

Increased population and government rules prohibiting the development of new fields in forest regions will ultimately induce intensification of cropping on existing agricultural fields, resulting in a shorter fallow period. Is there any type of farming system that could continue to produce optimally under such conditions? Even though soil fertility may not be fully restored after a short fallow period, the nutrients accumulated in the soil during short fallow period are almost sufficient to support one year of cultivation. The weed population, on the other hand, increases as the fallow time shortens. As a result, fallow plays a major role in weed management throughout the cropping season, but its significance in soil fertility recovery is limited.

5.3 IMPACT OF VARIOUS FACTORS

5.3.1 Variation in rainfall

Litter production in the southern Yucatan peninsula is totally dependent on the timing and amount of precipitation, similar to the dry forests of Western Mexico and portions of the Northern Yucatan, and in the wetter forests of Guatemala south (Whigham *et al.*, 1990). This evidence indicates that the trees re-engrossed needed nutrients prior to enormous leaf loss (Killingbeck, 1996). N concentrations of litter were reduced by only 33%. Lower nutrient concentrations have been seen in both wetter and drier woods during the dry season (Wieder and Wright, 1995, McGrath *et al.*, 2001).

5.3.2 Impact of environmental gradients on biomass and nutrient cycling

With increasing precipitation, there is increase in forest floor biomass as well as its decomposition, resulting in lower forest-floor biomass if litter yield is constant. The increase in forest floor biomass indicate an accompanying increase in litter production. The availability of water determines the display of nutrient limitation. Campo *et al.* (2001) observed that P did not appear to be limiting in the dry forests, although small amounts of P were retained in the system. As a result, P cycling changes dramatically during the forest recovery process, with P use efficiency increasing as a function of age (Brown and Lugo 1990). Because older trees produced substantially more litter while cycling the same amount of P as younger forests, litter P concentration decreased with age. In older forests, on the other hand, the pool of N cycled through litter rose in lockstep with increased litter output. Although there was a tendency for slightly greater concentrations in older forests, litter N concentrations did not increase appreciably. The nutrient studied determines how changes in nutrient cycle occur during forest recovery. Furthermore, knowing the interplay between water and nutrient restriction is necessary for evaluating nutritional limitations on ecological activities. When water is scarce, production appears to be constrained by the supply of nitrogen. Higher litter output was linked to more N cycling through the system. Regardless of age, all forests cycled the same amount of P. Changes in soil parameters with forest age show that during forest regeneration, N availability may rise. Soil organic matter increased dramatically with age, despite the lack of data on N and P content. Organic matter in the top 15 cm soil increased by 10 to 11% in young secondary forests and up to 15% in mature forests. Litter concentrations would remain steady while productivity increased if N availability was increased in tandem with organic matter. This increase in soil organic matter is most likely due to a slowing of soil mineralization rates, as well as an increase in litter inputs. Surprisingly, both decomposition tests and field observations reveal that P may restrict decomposition and mineralization changes with forest age, despite the fact that N appears to limit plant output when water is scarce. N fixation may be hampered by P. Similar observations have been made by Crews *et al.*, (2000) and Uliassi *et al.*, (2000). Phosphorus may have a crucial role in reducing N supply to regenerating vegetation by decreasing decomposition, mineralization, and N fixation (Ewel, 1986). It's vital to note that determining whether P or N is the limiting nutrient in this system is challenging, especially if P mediates N limitation. Key ecosystem processes appear to be hampered by both nutrients.

Regional precipitation gradients had a significant impact on nitrogen cycling via soil organic matter and litter, but only a little impact on plant productivity and biomass. More phosphorus entered the system as a result of increased precipitation. Although forest age had a significant impact on many components of nitrogen cycling, it had the biggest impact on litter production and aboveground biomass. With age, P use efficiency increased significantly while N use efficiency decreased slightly. In these dry forests, water scarcity appears to be the most significant environmental factor affecting ecological processes, with phosphorus perhaps being the most critical limiting nutrient.

5.3.3 Management options

The cultivators generally look for low-cost techniques to raise crop productivity as use of lime or fertilizer may be out of their reach. Under such circumstances, agroforestry can be a more appropriate alternative. The first stage in low-input soil management is to employ acid-tolerant species or cultivars (Reddell 1993, Sanchez and Salinas 1981). Agroforestry is also regarded to be particularly suitable for marginal soils that are subject to severe physical, chemical, or drought constraints (Sanchez 1987). In tropical acid soils, the main challenge is to recycle the few nutrients available in soil-plant systems (Sutherland and Sprent 1993). One low-input method for sustaining nitrogen levels, which is a key nutrient for plant growth, is to plant nitrogen-fixing trees and crops in an agroforestry system.

5.4 TROPICAL SOILS CHARACTERISTICS

Characterized by a hot and humid climate, the tropics are subjected to rapid weathering processes, thereby increasing the amount of iron oxides in the soil. The colour of the soils varies with the location and soil properties (Troeh and Thompson; 2005). The organic matter (OM) is low in these soils due to high rate of decomposition. It is necessary to keep vegetation cover in tropical soils as the nutrients are contained in the living OM (Troeh and Thompson; 2005). Burning of vegetation affects weathering as well as increases all chemical soil processes. High temperatures, in addition to heavy rain, produce significant chemical weathering and soil formation. The amount of humidity in the air has a big impact on nutrient leaching and the acidification process in the soil. The soil type is greatly influenced by the natural vegetation in a given location, and vice versa. The degree and kind of vegetation, as well as root distribution and litter composition, all influence the soil formation process (Eriksson et.al.; 2005). Infiltration and leaching can be lower on hills than on plains due to increased surface runoff from steep slopes (Wood & Porro; 2002). The texture and mineral composition of the soil determine its fertility. P is shown to be the greatest limiting element in the Amazon sois. According to Wood and Porro

(2002), 90 percent of Amazon soils are deficient in P, 50 percent are deficient in K, and 73 percent have aluminium toxicity. Brady (2002) attributed this to phosphorus and potassium adsorption on Fe- and Al-oxides. Fe, Mn, Cu, Zn, Cl, B, Ni, and Mo are the micronutrients (Havlin; 2005), with a 1 kg-1 (Fe10 kg-1) absorption (Eriksson *et al.*; 2005).

Bryant *et al.* (1997) mentioned that about 50% of the global forests have been converted for agriculture and other uses and about 78% of the remaining are largely changed. Almost similar figures were quoted by FAO (1997), mentioning that about 72% of the original 1450 million ha of tropical forests have been altered to other uses. Marginal cultivators often are seen as the primary drivers of deforestation in the tropics (Hauck 1974), causing about 96% of forest losses (Myers, 1994). Increased soil nutrient availability after burning in shifting cultivation has been reported by many scholars (De Rouw, 1994) and, Maass (1995) ascribed this increase in soil nutrients to the addition of nutrient-rich ash post burning. Two thirds of human carbon dioxide emissions are caused by burning fossil fuels, while the remaining third is produced by biomass, such as grasslands, forests, and agricultural crops (Levy, 2004). It was quoted by a Scientist that, "When the soil is burnt, its nutrients are destroyed and the soil is exhausted. And, when the grass is burnt, it releases carbon dioxide into the atmosphere thus contributing to the depletion of the ozone layer and to climate change. slash and burn' farming techniques are bad for both local agriculture and the environment and why the Durban conference must provide alternatives for small scale farmers" (Wendi, 2011).

5.4.1 Shifting cultivation and soils

Shifting cultivation is environmentally sustainable and meets a wide range of human needs, particularly in terms of labour and other agricultural inputs, in a cost-effective manner (Christanty 1986). Despite the fact that the humid tropics have a wide range of soil types, the vast majority of the region's soils are nutrient poor (Jordan 1985). According to Sanchez (1987), Lal (1989), and Moorman and Kang (1978), phosphorus shortage, aluminium toxicity, drought stress, and low inherent fertility are all common and well-documented concerns in the humid tropics of Africa, Southeast Asia, and the Amazon. The amount of rainfall in the region appears to be the reason of the region's poor soils, as when rainfall exceeds 1000 mm, the soils become acidic and leached (Sanchez 1987). The nutrient inadequacies of tropical soils, according to Lal (1987), are the major limiting factor in tropical productivity. Tropical rainforests are supported by these "old, severely weathered, and excessively leached" soils, although the woods do not rely on the soil for nutrition. Instead of using soil, the tropical forest ecosystem uses its own biomass to establish a nutrient cycle. However, if a very disruptive form of clearing is adopted, even a tiny area might experience considerable runoff and erosion. Traditional and human forest clearing results in less severe soil erosion than land cleared by automated processes, particularly tree pushers. Lal (1987) reported a runoff of 3, 35, 86 and 202 mm per year resulting in soil loss of 0.01, 2.5, 3.8 and 17.5 t ha-1 yr-1, when the clearing was done by selective cutting (traditional), manual, sheer blade and tree pushers/root rake, respectively. For the Kanchanjunga Conservation Area's (KCA) shifting cultivators, soil fertility is a crucial concern. The fallows, as well as the controlled burning and spontaneous decomposition of their leaf litter, are the primary source of soil nutrients. On their shifting agricultural plots, none of the people in the study villages utilise farm yard manure or other fertilisers (Aryal *et al.* 2010). The labor-intensive process of producing and distributing farmyard manure is not worth it. Other ways for conserving soil nutrients throughout the cropping phase are used by certain people. These include eradicating the weeds at the base of the maize stalk, intercropping legumes (mainly beans) between the other crops, mulching, and plot rotation, among others. In India, the shifting cultivation is mainly practiced in the northeastern states of the country and Odisha state of eastern India. The total population engaged in shifting cultivation is shown in Fig. 5.1.

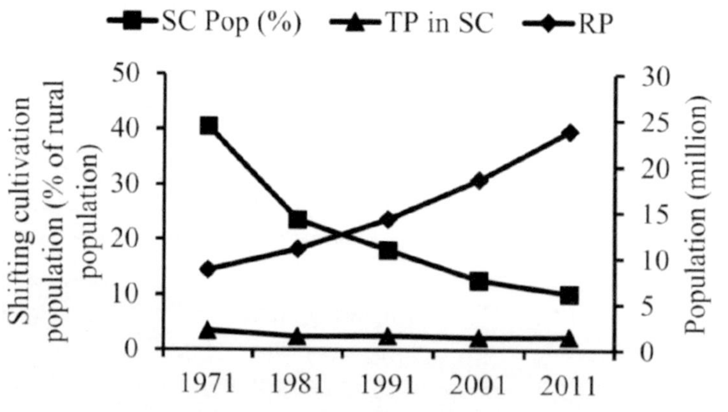

Fig. 5.1. Population engaged in shifting cultivation in northeastern states of India
SC Pop % (% rural population practicing shifting cultivation); TP in SC (Total rural population in shifting cultivation); RP (Total rural population).

5.4.2 Nutrient cycling

The work of Bartholomew, *et al.* (1953) is one of the few early attempts to understand the pattern of element cycling in tropical secondary successional fallows after slash and burn agriculture in Belgian Congo. They concluded that storage capacity of leaves and litter is saturated at an early stage in succession, and thereafter ,the total storage capacity increases more slowly and only in the woody perennials only. Another comprehensive study carried out at Burnihat in north-eastern India (Toky and Ramkrishnan,1983b) showed that leaves had the highest concentration of elements and the boles had the least. The concentration of nitrogen, phosphorus and potassium was much higher in the branches and leaves than in the litter, but the concentration of calcium and magnesium was the same in all these three compartments. The concentration of calcium and magnesium was much higher in trees and shrubs than the bamboos, while the later one had a much higher concentration of potassium. The weeds colonizing the field immediately after abandonment were found to be rich with phosphorous. With the development of secondary vegetation, the weight of elements in the living biomass increased due to linear increase in the biomass with age. The high enrichment quotient (concentration in the vegetation/uptake) for potassium in fallows 10 to 20 years old indicates a maximum accumulation of this element in the living biomass, and therefore, conservation of this vital element in the vegetation and releasing it slowly through litterfall. Greenland and Kowal (1960) reported almost three times more calcium than potassium in the above-ground biomass of 40-yera old forest in Ghana. Edward and Grubb (1982) also showed higher budget of calcium and magnesium in the trunks of large trees in the forest of New Guinea.

After the destruction of element cycling by slash and burn agriculture, the system recovers rapidly through a quick succession due to high rate of net primary productivity. Predominant weeds such as Eupatorium odoratum, Imperata cylindrica, Mikania macrantha, Borreria hispida and Sachharum spontaneum play important role in conservation of nutrients. The soil erosion is almost reduced to zero. Potassium which is one of the important elements and highly susceptible to leaching and run-off losses during cropping and early growth of fallows, is rapidly conserved in the living biomass of bamboos during the first 20 years or so. Any disturbance as in the short agricultural cycles of 5 years, as is now common is likely to adversely affect the recovery of soil fertility and elements conservation by the system as a whole. Biomass in shifting cultivation under differet cycles is given in Table 5.4.

Table 5.4. Biomass in shifting cultivation under short, medium and long shifting cycles at Burnihat, Meghalaya in north-eastern India (Toky and Ramakrishnan,1983a).

Biomass	**1 Year**	**5 Years**	**10 Years**	**15 Years**	**20 Years**
Standing biomass (t ha^{-1})	5	23	57	104	148
Litterfall (t ha^{-1} yr^{-1})	1.2	4.8	7.1	7.7	9.7
Net primary productivity (t ha^{-1} yr^{-1})	5	8	14	17	18

Shifting agriculture with a fallow period of more than 10 years is sustainable from ecological, economical and energy efficiency points of views; and it should continue. The germplasm of native crops should be preserved for a long time preferably by cryopreservation. Home gardens, which contain a wide variety of native trees, shrubs, and herbs, should be bolstered. Species diversity of the region which at presentis threatened due to heavy pressure on the natural forests should be preserved for future generations. To take the pressure off the lands under short cycles of 5 years or so, the valley cultivation should be strengthened for cereal production. Apatani way of wet rice cultivation as prevalent in Arunachal Pradesh, effectively recycling the domestic wastes, should be strengthened in other areas also. Irrigation of crops during dry periods using native technology of bamboo irrigation should be used extensively. Short-cycle degraded lands should be repaired with cash-producing crops like fruit, spices, and short-rotation woods. Cash crops such as areca nuts, coconuts, bananas, citrus, medicinal and fragrant plants should be integrated into multistory agroforestry systems. Nitrogen-fixing trees like alders (*Alnus nepalensis*), which coppice well and may be managed as a single bole, should be planted in large numbers in agroforestry systems. Bamboos (*Dendrocalamus hamiltonii*) should be planted because they are incredibly beneficial and successfully conserve potassium during the early stages of secondary succession. Superior vegetative produced bamboo saplings should be provided to the farmers at a subsidized rate. Agroforestry systems should be extensively managed using bio-fertilizers, bio-pesticides, and mulch made from in-situ created litter to conserve nutrients and moisture. For the development of bio-fertilizers and bio-fungicides, native stains of beneficial bacteria and fungi should be utilised. The Taungya method, which allots lands with Forest Department-grown plantation crops, should be implemented because it has shown to be incredibly successful in the past and is still in use in various tropical countries.

Fire is used as a tool to modify the acidic pH of the soils heavy rainfall zones. Terracing the land with the hope of making permanent cultivation should not be encouraged as it is not environment friendly and involves high inputs. Development of essential technological packages to encourage farmers to embrace eco-friendly and environmentally sound technologies. Diversifying land use improves the living conditions of those who live near forests. By trapping carbon in the soil, soil productivity can be increased while greenhouse gas emissions are reduced. One such example is bio-char. Traditional techniques can be modified to improve soil organic matter and plant biomass. Involve local people in all research and technology development endeavors. Develop appropriate agriculture marketing and subsidy methods. Coordination of numerous agencies for the farmers' benefit and to increase their purchasing power.

5.4.3 Improving soil and water conservation

For soil and water conservation in the tropics, improvement in infiltrarion rates is of great importance (Okigbo and Lal, 1979), and for this mulching has been found to be a very successful techniqueis in shifting cultivation areas. Direct impact of raindrops is absorbed by the mulches, which reduces soil detachment. Under such condition the infiltration rate of the soil is at its maximum level,. and therefore, runoff losses decline sharply. So, mulching helps in the improvement of crop yields by favourable soil physical condition as well as, adequate supply of soil moisture to the standing crop. Sufficient cover crops save the soil from wind and water erosion and modify soil bulk density and infiltration rates. Okigbo and Lal (1979) reported that mixed cropping systems are highly productive and conserve soil and that, mixed cropping of maize and cassava was better in reducing runoff, conserving soil and crop productivity as compared to monoculture cassava. Increase in plant density has been found an effective method for reduction of soil loss via erosion (Hudson 1973). Using a mixture of little tillage, mulching, and mixed cropping, farmers have been able to consistently produce high yields (Sanchez, 1976).

5.4.4 Soil nutrient dynamics

When total nutrient stock at farms is compared to annual production through soil dynamics without human activity, the increase in P and K is less than 1%, and the loss of 1.1 percent in the case of N demonstrates a considerable shift in nutritional status. Other studies have found decreased nitrogen (Lehmann *et al.,* 2003b) and phosphorus (Schulz and Glaser, 2012) leaching in the first year after compost and biochar application, while Schulz and Glaser (2012) found increased phosphorus and nitrogen leaching in the first year after compost and biochar application (Agegnehu *et al.* 2015). The hardy biochar could last for decades in the soil, enhancing nutrient and water storage capacity (Verheijen *et al.,* 2010).

5.4.5 Development and management of soil

Expanding the quantity of land under cultivation or using a number of other technologies that produce higher productivity per unit area can help increase food and agricultural output. The former is the most significant technique of increasing crop productivity in tropical Africa. The amount of land and labour available each year to clear new area will determine the strategy's effectiveness. The size of land holdings has reduced in tropical West Africa, as against other parts of Africa with lower population km^{-2}. Land demand for various development operations is increasing at the same time. As a result, measures for retaining soil fertility and removing or severely lowering the time spent in bush fallows as part of permanent cropping systems must be developed if farm sizes are to be raised and the current trend of shrinking farm sizes is to be avoided. A variety of competing objectives need the use of land-saving technologies.

Donahue (1970) mentioned that the soils of the tropical areas where shifting cultivation is practiced are generally low in inherent fertility due to the following reasons:

- The pedons are highly weathered;
- Silicon presence is extremely low;
- Iron and aluminium are present inhigher amounts;
- The presence of montmorillonite is low but kaolinite percent is high;
- Higher anion exchange capacity and low CEC;
- Buffer capacity is low;
- Lower WHC;
- Laterite layer in some soils hardening due to wetting and drying as in the case of continuous cultivation;
- Higher friability;
- Due to fast decomposition there is low accumulation of leaf litter;
- The soils are poor in total and available nutrients.

On hill slopes in Bangladesh's Chittagong Hill Tracts (CHT), soil fertility has diminished due to improper agricultural practises such as shifting cultivation, land use change, and land cover degradation (Hossain 2011). Hills cover around 12% of Bangladesh's total land area and are primarily found in the country's south and southeast. The alternating strata of little-consolidated sands and shale in these hills provide as a foundation for the production of diverse combinations of deep and shallow soils. Approximately 2.5 million tonnes of soil are carried from the hills to the piedmont, drainage canals, and finally floodplains and the sea (Shoaib *et al.*, 1998). Shifting cultivation is the major land use, covering an annual area of 32 thousand ha. Mean soil loss from shifting cultivation was 40-45 t $ha^{-1}yr^{-1}$, the highest being observed in steep (32-50%) slopes and the lowest in gentle slopes (15-20%) (Shoaib *et al.*, 1998, Sharma and Sharma 2003).

5.4.6 N, P and K use efficiency

Because of the high rate of phosphate-fixation, significant volumes of P fertiliser are needed to overcome the limitations of P-fixation. In contrast, most tropical farmers find mineral phosphate fertiliser to be an expensive input, and many do not have the resources to purchase phosphorus fertilisers (Yang and Post 2011). In Amazonian soils that are prone to cohesion and nutrient leaching, the effective management of low-input agrosystems depends heavily on how well nitrogen and potassium are used. Using exclusively inorganic potassium (K) and nitrogen (N) fertilisers is not advised, unlike in other parts of Brazil, as it keeps crops from producing to their maximum capacity (Moura *et al.* 2012). Due to the high phosphate-fixing rate, significant volumes of P fertiliser must be administered to overcome the limits of P-fixation. Mineral phosphate fertiliser, on the other hand, is an expensive input for most farmers in tropical locations, and many farmers do not have the financial means to obtain it (Yang and Post 2011). In Amazonian soils prone to cohesion and nutrient leaching, the efficiency with which nitrogen and potassium are utilised is a significant component in the efficient management of low-input agrosystems. Using exclusively inorganic potassium (K) and nitrogen (N) fertilisers, unlike in other parts of Brazil, is not recommended since it hinders crops from attaining their full productive potential (Moura *et al.* 2012).

Out of a total loss of 601 million tones of soil from Northeast Indian states through water erosion, 88.3 million tones (about 14.7% of total loss from the region) is contributed by shifting cultivation, though area affected by the practice of shifting cultivation is only 5.65% of the total geographical area (Sharma and Prasad 1995). The amount nutrient transfer causing environmental degradation from the region is estimated at 685, 99, 511.1, 22.6, 14.0, 57.1, and 43.0 thousand tonnes of N, P, K, Mn, Zn, Ca, and Mg, respectively (Sharma 2003).

5.5 SOIL PROPERTIES UNDER SHIFTING CULTIVATION

5.5.1 Effect of burning and fallow period on soil attributes

Values of BD, pH, EC, avail-P and avail-K in soils where vegetation was burnt, were found higher as compared to their values in plots where no biomass was burnt irrespective of fallow lengths in both Nagaland and Mizoram states. On the other hand the contents of SOC and avail-N under burnt sites were found lesser relative to their values in unburnt sites. The length of fallow period showed a distinct influence on soil BD, pH, EC, SOC, avail-N, avail-P and avail-K. Except BD, values of all above soil properties were found significantly higher in the longer fallow phase compared to that in shorter fallow phase. Soil BD decreased with increasing the length of the fallow phase. Jia *et al.* (2005) found similar trend of decreasing BD (1.44 g cm^{-3} in one-year secondary forest to 1.16 g cm^{-3} in seventeenyear secondary forest) with increasing age of secondary forest stand. Burning had significant impact on BD and was observed to increase after burning (1.58 to 1.6 Mg m^{-3}). However, significant interaction was not found between burning effect and fallow length in soils of Mizoram. Probably, higher organic matter content in unburnt situation lowered the bulk density and the finding was also in harmony with the report of Biswas *et al.* (2012) who also found that BD increased from forested site to slashed and burnt situation (from 1.38 to 1.43 Mg m^{-3}) in the Chittagong hill tracts of Bangladesh. In Peru, it was reported that BD increased from 1.16 Mg m^{-1} before clearing to 1.27 Mg m^{-1} after burning (Alegre and Cassel 1996). Another study on soil properties and vegetation cover in a Mediterranean Heathland after experimental burning also shown that BD increased after burning (from 1.4 to 1.5 g cm^{-3}) (Granged *et al.* 2011). Soil pH increased with the increasing length of fallow period and burning activity. Within each individual situation, the increase in pH was significant in most situations except for unburnt situation of Nagaland where it had insignificant increment. The interaction between burning and fallow length on pH was insignificant in Nagaland. The mature fallow phase accumulates more quantity of FFLs than younger fallow phase. The burning of more quantity of FFLs in mature fallow phase produces more alkaline ash materials and this might be the possible reason for higher pH in longer fallow phase than shorter one under burnt situation. Burning activity significantly increased EC of jhum soils (from 0.38 dS m^{-1} to 0.51 dS m^{-1} in Nagaland and from 0.34 dS m^{-1} to 0.37 dS m^{-1} in Mizoram). However, fallow types and the interaction between burning and fallow did not have significant differences with EC. It was portrayed that EC of the soil increased from 0.2 dS m^{-1} before pre-fire conditions to 1.4 dS m^{-1} immediately after burning (Granged *et al.* 2011).

Soil heating due to burning activity alters the content of SOC. The SOC content decreased significantly after burning and the extent of decrease ranges from 17.3 to 22% both in Nagaland and Mizoram. Our findings corroborated the past findings of Lenka *et al.* (2012). Higher topsoil temperature causing faster decomposition rates of organic matter and limited incorporation of litter materials in burnt plots were also accounted for decline in SOC. The content of avail-N in unburnt soil was higher than in burnt soil and the longer length of fallow phase contained higher content of avail-N than that in shorter fallow phase. This is in agreement with Neff *et al.* (2005). The effect of burning and the interaction effect between fallow period and burning situations were significant on the content of avail-N, but the length of fallow phase alone did not show significant difference. Decrease in the amount of NO_3-N after burning was previously reported, where the fallow period did not have much effect on the concentration of $NO_{3-}N$ (Ramakrishnan and Toky 1981). Xue *et al.* (2014) reported about a significant decrease in available N after burning as compared with unburnt soils while investigating the effect of wildfire in South China. Though the content of avail- P increased after burning, the extent of change in avail-P content was non-significant. The longer fallow length supported higher content of avail-P as compared to that in shorter fallow length and however, such effect was non-significant. Ramakrishnan and Toky (1981) described that available P shows an increasing trend with the lengthening of jhum cycle. Adeyolanu *et al.* (2014) also cited an insignificant increase of available P from 27.5 to 30.0 mg kg-1 while studying a slash and burn situation in tropical rainforest of Nigeria. The increase in available P can be attributed to the incorporation of P from the slashed biomass in the form of ash as indicated by Raison *et al.* (1985). The content of avail-K also increased insignificantly with the increasing length of fallow period. However, burning increased the content of avail-K significantly (14.9 and 7.8% increase in Nagaland and Mizoram, respectively). In similar situation, Ramakrishnan and Toky (1981) observed similar phenomenon of increase in K concentration after burning as well as an increasing trend with increase in fallow length and summarized the phenomenon to be controlled by the above-ground species. Neff *et al.* (2005) have however reported about a mild shift in the K content after burning.

The soils of the various land use management methods were classified as sandy loam (Table 5.5). All of the land use systems had different soil bulk density. The continuous cropping area had the highest soil bulk density, with the highest value of 1.76gcm-1 occurring at the 15–30cm soil depth. In this investigation, it was discovered that soil bulk density increased as soil depth increased. The increased bulk density of continuous cropping land compared to other land use regimes may be due to organic matter and tillage levels. According to Anikwe (2000), trafficking during field activities could increase soil bulk density since tillage procedures can create soil compaction.

Table 5.5. Soil properties at the different land use management systems before clearing.

Soil properties	Forest		Bush	fallow	Continuous	cropping	Alley	cropping
	0-15 cm	**15-30 cm**	**0-15 cm**	**15-30 cm**	**0-15 cm**	**15-30 cm**	**0-15 cm**	**15-30 cm**
Texture	SL	SL	SL	SL	SL	SL	SL	SL
BD (g cm^3)	1.32	1.43	1.02	1.25	1.52	1.76	1.26	1.35
pH	4.60	4.5	5.40	4.90	4.40	4.20	5.80	5.60
OC (g kg^{-1})	2.18	1.39	2.94	1.98	1.98	1.10	2.10	1.17
OM (g kg^{-1})	3.76	2.40	5,07	3.41	3.41	1.90	3.52	2.02
Total N (g kg^{-1})	0.64	0.39	0.84	0.46	0.18	0.08	1.84	1.24
Av. P (mg kg^{-1})	14.86	10.29	16.47	13.09	10.92	8.75	19.58	13.47
Ca (c mol kg^{-1})	2.07	1.02	2.37	1.17	1.41	0.95	2.97	1.64
Mg (c mol kg^{-1})	1.17	0.93	1.21	1.01	0.86	0.38	1.46	1.23
K (c mol kg^{-1})	0.36	0.12	0.49	0.16	020	0.06	0.72	0.31
Al + H(c mol kg^{-1})	2.69	3.72	2.04	2.89	3.06	4.12	1.98	2.17
CEC (c mol kg^{-1})	3.60	2.07	4.07	2.34	2.47	1.39	5.15	3.18

Source: Page et al. (1982)

5.5.2 Nutrient cycling

The work of Bartholomew, *et al.*(1953) is one of the few early attempts to understand the pattern of element cycling in tropical secondary successional fallows after slash and burn agriculture in Belgian Congo. They concluded that storage capacity of leaves and litter is saturated at an early stage in succession, and thereafter ,the total storage capacity increases more slowly and only in the woody perennials only. Another comprehensive study carried out at Burnihat in north-eastern India (Toky and Ramkrishnan,1983b) showed that leaves had the highest concentration of elements and the boles had the least. The concentration of nitrogen, phosphorus and potassium was much higher in the branches and leaves than in the litter, but the concentration of calcium and magnesium was the same in all these three compartments. The concentration of calcium and magnesium was much higher in trees and shrubs than the bamboos, while the later one had a much higher concentration of potassium. The weeds colonizing the field immediately after abandonment were found to be rich with phosphorous. With the development of secondary vegetation, the weight of elements in the living biomass increased due to linear increase in the biomass with age. The high enrichment quotient (concentration in the vegetation/uptake) for potassium in fallows 10 to 20 years old indicates a maximum accumulation of this element in the living biomass, and therefore, conservation of this vital element in the vegetation and releasing it slowly through litterfall.Greenland and Kowal (1960) reported almost three times more calcium than potassium in the above-ground biomass of 40-yera old forest in Ghana. Edward and Grubb (1982) also showed higher budget of calcium and magnesium in the trunks of large trees in the forest of New Guinea.

After the destruction of element cycling by slash and burn agriculture, the system recovers rapidly through a quick succession due to high rate of net primary productivity. Predominant weeds such as Eupatorium odoratum, Imperata cylindrica, Mikania macrantha, Borreria hispida and Sachharum spontaneum play important role in conservation of nutrients. The soil erosion is almost reduced to zero. Potassium which is one of the important elements and highly susceptible to leaching and run-off losses during cropping and early growth of fallows, is rapidly conserved in the living biomass of bamboos during the first 20 years or so. Any disturbance as in the short agricultural cycles of 5 years, as is now common is likely to adversely affect the recovery of soil fertility and elements conservation by the system as a whole.

Ramakrishnan and Toky (1981) reported that the carbon percent in 5, 10 and 30 years shifting cycle decreased from 1.82 to 1.78, 2.65 to 2.08 and 249 to 2.12, respectively, after 1 year of cropping. Other parameters are given in Table 5.6.

Table 5.6. Biomass at different cycles of shifting cultivation

Biomass Stage	1 Year	5 Years	10 Years	15 Years	20 Years
Standing biomass (t/ha)	5	23	57	104	148
Litterfall (t/ha/year)	1.2	4.8	7.1	7.7	9.7
Net Primary Productivity (t/ha/year)	5	8	14	17	18

Source: Ramakrishnan and Toky (1981)

5.6 NUTRIENT DYNAMICS DURING LAND CLEARING

Shifting cultivation can be broken down into three stages: (1) conversion; (2) cropping; and (3) fallow. Clearing and burning native vegetation removes shade canopies, reduces insect competition, and removes other physical barriers to cropping in the conversion process. Some of the nutrients held in plant biomass are also liberated during clearing and burning, making them available for crop use (Zinke *et al.*, 1978). Cropping practises differ greatly, as do their environmental consequences. Soil conservation methods are common in traditional agroecosystems, presumably contributing to their long-term viability, while they are few in shifting agroecosystems (Myers, 1992). In reaction to decreasing crop yields and increased insect pressure, fallows are used. Fallows help to restore soil fertility and other soil conditions that have been harmed by cultivation and conversion (Kalpagr, 1974; Pimentel and Heichel, 1991).

During combustion, sulphur and carbon are also released (Christanty, 1986). Organic material in the first few centimetres of the soil is also destroyed at higher burning temperatures (Andriesse and Koopmans, 1984). Desiccation of the soil by evaporation (Uhl *et al.*, 1981), changing of soil texture (Ahn, 1974), and deterioration of soil structure are all examples of physical changes (Christanty, 1986). Finally, by injuring or destroying seeds and self-propagating root systems and producing fertile conditions favourable to the recovery of invasive species such as Imperata cylindrical and Chasmopodium caudatum, fire affects the character of ecological succession during fallow (Ewel *et al.*, 1981; Nyegers, 1989; Seibert, 1990). Several strategies exist to mitigate the edaphic effects of clearance. Although clearance does not normally result in significant soil degradation, influences on structural qualities and bulk density can be reduced by minimising the use of heavy gear and strategically cutting trees such that some are coppiced or simply left alone (Lal, 1985; Nyegers, 1989).

Many investigations have been undertaken in various agro-ecosystems to assess the impacts of deforestation, afforestation, and agriculture on physical and chemical properties of the soil (Paix *et al.*, 2013; Belay *et al.*, 2015). Large cow grazing (pasture), coffee plantations, and secondary forest have been the main crop systems in the Amazon. Farmers in certain instances intercrop annual agricultural species with secondary forest after slashing and burning wooded areas. This strategy is based on an indigenous agricultural system that includes long fallow periods after a 1-3-year phase of cultivation (Comte *et al.*, 2012; Reichert *et al.*, 2016). Deforestation practises such as burning vegetation have had considerable effects on the physical (Comte *et al.*, 2012; Cruz *et al.*, 2014) and chemical (Lindell *et al.*, 2010) features of various soil classes. Ultisols and Oxisols under pastures in the Amazon had higher SBD than the same soil types in woody plots (Comte *et al.*, 2012). After cutting and burning of forest vegetation, the surfsce soil layer becomes compact which can be loosened by ploughings, though effect may be short lived (Reichert *et al.*, 2014), since soil integrate fast with reduced effect of organic matter on soil structure and system equilibrium.

Braz *et al.* (2013) reported that the Typic Hapludox soils in the Brazilian Amazon have, low levels of available P, Ca^{2+}, and Mg^{2+}, and high acidity and Al^{3+} saturation. The parent material with with more loosened clayey sediments, deposited during Holocene period, is the cause of chemical fragility in this environment. Vegetation cutting and burning method of forest clearing results in high nutrient losses due to volatilization, leaching, and erosion of ash, as reported by van Reuler and Janssen (1993) in an Ultisol from Africa and in different soil categories globally (Juo and Manu, 1996). At a deeper level, the pH difference between pasture and natural woodland was considerable (B1 horizon). Up to the B1 horizon, secondary forest and grassland had higher Zn values than native forest. In comparison to soil under native forest, coffee farming resulted in a considerable fall in Zn levels. In comparison to the other soils, soil under native woodland has a higher total porosity (STP). The most physical harm was caused by the conversion to grazing. After nine years of fallow, the secondary forest system demonstrated more sustainable use and good resilience. Coffee farming is one of the most prevalent strategies employed by farmers in the southern Amazon to reduce soil degradation. In this management approach, the farmer must fertilise to replace the exported nutrients in order to maintain soil fertility.

Under ever-increasing demographic pressure, the conventional farming methods' productivity and sustainability have been threatened by longer crop periods and shorter fallow periods. In the move from traditional shifting farming to semi-permanent systems using fertilisers, viable alternatives to these methods are necessary. Soil fertility management is essential for the long-term viability of semi-permanent agricultural systems that allow for crop intensification. In tropical regions, deforestation and forest degradation are serious challenges in terms of environmental degradation, biodiversity loss, greenhouse gas emissions, and ecosystem service reduction (Chazdon 2008). Deforestation and forest degradation are exacerbated by forest removal and land reclamation for agriculture (Don *et al.* 2011). Afforestation and reforestation, on the other hand, are proven methods for rehabilitating degraded and abandoned lands. Because it consists of various plant species with diverse root systems, natural bush regrowth is now regarded the most efficient type of fallow for nutrient recycling and biomass accumulation (Ewel 1986, Jaiyebo & Moore 1964).

5.6.1 Effects of logging

The mining and logging operations have been found to cause serious erosion problems, increasing the sediment yields at the catchment scale by 10 to 20 times, sediment loads of rivers, and aggravating flooding (Pushparajah, 1985), due to rapid economic growth in Southeast Asia. But when roads and log extraction (snig) tracks are carefully built and sediment yield to streams is limited by buffer strips, the sediment yield is much less. As illustrated in the Huay Ma Feung stream in Thailand during the rainy season of 1983, roads and snig tracks play a vital influence in enhancing sediment output. As a common precondition for shifting agriculture, cutting and burning vegetation exacerbates soil erosion, land degradation, and degradation of fauna and microbiology (Nair 1993; Marafa and Chau 1999, Sharma and Sharma 2004).

5.7 NUTRIENT DYNAMICS DUE TO BURNING

5.7.1 Fire and soil properties

Saharjo (1999) conducted a study to see the effects of fire on soil chemical and physical properties of young Acacia mangium plantations aged 1, 2, 3 and 4 years in South Sumatra, Indonesia. Fire was allowed to propagate naturally in a 300 m2 (15 x 20 m) plot in each aged plantation. Soil samples were taken before, immediately after and one year after burning. The results show that in the 1-y-old plantation there were significant increases in pH, available - P, Mg, K, Na, total bases and base saturation, and non-significant changes in the other chemical parameters in the soil immediately after burning. At one year after burning, relative to pre-burning levels, all the parameters showed significant decreases except for total bases, where the decrease was not significant, and available - P, Ca and base saturation, which gave significant increases. This trend was not repeated in the other aged plantations. Only C-organic and total-N decreased immediately (though only significantly in the two older plantations), and at one year after burning (significantly relative to pre-burning levels) in all plantations. Another common trend in all plantations was the significant increase in available -P immediately and at one year after burning (over the pre-burning levels). Soil permeability, porosity, and soil moisture content generally decreased immediately after burning, due to the destruction of litter and vegetation protecting the surface layer, but at one year after burning little destruction was found in the physical properties of soil. It can be concluded that some changes that occur immediately after burning are a temporary effect.

5.7.2 Heat effects

The greatest part of the total organic matter was Imperata cylindrica and shrubs with a diameter of less than 1 cm. The highest fuel load of 21.2 t ha-1 was found in the 2-y-old plantation, and the lowest of 14.5 t ha^{-1} in the 4-y-old plantation. Dead fine fuel moisture content was about 5%. This condition was supposedly caused by the dry season from July until October, especially in 1994. Fires spread in the plot naturally with the wind blowing irregularly. The maximum temperature reached at 0 m above the soil in the 1-y-old plantation was 343 °C, 454 °C in the 2-y-old plantation, 302 °C in the 3-y-old plantation and 302 °C in the 4-y-old plantation. The highest fire intensity of 863.5 kW m^{-1} was reached in the 2-y-old plantation. Immediately afterburning, the soil surface was covered with unburned materials and ash resembling snow with white and black colour. Average ash thickness was around 10 cm and in the zone with high fuel load it was around 15 to 20 cm. Shifting cultivation results in deforestation, drying-up of springs, wild animals' shelter, flooding, soil erosion, siltation of river beds and ecological imbalance. There are changes in physical, chemical and biological properties of the soil after burning. Immediately after burning, the soil properties undergo a complete change. The pH rises initially, favouring the release of soil nutrients for the benefit of the succeeding crop (Table 5.7).

Table 5.7. Changes in soil properties after burning of vegetation.

Stage	pH	OC (%)	Av. P (kg ha^{-1})	Av. K (kg ha^{-1})	Exch. Ca (me/ 100g)
Before burning	4.9	0.51	5.2	185	5.6
After burning	5.6	0.82	9.6	530	7.2

Source: Sharma, Datta and Sharma

5.7.3 Burning and soil properties

Because of oxidation loss, the organic carbon content of soil reduced dramatically after burning. After fire, the soil's pH, temperature, and bases may have boosted microbial activity, resulting in faster reduction of organic nitrogen to inorganic forms (Ahlgren and Ahlgren 1965). Table 5.8 depicts soil physical properties as affected with kind of MPTs.

Table 5.8. Soil physical properties as affected with kind of MPTs.

MPTs	OC (g kg^{-1})	Aggregate stability	Available water (m3m^{-3})	Infiltration rate (mm hr^{-1})	Erosion ratio
Alnus nepalensis	32.2	72.1	0.201	7.28	0.23
Gmelina arboria	28.6	67.9	0.183	5.36	0.24
Michelia oblonga	33.6	73.2	0.210	6.10	0.22
Parkia roxburghii	23.1	63.4	0.192	4.85	0.30
Pinus kesiya	35.4	75.6	0.220	8.04	0.20
Control (No tree)	15.6	56.8	0.151	3.84	0.39

5.7.4 Physical properties

(a) Impact of conversion

The opening of an area in which cultivation will take place is referred to as the conversion phase. To make room for the crop, a clearing is created, usually inside a vegetable mosaic with plants at various stages of succession. This opening is often cleaned using hand tools, and when the trees have been destroyed, fire is used to assist clean the area and boost the crop's productivity (McGrath, 1987; Kleinman *et al.*, 1995). Table 5.9 shows the main impacts on physical properties during the conversion phase, as well as their soil implications (Filho *et al.* 2013). It is demonstrated that the majority of the affects on the physical parameters of the system occur during this period.

Table 5.9. Shifting cultivation and soil physical properties during the conversion period.

Physical property of soil	Impact during conversion phase
Texture	The soil texture has a big influence on how the soil behaves during the conversion process. When the soil surface is exposed after clearing plants, the surface soil layer is eroded if the texture is light. Increased runoff causes significant soil erosion during rainstorms. The loss of soil also involves the loss of nutrients, making the soil less fertile, especially in terms of organic matter and some micronutrients such as zinc. Because of the usage of fire, the fine fraction of the soil changes, causing compaction. Additionally, higher granulometry increases the potential for nutrient absorption (CEC).

Soil structure	With clearing of biomass and its burning, a change in macroaggregates is possible. With burning of vegetation on soil surface, there is incorporation of carbon and carbonated matter in the humus. Fire also results in the loss of some favourable microorganisms, making the soil poorer of microbial communities.
Soil water	The air temperature in the tropics is often very high during the day and night. The soil surface is exposed to direct sunlight when the ground is stripped of vegetation for shifting cultivation, causing the soil to lose a lot of moisture. If timely rains are not received, dibbled seeds will not germinate, resulting in a poor crop stand. It is vital to maintain proper and adequate soil moisture. The burning of biomass lowers the surface layer's humidity, limiting biological activity and jeopardising the structure.
Bulk density	The absorption capacity of the soil is reduced when forest vegetation is cleared. The process also results in a drop in soil organic matter content, an increase in leeching, erosion, and a change in water conductivity and infiltration rate.
Temperature	When vegetation is cleared there is increase in decomposition process and a decrease in organic matter content of the soil.
Colour	The colour of the soil after burning process may be changed; however, there appears an Uncertainty with a density above 600 °C.

Improved over Filho et al. 2013.

5.7.5 Post-burn soil Properties

Ash improved soil pH and nutrient availability in Southwest Nigerian soils, according to Owolabi at al (2003) and Odedina *et al.* (2003). Ash caused a rise in soil pH and exchangeable cations, indicating that it had a liming impact. The CEC of forest land grew by 13% after the burn, whereas bush fallow and continuous cropping increased by 11% and 4%, respectively, over the baseline CEC levels. As a result of the burning, the total acidity (Al +H) decreased. Post-burn soil chemical properties are shown in Table 5.10.

Table 5.10. Soil chemical properties at 50 days post – burn (0 – 15cm)

Soil properties	Forest	Bush fallow	Continuous cropping	Alley cropping
pH	0.48	0.50	0.43	0.56
OC (g kg^{-1})	1.36	1.22	0.77	1.08
OM (g kg^{-1})	2.85	2.10	1.34	1.86
Total N (g kg^{-1})	1.24	1.89	0.56	2.97
Av. P (mg kg^{-1})	16.32	19.06	10.48	21.76
Ca (c mol kg^{-1})	1.41	1.62	1.18	2.49

Mg (c mol kg^{-1})	0.32	0.94	0.10	1.06
K (c mol kg^{-1})	0.19	0.14	0.07	0.99
Al + H(c mol kg^{-1})	1.31	0.92	1.97	0.76
CEC (c mol kg^{-1})	1.92	2.70	1.35	4.54

Source: Page *et al.* (1982)

Cultivation increased the bulk density of the soil. Prunings were used to reduce the bulk density of the soil. This shows that prunings may have lowered erodobility, which is one of the variables that causes soil bulk density to increase. The moisture content and overall porosity were both lowered. The pH of the soil, which had raised as a result of burning, had lowered as a result of farming. The soil pH of continuous cropping land became exceedingly acidic, whereas forest and bush fallow land reverted to a very strongly acidic state. The soil in the alley cropping scheme, on the other hand, remained mildly acidic. The elemental nutrients, such as organic C, total N, accessible P, and exchangeable cations, also decreased. The continuous cropping soils had the lowest nutrient levels. Ayanaba *et al.* (1976) and Juo and Lal (1977) obtained similar results after removing a secondary forest in Southwest Nigeria with a continuous farmed Alfisol.

5.7.6 Soil physical properties and fire

Immediately after burning, soil porosity, moisture content, poor pore drainage and permeability decreased significantly in the 1-y-old plantation, but available water increased. However, soil porosity and permeability decreased significantly while available water was still high relative to the pre-burning levels at one year after burning. In the 2-y-old plantation, soil porosity, soil moisture content, poor pore drainage and permeability decreased significantly immediately after burning whereas good pore drainage and available water increased. At one year after burning, poor pore drainage and permeability decreased, but soil porosity, good pore drainage and available water increased significantly. Immediately after burning, poor pore drainage, available water and permeability in the 3-y-old plantation increased significantly but soil porosity and good pore drainage decreased. At one year after burning, soil moisture content and permeability increased but soil porosity, good pore drainage and available-water decreased significantly to less than the pre-burning levels. Available water and permeability in the 4-y-old plantation increased significantly immediately after burning but soil porosity, soil moisture content and pore drainage (good and poor) decreased. At one year after burning, available water and permeability increased while soil porosity decreased significantly relative to pre-burning levels.

5.7.7 Chemical soil properties

pH

Plant uptake has a net acidifying effect on soil because cation uptake is usually greater than anion uptake. If biomass is removed from the earth or trees are cut down, this will only result in acidification in the long run. If the biomass is not removed, the litter will be broken down by the acidity. Processes like as nitrification, for example, can lead to a drop in pH. The hydrogen ions are neutralized by mineralization and weathering. CaCl2 has a pH that is typically 0.5 pH units lower than de-ionized water. If soils contain large quantities of iron or aluminium oxides, observed pH in salt solution is higher than in water because oxides have more positive than negative ions. The latter is true in many tropical soils (such as oxisols) (Eriksson *et al.*; 2005). The availability of P and B diminishes as pH rises, resulting in deficits at higher pH values. The probability of Co, Cu, Fe, Mn, and Zn shortage increases when the pH goes over 7.0. Most crops enjoy a pH range of 5.5-7.0, however the lower end of this range is too acidic for some crops and plants. Acid soils with a pH below 5.5 include issues such as probable Al toxicity, excess Co, Cu, Fe, Mn, and Zn, and a deficit of Ca, K, N, Mg, Mo, P, and S. (and B below pH 5). With the exception of iron, the availability of micronutrients decreases as pH rises, except for Mo (Landon, 1991)

CEC

The capacity of the soil to bind exchangeable cations is measured as CEC (cation exchange capacity). It's the number of negative charges on the mineral's surface divided by the number of positive charges. Because the amount of changeable negative charges grows with pH, CEC is affected by it. CEC is limited in a soil with a low pH because some sites are not dissociated. Effective CEC is the actual CEC in that type of soil (Eriksson *et al.*; 2005). In heavily worn soils, such as many tropic soils, the contribution to CEC is extremely crucial. CEC is influenced by the amount of organic matter and clay in the soil. The type of clay is more essential than the amount of clay, because weathered clays do not contribute to the CEC. Weathered clays are widespread in warm and humid climates, and CEC is frequently found to be low in these places. CEC varies significantly with soil order and is frequently used to determine soil order (Brady; 2002).

BS

The fraction of base cations in respect to the total CEC is shown by BS (base saturation). A low BS implies that the soil is acidic (Eriksson *et al.*; 2005). At pH 5.0, some tropical soils have a high BS, which may explain why some acid-sensitive crops may be cultivated on acid tropical soils and why liming has little effect on their yield. Soils with a high BS are regarded fertile (eutric), while soils with a low BS are deemed less fertile (dystric) (Landon; 1991).

OM

When organic nitrogen, phosphorus, and sulphur become available when organic matter decomposes, OM (organic matter) is an important source. Organic humified matter has a carbon content of 58 percent, a nitrogen content of 5-6 percent, and a phosphorus content of 0.04-0.20 percent. Humus binds cations and is a significant source of plant-available cations. In coarse soils or severely worn soils with iron and aluminium oxides, humus makes a significant contribution to CEC. Because the components of humus have a poor solubility, they immobilise micronutrients and heavy metals. With hydrogen ions, organic matter contributes to the acidifying effect in the soil (Eriksson *et al.*; 2005).

EC

A soil with an EC (electrical conductivity) of 0-2 dS/m is classed as insaline since it has no salt effects (Landon; 1991). Most crops grow best in loam and silt loam soils, which contain just sufficient clay to hold water and nutrients for normal crop production but not enough clay to produce poor aeration and make the soils difficult to handle. A loamy texture describes a soil that contains 7-27% clay and an equal quantity of silt and sand. If the soil also has a lot of organic stuff, it's ideal for growing plants (Troeh & Thompson; 2005).

Exchangeable acidity (Al3+/H+)

The more acidic the soil is, the closer the exchangeable acidity (in number) is to the CEC (Clemson University Extension Service (1); 2008). Al-ion is liberated from clay lattices at pH levels below 5.5, putting soils with low pH at risk of Al toxicity (Landon; 1991). Aluminum is rarely compared to recommended values as an exchangeable cation. Because Al3+ and H+ are acid ions, determining pH and base saturation is more useful (Berggren (1995).

Phosphorus

Organically bound phosphorus accounts for 30-60% of the total phosphorus content in top soil, making it a significant source of phosphorus for plants. Mineralization allows it to become available to plants. The amount of phosphorus available to plants is directly proportional to the pH of the soil. Between pH 5.5 and 8.0, phosphorus is most soluble; at lower pH levels, it forms low-solubility compounds with iron and aluminium. At high pH, it can precipitate with calcium. Because phosphorus is immobilised by precipitation with Fe, Al, and Ca, the effects of P-fertilizers may be absent. Organic compounds can increase solubility by pushing phosphate ions away from colloid surfaces with negatively charged humic acids. The distribution of phosphorus in agriculture soils is frequently irregular in the soil profile. It prefers to stay in the top soil due to its poor motility. Plants also take up phosphorus from the subsoil, and because of the low mobility, phosphorus levels in the subsoil are extremely low (Eriksson *et al.*; 2005). Fertilizer reaction is unlikely at high levels, possible at medium levels, and most likely at low levels. Maize has a low P demand, with a shortfall of 4 ppm compared to >8 ppm for a sufficient level (Landon; 1991).

Nitrogen

Inorganic nitrogen (NO_3^-) and ammonia ions (NH_4+) are both present in soil solution. In clay soils, NH4+ leaching is minimized because it binds strongly to the negative charges of the clays, but NO3- is considerably more mobile and poses a greater risk of leaching (Eriksson *et al.*; 2005). Nitrate leaching, on the other hand, has been discovered to be linked to clays quite deeply, as seen in Ultisol profiles, rather than seeping into the groundwater. This is beneficial because trees with deep root systems can be planted and so integrated into a crop cycle, facilitating nitrogen availability for crops (Brady; 2005). The ammonia ion, like the potassium ion, is adsorbed in exchangeable form and fixed in clay minerals. In humus components, nitrogen is also found in NH2 groups and related nitrogen compounds. Humus contains about 6% nitrogen and hence represents the actual nitrogen concentration of the soil. Microbial activities and mineralization make it available to the plants. Many tropical bush and tree species rely on bacteria and root nodules for symbiotic nitrogen fixation. Bacteria that produce root nodules on the roots of legume plants can fix nitrogen (Eriksson *et al.*; 2005). The levels of Ntot are categorised as shown in table X when using the Kjeldahl technique (Landon; 1991).

Calcium

The chemical and physical qualities of the soil are influenced by Ca^{2+} (Eriksson *et al.*; 2005). In both the soil solution and the exchangeable fraction, Ca^{2+} is the most common type of calcium. Exchangeable calcium can make about 50-90 percent of the total base cation concentration in agricultural soils. Rapid losses of calcium and magnesium from forest ecosystems are caused by timber harvesting methods that expose the soil to erosion and nutrient leaching. Mineral weathering's release of calcium will not be able to keep up with the losses. In the most weakly buffered watersheds, acid rain mixed with extensive timber harvesting may be reducing the Ca stocks. Forest growth has rarely responded positively to calcium application (Brady; 2002). Ca shortage is only found in soils with a low CEC and a pH of 5.5 or less. However, in soils with a more neutral response, high K levels may impede plant uptake of Ca (Landon; 1991).

Magnesium

Minerals often have more magnesium than calcium, but more Caex than magnesium. Mg^{2+} makes up 10-30% of base cations in most soils, and the amount rises with clay content, and is generally higher in subsoil than in topsoil. Plant Mg absorption is influenced by the proportions of K^{+} and Mg^{2+} in the soil, as well as soil pH. (Eriksson *et al.*; 2005). Mg is less plant accessible when Ca levels are high. On coarse, acidic soils, magnesium shortage is more likely. Mg levels >4.0 meq/100g are regarded high in tropical soils, while 0.5 meq/100g is considered low (Landon; 1991).

Potassium

When K is present in the soil solution and adsorbed to soil particles, it is available to plants. Apart from that, it can be found in biological matter in exchangeable form, albeit in little amounts due to K+'s low binding capability. Plants cannot absorb K because it is linked to K minerals or attached to clay minerals (Eriksson *et al.*; 2005). When K fertiliser levels are regarded excessive, a response is improbable, but it is likely when levels are less than 0.2 meq/100g. Between 0.07 and 0.20 meq/100g is the absolute minimum level of K. A relative level of at least 2% of the sum of all exchangeable bases is required (Landon; 1991).

Micronutrients

Plant micronutrients are nutrients that are required for plant growth and development but are only present in trace amounts. In acid soils, where it is frequently present at dangerous amounts, Al is a major limiting factor for growth. Because Al toxicity cannot be directly attributed to exchangeable, accessible, or total aluminum, it is difficult to find reference values. Al toxicity varies depending on the environment at a given location and the plant species (lvarez *et al.*, 2005). Weathering releases Cu from mineral rocks, which is then adsorbed by cation exchange. Copper shortage is more common in organic soils than mineral soils because copper derives from mineral weathering. Because its solubility declines with increasing pH, deficiency can cause issues for plant growth in alkaline soils and at very low pH levels (Troeh & Thompson; 2005). Copper has a usual range of 2-250 ppm, with a mean value of 30 ppm (Landon; 1991). 5-60 ppm is a common range for agricultural soils (Stevenson; 1986). Copper deficiency is particularly harmful to cereals (such as maize) and vegetables. Copper levels above 100 ppm are considered high, but hazardous amounts are uncommon (Landon; 1991). When copper content in mineral soils is less than 4 ppm and 20-30 ppm in organic soils, copper shortage is likely (Stevenson; 1986). Iron is the micronutrient that is required in the most significant levels. Even when huge levels of iron are available in the soil, deficiency can emerge because iron is commonly found in insoluble forms (Troeh & Thompson; 2005). Iron is the micronutrient with the highest rate of deficiency in soil (Stevenson; 1986). Because iron ions precipitate in low-solubility compounds at high pH, iron is more accessible at low pH. Mn in soil ranges from 20 to 3000 ppm and can be poisonous at high levels. Iron can become toxic at pH levels below 4-5. Low pH raises solubility and the likelihood of leaching. There is a potential of Mn poisoning in acidic environments and poorly drained soils (Troeh & Thompson; 2005). Zinc deficiency can be seen in heavily worn soils, but it is uncommon. Zinc solubility increases with decreased pH, and mineral soils commonly have zinc concentrations of 10-300 ppm, but only a small percentage is in solution. The zinc ion is strongly adsorbed to clays, carbonates, and other soil minerals, which is one reason for this (Troeh & Thompson; 2005). Sandy soils, degraded soils, and peat and muck soils all have low zinc levels. When phosphate levels are high, zinc deficiency is also on the rise (Stevenson; 1986. p.326-327). Zinc shortages are uncommon in acid soils because zinc availability increases as pH rises (Landon; 1991).

5.7.8 Soil nutrient status of burnt and unburnt soils

Burning is an important aspect of the shifting cultivation in the southern portion of Nigeria, where it is commonly utilised to prepare land for ploughing (Jamala *et al.*, 2012 ; Edem *et al.*, 2013). Burning has been the easiest and most practical way of clearing and preparing land for planting in the humid tropics for food production as more land is cleared and prepared each year (Ruddiman, 2003; Edem *et al.*, 2012). Slash fires have some of the highest rates of nutrient loss of any fire (Kauffman *et al.*, 1995), and maintaining soil health necessitates a full knowledge of nutrients gain and losses. The average sand, silt, and clay (0-15 cm) of an unburned plot were 73.0%, 8%, and 19%, respectively, while the burnt plot had 67.2, 6.7, and 26.1 percent (Table). Kattering *et al.* (2000) also noticed a decrease in silt in the fire area. The other parameters of burnt and unburnt soils are also given in the Table 5.11.

Table 5.11. Soil quality traits for burnt and un-burnt soil at different soil depths

Soil indicator	Unburnt plot	Burnt plot	Soil indicator	Unburnt plot	Burnt plot
Soil depth	0-15 cm	0-15 cm	Soil depth	0-15 cm	0-15 cm
pH (H2O)	5.58	7.22	Av. P (ppm)	2.54	3.45
pH (KCl)	4.86	6.50	Ca (cmol kg^{-1})	2.53	2.66
Sand (%)	73.00	67.2	Mg (cmol kg^{-1})	1.60	1.96
Silt (%)	8.00	6.70	K (cmol kg^{-1})	1.85	0.59
Clay (%)	19.00	26.10	Na (cmol kg^{-1})	0.13	0.15
SOC (%)	2.31	2.73	EC (cmol kg^{-1})	5.02	5.38
SOM (%)	3.99	4.72	BS (%)	82.5	82.2
Total N (%)	0.19	0.66			

5.7.9 Soil pH

The pH (KCL) of slash and unburnt samples ranged from pH4.9 to 5.9 (0-30cm), while slash and burnt samples ranged from pH5.9 to 6.5. This suggested that ashes were released throughout the burning process. As a result, the ash acts as a liming and fertilising ingredient while also lowering soil acidity (Adeyolanu *et al.* 2013). During the fractionation process, the early formation of K and Na oxides, hydroxides, and carbonates increased Ca^{2+} affinity for P and the potential for precipitation of Ca phosphate minerals (Arocene and Opio, 2003). Higher Ca^{2+} affinity for P and the possibility for Ca phosphate mineral precipitation during the fractionation process were also enhanced by increased soil pH. (Arocene and Opio, 2003). Giardina and colleagues (Giardina *et al.*, 2000).

5.7.10 Soil SOC and SOM

The subsurface SOC ranged between 1.56-1.96 g.kg^{-1}, and it was higher in the fields where the the vegetation was burnt than the fields where slash was not burnt but simply incorporated, possibly because the soils were sandy in texture and more leaching took place. These values ranged between 2.31-2.7 g.kg^{-1} in surface soils, with higher SOC mean content in burnt fields than slash and unburnt fields. Edem and Alphonsus (2016) found an increase in soil SOC in the soil surface after burning, which supports this discovery but contradicts Pyne and Goldammer's (1997) finding that organic carbon loss in soil happens as a result of fire depleting litter on the surface. According to Ubuoh *et al.* (2016), adding organic matter to the soil improves structure stability, rainfall resistance, infiltration rate, and faunal activity.

5.7.11 Soil carbon

Suarez de Castro and Rodriguez (1955) found very little change in nitrogen in volcanic ash soils in Colombia and Guatemala, and that it decreased at a far slower rate than organic carbon. Shifting agriculture at lower elevations involves high intensity burn depending upon quantity of slash under different length of fallow period from 5 years to sometimes 30 years. In comparison to medium and longer cycles, the short jhum cycle of 5 years starts with a low carbon budget (Ramakrishnan and Toky, 1981). Because of poor litter return (1.2 t $ha^{-1}yr^{-1}$) and increased oxidation at the surface soil, soil carbon depletion occurs for up to 22 percent during the first year of cropping and for up to 5 years during the early stages of succession. One of the strongest arguments against a short cycle is this. Because of the large litter fall (7.1 t $ha^{-1}yr^{-1}$), which bamboos (*Dendrocalamus hamiltonii*) are the main source, soil carbon improvement is greatest at 10-year age fallow (Toky and Ramakrishnan, 1983). This is one of the strong reasons that soil fertility is replenished in a fallow of 10-year which is the optimum cycle for the yields of cereals and legume against a short cycle of 5-year where the yields of rice and legumes is much low. Higher level of organic carbon maintained after one year of cropping in medium and longer cycles may be able to support another year of cropping. However, there was a net loss of carbon throughout cropping phases and fallow phases up to 10 years irrespective of one or two years of cropping or the types of fallows (Nye and Greenland, 1964; Greenland, 1975; Ramakrishnan and Toky, 1981). The keystone tree species such as bamboos (*Dendrocalamus hamiltonii*) in the lower elevations and alders (Alnus nepalensis) in the higher elevations are extremely important for conservation of potassium and fixation of atmospheric nitrogen in shifting fallows fallows.

Higher level of organic carbon maintained after one year of cropping in medium and longer cycles may be able to support another year of cropping. However, there was a net loss of carbon throughout cropping phases and fallow phases up to 10 years irrespective of one or two years of cropping or the types of fallows (Nye and Greenland, 1964; Greenland, 1975; Ramakrishnan and Toky, 1981). The keystone tree species such as bamboos in the lower elevations and alders (Alnus nepalensis) in the higher elevations are extremely important. Changes in carbon (%) in slash and burn agriculture (jhum)under short, medium and long cycles in north-eastern India (Ramakrishnan and Toky,1981). In some parts of Meghalaya around Shillong, India, the farmers are not burning the slash completely. The slash of branches and needles of Pinus kesiya trees and other herbaceous weeds are pulled in rows, the soil is put on the slash and then it is slowly charred to retain maximum carbon in the soil. Instead of slash and burn agriculture, it is slash and char. They also add some inorganic fertilizers to raise the potato crops. These are more and less permanent systems (Mishra and Ramakrishnan,1983). Amendment of soil with bio-char has been observed to increase the activity of mycorrhizal fungi and microbial growth. It also increase the soil cation exchange capacity leading to improved plant nutrient uptake.

5.7.12 Effect on soil chemical properties

Under the influence of land-use schemes, all of the measured soil chemical characteristics changed (Table 5.12). The land-use systems had a substantial impact on soil pH, total nitrogen, organic carbon, accessible P, exchangeable Ca, exchangeable Al, base saturation, Al saturation, and ECEC (Effective Cation Exchange Capacity).

Table 5.12. Effects of land-use systems on chemical properties of the soil

LUS	pH	Total N (%)	OC (%)	Logav P mg kg^{-1}	Ca	Mg	K	Al	ECEC
						cmol.	kg^{-1}		
Sfor	4.13c	0.17a	2.24a	0.98a	0.35b	0.27a	0.09a	4.39a	5.10a
Coco	4.30b	0.10b	1.35c	0.65b	0.32b	0.24a	0.07a	3.53b	4.16b
Yfor	4.59a	0.12b	1.72b	0.75ba	0.82b	0.50a	0.07a	1.90d	3.29c
Chro	4.75a	0.11b	1.48cb	047b	1.42a	043a	0.10a	2.70c	4.66ba
Crop	4.75a	0.10b	1.41c	0.74ba	075b	0.32a	0.08a	1.96d	3.12c
P	<0.0001	<0.0001	<0.0001	=0.0008	=0.0008			<0.0001	<0.0001
R2 (%)	85	88	87	89	71			91	79
CV (%)	4.17	21.26	19.63	41.88	97.80			17.43	16.70
Mean	4.50	0.12	1.64	072	0.72			2.94	4.09

Source: Agoume and Birang, 2009

The pH of the soil and the amount of nutrients in it were both low (Table 1). The higher degree of weathering of rock component minerals caused this. Soil organic matter is responsible for a major portion of plant nutrients and roughly 90% of the soil nutrient retention capacity in such soils (Kauffmann 1998). The soils with ECEC of 4 me/100g or low had little ability to retain nutrient cations (Buol *et al.* 1975).

5.7.13 Changes in soil chemical properties due to fire

Immediately after burning, pH, available-phosphorus, magnesium, potassium, sodium, total bases and base saturation increased significantly in the 1-y-old plantation, while one year later, pH, C-organic, total nitrogen, magnesium, potassium, sodium and CEC decreased significantly to less than the pre-burning levels. In the 2-y-old plantation, pH, calcium, magnesium, sodium, total bases and base saturation decreased, but available phosphorus increased significantly. In addition, at one year after burning, all parameters except available phosphorus decreased significantly relative to levels before burning. In the 3-y-old plantation, pH, available phosphorus, calcium, total bases and base saturation increased, although C-organic, total-nitrogen, sodium and CEC decreased significantly. At one year after burning, C-organic, total-nitrogen, calcium, magnesium, potassium, sodium, total bases, CEC and base saturation decreased significantly to less than the pre-burning levels. In the 4-y-old plantation, C-organic, total-nitrogen, calcium, total bases, CEC and base saturation decreased significantly immediately after burning, but available-phosphorus and potassium increased. At one year after burning, all parameters except available-phosphorus decreased significantly relative to pre-burning levels.

The increase of pH immediately after burning in the 1-y-old and 3-y-old plantations may be due to the increase of ash left above the soil that is rich in exchangeable cations, including available phosphorus and calcium in the 3-y-old plantation and magnesium, potassium, and sodium in the one year old plantation. Ash high in such exchangeable cations tends to raise the pH (Chandler *et al.* 1983). An increase of these cations causes total bases and base saturation to increase. The decrease of pH immediately after burning in the 2-y-old plantation may be due to the decrease of calcium, magnesium and sodium in the forest soil. In any case, ashes are beneficial to plants, with superficial roots absorbing the released nutrients. Nutrient concentration is often highest in early burnt plots and lowest in later burnt ones, with intermediate values found in protected plots (Menaut *et al.* 1993). At one year after burning pH in all but the 3-y-old plantation decreased. This may be due to the lack of cations such as calcium (except in the 1-y-old plantation), magnesium potassium and sodium because of leaching and runoff (Toky & Ramakrishnan 1981). A decrease in these cations causes a decrease in total bases and base saturation (except in the 1st year) and CEC is commonly reduced by fires as the organic matter content of soil is reduced (Trabaud 1990), since organic matter can account for a sizeable portion of the CEC (Brady 1974). CEC also tends to decrease after burning (Scotter 1963, Trabaud 1994). The significant decrease of total nitrogen immediately after burning, especially in the 3-y-old and 4-y-old plantations, may be caused by volatilisation during burning. Nitrogen will volatilise at 200 °C (White *et al.* 1973). Burning releases cations from forest vegetation, but it results in volatile losses of a significant portion of ecosystem nitrogen and sulphur. A decrease in the organic matter composition of the forest floor is linked to nitrogen loss and the release of other nutrients (Chandler *et al.* 1983).

5.7.14 Properties of soil

The physicochemical parameters of the soil in the four study sites are summarised in Table 5.13. The soils are sandy loam in texture in all four study sites, with a blackish grey colour in sites I, II, and III and a reddish colour in site IV. Across the sites, the soil texture was constituted of sand (69.00–72.50 percent), silt (16.50– 17.50 percent), and clay (11.00–13.50 percent). The soil at all four study locations is acidic, with a pH range of 4.96-5.72.

Table 5.13. Physico-chemical properties of soil at different sites

Parameter	Site I (Recent slash and burn)	Site II (3 years old fallow)	Site III (7 years old fallow)	Site IV (Protected forest fallow)
pH	5.72	5.22	5.21	4.96
Soil moisture (%)	15.67	16.67	18.67	20.00
Soil temperature	22.87	20.33	20.22	20.00
OC (%)	2.38	0.54	0.65	1.20
Total N (%)	0.28	0.13	0.14	0.19
Total P ($\mu g g^{-1}$ soil)	411.30	368.00	376.70	404.30

Significant human habitation of the Brazilian Amazon has resulted in increased deforestation over the previous 50 years. Despite the fact that some nutrients (such as exchangeable phosphorus) grew at the soil surface, total carbon and nitrogen (N) pools remained unchanged. The buildup of both forms of accessible nitrogen (NO3 and NH4) as well as potassium in subsoil horizons was also a sign of nutrient leaching. In addition, there were indicators of erosion, with a considerable rise in surface density and up to 25% fine particle loss at the surface (Beliveau *et al.* 2015). The dynamics of soil nutrients throughout the cropping period of shifting cultivation were investigated. Numerous loss processes, such as erosion, leaching, and crop uptake, kick in after the slash and burn practice's initial enrichment of soil with key nutrient components. The most severe nutrient loss happened as a result of premonsoon showers, which occurred even before the first corn crop was sowed (Tulaphitak *et al.* 1985). As a result, in monsoon tropics with distinct wet and dry seasons, the initial enrichment was followed by a considerable initial loss, which appeared to be inescapable. The process of nutrient loss, on the other hand, is anything but simple and uniform. One part of the process that complicates it is the release of nutrients from decomposing organic materials.

5.7.15 Biological properties

The main biological properties of the soil affected by the shifting cultivation are the following:

a) Impacts of conversion and cultivation

The action of fire and the repetition of farming cycles have a negative impact on the biological qualities of the soil. The structure, composition, and variety of the soil biota (micro-, meso-, and macrofauna) are influenced by the decrease of organic matter induced by fire. The soil's exposure raises its temperature by encouraging desiccation, which reduces the activity of micro-, meso-, and macrofauna. However, when the temperature rises, the rate of decomposition accelerates, increasing the biomass of fauna. Fire has a significant impact on the seed bank, which has been emphasised as a crucial component in various studies due to its effect on ecological succession. The emergence of invasive species is aided by the decrease of seed diversity and density, and the soil's resistance is jeopardised.

Effects of Shifting Cultivation on soil nutrient status

Rare plant species, shrubs, medicinal plants, and minor forest products are all lost as a result of shifting farming. The practice results in deforestation, drying-up of springs, wild animals' shelter, flooding, soil erosion, siltation of river beds and ecological imbalance. There are changes in physical, chemical and biological properties of the soil after burning. Immediately after burning, the soil properties undergo a complete change. The pH rises initially, favouring the release of soil nutrients for the benefit of the succeeding crop (Table 5.14).

Table 5.14. Effect of biomass burning on soil properties.

Soil characteristics	Before burning	After burning	End of crop cycle
pH	5.1	5.5	4.2
OC (%)	0.13	0.25	0.05
Available P (kg ha^{-1})	3.3	5.5	1.2
Available K (kg ha^{-1})	210	570	40
Exch. Ca (meq $100g^{-1}$)	7.1	9.5	2.8

Source: "Waste land management" by V.V. Dhrubanarayan (ICAR).

The abandoned plot of land regenerates and becomes fertile after the cropping phase, but the original forest flora, animals, and ecosystem are not fully restored. After 20-30 years, the cultivators return to this spot. In northeastern India, where shifting farming is practised on a large scale, this interval has been reduced to 3-4 years due to demographic pressure. There is substantial increase in organic carbon and, available P and K. However, this increase depends on the amount of vegetation burnt (Sharma and Bujarbaruah 2016) (Table 5.15). Therefore, sufficient time has to be given between two cropping phases to help rejuvenation of sufficient vegetation for the success of the practice. Much of the microbial population may perish during burning due to intense heat and high temperature in the surface layers of soil. Some species of microbes may resist this heat. The perished microbes are likely to be there in deep layers of the soil and may re-establish in the surface soil with the passage of time. Detailed studies are required on this aspect.

Table 5.15. Changes in soil properties after burning of vegetation.

Stage	pH	OC (%)	Av. P (kg ha^{-1})	Av. K (kg ha^{-1})	Exch. Ca (me/ 100g)
Before burning	4.9	0.51	5.2	185	5.6
After burning	5.6	0.82	9.6	530	7.2

To improve the places where shifting agriculture has been practised for generations, changes in the soil qualities of shifting cultivation sites must be documented and constantly monitored. As a result, soil samples were gathered over a period of one to three years from two sites in India's Tripura state and analysed (Datta *et al.*,2001, 2004). Table 5.16 shows that the soils were acidic, with pH ranging from 4.5 to 5.1.

Table 5.16. Impact of shifting cultivation on soil properties.

Properties	West Tripura			South Tripura		
	1st yr	**2nd yr**	**3rd yr**	**1st yr**	**2nd yr**	**3rd yr**
pH (H2O)	4.70	4.60	4.50	5.10	5.10	5.00
Exchangeable acidity [c mol (p+) kg^{-1}]	2.33	2.30	2.46	1.95	2.31	2.51
Org. Carbon (g kg^{-1})	7.30	6.70	6.50	8.30	7.00	6.10
Bulk density (mg m^{-3})	1.30	1.30	1.30	1.20	1.30	1.30
Water holding capacity (%)	34	32	32	40	35	35
Cation exchange capacity [c mol (p+) kg^{-1}]	4.22	3.86	3.38	4.29	4.20	3.33
Exchangeable cations [c mole (p+) kg^{-1}]	0.84	0.76	0.75	1.10	1.37	1.06
Base saturation (%)	20.40	19.90	22.40	25.80	32.10	34.40

Source: Datta et al. 2004

Following changes were noted:

i) Acidic pH (4.5 to 5.8) noted in shifting cultivation sites.

ii) A decline in soil pH by 0.1 to 0.2 units noted with the rise in shifting cycle.

iii) Dominance of exchangeable acidity in soils under shifting cultivation.

iv) A decline in soil organic carbon by 0.8 to 2.2 g kg^{-1}.

v) Low aromatic condensation of humic acids after 2nd and 3rd year of shifting cultivation.

vi) Low cation exchange capacity in soils and decreasing trend over shifting cycle.

vii) A decline in exchangeable cations in soils.

viii) An average of 20.4 to 22.4% and 25.8 to 34.4% base saturation in soils.

ix) 9.8 to 11.4% decline in nitrogen availability in soils under study.

x) Very low phosphorus availability.

xi) A decline in available potassium from 150 to 125 kg ha^{-1} and 319 to 203 kg ha^{-1} in soils.

5.7.16 Soil carbon

Higher level of organic carbon maintained after one year of cropping in medium and longer cycles may be able to support another year of cropping. However, there was a net loss of carbon throughout cropping phases and fallow phases up to 10 years irrespective of one or two years of cropping or the types of fallows (Nye and Greenland, 1964; Greenland, 1975; Ramakrishnan and Toky, 1981). The keystone tree species such as bamboos (Dendrocalamus hamiltonii) in the lower elevations and alders (Alnus nepalensis) in the higher elevations are extremely important for conservation of potassium and fixation of atmospheric nitrogen in jhum fallows.

In some parts of Meghalaya around Shillong, the farmers are not burning the slash completely. The slash of branches and needles of Pinus kesiya trees and other herbaceous weeds are pulled in rows, the soil is put on the slash and then it is slowly charred to retain maximum carbon in the soil. Instead of slash and burn agriculture, it is slash and char. They also add some inorganic fertilizers to raise the potato crops. These are more and less permanent systems (Mishra and Ramakrishnan,1983). The addition of bio-char to soil has been shown to boost mycorrhizal fungus activity and microbial development. It also improves plant nitrogen uptake by increasing soil cation exchange capacity. Since of its somewhat alkaline composition, it is beneficial in acidic tropical soils because it raises pH. If bio-char is widely employed for soil improvement, it will result in massive carbon sequestration on a worldwide scale, assisting in the mitigation of global warming. The increase in crop yield is directly proportional to the increase in soil carbon. Crop yields would increase by 20 to 40 kg/ha of wheat, 10 to 20 kg/ha of maize, and 1 kg/ha of cowpeas for every tonne of carbon added to the soil (Lal, 2004).

5.7.17 Soil nutrient changes after clearing and burning

One of the most prominent reasons for surrendering the field to forest regrowth, according to Sanchez (1976), is the loss in accessible phosphorus. Although many tropical soils are poor in phosphorus, mature tropical forests show no signs of phosphorus deficit, presumably due to the closed nutrient cycle. The soil microflora is also altered as a result of clearing and burning. After a "flush" of microbial population, followed by a drop nearing new equilibrium levels, burning induces partial sterilisation of the soil (Laudelot 1961). After clearance, a higher number of bacteria (as opposed to fungus) is discovered. During the dry season, the overall microbial population falls, but increases during the rainy season, as well as with mulching and fertilising. Burning has little effect on nitrogen-fixing microorganisms, although pH variations undoubtedly alter their number.

After clearing and burning, quick mineralization of organic matter and inputs of ash result in a large increase in available nutrients for the first crop seeded in most tropical soils (Sanchez 1973). With repeated croppings, yields decrease over time, which is one of the main reasons growers relocate. Decreased yields are caused by an increase in pests, diseases, and weeds, topsoil erosion, deterioration in the physical state or nutrient status of the soil, and changes in the number and composition of soil fauna and flora (Nye and Greenland 1960). Lower crop yields are frequently caused by a decline in soil fertility, which can be caused by a loss of humus in the soil, changes in the physical condition of the soil, or changes in the nutritional status. During the cropping season, the loss of humus content is significant in the first year and then gradually slows. Nye and Greenland (1960) considered that humus loss is not always a severe concern, even during a ten-year cropping cycle, and that the change in humus quality is more important than the change in quantity. The characteristics of humus, as well as its microfauna, are affected (Ahn 1974). As a result, humus changes are frequently regarded as basic, causing most other soil changes to some extent.

All of these factors might contribute to a decrease in soil production when shifting cultivation is used. The phrase "soil constitution" refers to the physical qualities of the soil that impact its fertility and may deteriorate with cropping (Nye and Greenland 1960). The amount of water held for crops when drainage is no longer possible, as well as the exchange of carbon dioxide and oxygen in the root zone, are all determined by the size and continuity of the pores in the soil. The unhumified organic stuff on the immediate surface decays quickly, keeping the pores open. Further deterioration in soil constitution is minimal if an effective vegetative cover has grown. Natural soil aggregation, as opposed to humus-induced aggregation, varies greatly from soil to soil. In Brazil, for example, there was less reduction in soil structure on Oxisols than on Ultisols under changing cultivation cropping (Seubert 1975).

The investigation's conclusions showed that various land use regimes had varying nutrient levels. Agriculture and burning altered the soil's chemical and physical properties. Burning increased the pH, accessible P, exchangeable K, Ca, and Mg levels but decreased the bulk density of the soil. Nutrient accessibility was decreased and bulk density rose as a result of cultivation. The results of this investigation show that although burning could lead to a brief drop in the quantity of microorganisms, enhancements in physico-chemical properties might promote microbe growth.

5.8 NUTRIENT DYNAMICS DURING CROPPING

5.8.1 Soil properties

Cropping has yet another set of soil consequences. Cropping practices differ greatly in terms of relative benefit and agroecosystem. How much of the soil surface is exposed to incoming precipitation, wind, and sunshine, and thus how much runoff and erosion occurs, is determined by the rate of development and the degree of canopy cover (Renard *et al.*, 1991). Furthermore, the types and combinations of crops planted in a swidden have an impact on soil fertility and erosion susceptibility. Tillage using hoes may enhance short-term agronomic conditions, but it also disturbs the soil surface, breaking up aggregates and increasing the area exposed to rain and wind erosion (Lal, 1987, Pimentel and colleagues 1987). In fact, under these circumstances, sediment loss throughout the cropping period could skyrocket (Lal, 1987). Because of increasing decomposition and restricted replacement, soil organic matter normally decreases during the cropping season. According to Driessen *et al.* (1976), the length of continuous cropping has a negative connection with organic matter concentration. Decreased CEC, higher acidity, lower nutrient levels (particularly nitrogen and phosphorus), deterioration of soil structure, reduced porosity and aeration, and reduced water infiltration capacity may all result from a drop in soil organic matter levels (Ahn, 1974).

In the upper Amazon in Peru, migration and changes in population and land use, have affected the ecosystem and environment. In a study conducted with five different land uses such as; primary vegetation, secondary vegetation, pasture, plantation and burned fields, showed that the differences in N content, base cations and other soil properties among the land uses were not as consipicuous as thought to be. However, nitrate was observed to be higher level in the burned fields. This is likely due to increased nitrification as a result of soil disturbance and the focused input of organic matter. The altering geology in the area, as opposed to the more homogeneous geology of the lowlands, could be the explanation for not observing clearly dropping fertility numbers (Olsson 2008).

As a result of leaching beyond the shallow root zone, there may be net nutrient losses during cropping. Many forest soils lack exchangeable potassium, which carbohydrate crops like cassava and plantain can readily absorb. A quick reduction in yields as a result of insufficient potash is possible (Nye and Greenland 1960). The content of nitrate nitrogen (NO3-N) followed a distinct pattern. Despite the fact that fire increased ammonification, the concentration did not increase much shortly after burning. However, the concentration of NOs-N had increased dramatically by the time of planting and then declined during the cropping period. The increase in pH caused by the addition of ash and the better aeration of the soil after litter removal could be factors favouring the growth of nitrifying bacteria. The Thai experiment should be treated similarly. At the time of seeding the second crop, nitrate nitrogen reached a new high. This could be the result of soil drying during the previous lengthy dry season.

5.8.2 Changes in soil during the cropping season

Leaching causes nitrate losses throughout the cropping period, especially in wetter forest sites with persistent rainfall and before a full crop cover is established. Crop uptake and a decrease in availability cause soil phosphate loss throughout the cropping period, especially when iron and aluminium oxides are present and the soil pH lowers. Leaching only wastes a little quantity of water. According to Nye and Greenland (1960), a ten-year forest fallow adds roughly 27 kilogramme phosphorus per ha, while humus mineralization contributes about 9 kg phosphorus per ha annually throughout the cropping phase.

Despite the fact that the first rain washes away some of the ash, a forest fallow contributes a large amount of nutritional cations to the soil's exchange complex (Webster and Wilson 1966). Due to leaching and removal by the crop, the quantity of exchangeable cations decreases during cropping, but this is somewhat countered by cation release from nonexchangeable forms. Cation loss reduces their percentage saturation and availability in the exchange complex, however this is offset by a reduction in exchange capacity due to humus oxidation (Nye and Greenland 1960). There are usually no serious nutritional deficits over a cropping period of one to three years, with the exception of potassium. Longer farming periods or several cropping cycles with short fallow periods might result in deficiencies. Field crops, unlike trees in a natural forest, do not carry nutrients from the subsoil to the surface. As a result of leaching beyond the shallow root zone, there may be net nutrient losses during cropping. Many forest soils lack exchangeable potassium, which carbohydrate crops like cassava and plantain can readily absorb. Insufficient potash can cause a rapid decline in crops (Nye and Greenland 1960).

Sanchez *et al.* (1973) mentioned that the additions of ash after clearing and burning and fast mineralization of organic matter make available the required nutrients to the first crop planted in most tropical soils (Sanchez 1973). With each cropping, yields eventually decrease. Deterioration in the physical state or nutrient status of the soil; and changes in the numbers and composition of soil fauna and flora are all factors that contribute to falling crop yields (Nye and Greenland 1960). However, a drop in soil fertility, which can be caused by a loss of humus in the soil, changes in the physical condition of the soil, or changes in the nutritional status, is often the cause of lower crop yields. Despite the fact that the first rain washes away some of the ash, a forest fallow contributes a large amount of nutritional cations to the soil's exchange complex (Webster and Wilson 1966). Due to leaching and removal by the crop, the quantity of exchangeable cations decreases during cropping, but this is somewhat countered by cation release from nonexchangeable forms. Cation loss reduces percentage saturation and availability in the exchange complex, however this may be countered by a decrease in exchange capacity due to humus oxidation (Nye and Greenland 1960). Field crops, unlike trees in the natural forest, do not transmit nutrients from the subsurface to the surface soil. There can be net loss of nutrients due to leaching as the nutrients become unavailable to shallow rooted crops. Insufficient supply of nutrients will cause decline in crop yields (Nye and Greenland 1960).

The oldest fallow sites had higher soil carbon and P concentrations, while the youngest fallow sites had higher N concentrations (Mukul *et al.* 2016). One of the most important environmental qualities influenced by shifting agriculture is soil quality, and any changes in shifting cultivation practises are likely to have an effect (Paudel *et al.*, 2015; Bruun *et al.*, 2013, Lawrence *et al.*, 2010; Kotto-Same *et al.*, 1998). The ability of a soil to enable forest development without deteriorating the soil or hurting the ecosystem is referred to as soil quality (Lal, 1997). Soil resilience, or the ability of the soil to recover from disturbances and resume its functions, is also linked to soil quality (Estes *et al.*, 2011; Lal, 1997). The ecological features of shifting cultivation, on the other hand, are significantly more involved than commonly depicted in the literature, and converting shifting agriculture systems to other farming systems has a varied type of ecological effects, locally and worldwide (Mukul and Herbohn, 2016; van Vliet *et al.*, 2012).

5.8.3 Use and soil chemical properties interrelationship

Within the first 10 years, the pH rises as the fallow period lengthens, owing to the accumulation of alkaline elements during the fallow period and the deposition of ash during cultivation periods. The pH values, on the other hand, dropped after ten years of fallow, presumably due to the significant degradation of these areas prior to abandonment. In other words, for many years, farmers did not cultivate these plots because the soil qualities were too inadequate to sustain ongoing cultivation. As a result, there was a protracted period of fallow. According to Tanaka *et al.* (1997), continuous cropping and fallow forest soils were more acidic than newly recovered shifting cultivation soils. Kyuma *et al.* (1985) found that after 2 years of farming, the pH value was higher than before cultivation. Previous research has also shown that after burning, the amount of exchangeable Al decreases (Tanaka *et al.* 1997). Even in the farmed area on the eastern side, however, the amount of exchangeable Al was substantial. The higher the Ca content, the higher the pH value. The values grew as the length of cultivation and fallow periods grew longer, up to ten years. The pattern of exchangeable Mg distribution was comparable to that of exchangeable Ca. However, there were no correlations between exchangeable K and Na and agricultural or fallow periods. The amount of exchangeable Na in the eastern half was slightly higher than in the western side. Ca, Mg, and K levels dropped fast after abandonment, according to Jordan (1987), due to increased soil erosion. This observation differs from our findings, most likely because soil loss is less severe at our sites due to the high clay content.

5.9 NUTRIENT DYNAMICS DURING FALLOW PHASE

Although traditional agroecosystems are increasingly using fallow to suppress weeds, it also serves to improve soil conditions for future farming periods by raising soil organic matter levels (Van Wambeke, 1992). Fallow times, first and foremost, signify the end of the cropping season and the start of natural or managed plant succession. Sabhasri (1978) concluded that trees are "soil builders" in the case of forest fallow. According to Zinke *et al.* (1978), "too short a fallow time could eventually result in fertility depletion to the point where land abandonment would be essential." They discovered that whereas residual ash returns the majority of calcium, phosphate, and potassium, forest fallow adds nitrogen and organic matter. In slash-and-bum agroecosystems, managing the fallow is critical for maintaining agronomically beneficial soil conditions.

5.9.1 Effects of fallow

The fallow component of the SCS symbolises the restoration of the forest ecosystem's initial conditions prior to agricultural management. Natural ecological mechanisms associated to ecological succession are responsible for this ecosystem recovery. The more relevant matter in this phase is the period of time for which the ground is left to rest. The unfavourable effects are only present at the start of the fallow period. The variables that make up the chemical characteristics of the soil alter as ecological succession progresses (e.g., become less acidic). These modifications have good consequences, restoring the soil's original fertility. With an insufficient rest period, the fallow season would only have negative effects on the soil system, reducing its resilience. To avoid soil degradation, a ten-year term is considered the absolute minimum. Kleinman *et al.* (1995) discovered that the ratios range from 1:3 (Nye and Greenland, 1960) to 1:9 (Kleinman *et al.*, 1995). (Kleinman *et al.*, 1995). (Zinke *et al.* 1978).

5.9.2 Influence of burning and fallow period on soil biochemical properties

The following factors may contribute to the negative impact of burning on soil enzyme activities: (1) a sudden decrease in soil biota population and activities due to the loss of FFLs substrates from surface layers, (2) depletion of hydrolytic enzyme pools due to the breakdown of above- and below-ground community linkages, and (3) nutrient enrichment in soils following burning reduce crop plant reliance on enzyme activities. The decrease in microbial biomass is a sign of reduced abundance of microbial population as observed in burnt soils, which may be linked to the drop in DHA in burnt soils as compared to that of unburnt soils. The negative impact of burning on the activity of Docosahexaenoic acid (DHA) was previously reported (Ajwa *et al.* 1999). As DHA has close linkage with the dynamics of microbial activity and therefore would get reduced after burning where oxidoreductive microbial population is low. This study also indicated lower values of MBC, MBN and MBP in shorter fallow phase due to burning at frequent intervals as compared to longer fallow phase. The activity of GSA increased with the increase in length of the fallow phase in both Mizoram and Nagaland. The higher activity of GSA is thought to be closely linked with the greater quantity and more complexity of substrates available in the longer fallow phase. Decrease in GSA activity was also reported to decline for over a month after burning (Ajwa *et al.* 1999). López-Poma and Bautista (2014) reported a decline in GSA activity for over a year after burning. In another observation at Southern Ohio, USA, the activity of GSA has decreased with the increase in frequency of fire while comparing with unburnt plots of oak-hickory forest (Boerner and Brinkman 2003). The activity of PHA decreased (25 to 59%) after burning and the cause of decrease may be related to discontinuation of PHA secretion by plant root after burning. The quantity of PHA in soil is usually contributed by both plant as well as soil microorganisms. The activity of PHA increased with increase in the length of fallow phase. In Eastern Spain, López-Poma and Bautista (2014) found that the activity of acid phosphatase continued to decrease for over a year after burning. Ajwa *et al.* (1999) has also reported on the decrease of PHA after burning. Boerner and Brinkman (2003) also described that burning has reduced the acid phosphatase activity in the soil while studying the effect of burning on soil enzyme activity in a southern Ohio hardwood forest. The activity of phosphatase enzyme is involved in P-cycling and has also been reported to be governed by soil microclimate, SOC and the availability of P in the soil (Hamman *et al.* 2008). Compared to unburnt condition, the activity of PRO decreased to an extent of 24 to 49% under burnt condition. However, Oseni *et al.* (2009) did not find any change in the proteolytic activity while comparing unburnt and recently burnt plots of teak plantations in Nigeria. Yuan and Yue (2012) found that the activity of protease has peaked up linearly in a plantation forest in a succession pattern 3 years >7 years >13 years >21 years > 28 years. Malcheva *et al.* (2015) also mentioned that the activity of PRO is reduced at higher temperature, which in our case is brought about by high intensity burning of biomass. The increase in PRO activity with increasing fallow length could be attributed to the accumulation of non cellulosic glucose in higher succession tree species which may have been lesser in composition in shrubs and taller grass species of shorter fallows. Srilakshmi *et al.* (2012) while comparing littered forest soil with adjacent non-forest soil had the opinion that increase in PRO activity could be due to the availability of casein based substrates in the litter biomass, the prevailing low pH and abundance of proteolytic microbes in the soil. The extent of increase in the activity of ASA ranged from 15.6 to 45.5% due to burning of slashed biomass irrespective of fallow phases. The activity of ASA also increased with increasing length of fallow phase. The different rates of S-immobilization at different lengths of fallow phases as well as the difference caused by burning could be a reason for change in ASA activity.

5.9.3 Biological properties

b) Impacts of fallow

The quantity of organic matter in the soil grows with ecological succession, as does the pace of fauna activity, biomass, and variety. The fallow period allows the seed bank to germinate and seedlings to enter, promoting the environmental succession's evolution. The re-establishment of the soil conditions at the initial level is favoured as a result of this. With the help of model forecasts and quantum indicators, the dynamics of soil qualities under various land uses have been monitored (Yemefack *et al.*, 2006a, 2006b). The soil quality index is another tool for analysing the quality, state, and productivity potential of soils in agricultural systems (SQI). Andrews and Carroll (2001) created an index that evaluated the soil's functional capabilities using physical, chemical, and biological characteristics.

5.9.4 Influence of fire and fallow period on microorganisms

Microbial biomass in soils was found to be negatively impacted by the burning activity in each fallow phase. Microbial Biomass Carbon (MBC), Microbial Biomass Nitrogen (MBN), and Microbial Biomass Phosphorus (MBP) were significantly higher in the longer fallow phase compared to the short fallow phase in both unburned and burned situations, and the higher values of MBC, MBN, and MBP were in the order F3 > F2 > F1. Besides, the C:N:P within microbial biomass was found to vary with the length of fallow phase and also with burning activities. The longer fallow phase supported narrower C:N:P within microbial biomass, whereas wider C:N:P was observed in shorter fallow phase. The narrower C:N:P within microbial biomass was in the order

F3 > F2 > F1. The reduction in MBC and MBN caused by the after effect of burning was reported elsewhere (Ajwa *et al.* 1999). Moreover, MBC, MBN and MBP have also increased with the increase in fallow length and the same phenomena were observed both for Mizoram and Nagaland situations. Jia *et al.* (2005) reported that MBC and MBN increased gradually from the first year to the seventeenth year of secondary forest succession. Many of the soil decomposer community's numbers would be diminished or eliminated as a result of the fire effect, according to previous research. The dramatic changes in the environment generated by the burning activities, such as changes in pH, temperature, and low soil moisture content, would further impede the surviving species that may have survived. Microbial activity may be reduced as a result of the loss of SOC and N following burning activities. Our findings clearly indicated that wider C:N:P within microbial biomass is an good indicator of the degree of anthropogenic disturbances in an ecosystem.

This study showed that a longer fallow period not only supports a bigger quantity of accumulated forest floor litters, but also maintains higher soil nutrient availability and a better state of C, N, and P cycling, as evidenced by increased soil enzyme activity. Under both burned and unburned conditions, the length of the fallow period had a substantial positive connection with GSA, DHA, ASA, PHA, and PRO activity. However, the burning practice imposed a retarding influence on the positive relationship of the length of fallow phase and the activities of soil enzymes except AMY. In terms of increasing physicochemical and biochemical qualities of agricultural cultivated soils, it can be stated that less frequent biomass burning at longer intervals is more advantageous than slashing biomass burning at frequent intervals.

The aveage pH was found to be higher in the soils of grasslands of secondary forest, followed by secondary forests, grassland soils and primary forest soils. However, CEC was significantly higher in primary forest soils (Table 5.17). Calcium was more in the soils of secondary forest and Mg was higher in the the grassland secondary forest soils. Total N, Na and OC was significantly higher in primary forest soils. Exchangeable K was found to be higher in grassland soils.

Table 5.17. Summary of analyses for surface soils under various vegetation formations

Soil No	pH	CEC	Ca	Mg	K	Na	OC (%)	Total N (%)
		meq.100g-1						
Primary forest soils								
580-2	5.40	65.3	1.96	1.14	0.50	0.96	17.3	0.91
581-2	5.50	69.8	6.74	2.08	0.76	0.40	20.9	1.22
582-2	5.45	61.8	1.17	0.87	0.40	0.14	15.1	0.67
Average	5.45	65.6	3.29	1.36	0.55	0.50	17.7	0.93
Secondary forest soils								
161267T-1	5.62	47.8	14.23	3.92	0.59	0.27	11.6	0. 77
161267S-1	5.94	47.4	29.31	3.36	0.82	0.42	7.9	0.79
31067-1	5.58	53.8	9.93	2.22	0.40	0.24	7.8	0.83
Average	5.71	49.6	17.28	3.17	0.60	0.31	9.1	0.80
Grassland soils								
121567-1	5.82	33.7	7.08	2…63	0.68	0.17	8.01	0.51
121567-2	4.90	30.7	6.17	2.69	0.79	0.17	6.49	0.47
Ming-31	5.75	45.8	7.64	2.42	0.99	0.18	7.78	0.70
Ming-38	5.50	48.1	4.51	1.60	0.70	0.18	12.68	-
Ming-52	5.69	44.4	5.69	1.79	071	0.28	7.86	0.73
Average	5.53	45.5	6.24	2.23	0.77	0.20	8.56	0.60
Grassland secondary forest soils								
299678-1	5..90	42.8	10.54	4.14	0.75	0.29	6.87	0.42

Sourcs: Manner, 1969

5.9.5 Soil erosion and nutrient loss

Because vegetation stores a major amount of the mineral resources in the tropical ecosystem (Richards 1952, Nye and Greenland 1960), slashing and burning appear to be the only option to prepare the land for cultivation in order to absorb nutrients gathered in vegetation into the soil. Changes in soil physical and chemical qualities, as well as nutrient status, are all results of cutting and burning (Christanty 1986, Khresat *et al.* 2008). Drastic changes occur in soil macro- (>250 μm) and microaggregates (<250 μm) and porosity because of densification under short fallow period. Soil compaction reduces water in soil (θ) and hydraulic conductivity. Short fallow causes higher soil erosion (60–80 percent more) and surface runoff (80–100 percent more) than natural woodland. As a result, slicing and burning vegetation with brief fallows degrades soil physical characteristics by dissolving soil stable aggregates and interrupting the continuity of pores from the surface to the subsurface soil, reducing water infiltration and percolation and increasing surface runoff. Similar to NEI, a study from Zambia indicated decline in maize yield (*Zea mays*) under short fallows (Ando *et al.* 2014). The magnitude of annual losses (Mg ha^{-1} yr^{-1}) of top soil, N, and K under shifting cultivation area in NEI may be as high as 58.9, 7.1, and 4, respectively (Sharma 1998). Decline in soil fertility leads to decline in crop yields (Tinker *et al.* 1996, Gafur *et al.* 2003) and forces cultivator to abandon the land and shift to another forest patch and begin the new cycle. Therefore, shifting cultivation with short fallow is the prime driver of poor socioeconomic status of the hill farmers and of the degraded biophysical environment. In contrast, long fallow period improves the soil quality by improving soil structure (Fig. 2). Changes in soil structure are directly linked to increases in soil organic matter (SOM) and soil humus contents throughout the extended fallow period (Ziegler 2009, Nye and Greenlang 1960). In NEI, litterfall increased from 1.2 Mg ha^{-1} yr^{-1} during the first-year fallow to 9.7 Mg ha^{-1} yr^{-1} in the 20-year-old fallow (Toky and Ramakrishnan 1983), almost equivalent to the mature forests (Swamy *et al.* 2010), implying a critical role of biomass accumulation in long fallows to soil restoration. In case of older fallow, organic inputs through above- and belowground litterfall and development of macro- and microfaunal population (Reza *et al.* 2014) accelerate the decomposition process and subsequently improve soil quality, and the system slowly approaches to the level of climax forest.

5.9.6 Soil organic carbon (SOC) and shifting cultivation

The SOC derived from the partially combusted biomass is the basis for maintenance of productivity of soil under shifting cultivation. On the exchange sites of SOC, nutrients generated from biomass combustion are stored in organic form and held in mineral form (Palm *et al.* 1996, Bahr *et al.* 2014). SOC created in the first and second years of cultivation is vulnerable to rapid erosion in the absence of plant cover. During the cropping period of shifting agriculture, a loss of 14–20 percent of SOC in top soil has been documented (Bahr *et al.* 2014), and regaining content comparable to that under a primary forest could take up to 35 years (Palm *et al.* 1996). The oxidative loss of SOC during biomass burning releases CO_2 and, therefore, has positive feedback on global climate change. However, fallow systems apparently regain the SOC content.

5.9.7 Alternatives to shifting cultivation in Eastern India Himalayas

Synthesis of data on soil erosion, loss of nutrients, and SOC reveals that a short fallow cannot fulfill the livelihood requirements of the increasing hill populace due to severe decline in agronomic yield. Poverty and a lack of alternative sources of income force hill farming communities to use even marginal lands in unsustainable ways. The result is a loss of soil resilience, which makes the shifting agricultural system unsustainable. Sustainability of the fragile ecoregion can be managed through maintaining longer fallow periods, but this also seems to be unachievable due to rise in population and their subsequent livelihood requirement in the form of fuel, food, and fodder. Further, maximizing yield from short-term fallow is not a viable option considering soil degradation and its attendant socioenvironmental problems. Thus, there is a need to develop a comprehensive plan so that forest can be managed on long-term basis and, at the same time, the livelihood requirement of hill farmers is fulfilled. In this context, we propose the introduction of "grains for forest management program" as an alternative to shifting cultivation, where farmers can be provided with food grains that they otherwise grow by shifting cultivation. Rather than a subsidy, food grains provided are an inkind payment for ecosystem services (PES) for restoration of degraded lands. India has made commendable success in food grain production with an annual production rate of 264 million metric ton. However, 10–20% of the total production is wasted annually due to lack of adequate storage facility (Nellemann *et al.* 2009). Therefore, implementing "grains for forest management program" in NEI and provisioning food grains are appropriate use of surplus food grains, while restoring degraded soils, strengthening ecosystem services, and earning C credits through PES from soil and ecosystem C sequestration. The establishment of such program will also enable the hill farmers to pursue their traditional semiperennial and perennial crop cultivation understory of the restoring vegetation. This program will fulfill their food, fodder, and other daily life requirements. Considering the soil C sequestration rate of 0.61–0.64 Mg ha^{-1} yr^{-1} with conversion of shifting cultivated area to a fallow land in India, total C sequestration in 1.5 Mha of jhum land in NEI should be around 1 Tg yr^{-1}.

Considering the socioeconomic condition of hill farmers, increasing population pressure and decreasing dense forest cover in NEI proposed "grains for forest management program" may be a viable strategy to restore degraded fragile ecoregion and enhance livelihood security. Implementation of the program will also strengthen the ecosystem resilience of this biodiversity hotspot to offer critical ecosystem services. In addition to SOC, C sequestered in the biomass in successional fallows of restored shifting cultivation landscape can bring additional income to the land managers of NEI in the form of offsetting C credits under climate negotiations. Fig. 5.2. showing schematic diagram of shifting cultivation, Long and short phases, and their impact on soil fertility and socio-economic conditions is presented below.

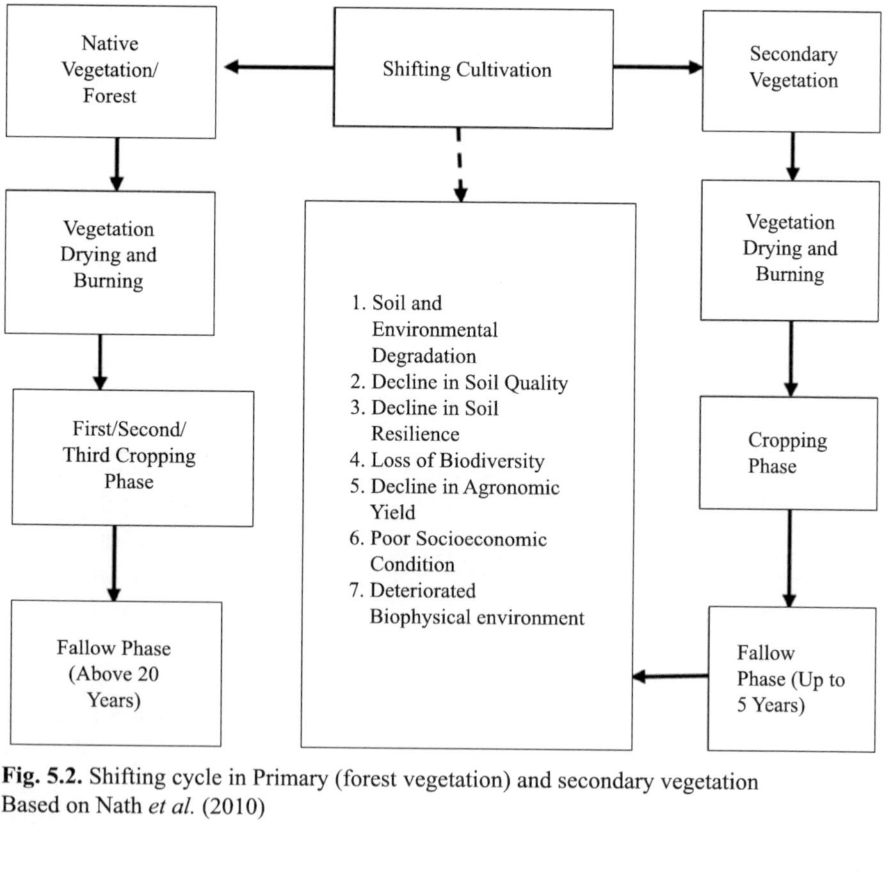

Fig. 5.2. Shifting cycle in Primary (forest vegetation) and secondary vegetation Based on Nath *et al.* (2010)

5.10 ECORESTORATION OF SHIFTING CULTIVATION AGROECOSYSTEM

Soil biota communities are essential for soil processes such nutrient cycling, organic matter breakdown, and soil structure formation, and hence play an important role in ecosystem productivity and community dynamics (Bardgett, 2008). These roles are critical in ecosystem restoration because they help to restore function and biodiversity. The condition of a system in respect to restoration targets and the success of management interventions may be determined by measuring the community, and manipulating the community has shown promise in improving the rate of recovery of degraded systems. As a result, assessing functional connections between biological inputs (above and below ground) could be a critical component of any programme aimed at soil restoration under shifting cultiation (Thakuria and Sharma 2014). Clearing of forest cover due to shifting cultivation practices and subsequent cultivation of crops can modify the stability of the existing ecosystem through the processes like disturbance in linkages between above- and below-ground communities, alteration of decomposer community composition, and addition of nutrient through ash material, etc. Previous research has shown that the shifting cultivation agroecosystem's stability is mostly dependent on the fallow period, which permits secondary forest to re-grow (Ramakrishnan, 1998). Above-ground biomass is contributed by regenerated forest cover in the form of litter falling on the soil surface, as well as below-ground substrates such as roots and their exudates. These biological inputs help to rejuvenate soil biota components and their activity throughout time during the fallow period. Thus, in the disturbed shifting cultivation agroecosystem, soil restoration programmes must take into account the philosophy of functional relationships between above-ground biomass and below-ground biota populations, as well as the impact of such interactions on soil processes (Thakuria and Sharma 2014).

5.11 SHIFTING CULTIVATION AND SOIL BIOTA

Out of 5.5 million hectare of net sown areas, about 1.6 million hectare area is under shifting cultivation in Northeast India. Because of the hilly terrains and prevailing meteorological conditions, as well as the different bio-resources accessible, hill farmers can practise shifting agriculture. Shifting cultivation is a farming approach that involves balancing agricultural practises with biodiversity by protecting, promoting, and defending agriculture and hill forest ecosystem coexistence (Thakuria and Sharma 2014). Traditional shifting cultivation is considered viable in the literature within a reasonable range of fallow length and accompanying demographic conditions. Fallow periods of 10-20 years or longer are regarded to be necessary to minimise soil erosion, loss of soil fertility, and water balance, as well as to allow for forest regeneration and ecosystem stability. These circumstances are usually linked with population densities of less than 20 people per square kilometre (Kumar, 2008). However, because to the ever-rising population and their increasing food demands, the length of the fallow period has been reduced to 3 to 10 years in recent years. The greatest serious limitation to improved production from changing agricultural areas is land degradation. Deforestation, loss of biodiversity, loss of fertile top soil (46.0 t ha^{-1} $year^{-1}$ compared to 16.0 t ha^{-1} $year^{-1}$ in India), reduction of water storage capacity, and exposure of unfertile acidic subsoil (soil pH 5.5) lead to impairment in carbon mineralization dynamics and destruction of soil structure on-site when hill slopes are used for shifting cultivation. In the valley regions, shifting farming has off-site effects such as siltation and associated environmental problems. Soil is one of the most important natural resource qualities that determines the productivity of terrestrial ecosystems. The control of soil functions and associated processes, as well as the optimal use of bioresources, must be at the forefront of future plans for boosting agricultural yield from shifting cultivation fields (Thakuria and Sharma 2014).

5.12 DEGRADATION OF SOIL FERTILITY

Soils are a critical natural resource on which the life-supporting system and socioeconomic development rely. A wide range of human activities, including overgrazing, bad farming methods, land-use change, and forest clearing, are threatening the stability and future of many soils. Shifting cultivation leads in a massive loss of soil and soil fertility, as well as a considerable amount of carbon being released into the atmosphere. It is eventually supported by spontaneous soil fertility restoration. It is a critical time since natural processes only take a short time to return cultivated land to its former fertility level. The Jhum cycle is maintained differently in each hamlet in the Mokokchung district. Jhum cycles have been seen to have dramatically narrowed as a result of regular practise in some areas. This period fluctuates depending on the area's biophysical and agroclimatic circumstances, and is commonly referred to as the critical phase (Sharma, 1984). Shifting cultivation cycles have been seen to have been considerably reduced. Crop production productivity declines as soil nutrients are depleted. It is true that in shifting cultivation, productivity is low in terms of land utilisation. During shifting cultivation, larger forest tracts are destroyed and burned away. Nearly 3.7 tons/hectare of soil materials are slid down to hillsides as a result of clearing jungles, burning, and seed dibbling (Singh and Singh, 1978). Organic matter in the surface soil is likely to be oxidised as well, resulting in lower levels of organic matter and other nutrients in the soil. The high amount of mineral-rich ash added to the soil kills large amounts of organic matter and bacteria, affecting soil formation processes and soil microorganisms.

Due to variations in the quality and quantity of organic carbon inputs into the soil, shifting cultivation alters the physical, chemical, and biological aspects of the soil (Yang *et al.* 2004). Forest conversion to jhum field/cropland often results in a decrease in organic carbon content (Guo and Gifford 2002). The collection and decay of organic materials in the area could have an impact on carbon stores in the soil. As a result, the high level of SOC could be ascribed to the burning of slashed vegetation in March, which leaves ashes and partially combusted particles depending on the severity of the burn (Andriesse and Schelhaas 1987). Furthermore, Nye and Greenland (1960) suggested that decaying roots from cut and burned forest plants can contribute soil organic matter during forest conversion to shifting cultivation land.

5.13 STRATEGIES FOR IMPROVING SOIL HEALTH

Several agroforestry treatments for crop sustainability include MPTs (multipurpose tree species). MPTs contribute to the enhancement of soil health and other environmental factors while also producing fuel, fodder, lumber, and other commodities. Farmers use a range of tree species for varied land uses, although priority species differ by area due to ethnic diversity and tribal feeding habits. Tree density per hectare is also essential on sloppy grounds. The optimum tree density in an agri-horticulture system is 400 trees per hectare, however in an agri-silviculture system, it is 200 plants per hectare, to limit the impact of shadow and biochemical interactions on agricultural crop growth and output (Bhatt 2003). The long-term effects of various multifunctional tree species on soil physical parameters were investigated by Saha *et al.* (2007). Soil organic C increased by 96.2 percent, porosity improved by 10.9 percent, aggregate stability increased by 24.0 percent, and accessible soil moisture increased by 33.2 percent in multipurpose tree species with robust surface cover, litter fall, and a large root system. Tree-based land use systems are resilient, productive, and beneficial to soil health. Under a jackfruit-based agroforestry system (AFS), Singh and Dhyani (1995) found that MPTs increased organic carbon, available P, and exchangeable cations contents in surface soil by 2.0–2.5 percent, 10.4–13.2 ppm, and 5.9–8.4 cmol (p+) kg1, respectively, and 1.5–1.8 percent organic carbon, 3.8–6.7 ppm available P, and 3.9–5.9 cmol (p+) kg. In comparison to the baseline state, all AFS showed a large increase in organic carbon (1.17–1.65 fold), with the silvi-horti-pastoral AFS contributing the most. In comparison to natural forest, the same system recorded 43.2 percent higher exchangeable Al, resulting in a pH drop of up to 0.50 units (Table 5.18).

Table 5.18. Soil properties as affected by agroforestry

Soil properties	**Agri-sil viculture**	**Khasi mandarin and crops**	**Assam lemon + crops**	**Silvi-horti-pastoral**	**Alder + tea+ black pepper**	**Natural forest**
pH	4.65	4.62	4.80	4.25	4.61	4.62
OC (%)	1.62	1.55	2..02	2.19	1.91	1.92
Exch. Ca [cmol (p+) kg^{-1}]	0.40	0.86	0.74	0.31	0.65	0.26
Exch. Mg [cmol (p+) kg^{-1}]	0.75	0.51	0.33	0.48	0.71	0.16
Exch. K [cmol (p+) kg^{-1}]	0.232	0.244	0.249	0.238	0.201	0.169
Exch. Na [cmol (p+) kg^{-1}]	0.201	0.220	0.194	0.195	0.197	0.196
Exch. Al [cmol (p+) kg^{-1}]	2.65	2.70	2.20	3.15	2.05	2.20
Av. N (ppm)	190.1	180.8	203..6	199.4	216.9	167.2
Av. P (ppm)	2.75	4.10	5.36	0.94	3.36	0.63
Av. Fe (ppm)	8.9	10.4	12.8	10.9	13.9	7.3
Av. Mn (ppm)	0.58	0.92	0.79	0.83	1.04	0.04
Av. Zn (ppm)	0.08	0.05	0.07	0.06	0.08	0.02
Av. Cu (ppm)	0.21	0.23	0.37	0.30	0.27	0.10

Source: (Majumdar et al. 2004)

The effect of various Agroforestry systems (AFS) such as silvi-pastoral, silvi-horticulture, and agri-horti-silvipastoral on soil fertility characteristics was studied in a study done at ICAR Research Complex, Barapani, in the Indian state of Meghalaya (Majumdar *et al.* 2002). For comparison, natural fallow and abandoned shifting agriculture land were used. Organic carbon content increased in all AFS, including natural fallow, but the amount varied significantly depending on the vegetation type in each system. Different cropping patterns in various AFS considerably changed the exchangeable Ca, Mg, and K concentration in the soil. Agri-horti-silvipastoral and silvihorticulture AFS had the highest cation accumulation, followed by natural fallow and silvi-pastoral systems. Silvi-horticulture accumulated the most exchangeable K, followed by agri-horti-silvipastoral.

5.13.1 Effects of shifting cultivation

Sahu *et al.* (2005) mentioned about some bad effects of shifting cultivation:

i) The undomesticated creatures are deprived of a safe haven.

ii) Drying up of springs.

iii) Flooding occurs in the rivers below.

iv) Increases the shortage of water and moisture for plants, animals, and humans.

v) Tanks, nallas, rivers, and reservoirs are all affected by siltation.

vi) Ecology has been disrupted and degraded, and it has never been restored at many places.

vii) Soil erosin due to shifting cultivation is a major cause of concern.

5.13.2 Control measures

The shifting cultivation needs to be discouraged if can not be stopped even after persuation. The should be made aware of the disadvantages of the practice in areas population densities are higher and availability of land low. Sahu *et al.* (2005) advocated for the physical, agronomic and socio-economic approaches to be adopted to minimize the bad effects of shiting cultivation.

Physical

The objective is to control soil erosion through soil conservation measures and includes;

- Contour bunding (up to 30% slope)
- Graded bunding if the slope is 30-60%.
- Terraces on 60-100% slope

- Contouring (>1500 mm annual rainfall)
- Grasses on drainage channels.
- Water reservoirs.s

Agronomical

In order to get a good harvest of the crops for food security, the following agronomic management practices are essential:

- Ploughing across the slop
- Planting of suitable crop and variety
- More use o organic manures and bio-fertilizers, integrated nutrient management
- Mixed farming; such as livestock, crops, medicinal plants and mushrooms etc.

Socio-economic

The system's survival was based in part on its compatibility with the tribal people's physio-social condition, which included low population, land tenure, poverty, poor resources, illiteracy, food habits, social taboos, and so on.

References

Adams, C.; Murrieta, R. S. S.; Sanchez, R. A. 2005. Agricultura e alimentação em populações ribeirinhas das várzeas do Amazonas: novas perspectivas. Ambiente & Sociedade, v. 8, n. 1, p. 1-22, .

Adedeji, F.O., 1984. Nutrient cycles and successional changes following shifting cultivation practice in moist semi-deciduous forests in Nigeria. For. Ecol. Manag. 9, 87–99.

Adeyolanu, O.D., Are, K.S., Oluwatosin, G.A., Ayoola, O.T. and Adelana, A.O. 2014. Evaluation of two methods of soil quality assessment as influenced by slash and burn in tropical rainforest ecology of Nigeria. Archives of Agronomy and Soil Science 59, 1725–1742.

Adeyolanu, OD; Oluwatosin, KS; Ayoola, OT; Adelana, AO 2013. Evaluation of two methods of soil quality assessment as influenced by slash and burn in tropical rainforest ecology of Nigeria. Archives of Agronomy and Soil Science 59(12): 37-41, DOI:10.1080/03650340.2012.760037

Agegnehu, G., M. I. Bird, P. N. Nelson, and A. M. Bass. 2015. The ameliorating effects of biochar and compost on soil quality and plant growth on a Ferralsol. Soil Research 53(1):1-12.https://doi.org/10.1071/SR14118

Agoume V and Birang AM, 2009. Impact of Land-use Systems on some Physical and Chemical Soil Properties of an Oxisol in the Humid Forest Zone of Southern Cameroon. Tropicultura, 27: 15-20.

Aguiar, A.C.F.; Candido, C.S.; Carvalho, C.S.; Monroe, P.H.M.; Moura, E.G. (2013) Organic matter fraction and pools of phosphorus as indicators of the impact of land use in the Amazonian periphery. Ecol. Indic. 2013, 30, 158–164.

Aguiar, A.C.F.; Freitas, I.C.; Carvalho, C.S.; Monroe, P.H.M.; Moura, E.G. (2011). Efficiency of an agrosystem designed for family farming in pre-Amazon region. Renew. Agric. Food Syst. 2011, 26, 24–30.

Ahlgren IF and Ahlgren CE, 1965. Effect of prescribed burning on soil microorganisms in a Minnesota jack pine forest," Ecology, vol. 46, no. 3, pp. 304–310, 1965.

Ahmed S, Gaby S. 1996. Bio pesticides. In: Joske B. (eds.), Biotechnology Building on Farmers Knowledge ETC, Netherlands. pp.52-79.

Ahn, P. M. 1974. Some Observations on Basic and Applied Research in Shifting Cultivation. In: Food and Agricultural Organization of the United Nations (FAO) (ed.) Shifting Cultivation and Soil Conservation in Africa, Papers presented at the FAO/SIDA/ARCN Regional Seminar, held at Ibadan, Nigeria (July 2-21, 1973), pp. 123-154; available at: http://pdf.usaid.gov/pdf_docs/PCAAA007.pdf.

Ajwa, H.A., Dell, C.J. and Rice, C.W. 1999. Changes in enzyme activities and microbial biomass of tall grass prairie soil as related to burning and nitrogen fertilization. Soil Biology and Biochemistry 31, 769– 777.

Alegre, J.C. and Cassel, D.K. 1996. Dynamics of soil physical properties under alternative systems to slash– and–burn. Agriculture, Ecosystems and Environment 58, 39–48.

Alexander, M. (1997). Introduction of Soil Microbiology (New York: John Wiley and Sons, 2nd edition).

Allan, W. 1965. The African Husbandman. London: Oliver and Boyd.

Allan, W. 1972. Ecology, techniques and settlement patterns. In Ucko, P.J., Tringham, R. and Dimbleby, G. W. Man Settlement and Urbanism. London: Duckworth & Co. Ltd. pp. 211 - 226.

Altieri, M. A. 1999. The ecological role of biodiversity in agroecosystems. Agriculture, Ecosystems & Environment, v. 74, n. 1-3, p. 19-31.

Altieri, M. A.; Anderson, M. K.; Merrick, L. C. 1987. Peasant agriculture and the conservation of crop and wild plant resources. Conservation Biology, v. 1, n. 1, p. 49-58,.

Allison, F. E. (1973) Soil Organic Matter and Its Role in Crop Production (Amsterdam, London and New York: Elsevier Scientific Publishing Company).

Álvarez, E *et al.* 2005. Application of aluminum toxicity indices to soils under various forest species. Forest Ecology and Management. 211 (p.227-239) Elsevier

Ambe, BA; Eja, lE; Agbor, CE 2015. Assessment of the Impacts and People'S Perception of Bush Burning on the Grasslands and Montane Ecosystems of the Obanliku Hills/ Plateau, Cross River State, Nigeria. Journal of Natural Sciences Research, 5, (6): 12-20.

Ando, K.; Shinjo, H., Noro, Y., Takenaka, S., Miura, R., Sokotela, S.B., Funakawa, S. 2014. Short-term effects of fire intensity on soil organic matter and nutrient release after slash-and-burn in Eastern Province, Zambia. Soil Sci. Plant Nutr. 2014, 60, 173–182.

Andrews, S. S.; Carroll 2001., C. R. Designing a soil quality assessment tool for sustainable agroecosystem management. Ecological Applications, v. 11, n. 6, p. 1573-1585.

Andriesse JP & Schelhaas RM 1987. A monitoring study on nutrient cycles in soils used for shifting cultivation under various climatic conditions in tropical Asia. III. The effects of land clearing through burning on fertility level. Agric Ecosys Environ 19(4), 311-332.

Andriesse JP. 1989. Nutrient management through shifting cultivation. In: van der Heide, J. (Ed.), Nutrient Management for Food Crop Production in Tropical Farming Systems. Institute for Soil Fertility, Haren, The Netherlands, 1989, 29-62.

Andriesse, J. P., and Koopmans, T.T. 1984.. A Monitoring Study on Nutrient Cycles in Soils Used for Shifting Cultivation under Various Climatic Conditions in Tropical Asia. I. The Influence of Simulated Burning on Form and Availability of Plant Nutrients. Agriculture, Ecosystems & Environment 12: 1–16. doi:10.1016/ 0167-8809(84)90057-4.

Andriesse, J. P.; Schelhaas, R. M. A monitoring study on nutrient cycles in soils used for shifting cultivation under various climatic conditions in tropical Asia. II. Nutrient stores in biomass and soil. Agriculture, Ecosystems and Environment, v. 19, p. 285-310, 1987a.

Andriesse, J. P.; Schelhaas, R. M. A monitoring study on nutrient cycles in soils used for shifting cultivation under various climatic conditions in tropical Asia. III. The effects of land clearing through burning on fertility level. Agriculture, Ecosystems and Environment, v. 19, p. 311-332, 1987b.

Andriesse, J.P., 1987. Monitoring project of nutrient cycling in soils used for shifting cultivation under various climatic conditions in Asia, Part I, Text. Research Report, Royal Tropical Institute, Amsterdam, 141 pp.

Anikwe, M.A.N., 2000. Amelioration of a heavy clay loam soil with rice husk dust and its effect on soil physical properties and maize yield. Bioresource Technology, 74: 169-173.

Arocenaa JM and Opio C 2003.. Prescribed fire-induced changes in properties of sub-boreal forest soils. Geoderma, 113: 1-16.

Arunachalam, A.. Dynamics of soil nutrients and microbial biomass during first year cropping in an 8-year jhum cycle. Nutrient Cycling in Agroecosystems. 64, 2002, 283-291.

Aweke, M.; Gelaw, R.; Singh, R.; Lal, R. Soil Quality indices for evaluating smallholder agricultural land uses in Notthern Ethiopia. Sustainability 2015, 7, 2322–2337.

Aryal KP, Kerkhoff EE, Maskey N and Sherchan R, 2010. Shifting cultivation in sacred Himalayan landscape: A case study in Kengchenjunga Conservation Area, WWF Nepal, First Edition.

Awodon, M.A., 2007. Influence of saw dust ash on soil chemical properties and Cowpea performance in Southwest Nigeria. International Journal of Soil Science, 2(1): 78-81.

Ayanaba, A. and A.P.O. Omayuli, 1975. Microbiology of acid tropical soils – a preliminary report.

Ayanaba, A.S.B. Tuckwell and D.S. Jen Kinson, 1976. The effects of clearing and cropping on organic reserves and biomass of tropical forest soils. Soil Biology and Biochemistry, 8: 519-525.

Bahr, E.; Zaragocin, D.C.; Makeschin, F. 2014.. Soil nutrient stock dynamics and land-use management of annuals, perennials and pastures after slash-and-burn in the Southern Ecuadorian Andes. Agr. Ecosyst. Environ. 2014, 188, 275–288.

Baker, K. F. and R. J. Cook 1974. Biological Control of Plant Pathogens (San Francisco, CA: W. H. Freeman and Company).

Balee, W. 1989. Cultura na vegetação da Amazônia brasileira. In: NEVES, W. A. (Org.). Biologia e ecologia humana na Amazônia: avaliação e perspectivas. Belém: Museu Paraense Emílio Goeldi/CNPq, 1989. p. 95-109.

Bandy, D. E.; Garrity, D. P.; Sanchez, P.A. 1993. The worldwide problem of slash-and-burn agriculture. Agroforestry Today, v. 5, n. 3, p. 2-6.

Bardgett, R.D. 2008. The Biology of Soil: a Community and Ecosystem Approach. Oxford University Press.

Bartholomew, W.V., Meyer, J. and Laudelout, H. 1953. Mineral nutrient immobilization under forest and grass fallow in the Yangambi (Belgian Congo) region. Publications de Institut National pour I Etude Agronomique du Congo Belge.SerieScientifique 57: 1-27.

Başaran M, Erpul G, Tercan AE, Çanga MR. 2008. The effects of land use changes on some soil properties in İndağı Mountain Pass –Çankırı, Turkey, Environmental Monitoring and Assessment 136: 101–119.

Belay KT, Van Rompaey A, Poesen J, Van Bruyssel S, Deckers J, Amare K. Spatial analysis of land cover changes in eastern Tigray (Ethiopia) from 1965 to 2007: are there signs of a forest transition? Land Degrad Develop. 2015;26:680-9. https://doi.org/10.1002/ldr.2275

Beliveau A, Davidson R, Lucotte M, Canto Lopes LOD, Paquet S and Vsseur C, 2015. Early effects of slash-and-burn cultivation on soil physicochemical properties of small-scale farms in the Tapajós region, Brazilian Amazon. The Journal of Agricultural Science, 153: 205-221. DOI: https://doi.org/10.1017/S0021859613000968

Bellwood P, 2005. First Farmers: The Origins of Agricultural Societies. Blackwell Publishers

Benneh, G. 1972. Systems of agriculture in tropical Africa. Economic Geography 48(3): 244-257.

Berggren D, Mulder J 1995. The role of organic matter in controlling aluminum solubility in acidic mineral soil horizons. Geochimica et Cosmochimica Acta, 59(20): 4167-4180.

Bhatt BP 2003. "Agroforestry for sustainable mountain development in N.E.H. region," in Central Himalaya Environment and Development: Potentials, Actions and Challenges, M. S. S. Rawat, Ed., pp. 206–223, Transmedia Publisher, Uttaranchal, India.

Birang A.M., Hauser S., Brussaard L. & Norgrove L., 2003. Earthworm surface-casting activity on slash-and-burn cropped land and in undisturbed Chromolaena odorata and young forest fallow in south Cameroon. Pedobiologia, 47, 811-818. 5.

Biswas A, Alamgir M, Haque SMS, Osman KT. 2012. Study on soils under shifting cultivation and other land use categories in Chittagong Hill Tracts, Bangladesh, Journal of Forestry Research, 23(2): 261–265.

Biswas S, Swanson ME, Shoaib JUM, Haque SMS. 2010. Soil Chemical properties under modern and traditional farming systems at Khagrachari, Chittagong Hill District, Bangladesh. Journal of Forestry Research, 21(4): 451–456.

Blanco-canqui, H.; Shapiro, C.A.; Wortmann, C.S.; Drijber, R.A.; Mamo, M.; Shaver, T.M.; Ferguson, R.B. Soil organic carbon: The value to soil properties. J. Soil Water Conserv. 2013, 68, 129A–134A.

Blume, H.P.; Brümmer, G.W.; Fleige, H.; Horn, R.; Kandeler, E.; Kögel-Knabner, I.; Kretzschmar, R.; Stahr, K.; Wilke, B.M. Soil Organic Matter. In Scheffer/Schachtschabel Soil Science; Springer: Berlin, Germany; Heidelberg, Germany, 2016; pp. 55–86.

Boerner, R.E.J and Brinkman, J.A. (2003) Fire frequency and soil enzyme activity in southern Ohio oak –hickory forests. Applied Soil Ecology 23, 137–146.

Borggaard O.K. 2003. Sustainability appraisal of shifting cultivation in the Chittagong hill tracts of Bangladesh. AMBIO : A journal of the human environment. Volume 32, Issue 2, 118 – 123.

Brady NC and Well RR 2005. The Nature and Properties of Soil. 13th Edition, New Jersey: Prentice-Hall, Inc.

Brand J, Pfund JL. 1998. Site-and watershed-level assessment of nutrient dynamics under shifting cultivation in eastern Madagascar. Agriculture, Ecosystems & Environment, 71(1-3): 169–183.

Braz AMS, Fernandes AR, Alleoni LRF. Soil attributes after the conversion from forest to pasture in Amazon. Land Degrad Develop. 2013;24:33-8. https://doi.org/10.1002/ldr.1100

Brown S, Lugo AE 1990. Tropical secondary forests. J. Tropical Ecol. 6: 1-32.

Brams, A. 1971. Continuous cultivation of West African soils: Organic matter diminution and effects of applied lime and phosphorus. Plant and Soil 35:40114.

Brown, A. A. & Davis, K. P. 1973. Forest Fire: Control and Use. 2nd edition. McGraw-Hill Book Company, New York. 658 pp.

Bruun, T.B., de Neergaard, A., Lawrence, D., Ziegler, A.D., 2009. Environmental consequences of the demise in swidden cultivation in Southeast Asia: Carbon storage and soil quality. Human Ecology, 37, 375–388.

Bruun, T.B., Egay, K., Mertz, O., Magid, J., 2013. Improved sampling methods document decline in soil organic carbon stocks and concentrations of permanganate oxidizable carbon after transition from swidden to oil palm cultivation. Agriculture, Ecosystems & Environment, 178, 127–134.

Bruun, T.B., Mertz, O. and Elberling, B. 2006. Linking yields of upland rice in shifting cultivation to fallow length and soil properties. Agriculture, Ecosystems and Environment. 113,139-149.

Bryant, DD; Nielsen, BN; Tangley, L 1997. The last frontier forests: Ecosystems & economies on the edge. World Resources Inst. Washington, DC.pp67-98.

Buol S.W., Sanchez P.A., Cate R.B. Jr. and Granger M.A., 1975, Soil fertility capability and classification. A technical soil classification system of soil fertility management. In: Bornemisza E. and Alvarado A. Soil Management in Tropical America (eds) (pp. 126-145). North Carolina University, Raleigh, NC.

Campo J, Maass JM, Jaramillo VJ, Martínez- Yrizar A, Sarukhan J (2001) Phosphorus cycling in a Mexican tropical dry forest ecosystem. Biogeochemistry 53: 161-179.

Chandler, C., Cheney, P., Thomas, P., Trabaud, L. and Williams, D. 1983. Fire in Forestry. Volume 1: Forest Fire Behavior and Effects. John Willey and Sons, Inc. New York. 450 pp.

Charrau, C. and R. Nicou, 1971. L'amelioration due profile cultural dans les sols sableux et sabloargileux de La Zone tropicale secha Ouest – adicane et Ses incidence agronomique. Agron. Trop., 26: 565-631.

Chaudhari, P.R.; Ahire, D.V.; Ahire, V.D.; Chkravarty, M.; Maity, S. 2013. Soil bulk density as related to soil texture, organic matter content and available total nutrients of Coimbatore soil. Int. J. Sci. Res. Publ. 2013, 3, 1–8.

Chidumayo EN and Kwibisa L 2003. Effects of deforestation on grass biomass and soil nutrient status in miombo woodland, Zambia. Agriculture, Ecosystems & Environment, 96: 97-106.

Christanty L, 1986. Shifting Cultivation and Tropical Soils: Patterns, Problems, and Possible Improvements.. In Marten GG (edn) Traditional Agriculture in South East Asia: A Human Ecology Perspective. Westview Press, Boulder, Colorado. P-226-240.

Cobo J.G. et al. 2002. Nitrogen mineralization and crop uptake from surface-applied leaves of green manure species on a tropical volcanic-ash soil. Biol Fertil Soils, 36, 87 - 92.

Compton JE, Boone, RD. 2000. Long-term impacts of agriculture on soil carbon and nitrogen in New England forests. Ecology 81: 2314–2330.

Comte I, Davidson R, Lucotte M, Carvalho CJR, Oliveira FA, Silva BP, Rousseau GX. Physicochemical properties of soils in the Brazilian Amazon following fire-free land preparation and slash-and-burn practices. Agric Ecosyst Environ. 2012;156:108-15. https://doi.org/10.1016/j.agee.2012.05.004

Conklin. H C 196). 'The Study of Shifting Cultivation'. Current Anthropology. Vol. 2. p. 27-61.

Cornwell, W.K., Cornelissen, J.H.C., Amatangelo, K., Dorrepaal, E., Eviner, V.T., Godoy, O., Hobbie, S.E., Hoorens, B., Kurokaw, H., *et al.* (2008) Plant species traits are the predominant control on litter decomposition rates within biomes worldwide. Ecology Letters 11, 1065– 71.

Cotrufo, M.F.; Wallenstein, M.D.; Boot, C.M.; Denef, K.; Paul, E. 2014. The microbial efficiency-matrix stabilization (mems) framework integrates plant litter decomposition with soil organic matter stabilization: Do labile plant inputs form stable soil organic matter? Glob. Chang. Biol., 19, 988–995.

Crews TE, Farrington H, Vitousek PM (2000) Changes in asymbiotic, heterotrophic nitrogen fixation on leaf litter of Metrosideros polymorpha with long-term ecosystem development in Hawaii. Ecosystems 3: 386-395.

Cristancho, R.J.A., Hanafi, M.M., Syed Omar, S.R., Rafii, M.Y. 2014. Aluminum speciation of amended acid tropical soil and its effects on plant root growth. J. Plant Nut., 37, 811–827.

Cruz DLS, Vale Júnior JF, Cruz PLS, Cruz ABS, Nascimento PPRR. 2014. Atributos físico-hídricos de um Argissolo Amarelo sob floresta e savana naturais convertidas para pastagem em Roraima. Rev Bras Cienc Solo.;38:307-14. https://doi.org/10.1590/S0100-06832014000100031

Cunningham, R.K. 1963. The effect of clearing a tropical forest soil. Journal of Soil Science 14:334-45.

Dabiņ B., 1984, Les sols acides tropicaux. Cahiers ORSTOM, série Pédologie, 21, 7-19.

Datta,M.,Bhattacharyya, B.K. and Saikh, S. 2001. Soil fertility - a case study of shifting cultivation sites in Tripura. Journal of Indian Society of Soil Science, 49 (1),104-109.

Davidson, E. A.; Carvalho, C. J. R.; Figueiral, A. M.; Ishida, F. Y.; *et al.* 2007. Recuperation of nitrogen cycling in Amazonian forests following agricultural abandonment. Nature, v. 447, n. 7147, p. 995-997.

de Moura, E.G., Christoph Gehring, C., Braun H., Ferraz Junior, A.S.L., Reis, F.O. and Aguiar, A.C.F. 2016. Improving Farming Practices for Sustainable Soil Use in the Humid Tropics and Rainforest Ecosystem Health. Sustainability, 8: 841; doi:10.3390/su8090841.

De Rouw, A.1994. Effect of fire on soil, rice, weeds and forest regrowth in a rain forest zone (Cote d'Ivoire). Catena 22 (3):133–152.

DeBano LF, Neary DG, Ffolliott PF. 1998. Fires Effects on Ecosystems. John Wiley and Sons, New York. 203-208.

Demeyer, A., J.C. Voundi Nkana and M.G. Verloo, 2001. Characteristics of wood ash and influence on soil properties and nutrient uptake: An overview. Bioresource Technology, 77: 287- 295.

Denevan, W. M. 1996. A bluff model of riverine settlement in prehistoric Amazonia. Annals of the Association of American Geographers 86(4):654-681. https://doi.org/10.1111/j.1467-8306.1996.tb01771.x

Desjardins, T.; Barros, E.; Sarrazin, M.; Girardin, C.; Mariotti, A. Effects of forest conversion to pasture on soil carbon content and dynamics in Brazilian Amazonia. Agric. Ecosyst. Environ. 2004, 103, 365–373

Devi NL and Choudhury, B.U. 2013. Soil fertility status in relation to fallow cycles and landuse practices in shifting cultivated areas of Chandel district Manipur, India. IOSR Journal of Agriculture and Veterinary Science, 4: 2319-2372, www.iosrjournals.org www.iosrjournals.org,

Dhyani, S.K. and Chauhan, D.S., 1994. Agroforestry practices of North-Eastern hill region of India. In: Agroforestry traditions and innovations. CSWCR&TI, Dehradun. ,19-23.

Dias, T.; Oakley, S.; Alarcón-Gutiérrez, E.; Ziarelli, F.; Trindade, H.; Martins-Loução, M.A.; Sheppard, L.; Ostle, N.; Cruz, C. N-driven changes in a plant community affect leaf-litter traits and may delay organic matter decomposition in a Mediterranean maquis. Soil Biol. Biochem. 2013, 58, 163–171.

Don, A. et al. (2011) Impact of tropical land-use change on soil organic carbon stocks - a meta-analysis. Glob. Change Biol.,17, 1658-1670.

Donahue, R.L. Soils of equatorial Africa and their relevance to 1970 rational agriculture. East Lansing, Dept. of Crop and Soil Sciences, Michigan State University.

Doré, T.; Makowski, D.; Malézieux, E.; Munier-Jolain, N.; Tchamitchian, M.; Tittonell, P. (2011) Facing up to the paradigm of ecological intensification in agronomy: Revisiting methods, concepts and knowlegde. Eur. J. Agron., 34, 197–210.

Dove M, 1983. "Theories of Swidden Agriculture, and the Political Economy of Ignorance." Agroforestry Systems 1:85-99.

Driessen, P. M., P. Buurman, and Permadhy. 1977. The influence of shifting cultivation on a ''podzolic'' soil from Central Kalimantan. Pages 95–115 in Proceedings of peat and podzolic soils and their potential for agriculture in Indonesia. October 13–14, 1976. Soil Research Institute, Bogor, Indonesia.

Dung, N.V., Vien, T.D., Lam, N.T., Tuong, T.M., Cadisch, G., 2008. Analysis of the sustainability within the composite swidden agroecosystem in northern Vietnam: 1. Partial nutrient balances and recovery times of upland fields. Agric. Ecosyst. Environ. 128, 37–51.

Dungait, J.A.J.; Hopkins, D.W.; Gregory, A.S.; Whitmore, A.P. Soil organic matter turnover is governed by accessibility not recalcitrance. Glob. Chang. Biol. 2012, 18, 1781–1796.

Eckholm, E.P. 1976. Losing Ground. New York, N.Y.: Norton and Company. Ecol. Management 154: 353-370.

Edem, ID; Alphonsus, DU (2013).soil properties dynamics induced by passage of fire during agricultural burning. International of Advance Agricultural Research, 5(1) : 43-52

Edem, ID; Uduak, C; Udo, I; Ifiok, RI (2012) .Erodibility of Slash-and-Burn Soils along a Toposequence in Relation to Four Determinant Soil Characteristics. Journal of Biology, Agriculture and Healthcare (2)5:93-102.

Eden, M. J.; Andrade, A. Ecological aspects of swidden cultivation among the Andoke and Witoto Indians of the Colombian Amazon. Human Ecology, v. 15, n. 3, p. 339-359, 1987.

Edem, ID; Alphonsus, DU (2016).soil properties dynamics induced by passage of fire during agricultural burning. International of Advance Agricultural Research, 5(1) : 43-52

Edwards, P.J. and Grubb, P.J. (1982). Studies of mineral cycling in a montane rain forest in New Guinea. IV. Soil characteristics and the division of mineral elements between vegetation and soil. Journal of Ecology 70:649-666.

Erenstein O. (2002) Crop residue mulching in tropical and semi-tropical countries: an evaluation of residue availability and other technological implications. Soil and Tillage Res.; 67:115-133.

Eriksson *et al.* (2005) Wiklanders marklära. Studentlitteratur, Lund. FAO. (1993) Management and conservation of closed forests in Tropical America. 36

Estes, J.A., Terborgh, J., Brashares, J.S., Power, M.E., Berger, J., *et al.* (2011) Trophic downgrading of planet earth. Science, 333, 301–306.

Ewel JJ (1986) Designing agricultural systems for the humid tropics. Annual Rev. Ecol. Systematics 17: 245-271.

Ewel, J. J., C. Berish, B. Brown, N. Price, and J. Raich. 1981. Slash and burn impacts on a Costa Rican wet forest site. Ecology 62:816–829.

Fahey, T.J., Hughes, J.W., Pu, M. and Arthur, M.A., 1988. Root decomposition and nutrient flux following wholetree harvest of northern hardwood forest. For. Sci., 34(3): 744-768.

Fanning DS, Fanning MCB. 1989. Soil Morphology, Genesis, and Classification. John Wiley and Sons, New York.

FAO. 1957. Shifting cultivation. Unasylva 11: 1.

FAO. 1985. Changes in Shifting Cultivation in Africa: Seven Case Studies. FAO Forestry Paper 50/1. Rome: FAO.

FAO 1998. World Reference Base for Soil Resources. World Resources Report 84, Rome.

Fearnside, P. M. 2001. Land-tenure issues as factors in environmental destruction in Brazilian Amazonia: The case of Southern Para´ . World Dev. 29:1361–1372.

Fearnside, P. M. (2005) Deforestation in Brazilian Amazonia: history, rates, and consequences. Conservation Biology, v. 19, n. 3, p. 680-688.

Feller, C.; Beare, M.H. Physical control of soil organic matter dynamics in the tropics. Geoderma 1997, 79, 69–116.

Feng, W.; Plante, A.F.; Aufdenkampe, A.K.; Six, J. Soil organic matter stability in organo-mineral complexes as a function of increasing C loading. Soil Biol. Biochem. 2014, 69, 398–405. 133.

Filho, A.A.R., Adams, C. and Murrieta, 2013. The impacts of shifting cultivation on tropical forest soil: a review. Bol. Mus. Para. Emílio Goeldi. Cienc. Hum., Belém, 8: 693-727.

Folester, H. (1986) Nutrient loss during forest clearing. In: Land Clearing and Development in the Tropics (eds. Lal, R., Sanchez, P.A. and Cummings, Jr. R.W.), pp. 241-246. Balkema, Rotterdam.

Frizano, J.; Vann, D. R.; Johnson, A. H.; Johnson, C. M.; Vieira, I. C. G.; Zarin, D. J. (2003). Labile phosphorus in soils of forest fallows and primary forest in the Bragantina Region, Brazil. Biotropica, v. 35, n. 1, p. 2-11.

Funakawa, S., Makhrawie, M., Pulunggono, H.B., 2009. Soil fertility status under shifting cultivation in East Kalimantan with special reference to mineralization patterns of labile organic matter. Plant Soil 319, 57–66.

Funakawa, S., Tanaka, S., Kaewkhongkha, T., Hattori, T. and Yonebayashi,K.. (1997) Soil ecological study on dynamics of K, Mg and Ca and soil acidity in shifting cultivation in Northern Thailand. Soil Science and Plant Nutrition. 43 (3), 695-70

Gafur A, Borggaard OK, Jensen JR, Petersen L. 2000. Changes in soil nutrient content under shifting cultivation in the Chittagong Hill tracts of Bangladesh. Danish Journal of Geography, 100: 37–46

Gafur A, Jensen JR, Borggaard OK, Petersen L. 2003. Runoff and losses of soil and nutrients from small watersheds under shifting cultivation (Jhum) in the Chittagong Hill Tracts of Bangladesh. Journal of Hydrology, 279: 293–309.

Gafur, A., Koch, C.B., Borggaard, O.K., 2004. Weathering intensity controlling sustainability of ultisols under shifting cultivation in the Chittagong Hill Tracts of Bangladesh. Soil Sci. 169, 663–674.

Gain P. 1998. Bangladesh: Land Forest and Forest People (second edition).Dhaka: Society for Environment and Human Development (SHED).

Gama-Rodrigues, A.C.; Sales, M.V.S.; Silva, P.S.D.; Comerford, N.B.; Cropper,W.P.; Gama-Rodrigues, E.F. An exploratory analysis of phosphorus transformations in tropical soils using structural equation modeling. Biogeochemistry 2014, 118, 453–469.

Garcı´a-Oliva, F., Sanford Jr., R.L., Kelly, E., 1999. Effects of slash-and-burn management on soil aggregate organic C and N in a tropical deciduous forest. Geoderma 88, 1–12.

Garcia G. Hernandez T. Roldan A. Martin A. 2002. Effect of plant cover decline on chemical and microbiological parameters under Mediterranean climate. Soil Biol. Biochem. 34: 635-642.

Gentile, R.; Vanlauwe, B.; Six, J. Litter quality impacts short- but not long-term soil carbon dynamics in soil aggregate fractions. Ecol. Appl. 2011, 21, 695–703.

George, M. & Buvaneswaran, C. (2001). Productivity and nutrient cycling in teak plantations. Pp.73–91 in Mandal, A. K. & Ansari, S. A. (Eds.) Genetics and Silviculture of Teak. International Book Distributors, Dehra Dun

Gessner, M.O., Swan, C.M., Dang, C.K., McKie, B.G., Bardgett, R.D., Wall, D.H. and Haettenschwiler, S. (2010) Diversity meets decomposition. Trends in Ecology and Evolution 25, 372–380.

Giardina CP, Sanford Jr. RL, and Døckersmith IC, 2000. Changes in soil phosphorus and nitrogen during slash-and-burn clearing of a dry tropical forest. Soil Sci. Soc. Am. J. 64:399–405.

Glaser, B. 2006. Prehistorically modified soils of central Amazonia: a model for sustainable agriculture in the twenty-first century. Philosophical Transactions of the Royal Society B: Biological Sciences 362(1478):187-196. https://doi.org/10.1098/rstb.2006.1978

Glaser, B., and J. J. Birk. 2012. State of the scientific knowledge on properties and genesis of anthropogenic dark earths in Central Amazonia (terra preta de índio). Geochimica et Cosmochimica Acta 82:39-51. https://doi.org/10.1016/j.gca.2010.11.029

Goudie, A. 1984. The Human Impact: Man's role in Environmental Change . Oxford: Blackwell.

Grandstaff, T.B. 1981. Shifting cultivation. Ceres 4:28-30.

Grange, I., Kansuntisukmongkol, K., 2003. Effect of fallow length on soil structure, hydraulic properties, and soil organic C in a swidden cultivation system of western Thailand. Trop. Agric. 80, 246–251.

Granged, A.J.P., Zavala, L.M., Jordán, A. and Bárcenas– Moreno, G. (2011) Post–fire evolution of soil properties and vegetation cover in a Mediterranean heathland after experimental burning: A 3–year study. Geoderma 162, 85–94.

Grayston S.J. Campbell C.D. Bardgett R.D. Mawdsley J.L. Clegg C.D. Ritz K. Griffiths B.S. Rodwell J.S. Edwards S.J. Davies W.J. Elston D.J. Millard P. 2004. Assessing shifts in microbial community structure across a range of grasslands of different management intensity using CLPP, PLFA and community DNA techniques. Applied Soil Ecology. 25: 63-84.

Greenland, D. J. 1974. Evolution and development of different types of shifting cultivation. In Shifting Cultivation and Soil Conservation in Africa. Soil Bulletin 24. Rome: FAO.

Greenland, D.J. (1975). Bringing the green revolution to the shifting cultivator. Science 190: 841-4.

Greenland, D.J. and Kowal, J.M.L. (1960). Nutrient content of the moist tropical forest of Ghana. Plant and Soil12:154-174.

Griffin, D. M. (1972). Ecology of Soil Fungi (London: Chapman and Hall).

Guo LB & Gifford RM (2002). Soil carbon stocks and land use change: a meta analysis. Global Change Biol 8, 345-360.

Gupta PK, Sahai S, Singh N, Dixit CK, Singh DP, Sharma C *et al.* Residue burning in rice-wheat cropping system: causes and implications. Curr. Sci. 2004; 87(2):1713-1717.

Gupta, P.C. and O'Toole, J.C., 1986. Upland Rice: A Global Perspective. International Rice Research Institute, Los Banos, 360 pp.

Hagvall, K.; Persson, P.; Karlsson, T. Speciation of aluminum in soils and stream waters: The importance of organic matter. Chem. Geol. 2015, 417, 32–43.

Haider MZ. Options and determinants of rice residue management practices in the south–west region of Bangladesh. SANDEE Working Paper no 71-12, South Asian Network for Development and Environmental Economics (SANDEE) EPC 1056, Kathmandu, Nepal, 2012.

Hamman, S.T., Burke, I.C. and Knapp, E.E. (2008) Soil nutrients and microbial activity after early and late season prescribed burns in a Sierra Nevada mixed conifer forest. Forest Ecology and Management 256, 367–374.

Hargreaves P.R. Brookes P.C. Ross G.J.S. Poultion P.R. 2003. Evaluating soil microbial biomass carbon as an indicator of long-term environmental changes. Soil Biology & Biochemistry. 35:401-407.

Harper, C.W., Blair, J.M., Fay, P.A., Knapp, A.K. and Carlisle, J.D. (2005) Increased rainfall variability and reduced rainfall amount decreases soil CO2 flux in a grassland ecosystem. Global Change Biology 11, 322–334.

Harris, D. R. The ecology of swidden cultivation in the upper Orinoco rain forest, Venezuela. The Geographical Review, v. 61, n. 4, p. 475-495, 1971.

Hattori, D., Sabang, J., Tanaka, S., Kendawang, J.J., Ninomiya, I., Sakurai, K., 2005. Soil characteristics under three vegetation types associated with shifting cultivation in a mixed dipterocarp forest in Sarawak, Malaysia. Soil Sci. Plant Nutr. 51, 231–241.

Hauck, FW (1974. Shifting cultivation and soil conservation in Africa. FAO Soils Bull. 24: 1–4.

Havlin et al. (2005) Soil Fertility and Fertilizers (7th ed.). Pearson Education Inc., New Jersey, USA.

Haws, L.D., H. Inoue, A. Tanaka, and S. Yoshida. 1983. Comparison of crop productivity in the tropics and temperate zone. In Potential Productivity of Field Crops Under Different Environments International Rice Research Institute. Los Banos, Philippines. pp. 403-413.

Heard J, Cavers C, Adrian G. Up in smoke-nutrient loss with straw burning. Better crops. 2006; 90:10-11.

Hinsinger, P.; Bengough, A.G.; Vetterlein, D.; Young, I. Rhizosphere: Biophysics, biogeochemistry and ecological relevance. Plant Soil 2009, 321, 117–152.

Hoffman J. Bezchlcbova J. Dusck L. Dolczal L. Holoubek I. Andel P. Ansorgova Maly S. 2003. Novel approach to monitoring of soil biological quality. Environmental International. 28:771-778.

Hossain, M.A. 2011. An overview on shifting cultivation with reference to Bangladesh. Scientific Research and Essays, 6: 6509-6514. DOI: 10.5897/SRE11.1282.

Hudson, N. 1973. Soil Conservation. London: B.T. Batsford Ltd.

IBSRAM (International Board for Soil Research and Management). (1987). Report offhe First Regional Seminar on Management of Vertisols under Semi-Arid Conditions in Africa and Southwest Asia, Nairobi, Kenya, 1-6 December 1986. Bangkok, Thailand: IBSRAM.

IFAD (International Fund for Agricultural Development (2001). International Development Research Centre (IDRC); Cornell International Institute for Food, Agriculture and Development (CIIFAD); International Centre for Research in Agroforestry (ICRAF); International Institute of Rural Reconstruction (IIRR). Shifting cultivation: towards sustainability and resource conservation in Asia. Silang/Cavite/Philippines: International Institute of Rural Reconstruction.

Innes, L., Hobbs, P.J., Bardgett, R.D. (2004). The impacts of individual plant species on rhizosphere microbial communities varies in soils of different fertility. Biol. Fertil. Soils, 40:7–13.

Ishaq M. 1971. Bangladesh District Gazetteers Chittagong Hill Tracts. Govt. of Bangladesh, Dhaka. Meah F, Recreational possibilities in forests of Bangladesh. Review Paper No.110.

Jackson, R.B. *et al.* (2005) Trading water for carbon with biological carbon sequestration. Science, 310, 1944-1947.

Jamala, GY; Boni, PG; Abraham, P; Teru, CP (2012). Evaluation of Environmental and Vulnerability Impact of Bush Burning in Southern Guinea Savanna of Adamawa State, Nigeria. American Journal of Experimental Agriculture 2(3): 359-369.

Jendoubi, D., Hodel, E., Liniger, H. P., and Subhatu, A. T.: (2019). Land degradation assessment using landscape unit approach and normalized difference vegetation index in Northwest of Tunisia, Journal of Mediterranean Ecology, 17, 67–79, 2019.

Jenkinson, D.S. & Ayanaba A. (1977) Decomposition of carbon- 14 labeled plant material under tropical conditions. Soil Sci. Soc. Am. J., 41, 912-915.

Jenny, H. (1941) Factors of soil formation. New York, McGraw-Hill. 1941

Jia, G., Cao, J., Wang, C. and Wang, G. (2005) Microbial biomass and nutrients in soil at the different stages of secondary forest succession in Ziwulin, northwest China. Forest Ecology and Management 217, 117–125.

Jiao, F.; Shi, X.-R.; Han, F.-P.; Yuan, Z.-Y. Increasing aridity, temperature and soil pH induce soil C-N-P imbalance in grasslands. Nature 2016, 6, 1–9.

Johan, A.; Hakansson, I. Response of different crops to soil compaction—Short-term effects in Swedish field experiments. Soil Tillage Res. 2014, 138, 56–63.

Johnson, C. M.; Vieira, I. C. G.; Zarin, D. J.; Frizano, J.; Johnson, A. H.(2001) Carbon and nutrient storage in primary and secondary forests in eastern Amazonia. Forest Ecology and Management, v. 147, n. 2-3, p. 245-252, 2001.

Jones, D.L.; Nguyen, C.; Finlay, R.D. Carbon flow in the rhizosphere: Carbon trading at the soil–root interface. Plant Soil 2009, 321, 5–33.

Jordan C.F., 1985, Nutrient cycling in forest ecosystems. Wiley: New York, NY.

Jordan, C. F. 1987. Shifting Cultivation, Case Study No. 1: Slash and Burn Agriculture near San Carlos de Rio Negra, Venezuela, In Amazonian Rain Forest, Ecosystem Disturbance and Recovery, edited by C. F. Jordan.

Jordan, C. F.1982. Amazon Rain Forests. American Scientist 70: 394 - 401.

Jordan CF (1989). An Amazonian Rain Forest. The Structure and Function of a Nutrient Stressed Ecosystem and the Impact of Slashand- Burn Agriculture. Man and Biosphere Series. The Parthenon Publishing Group, Lancs, UK.

Juo A.S.R. & Manu A., 1996, Chemical dynamics in slash-and-burn agriculture. Agriculture, Ecosystems and Environment, 58, 49-60.

Juo, A.S.R. and E.J. Kamprath, 1978. Copper Chloride as an extractant for estimating the reactive Al pool in acid soils. Soil Science Society of America Journal, 18: 525-531.

Juo, A.S.R. and R. Lal, 1977. The effects of fallow and continuous cultivation on the chemical and physical properties of an Alfisol in Western Nigeria. Plant and Soil, 47: 567-587.

Juo, A.S.R. et al. (1995a) Acidification of a kaolinitic Alfisol under continuous cropping with N fertilization in West Africa. Plant Soil, 171, 245-253.

Juo, A.S.R. et al. (1995b) Changes in soil properties during longterm fallow and continuous cultivation after forest clearing in Nigeria. Agric. Ecosyst. Environ., 56, 9-18.

Juo, A.S.R. et al. (1996) Soil properties and crop performance on a kaolinitic Alfisol after 15 years of fallow and continuous cultivation. Plant Soil, 180, 209-217.

Kalpage, F.S.C.P. 1974. Tropical Soils Classification, Fertility and Management. London: MacMillan Press.

Kamada, M.; Nakagoshi, N.; and Takahashi, F. 1987. Effects of Burning on Germination of Viable Seed in a Slash and Burn Agriculture. Jpn. J. Ecol. 37: 91–100. (in Japanese with English summary)

Kang, B.T., Wilson, G.F., Lawson, T.L.. (1984) Alley Cropping: A Stable Alternative to Shifting Cultivation. International Institute of Tropical Agriculture, Ibadan, Syria.

Kamarck, A.M. (1976). The tropics and economic development: a provocative enquiry into the poverty of nations. Baltimore: John Hopkins University Press.

Kardol, P. and Wardle, D.A. (2010) How understanding aboveground – belowground linkages can assist restoration ecology. Trends in Ecology and Evolution 25, 670–679.

Kato MSA, Kato OR, Denich M, Vlek PLG. Fire free alternatives to slash-and-burn for shifting cultivation in the eastern Amazon region: the role of fertilizers. Field Crops Res. 1999; 62:225-237.

Kauffman, JD; Cummings, DW; Babbitt, R (1995). Fire in the Brazilian Amazon: I. Biomass, nutrient pools, and losses in slashed primary forests. Oecologia 104:397–408.

Kauffman, J. B.; Hughes, R. F.; Heider, C. (2009) Carbon pool and biomass dynamics associated with deforestation, land use, and agricultural abandonment in the Neotropics. Ecological Applications, v. 19, n. 5, p. 1211-1222.

Kauffmann S., Sombrooek W. & Mantel S., 1998, Soils of rainforests: characterization and major constraints of dominant forest soils in the humid tropics. In: Schulte A. and Ruhiyat D. Soils of Tropical Forest Ecosystems (eds): Characteristics, Ecology and Management, Springer- Verlag, Berlin, pp. 9-20.

Kellman M, Miyanishi K, and Hiebert P, "Nutrient retention by savanna ecosystems II. Retention after fire," Journal of Ecology, vol. 73, no. 3, pp. 953–962, 1985.

Ketterings,Q. M. Bigham J.M., Lapeche V. (2000). Changes in Soil Mineralogy and Texture Caused by Slash and Burn Fires in Samatra, Indonesia soil sci. Soc. Am 64, 1108-1117.

Khresat, S.; Al-Bakri, J.; Al-Tahhan, R. 2008. Impacts of land use/ cover change on soil properties in the Mediterranean region of northwestern Jordan. Land Degrad. Dev., 19, 397–407.

Kidd, C.V. and Pimentel, D. (Editors), 1992. Integrated Resource Management: Agroforestry for Development. Academic Press, San Diego, CA, 223 pp.

Killingbeck KT (1996) Nutrients in senesced leaves: Keys to the search for potential resorption

Kleinman, P. J. A., R. B. Bryant, and D. Pimentel. 1996. Assessing ecological sustainability of slash-and-burn agriculture through soil fertility indicators. Agronomy Journal 88(2):122-127. https://doi.org/10.2134/agronj1996.00021962008800020002x

Kleinman, P. J.; Pimeentel, D.; Bryant, R. B. The ecological sustainability of slash-and-burn agriculture. Agriculture, Ecosystems & Environment, v. 52, n. 2-3, p. 235-249, 1995.

Kotto-Same, J., Woomer, P.L., Appolinaire, M., Louis, Z., 1998. Carbon dynamics in slash-andbum agriculture and land use alternatives of the humid forest zone in Cameroon. Agriculture, Ecosystems & Environment, 65, 245-256.

Kumar, S. (2008). Report of the Inter-Ministerial National Task Force on Rehabilitation of Shifting Cultivation Areas.

Kyuma, K.; Tulaphitak, T.; and Pairintra, C. 1985. Changes in Soil Fertility and Tilth under Shifting Cultivation: I. General Description of Soil and Effect of Burning on the Soil Characteristics. Soil Sci. Plant Nutr. 31 (2): 227–238.

Lal, R., 1985. Need for, approaches to, and consequences of land clearing and development in the tropics. In: IBSRAM, Tropical Land Clearing for Sustainable Agriculture: Proc. IBSRAM Inaugural Workshop, Bangkok, IBSRAM, Bangkok, pp. 15-27.

Lal R., 1996, Deforestation and land-use effects on soil degradation and rehabilitation in Western Nigeria. I. Soil physical and hydrological properties. Land Degradation & Development, 7, 19-45.

Lal, R. (1975). Role of Mulching Techniques in Tropical Soil and Water Management. Ibadan, Nigeria: International Institute for Tropical Agriculture, Technical Bulletin, No. 1.

Lal, R. (1987) Managing the soils of sub-Saharan Africa. Science, 236, 1069-1076.

Lal, R. (2004).Soil carbon sequestration impacts on global climate change and food securityScience 304: 1623-1627.

Lal, R. Challenges and opportunities in soil organic matter research. Eur. J. Soil Sci. 2009, 60, 158–169.

Lal, R. 1989. Agroforestry systems and soil surface managment of a tropical alfisol. Agroforestry systems 8:1-6.

Lal, R., 1997. Degradation and resilience of soils. Philosophical Transactions of the Royal Society B, 352, 869-889.

Lal, R.; Kimble, J.; Levine, E.; Whitman, C. World soils and greenhouse effect: an overview. In: Soils and global change. Boca Raton: Lewis Publishers, 1995. p. 1-7.

Laudelot, H. (1961). Dynamics of Tropical Soils in Relation to Their Fallowing Techniques. Rome, Italy: Food and Agricultural Organization (FAO), Paper 11266/E.

Landon, J. R. (1991). Booker Tropical Soil Manual. A handbook for soil survey and agricultural land evaluation in the tropics and sub tropics. Longman Scientific & Technical. Hong Kong.

Lawrence D and Schlesinger WH, 2001. Changes in soil phosphorus during 200 years of shifting cultivation in Indonesia. Ecology, 82: 2769–2780.

Lawrence, D., Radel, C., Tully, K., Schmook, B., Schneide, L., 2010. Untangling a decline in tropical forest resilience: constraints on the sustainability of shifting cultivation across the globe. Biotropica 42, 21–30.

Lee, D.H.K. (1957). Climate and economic development in the tropics. Harper, New York.

Lehmann, J., D. Kern, L. German, J. McCann, G. C. Martins, and A. Moreira. 2003a. Soil fertility and production potential. Pages 105-124 in J. Lehmann, D. C. Kern, B. Glaser, and W. I. Woods, editors. Amazonian dark earths: origin, properties, management. Kluwer Academic, Dordrecht, The Netherlands. https://doi.org/10.1007/1-4020-2597-1_6

Lehmann, J., J. Pereira da Silva Jr., C. Steiner, T. Nehls, W. Zech, and B. Glaser. 2003b. Nutrient availability and leaching in an archaeological Anthrosol and a Ferralsol of the Central Amazon Basin: fertilizer, manure and charcoal amendments. Plant and Soil 249:343-357. https://doi.org/10.1023/a:1022833116184

Lenka, N.K., Choudhury, P.K.R., Sudhishri, S., Dass, A. and Patnaik, U.S. (2012) Soil aggregation, carbon build up and root zone soil moisture in degraded sloping lands under selected agroforestry based rehabilitation systems in eastern India. Agriculture, Ecosystems and Environment 150, 54-62.

Levy, D (2004).Forest burning is a net contributor to global warning Scientist say, August 12,News Service.

Liliensiek, A.K., Thakuria, D., Clipson, N. (2013) Influences of plant species composition, fertilisation and Lolium perenne ingression on soil microbial community structure in three Irish grasslands. Microbial Ecology, 63:509-521.

Lindell L, Åström M, Öberg T. (2010) Land-use versus natural controls on soil fertility in the Subandean Amazon, Peru. Sci Total Environ.;408:965-75. https://doi.org/10.1016/j.scitotenv.2009.10.039

Lôpez–Poma, R. and Bautista, S. (2014) Plant regeneration functional groups modulate the response to fire of soil enzyme activities in a Mediterranean shrubland. Soil Biology and Biochemistry 79, 5–13.

Luna-Orea, P., and M.G. Wagger. 1996. Management of tropical legume cover crops in the Bolivian Amazon to sustain crop yields and soil productivity. Agron. J. 88:765–776.

Lutz, H. J. and R. F. Chandler (1946). Forest Soil (New York: John Wiley and Sons).

Manns, H.R.; Berg, A.A. Importance of soil organic carbon on surface soil water content variability among agricultural fields. J. Hydrol. 2014, 516, 297–303.

Maass, J (1995).Conversion of tropical dry forest to pasture and agriculture. In S.H. Bullock, H.A. Mooney, and E. Medina. Cambridge Univ. Press, p. 399–422.

Mahmood, S.R.D. Finlay, A.M. Fransson and H. Wallander, 2003. Effect of hardened wood ash on microbial activity, plant growth and nutrient uptake by ectomycorrhizal spruce seedlings. FEMS Microbial. Ecol., 43: 121-131.

Majumdar B, Kumar K, Venkatesh MS, Patiram, and Bhatt BP, "Effect of different agroforestry systems on soil properties in acid Alfisols of Meghalaya," Journal Hill Research, vol. 17, no. 1, pp. 1–5, 2004.

Malcheva, B., Velizarova, E., Nustorova, M. and Ibrahim, M. (2015) Enzyme activity of forest fire–influenced soils from north slopes of Rila mountain (region of Dolna Bania). Forestry Ideas, 21, 55–66.

Manner, H.I. 1969. Effect of shifting cultivation on some soil properties of the Bismarck mountains territory of Papua and New Guinea. A theses submitted to the Graduate Division of the University of Hawaii for the partial fulfillment of the requirements of the degree of Master of Arts in Geography.

Marafa L.K. and Chau K.C. 1999. Effect of hill fire on upland soil in Hong Kong. Forest Ecol. Mgmt. 120: 97-104.

Martins, P. S. (2005). Dinâmica evolutiva em roças de caboclos amazônicos. Estudos Avançados, v. 53, n. 19, p. 209-220, 2005.

Mbagwu, J.S.C., 1992. Improving the productivity of a degraded Ultisol in Nigeria using organic and inorganic amendments. Part II. Changes in physical properties. Bioresource Tech., 42: 167-175.

McDonald, M. A.; Heleay, J. R.; Stevens, P. A. (2002) The effects of secondary forest clearance and subsequent land-use on erosion losses and soil properties in the Blue Mountains of Jamaica. Agriculture, Ecosystems & Environment, v. 92, n. 1, p. 1-19.

McGrath DA, Comerford NB, Duryea ML (2001) Litter dynamics and monthly fluctuations in soil phosphorus availability in an Amazonian agroforest. Forest Ecol. Management 131: 167-181.

McGrath, D. G. 1987. The role of biomass in shifting cultivation. Human Ecology, v. 15, n. 2, p. 221-242.

McKnight, T L (1992). Essentials of Physical Geography. New Jersey: USA.

Menaut JC, Abbadie L and Vitousek PM (1993). Nutrient and organic matter dynamics in tropical ecosystems. Pages 215–231 in P. J. Crutzen and J. G. Goldammer, editors. Fire in the environment: the ecological, atmospheric, and climatic importance of vegetation fires. John Wiley and Sons, New York, New York, USA.

Mertz, O., Leisz, S.J., Heinimann, A., Rerkasem, K., Thiha., Dressler, W., Pham, V.C., Vu, K.C., Schmidt-Vogt, D., Colfer, C.J.P., Epprecht, M., Padoch, C. and Potter, L., 2009b. Who Counts? Demography of swidden cultivators in Southeast Asia. Human Ecology, 37, 281-289.

Miah S, Dey S, Haque SMS. 2010. Shifting cultivation effects on soil fungi and bacterial population in Chittagong Hill Tracts, Bangladesh. Journal of Forestry Research, 21(3): 311–318.

Mieville, A; Granier, C; Liousse, C; Guillaume, B; Mouillot, F; Lamarque, JF; Grégoire, JM; Pétron, G (2010). Emissions of gases and particles from biomass burning during the 20th century using satellite data and an historical reconstruction. Atmos. Environ. 44: 1469–1477.

Mishra, A.K., Aggarwal, P., Bhattacharyya, R., Das, T.K., Sharma, A.R., Singh, R. Least limiting water range for two conservation agriculture cropping systems in India. Soil Tillage Res. 2015, 150, 43–56.

Mishra, B.K. and Ramakrishnan, P.S. (1983). Slash and burn agriculture at higher elevations in north-eastern India. ii. Soil fertility changes. Agriculture, Ecosystem and Environment, 9: 83-96.

Moura, E.G.; Aguiar, A.C.F.; Piedade, A.R.; Rousseau, G.X. The contribution of legume tree residues and macrofauna to the improvement of abiotic soil properties in the eastern Amazon. Appl. Soil Ecol. 2015, 86, 91–99.

Moura, E.G.; Oliveira, A.K.C.; Pinheiro, K.M.; Aguiar, A.C.F. Management of a cohesive tropical soil enhance rootability and increase the efficiency of nitrogen and potassium use. Soil Use Manag. 2012, 28, 370–377.

Moura, E.G.; Sena, V.G.L.; Correa, M.S.; Aguiar, A.F.C. The Importance of an Alternative for Sustainability of Agriculture around the Periphery of the Amazon Rainforest. Recent Pat. Food Nutr. Agric. 2013, 5, 70–78.

Moura, E.G.; Sena, V.G.L.; Sousa, C.C.M.; Silva, F.R.; Coelho, M.J.A.; Macedo, V.R.A.; Aguiar, A.C.F. Enhancement of the rootability of a structurally fragile tropical soil using gypsum and leguminous residues to increase the maize yield. Soil Use Manag. 2016, 32, 118–126.

Mukul S.A. and Herbohn J. 2016. The impacts of shifting cultivation on secondary forests dynamics in tropics: A synthesis of the key findings and spatio temporal distribution of research. Environmental Science & Policy 55: 167–177

Mukul, S.A., Herbohn, J., Firn, J., Ferraren, A., Congdon, R.(2016). In review. Soil carbon and nutrients status in relation to shifting cultivation in regenerating secondary forests in the Philippines uplands. Land Degradation & Development.

Mulumba, L.N.; Lal, R. Mulching effects on selected soil physical properties. Soil Tillage Res. 2008, 98, 106–111

Myers, N (1994). Tropical deforestation: Rates and patterns. In K. Brown and D.W. Pearce (eds.). The causes of tropical deforestation. UCL Press, London.

Myers. Norman (1995). 'Tropical deforestation: rates and patterns' in Katrina Brown & David W Pearce (ed.) The Causes of Tropical Deforestation, The Economic and Statistical Anajysis of Factors giving Rise to the Loss of Tropical Forests. UCL Press London & Research Press New Delhi.

Nair PKR (1990). The prospects for agroforestry in the tropics. World Bank Technical paper no.131, World Bank, Washington D.C.

Nair P.K.R. 1993. An introduction to Agroforestry. Kluwer Academkic Publishers, Dordrecht, The Netherlands. pp. 785-790.

Namgyel, U.; Siebert, S. F.; Wang, S. (2008) Shifting cultivation and biodiversity conservation in Bhutan. Conservation Biology, v. 22, n. 5, p. 1349-1351, 2008.

Nath AJ, Brahma B, Lal R and Dass AK, 2016. Soils under jhum cultivation in Eastern Himalayas, Encyclopedia of Soil Science, Third Edition DOI: 10.1081/E-ESS3-120053748

Neff, JC; Harden, JW; Gleixner, G (2005). Fire effects on soil organic matter content composition, and Nutrients in boreal interior Alaska. Can. J. Forest Res. 35(9- 10): 2178-2187.

Nellemann, C.; MacDevette, M.; Manders, T.; Eickhout, B.; Svihus, B.; Prins, A.G.; Kaltenborn, B.P., (2009) Eds. The Environmental Food Crisis – The Environment's Role Inaverting Future Food Crises, A UNEP Rapid Response Assessment; United Nations Environment Programme, GRID-Arendal, 2009. Q2

Nnabude, P.C. and J.S.C. Mbagwu, 1999. Soil water relations of a Nigerian typic haplustult amended with fresh and burnt rice mill wastes. Soil and Tillage Research, 50: 207-214.

Novotny, E. H., M. H. Hayes, B. E. Madari, T. J. Bonagamba, E. R. de Azevedo, A. A. de Souza, G. Song, C. M. Nogueira, and A. S. Mangrich. 2009. Lessons from the Terra Preta de Índios of the Amazon region for the utilisation of charcoal for soil amendment. Journal of the Brazilian Chemical Society 20(6):1003-1010. https://doi.org/10.1590/S0103-50532009000600002

Nye P.H. & Greenland D.J., 1964, Changing in the soil after clearing a tropical forest. Plant and Soil, 21, 101-112.

Nye PH and Greenland DJ 1960. Soil under shifting cultivation. Technical Communication. Commonwealth Bureau of Soils, Harpenden 51, 73-126.

Nyegers, A.E., 1989. Coppice swidden fallows in tropical deciduous forest: biological, technological, and socioculturat determinants of secondary forest successions. Hum. Ecol., 17(4): 379-400.

Odedina, S.A., J.N. Odedina, L.S. Ayeni, S.A. Arowojoh, S.D. Adeyeye and S.O. Ojeniyi, 2003. Effects of types of ash on soil fertility, nutrient availability and yield of tomato and pepper. Nig. J. Soil Sci., 13: 61-67.

Okigbo, B.N., and R. Lal. 1979. SoU fertility maintenance and conservation for improved agroforestry systems in the lowland humid tropics. In Soil Research in Agroforestry, eds. H.O. Mongi and P.A. Huxley, pp. 41-77. International Council for Research in Agroforestry, Report Number ICRAF..001e, Nairobi, Kenya.

Okonkwo, C.I. 2010. Effect of Burning and Cultivation on Soil Properties and Microbial Population of Four Different Land Use Systems in Abakaliki. Research Journal of Agriculture and Biological Sciences, 6(6): 1007-1014, 2010.

Oliver, D.P.; Bramley, R.G.V.; Riches, D.; Porter, I.; Edwards, J. Soil physical and chemical properties as indicators of soil quality in Australian viticulture. Aust. J. Grape Wine Res. 2013, 19, 29–139.

Olsson K. 2008. Soil fertility and changes caused by slash and burn agriculture in two watersheds of the upper Amazon, Peru. Thesis submitted to the University of Kalmar, Sweden.

Oseni, O.A., Ekperigin, M.M., Oboh, G., Akindahunsi, A.A. and Adekunle, V.A.J. (2009) Effects of fire on physico–chemical properties and proteolytic enzyme activities of soil from teak (Tectona grandis L.F) plantations. Journal of Agricultural Biotechnology and Sustainable Development 1, 079–085.

Osman, K.S., Jashimuddin, M, Haque, S.M.S and Miah, S. 2013. Effect of shifting cultivation on soil physical and chemical properties in Bandarban hill district, Bangladesh. Journal of Forestry Research, 24: 791-795, DOI 10.1007/s11676-013-0368-3

Owolabi, O., A. Adeyeye, B.T. Oladgo and S.O. Ojeniyi, 2003. Effect of wood ash on Soil fertility and crop yield in Southwest Nigeria. Nig. J. Soil Sci., 13: 56-60.

Padoch, C. 1985. Labor efficiency and intensity of land use in rice production: an example from Kalimantan. Human Ecology 13: 271-289.

Page, A.L., Miller, R.H. and Keeney, D.R. (1982) Methods of Soil Analysis, Part II- Chemical and Microbiological Properties, Second Edition, Agronomy Monograph No.9, American Society of Agronomy, Inc., Soil Science Society of America, Inc., Madison, Wisconsin, USA. 1159 p.

Paix MJ, Lanhai L, Xi C, Ahmed S, Varenyam A. Soil degradation and altered flood risk as a consequence of deforestation. Land Degrad Dev. 2013;24:478-85. https://doi.org/10.1002/ldr.1147

Palm, C. A., M. J. Swift, and P. L. Woomer. 1996. Soil biological dynamics in slash-and-burn agriculture. Agriculture, Ecosystems, and Environment 58: 61–74.

Paudel, E., Dossa, G.G.O., Xu, J., Harrison, R.D., 2015. Litterfall and nutrient return along a disturbance gradient in a tropical montane forest. Forest Ecology and Management, 353, 97–106.

Pausas, J.G., Bradstock, R.A., Keith, D.A., Keeley, J.E. and the GCTE (Global Change of Terrestrial Ecosystems) Fire Network (2004) Plant functional traits in relation to fire in crown-fire ecosystems. Ecology 85, 1085– 1100.

Pedroso-Junior, N. N.; Murrieta, R. S. S.; Adams, C. Slashand- burn agriculture: a system in transformation. In: BEGOSSI, A.; LOPES, P. (Orgs.). Current trends in human ecology. Cambridge: Cambridge Scholars Publishing, 2009. p. 12-34.

Pedroso-Junior, N. N.; Murrieta, R. S. S.; Rasul, G.; Thapa, G. B.; Zoebisch, M. A. Determinants of land-use changes in the Chittagong Hill Tracts of Bangladesh. Applied Geography, v. 24, n. 3, p. 217-240, 2004.

Peters, J.W and Neuenschwander, L.F. (1988). Slash and Burn: Farming in the Third World Forest. University of Idaho Press. USA.

Pimentel D, Allen J, Beers A, Guinand L, Linder R, P. McLaughlin, Meer B *et al.* (1987). World agriculture and soil erosion. Bioscience, 37: 277-283.

Pimentel, D. and Heichel, G.H., 1991. Energy efficiency and sustainability of farming systems. In: R. Lal and F.J. Pierce (Editors), Soil Management for Sustainability. Soil and Water Conservation Society, Ankeny, IA, pp. 113-123.

Popenoe, H.L. 1960. The influence of shifting cultivation on natural soil constituents in Central America. Ph.D. thesis, University of Florida, Gainesville, Florida.

Posey, D. A. l983. Indigenous ecological knowledge and development of the Amazon. In Moran, E. F. (ed.), Dilemma of Amazonian Development . Boulder, Colorado: Westview, pp. 225 - 257.

Pritchet, W. L., & Fisher, R. F. 1987. Properties and Management of Forest Soils. 2nd edition. John Willey and Sons, Inc., New York. 494 pp.

Pushparajah, E. (1985) Development induced soil erosion and flash floods in Malaysia. Ecologist 15(1-2), 19-20.

Pyne, SJ; Goldammer, JG (1997).The culture of fire: An Introduction to anthropogenic fire history: in: Clark J. S., Cachier H. Goldammer, J. G. stocks B. (ed.) Sediment Records of Biomass Burning and global change., NATO ASI series 1 Vol. 51. Springer, Pp. 71-14.

Raharimalala, O., Buttler, A., Ramohavelo, C.D., Razanaka, S., Sorge, J.P., Gobat, J.M., 2010. Soil–vegetation patterns in secondary slash and burn successions in Central Menabe, Madagascar. Agric. Ecosyst. Environ. 139, 150–158.

Raison, R.J., Khanna, P.K. and Woods, P.V. (1985) Transfer of element to the atmosphere during the low intensity prescribed fire in three Australian subalpine eucalypt forests. Canadian Journal of Forest Research 15, 657–664.

Ralston, C. W. & Hatchell, G. E. 1971. Effects of prescribed burning on physical properties of soil. Pp. 68-84 in Proceedings Prescribed Burning Symposium. USDA Forest Service. Southeastern Forest Experimental Station.

Ram SC and Ramakrishnan PS, "Hydrology and soil fertility of degraded grasslands at Cherrapunji in North Eastern India," Environment Conservation, vol. 15, pp. 29–35, 1988.

Ramakrishnan PS, Toky OP. Soil nutrient status of hill agro ecosystems and recovery pattern after slash and burn agriculture (Jhum) in north-eastern India. Plant Soil. 1981; 60:41-64.

Ramakrishnan, P.S. (1992). Shifting Agriculture and Sustainable Development: An Interdisciplinary Study from North-Eastern India. Man and the Biosphere Series No. 10. UNESCO, Paris. pp 424.

Ramakrishnan, P.S. (1998). Ecology in Developing countries. The Emerging Paradigms. PINSA, 65 (1&2): 51-82.

Rambo, A. T. l983. Why shifting cultivators keep shifting: understanding farmer decision-making in traditionalagroforestry systems. Paper prepared for the Joint East-West Environment and Policy Institute/ United Nations Food and Agriculture Organization

Regional Worksop on Community Forestry. Nakorn Rachasima, Thailand, August 22-29, 1983.

Randrianarison, A., Schlaepfer, R., Mills, R., Hervé, D., Razanaka, S., Rakotoarimanana, V., Buttler, A. (2016). Linking historical land use to present vegetation and soil characteristics under slash-and-burn cultivation in Madagascar. Applied Vegetation Science, 19(1), 40–52. https://doi.org/10.1111/ avsc.12202

Rasul G, Thapa GB. 2003. Shifting cultivation in the mountains of south and Southeast Asia: Regional patterns and factors influencing the change, Land Degradation and Development, 14: 495–508.

Rasul, G; Thapa, GB; Zoebisch, MA (2004) 'Determinants of land-use changes in the Chittagong Hill Tracts of Bangladesh.' Applied Geography 24(3): 217–240

Reddell P., 1993, Soil constraints to growth of nitrogen-fixing trees in tropical environments. In: N.S. Subba Rao and C. Rodriguez-Barrueco (eds), Symbiosis in nitrogen-fixing trees. New Delhi: Oxford and IBH Publishing Co. pp. 65-79.

Reichert JM, Bervald CMP, Rodrigues MF, Kato OR, Reinert DJ. Mechanized land preparation in eastern Amazon in fire-free forest-based fallow systems as alternatives to slash-and-burn practices: hydraulic and mechanical soil properties. Agric Ecosyst Environ. 2014;192:47-60. https://doi.org/10.1016/j.agee.2014.03.046

Reichert JM, Rodrigues MF, Bervald CMP, Kato OR. Fire-free fallow management by mechanized chopping of biomass for sustainable agriculture in eastern Amazon: effects on soil compactness, porosity, and water retention and availability. Land Degrad Dev. 2016;27:1403-12. https://doi.org/10.1002/ldr.2395

Renard, K.G., Foster, G.R., Weesies, G.A. and Porter, J.P., 1991. Rusle: revised universal soil loss equation. J. Soil Water Conserv., 46(1): 30-33.

Rerkasem, K.; Lawrence, D.; Padoch, C.; Schmidtvogt, D.; Ziegler, A. D.; Bruun, T. B. Consequences of swidden transitions for crop and fallow biodiversity in Southeast Asia. Human Ecology, v. 37, n. 3, p. 347-360, 2009.

Reza, S.K.; Baruah, U.; Nath, D.J.; Sarkar, D.; Gogoi, D. Microbial biomass and enzyme activity in relation to shifting cultivation and horticultural practices in humid subtropical North-Eastern India. Range Manag. Agrofor. 2014, 35, 78–84.

Reuler HV and Janssen BH (1993) Nutrient fluxes in the shifting cultivation system of south-west Côte d'Ivoire. Plant and Soil, 154: 179-188.

Richards, P. W. 1952. The Tropical Rain Forest. Cambridge: the Cambridge University Press.

Rodenburg, J., Stein, A., Noordwijk, M., Ketterings, Q.M., 2003. Spatial variability of soil pH and phosphorus in relation to soil run-off following slash-andburn land clearing in Sumatra, Indonesia. Soil Till. Res. 71, 1–14.

Roder W, Phengchanh S, Keoboulapha B and Maniphone (1995) Relationships between soil, fallow period, weeds and rice yield in slash-and-burn systems of Laos. Plant Soil. 1995; 176:27-36.

Roder, W.; Phengchanh, S.; Keoboulapha, B.; and Maniphone, S. 1994. Upland Agriculture-activities by Lao-IRRI Project. In Shifting Cultivation and Rural Development in the LAO PDR, Report of the Nabong Technical Meeting, pp.152–169. Vientiane: Nabon College.

Roder W, Phengchanh S, Maniphone S (1997) Dynamics of soil and vegetation during crop and fallow period in slash-and-burn fields of northern Laos. Geoderma 76:131–144

Ruddiman, WF (2003). The anthropogenic greenhouse era began thousands of years ago. Clim. Change., 61: 261-293.

Russell, A.E.; Raich, J.W. Rapidly growing tropical trees mobilize remarkable amounts of nitrogen, in ways that differ surprisingly among species. Proc. Nat. Acad. Sci. USA 2012, 109, 10398–10402.

Sabhasri, S., 1978. Effects of forest fallow cultivation on forest production and soil. In: P. Kunstadter, E.C. Chapman and S. Sabhasri (Editors), Farmers in the Forest: Economic Development and Marginal Agriculture in Northern Thailand. East-West Center, Honolulu, HI, pp. 160-184.

Saha R, Tomar JMS, Ghosh PK (2007) Evaluation and selection of multipurpose tree for improving soil hydro-physical behaviour under hilly eco-system of north east India. Agroforestry Systems 69(3): 239.247.

Saharjo BH, 1999. The effects of fire on the properties of soil in Acacia Mangium plantations in South Sumatra, Indonesia. Journal of Tropical Forest Science ll :459-469.

Saharjo, B. H. 1995. The changes in soil chemical properties following burning in a shifting cultivation area in South Sumatra. Wallaceana 75: 23-26.

Sahrawat, K.L. *et al.* (2001) Application of inorganic phosphorus fertilizer. In Sustaining soil fertility in West Africa. SSSA Spec. Publ. No. 58, eds. Tian, G., Ishida, F., Keating, D., SSSA, Madison, Wisconsin, 225-246. 331

Sanchez, P.A. (2002) Soil fertility and hunger in Africa. Science, 295, 2019-2020.

Sanchez, P.A. 1973. Soil management under shifting cultivation. In A Review of Soils Research in Tropical Latin America, ed. P.A. Sanchez, pp. 46-67. North Carolina Agricultural Experiment Station Technical Bulletin 219, Raleigh, North Carolina.

Sanchez, P.A. 1982. Nitrogen in shifting cultivation systems of Latin America. Plant and Soil, 67: 91-103. 0032-079X/82/0671-0091 $01.95.

Sanchez P.A. 1976. Properties and management of soils in the tropics. New York: Wiley.

Sanchez P.A. 1987. Management of acid soil in the humid tropics of Latin America. In: Sanchez P.A., Stoner E.R. and Pushparajah E. eds. Management of acid tropical soils for sustainable agriculture. IBSRAM Proceedings 2:63-107. Bangkok, Thailand: IBSRAM.

Sanchez, P. A., & Cochrane, T. T. (1980). Soil constraints in relation to major farming systems of tropical America. Soil related constrain to food production in the tropics. Los Baños, Filipinas, 107-139.

Sanchez P A, Couto W and Buol S W (1982) The fertility capability soil classification system: interpretation, applicability and modification. Goederema 27: 283–309

Sanchez PA and Hailu M (1996). Alternatives to slash-and-burn agriculture. Agriculture, ecosystems and environment, no. 1, Elsevier Publications.

Sanchez, P. A. & Leakey, R. R. B. 1997 Land-use transPhil. Transformation in Africa: Three determinants for balancing food security with natural resource conservation. Fourth Congress, European Society of Agronomy, 7: 15-23'.

Sanchez P.A. & Salinas J.G., 1981, Low input technology for managing Oxisols and Ultisols in Tropical America. Adv. Agron. 34, 279-406.

Sanchez, P.A. 1973. Soil management under shifting cultivation. In A Review of Soils Research in Tropical Latin America, ed. P.A. Sanchez, pp. 46-67. North Carolina Agricultural Experiment Station Technical Bulletin 219, Raleigh, North Carolina.

Sanchez, P.A., D.A. Bandy, J.H. Villachica, and J.J. Nicholaides (1982), 'Amazon Basin soils: management for continuous crop production', Science 216: 821–827.

Sanchez, P.A., Villachica, J.H. and Bandy, D.E., 1983. Soil fertility dynamics after clearing a tropical rainforest in Peru. Soil Sci Soc. Am. J., 47:1171-1178.

Sánchez, P.A., C.A. Palm, S.A. Vosti, T.P. Tomich, and J. Kasyoki. 2005. Alternatives to slash and burn: Challenge and approaches of an international consortium. In Slash-and-burn agriculture: The search for alternatives, edited by C.A. Palm, S.A. Vosti, P.A. Sanchez, and P.J. Ericksen. New York: Columbia University Press, 3-37.

Saplalrinliana, H, Thakuria, D, Sapu Changkija, D.S. and Hazarika, S. 2016. Impact of Shifting Cultivation on Litter Accumulation and Properties of Jhum Soils of North East India. Journal of the Indian Society of Soil Science, 64, No. 4: 402-413, DOI: 10.5958/0974-0228.2016.00052.9

Schipper L.A. & Sparling G.P., 2000, Performance of soil condition indicators across taxonomic groups and land uses. Soil Science Society of America Journal, 64, 300-311.

Schlesinger WH, "Carbon sequestration in soils," Science, vol. 284, no. 5423, p. 2095, 1999.

Schritt, H., C. Beusch, P. Ríos Guayasamín, and M. Kaupenjohann. 2020. Transformation of traditional shifting cultivation into permanent cropping systems: a case study in Sarayaku, Ecuador. Ecology and Society 25(1):10. https://doi.org/10.5751/ES-11252-250110.

Schulz, H., and B. Glaser. 2012. Effects of biochar compared to organic and inorganic fertilizers on soil quality and plant growth in a greenhouse experiment. Journal of Plant Nutrition and Soil Science 175(3):410-422. https://doi.org/10.1002/jpln.201100143

Scotter, G. W. 1963. Effects of fire on soil properties in northern Saskatchewan. Forest Chronicle 39: 412-421.

Seibert, S.F., 1990. Hillside farming, soil erosion, and forest conservation in two Southeast Asian national parks. Mt. Res. Dev., 10(1): 64-72.

Seubert, C.E. 1975. Effect of land clearing methods on crop performance and changes in soU properties in an Ultisol of the Amazon Jungle of Peru. M.S. thesis, North Carolina State University, Raleigh, North Carolina.

Sharma UC, (1998) "Methods of selecting suitable land use system with reference to shifting cultivation in NEH region," Indian Journal of Soil Conservation, vol. 26, no. 3, pp. 234–238, 1998.

Sharma UC and Bujarbaruah KM 2016. Issues and Challenges in Shifting Cultivation and its Relevence in the Present Context, National Academy of Agricultural Sciences, Policy Paper No. 83.

Sharma, U. C. & Prasad, R. N. (1995) Socio-economic aspects of acid sooil management and alternate land use systems for north-eastern states of India. In: Plant-Soil Interactions at Low pH. (ed. by R. A., Date; N. J. Grundon.; G. E. Rayment, . & M. E. Probert), 689-696, Kluwer Academic Press, The Netherlands.

Sharma, U. C. (2003) Impact of population growth and climate change on the quantity and quality of water resources in the northeast region of India. In: Water Resources System - Hydrological Risks, Management and Development. (ed. by G. Bloschl, S. Franks, M. Kumagia, K. Musiake & D. Rosbjerg), IAHS Publ. No. 281, 349-357, IAHS Press, Wallingford, U.K.

Sharma U. C. and Sharma, V. (2004). Implications of nutrient and soil transfer with runoff in the northeastern reghion of India. In: Sediment Transfer through the Fluvial System (ed. by V. Golosov, V. Belyaev and D. E. Walling). IAHS Publ. No. 288, 488-493, IAHS Press, Wallingford, U.K.

Sharma, U. C., Datta, M. and Sharma, V. (2014). Soil fertility, erosion, runoff and crop productivity affected by different farming systems. Ecopersia, 2: 629-650.

Sharma E, Ambasht RS (1984) Seasonal variation in nitrogen fixation by different ages of root nodules of Alnus nepalensis plantations in the Eastern Himalayas. Journal of Applied Ecology 21: 265-270.

Shepherd G., Bureh R.J. & Gregory P.J., 2000, Land use affects the distribution of soil inorganic nitrogen in smallholder production systems in Kenya. Biology and Fertility of Soils, 31, 348-355.

Shoaib, J. U.; G. Mostafa; and M. Rahman (1998). Soil Erosion Hazard in Chittagong Hill Tracts: A Case Study. Dhaka, Bangladesh: Soil Resources Development Institute, Annual Report, 1998.

Singh BP and Dhyani SK, "Significance of jackfruit in restoration of soil fertility," Annual Report, ICAR Research Complex, Umiam, Meghalaya, 1995.

Singh A and Singh MD, "Effect of various stages of shifting cultivation on soil erosion from steep hill slopes," Indian Forester, vol. 106, no. 2, pp. 115–121, 1981.

Singh, B.P., Madhumita Das and Prasad, R.N. Effect of different landuse systems on nutrient availability in hill terrains of Meghalaya (India). Proc. of the Inter. Symp. on Nutrient Management for sustained productivity. Vol.II. 1992, 60-62.

Singh A and Singh, M. D. (1978). Effect of various stages of shifting cultivation on soil erosion from steep hill slopes. Indian Forester, Vol.106:pp 115-121.

Srilakshmi, A., Dakshayani, R., Saigopal, D.V.R. and Narasimha, G. (2012) Influence of forest litter on soil enzyme activities. Ecology, Environment and Conservation 18, 105–112.

Stark, S.; Männistö, M.K.; Eskelinen, A. Nutrient availability and pH jointly constrain microbial extracellular enzyme activities in nutrient-poor tundra soils. Plant Soil 2014, 383, 373–385.

Steiner, C., W. G. Teixeira, J. Lehmann, T. Nehls, J. L. V. de Macêdo, W. E. H. Blum, and W. Zech. 2007. Long term effects of manure, charcoal and mineral fertilization on crop production and fertility on a highly weathered Central Amazonian upland soil. Plant and Soil 291(1-2):275-290. https://doi.org/10.1007/s11104-007-9193-9

Stevenson, F.J. (1986) Cycles of soil C, N, P, S, Micronutrients. John Wiley & Sons, Canada. Troeh, Frederick R. and Thompson, Louis M. (2005) Soils and Soil Fertility (6th ed.). Blackwell Publishing, UK.

Street, J. M. 1969. An evaluation of the concept of carrying capacity. The Professional Geographer 21(2): 104 - 107.

Suarez, R.K., Sajise, P.E., 2010. Deforestation, swidden agriculture and Philippine biodiversity. Philippine Science Letters, 3, 91-99.

Sutherland J.M. & Sprent J.I., 1993, Nitrogen fixation by legume trees. In: N.S. Subba Rao and C. Rodriguez-Barrueco (eds). Symbiosis in nitrogen- fixing trees. New Delhi: Oxford and BH Publishing Co, pp. 33-54.

Swamy PS, Ramakrishnan PS (1988). Nutrient budget under slash and burn agriculture (Jhum) with different weeding regimes in north-eastern India. Acta Oecologica. 1988; 9:85-102.

Swamy, S.L.; Dutt, C.B.S.; Murthy, M.S.R.; Mishra, A.; Bargali, S.S. Floristics and dry matter dynamics of tropical wet evergreen forests of Western Ghats, India. Curr. Sci. 2010, 99, 353–364.

Talbot, J.M.; Yelle, D.J.; Nowick, J.; Treseder, K.K. Litter decay rates are determined by lignin chemistry. Biogeochemistry 2012, 108, 279–295.

Tanaka S. Funakawa S. Kawekhangkha T. Hattori T. Yonebayashi K. 1997. Soil ecological study on dynamics of K, Mg and Ca and soil acidity in shifting cultivation in northern Thailand. Soil Sci. Plant Nutr. 43: 695-708.

Tanaka, S., Andoa, T., Funakawa, S., Sukhrun, C., Kaewkhongkha, T., Sakurai, K., 2001. Effect of burning on soil organic matter content and N mineralization under shifting cultivation system of Karen people in Northern Thailand. Soil Sci. Plant Nutr. 47, 547–558.

Tanaka, S., Kendawang, J.J., Yoshida, N., Shibata, K., Jee, A., Tanaka, K., Ninomiya, I. and Sakurai, K.. Effects of Shifting Cultivation on Soil Ecosystems in Sarawak, Malaysia II. Changes in soil chemical properties and runoff water at Balai Ringin and Sabal experimental sites. Soil Science and Plant Nutrition. 50 (5). 2004, 689-699.

Thakuria D and Sharma UC, 2016. Shifting Cultivation : Coexistence of Forest and Agriculture, National Academy of Agricultural Sciences News, New Delhi, India.

Thomas ML, Lal R, Logan T and Fausey NR (1992) Land Use and Management Effects on Nonpoint Loading from Miamian Soil. Soil Sci. Soc. Am. J., 56, 1871-1875.

Thomaz, E.L., 2013. Slash-and-burn agriculture: establishing scenarios of runoff and soil loss for a five-year cycle. Agric. Ecosyst. Environ. 168, 1–6.

Tinker, P.B.; Ingram, J.S.I.; Struwe, S. Effects of slash-andburn agriculture and deforestation on climate change. Agr. Ecosyst. Environ. 1996, 58, 13–22.

Toky, O.P. and Ramakrishnan, P.S. (1982). A comparative study of the energy budget of hill agro-ecosystems with emphasis on the slash and burn system (Jhum) at lower elevations of north-eastern India. Agricultural Systems 9: 143-154.

Toky, O.P. and Ramakrishnan, P.S. (1983a). Secondary succession following slash and burn agriculture in north-eastern India.1. Biomass, litter fall and productivity. Journal of Ecology 71, 735-745.

Toky, O.P. and Ramakrishnan, P.S. (1983b). Secondary succession following slash and burn agriculture in north-eastern India. II. Nutrient cycling. Journal of Ecology 71, 747- 757.

Tomich TP, Fagi AM, De Foresta H, Michon G, Murdiyarso D, Stolle F, Van Noordwijk M. 1998. Indonesian fires: smoke as a problem, smoke as a symptom. Agroforestry Today, 10: 4–7.

TPR. 1987. The third party preview of the upland settlement component of Chittagong hill tracts development project (multi-sectoral). Mid-term evaluation, May 1987. Cabinet Division, Bangladesh Secretariat, Dhaka.p.246

Trabaud, L. 1987. Fire and survival traits of plants. – In: Trabaud, L. (ed.), The role of fire in ecological systems. SPB Academic Publishing, pp. 65–89.

Trabaud, L. 1994. Post-fire plant community dynamics in the Mediterranean basin. – In: Moreno, J. M. and Oechel, W. C. (eds), The role of fire in Mediterranean type ecosystems. Ecological Studies. Springer-Verlag. Vol. 107, pp. 1–15.

Troeh, F.R. and Thompson, L.M. (2005) Soils and Soil Fertility. Sixth Edition, Blackwell, Ames, Iowa, 489. Troeh & Thompson; 2005

Tulaphitak, T.; and Pairintra, C.; and Kyuma, K. 1985. Changes in Soil Fertility and Tilth under Shifting Cultivation: II. Change in Soil Nutrient Status. Soil Science and Plant Nutrition 31: 239–249.

Ubuoh, EA; Nwankama, C; Ejekwolu, CC (2016). The Effect of Selected Land Uses on Soil Organic Carbon (Soc) and Total Nitrogen (Tn) Dynamics in Imo State, Nigeria (Carbon Sequestration Implications) Academia Journal of Environmental Science 4(11): 220-226.

Uhl, C., Clark, K., Clark, H. and Murphy, P., 1981. Early plant succession after cutting and burning in the Upper Rio Negro of the Amazon Basin. J. Ecol., 69:631-649.

Uhl, C. (1987). Factors Controlling Succession Following Slash-and-Burn Agriculture in Amazonia. Journal of Ecology 75: 377–407. doi:10.2307/2260425.

Van der Kruijs A.C.B.M. *et al.* (1988) Recovery of 15N-labelled fertilizer in cops, drainage water and soil using monolith lysimeters in south-east Nigeria. Eur. J. Soil Sci., 39, 483-492.

Van Reuler H, Janssen BH. Nutrient fluxes in the shifting cultivation system of South-West Cote D'Ivoire. I. Dry matter production, nutrient contents and nutrient release after slash and burn for two fallow vegetations. Plant Soil. 1993; 154:169-177.

van Vliet, N., Mertz, O., Heinimann, A., Langanke, T., Pascual, U., Schmook, B., Adams, C., Schmidt-Vogt, D., Messerli, P., Leisz, S., Castella, J.-C., Jørgensen, L., BirchThomsen, T., Hett, C., Bech-Bruun, T., Ickowitz, A., Vu, K.C., Yasuyuki, K., Fox, J., Padoch, C., Dressler, W., Ziegler, A.D., 2012. Trends, drivers and impacts of changes in swidden cultivation in tropical forest-agriculture frontiers: a global assessment. Glob. Environ. Change 22, 418–429.

Van Wambeke, A., t 992. Soils of the Tropics: Properties and Appraisal. McGraw-Hill, New York, 343 pp.

Watters, R.F. 1971. Shifting Cultivation in Latin America. FAO Forestry Development Paper, Rome, Italy.

Webster, C.C. and Wilson, P.N. (Ed.). 1998 Agriculture in the Tropics (3rd ed.). Blackwell Science Ltd, UK.

Webster, C.C., and P.N. WUson. 1966. Agriculture in the Tropics. London: Longmans Green and Co., Ltd.

Wendi , B. (2011). Rural women want alternatives to 'slash and burn. conservation, environment, farming, livelihoods, Panos London achieved visit

Whalen, J.K. Managing soil biota-mediated decomposition and nutrient mineralization in sustainable agroecosystems. Adv. Agric. 2014, 2014, 1–13.

White, E. M., Thompson, W. W. & Gartner, F. R. 1973. Heat effects of nutrient release from soils under Ponderosa pine. Journal of Range Management 26:22-24

Whigham DF, Zugasty Towle P, Cabrera Cano E, O'Neill J, Ley E (1990) The effect of annual variation in precipitation on growth and litter production in a tropical dry forest in the Yucatan of Mexico. J. Tropical Ecol. 31: 23- 34.

Wieder RK, Wright SJ 1995 Tropical forest litter dynamics and dry season irrigation on Barro

Wiedinmyer, C; Akagi, SK; Yokelson, RJ; Emmons, LK; Al-Saadi, JA; Orlando, JJ; Soja, AJ (2011). The Fire INventory from NCAR (FINN): a high resolution global model to estimate the emissions from open burning. Geosci. Model Develop. 4 (3): 625–641

Willis, K. J.; Gillson, L.; Brncic, T. M. How "virgin" is virgin rainforest? Science, v. 304, n. 5669, p. 402-403, 2004.

Wilson, G.F. and Akapa, K.L., 1983. Providing mulches for no-tillage in the tropics. In: I.O. Akobundu and A.E. Deutsch (Editors), No-Tillage Crop Production in the Tropics. International Plant Protection Center, Oregon State University, Corvallis, pp. 51-65.

Wood, Charles H. and Porro, Roberto (Ed.). 2002 Deforestation and Land Use in the Amazon. University Press of Florida, FL, USA.

Xue, L., Li, Q. and Chen, H. 2014 Effects of a wildfire on selected physical, chemical and biochemical soil properties in a Pinus massoniana forest in South China. Forests 5, 2947–2966.

Yang JC, Huang JH, Pan QM, Tang JW & Han XG. 2004. Long-term impacts of land-use change on dynam-ics of tropical soil carbon and nitrogen pools. J Environ Sci 16, 256-261.

Yang, F.; Zhang, G.L.; Yang, J.L.; Li, D.C.; Zhao, Y.G.; Liu, F.; Yang, R.M.; Yang, F. Organic matter controls of soil water retention in an alpine grassland and its significance for hydrological processes. J. Hydrol. 2014, 519, 3086–3093.

Yang, X.; Post,W.M. Phosphorus transformation as a function of pedogenesis: A synthesis of soil phosphorus data using Hedley fractionation method. Biogeosciences 2011, 8, 2907–2916.

Yang, Y., Guo, J.F., Chen, G.S., He, Z.M., Xie, J.S., 2003. Effect of slash burning on nutrient removal and soil fertility in Chinese fir and evergreen broadleaved forests of mid-subtropical China. Pedosphere 13, 87–96.

Yemefack M., Njomgang R., Nounamo L. and Rossiter, D.G. 2010. Quantified soil dynamics and spatial fragmentation within the shifting agricultural landscape in southern Cameroon.19th World Congress of Soil Science, Soil Solutions for a Changing World, 1 – 6 August 2010, Brisbane, Australia.

Yemefack M, Bijker W, De Jong SM (2006a) Investigating relationships between Landsat-7 ETM+ data and spatial segregation of LULC types under shifting agriculture in southern Cameroon. International Journal of Applied Earth Observation & Geoinformation 8, 96-112.

Yemefack M, Jetten VG, Rossiter DG (2006b) Developing a minimum data set for characterizing soil dynamics under shifting cultivation systems. Soil & Tillage Research 86, 84-98.

Yuan, B.C. and Yue, D.X. (2012) Soil microbial and enzymatic activities across a chronosequence of Chinese pine plantation development on the Loess Plateau of China. Pedosphere 22, 1–12.

Zaffar, M.; Lu, S.G. Pore size distribution of clayey soils and its correlation with soil organic matter. Pedosphere 2015, 25, 240–249.

Zarin, D. J.; Duchesne, A. L.; Hiraoka, M. Shifting cultivation on the tidal floodplains of Amazonia: impacts on soil nutrient status. Agroforestry Systems, v. 41, n. 3, p. 307-311, 1998.

Zhao, C.; Fu, S.; Mathew, R.P.; Lawrence, K.S.; Feng, Y. Soil microbial community structure and activity in a 100-year-old fertilization and crop rotation experiment. J. Plant Ecol. 2015, 1–10.

Zhao, J.; Dong, Y.; Xie, X.; Li, X.; Zhang, X.; Shen, X. Effect of annual variation in soil pH on available soil nutrients in pear orchards. Acta Ecol. Sin. 2011, 31, 212–216.

Ziegler, A. D., T. B. Bruun, M. Guardiola-Claramonte, T. W. Giambelluca, D. Lawrence, and Nguyen Thanh Lam. 2009b. Environmental consequences of the demise in swidden agriculture in Southeast Asia: hydrology and geomorphology. Human Ecology 37:361–373.

Ziegler A. D., Bruun T. B. and Guardiola-Claramonte M, Giambelluca T. W., Lawrence D. and Lam N. T., 2009a. Environmental Consequences of the Demise in Swidden Cultivation in Montane Mainland Southeast Asia: Hydrology and Geomorphology. Hum Ecol., 37:361–373 DOI 10.1007/s10745-009-9258-x

Zinke, P. J.; S. Sabhasri; and P. Kunstadter (1978). Soil Fertility Aspects of the Lua' Forest Fallow System of Shifting Cultivation. In: P. E. Kunstadter, E. C. Chapman and S. Sabhasri (eds.) Farmers in the Forest: Economic Development and Marginal Agriculture in Northern Thailand (Honolulu, Hawaii: University Press of Hawaii), pp. 134-159.

6

Deforestation and Degradation

6.1 GENERAL DESCRIPTION

Between 13,000 and 3,000 BC, Neolithic period, mountain people in Asia, Africa, and Latin America have practiced shifting cultivation (Mazoyer and Roudart 2006). A cultivation phase, consisting of removing forest and crop production for 1 to 3 years, is followed by a fallow phase, in which agriculture is discontinued to allow soil nutrients to recoup (Conklin 1961). Conklin (1961) distinguished between total and partial shifting cultivation. Pioneer shifting cultivation, in which primary or apex forest is cleared annually, and replaced by shifting cultivation, which entails clearing mostly secondary forests, are both essential forms of cultivation that are primarily practised for farmers' subsistence. Partial shifting cultivation, on the other hand, is done as a supplemental or temporary activity (Conklin 1961). Throughout history, indigenous peoples have used fire to manage their land and natural resources. Pastoralists and hunters-gatherers in Africa's savannas, for example, have used fire to preserve the ecosystem's productivity for livestock and game for millennia (McClanahan and Young 1996). In pre-colonial times, indigenous peoples in North America also significantly altered the landscape, establishing and preserving pastures and open forest savannas through regulated burning (Pyne 1982, cited in: Schneider 2000).

Forests cover 27 percent of the earth's land surface (Sanchez, 1994). Forests are a major source of energy and building materials, as well as a source of off-farm revenue and employment. They help to maintain environmental stability, produce items that promote food security, and provide a source of off-farm income and employment. Forests are also an important part of the biosphere, as they help to stabilize natural processes. Forest ecosystems provide a wide range of environmental services, some of which are required for the system's operation and some of which are directly beneficial to humans (Learman and Sedjo, 1994). In recent years, preserving biological diversity in the tropics has become a more pressing issue. Tropical forests are very diverse ecosystems that support a disproportionately large number of plant and animal species around the planet (Kahn and McDonald, 1995). Tropical forests once covered roughly fifteen million square kilometers of land in Asia, Africa, and Latin America, accounting for around 12% of the earth's geographical surface. Tropical forest cover has fallen to half its original size, or 7,5 million square kilometers (Gillis, 1996). Tropical deforestation is one of the world's most pressing environmental issues today, and it is expected to remain so in the first decades of the twenty-first century (Sanchez, 1994). Deforestation occurs when land is permanently transferred from forest to non-forest use. Human activity can affect the tropical forest in a variety of ways and on a variety of scales. Agriculture and livestock expansion, fuelwood harvesting, and rising demand for commercial and noncommercial forest products are the main causes of deforestation. Rural communities are the first to suffer the effects of deforestation when trees and other forest products become scarce. Deforestation is a serious problem not just because it depletes a vital natural resource for the region's population that cannot be replenished immediately, but also because of its far-reaching consequences (Sharon et al., 1988). The problem of poverty in rural regions is exacerbated by deforestation, land degradation, and loss of biological diversity. Other negative environmental and social repercussions, as well as a decline in agricultural production, are common. The gathering of fuelwood, commercial logging, deterioration through grazing and fire, and shifting cultivation are the main causes of deforestation in Asia. Conversion to agriculture is thought to be responsible for 94 percent of deforestation in tropical countries, split evenly between permanent agriculture and shifting cultivation (Siren, 1997). Deforestation in the tropics and other developing countries is part of a larger land-use change process that impacts both the tropics and the rest of the world (Nilsson, 1996). In the debate about tropical deforestation, shifting cultivation and agricultural growth are at the forefront.

While the development of cultivated land and pasture is the most visible proximal cause of tropical deforestation, changing farmers are not usually the principal deforestation agents: Traditional shifting cultivation (swidden-fallow farming) is generally found in upland and foothill Asian cases, whereas colonist shifting cultivation (in-migrant slash-and-burn agriculture) is mostly found in humid lowland Latin American examples, with many of the latter driven by poverty. Agricultural expansion is dominated by smallholders' growth of persistently farmed land for food, rather than changing cultivation, leading in deforestation. The disparity between "traditional shifting farming" and "slash and burn agriculture" practised by migrant settler colonialism is crucial. Not only are they essentially different types of land use, but they are also carried out by different persons. Traditional shifting agriculture, as defined by Geist and Lambin (2002), is practised by indigenous peoples in Southeast Asia-Asia, which we are primarily concerned with here. Traditional shifting cultivation's tangible expressions, on the other hand, are as varied as the people who practise it, making it difficult to characterise. Mertz *et al.* (2009) who "decided to define swidden cultivation in Southeast Asia as a land use system that employs a natural or improved fallow phase, which is longer than the cultivation phase of annual crops, sufficiently long to be dominated by woody vegetation, and cleared by means of fire".

Despite disagreements over the root reasons of forest conversion, shifting agriculture is commonly cited as the primary cause of deforestation in the tropics (Myers 1993; Riswan and Hartanti 1996). Because it is carried out by people who live in or near the forest, shifting agriculture will continue to have an impact on rainforests. Every year, Indonesia deforests 1.2 million hectares (m ha) of tropical rainforest (FAO 1993). Each year, selective logging affects an additional 1.2 million hectares of primary forest, but this is not deforestation or permanent forest alienation (Grainger 1984; FAO 1993). Because they are assumed to exploit logging roads to grow into previously uncultivated areas, shifting cultivators are sometimes blamed for deforestation after logging. Kummer (1992) shown that in the Philippines, weak economic growth and massive timber extraction aided the expansion of non-shifting, small-scale agriculture. Large-scale industrial agriculture replaced logged forest in Malaysia (Repetto 1988) and Thailand (Feeny 1984). Much of the deforestation in Indonesia, like in the Philippines, has been attributed mistakenly to non-migrant hill farmers changing their agricultural practises (Collins *et al.* 1991; Riswan and Hartanti 1996). People practising shifting cultivation in the tropics typically use standing primary forest for game, wood, and other resources while clearing it for agriculture (Denslow and Padoch 1988; Poffenberger 1990; Anderson 1990). Shifting farming is distinct from other "slash-and-burn" land-use systems, which include a variety of land-uses that depend on fire to clear rainforest for cultivation or further development. In shifting agriculture, farmers return to an area after a certain period of time to clean and cultivate it, while in slash-and-burn, they might never return. In addition to upland rice cultivation, shifting growers in Indonesia cultivate and manage many types of secondary forest (Michon 1991; Padoch and Peters 1993; Salafsky 1994; Lawrence *et al.* 1995). Only a few households are able to establish all of their land holdings within walking distance of the village. Travel time reduces productive work time since home labor is limited. As a result, even if the effort required to avoid vertebrate pest harm increases with distance, distant land is maintained less actively. Rubber has had a significant impact on the landscape's spatial layout, initially reflecting trade-offs between employment and travel time. Rubber's growing dominance will eventually dictate the dynamics of dry-rice agriculture, which was once the sole factor shaping the terrain. The survival of primary forest species whose native habitat is becoming increasingly threatened may be harmed as a result of these changes in landscape structure and dynamics. To further understand how changes in the controlled landscape affect forest regeneration processes, more study is needed.

Shifting cultivation is practiced on a diverse range of soil types, slashing a diverse range of vegetational types, and by people of widely disparate origins and cultures, resulting in significant differences in the types of crops grown, the length of the cropping and fallow periods, and the method of cultivation itself. However, the method is universally defined by (1) partial or complete removal of vegetation by cutting and burning (2) cultivation of pure or mixed crops (3) abandonment of the plot to fallow regeneration when the soil fertility has been depleted. A brief period of cropping is usually alternated with a longer period of fallow, with human energy being used exclusively for all agricultural activities. It is one of the oldest agricultural systems, implying that it must be, or has been, more or less in balance with the environment until recently (Om Prakash 1980). Although some anthropologists prefer the term 'swidden,' it is commonly referred to as'slash and burn' agriculture (Exwall, 1955). The technique is also known by a variety of regional names, including *milpa' in Central America, 'chena* in Ceylon, and 'xaingin' in the Philippines. Conklin has gone over the system's terminology in great detail (1957).

Efforts to alleviate climate change and conserve living variety continue to focus on reducing deforestation in tropical regions. Population development, greater climate unpredictability, inadequate preservation of forest reserves, and, in certain cases, the loss of local eco-fiendly information are all putting strain on tropical agro-ecosystems and their surrounding landscapes (Berkes *et al.* 2000). As a result of such stresses, a growing number of ecological collapses are being observed in tropical regions (Davidson *et al.* 2012). Despite this trend, the causes of deforestation, low productivity, and low diversity stable states in tropical forest landscapes are unknown. This unpredictability makes it challenging to manage for human needs while protecting ecological services and cultural value in woody environments.

In Bangladesh, ethnic minorities' shifting cultivation is to blame for much of the deforestation and subsequent land degradation in the hills, despite the fact that it is their sole source of income. In CHT, there has been substantial deforestation. The Reinkhyong, Kassalong, Sangu, and Matamuhuri reserve forests have been the most severely impacted by plains people settling in the highlands, encroachments by displaced shifting cultivators, and illicit logging between 1979 and 1984. Many researchers believe that shifting cultivation contributes to land deterioration by promoting loss of fertile topsoil, exposing rocks (Kleinman *et al.* 1995). As a result of shifting cultivation, a significant proportion of nutrients is carried away from the upper 10 cm soil with run off sediments (Gafur, 2001; Gafur *et al.* 2003).

As in many other regions of the world, deforestation and forest degradation are major problems in Indonesia. This is mainly because of demographic pressure, bringing in more area under cultivation, increasing demand for food and food products, and fast economic growth. United Nations' Food and Agriculture Organization mentioned about numerous causes of deforestation (FAO, 2005). It was further elaborated that, diversifying tenure arrangements, such as giving secure tenure rights to local stakeholders, was important mechanism for increasing local forestry operations accountability and control, and so evolving better circumstances for long-term forest management (Wulandari 2017).

6.1.1 Shifting cultivation and related issues

Shifting cultivation is a natural resource management approach in Central Africa's forest zones and beyond that supports food production and livelihoods. In primary and secondary forests, shifting cultivation is the most common type of agriculture. Although the shifting agricultural techniques outlined here are generally used to grow subsistence crops, households also plant commercial crops and long-term income crops like cocoa and coffee. In general, women do the majority of agricultural labour, while men in some areas clear and burn sites in preparation for planting. While women play an important role in agriculture in these areas, their involvement in REDD$^+$ has been limited thus far. Cocoa agroforestry produces major income at the locations in Cameroon, although plantations are distinct from food-crop areas and are mainly overseen by men. People are drawn to the forest frontier by the success of smallholder cocoa farming. Farmers abandon sites with increasing population density, falling fertility, and aged cocoa plants in favor of establishing new cocoa plantations in the woods (Pollini Jacques 2015). Larger investors are drawn to forest areas by the construction of road networks to establish large-scale oil palm and rubber tree plantations.

Since the 1990s, the cultivation of continual commercial crops such as, cocoa, coffee and plantains has been diminishing in the Democratic Republic of Congo (DRC). This tendency has resulted in lower household cash income and jeopardised household food security. Weak markets, increasing crop loss from disease, and falling hunting and fishing returns have all added to household vulnerability. According to the Global Food Security Index, the DRC is currently the world's most food-insecure country. Smallholders' security of tenure is often lacking, and large-scale agricultural, forestry, and mining concessions are the most common land uses. Obtaining tenure rights and assisting smallholder farmers and other shifting cultivators who engage in shifting cultivation and other economic activities on modest scale may be able to offset the risk of larger-scale deforestation by outside loggers and large-scale plantation or other investors. It has the potential to help with poverty reduction. However, if other fundamental constraints remain in place, there is no certainty that this course of action will minimise deforestation. Shifting cultivation has been regarded an unsustainable land use in various REDD$^+$ and Payment for Ecosystem Services (PES) initiatives. Since the colonial era, this viewpoint has been widely held (Colfer *et al.*, 2015). Crop rotation, fire control, ploughing, agroforestry, conservation agriculture, perennial crops, and income-generating activities including fish farming, poultry farming, and apiculture are all recommended in some programmes. The alternatives' goal is to increase farmers' income and food security by diversifying and, in certain cases, eliminating shifting farming. Many of the techniques offered required more effort, which many women were unwilling or unable to provide. A bright spot was support for the marketing of non-timber forest products. Villagers also stated that the programmes primarily involved men in the community, with very little labour including women.

Farmers may develop anti-conservation and/or anti-nongovernmental organization (NGO) sentiments as a result of the unpopularity of shifting agricultural alternatives among some intended project beneficiaries, as well as the inability of certain recommended alternatives to suit local demands. Overall weak ties between local and state players are likely exacerbating these negative sentiments. According to some farmers contacted, forest rangers, sometimes known as eco-guards, frequently disregard villages' human rights and make no distinction between subsistence and commercial hunting (Pollini Jacques 2015). Tenure, usage, and/or access rights for land and resources must be specified and guaranteed so that people living in forest regions can invest in their land over time. Forest-dependent communities must be given both true and concrete incentives to avoid deforestation, as well as a fair share of the benefits under the REDD+ and PES benefit-sharing mechanisms. Management, access, and usage rights, as well as ownership rights, should all be protected by tenure security. Increased de jure security would provide more flexibility and minimise transaction costs if carefully constructed to avoid mass-scale property transactions, as local institutions have maintained these types of rights for generations de facto. To avoid conflict and enable investment, tenure regimes must contain individual and family rights and obligations. Women are the primary agricultural labour force and the driving force behind agricultural system development. This is partly due to the fact that food crop farming is largely unprofitable, and women's labour is undervalued, despite the fact that household food security is dependent on these efforts. In social interactions with projects and other external stakeholders, men are more involved than women. It is necessary to identify pathways to ensure that women can participate effectively in REDD$^+$ decision-making.

Policymakers should create and adopt new analytical methods, including existing bottom-up approaches, to examine the drivers of deforestation. Local dynamics, such as context-specific drivers and biophysical, socioeconomic, and cultural aspects, should be considered in REDD$^+$ feasibility studies at the site level, as demonstrated by active REDD$^+$ pilot projects in Cameroon and the Democratic Republic of the Congo. REDD$^+$ projects and social and environmental protections are at risk without such fine-grained data.

6.2 HUMID TROPICS AND ITS RESOURCES

6.2.1 Forests

Forests are the natural vegetation of the planet earth (Hadly and Lanly 1983). There are two sorts of forests: closed forests and open forests (Hadly and Lanly 1983). Closed forests thrive in areas with an average annual rainfall of more than 1600 mm. Closed forests have a continuous canopy, are multi-layered, and have a lot of undergrowth. The area under humid tropics is presented in Table 6.1.

Table 6.1. Extent of warm humid tropics (million ha)

Region	Africa	South America	Central America	Southeast Asia	Total
Extent of warm humid tropics	911.7	1001.5	76.3	491.8	2481.3
% of total area in the region	31.7	56.5	28.1	54.8	42.6s

Source: FAO 1981; 1982

The ability of the tropical forest ecosystem to "withstand climatic and other natural disasters" contributes to its stability (Richards 1977). These are the characteristics:

1. The humid tropical forest is home to a diverse range of plant and animal species. The forest ecosystem's stability is based on its high level of species diversity.
2. Tropical forests are the most complex terrestrial ecosystems on the planet (Connell 1978). Within the tropical forest ecosystem, plants and animals are inextricably intertwined. Animals in the tropical forest provide the same seed distribution and pollination functions as wind in the temperate forest (Hadly and Lanly 1983). Because tropical forests

have significantly more species diversity, this helps to reestablish and preserve local diversity.

3. Due to the lack of nutrients in tropical soils, the ecosystem of tropical forest relies on a self-supporting, nearly closed nutrient cycle.

6.2.2 Deforestation and soils

Despite the fact that the humid tropics have a wide range of soil types, the vast majority of the region's soils are nutrient poor (Jordan 1985). Phosphorus deficiency, aluminium toxicity, and low intrinsic fertility are all typical concerns in the humid tropics (Sanchez 1987; Lal 1989). The rainfall may be the cause of the region's bad soils, since if an area's rainfall surpasses 1000 mm, the soils are frequently acidic and leached (Sanchez 1987). Tropical soil nutrient shortages are a major limiting factor in tropical productivity. According to a survey of 70 studies from throughout the world, the impact of population growth on tropical hillside resources has been highly variable. Tree cover grew with population development, according to aerial pictures and ground surveys from mountainous areas of Kenya, Nepal, and Rwanda. The land under shifting agriculture loses nutrients and top soil when the shifting cycle is reduced from 20–30 years to 3-5 years. Families begin to migrate to other virgin places as agricultural yields decline. Frequent moving from one land to another has had an impact on the ecology of these areas, reducing the amount of natural forest cover, fragmenting habitat, and causing the extinction of native species, as well as the invasion of alien weeds.

Despite the fact that shifting cultivation is seen as a harmful agricultural practice, a cost-benefit analysis demonstrates that the livelihood, food security, and socio-cultural events of the affected communities outweigh the environmental concerns. The term'sustainability' refers not just to environmental management, but also to other aspects of a community's survival. The term "sustainable development" refers to a situation in which resources are extracted but not depleted. It manages economic growth in such a way that it does not cause irreversible environmental damage. It is founded on the idea that given enough time, nature can recycle its renewable resources. According to Swaminathan (1991), there are some basic principles of sustainability, such as; water, land, nutriet supply, energy, genetic diversity, pest management, post harvest systems, systems approach, location-specific research and development, evaluation, equity in conservation, and no less pivotal and vitally important areas of biodiversity, which should be pursued for establishing ecologically strong agriculture. Effect of shifting cultivation on social and ecological aspects is shown in Fig. 6.1.

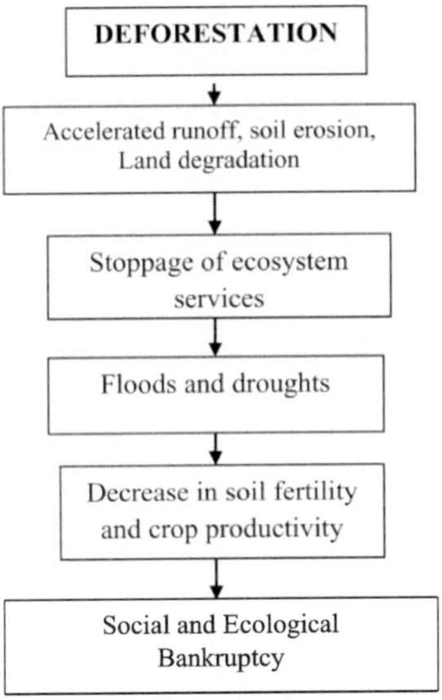

Fig. 6.1. Effect of deforestation on the social and ecological aspects

6.2.3 Preservation of rainforests

Although primary forests are being deforested in every country with humid tropics, 12 countries account for more than 80% of the total. Deforestation rates are predicted to rise in the coming decades, and by the second or third decade of the twenty-first century, their overall contribution to global warming will equal or exceed that of fossil fuel use. If current trends continue, much of the remaining tropical forest will be gone by the twenty-first century. Destruction of the world's largest repository of plant and animal genetic variety is also a result of deforestation (Myers 1989). Recent discoveries of rainforest plants as new sources of food or chemicals in cancer chemotherapy highlight the need of preserving rainforest biodiversity. As a result, one of the most pressing worldwide environmental challenges of our day is finding practical solutions to protect tropical rainforests.

6.2.4 Land of plenty but hunger persistse

Cameroon, Central African Republic, Congo, Democratic Republic, Congo, Republic, Equatorial Guinea, and Gabon are among the Congo basin countries. Despite the fact that most Congo Basin countries have significant natural resources and rainfall, hunger is a major to extremely frightening concern in all but Gabon (IFPRI Global Hunger Index 2011). Traditional low-input, low-output subsistence farming methods still dominate agriculture, with huge yield disparities between actual and prospective yields. Farmers are shut off from potential markets and growth prospects due to poor infrastructure, which isolates a large percentage of the Congo Basin's people from the global economy. Fig. 6.2 shows the net deforestation and degradation during 1990 and 2000, and 2000 and 2005 in Congo basin.

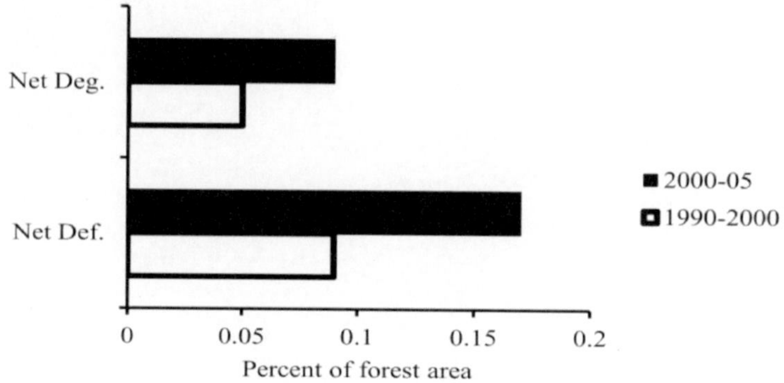

Fig. 6.2. Average Annual Net Deforestation and Net Degradation Rates, Congo Basin, 1990–2000 and 2000–05 (Source: de Wasseige *et al.* 2012.)

6.2.5 Loss of forest area

The net loss of forestland fell from 7.8 million hectares per year in the 1990s to 4.7 million hectares per year between 2010 and 2020. (Fig. 6.3). While some areas are being deforested, others are being reforested either naturally or via deliberate efforts. As a result, deforestation is occurring at a slower rate than forest loss. The worldwide forest area decreased by 178 million hectares between 1990 and 2020.

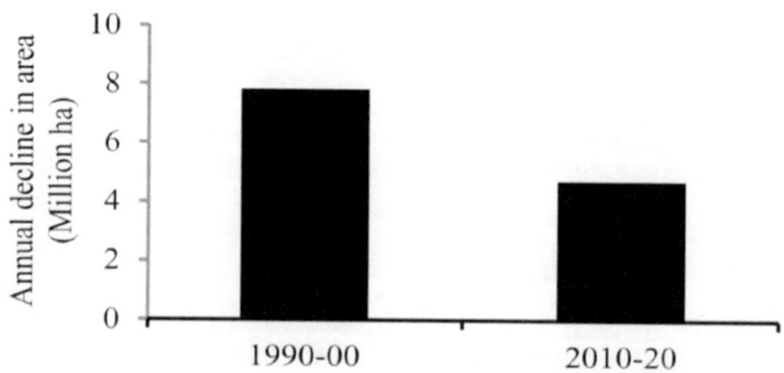

Fig. 6.3. Global annual decline in forest area during 1990's and 2010-20.

Decrease in area of forest cover in the world was 7.84, 5.17 and 4.74 million ha per year during 1990-2000, 2000-2010 and 2010-2020, respectively. The net change rate was 0.19%, 0.13% and 0.12% per year during the above periods, respectively. Forest expansion and deforestation rates are shown in Fig. 6.4.

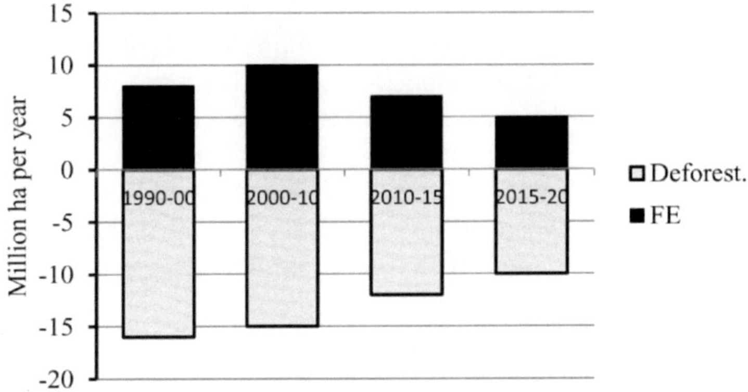

Fig. 6.4. Forest expansion (FE) and deforestation in the World between 1990 and 2020.

6.2.6 Area under forests and its trends

Forest ecosystems are more biodiverse than other ecosystems, making them a crucial component of global biodiversity. Forests encompass about 4.06 billion hectares, however they are not distributed uniformly over the World. Forest loss is mostly caused by agricultural growth, but it can also happen naturally, such as on abandoned agricultural land, or as a result of reforestation (including assisted natural regeneration) or afforestation. Table 6.2 shows the forest area in 10 major countries of the world.

Table 6.2. Forests in ten major countries of the World during 2020.

Country	Forest area (Mha)	Percent of Global area	Country	Forest area	Percent of Global area (Mha)
Russia	815	20.0	DRC*	126	3.1
Brazil	497	12.2	Indonesia	92	2.3
Canada	347	8.5	Peru	72	1.8
USA	310	7.6	India	72	1.8
China	220	5.4	Rest of the World	1,375	33.9
Australia	134	3.3	Total	4,060	100.0

*Democratic Republic of Congo

Source: FAO 2020.

6.2.7 Forests and climate

The tropics contain the majority of the forest (45 percent), followed by the boreal, temperate, and subtropical domains. There are five main climate zones on Earth: tropical, subtropical, temperate, arctic, and boreal. Globally, tropical regions have the highest percentage of forest land (45%), followed by sub-tropical regions (11%) and arctic regions (27%) and temperate regions (16%)

(FAO 2020). Normal and changes caused by human factors change the capacity of forests to retain biodiversity, sequestration of carbon, and promote human welfare, causing this heterogeneity (Hansen *et al.* 2020). Despite the fact that international environmental agreements require the protection and restoration of forests, forest quality is often overlooked. As a result, the sites and rates of forest loss of high biological significance are unclear, and coordinated conservation strategies for these forests have yet to be devised. The majority of these forests are unprotected and, given recent loss rates, are under grave danger.

6.2.8 Rates and extent of deforestation

Although some of the deforestation stretches back over 10,000 years, credible data on rates of deforestation has only lately been available (Houghton 2016). Prior to 1950, the majority of deforestation happened when these countries built agricultural areas in Europe, Russia, China, North America, and Australia. In Amazonia, Brazil, the annual deforestation has decreased by almost 80% since 2004. Across the rest of the tropics, deforestation is still happening at a rate of about 9 million hectares per year. Plantations or planted forests, such as those found in Brazil, India, and Vietnam, have helped to counteract this trend (Houghton 2016). Concern about tropical deforestation has risen dramatically in recent decades, as more people have realised that deforestation, along with linked processes like as global warming and biodiversity loss, poses a serious threat to our future (Metz 2009). The destruction of tropical rain forests is more than a cosmetic issue. The world's rain forests will be finished in 100 years if present forest cutting rates continue, causing devastation in the world climate and vanishing the many of plant and animal species.

During the 1980s, the Food and Agriculture Organization (FAO) estimates that 137,270 km^2 of tropical forest (rain forest and other) were destroyed annually. They estimate that 54,390 km^2 of land in South America are deforested each year, with the Amazon Basin accounting for the majority of this. The rate of deforestation varies from one location to the next. According to a recent research, deforestation in the Brazilian Amazon averaged 16,058 km^2 per year between 1978 and 1986, but dropped to 4800 km^2 per year between 1986 and 1993. By 1988, 6% (233,100 km^2) of the Brazilian Amazon had been logged (an area the size of New England). However, due to fragment isolation and the widening of forest/clearing boundaries, deforestation damaged 16.5 percent of the forest (595,700 km^2) (595,700 km^2). Cambodia, Indonesia, Laos, Malaysia, Myanmar, Thailand, and Vietnam lost nearly as much forest every year as the Brazilian Amazon between 1975 and 1985, with roughly 12500 km^2 per year converted to cropland or deforested for lumber.

6.2.9 Is shifting cultivation responsible for deforestation?

One of the most essential differences to draw in the subject of shifting agriculture and deforestation is between established, rotational secondary forest systems and pioneer systems that open up primary forest. The shifting cultivation of short cultivation and long fallow, does not result in 'deforestation' as per the FAO definition of 'forest' as argued by Van Noordwijk *et al.* (2008).

Forests cover 31% of all surface area on the planet, yet they are not evenly distributed. Almost half of the forest is largely undisturbed, and over a third is primary forest (FAO, UNEP 2020). Shifting cultivation is thought to be a major cause of deforestation before 1991, this practice was responsible for 61 percent of all tropical forest degradation worldwide (Myers 1991). Nonetheless, the practice continues because it provides subsistence income to between 300 and 500 million people worldwide (Brady 1996) and is closely linked to communities' cultural, ecological, and economic features (Ramakrishnan 1992). While some ecologists dispute the practice's long-term viability because it entails destroying primary and secondary forests (Bandy *et al.* 1993), others recognise that timber harvesting, monoculture plantations, and other commercial aims are also key drivers of deforestation (Dressler 2005). Fallow cycles fall below a threshold time period due to rising human population, resulting in land unavailability, which has a substantial impact on plot productivity and forest regeneration (Mazoyer and Roudart 2006, Raman *et al.* 1998). Whitmore (1984) estimated that the maximum density of human population shifting cultivation can support is 7 per km^2, which is significantly lower than current concentrations in tropical SC settings. Even in several of the tropics' biodiversity hotspots in 1995, human population density was substantially greater than the world average of 42 persons per km^2 (Cincotta *et al.* 2000).

The spatial scales of sampling, species focused on, and sample methods employed in studies conducted in diverse tropical and sub-tropical locations vary. Teegalapalli *et al.* (2009) compiled the findings of 33 studies on plants, birds, and animals in South Carolina landscapes, as well as four reviews on birds and vegetation in anthropogenically damaged ecosystems. The bulk of investigations were conducted in South and Central America (Blake and Loiselle 2001), with fewer research conducted in Africa (Myers 1991) and Asia (Cincotta *et al.* 2000). In comparison to studies on avifauna (Raman *et al.* 1998) and vegetation (Blake and Loiselle 2001), few studies (Ramakrishnan 1992) have concentrated on mammals in SC settings. Furthermore, only six studies have been conducted in sites adjacent to tropical dry and semi-deciduous forests, and the majority of knowledge about recovery after SC comes from sites adjacent to humid and moist forests.

6.3 DEFORESTATION AND SHIFTING CULTIVATION

Another source of worry is determining the extent of deforestation caused by shifting cultivation. There is no precise definition of deforestation, no solid estimate of its scale, and no evidence that shifting cultivation is the primary cause. WRI (World Resource Institute) offers the following definition; Deforestation refers to three distinct changes in land use from forest to agriculture, such as shifting cultivation, grazing, or urban use. It excludes logged woodland that has been allowed to regenerate, even if it is obvious (WRI. 1996 and Angelson. 1995). Because forest opened by shifting cultivation is frequently secondary forest, the concept contains a contradiction. As a result, temporary logging clearing is not considered deforestation, but temporary clearing by shifting agriculture is. As a result, policymakers in developing countries should acknowledge that asset heterogeneity is crucial when analyzing technology uptake (Bellon and Taylor, 1993).

Biomass recovery follows a similar pattern to that of species diversity recovery, with a fast start and then a slowdown. The accumulation of biomass grows with the age of abandonment (Raharimalala *et al.*, 2012). Similarly, Gehring *et al.* (2005) showed that in the early stages of succession, biomass recovery is swift, with approximately half of the biomass returned within 25 years of the previous use. According to Read and Lawrence (2003) from Mexico, biomass recovery after shifting cultivation could take up to 55–95 years. Do *et al.* (2010) discovered that in an undisturbed forest in Vietnam, obtaining 80 percent biomass can take up to 60 years. Significant biomass recovery has been recorded following recurrent fire events in Brazil, with recovery to pre-fire AGB levels occurring within 30 years (d'Oliveira *et al.*, 2011). According to Lawrence *et al.*, seed rain decreased as distance from the undisturbed forest increased (2005). Surprisingly, the largest seed banks were discovered in the Brazilian Amazon's young, desolate regions (Vieira and Proctor, 2007).

6.3.1 Does shifting cultivation leads to deforestation?

In its overview graph on causes of deforestation, the UN REDD Framework Document (FAO, UNDP, UNEP 2008) makes a distinction between "extension of shifting cultivation into undisturbed forests" and "expansion of shifting cultivation into disturbed forests. "It also makes a distinction between "intensification of agriculture in shifting cultivation areas," "direct conversion of forest area" to small- or large-scale permanent agriculture. It's worth noting that converting shifting cultivation to a more intense kind of agriculture is referred to as deforestation. As a result, it appears that shifting cultivation is not included, i.e., removing land as part of shifting cultivation is not deemed "deforestation," as the FAO definition suggests. However, if that were the case, shifting cultivation extension into "undisturbed forest" for the purpose of sustaining rotating shifting agriculture could not be considered "deforestation." It might be classified as "forest modification," much as logging companies' temporary clear-cutting of forest is not classified as deforestation by the FAO. Clearly, there is still some disagreement between the concept of forests and the statistics reported, which is likely attributable in part to a lack of a clear definition and difference between different types of shifting agriculture.

6.4 DEFORESTATION

Shifting cultivation, which was once thought to be a major cause of deforestation, is now being seen as a kind of forest management and even conservation. Under historic conditions, it is currently assumed that this approach was sustainable, and that agricultural cycles were long enough to promote forest regeneration throughout the fallow season. These allegations are investigated using historical data from Sulawesi (Indonesia) from 1820 to 1950. According to the statistics, intensive bush-fallow shifting cultivation systems with five- to six-year fallow periods were already the norm in the early 1800s, when average population densities were still low and commercial production was limited (Henley 2011). In most cases, conventional short-fallow systems were sustainable because they did not require more deforestation beyond a pre-existing swidden-fallow complex. The natural forest, however, was permanently replaced with human vegetation that was significantly less abundant and diverse within that complex. Furthermore, in some regions, shifting cultivation has taken on an unsustainable, itinerant character, resulting in the establishment of fire-climax grassland. This, too, appears to be a long-standing tendency, with no evidence that it was triggered by population growth or external causes such as migration or commerce. Traditional shifting cultivation's image as an environmentally friendly method is idealized, and it should not be permitted to mask agriculture's basic incompatibility with nature protection (Henley 2011). Farmers also preferred to open their swiddens in secondary vegetation that was still young and hence reasonably easy to cut and clear (Adriani and Kruyt. 1912–14). As Seavoy has pointed out, fallow vegetation that is more than eight years old may no longer be felled with a bush knife and must be felled with an axe (Seavoy 1973). Older forest, in addition to being more difficult to fell, is also more difficult to burn, posing the risk that the cleared plot will have to be abandoned until the large trunks have rotted (Li 1991). The fact that short-fallowed shifting cultivation productivity did not offset the drawbacks of protracted fallowing is consistent with recent empirical evidence that fallow length is a poor predictor of crop yields in swidden farming (Mertz *et al.*2008).

With shifting cultivation fallow intervals averaging only five to six years from the start, it's clear that these environment never resembled the natural rainforest it replaced within populated regions, where canopy trees were generally centuries old and over thirty metres tall. The fallow vegetation's man-made character was particularly apparent when it included some intentionally planted fruit trees and palms. If human population grew so rare or far away that the burning stopped all together, the forest cover would gradually rebound, making the region ideal for settlement and farming once more (Adriani and Kruyt 1950–51).

6.4.1 Tenure system and deforestation

Land-tenure issues are often cited as a root cause of communal conflict and even separatist violence. More generally, land and natural-resources issues are widely believed to be the main causes of conflict. Herrera and Da Passano (2006) placed the causes of land-tenure conflicts into three categories: the first was political influence, a factor present in almost every land-tenure conflict; second was legal factors arising when actors were unaware of their legal rights or the different legal frameworks that regulated access to areas and the use of natural resources in different or opposing ways; and the third major cause was economic factors. Because their land rights are weak and their tenure is uncertain, the majority of impoverished people in rural areas remain poor (Bruce, 2004). Forests and forestry reform could help them, but an effective rights-based approach to forestry reform that assures justice and poverty alleviation requires that more than simply property rights be considered. Forest governance systems, according to Colchester (2007), must secure a broad spectrum of rights (e.g., access to markets, the ability to commercialise forest products, and the ability to give or withhold prior and informed consent for activities affecting people's land) if forest peoples are to benefit from forests and forestry reform. There is much potential for solving forest degradation and destruction by implementing effective tenure arrangements. If this potential is to be fulfilled, greater focus should be placed on developing and implementing more effective and diverse tenure systems that benefit local users, particularly disadvantaged populations, as well as enacting supportive laws.

Diversification of tenure systems through procedures that expand and ensure poor people's access to natural resources is also critical for poverty reduction and human rights fulfilment (FAO, 2004; SIDA, 2009). In these circumstances, land tenure might be characterized as a legal or customary relationship between people, as individuals or groups, and land. As a result, a tenure assessment would be required to see if a program will jeopardize other people's rights to resources, livelihood, and security. Freudenberger (1994) outlined three reasons why tenure is important in natural-resources management programmes: first, it influences who has access to resources; second, it influences whether people are willing to participate in project activities; and third, it influences how programme benefits are distributed. Land tenure concerns that affect long-term forest, range, and agriculture management, as well as tree resources and minor forest product sources, must be investigated (Otsuka and Place, 2001). Given these three considerations, it is clear that a thorough grasp of tenure is critical to a community's long-term viability. This is also true for communities living on the outskirts of the forest. Because of their long-term impact and relationship to security in land management, communities that use their land for shifting cultivation must comprehend local land-tenure rules (Wulandari and Cahyaningsih, 2010). Legal considerations of restricting access to certain regions (tenure rights systems), as mentioned by Herrera and Da Passano (2006) and Colchester (2007), are also vital in securing a broad spectrum of community rights.

6.4.2 Forests and policies relevant to shifting cultivation

Indonesia's problems of deforestation and forest degradation are the consequence of encroachment, either by local communities, outsiders, or indigenous people. In the 20 years from 1990 to 2010 there were 1065 forest conflicts involving communities and indigenous people in 27 provinces (Purba *et al.*, 2014). On this basis alone, the government should implement programmes of community-based forest management in an attempt to reduce the incidence of encroachment while at the same time increasing community incomes. In 1986, Linda Christanty stated that shifting cultivation could refer not only to actual changes that occurred on a cropping site (or swidden), but also to the system of land tenure that applied to the site. Although the government has imposed a zero burning programme, the burning of slashed vegetation for land clearing is still widely practised by farmers (Wulandari and Zakaria, 2010). The research of Ketterings *et al.* (2002) in Sumatra found that farmers still burned slashed vegetation to clear the land for the following reasons:

1. It is the fastest and most effective method of clearing the land;
2. It can suppress the growth of weeds and other wild vegetation such as Imperata cylindrica, an invasive species that dominates a large area of land and forest in Sumatra;
3. It turns biomass into natural fertilizer that is beneficial to both plants and soil;
4. It loosens the soil so that seeds grow more rapidly; and
5. It is an effective way to kill pests and pathogens.

6.4.3 Management of protected and conservation forests in Indonesia

In principle, sustainable forest management in Indonesia is based on Act no. 41 of 1999 (Government of Indonesia, 1999) which covers the management of all types of forests, i.e. production, protected and conservation forests. Conservation forests are defined as forest areas with particular characteristics. Their principal function is preserving the diversity of flora and fauna and their ecosystems. Within the nature preservation areas category are national park areas, which are described as forests with certain characteristics whose main function is the preservation of biodiversity of flora and fauna and protection of ecosystems that serve as life-support systems for the whole region. Nature reserves areas are forests with certain characteristics whose main function is to protect the region's life-support systems by preserving the biodiversity of flora and fauna and supporting the sustainable use of natural resources and ecosystems.

6.4.4 Transformation of shifting cultivation due to external factors

External influences, mainly government policies and fluctuations in prices for agricultural and agroforestry products, have brought about fundamental changes to traditional systems of shifting cultivation in Indonesia (Mulyoutami *et al.*, 2010). The basic patterns of shifting cultivation in Indonesia are now no longer recognizable. As well as government policies, the transition of shifting cultivation was driven by the increasing market integration of NTFPs and agroforestry crops such as coffee, rubber and cacao, which mature after the annual swidden crops and ensure that the land still provides income in the fallow period. In Lampung, the changes to shifting cultivation systems were also due to migration by various ethnic groups who brought horticulture with them (Wulandari, 2007).

The secondary forest grows quickly, drawing nutrients from the soil, including those slowly released by unburned decomposing forest biomass, and storing them above ground for 10 to 20 years, until the cycle repeats itself (Nye and Greenland 1960, Szott and Palm 1986). Another distinguishing element of shifting farming is its biodiversity. Complex polycultures make up the majority of indigenous shifting farming systems. In 'conventional' agroforestry systems, they feature a variety of crops and species, including trees and food crops. This diversity helps to lower the danger of illness and pests while also providing families with a broad range of foods and goods. Furthermore, indigenous systems are frequently linked to complex cultural norms and socioeconomic practices, as well as unique species-specific cultural knowledge. In addition, the fallow stage vegetation serves as a genetic reservoir for many important plants as well as a sanctuary for invertebrate and vertebrate creatures. In terms of the environmental issues caused by slash-and-burn farming, burning releases half of the nitrogen, as well as the majority of the sulphur and carbon, into the atmosphere, contributing to greenhouse gas emissions (Bandy 1994).

6.4.5 Environmental context

According to recent estimates, the clearance of tropical rainforests, which is presently occurring at a rate of 14 million hectares of primary forest per year, is responsible for around 18 percent of global warming (EPA 1990). Tropical America and tropical Asia currently account for the majority of deforestation, accounting for 40% and 37% of estimated net carbon emissions from deforestation in 1980, respectively. Third place goes to tropical Africa, which accounts for 23% of world emissions (Houghton *et al.* 1987). Between 1979 and 1989, deforestation rates nearly quadrupled, from 7.6 million hectares per year to 13.9 million hectares per year (Myers 1989). Slash-and-burn agriculture can cause soil erosion and degradation, leaching, watershed degradation, and biodiversity loss in an unbalanced ecosystem, among other things. These implications reflect resource depletion and productivity losses at the local level. A large range of forest products are also underutilised. Changes in rainfall patterns and climate change are also influenced by the loss of forest cover. Soil organic matter may act as an extra supply of carbon dioxide to the atmosphere or as a sink where carbon dioxide can be absorbed, depending on how the land is maintained after deforestation.

6.4.6 Socioeconomic, policy and equity context

Because shifting cultivation is generally practiced by the poorest, most displaced rural inhabitants of the tropics, tropical deforestation through shifting cultivation is also a major human equality concern. Furthermore, the poor are frequently the ones that face the brunt of environmental degradation. A complex collection of demographic, biological, social, geopolitical, and economic causes, as well as governmental demands, promote deforestation (Sanchez *et al.* 1990). The rate of population increase in emerging countries remains high, despite the fact that most of the productive and accessible lands have already been heavily exploited. Various government actions frequently worsen and contribute to resource depletion, land scarcity, and unsuitable land usage. Gross disparities in land tenure are allowed by land-use rules. As a result of these circumstances, there are an increasing number of landless rural people who have three options: stay-put or remain, relocate to the metropolis, or migrate to the rainforests, which form the border of many emerging countries. Deforestation is occurring at an alarming rate in practically all humid tropic nations, and shifting cultivation is one of the main causes. Table 6.3 shows the rate of deforestation in a few selected nations around the world (Myres 1989).

Table 6.3. Rates of deforestation and carbon emissions for leading countries

Country	Annual deforestation (km^2 $year^{-1}$)	Rate (% $year^{-1}$)	CO2 emitted (million t $year^{-1}$)
Brazil	50000	2.1	46
Indonesia	12000	1.4	12
Myanmar	8000	3.3	8
Mexico	7000	4.2	6
Colombia	6500	2.3	6
Thailand	6000	8.4	6
Malaysia	4000	3.1	5
Zaire	4000	0.4	5
India	4000	2.4	4
Nigeria	3500	14.3	6
Peru	3500	0.7	3
Vietnam	3500	5.8	3
World total	111500	1.8	140

Source: Myers 1989, estimates for 1989

Many scholars have stated that shifting agriculture does not promote deforestation because abandoned plots are allowed to recuperate after 2-3 years of farming and forest flora grows back in these plots. However, we believe that this is not the case in most cases, as abandoned plots are rarely given the attention they require for full regeneration. Due to the high slope, the top layer of the slope is eroded, resulting in irreversible damage. Similarly, land-tenure rules frequently do not provide individuals with secure title to land, which may deter producers from adopting long-term sustainable practises. Although urban migrations are often driven by geopolitical factors, national initiatives in important countries such as Brazil, Peru, and Indonesia include colonisation projects that encourage the occupation of their tropical rainforests. Deforestation-related environmental degradation has a significant impact on rural production and subsistence systems. Agricultural production is harmed by erosion, flooding, groundwater depletion, and silting, which reduces food supply, income, and employment. Forests also act as 'food banks' for poor communities and are frequently the primary source of household energy for cooking; however, these resources are depleted as the environment deteriorates. Reduce the rate of tropical deforestation caused by slash-and-burn agriculture and related processes, as well as increase the well-being of forest and forest edge residents, through a joint technology-policy strategy focused at generating viable land-use alternatives. Agriculture will serve as a foundation for the development of biologically-based management systems.

6.4.7 Quantifying environmental, agronomic and socio-economic parameters

Climate Change

Despite the fact that tropical deforestation and land use change contribute up to 25% of yearly CO_2 fluxes to the atmosphere (IPCC 2001), there is still a lot of disagreement on the subject due to uncertainty in biomass estimates, deforestation rates, and land use change sequences. The majority of previous research has focused on forest and grassland data, as well as continuous cropping systems (Wood *et al.* 2000).

The amount of deforestation caused by shifting cultivation varies. Between 1850 and 1985, around 10% of Latin American forests were deforested, according to Houghton and associates (1991), whereas De Jong (1997) estimates 50% for Indonesia. Serro and associates (1996) calculated a figure of between 30 and 35 percent in the Amazon. Because no distinction is made between permanent and temporary conversions, conversion and changes, or deforestation and forest degradation, there is much uncertainty concerning the significance of shifting cultivation agricultural systems (Angelsen 1995). The scholars also emphasizes the importance of distinguishing between different forms of forest conversions or adjustments that fall under the deforestation umbrella because the environmental and social impacts might vary greatly. Many international organisations, like the World Bank, now recognise that the reasons of deforestation are far more complex than previously imagined, and that blaming small farmers and the extension of traditional shifting farming systems is a mistake. Market and legislative factors emerging outside the conventional agricultural economy are far more likely to cause deforestation in modern times (Brown and Schreckenberg) (1998). Public policies, economic incentives, and the agricultural credit system, for example, have exacerbated deforestation in the Amazon region (Fearnside 2005). Other factors include institutional changes related to the agrarian system and macroeconomic change (Fearnside 2001; Malhi *et al.* 2008), land and resource ownership, horticulturist experience and technologies (Brondzio 2005, 2006), microeconomic processes at the household level, and market opportunities (Brondzio 2005, 2006; Homma *et al.* 1993). Historically, this land use was seen as exploiting rather than managing the humid tropics' natural resources. This viewpoint has now shifted. The integral shifting cultivator's natural resource management focuses on preserving the forest ecosystem's highly important diversity (Warner 1991). Integral shifting cultivation has been used in the tropics for centuries and is still used today.

6.4.8 Influence of shifting cultivation on tropical forest

The Lao People's Democratic Republic faces a number of environmental issues, the most serious of which being deforestation. The ecological situation in Lao PDR is quickly deteriorating. Natural vegetation cover has been removed or altered to an alarming amount in several parts of the country due to human impact. The dry dipterocarp forest, bushland, and savannah, for example, are frequently burned by the local inhabitants at the end of the dry season (Persson, 1983). In Lao PDR, the rate and scope of commercial logging has been quite high. It has been limited to the most accessible locations. Felling should be done with caution, taking into account the species and diameter guidelines. However, it has been inadequately managed and operated in practice due to a lack of proper forest management plans (Young and Hyde, 1988).

6.5 CONGO BASIN AND FORESTS

The Congo Basin forests, after the Amazon, are the world's second-largest tropical forest blocs, with an estimated 300 million hectares (ha) of forest. These woods cover six countries and account for 70% of Africa's forest cover. They are important for biodiversity and ecosystem services on a global scale, and they provide crucial habitat for endangered species. They are responsible for around 25% of global forest carbon sequestration. Forests are home to thirty million people, and more than 75 million people rely on them for their livelihood (Megevand *et al.*, 2013). In comparison to the rest of the world, the Congo Basin has a lower rate of deforestation and forest degradation. Only 5.4 percent of the total worldwide loss of humid tropical forest happened in the Congo Basin between 2000 and 2005. (Hansen *et al.*, 2008, cited in Megevand *et al.*, 2013). "The Congo Basin forests may be at a turning point, headed toward increased deforestation and forest degradation rates," according to Megevand (2013). The Congo Basin woods have been primarily "passively" conserved due to chronic political instability and warfare, insufficient infrastructure, and poor governance. As a result, the Congo Basin countries continue to meet the High Forest Cover and Low Deforestation criterion (HFLD). "Eutrophication and chemical pollution, as well as water overuse, are all environmental consequences of modern agriculture.

Land use change because of forest clearing and the consequent carbon release, surface and groundwater pollution due to pesticides and fertilizers, and water misuse are all environmental repercussions of present agriculture. Between 2005 and 2050, global agriculture demand is predicted to treble, putting even more strain on these systems (Tilman *et al.*, 2011). In the tropics, these challenges are of special concern. The risk of biodiversity loss is substantially higher than in other areas (Wiens and Donoghue, 2004). Small-holder farming is a major cause of deforestation (Southgate, 1990), but policy implementation is problematic due to the large numerals of marginal farmers and fragile governments (Shah *et al.*, 2003). Poor soil health and shifting cultivation methods employed to address it, have entrapped many people in harmful habits and poverty. Slash-and-burn agriculture is practised by 240 to 300 million subsistence farmers worldwide on an estimated 12.4 million km^2 (Dove, 1983). Over 1.5 million km^2 of this is being collected in an unsustainable manner, resulting in deforestation and its numerous consequences.

6.5.1 Shifting cultivation is not the only factor for deforestation

Agricultural intensification has mostly replaced shifting cultivation, which can be sustainable in areas with low population density and large land areas (Chomitz 2007; Sanchez *et al.* 2005). The system becomes unsustainable as human density and land demands increase, but soil nutrients and land productivity decline (Sanchez *et al.* 2005; Giller and Palm 2004). In the past, shifting cultivators were thought to be the primary cause of deforestation in the tropics. Recent studies, however, suggest that shifting cultivators are not responsible for the majority of deforestation, particularly in the most deforested tropical areas. There are regional variations in this trend, but small farmers are most important in areas with the lowest deforestation rates (Rudel *et al.* 2009; Geist and Lambin 2002). Between 2000 and 2005, the majority of humid tropical deforestation took place in Latin America, accounting for three-fifths of total forest cover loss, and Asia, accounting for one-third of total forest cover loss. According to Hanen *et al.* (2008), Brazil (47.8%) and Indonesia (12.8%) had the highest percentages of humid tropical forest loss relative to overall loss. Brazil had almost four times higher forest loss than the second ranked country, Indonesia. Small farmers who responded to government initiatives encouraging colonization and migration to forest frontier areas were usually cited as the principal cause of deforestation in areas with substantial deforestation before roughly 1990, according to research. Because they do not manage huge areas due to demographic and political pressures, as well as economic needs in tropical climates, only a tiny percentage of shifting cultivators can allow for long fallow periods and forest regeneration. According to Rudel *et al.* (2009), deforestation in the Amazon and Southeast Asia grew more "enterprise-driven" around 1990. In Southeast Asia, the ancient practice of shifting cultivation is dwindling, and permanent cropping and commercial plantations are taking its place (Padoch *et al.* 2007). Unlike subsistence farmers, these commercial operations are linked to metropolitan markets and global demand for agricultural commodities (Rudel *et al.* 2009). Increased tropical deforestation has been impacted by migration to cities, urban population expansion, and consequent urban demand for agricultural products, mainly in Asia and Latin

America. Increases in agricultural exports have also been linked to more forest destruction. While small-scale agricultural operations continue to contribute to deforestation in the Amazon and Southeast Asia, large-scale commercial enterprises currently predominate (De Fries *et al.* 2010).

Shifting cultivation, which can be sustainable in areas with low population densities and large land areas, is now uncommon and has been largely replaced by agricultural intensification (Chomitz 2007). Fallow times have been shortened as a result of increasing population pressure and declining land availability, leading in a loss in soil nutrients and land productivity, rendering the system unsustainable (Sanchez *et al.* 2005; Giller and Palm 2004).

6.5.2 Forest degradation and deforestation

In the last two and four years, Brazil and Indonesia have lost more forest than the Congo Basin has lost in the previous 15 years. Contributions of major blocks to the global loss in humid tropical forest cover between 2000 and 2005 (Fig. 6.5). Africa only contributed 5.4 percent of the expected global loss of humid tropical forest cover from 2000 to 2005, compared to 12.8 percent in Indonesia and 47.8% in Brazil alone. In Sub-Saharan Africa, Central African countries have the lowest deforestation rates. The results of a global analysis are summarised in Table 1.7. Deforestation rates in Central Africa are not only lower than those of the world's major negative contributors to total forest area, but also lower than those of most other African regions, according to estimates.

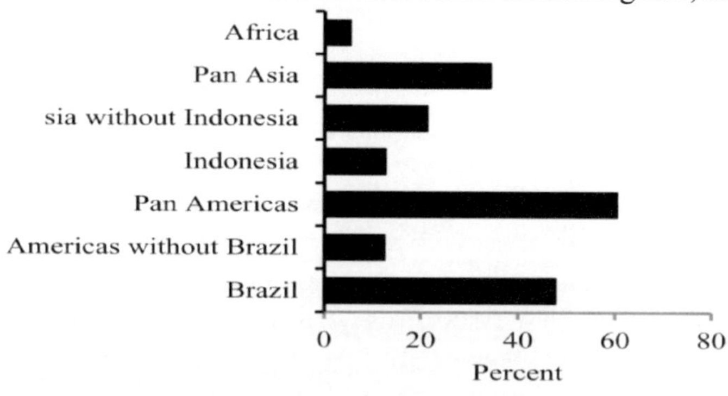

Fig. 6.5. Humid forest loss in various regions

Source: Authors, derived from Hansen *et al.* 2008.

The Congo Basin countries, which are heavily wooded, have significantly lower deforestation rates. During the 1990–2000 period, the overall yearly net deforestation rate in the Basin rainforest was projected to be 0.09 percent. This pace nearly doubled between 2000 and 2005, resulting in a net loss of around 300,000 square kilometers each year. The forest area in Africa and America, annual change and annual change precent is given in Table 6.4.

Table 6.4. Changes in Forest Area in Africa and America (1990–2010)

Sub-region	Forest area ('000 ha)			Annual ('000 Change ha)		Annual rate Change (%)	
	1990	2000	2010	1990-00	2000-10	1990-00	2000-10
Central Africa	268,214	261,455	254,854	-676	-660	-0.25	-0.26
East Africa	88,865	81,027	73,197	-784	-783	-0.92	-1.01
North Africa	85,123	79,224	78,814	-590	-41	-0.72	-0.05
Southern Africa	215,447	204,879	194,320	-1,057	-1,056	-0.50	-0.53
West Africa	91,589	81,979	73,234	-961	-875	-1.10	-1.12
Total Africa	749,238	708,564	674,419	-4,067	-3,414	-0.56	-0.49
Southeast Asia	247,260	223,045	214,064	-2,422	-898	-1.03	-0.41
Oceania	198,744	198,581	191,384	-36	-700	-0.02	-0.36
Central America	96,008	88,731	84,301	-728	-443	-0.79	-0.51
South America	946,454	904,322	864,351	-4,213	-3,997	-0.45	-.0.45
World	4,168,399	4,085,063	4,032,905	-8,334	-5,216	-0.20	-016

Source: FAO 2011.

Deforestation rates have risen considerably in the Republic of Congo and the Democratic Republic of Congo, whereas they have remained consistent in the Central African Republic and even declined in Gabon, Equatorial Guinea, and Cameroon (Ernst *et al.* 2010). Forest degradation is also an important driver of change in Congo Basin forests, albeit it is more difficult to quantify. In terms of deforestation, the Basin's worldwide forest degradation has accelerated in recent years. The Democratic Republic of Congo is primarily responsible for this trend, as degradation rates in other nations have stabilised globally. Gabon even has a negative rate of degradation, indicating that forest density is increasing globally. The rate of net deterioration in dense forests is a mix of gross degradation and forest regrowth.

6.5.3 Deforestation and its impact in the Congo Basin

Farming

In the Congo Basin, agriculture is a critical but underappreciated sector. Agriculture remains the region's major employer by a long shot. Agriculture still employs more than half of Cameroon's, Central African Republic's, Democratic Republic of Congo's, and Equatorial Guinea's economically active population. In comparison to other tropical countries, agricultural productivity in the region is quite poor, and fertiliser use is also very low. As a result, there is a significant and growing reliance on food imports. The Congo Basin's agricultural growth potential is significant for a variety of reasons. First, Congo Basin countries account for roughly 40% of noncultivated land not suitable for cultivation. If just appropriate nonforested regions are taken into consideration, the Congo Basin accounts for roughly 20% of land available for agricultural expansion in Sub-Saharan Africa and 9% of land available worldwide (Deininger *et al.* 2011). Congo In comparison to other tropical countries, agricultural productivity in the region is quite poor, and fertilizer use is also very low. As a result, there is a significant and growing reliance on food imports. For various reasons, the Congo Basin's agricultural growth potential is substantial. The location has limitless water supplies, placing it ahead of other areas that may encounter water scarcity due to climate change. The Congo Basin countries have some of the world's biggest output growth potential. Finally, the Congo Basin's agricultural products may be in high demand due to increasingly urbanizing people and rising international food and energy demands. Agriculture is a very promising sector as a result of all of these characteristics. Future agricultural advances, on the other hand, could put forests at risk.

Congo Basin forests and their biodiversity

The Congo Basin woods are home to a staggering amount of biodiversity and endemism. Although the Congo's coastal forests are likely the most diverse in the Afrotropics, data for some ecoregions in the central Congo Basin is limited (Billand 2012). There are around 10,000 species of higher plants in low-altitude forests, with 3,000 of them being indigenous. There are 4,000 species in montane forests, with at least 70% of them being indigenous. African elephants and buffalo live in the Congo Basin woods, as do indigenous species such the okapi, bongo, bonobo, and gorilla, as well as numerous endemic bird species.

Deforestation in the Congo Basin

Population variables, which are significantly correlated with demographic patterns and are largely associated with the increase of subsistence activities, are a primary underlying cause of deforestation. As a result, forest degradation have primarily occurred in urban areas and densely populated areas. Despite the fact that the Congo Basin countries' population densities are still low in contrast to the rest of the world, urbanisation trends are emerging.

The rise of urban centres with populations of at least 100,000 people shows that rural places in the humid forests are turning more and more populated. Zhang *et al.* (2006) used a GIS-based analysis of the Congo Basin's tropical forests' susceptibility and future extent to show that annual dense forest clearance is highly linked to rural population density. It also showed that the deteriorated forest area in the 1980s had a close link to the damaged thick forest in the 1980s–1990s. In the transition zones between rainforest and savanna, deforestation and forest degradation are prevalent, and population density can reach 150 people per square kilometre.

6.5.4 How shifting cultivators cause deforestation?

Small-scale and subsistence farmers may be forced to move into forested regions to grow food and earn a living due to a lack of resources or secure land tenure in some areas, resulting in deforestation (Sanchez *et al.* 2005; Geist and Lambin 2002). Small-scale farmers, particularly in Sub-Saharan Africa, who lack the resources to boost crop yield on nutrient-depleted soils may resort to more forested land to protect their livelihoods and productivity (Palm *et al.* 2010). Wealthier farmers are better suited to deforest land, especially large tracts of land, because deforestation may be costly and complex (Chomitz 2007). Although shifting farmers are not the primary cause of deforestation in the areas where it occurs, they do contribute to it. It may be more difficult to reduce their deforestation impacts than it is to reduce deforestation caused by large-scale commercial agricultural or logging enterprises (Shearman *et al.* 2009). Community forestry entails a group of people working together to manage forests in a way that benefits them socially and economically. If the right incentives are in place, small-scale agriculture can also help to limit agricultural development into wooded regions (Palm *et al.* 2010). REDD+ is an international policy that can value standing forests and provide financial incentives for (i) lowering emission of CO_2 from deforestation and (ii) increase in capturing and storing carbon through forestry practises. For small farmers, indigenous peoples, and other stakeholder groups like women to engage in a REDD+ system in these programmes, land tenure and other rights must be developed. Deforestation is less frequent among small-scale subsistence farmers with restricted access to markets, highlighting the importance of commercial markets, as well as urban and global demand, as underlying causes of deforestation (DeFries *et al.* 2010; Pacheco 2009).

The rise in human population, especially in emerging countries, has put enormous strain on land resources. The expansion of crop lands to increase food production has resulted in a direct decline in the area covered by woods and grasslands. By the early 1990s, about 40% of the earth's land surface had been turned to crop lands and permanent pastures, according to one estimate (World Resource Institute 1996). Between 1980 and 1990, more than 6% of tropical forest land was converted to shifting agriculture in all tropical countries. During the time period mentioned, almost 10% of forest land in Asia was converted to shifting cultivation (Singh and Marzoli 1988).

Tropical countries are the most common places to see shifting cultivation. Shifting cultivation on hill slopes is practiced by people in India's eastern and north-eastern regions. Shifting farming accounts for 85 percent of overall cultivation in northeast India (Singh and Singh 1992). Due to the increased demand for land, the cycle of agriculture followed by fallow land has been decreased from 25–30 years to 2–3 years. Previously, the fallow period lasted 20–30 years, allowing the land to revert to its original state. The ecosystem's resilience has been broken down as a result of the decrease of the cycle to 2–3 years, and the land is rapidly deteriorating.

6.5.5 REDD+ planning Congo and Cameroon

If REDD+ is not implemented in the DRC, the R-PP (Readiness Preparation Plan) and National REDD+ planning predict that shifting cropping will be responsible for 20% of additional deforestation by 2030. To solve this problem, the DRC intends to boost subsistence agriculture productivity. Moving farmers are unfairly blamed for deforestation, according to critics, and the catalytic role of logging and industrial plantations is undervalued or ignored.

Although no National REDD+ Strategy had been formed at the time of the research, an R-PP was filed to the FCPF (Forest Carbon Partnership Facility) in Cameroon. The R-PP cites four elements as direct causes of deforestation and degradation, but warns that due to the country's different agro-ecological zones, prioritising direct causes of deforestation at the national level is challenging. Forest conversion for a range of land uses, such as shifting farming (practised by over 80% of rural Cameroonian populations), cash crops such as coffee and cocoa, and industrial plantations such as oil palm and cow grazing, is one of these factors. The main perpetrators include intensive fuel wood consumption and inefficiency, noncompliance with forest concession management plans, and anarchic exploitation of timber, mining, and infrastructure. Various agricultural intensification measures, enhancing the efficiency of fuel wood utilisation, and various forest management approaches were highlighted as potential solutions to these problems.

According to this analysis, the current REDD+ market in the Congo Basin is dominated by a few groups that appear to work more closely with governments and regional elites than with farmers. The existence of a few intermediaries, primarily foreign non-governmental organisations, is essential for the provision of environmental services (NGOs). Only educated elites are aware of how carbon markets work, providing considerable impediments to participation. REDD+ is a completely new tool. If and when sufficient funds for REDD+ become available, it is projected to have a considerable impact on land use and alterations.

6.5.6 FAO view on deforestation

It is estimated that nearly 90% of the land area of the Philippines was previously forested. This has already been reduced to 20%, and if the current rate of deforestation continues, no forest cover is expected to remain within the next decade (Suarez and Sajise 2010). Forest devastation has been accomplished in two stages: first, logging, and then, various forms of swidden cultivation. Forest flora and animals have such high richness and endemism that, in most situations, the quantity and types of species in the swidden are insufficient to compensate for what is lost as a result of forest degradation. Studies comparing forest and swidden biodiversity, on the other hand, have proven insufficient, failing to account for ecosystem function and services. Because many indigenous and unique species in the Philippines developed as forest specialists, continuous destruction will almost certainly result in extinction. The value of ecosystem services provided by Philippine forests may yet reveal that the benefits derived from their preservation significantly outnumber those derived from their destruction (Suarez and Sajise 2010).

The decrease in forest cover in the Philippines is linked to an increase in agricultural land, implying that most of the deforested land has been turned to agricultural production (Dobson *et al.* 1997, Sajise *et al.* 1992). The work of Kummer (1992a), which records how the majority of Filipinos remained impoverished and did not profit from economic progress during the postwar period, is a landmark in studies of Philippine deforestation and kaingin. The poor relocated to the uplands, where they cut down secondary forests and practiced kaingin, due to a lack of economic possibilities in the lowlands. There were times when the government promoted such migration, allowing kaingin to act as a "safety valve" that relieved pressure on the government to implement much-needed socioeconomic reform. There is evidence that forestry data on the extent and rate of deforestation has been manipulated (Kummer 1995). In a comprehensive overview, Cuevas (1991) examines the human elements of kaingin in the Philippines, concluding, as Kummer (1992a) does, that Philippine deforestation and kaingin can only be completely understood in the context of the country's socioeconomic and political realities. The socioeconomic marginalization and detribalization of indigenous populations is one of the consequences of commercial logging and migration of lowlanders to upland areas (Cuevas 1991, Eder 1990). The decline in forest area in the Philippines between 1900 and 2000 is shown through Fig, 6.6.

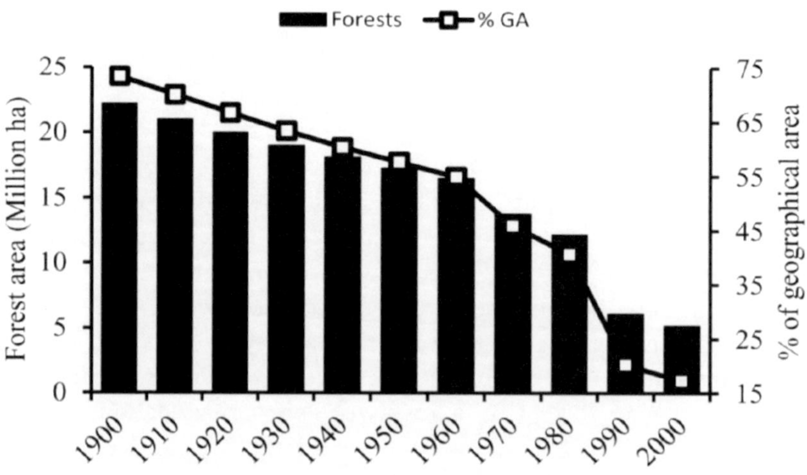

Fig. 6.6 Trend of decline in area under forests in the Philippines

Source: Suarez R.K. and Sajise P. E. (2010)

According to Hansen, between the 2010s and the 2000s, the average yearly rate of primary forest loss in the tropics increased by 30% to 3.7 million hectares from 2.9 million hectares. Tropical forest loss contributes more to global emissions than forest loss in temperate climates, owing to the fact that tropical trees store more carbon in their biomass than soil. Conversion to agriculture and pasture, primarily for subsistence, is the single most important direct driver of tropical deforestation (producing crops or keeping livestock to meet daily needs). A number of key factors frequently influence the conversion of land to agricultural use. Tropical forests make up the majority of the world's forests (45 percent), with boreal, temperate, and subtropical forests following closely after. The United States' tropical humid woods are found below 1000 metres in elevation, with average year-round temperatures ranging from 20 to 26 degrees Celsius, more than 1500 mm of precipitation annually, and less than three dry months per year. In the humid tropics, the main forest conversion process was the conversion of closed, open, or fragmented forests to farmland. The big forest alterations were mostly localised to a few hotspot locations with frighteningly high change rates (Achard *et al.* 2002).

Shifting cultivation, as it has been done for centuries around the world, entails chopping down and burning plant growth before planting and collecting crops. Farming continues until soil fertility is depleted, at which point the shiting cultivators move on to new pastures. The fields are left fallow for a suitable period of time until soil fertility and vegetative growth resume in this condition. This allows for a significant return to its previous state and the "slash and burn" cycle to be repeated (Noble and Dirzo 1997, Russell 1988). Indigenous peoples in the Philippines have been described as practicing relatively mild types of kaingin, such as the Igorot in the Cordillera (Kowal 1966), T'boli in Cotabato (Hyndman *et al.* 1994), and Hanunoo in Mindoro (Russell 1988). Russell (1988) says "The system is often practiced with great sophistication. The Hanunoo people, for instance, of Mindoro Island in the Philippines are expert botanists and ecologists (Conklin 1957). Their soil classification stands up to modern scientific analysis. They know all about slopes, erosion, and the value of litter as mulch. They can recognize 1600 different kinds of plants (including varieties as well as species) and treat them all differently and appropriately, and they cultivate more than 400 kinds of plants in the swidden, a veritable botanic garden." The Hanunoo are claimed to only till a plot for 2-4 years before leaving it fallow for 8-10 years to allow soil and forest plants to regenerate. When done this manner, kaingin is superior to traditional agriculture because, despite the poor yield per unit area, it maintains the soil and does not require fossil energy-based inputs like commercial fertiliser, herbicide, or pesticide. However, because of the need for unfettered access to wide amounts of land, kaingin can only be sustained at low population densities. Beyond a threshold population density, increasing the period of agriculture and decreasing the time of fallow becomes necessary.

Because of the subsequent progressive decreases in soil quality and crop output, this sets off a vicious cycle. In order to maintain total output, additional land is cultivated, resulting in the deterioration of even more land. The need to increase production in response to rising population sizes has resulted in the kaingin of several indigenous groups being short-lived (Cuevas 1991). Most forms of upland agriculture practised in the postwar period, loosely referred to as kaingin, are basically sedentary agriculture, according to Kummer (1992b). Now that the prime forests have all but vanished, the only forest left is primarily secondary growth, which is where the majority of kaingin takes place (Lasco *et al.* 2001, Viloria *et al.* 2005). Others, such as Viloria *et al.* (2005) in Mindanao and Lawrence (1997) in Leyte and Bohol, agree that most migrants to the uplands practise sedentary (rather than shifting) agriculture. Despite variations in how it is carried out, kaingin has been and continues to be an important part of the Philippines' forest degradation process.

Russell (1988) describes the process of biodiversity loss as follows: in sustainable shifting cultivation, periods of cultivation last only 10-20% of the whole cycle (in some cases, fallow periods last 30 years), Because many of these species are endemic, or found nowhere else on the planet, forest loss is thought to be a likely cause of worldwide extinction. Long periods of cultivation and insufficient times of fallow result in significant soil erosion and nutrient depletion, to the point where forest regrowth is impossible when the site is abandoned (due to a lack of production). Instead, a common scenario involves grasses invading the area. In 1966, two tough grass genera, *Imperata* and *Hyparrhenia*, had already covered 40% of the Philippines (Russell, 1988). Because grasses are less effective than forests at keeping soil and recovering fertility, the traditional, sustainable shifting cultivation cycle comes to an end. Wallace (1996) estimated an average per capita consumption of 1.53 percent of a hectare of secondary forest per year based on research of shifting cultivation villages in Northern Luzon. The invasion of *Imperata cylindrical* was followed by deforestation (commonly known as cogon). He claims that if kaingin is practiced in secondary forest and enough fallow time is given, the forest can revive. Imperata-dominated areas become fire-prone, according to Sajise *et al.* (1976), and fire increases Imperata's competitive dominance.

Deforestation and shifting cultivation are considered as fundamental aspects of the process of forest loss, best understood in the context of the country's socioeconomic realities and politics, according to social scientists. The shifting cultivation of indigenous people, which may be sustained at low population density, is at the most benign end of the spectrum of activities. They practice various forms of shifting cultivation at population densities over critical limits, causing significant erosion, loss of soil nutrients, harm to watersheds, and loss of floral and faunal biodiversity. More benign agriculture practises, which are designed to improve plant biodiversity, stimulate native tree growth, prevent erosion, and conserve watersheds, fall somewhere between these two extremes. If the current rate of habitat loss through deforestation, followed by shifting cultivation, continues, a massive loss of many forest species is projected in the future. However, it's vital to distinguish between "global extinction," which the remaining Philippine endemic species have so far avoided, and "local extinctions," which are most likely the result of extensive habitat degradation across the country. The concept of "ecological extinction" (Estes *et al.* 1989), which states that a drop in population size may render a species ineffectual in its ecological interactions with other species in a community, is of major importance. Because the relationships between population increase, culture, socioeconomic issues, politics, and biology are complicated, preventing biodiversity loss is a challenging task.

The determination of the level of deforestation induced by shifting agriculture is a major concern when it comes to deforestation. There is no reliable estimate of deforestation caused by shifting agriculture, nor of its scope in the many nations where it is practised (Jyotishi 2000). It's also unclear whether shifting farming is the leading source of deforestation. 'The word deforestation describes a total transition in land use from forest to agriculture- including shifting cultivation and grazing or urban use,' according to WRI (World Resource Institute). Even if it is clear-cut, it does not contain forest that has been logged and left to recover' (WRI. 1996 and Angelson. 1995). The term is contradictory, because forest opened through shifting agriculture is frequently secondary forest, employed for the same purpose and then abandoned. Thus. 'Temporary clearance by logging is not categorized as deforestation, but temporary clearing by shifting agriculture is' is a key aspect to consider. There is a lot of misunderstanding because there isn't a clear distinction between permanent and temporary conversions, conversions and alterations, or deforestation and forest degradation. Deforestation, on the other hand, is defined by the FAO as "the transfer of forest land to non-forest uses and includes all land where the forest cover has been stripped and the land converted to such uses as: permanent cultivation, shifting cultivation, human settlements, mining, dam construction, and so on." Degradation, on the other hand, is defined as a "decrease in the area and quality of forest cover as a result of factors such as indiscriminate logging, improper road-making practices, forest fires, and other reasons." (Rao, 1989, as referenced by Fox *et al.*, 1999) Traditional shifting farming, according to studies, does not imply permanent conversion, but rather the transitory use of forestland (Fox *et al.* 1999; Xu *et al.* 1999, Angelson. 1995). As a result, counting cleared land that regenerates into secondary forests following shifting cultivation as deforestation will overestimate deforested land. According to some estimates, the area under shifting agriculture has been relatively stable over the last 40 years, and shifting cultivation's contribution to deforestation is small (Fox *et al.* 1999: Xu *et al.* 1999). As a result, there's a good probability that ambiguous definitions and estimations may increase the extent of biases against shifting cultivation.

6.5.7 Deforestation and Biodiversity

The "biodiversity" of planet Earth consists of 5 to 80 million plant and animal species worldwide. Over half of all these species are found in tropical rain forests, which cover only 7% of the Earth's dry area. Only roughly 1.5 million of the tens of millions of species thought to exist on Earth have been named by scientists, and even fewer have been researched in depth. Because they require a specific ecosystem to survive, many rain forest plants and animals can only be found in tiny areas. As a result, they are extremely sensitive to deforestation. They may become extinct if their habitat is eliminated. As tropical rain forests are removed, species are disappearing on a daily basis. According to estimates, up to 137 species perish every day on the planet. Species extinction will have a significant influence on the earth.

6.6 GREENHOUSE GAS EMISSION DUE TO DEFORESTATION

6.6.1 Greenhouse Gases and Global Warming

Carbon dioxide is released not just during combustion, but also during the oxidation of plants that have been destroyed and left to degrade on deforested land, which accounts for up to a quarter of all carbon released into the atmosphere (Detwiler and Hall 1988). Secondary forest sprouting on abandoned plots, on the other hand, absorbs a portion of the carbon emitted, making it a major contribution to the carbon balance in deforested tropical areas (Fearnside 1996). Because shifting cultiation is one of the drivers of forest conversion, numerous research have tried to figure out how it affects carbon and other greenhouse gas emissions, and hence global warming. Sharma and Sharma (2018) estimated that 1237.065 Gg of CO^2e was produced from an area of 386 thousana ha under shifting cultivation in northeast India.

The shifting cultivation has been responsible for feeding millions of people, particularly the marginalized and small farming communities and, causing large scale deforestation. Even so, let's accept that "changing agriculture expansion into undisturbed forests" qualifies as deforestation. The global contribution of shifting farming to deforestation is estimated to be roughly 8%. If deforestation accounts for 17 percent of carbon emissions and shifting cultivation accounts for 8% of deforestation, shifting cultivation's proportion of global carbon emissions is 1.36 percent. If the entire forest was destroyed in the process, this would be the situation. However, this is not the case in shifting cultivation, where forest regeneration during the fallow period is an essential, if not critical, component of the system. The fact that shifting cultivation does not cause deforestation, at least not to the extent that is commonly claimed, appears to be increasingly recognised by international organisations such as the FAO, UNDP, and UNEP, and shifting cultivation has become associated with forest degradation rather than deforestation in the global climate change discourse. In other words, when a primary forest is opened for shifting cultivation, just a small quantity of carbon is released into the atmosphere owing to decaying soil carbon, however when shifting cultivation is converted to permanent agriculture, a significantly larger amount is released. They discovered that the trend in carbon stock above ground was positive, "is similar to that below ground, except that the magnitude of the decrease is much higher as forest is converted to shifting cultiation and shifting cultiation converted to permanent cropping".

6.7 SOCIO-ECONOMIC AND EQUITY ISSUES

6.7.1 Socio-economic, policy and equity context

Shifting cultivation deforestation in the tropics is also a huge human equality concern, as slash-and-burn is mostly practiced by the poorest, most displaced rural inhabitants in the tropics. Furthermore, the poor are frequently the ones that face the brunt of environmental degradation. A complex collection of demographic, biological, social, geopolitical, and economic causes, as well as governmental demands, promote deforestation (Sanchez *et al.* 1990). The rate of population increase in emerging countries remains high, despite the fact that most of the productive and accessible lands have already been heavily exploited. Various government actions frequently worsen and contribute to resource depletion, land scarcity, and unsuitable land usage. Land-use policies, for example, allow for massive disparities in land tenure. As a result of these circumstances, there are an increasing number of landless rural people who have three options: stay put, relocate to the metropolis, or migrate to the rainforests, which form the border of many emerging countries. Similarly, land-tenure rules frequently do not provide individuals with secure title to land, which may deter producers from adopting long-term sustainable practices. Although urban migrations are spontaneous, national strategies in significant countries such as Brazil, Peru, and Indonesia involve colonization programs that encourage the occupation of their tropical rainforests, which are frequently geopolitically motivated.

While regulations that promote these critical techniques should be maintained, they will not be enough to stop deforestation. Linking environmental and economic policies gives a viable and realistic strategy. New initiatives in this manner are gaining traction (Bouwman 1990, Gillis and Repetto 1988) and have sparked passionate debates. In response to national and international environmentalist demands, a few deforestation-prone countries have created legislation to limit deforestation. Some of them are much too far-fetched to be practical. In certain circumstances, they backfire, causing severe economic consequences and encouraging deforestation in neighboring countries. Policy research on tropical deforestation is just as important as biophysical studies in this regard. This should be done in conjunction with and in conjunction with study on technological and socioeconomic choices.

6.8 FOREST RECOVERY FOLLOWING SHIFTING CULTIVATION

In the majority of tropical hilly regions, shifting cultivation is the norm. Only a few studies have looked at forest recovery after shifting farming. Most studies show that, whereas pioneer tree species recover more quickly, mature forest tree woody biomass recovers over several decades after cultivation is stopped. According to a review of bird-species inventories from ten studies, successional sites can recover up to 70% of mature forest birds, and the species composition in successional sites is roughly 55% comparable to mature forests after 25 years (Teegalapalli *et al.* 2009). When relatively large forest tracts adjoin a dynamic agriculture landscape, community recovery should be accelerated compared to recovery at sites with shorter fallow cycles and no adjoining woods, which act as sources for fauna and flora recolonization.

6.8.1 Alternatives to shifting cultivation

Agricultural intensification, agroforestry, and other "green" technologies were advocated by project developers at the locations visited, as well as encouraging agriculture outside of forests in nearby savanna areas. In conservation and development circles, there may be bias against shifting cultivation. The majority of official papers, such as the Cameroon 2035 vision, stress the promotion of mid- to large-scale farms as a means of advancing the country to "agriculture de seconde génération" status (second generation agriculture). Shifting cultivation and other traditional farming methods are frequently considered as impediments to accomplishing this transformation. Marginal cultivators are motivated to enhance productivity and revenue, and can make better use of land than affluent cultivators (Pedelahore, 2012), if supplied with proper tools and support, according to observations made throughout the study. Indigenous peoples are still underserved since their livelihoods were based on hunting, fishing, and gathering until one or two generations ago, and they are still attached to and reliant on this way of life. Despite the fact that women are nonetheless responsible for household food security, they are essentially absent from decision-making. The global areas under humid tropics are shown in Fig. 6.7.

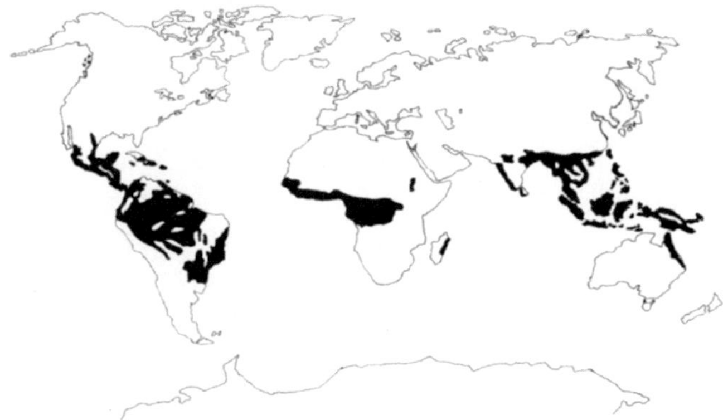

Fig. 6.7. Shifting cultivation in the humid tropics (Warner 1991)

6.8.2 Consequences of tropical deforestation

Tropical deforestation is becoming a growing source of international concern. According to current forecasts, large-scale deforestation in the tropics will result in flooding, drought, and widespread erosion, rendering broad areas uninhabitable. Recent research on the worldwide impact of tropical deforestation has revealed that the world will get warmer. Whereas tropical forests were formerly seen of as a national resource to be controlled, there is now a growing consensus that they are a global resource whose management is of worldwide significance. As a result of this new belief, when the bleak forecasts are revealed, the question becomes: What has been done to conserve the forests? Who is causing the forest to be destroyed? Why aren't they being detained? Shifting cultivators were traditionally held responsible for deforestation in the tropics (and still are in certain cases now). Governments and international organisations attempted to stop them, believing they were indiscriminately destroying nations' natural resources. It was simple to blame them and pass laws against forest cutting and burning, but it was more difficult to eliminate shifting cultivation. Shifting cultivators are still in use today and will be for a long time. According to recent analyses, much of the blame was misplaced. To succeed, you'll need a thorough understanding of the tropical environment as well as a high level of managerial ability. The failure of agricultural development efforts in the tropics has reinforced this new perspective on shifting agriculture.

When land becomes unsuitable for crop cultivation, natural vegetation is allowed to recover it. The ecological repercussions are frequently negative, but they can be alleviated to some extent if fresh woods are not invaded (Lal, 1975). Slash-and-burn farming is used by many cultivators, while others use land clearing without burning as part of their farming cycle (Serajuddin, 1971). Forestal (1966) estimates that 73 percent of CHT land is suitable only for forest, 15% for horticulture, and only 3% for intensive terraced agriculture. Locals in CHTs have been practising shifting cultivation, or jhum, for hundreds of years. Sedentary agriculture has nearly completely replaced shifting cultivation in Europe and Nepal. In Thailand, it has been largely replaced, and in Indonesia and Malaysia, it has been replaced to a lesser extent. In Bangladesh, Laos, and northeastern India, however, shifting farming is still commonly used (Mohsin, 1997). Nearly two-thirds of the formerly present CHT woodlands were destroyed by shifting cultivation and accompanying fires (Farid and Husain, 1988), hastening soil erosion (Shoaib *et al.*, 1998). While it was an environmentally suitable land use system in the past when population pressure was low (Nye and Greenland, 1960), as the fallow period has shrunk due to increasing population pressure, low agricultural investment, and poor forest management rights in Bangladesh's Chittagong Hill Tracts, it has gradually become an environmentally incompatible land use system (Knudsen and Khan, 2002). Shifting cultivation practices have an impact not only on the soil of the farmed field, but also on the surrounding environment. Shifting agriculture is viewed by many experts and the general public as primitive, backward, inefficient, unproductive, and exploitative, as well as the source of severe environmental degradation (Thomas, 1956).

6.9 SOIL QUALITY AND DEFORESTATION

Policymakers face a difficult problem in developing regional development strategies for the Amazon Basin. Forests are continuing to disappear, and agriculture is moving into a region with nutrient-poor soils that are patchy (in terms of agronomic potential). At finer sizes relevant to agriculture, the geographical division of soil types is not clearly mapped (Witcover *et al.* 2006). Furthermore, there is little information in this frontier environment about how agricultural land usage or farm household welfare varies with soil quality. "Shifting cultivation is not one system, but hundreds or thousands of systems," Brookfield and Padoch (1994) mentioned. Ruthenberg (1980) claims that “the forms assumed by shifting cultivation are more varied than in practically any other land-use system.” According to him, shifting cultivation is grouped into the following systems: vegetation, migration, rotation, clearance, cropping, and tool.

The amount and type of degradation that may take place, as well as the potential productivity of low-input agroecosystems, are determined by inherent soil features, according to Kleinman *et al.* (1995). They cite a claim by Andriesse (1980) "that the natural variability within one specific soil series may be greater thandifferences in nutrient level caused by the bush/fallow stage." And according to Szott *et al.* (1999), "(w)ithin a climatic regime, fallow growth can vary greatly as a result of interactions among vegetation vegetation, soil, and management." A study by Mapa and Kumaragamage (1996) shows that different soil characteristics can vary widely even within the same immediate area: "differences may be attributed to the previous management practice of shifting cultivation or due to natural variation in soil formation (Mapa and Kumaragamage 1996)."

6.9.1 Changing forests in Myanmar

Total forest area in Myanmar decreased from 442 to 430 thousand km2 between 1990 and 2000. Myanmar lost 12000 km of forest due to human activities in the ten years between these forest observations, with an annual rate of forest loss of 0.3 percent. Around 3000 km2 of forest regenerated at the same time period, lowering the annual net deforestation rate to 0.2 percent. Ayeyarwady, Mandalay, and Sagaing Divisions had the greatest clearing rates, ranging from 0.4 percent to 1.2 percent every year. The middle dry zone, where the bulk of Myanmar's inhabitants live, is covered by all three divisions. In just ten years, the Ayeyarwady delta region had lost around 12% of its remaining forest cover. Only four of the country's Divisions and States, including the Ayeyarwady and Mandalay Divisions, have less than 35 percent forest cover. Forest cover was greater than 70% in half of the Divisions and States, and annual forest losses were typically lower than the national average.

6.10 DEGRADATION OF FORESTS

The term "forest degradation" refers to the reduction or loss of the biological and/or economic productivity of forest ecosystems, which causes an ongoing loss in the total amount of benefits that forests provide, including biomass, fuel wood, lumber, and other products and services. But there isn't a single, agreed-upon definition of what constitutes deforestation. FAO inquired from different countries if they were evaluating forest degradation and, if so, what trchniques they were employing to help future information on relevant objectives and goals in relation to forest degradation. A total of 58 countries (representing 38% of global forest area) responded, implying that they were attempting to track the rate of forest degradation. On the other hand, many of those countries just rated one or a few crucial elements (FAO 2020).

Forest deterioration, like the term "forest," is a difficult concept to define. The United Nations Food and Agriculture Organization defines forest as a unique class of land use type, which has a lengthy institutional history (FAO). As long as the canopy cover remains over 10%, forest degradation is thought to occur when the forest's canopy cover and/or stocking are reduced due to logging, fire, windfelling, or other occurrences. Forest degradation, in a larger sense, refers to the long-term loss of forest advantages such as timber, biodiversity, and other products or services (FAO 2000). Forest degradation is defined by the UNFCCC as "direct human-induced long-term loss of at least Y percent of forest carbon stocks (and forest values) since time (T) and not qualifying as deforestation" (Penman *et al.*, 2003). As a result, even a brief period of time without trees, such as areas in shifting agriculture landscapes that are later fallowed, may qualify as degraded forest if the trees are projected to regenerate. However, when forest is converted to agricultural purposes, shifting cultivation is considered deforestation under the FAO forest definition, which is frequently used. The FAO definition is insufficient to accurately identify mosaic landscapes that include both forest and non-forest elements, as well as shifting agriculture.

6.10.1 Forest degradation and shifting cultivation

During the clearing of shifting cultivation plots, the Amazon Basin's indigenous peoples purposefully keep commercially valuable species. These species are fire-resistant and contribute not only to sustenance but also to shaping the species content of subsequent successions (Barton 1994). Pollarding may be done to leave trunks for cultivated climbing plants, as well as coppicing. Many times, valuable tree seeds are planted. These comprise both wild and semi-domesticated plants, which help the forest regenerate more quickly and with a greater diversity of species during the fallow time. The shifting cultiation plot's care is reduced to periodic weeding around valuable trees after the annual or semi-perennial crops have stopped producing during the second or third year. The plot, on the other hand, might provide valuable crops for up to 25-30 years. After this period, the field may be harvested and burned again, or high woodland may outcompete it.

There are a few things that these Latin American indigenous agroforestry systems have in common (Alcorn, 1990). To begin with, fallows are viewed as a beneficial stage in the production process rather than merely a response to deteriorating fertility, weed, and pest concerns. Native plants, which often exist naturally in primary or secondary forest and are chosen for its numerous uses, are incorporated into farmers' managed £allows. Farmers, on the other hand, make the most of natural environmental variety. They recognize a variety of biological zones, soil types, and topographical units, which they apply to various species depending on their demands. Finally, producers control pests and reduce risk by increasing diversity in the field by introducing a mix of native and foreign tree and crop species.

6.10.2 Causes of forest degradation

For millennia, indigenous knowledge and traditional agricultural practices have been practiced in the HKH region. Farmers have devised many agricultural strategies that are suitable to marginal and vulnerable alpine areas using their practical expertise. Even in such difficult circumstances, eco-friendly agriculture and sustainable soil management are conducted through terracing composting, mulching, mixed copping, mixed farming, and other methods. These farming communities are also well-versed in the many agro-ecosystems that make up a watershed. Because of specialized prospects for limited volume high value crops, the hills and mountains have a comparative advantage. They have enormous production potential to contribute to the sustainable use of varied mountain ecosystems on a global scale. Farming communities are aware of micro niche variability and have knowledge of how to manage and utilise niche-specific diversity. Even in such difficult circumstances, eco-friendly agriculture and sustainable soil management are conducted through terracing, composting, mulching, mixed copping, mixed farming, and other methods. These farming communities are also well-versed in the many agro-ecosystems that make up a watershed. Because of specialised prospects for limited volume high value crops, the hills and mountains have a comparative advantage. They have enormous production potential to contribute to the sustainable use of varied mountain ecosystems on a global scale. Farming communities are aware of micro niche variability and have knowledge of how to manage and utilise niche-specific diversity. Causes of deforestation in developing countries are depicted in Fig. 6.8.

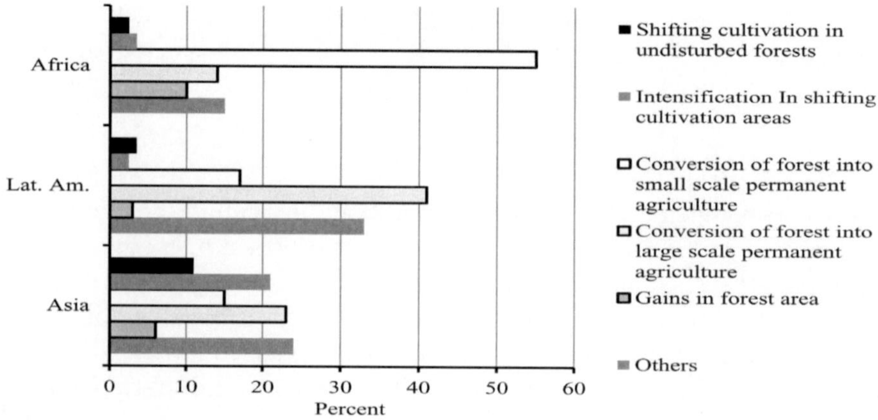

Fig. 6.8. Causes of deforestation in developing countries, by region (*Source*: FAO, UNDP, UNEP 2008: p.3)(1980-2000)

Even so, let's accept that "changing agriculture expansion into undisturbed woodland" qualifies as deforestation. According to the graph, its global contribution to deforestation is estimated to be roughly 8%. If deforestation accounts for 17 percent of carbon emissions and shifting cultivation accounts for 8% of deforestation, shifting cultivation's proportion of global carbon emissions is 1.36 percent. If the entire forest was destroyed in the process, this would be the case. However, this is not the case in shifting cultivation, where forest regeneration during the fallow period is an essential, if not critical, component of the system. The fact that shifting cultivation does not cause deforestation, at least not to the extent that is commonly claimed, appears to be increasingly recognised by international organisations such as the FAO, UNDP, and UNEP, and shifting cultivation has become associated with forest degradation rather than deforestation in the global climate change discourse.

6.10.3 Depletion of forest resources in Northeast India

The total forest area of the North-Eastern Region of India is 167.4 thousand km^2, with thick forest cover covering 91.2 thousand km^2 (54.4%) and open forests covering 76.2 thousand hectares (45.6%). Arunachal Pradesh has the most forest cover, accounting for 68.6 thousand acres or 82.0 percent of its total land area. Assam contains 24,000 hectares of forest, yet it only makes up 30.6 percent of the state's total land area. The region as a whole has 63.9 percent of its geographical area covered by forest, which is significantly above the government's requirements of 60 percent forest cover in hills. Between 1991 and 1993, 635 km^2 of forest were lost, and between 1993 and 1995, 791 ha were reduced. Assam experienced the greatest decline in forest area, with a 0.494 percent decrease in 2 years.

Although primary forests are being deforested in every country with humid tropics, 12 countries account for more than 80% of the total. Deforestation rates are predicted to rise in the coming decades, and by the second or third decade of the twenty-first century, their overall contribution to global warming will equal or exceed that of fossil fuel use. If current trends continue, much of the remaining tropical forest will be gone by the twenty-first century. Destruction of the world's largest repository of plant and animal genetic variety is also a result of deforestation (Myers 1989). Recent discoveries of rainforest plants as new sources of food or chemicals in cancer chemotherapy highlight the need of preserving rainforest biodiversity. As a result, one of the most pressing worldwide environmental challenges of our day is finding practical solutions to protect tropical rainforests.

6.10.4 Forest degradation in Amazon

The Peruvian Amazon covers a total area of 775 650 km2 (FAO; 1993). Selva alta is separated into two sub zones based on precipitation differences, wet tropical sub-montane forest and very moist tropical sub-montane forest (FAO; 1993). In the Amazonian rainforest, deforestation has become a serious issue. Peru had 68.5 million hectares of amazonian forest cover in 1990. Deforestation is expected to be 260 000 ha each year. The region of San Martn, is one of Peru's most deforested. Every year, an estimated 23 260 ha of land in San Martn is deforested. This equates to almost 160 hectares per day (Predes; 2008). Deforestation is mostly caused by logging and forestry businesses, as well as small-scale farming. In the tropics, slash-and-burn agriculture is the customary method of clearing land for cultivation. It is left to recover for a length of time after a couple of years of use. When population pressure is too high, as it is in the San Martn region, this sort of agriculture is unsustainable (Troeh & Thompson; 2005). Every year, over 21.000 people come to San Martn, which equates to nearly seven families per day (Paitán; 2006). Increases in population put more pressure on the soil because it is no longer given enough time to recuperate (Troeh & Thompson; 2005). The Peruvian forest law was enacted in 1975 (and ratified in 1979) and covers all natural resources. The law contains enough information to support sustainable forest management, however it is not fully implemented. According to an FAO research from 1989, just 40% of people follow the legislation. Conflicts between forest conservation and industrial production are caused by economic considerations (FAO; 1993). Deforestation causes a slew of issues, including reduced water flow, soil degradation, and biodiversity loss. Because there is insufficient land cover, nutrients collected in the plants are released to the soil without cycling back, resulting in nutrient leakage. Nutrients can thus be lost to ground water, stream water, and other locations, and will not be naturally returned to the original area. The nutrient status often deteriorates with time, and after a few years, the nutrient levels are too low to support agricultural activities such as pasture production, and new land must be removed to maintain the agricultural activity (Wood & Porro; 2002). Improving the long-term viability of natural resource usage and biodiversity conservation in Latin America is a top objective for reducing tropical rainforest loss.

6.10.5 Environmental degradation

The settlement of the shifting cultivators in the Amazon Basin forests has been fueled by a mix of government policy incentives and socioeconomic conditions that result in a scarcity of appropriate land for cultivation in the settlers' native regions. Forest colonization has been encouraged with few, if any, accompanying programs to promote sustainable agriculture methods. Indeed, policies have fostered the use of unsustainable methods, as the examples from Brazil, Ecuador, Peru, and Bolivia demonstrate. Although it is possible to argue that authorities were unaware of the issues linked with the farming and livestock rearing activities that they favoured in the humid forest zone during colonization, this is not the case now.

The Brazilian government has supported migration as a way to assert national sovereignty over the Amazon and avoid land distribution disparities in the migrants' home states (Mahar, 1990). Massive road-building projects in the 1960s and 1980s made significant portions of the Amazon accessible for the first time, as well as direct incentives like subsistence handouts (Binswanger, 1991). Because impoverished settlers in their native regions have little economic options, they have developed farming strategies based on cutting forest, mining the land's resources, earning from converting land to pasture, and moving on (Mahar, 1990). As a result, enormous swaths of forest have been cleared for unsustainable agriculture.

Settlers to the forest zone in northern Ecuador were required to pay for the land and its survey, and they could not obtain full ownership until they fulfilled their debt (Collins, 1986). They couldn't get credit because they didn't have a clear title. They didn't have much of a choice except to maximize short-term output and profits by converting forest land to farmland and extracting forest resources. The majority of farmers sought to create pasture, but this failed due to a lack of long-term soil fertility management strategies. They had little choice but to sell out to wealthy ranchers and move on, putting even more strain on the colonizing frontier.

A smallholder coffee production initiative in southern Peru failed due to a combination of low access to credit and issues acquiring clear title to land

(Collins, 1986). Without external inputs, which were not affordable without loans, coffee did not yield well enough. As a result, a seasonal migration pattern emerged, with settlers from the highlands moving back and forth between upland and forest zones to keep agricultural activities going in both. This has put strain on the forest boundary, resulting in poor management in both zones and a drop in agricultural production as a result. Colonist growth was supported in the Bolivian Amazon as early as the 1960s. For rice production, land is cleared of dense forest. After one or two years of farming, the area is abandoned to bush regrowth due to diminishing yields and weed encroachment (known locally as barbecho). This technique is repeated until the high forest on the farmer's property is depleted, at which point barbecho land must be cultivated. As yields decline and weed issues worsen, migrants are forced to migrate further into the forest, resulting in the 'barbecho catastrophe' (Maxwell, 1980; Stearman, 1983; Thiele, 1991).

Mechanization for arable crop production or shift to pasture for cattle or dairy production have both failed. In the long run, mechanized land will be converted to pasture. Those with money have become ranchers with large holdings, while the lesser growers have been moved to the frontier to restart the cycle. Herbicides and integrated weed management tactics have allowed for more extensive cropping of the land in recent years (Thiele, 1991), but falling soil fertility threatens its long-term viability.

6.10.6 Direct factors of forest deterioration

Expansion of agricultural land for food production, construction of hydro-electric dams, construction of infrastructure (such as roads and power lines), and mining operations are all well-established direct drivers of deforestation. The surviving woodlands were steadily deteriorated by cattle grazing, forest fires, and overexploitation for wood and non-wood forest products, even though these operations completely changed the usage of the land. In this setting, gathering and harvesting firewood remains a significant factor in the destruction of the forest. Poor silviculture practises in numerous forest sites contributed to the negative outcomes (Kant *et al.* 2007).

6.10.7 Indirect causes of forest degradation

In order to create ways for successful forest rehabilitation, the underlying reasons of forest degradation and include (Kant *et al.* 2007):

- Population pressure: As the population of both rural and urban regions grows, so does the requirement for forest products, such as timber extraction and livestock grazing, leading to forest degradation.

- Poverty: The majority of the population remained impoverished, forcing them to rely on the forests for fuelwood gathering and forest product extraction for subsistence.
- Unintended consequences of governmental interventions: Government policies have occasionally had harmful consequences on forests. The privatization of common property, mostly for communities, resulted in an increase in the usage of nearby residual forest areas for cattle grazing.
- Land rights: In many parts of South Asia, local customary rights have governed access to forest resources for hundreds of years. Often, governments do not recognize these de facto land tenure rights. Forest nationalisation, such as that occurring in Nepal, has resulted in faster deterioration as a result of rising demand and economic development.
- Market failures: Because forest departments still control a major portion of the timber and firewood markets, the true costs of production are not recognized, preventing adequate expenditures in forest restoration. The energy-producing sector in India, for example, prefers forest locations for new power plants since clearance is easier to come by. Similarly, in Sri Lanka and Bangladesh, agricultural land growth occurred nearly exclusively at the expense of forests.

6.10.8 Soil erosion and shifting cultivation

Shifting cultivation is one of the earliest land-use strategies, and its long-term viability is dependent on maintaining soil fertility. There will be no decline in crop productivity if the balance of nutrients lost or shifted throughout the cultivation phase of the cycle is maintained, and the system will continue to operate indefinitely (Xu *et al.* 1999: Ramakrishnan. 1992: Hong. 1990). The first monsoon rains of the season, in fact, have caused considerable erosion on the newly cleared plots. However, moist tropical woods are remarkably robust, and vegetation quickly covers the ground, reducing erosion for the remainder of the monsoon season. On the quantity of annual eroded soil from shifting cultivation, different scholars have different views. Singh and Singh (1981) reported a loss of 170 tha^{-1} from shifting cultivation on 70% slope in India during the first year of cultivation. On an average, Sharma amd Prasad (1994, 1995), Sharma and Sharma (2005) and Sharma *et al.* (1994) have reported a soil loss of 23.5-45.3 tha^{-1}, due to shifting cultivation, from seven northeastern states of India, depending on slope, amount of rainfall and crops grown intensity of soil cover. Other investigations found that, when compared to other current agricultural methods, shifting culture is the most efficient system of cultivation in the forest and on hill slopes (Hatch. 1980; Padoch 1980). The reason for this was invariably attributed to the swidden agriculture method, which involves only a few millimeters of soil disturbance when planting and weeding. Forest covcr is important for mass erosion because of its root system. It was due to the fact that in mixed crop farming, trees are frequently strategically positioned to prevent erosion (Nadkarni and Govindaru. 1995). The availability of nutrients would rise as ash deposition occurred. Increased soil pH would also reduce the toxicity of microelements such as aluminum and manganese (Ramakrishnan 1992). In a highly transitory ecology, shifting cultivation entails optimizing production by taking advantage of limited soil fertility.

6.10.9 Impact of slope and management practices on soil loss

Water erosion has been found to have a significant impact on soil loss in shifting cultivation plots, largely due to the slope of the area under cultivation (Sharma and Sharma 2003, Singh and Singh 1981). Level 1 management practises involve minimal soil cover and no efforts to conserve soil or water. Level 2 management practises involve trenches, contour planting, and sparse soil cover. Level 3 management practises involve implementing all feasible measures to conserve soil and water, such as bench terraces, contour planting, grassed waterways, etc., and full vegetative soil cover, including addressing localised issues. The interplay between these two variables affected the soil erosion resulting from moving crop plots (Fig. 6.9). The interactional impact was roughly exponential in character rather than linear. By using level 3 management practises, the soil loss up to 60% slope can be stopped and reduced to below the critical limit (< 10 t ha^{-1}). For level 2 management, there may be a slope of up to 30%, while for level 1 management, there may be no slope at all. It demonstrates how crucial it is to use suitable soil and water conservation measures along with effective crop management techniques in order to prevent soil erosion caused by shifting cultivation.

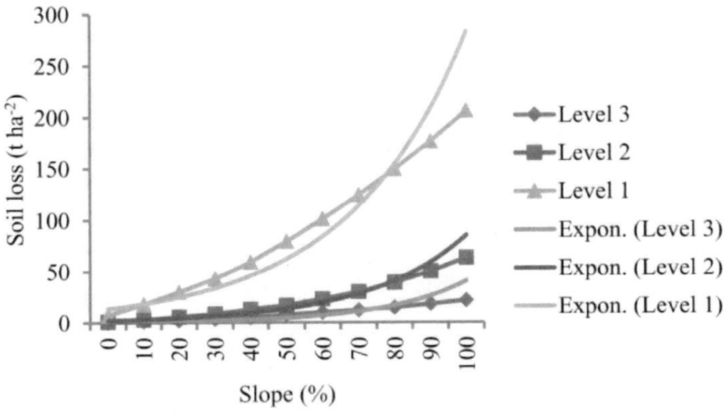

Fig. 6.9. Soil loss through erosion by water as affected by slope of cultivated land and management practices in shifting cultivation.

6.10.10 Shifting cultivation's advantages

Sustainability, productivity, biodiversity, soil conservation, adaptability, population density, and the use of native species are the main benefit characteristics of both indigenous and market-oriented systems.

Sustainability

The case studies demonstrate that indigenous people's agroforestry methods are both productive and potentially sustainable, as the forest is only destroyed in discrete chunks and can easily recover. The use of species-enhanced long-term fallows and biomass recycling keep production going. The systems also aid in the conservation of biodiversity and soil. They can respond to new conditions by using a mix of locally adapted and foreign species, and this allows them to support large population densities. It is now obvious that Amerindians make effective use of the natural resources at their disposal, both in terms of subsistence production and long-term sustainability. They have a thorough grasp of their surroundings, including micro-variations and crop potential (Clay, 1988; Hecht, 1989).

Productivity

In the eastern Amazon, Hecht (1990) studied three different land use systems. Over a five-year period, Kayapo yields (edible harvest) were around 200 percent greater than new colonist systems and 175 times higher than animal production systems.

Biodiversity

Amerindians have exploited natural resources so greatly, according to Balee (1989) and others, that they have altered much of Amazonia by exchanging valuable plant species between ethnic groups, increasing biological diversity throughout the region. Long-fallow swidden cultivation mimics the diversity, complexity, and utilisation of biomass for nutrients found in natural forests (Warner, 1991). To ensure long-term stability and productivity, swidden cultivators deliberately recreate the forest in their fields, selecting protecting beneficial species. As a result, native biology is disrupted as little as possible, and annual re-establishment insures future plant material and game production. Variety in kinds, crops, and the quantity and type of swiddens farmed decreases risk, expands possibilities, and ensures household security.

Soil conservation

Indigenous societies not only have technical understanding of which soil types are suitable for agriculture, but it is also claimed that their soil management is superior to that of new colonists and large-scale animal production (Hecht,

1989; Unruh, 1988; 1990). Because coppicing and selective protection of beneficial species are used in the swidden cycle, there is less loss of the nutrient cycling root mat.

Flexibility

Traditional swidden systems are dynamic, with new plant kinds being added as they become available. The Bora of Peru, for example, has introduced valuable wood tree species to forest £allows (Denevan *et al.*, 1984). Although cassava remains the staple in the Amazon region, maize, plantains, and rice are relatively new additions (Warner, 1991). These new additions have increased the system's diversity, which may have enhanced food security and increased productivity (Alcorn, 1990). Although swidden cultivation is commonly practiced in low-population areas (with the exception of Mexico's Huastec), this is not always a reflection of the land's carrying capacity.

6.10.11 Impacts of shifting cultivation on forest degradation

Shifting cultivation has resulted in land and forest degradation, posing a severe threat to the people of Bangladesh's CHTs. Attack on alien weeds in shifting cultivation areas, among other negative consequences, is critical. Obviously, the ecology of the target area has an impact on weed invasion (Karim *et al.*, 1999). The following are the effects of shifting agriculture:

- Crop productivity has dropped by 50% in the last decade, despite the use of fertilizers and pesticides (Mantel *et al.*, 2006).
- The food was sufficient only for 6 to 10 months of a year and during rest of the year, the cultivators faced food shortage.
- To supplement their income, shifting cultivators turned to alternative activities such as wage work, animal husbandry, yearly monocropping, and the extraction and sale of forest products.
- Deforestation for shifting cultivation has caused landslides soil erosion, which has resulted in siltation and flooding in nearby lakes.
- Large sums of money were required to protect and repair drainage basins in order to conserve ecological resources, particularly water (Uddin *et al.*, 2000).
- The area's ecology has been severely harmed as a result of constant shifting from one land to another. It has resulted in forest canopy gaps visible from barren slopes. Different weed species colonised the canopy gaps due to insufficient soil disturbance (Karim and Ahmed, 1997).
- The forest area decreased from 1215636 ha in 1980 to 828745 ha in 1990.

- Mutrient depletion and deterioration of soil physical qualities;
- Weed encroachment;
- A surge in illnesses, insects, and parasitic nematodes.
- The soils lose not just nutrients but also important physical properties like structural stability and bulk density due to leaching and erosion (Volkoff *et al.*, 1989).

6.10.12 Shifting cultivators: enemies or friends of forests

More and more scientists have disputed this logic by demonstrating that (i) shifting cultivators have proved, on many occasions, their ability to adapt their agricultural practices and their livelihoods to changing environment (Cramb, 2007; Hansen and Mertz, 2006; Michon, 2005) and (ii) the quick deforestation that occurred in the Tropics during the twentieth century resulted mainly from other causes and drivers than rotational shifting cultivation. The tropical forest covered around 15 per cent of the earth in the 1950s; this had decreased to 12 per cent by the mid-1970s and is at 6 per cent nowadays (Achard *et al.*, 2002; Rudel, 2005). Such a steady pace raises strong and legitimate concerns in terms of biodiversity loss and climate impact. If shifting cultivators are the sole or main perpetrators of the forest slaughter as widely accepted, the solution is easy: we just have to ban shifting cultivation and provide alternative livelihoods for those populations, with either the promotion of new farming systems or direct compensatory payments for environmental services (Gullison *et al.*, 2007; Palm *et al.*, 2005). However, the factors of tropical deforestation are not so simple. In the last decade, many researchers (Achard *et al.*, 2002; Boucher *et al.*, 2011; Lambin *et al.*, 2001; Rudel, 2005) have shown that the causes of the quick deforestation are multiple and shifting cultivation plays only a minor role, behind forest conversion to permanent agriculture – by either smallholders or large estates; mainly for soybean, oil palm, hevea and sugarcane cropping, or ranching – logging, migration settlements and urban expansion, encroachments due to roads and other large-scale infrastructures (mines, hydropower dams, etc.). When focusing on shifting cultivation, the environmental policies fail to reduce the progress of deforestation (Rudel, 2005) and they further marginalize the shifting cultivators (Fox *et al.*, 2009). Moreover, when a ban of shifting cultivation is promulgated and enforced the fallow forests become access-free for other economic agents; the ultimate conversion to an immediate deforestation is favoured instead of a progressive, long-term degradation with shifting cultivation. The wrong political path to protecting tropical forests relies on some misunderstandings: (i) timescale confusion, with a demographic crisis being only a mid- or long-term prospect; (ii) a confusion of techniques, when slashing and burning may be the first steps for converting forest to open land, as well as bulldozers, for example, or steps in the rotation pattern for shifting cultivation in forests (Ziegler *et al.*, 2011). Such misunderstandings lead to overestimations of the population involved in shifting cultivation (Mertz *et al.*, 2009; Thrupp *et al.*, 1997). They base inefficient policies that aim to protect the forest by restricting shifting cultivation, but that may trigger a demographic crisis when the land allowed for fallow is shrunk or when shifting cultivators' villages are resettled and consolidated on limited land (Lestrelin and Giordano, 2006). Moreover, these policies may serve the interests of powerful economic agents and lobbies, easing their access to forest resources and land (Hecht and Cockburn, 2010; Ziegler *et al.*, 2009).

6.11 SOIL DEGRADATION UNDER SHIFTING CULTIVATION

6.11.1 Forest degradation

Disturbances to deciduous forest vegetation and soil showed no significant impact, but disturbances to tropical evergreen forest vegetation and soil had a considerable impact, according to these two studies. For Mediterranean dry and semi-arid environments, land degradation is a serious issue (Hill *et al.*, 2008). Deforestation, overgrazing, the loss of natural vegetation, and farming methods that worsen soil erosion all contribute to land degradation in Tunisia (Sarraf *et al.*, 2004). Anthropogenic disturbance for longer periods from agricultural use (Kosmas *et al.*, 2015), alongwith abiotic pressures such as climatic and terrain variability (Scarascia-Mugnozza *et al.*, 2000), creates a variety of situations in which generalised assumptions about SOC distribution and determinant factors are difficult to make (Jobbagy and Jackson, 2000). The degradation of other land resources (such as water and plants) is linked to the degradation of soil quality (Karamesouti *et al.*, 2015). Soil degradation can be caused by both natural and human-made factors (Bhattacharyya *et al.*, 2015). While traditional pepper cultivation in rows produces sediment yields 200 times greater than those under natural forest, shifting cultivation does not significantly increase them, owing to the organic debris left on the ground, which prevents long rills or channels for sediment movement from developing despite the burning.

6.11.2 REDD+ in Myanmar

Myanmar's government is now working on a countrywide REDD+ initiative. Deforestation and environmental degradation are exacerbated by logging and monoculture plantings in Myanmar and Southeast Asia. In contrast to these other methods of agriculture, shifting cultivation does not result in deforestation and retains relatively large carbon reserves. As part of a fair and effective REDD+ national plan, supporting sustainable shifting cultivation by recognizing customary land ownership would cut carbon emissions. In two ways, tenure security for shifting farming will minimize deforestation and forest degradation emissions:

1. Protecting shifting cultivation from conversion to monoculture agriculture and excessive logging will reduce carbon emissions.
2. Tenure instability in shifting agriculture limits the amount of land accessible for shifting cultivation and makes protracted fallow periods more difficult. Securing tenure will aid in the long-term sustainability of shifting cultivation as a varied, carbon-rich, and productive agricultural system.

6.11.3 Forest degradation in Northeast India

The practice of shifting cultivation on hill slopes has led the elimination of water sources and increased the run-off water. One of the factors contributing to increased floods in the Brahmaputra lowlands, the extension of the delta and construction of islands in the Bay of Bengal, huge siltation, and a reduction in accessible water supplies is the destabilization of mountain slopes through shifting cultivation. In Manipur, Loktak lake is silting with sediment materials eroded from the vegetation less hills around it. The various adverse effect of shifting cultivation is presented below (Borthakur, 1992). Forest and soil degradation due to shiting cultivation in Northeast India is shown in Fig. 6.10.

Fig. 6.10. Forest and soil degradation due to shifting cultivation in Northeast India
Source: Datta et al. 2004 (See colour version on page 653)

6.11.4 Deterioration of Soil Quality

The shifting cultivation is considered to be the driving force behind soil degradation (Sharma and Prasad 1995) in many many areas of Northeastern states of India where this practice of cultivation is done on high slopes.

Forest degradation may lead to land degradation if appropriate restoration and conservation measures are not followed. Degraded lands are those whose quality has deteriorated to the point where they can no longer be put to any productive use, other as present fallow, due to a variety of limitations. The GLASOD has divided the degree of soil degradation to reflect the decline of productivity into four classes:

Light: Still appropriate for local farming systems, with good management allowing for complete productivity restoration.

Moderate: Significant improvements are necessary to re-establish productivity.
Strong: The original biotic functions have been entirely damaged, necessitating major engineering work.

Extreme: All biotic functions have been completely destroyed, and the situation is unsalvageable.

The major causes of the high rate of silt load in these rivers are the result of shifting agriculture on hill slopes. Land degradation processes may be slow and continuous or quick and short lived between various stages of ecological equilibrium. Soil erosion is a primary source of soil degradation on steeply sloped fields that lack vegetation and are prone to landslides. During the monsoon season, landslides disrupt utility services such as roadways, power generation, reservoirs, human settlements, trade, tourism, and other economic and development activities. Landslides are a devastating natural occurrence that affects not only the land/soil, but also biodiversity and human life. The following visual and morphological observations made in the field as a result of degradation can be identified:

i) Topsoil erosion, as demonstrated by the presence of rills and gullies on the exposed soil surface,;
ii) Water erosion, as evidenced by the presence of rills, gullies, exposed roots etc,
iii) Change in terrain due to mass movement,
iv) Acidification, biological and chemical deterioration, as shown by low crop yields
v) Physical degradation, and
vi) Unproductive grassland vegetation dominates the field.

There is a strong link (Lal, 1999) among the soil degradation, environmental quality, food security and nutrition (Fig. 6.11). Mausbach and Tugel defined soil potential as "the ability of a certain type of soil to function within natural or managed ecosystem bounds, to sustain plant and animal productivity, maintain or improve water and air quality, and support human health and habitation" (1995). The ability of a soil to function to its full potential is referred to as soil health. Soil quality refers to the state of the soil in relation to the needs of one or more species, as well as any human need or goal (Johnson *et al.*, 1997). Because of human usage and management, as well as rare natural events, soil quality and health change over time. The soil nutrients are one of the components of soil quality/ health.

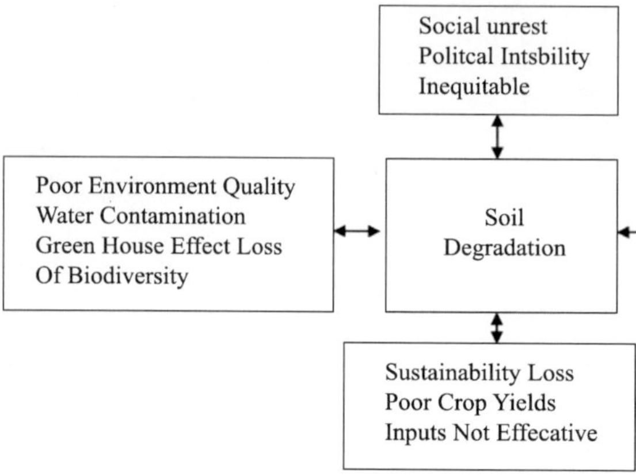

Fig. 6.11. Interrelationship among the soil degradation, environment quality, food and energy

Source: Lal 1999.

Many farmers have abandoned their traditionally cultivated lands in favour of marginal lands due to the loss of top rich soil from hill cultivation. The decline in yield after 2-3 consecutive cropping, which forces the farmer to abandon the area, is not always associated to nutrient depletion (Andriesse, 1989). Often, it is unfavourable soil physical conditions (e.g., compaction, reduction in porosity, restricted drainage), coupled with weed proliferation, that are decisive in limiting yields (Aweto, 1988). Structural deterioration of the soil is, in turn associated with a decrease in soil organic matter content (Andreux *et al.*, 1990). During fallow period, the lost soil fertility is stored, weed growth is suppressed, and soil organic carbon levels increase. The practices of jhum and hill agriculture accelerate the soil erosion by running water on hill slopes. Erosion badly influences the crop productivity as the nutrient enriched top soil gets depleted, thus ultimately restricting the rooting depth of plants.

Forest cutting, clearing, burning, and seed sowing on sloping fields cause a significant amount of loose soil material, ashes, earthworm casting, and unattached soil clods/stones to roll down the foothills, amounting to up to 3.7 tonnes soil material ha^{-1} (Borthakur, 1992). Erosion is minor in soils with high grasses or forest cover, but it is massive in steep, thinly covered shifting agriculture areas (Prasad *et al.*, 1986) that receive 1600mm of annual rainfall. From the standpoint of soil erosion, the second year of the shifting cropping cycle is more dangerous than the first. Table 6.5 shows the potential consequences of actual soil loss in terms of productivity loss (Higgins *et al.*, 1982).

Table 6.5. Soil erosion land productivity

Soil loss (t ha^{-1})	Loss in production
< 12	Crop production is not affected
12-15	About half of the area shows diminishing trend in productivity.
50-100	There is considerable decline in productivity from whole area.
101-200	About half of the area becomes unsuitable for cultivation, the rest shows a declining trend by one productivity class.
> 201	The whole area converted to non-suitable class.

Source: Higgins *et al.*, 1982

The forest vegetation undergoes a change from the continuous jhum cycle into semi-evergreen, moist deciduous, deciduous forest, grasses, bamboos and a number of aquatic and semi-aquatic plant species (Ghosh, 1992). In the course of time, a dominant stand of bamboos with the exclusion of tree growth, gradually succeeds these. It is particularly due to the extensive growth of bamboo rhizomes particularly in muli (*Melocanna baccifera*). The successive short jhum cycle (2-5 years) and lengthening of cropping periods at the same site have even eliminated the growth of bamboos. In some tracts, natural vegetation has been successively replaced by bamboo and thereafter by unproductive thatch grass (*Imperata cylindrica*) ,thus ,ultimately leading to land abandonment.

As a result, if more intense cropping is conducted on steep land, erosion may become a severe concern. Surface runoff can accelerate soil removal if the topsoil is exposed and disturbed during clearing, cropping, and weeding (Sharma and Bujarbaruah 2016). The studies conducted on the economics of shifting cultivation at ICAR Research Complex for NEH Region, have indicated that the practice is not economical. Against output/input ratio of 2.1, 1.9 and 1.8 in livestock based, horticulture and agriculture farming system, respectively, the shifting cultivation has an output/input ratio of only 0.6.

The issues involved in the problem of shifting cultivation are very complex and multifaceted. Over the years, various Working Groups and Committees have given considerable thought and recommendations. An open house seminar was held in Aizwal, Mizoram, in 1986; the task force on shifting cultivation in 1983, deliberation of National Board on Shifting Cultivation at Kohima in 1985, have suggested the extent of the problem and the means to tackle the problem. However, the intensity of the problem. The approach to solve these problems need different pathways, attitude and financial involvement (Sharma and Bujarbaruah 2016).

This problem is deepening because of additional areas for SF not being available to increase the cyclicity due to increased pressure on land on one hand and non-remunerative return through SF on the other. The immediate challenge of shifting cultivation area rehabilitation is thus two-fold: scientific management of the fallow cycle to ensure the land's long-term viability, and the creation of competent local institutions to support such scientific interventions for long-term development.

6.11.5 Linkages between soil and forest degradation and, poverty

The link between rural poverty, land degradation, and deforestation is well documented throughout Latin America and the Caribbean (Deininger and Minten 1999, McKinley and Alarcon 1995). What is often considered to be a direct link between poverty and environmental degradation turns out to be a complex interplay between a household's level of well-being and wealth, as well as its economic behaviour, such as labour allocation and other assets, after further investigation (Perrings 1989, Reardon and Vosti 1995). Cultivation shifts From forests to grasslands, savannas, and mountains, agriculture's diversity may be found at all levels and environments. The tropics have between 300 and 500 million shifting cultivators, according to Sanchez *et al.* (1996). Furthermore, it is believed that roughly 30% of the world's arable soils are used for shifting agriculture in some manner (Bandy, Garrity, and Sanchez 1993). Closed and open forests cover around 240 and 170 million hectares, respectively. The remaining closed forests in Latin America are made up of 16 percent fallow lands due to shifting farming (Brady 1996). Attempts to model the economic behaviour that determines shifting farming systems have recently been made (Angelsen 1999, DvoraK 1992, Holden 1993, Vosti and Witcover 1996) Furthermore, only recently has the impact of peasant household labour allocation decisions on rural resource degradation been investigated (Bluffstone 1995, Cooke 1998).

6.11.6 Shifting cultivation blamed for degradation

Shifting agriculture is regarded as destructive and disadvantageous because it not only harms the ecosystem but also has a negative economic impact. Some of the negative consequences of such cultivation include: helping springs dry up, causing soil erosion, destroying precious timber, creating very heavy floods, silting tanks and fields, and agricultural damage (Biswal and Kumar 2013). Forests, land, and water not only provide for one another, but also for the survival of life on Earth. Man can only live a healthy and peaceful life if these critical resources are used in a balanced manner. Several man-made factors/developments (such as mining, industrial use, timber extraction, and so on) have been blamed for destroying this harmonic balance between natural resources and human existence, with shifting cultivation being one of the most notable. It has a detrimental influence on land, water, and forests in the following ways: it limits the availability of floral resources, destroys rich forest cover, degrades land through soil erosion, and destroys rich forest cover. Waste that used to boost soil fertility owing to the presence of a decent forest cover is now impacting river and hill stream catchment areas in a number of ways, including decreased rainfall and ground water levels, increased siltification, and consequent burial of river channels due to soil erosion (Biswal and Kumar 2013).

6.11.7 Loss of Forest Cover and Biodiversity

The assessment made by Forest Survey of India (1997,2001) between 1995-97, observed that 1,875 km^2 forest areas were lost due to shifting cultivation. However, an area (1,700 km^2) under shifting cultivation came under forest cover as a regeneration (Table 6.6). It is noted that in most of the states, the forest cover tended to decline where shifting cultivation is practised. In Tripura state of India ,forest cover has a tendency to increase after 1997as a result of the settlement of shifting cultivators.

Table 6.6 Loss or gain of forest cover (ha) in the Northeastern states of India

State	Loss			Gain			Net change
	SC	Others	Total	NRSC	Others	Total	
Arunachal Pradesh	7500	-	7500	5600	-	5600	-1900
Assam	25700	15900	41600	16300	1600	17900	-23700
Manipur	60300	-	60300	46300	-	46300	-14000
Meghalaya	75000	200	7700	2000	-	2000	-5700
Mizoram	29200	-	29200	49100	-	49100	+19900
Nagaland	57300	-	57300	50300	-	50300	-7000
Tripura	-	300	3	400	700	1100	+800
Total	1,87500	16400	2,03900	170000	2300	1,72300	-31600

SC and NRSC = Shifting cultivation and natural regeneration in shifting cultivation.

6.11.8 Forest degradation in Northeast India

The conversion of forests into agriculture land use in shifting cultivation in northeast India has contributed towards soil loss through erosion and land degradation. The following photographs show land degradation after the cropping phase because fallow lands were not managed properly (Sharma 2010) (Fig. 6.12 to Fig. 6.15).

Fig. 6.12 Cropping phase (See colour version on page 654)

Fig. 6.13 Cropping phase (See colour version on page 654)

Fig. 6.14 Land degradation during fallow phase

Fallowing is a period of planned natural regeneration and forest enrichment that prepares the land for the following shifting cultivation cycle. This is a crucial time for reestablishing soil fertility. The fallow vegetation absorbs nutrients from a variety of soil depths. Fallowing is a period of planned natural regeneration and forest enrichment that prepares the land for the following shifting cultivation cycle. This is a crucial time for reestablishing soil fertility. Depending on their rooting zones, fallow plants take nutrients from a variety of soil depths. The foliage stores a portion of the nutrients, while the rooting zones are responsible for the rest. A portion of the nutrients are retained in plants, while the remainder is returned to the soil surface as litter and rainwash. The fallow period's ability to replenish or sustain soil fertility is critical to a shifting cultivation system's long-term success.

Fig. 6.15. Degradation of shifting cultivation areas if fallow phase is not managed properly. (See colour version on page 654)

6.12 POLICY OPTIONS AND RECOMMENDATIOS

Governments must implement policies and legal frameworks that recognise, protect, and reward the following farmer rights (Upadhayay 2004):

1. Farmers' right to prevent the misappropriation of their traditional knowledge related with plant genetic resources.
2. Farmers' rights to plant varieties and local knowledge, as opposed to corporate breeders' rights. 3. Rights of farmers enshrined in indigenous set-up to preserve, use, barter, sell, and proper care of crop seeds of any plant type.
4. Farmers' right to equitable benefit sharing in the production of new commercial varieties using plant genetic resources established and conserved by them, taking into consideration the economic value of plant and seed types developed by them over generations of evolution.
5. Farmers' right to be educated about market opportunities so that they can better assess the marketing choices for the products they produce using traditional knowledge.
6. Farmers' right to be protected from biopiracy and the theft of traditional knowledge.

Farmers' right to know about national and international agreements and legal systems that directly affect their livelihoods

Indigenous wisdom is enshrined in traditional farming practices. Traditional farming traditions and knowledge should be included into research and development efforts. It is necessary to explore forming a partnership with farmers on an equal footing in the management of agriculture and related natural resources. As a result, the following policy recommendations (Upadhayay 2004) are worth considering: Farmers' contributions to agricultural development and environmental conservation should be recognized and rewarded through institutional mechanisms that recognize and reward their indigenous farming expertise and practices. Provisions for officially registering farmers' varieties, landraces, and traditional knowledge would aid in creating community/ farmer ownership of resources and knowledge. Unauthorized piracy would be discouraged. Traditional farming practices use indigenous wisdom. Traditional farming knowledge and customs should be incorporated into research and development. It is vital to investigate the possibility of developing an equal-partnership with farmers in the management of agriculture and related natural resources. As a result, consider the following policy recommendations (Upadhayay 2004): Farmers should be recognized and rewarded for their contributions to agricultural development and environmental conservation through institutional systems that recognize and reward their indigenous farming expertise and traditions. Provisions for registering farmers' varieties, landraces, and traditional knowledge would help to establish community/ farmer ownership of resources and knowledge. Piracy that is not permitted would be discouraged.

6.12.1 Marginalisation of indigenous peoples

Indigenous peoples are the principal practitioners of integrated shifting cultivation, often known as rotating agroforestry. These organizations are often marginalized by mainstream society, and their techniques are disregarded as 'wrong in modern times' at best, and 'primitive' at worst, adding to the practice's misjudgment (Karkhoff and Sharma 2006). The profound perception of their surroundings is seldom appreciated unless it can be used for behefit, and it is commonly overlooked or dismissed. Farmers' entitlement to equitable monetary and non-monetary benefit sharing in the development of new commercial varieties using plant genetic resources they established and conserved, taking into account the economic value of plant and seed types developed by them over generations of evolution.

Several agro-forestry production strategies, developed with locally suitable trees and crops for varied slopes, have been discovered in Bangladesh's hills to optimize agro-forestry agricultural output and reduce environmental degradation (Paul and Hossain, 2001). Shifting cultivation causes the least amount of soil disturbance, generates natural fertility through residues of mixed cropping on moderate to steep slopes with minimum tillage under rainfed circumstances relying on local resources, and causes the least amount of soil disturbance (Sharma, 1976). In Meghalaya and other north-eastern Indian states, a team of experts led by P.S. Ramakrishnan investigated the economics and efficiency of shifting agriculture. These investigations indicated that, far from being primitive and inefficient, jhum is a sophisticated organic multiple cropping system that is well suited to the heavy rainfall zones of the hill tracts. Other methods of agriculture, such as terrace and valley farming, have worse economic and energy efficiency than Jhum (Ramakrishnan, 1992).

Shifting cultivation is still seen as a primitive agricultural method with little commercial potential, as well as a major cause of deforestation and environmental degradation, as well as a barrier to upland development. Despite an increasing quantity of scientific literature refuting the impression, this viewpoint remains. As a result of this misunderstanding, shifting cultivation management has remained a critical component of agricultural development in upland areas of South and Southeast Asia, with governments spending significant financial resources to replace the practise. There is a pressing need to reconsider past and current approaches to promoting the transition from shifting cultivation to established agriculture, as well as to investigate strategies to avoid the usual mistakes and harmful consequences of short-sighted interventions. Despite extensive and long-term efforts by the government to eliminate shifting cultivation in the eastern Himalayan region, the practise has remained entrenched in significant regions.

6.12.2 Principles

To arrive at the primary recommendations, the following principles were followed (Karkhoff and Sharma 2006).

1. There must be no prejudice towards shifting cultivation. Rather of being harmful to the environment or livelihoods, shifting cultivation could be viewed as a potentially beneficial tool for development.
2. Shifting cultivation communities will inevitably change. These adjustments, on the other hand, should originate from within the system. We must capitalize on and support technical and institutional breakthroughs and transformations.
3. To establish shifting cultivation, an adaptive management technique is required. This method emphasizes change rather than replacement of current methods, and it takes into account traditional knowledge and practices as well as social sensitivity.
4. To validate and document current excellent practices and traditional knowledge of shifting agriculture farmers, scientific study is required. Prioritize support for farmers' innovative processes. This will open

up practical opportunities for improving shifting cultivation and other farming methods in similar agro-ecological situations. These options should be the focus of agricultural extension and training.

5. Controlled burning by shifting cultivators should be recognized as a significant management strategy for fertility enhancement, weed and pest control, and weed and pest control.
6. A fundamental concept of shifting agriculture is the use of common property resources. All community members have access to the sources of production through customary institutions.
7. Alternative loan access options for shifting agriculture farmers should be developed. Local agencies that regulate common property, for example, could give bank guarantees on behalf of the borrower.
8. It is necessary to build, capacitate, and formalize local customary institutions. Rather than being supplanted by parallel government bodies at the local level, they should have proper synergy with the state and be integrated inside it. Gender concerns such as sensitization, representation, and equity must be addressed. In order to formulate and implement policies on shifting cultivation, they must work with local/state governments. To reduce disputes, cross-level cooperation should be fostered, and the policy process should be supported at the national level.
9. Within the eastern Himalayas, regional collaboration channels are critical for sharing and exchanging knowledge on excellent practices and innovations in shifting agriculture.

6.13 COMBATING LAND DEGRADATION

It is critical to assess, characterize, and classify different types of degradation concerns, as well as create appropriate technologies to recover degraded regions for productive use, in order to establish successful methods to combat land degradation. Following 'bottom up' participation approaches, participatory watershed management has been embraced in India as a tool for all developmental operations with an emphasis on socio-economic elements apart from biophysical qualities. Since April 1, 2008, all participating ministries have been following a set of common rules for watershed improvement projects. For effective natural resource management, the following challenges must be addressed first and foremost:

- To satisfy the predicted phytomass/biomass demand, an integrated land resource management program is required, accounting for the reclamation of degraded areas and incorporating all relevant ministries.

- Methodologies for effective land use planning at various scales using modern tools and methods must be developed for sustainable land use management.
- Integrated farming systems must be created and evaluated in different agro-ecological zones of the country to increase yields and profits. It would contain the best feasible combination of diverse operations for different types of farmers and agricultural scenarios, such as agriculture, horticulture, livestock, fishery, forestry, and so on, in order to maximise the use of land and water resources while avoiding overexploitation.
- Phytoremediation, bioremediation, altering microbial catabolic genes, and breeding resistant crops can all help to restore soils that have been contaminated by heavy metals or dangerous substances, as well as pesticide overuse.
- Soil management techniques such as residue integration, manuring, decreased tillage, and mulching help to sequester carbon in the soil and minimize CO_2 emissions.
- Soil quality indicators must be evolved for vartous production functions of the soil and prevent land forest degradation due to different factors.
- Restoring damaged soils and ecosystems using a watershed method can improve the terrestrial C pool, microbial population, and soil net C sinks, resulting in better and longer-term production.
- To develop an enabling policy framework, proper legislation must be enacted to allow for the restoration of damaged soils.

6.14 AGROFORESTRY SYSTEMS FOR DEGRADED LANDS

For improvement of the degraded lands some of the workers (Singh and Parihar 2015) on the basis of their studies have recommended the systems which had a profound impact on the different soil properties. Some of the practices are presented below:

6.14.1 Agroforestry and soil fertility of degraded soils

A significant increase in organic carbon was reported in all agroforestry systems when compared to starting state, with the silvihortipastoral system contributing the most. In comparison to natural forest, the same system recorded 43.2 percent higher exchangeable Al, resulting in a pH drop of up to 0.50 units. All of the systems had higher exchangeable Ca, Mg, Na, K, and Al content, as well as accessible N, P, and K content, than natural forest, and the content of these nutrients declined with increasing soil depth (Majumdar *et al.* 2004). After 17 years of implementation, the influence of various AFS such as silvipastoral, silvihorticulture, and agrihortisilvipastoral on soil fertility indices was studied (Majumdar *et al.* 2002). The amount of organic carbon in all of the systems increased, although the rate of increase was highly dependent on the type of vegetation. The exchangeable Ca, Mg, and K content in the soil was significantly altered by the use of different cropping patterns in various AFS. Agrihortisilvipastoral and silvihorticulture had the highest cation buildup, followed by natural fallow and silvi-pastoral systems. Silvihorticulture accumulated the most exchangeable K, followed by agrihortisilvipastoral. In agrihortisilvipastoral and silvihorticulture, available N, P, and S were higher than in natural fallow and silvipastoral.

Improvement of soil physical properties

Forest (17.6% loss in bulk density), agri-horti-silvi-pastoral (14.3%), livestock-based (13.4%), natural fallow, and agriculture are the land uses that have the greatest impact on soil physical attributes (12.6 percent). In the forest environment, there were also higher percentages of macro-aggregates (54.5%), organic carbon (2.95%), and biotic activity. Soil biota affects soil characteristics by forming stable aggregates, developing organo-mineral complexes, and enhancing macro-porosity and pore continuity from the surface to the sub-soil, all of which improve water conveyance and reduce run-off. In comparison to shifting-cultivated plots, modified land use systems had higher transmission and storage pore volume, as well as reduced residual pore volume. This demonstrated that the soil's pore shape was preserved under these conditions. Higher percentages of organic matter, clay content, and high levels of Al and Fe oxides in soil could explain enhanced soil aggregation in natural forest multi-storied agroforestry and silvihortipastoral systems that maintain broad vegetative cover throughout the year (Saha and Mishra 2007).

6.14.2 Loss of forest cover

The tree ecosystem, which includes a wide range of other trees, smaller plants, and animals from the forest environment. Forests are intertwined with various river and stream ecosystems, as well as nearby lowland or village ecosystems, in which visitors such as man or his domestic cattle invade into the forest and extract forest products such as timber, leaves, grass, and firewood. Forest ecosystems are thus immensely beneficial to civilization, particularly in the lives of rural poor and tribal people. Forests are important in our country for social, religious, cultural, and ethnic reasons in addition to their economic, ecological, and scientific values. A number of causes are threatening hill forest ecosystems, including (i) putting more forest land to agriculture, human settlements and other uses, (ii) forestland loss owing to multifunctional schemes, road, electric line, and other infrastructure construction, shifting cultivation etc., (iii) Degradation due to illegal cutting, overgrazing etc,, and (iv) Population increase and forest land encroachment.

References

Achard F, Eva HD, Stibig HJ, Mayaux P, Gallego J, Richards T and Malingreau JP 2002. Determination of deforestation rates of the world's humid tropical forests. Science, 297: 997-1003.

Adriani, N. and Alb. C. Kruyt. 1912–14. De Bare'e-sprekende Toradja's van Midden- Celebes. Batavia: Landsdrukkerij. 3 vols.

Adriani, N. and Alb. C. Kruyt. 1950–51. De Bare'e Sprekende Toradjas van Midden- Celebes (de Oost-Toradjas). Amsterdam: Noord-Hollandsche. 3 vols.

Alcorn JB. 1990. Indigenous agroforestry strategies meeting farmers' needs. In: Anderson A, editor. Alternatives to deforestation: steps towards sustainable use of the amazon rain forest. New York (NY): Columbia University Press.

Amanor, K., 1994, The New Frontier: farmers' response to land degradation, Zed Books,. London.

Andersen, L.E., C.W.J. Granger, E.J. Reis, D. Weinhold, and S. Wunder 2002. The Dynamics of Deforestation and Economic Growth in the Brazilian Amazon,Cambridge, UK: Cambridge University Press.

Anderson, A.B. 1990. Alternatives to deforestation: steps toward sustainable use of the Amazon rain forest. Columbia University Press, New York.

Andreux, F.G., Cerri, C.C., Eduardo, de P. and Chone, D. 1990. Science of total environment, 90:249-265.

Andriesse, J.P., 1980. Nutrient level changes during a 20-year shifting cultivation cycle in Sarawak (Malaysia). In: K.T. Joseph (Editor), Proc. Conf. on Classification and Management of Tropical Soils, 1977. Malaysian Society of Soil Science, Kuala Lumpur, pp. 479-49 I.

Andriesse, J.P. (1989) Nutrient management through shifting cultivation: A comparative study on cycling of nutrients in traditional farming systems of Malaysia and Sri Lanka. In : Nutrient Management for Food Crop Production in Tropical Farming System (eds. Van der Heide, J.), pp. 29-62. Institute of Soil fertility, Haren & University of Brawijaya, Malang, Indonesia.

Angelsen, A. 1995. Shifting cultivation and "deforestation": a study from Indonesia. World Development 23: 1713–1729. doi:10.1016/0305-750X(95)00070-S.

Angelsen, A. 1999. 'Agricultural expansion and deforestation: Modelling the impact of population, market forces and property regimes,' Journal of Development Economics, 58, 185–218.

Atamov, V.V., M. Cabbarov and E. Gurbanov, 2006. The pytosociological characteristics of ecosystems of mountain of talysh region of Azerbaijan. Asian J. Plant Sci., 5: 899-904.

Aweto, A.O. 1988. Effects of shifting cultivation on a tropical rain forest soil in southwestern Nigeria. Turrialba 38:19-22.

Balee, W. 1989. The culture of Amazonian forests . Advances in Economic Botany, 7: 1-21.

Bandy, D. E., D. P. Garrity, and P. Sanchez 1993. 'The world wide problem of slash-and-burn agriculture,' Agroforestry Today, 5, 2–6.

Bandy D, 1994. Alternatives to slash and burn – A global strategy (Report prepared by ICRAF, IFDC, IITA, IRRI,TSBF and CIAT), Alternatives to Slash and Burn Programme, Nairobi, Kenya, 37p

Barona, E., N. Ramankutty, G. Hyman, and O. T. Coomes. 2010. The role of pasture and soybean in deforestation of the Brazilian Amazon. Environ. Res. Lett. 5. http://iopscience. iop. org/1748-9326/5/2/024002/pdf/1748-9326_5_2_024002.pdf (accessed 30 March 2013).

Barrett, C. B. 1999. "Stochastic Food Prices and Slash-and-Burn Agriculture." Environment and Development Economics 4:161-176.

Barton, D. 1994. Indigenous Agroforestry in Latin America: A Blueprint for Sustainable Agriculture? NRI Socio-economic Series 6. Chatham, UK: Natural Resources Institute.

Beach, T., 1998. Soil catenas, tropical deforestation, and ancient and contemporary soil erosion in the Peten, Guatemala. Phys. Geogr. 19, 378–405.

Beckerman, S. 1987. Swidden in Amazonia and the Amazon Rim. In Turner, B. L. and Brush, S. B. (eds.. Comparative Farming Systems. New York: Guilford Press.

Bellon, M. R., and J. E. Taylor. 1993. "'Folk' Soil Taxonomy and the Partial Adoption of New Seed Varieties." Economic Development and Cultural Change 41:763-86.

Bene, J.G., Beall, H.W., and de J. Hart, R.A. 1977. Trees, Food, and People: Land Management in the Tropics. Ottawa: IDRC.

Berkes F, Colding J, Folke C. Rediscovery of traditional ecological knowledge as adaptive management. Ecol. Appl. 2000; 10:1251–1262.

Bhagawati, K., Bhagawati, G., Das, R, Bhagawati, R. and Ngachan, S.V. 2015. The Structure of Jhum (Traditional Shifting Cultivation System): Prospect or Threat to Climate. International Letters of Natural Sciences Online, 46: 16-30, doi:10.18052/www.scipress. com/ILNS.46.16

Bhattacharyya, R. Ghosh, B. N., Mishra, P. K., Mandal, B., Srinivasa Rao, C. H., Sarkar, D., Das, K., Anil, K. S., Lalitha, M., Hati, K. M., and Franzluebbers, A. J.: 2015. Soil degradation in India: challenges and potential solutions, Sustainability, 7, 3528–3570.

Billand, A. 2012. "Biodiversity in Central African Forests: An Overview of Knowledge, Challenges, and Conservation Measures." In The Forests of the Congo Basin—State of the Forest 2010, ed. de Wasseige *et al.* Luxembourg: Publications Office of the European Union.

Binswanger, H.P. (1991) Brazilian policies that encourage deforestation in the Amazon. World Development, 19(7): 821-829.

Biswal DK and Kumar S, 2013. Shifting cultivation and policies of sustainable development: a meaningless obsession Journal of Economic & Social Development, 9: 132-138.

Blake, J. G. and Loiselle, B. A. 2001. Bird assemblages in second-growth and old-growth forests, Costa Rica: perspectives from mist nets and point counts. The Auk 118: 304-326.

Bluffstone, R. A. (1995), 'The effect of labor market performance on deforestation in developing countries under open access: An example from rural Nepal,' Journal of Environmental Economics and Management, 29, 42–63.

Borthakur D N (1992) Agriculture of the North Eastern Region with Special Reference to Hill Agriculture, Bee Cee Prakashan, Guwahati, India.

Boucher D., Elias P., Lininger K., Roquemore S. and Saxon E. (2011), The Root of the problem: What is driving tropical deforestation today? Union of Concerned Scientists, Cambridge, US.

Bouwman A.F. ed. 1990. Soils and the greenhouse effect. New York: Wiley.

Brady, N. C. 1996. Alternatives to slash-and-burn: A global imperative. Agricultural Ecosystems and Environment 58: 3-11.

Brearley FQ, Prajadinata S, Kidd PS, Proctor J, Suriantata (2004) Structure and floristics of an old secondary rain forest in Central Kalimantan, Indonesia, and a comparison with adjacent primary forest. For Ecol Manag 195:385–397

Brokensha, D. and Riley, B. W. 1980. Mbeere knowledge of their vegetation, and its relevance for development (Kenya). In Brokensha, D. Warren, D. M. and Werner, O. (eds) Indigenous knowledge and development. Lanham, MD: University Press of America. pp. 111 - 128.

Brondízio, E. Intensificação Agrícola, Identidade Econômica e Invisibilidade entre Pequenos Produtores Rurais Amazônicos: caboclos e colonos numa perspectiva comparada. In: ADAMS, C.; MURRIETA, R. R. S., NEVES, W. (Ogs.). Sociedades Caboclas Amazônicas: modernidade e invisibilidade. São Paulo: Annablume/ FAPESP, 2006. p. 191-232.

Brondízio, E. Intraregional Analysis of Land-Use Change in the Amazon. In: MORAN, E. F.; OSTROM, E. (Eds.). Seeing the Forest and the Trees. Human Environment Interactions in Forest Ecosystems. Cambridge, Massachussets: MIT Press, 2005. p. 223-252.

Brookfield, Harold and Christine Padoch. "Appreciating Agrodiversity: a look at the dynamism and diversity of indigenous farming practices," Environment Vol. 36, No. 5 (June 1994): 6-21.

Brown, S. and Lugo, A.E. 1990. Tropical secondary forests, Journal of Tropical Ecology 6: 1-21.

Brown, D.; Schreckenberg, K. Shifting Cultivators as agents of deforestation: assessing the evidence. Natural Resource Perspectives, v. 29, p. 1-14, 1998.

Bruce, J. W. 2004, 'Strengthening property rights for the poor', in R. Meinzen-Dick and M. Di Gregorio (eds)Collective Action and Property Rights for Sustainable Development, International Food Policy Research Institute (IFPRI),Washington, DC

Bruun, T.B., Mertz, O., Elberling, B., 2006. Linking yields of upland rice in shifting cultivation to fallow lenght and soil properties. Agric. Ecosyst. Environ. 113, 139–149.

Castro-Luna, A.A., Castillo-Campos, G., Sosa, V.J., 2011. Effects of selective logging and shifting cultivation on the structure and diversity of a tropical evergreen forest in south-eastern Mexico. Journal of Tropical Forest Science, 23, 17-34.

Cerri, C.E.P., M. Bernoux, V. Chaplot, B. Volkoff, R.L. Victoria, J.M. Melillo, K. Paustian, and C.C. Cerri 2004, 'Assessment of soil property spatial variation in an Amazon pasture: basis for selecting an agronomic experimental area', Geoderma 123: 51–68.

Chazdon R. 1998. Tropical f orests (log 'em or leave 'em). Science 281: 1295–1296.

Chomitz, K.M., P. Buys, G. De Luca, T.S. Thomas, and S. Wertz-Kanounnikoff. 2007. At loggerheads? Agriculturalexpansion, poverty reduction, and environment in the tropicalforests. Washington, DC: The World Bank.

Christanty L, 1986. Shifting Cultivation and Tropical Soils: Patterns, Problems, and Possible Improvements. In: Traditional Agriculture in Southeast Asia – A Human Ecology Perspective, Martin GG (eds.) ,West View, Boulder Press, Colorado.

Cincotta, R. P., Wisnewski, J. and Engelman, R. 2000. Human population in the biodiversity hotspots. Nature 404: 990-992.

Clay, J.W. 1988, Indigenous peoples and tropical forests. Models of land use and management from Latin America. Cultural Survival Report 27. Cambridge, USA: Cultural Survival.

Colchester, M. 2007, Beyond Tenure.Rights-based Approaches to Peoples and Forests. Some Lessons from the Forest People Programme, Rights and Resource Initiative (RRI),Washington, DC

Colfer, C. J. Alcorn and Russell, D. 2015. Swiddens and Fallows: Reflections on the Global and Local Values of ' Slash and Burn'. pp 62-86 in Cairns, M. Shifting Cultivation and Environmental Change: Indigenous People, Agriculture and Forest Conservation. New York and Oxford UK: Routledge. Shifting Cultivation, Gender, and REDD+ in Cameroon and the Democratic Republic of Congo

Collins, M., J. Sayer, and T.C. Whitmore. 1991. The conservation atlas of tropical forests: Asia and the Pacific. Simon and Schuster, New York.

Collins, J.I. 1986. Smallholder settlement of tropical South America: the social causes of ecological destruction. Human Organisation, 25(1): 1-10.

Conklin HC 1957. Hanunoo Agriculture: A Report on an Integral System of Shifting Cultivation in the Philippines. FAO Forestry Development Paper No 12, Food and Agricultural Organization of the United Nations, Rome, Italy

Conklin, H. C. 1961. The study of shifting cultivation. Current Anthropology 2: 27-61

Connell, J. H. 1978. Diversity in tropical rain forests and coral reefs. Science 199: 1302 - 1310. Consult. Group Int. Agric. Res. (CGIAR). 1974. International Research in Agriculture. New York: CGIAR. Conservation Area (report series no.5)

Cooke, P. 1998. 'The effect of environmental good scarcity on own-farm labor allocation: The case of agricultural households in rural nepal,' Environment and Development Economics, 3, 443–469.

Costa, F. A. 1995. Agricultura familiar em transformac¸a˜o na Amazoˆ nia: O caso de Capita˜o Poc¸o e suas implicac¸o˜es para a polı´tica e o planejamento agrı´colas regionais. Working paper 49. Bele´m, Brazil: Nu´cleo de Altos Estudos Amazoˆnicos (NAEA). http://www. ufpa.br/naea/novosite/paper/91

Cramb RA. 2007. Land and Longhouse. Agrarian Transformations in the Uplands of Sarawak. Copenhagen: NIAS Press.

Cuevas VC. 1991. The continuing problem of shifting cultivation in the Philippines: a human ecology analysis. IESAM Bulletin; 11: 10-16 & 21-27.

Dalle, S. P. and de Bois, S. 2006. Shorter fallow cycles affect the availability of noncrop plant resources in a shifting cultivation system. Ecology and Society, Vol. 11, No. 2, 2.

Dalle, S.P. Pulido, M.A. and De Blois, S. 2011. Balancing shifting cultivation and forest conservation: lessons from a ''sustainable landscape'' in southeastern Mexico. Ecological Applications, 21: 1557–1572.

Darby, H. C. 1956. The Clearing of the Woodland of Europe. In: W. L. Thomas (ed.) Man's Role in Changing the Face of the Earth (Chicago and London: The University of Chicago Press), pp. 183-216.

Datta M, Singh NP 2007. Growth characteristics of multipurpose tree species, crop productivity and soil properties in agroforestry systems under subtropical humid climate in India. Journal of Forestry Research 18(4): 261-270.

Datta , M. Dhiman , K.R. and Singh , N. P. 2004. Shifting Cultivation – A case Study to Evaluate Soil Fertility ,Pub. No 6, ICAR Research Complex for NEH Region , Tripura, India

Davidson EA, de Abreu Sa TD, Reis Carvalho CJ, de Oliveria Figueiredo R, Kato M do SA, Kato OR, *et al.* 2008. An integrated greenhouse gas assessment of an alternative to slash-and-burn agriculture in eastern Amazonia. Glob Change Biol.;14: 998–1007.

Davidson EA, de Araújo AC, Artaxo P, Balch JK, Brown IF, Bustamante MM, *et al.* 2012.The Amazon basin in transition. Nature. 481:321–328. doi: 10.1038/nature10717 PMID: 22258611

Davis, S.D., V.H. Heywood, O. Herrera-MacBryde, J. Villa-Lobos, and A.C. Hamilton (eds) 1997. Centres of plant diversity: a guide and strategy for their conservation. Vol. 3: The Americas, WWF-World Wildlife Fund for Nature and IUCN–The World Conservation Union, Cambridge, UK: IUCN Publications Unit.

de Camino, R. (1999), 'Sustainable forest management in Latin America: relevant actors and policies', Discussion Paper, April, InterAmerican Development Bank, Washington, DC. Environment and Development Economics 367 de Jong, W. 1997. Developing swidden agriculture and the threat of biodiversity loss. Agriculture, Ecosystems and the Environment, 62: 187-197.

de Neergaard A, Magid J and Mertz O, 2008. SOIL erosion from shifting cultivation and other smallholder land use in Sarawak, Malaysia. Agriculture Ecosystems & Environment 125:182-190 DOI:10.1016/j.agee.2007.12.013

de Wasseige, C., D. Devers, P. de Marcken, R. Eba'a Atyi, R. Nasi, and P. Mayaux, eds. 2009. The Forests of the Congo Basin—State of the Forest 2008. Luxembourg: Publications Office of the European Union.

de Wasseige, C., P. de Marcken, F. Hiol-Hiol, P. Mayaux, B. Desclee, R. Nasi, A. Billand, P. Defourny, and R. Eba'a Atyi, eds. 2012. The Forests of the Congo Basin—State of the Forest 2010. Luxembourg: Publications Office of the European Union.

Defries, R.S., T. Rudel, M. Uriarte, and M. Hansen. 2010. Deforestation driven by urban population growth and agricultural trade in the twenty-first century. Nature Geoscience3: 178-181.

Deininger, K. W. and B. Minten 1999. 'Poverty, policies and deforestation: The case of mexico,' Economic Development and Cultural Change, 47(2), 313–344.

Deininger, K., D. Byerlee, J. Lindsay, A. Norton, H. Selod, and M. Stickler. 2011. Rising Global Interest in Farmland—Can It Yield Sustainable and Equitable Benefits? Washington, DC: World Bank

Denevan, W. M. (Ed). 2001. Cultivated landscapes of native Amazonia and the Andes. Oxford, Oxford University Press.

Denslow, J.S., and C. Padoch. 1988. People of the tropical rainforest. University of California Press, Berkeley.

Detwiler, R.P. and Hall, C.A.S., 1988. Tropical Forests and the Global Carbon Cycle. Science, 239: 42–47.

Ding, Y., Zang, R., Liu, S., He, F., Letcher, S.G., 2012. Recovery of woody plant diversity in tropical rain forests in southern China after logging and shifting cultivation. Biological Conservation, 145, 225–233.

Do, T.V., Osawa, A., Thang, N.T., 2010. Recovery process of a mountain forest after shifting cultivation in Northwestern Vietnam. For. Ecol. Manag. 259, 1650–1659.

Do, T.V., Osawa, A., Thang, N.T., Van, N.B., Hang, B.T., Khanh, C.Q., Thao, L.T., Tuan, D.X., 2011. Population changes of early successional forest species after shifting cultivation in Northwestern Vietnam. New Forests, 41, 247–262.

Dobson AP, Bradshaw AD, Baker AJM. Hopes for the future: restoration ecology and conservation biology. Science 1997; 277: 515-522.

Oliveira, A. H. de; Carneiro, M. S. de S.; Sales, R. de O.; Pereira, E. S.; Araújo Filho, J. M. de; Pinto, M. S. de C.; Magalhães, J. A.; Costa, N. de L., 2011. Value nutritive of bakery waste in the sheep feeding. Pubvet, 5 (8): 24

Dove MR. 1984. Government versus peasant beliefs c oncerning imperata and eupatorium:A structural analysis of knowledge, myth, and agricultural ecology in Indonesia. Honolulu: East–West Center. East–West Center Working Paper no. 54.

Dove, M. (1983). Theories of swidden agriculture, and the political economy of ignorance. Agroforestry systems, 1(2):85{99.

Dressler WH, Wilson D, Clendenning J, Cramb R, Rodney Keenan R, Mahanty S, Bruun TB, Ole Mertz O and Lasco RD, 2017. The impact of swidden decline on livelihoods and ecosystem services in Southeast Asia: A review of the evidence from 1990 to 2015. Ambio. 46: 291–310. doi: 10.1007/s13280-016-0836-z

Dressler, J. J., Teraoka, M., Espejo, G. L., Kishi, R., Takamuku, S., Gómez-García, C. J., ... & Haley, M. M. (2018). Thiophene and its sulfur inhibit indenoindenodibenzothiophene diradicals from low-energy lying thermal triplets. Nature Chemistry, 10(11), 1134-1140.

Dressler, W., Wilson, D., Clendenning, J., Cramb, R., Mahanty, S., Lasco, R., Keenan, R., To, P., Gevana, D., 2015. Examining how long fallow swidden systems impact upon livelihood and ecosystem services outcomes compared with alternative land-uses in the uplands of Southeast Asia. Journal of Development Effectiveness, 7, 210–229.

DvoraK, K. A. (1992), 'Resource management by West African farmers and the economics of shifting cultivation,' American journal of agricultural economics, 74(3), 809–815.

Eckholm, E.P. 1976. Losing Ground. New York, N.Y.: Norton and Company.

Eder JF. Deforestation and detribalization in the Philippines: the Palawan case. Population and Environment 1990; 12: 99-115.

Embrapa (1981), 'Servic¸o nacional de levantamento e conservac¸ao de solos', Mapa ˜de solos do Brasil: escala 1:5.000.000. Rio de Janeiro, Brazil.

Erni, C. (2009). Shifting the blame? Southeast Asia's indigenous peoples and shifting cultivation in the age of climate change. Paper presented at the seminar on "Adivasi/ ST Communities in India: Development and Change", Delhi, August 27-29, 2009, p. 6

EPA (Environmental Protection Agency). 1990. Greenhouse gas emission from agricultural systems, vols. 1 and 2. Washington, DC: EPA.

Eriksson, M; Jianchu, X; Shrestha, AB; Vaidya, RA; Nepal, S; Sandstorm, K (2008) The changing Himalayas: Impact of climate change on water resources and livelihoods in the Greater Himalayas. Kathmandu, Nepal: ICIMOD

Ernst, C., A. Verhegghen, C. Bodart, P. Mayaux, C. de Wasseige, A. Bararwandika, G. Begoto, F. Esono Mba, M. Ibara, *et al.* 2010. "Congo Basin Forest Cover Change Estimate for 1990, 2000 and 2005 by Landsat Interpretation Using an Automated Object-Based Processing Chain." The International Archives

Estes, J.A., Terborgh, J., Brashares, J.S., Power, M.E., Berger, J., Bond, W.J., Carpenter, S.R., *et al.*., 2011. Trophic downgrading of planet earth. Science, 333, 301–306.

FAO. 1957. Shifting cultivation. Unasylva 11: 1.

FAO/UNEP. 1981. Tropical Forest Resources Assessment Project(GEMS): Tropical Africa, Tropical Asia, Tropical America (4 volumes). Rome: FAO. Cited by OTA 1984.

FAO. 1980. Maximizing the efficiency of fertilizer use by grain crops. FAO Fertilizer and Plant Nutrition Bulletin 3. Food and Agriculture Organization of the United Nations, Rome (Reprinted in 1988). 30 p.

FAO. 1981. Crop production levels and fertilizer use. FAO Fertilizer and Plant Nutrition Bulletin 2. Food and Agriculture Organization of the United Nations, Rome. 69 p.

FAO. 1993. Agriculture: Towards 2010. FAO Conference report C93/24. Food and Agriculture Organization of the United Nations, Rome.

FAO (2000') Woodfuel production and marketing in Myanmar. Regional Wood Energy Development Programme Report No. 56, FAO, Bangkok, Thailand.

FAO and UNEP. 2020. The State of the World's Forests 2020. Forests, biodiversity and people. Rome. https://doi.org/10.4060/ca8642en.

FAO, 2005. The State of Food and Agriculture. Food and Agricultural Organizatio of the United Nations. Rome, Italy.

FAO 2004 Global Forest Resources Assessment Update 2005: Terms and Definitions, final revision, Food and Agriculture Organization of the United Nations, Rome

FAO and UNEP. 2020. The State of the World's Forests 2020. Forests, biodiversity and people. Rome. https://doi.org/10.4060/ca8642en.

FAO, 1982. Forest Resources of Tropical Asia. Rome, Italy.

FAO, UNDP, UNEP 2008. UN Collaborative Programme on Reducing Emissions from Deforestation and Forest Degradation in Developing Countries (UN-REDD). Framework Document 20 June 2008

FAO. 2011. State of the World's Forests 2011. Rome: FAO.

Farid, ATM; Hossain, SM (1988) Diagnosis of farming practices and their impact on soil resources loss and economic loss in the hill tracts areas of Bangladesh. Gazipur, Bangladesh: Bangladesh Agricultural Research Institute (BARI)

Fearnside, P.M. (1983) Land-use trends in the Brazilian Amazon region as factors in accelerating deforestation. Environmental Conservation 10: 141–148.CrossRefGoogle Scholar

Fearnside P.M., and W.M. Guimaraes. 1996. Carbon uptake by secondary forests in Brazilian Amazonia. For. Ecol. Manage. 80:35–46.

Fearnside, P. M. (2001). Land-tenure issues as factors in environmental destruction in Brazilian Amazonia: the case of Southern Pará. World Development, v. 29, n. 8, p. 1361-1372,.

Fearnside, P. M. (2005). Do hydroelectric dams mitigate global warming? The case of Brazil's Curuá-Una Dam. Mitigation and adaptation Strategies for Global change, 10(4), 675-691.

Feeny, D. 1984. Agricultural expansion and forest depletion in Thailand 1900–1975. Yale University, New Haven, Connecticut.

Fischer, G., G. Granat, and M. Makowski (1998), 'AEZWIN: An interactive multiple-criteria analysis tool for land resources appraisal', Report IR-98-051, International Institute for Applied Systems Analysis, Laxenburg, Austria.

Forest Survey of India (FSI). India State of Forest Report; FSI: Dehradun, 1997.

Forest Survey of India (FSI). India State of Forest Report; FSI: Dehradun, 2001.

Forestal (1966) Soil and land use survey of the Chittagong Hill Tracts. Vol 9. Vancouver, Canada: Forestal International

Forsyth, T.J., 1994. The use of Cesium-137 measurements of soil erosion and farmer's perceptions to indicate land degradation amongst shifting cultivators in northern Thailand. Mountain Res. Dev. 14, 229–244.

Fox J, Truong DM, Rambo AT, Tuyen NP, Cuc LT and Leisz S. 2000. Shifting Cultivation: A New Old Pa r a d i g m for Managing Tropical Forests. BioScience, 50: 521-528.

Fox J, Truing DM, A Terry Rambo AT, Tuyen NP. Cuc LT and Leisz S (1999). 'Shifting cultivation without deforestation: A case study in the Mountains of Northwestern Vietnam'. EWC Working Papers: Environmental Series No. 50. East-West Center. Honolulu. Hawaii.

Freudenberger, K. S. (1994) Tree and Land Tenure Rapid Appraisal Tools, Food and Agriculture Organization of the United Nations, Rome

Gade DW (1996) Deforestation and its effects in highland Madagascar. Mt Res Dev 16:101–116 Biodivers Conserv (2010) 19:187–204 203

Gafur A, Jensen JR, Borggaard OK, Peterson L. 2003. Run-off and losses of soil nutrients from small watershed and shifting cultivation (Jhum) in the Chittagong Hill Tracts of Bangladesh. Journal of Hydrology 274: 30–46.

Gafur, A (2001) Effects of shifting cultivation on soil properties, erosion, nutrient depletion, and hydrological responses in small watershed of the Chittagong Hill tracts of Bangladesh. PhD thesis, Chemistry Department, Copenhagen, Denmark

Gafur, A., Borggaard, O.K., Jensen, J.R., Petersen, L., 2000. Changes in soil nutrient content under shifting cultivation in the Chittagong Hill Tracts of Bangladesh. Geografisk Tidsskrift, Danish J. Geogr. 100, 37–46.

Gehring, C., Denich, M., Vlek, P.L.G., 2005. Resilience of secondary forest regrowth after slash-and-burn agriculture in central Amazonia. J. Trop. Ecol. 21, 519–527.

Geist HJ and Lambin EF (2001), What Drives Tropical Deforestation? (Louvain Univ., Louvain-Da-Neuve, Belgium.

Geist, H, and E. Lambin. 2002. Proximate causes and underlying driving forces of tropical deforestation. BioScience 52: 143-150.

Gerold, G. (1995) Soil differentiation and degradation by Agricultural and logging impacts in tropical forests of West Africa. In: Proceedings of International Congress on soils of tropical forest ecosystems. Third conference on forest soils (ISSS-AISS-IBG) Volume 2: Soils degradation and conservation. Mulawarman University Press, Samarinda, Indonesia.

Ghosh, A.K. (1992) Indian Forester., 118:754-762.

Giller, K., and C. Palm. 2004. Cropping systems: Slash-andburn cropping systems of the tropics. Encyclopedia of Plantand Crop Science: 363-366.

Gillis M. and Repetto R. 1988. Deforestation and wood, fodder and fibre, plus increased income and improved living standards. government policy. Cambridge: Cambridge University Press.

Gillis, M. (1996). Tropical Deforestation] 996. Rice environmental Conference, U.S.A. Shifting Cultivation Practices By Hmong, Khamu and Lao ethnic categories In The Nam Nane Watershed, Nane District, Luang Prabang Province, Lao PDR 20040530 64

Gilruth PT, Marsh S, Itami R (1995). A dynamic spatial model of shifting cultivation in the highlands of Guinea, West Africa. Ecol. Mod., 79: 179-197.

Government of Indonesia (1999) Act Number 41/1999 on Forestry, Ministry Secretariat, Jakarta

Grace, J. (2004). Understanding and Managing the Global Carbon Cycle. Journal of Ecology 92: 189–202. doi:10.1111/j.0022-0477.2004.00874.x.

Grainger, A. 1984. Quantifying changes in forest cover in the humid tropics: overcoming current limitations. The Journal of World Forest Resource Management 1: 3-64.

Grandstaff, T.B. 1981. Shifting cultivation. Ceres 4:28-30.

Grau, R. H., andM. Aide. 2008. Globalization and land-use transitions in Latin America. Ecol. Society 13. http://www.ecologyandsociety.org/vol13/iss2/art16 (accessed 30March 2013).

Gray, C. L., R. E. Bilsborrow, J. L. Bremner, and F. Lu. 2008. Indigenous land use in the Ecuadorian Amazon: A cross-cultural and multilevel analysis. Hum. Ecol. 36:97–109.

Guariguata MR, Ostertag R (2001) Neotropical secondary forest succession: changes in structural and functional characteristics. For Ecol Manag 148:185–206

Gullison, R. E., Frumhoff, P. C., Canadell, J. G., Field, C. B., Nepstad, D. C. and others (2007), 'Tropical forests and climate policy', Science, vol 316, pp. 985–986.

Hadley, M. and Lanly, J-P 1983. Tropical forest ecosystems: identifying differences seeking similarities. Nature and Resources 19: 1 (Jan-March): 2 - 19.

Hansen, M., S. Stehman, P. Potapov, T. Loveland, J. Townshend, R. Defries, K. Pittman, B. Arunarwati, F. Stolle, M. Steininger, M. Carroll, and C. DiMiceli. 2008. "Humid Tropical Forest Clearing from 2000 to 2005 Quantified by Using Multitemporal and Multiresolution Remotely Sensed Data." Proceedings of the National Academy of Sciences (PNAS) 105 (27): 9439–44.

Hansen P, Mertz O. (2006) Extinction or adaptation? Three decades of change in shifting cultivation in Sarawak, Malaysia. Land Degradation & Development.; 17:135–148. doi: 10.1002/ldr.720.

Hansen AJ, Burns P, Ervin J *et al.* (2020). A policy-driven framework for conserving the best of Earth's remaining moist tropical forests. Nature Ecology and Evolution, www.nature.com/natecolevol.

Hansen, M., S. Stehman, P. Potapov, T. Loveland, J. Townshend, R. Defries, K. Pittman, B. Arunarwati, F. Stolle, M. Steininger, M. Carroll, and C. DiMiceli. 2008. "Humid Tropical Forest Clearing from 2000 to 2005 Quantified by Using Multitemporal and Multiresolution Remotely Sensed Data." Proceedings of the National Academy of Sciences (PNAS) 105 (27): 9439–44.

Hatch, T., and Y.L. Tie. 1979. Shifting cultivation in Sarawak and its effects on soil fertility. Paper presented at a Seminar on the Fertility and Management of Deforested Land, Kota Kinabalu, Malaysia.

Hatch. T (1980). 'Shifting Cultivation in Sarawak: Past. Present and Future' in J L Furtado (ed.) Tropical Ecolog and Development. Proceeding of the Fifth International Symposium of Tropical Ecology. 16-21 April 1979. Kuala Lumpur. Malaysia. International Sociery of Tropical Ecology.

Hecht S. B. and Cockburn A. (2010), The fate of the forest: Developers, destroyers, and defenders of the Amazon. 2nd edition, University of Chicago Press, Chicago, IL, US.

HECHT, S.B. (1989) Indigenous soil management in the Amazon Basin. Some implications for development. In: Fragile Lands of Latin America. Browder, J.O. (ed.). Boulder, Colorado: Westview Press.

Hecht, S.B. (1993), 'The logic of livestock and deforestation in Amazonia', Bioscience 43: 687–695.

Heinimann A, Mertz O, Frolking S, Christensen AE, Humi K, Sedano F, Chini LP, Sahajpal R, Hansen M and Hurtt G, 2017. A global view of shifting cultivation: Recent, current, and future extent. PLoS ONE, 12 (9), https://doi.org/10.1371/journal.pone.0184479

Henley D. 2011. Swidden Farming as an Agent of Environmental Change: Ecological Myth and Historical Reality in Indonesia. Environment and History 17: 525–554.

Herrera, A. and Da Passano, M. G. (2006) Land Tenure Alternative Conflict Management, Food and Agriculture Organization of the United Nations, Rome

Higgins, G.M., Kassam, A.H., Naiken, L., Fisher, G. and Shah, M.M. (1982) Potential Population Supporting capacities of Land in Developing World.Technical Report (INT/75),p.13. FAO, Rome.

Hill, J., Stellmes, M., Udelhoven, T., Röder, A., and Sommer, S.(2008). Mediterranean desertification and land degradation: mapping related land use change syndromes based on satellite observations, Global Planet. Change, 64, 146–157, 2008.

Holden, S. T. (1993), 'The potential of agroforestry in the high rainfall areas of zambia: A peasant programming model approach,' Agroforestry systems, 24, 39–55.

Homma, A. K.; Walker, R. T.; Scatena, F. N.; Couto, A. J.; Carvalho, R. A.; Rocha, A. C. N.; Ferreira, C. A. P.; Santos, A. I. M. (1993). Dynamics of deforestation and burning in Amazonia: a microeconomic analysis. Rural Development Forestry Network, Network Paper 16c, p. 1-19.

Hong E (1990). 'Swidden as a Rational Agricultural System' in Return to the Good Earth, Damaging Effects of Modern Agriculture and the Case for Ecological Farming, Third World Network. Malaysia.

Houghton R.A., Boone R.D., Fruci J.R., Hobbie J.E., Melillo J.M., Palm C.A., Petersen B.J., Shaver G.R. and Woodwell G.M. 1987. The flux of carbon from terrestrial ecosystems to the atmosphere in 1980 due to changes in land use: geographic distribution of the global flux. Tellus 39B:122-139.

Houghton RA (2016). Deforestation. Biological and Environmental Hazards, Risks, and Disasters, 313-315p, https://doi.org/10.1016/B978-0-12-394847-2.00018-8.

Houghton, R. A., D. S. Lefkowitz,and D. L. Skole. 1991. Changes in the landscape of Latin America between 1850 and 1985. Progressive loss of forests. For. Ecol. Manage. 38:143–172.

Houghton, RA; Hackler, JL; Lawrence, KT (1999) _The U.S. carbon budget: Contributions from land-use change.' Science 285: 574–57

Houghton RA, Skole DL, Nobre CA, Hacker JL, Lawrence KT, Chometowski WH (2000) Annal fluxes of carbon from deforestation and regrowth in the Brazilian Amazon. Nature 403:301–304

Hunt, C.O., Rushworth, G., 2005. Cultivation and human impact at 6000 cal yr BP in tropical lowland forest at Niah, Sarawak, Malaysian Borneo. Quat. Res. 64, 460–468.

Hyndman D, Duhaylungsod L, Thomas B. To the last grain of rice: T'boli subsistence production. Dialectical Anthropology 1994; 19: 45-79.

Ickowitz, A. 2004. Shifting Cultivation and Deforestation: questioning the dominant narrative in tropical Africa, University of California, Riverside.

IFPRI (International Food Policy Research Institute). 2011. 2011 Global Hunger Index. Washington, DC: IFPRI. Available at http://www.ifpri.org/publication/2011-globalhunger- index.

IPCC (Intergovernmental Panel on Climate Change). 2001. Climate change 2001: The scientific basis. J.T. Houghton, Y. Ding, .D.J. Griggs, M. Nogues, P.J. van der Linden, K. Dai, *et al.* (eds.). Cambridge Univ. Press, Cambridge.

IPCC (2007a) Impacts, adaptation and vulnerability, Contribution of Working Group II to the Fourth Assessment Report of the Intergovernmental Panel on Climate Change. Cambridge, UK: Cambridge University Press

Itioka T, Takano K, Kishimoto K, Tzuchiya T, Ohshima Y, *et al.* (2014) Chronosequential changes in species richness of forest-edge-dwelling butterflies during forest restoration after swidden cultivation in a humid tropical rainforest region in Borneo. Journal of Forest Research, 20: 125-134, Doi, , 10.1007/s10310-014-0444-3

IWGIA 2009. The Indigenous World 2009. Copenhagen: IWGIA

Janssen, B.H. (1993). Integrated nutrient management: the use of organic and mineral fertilizers. In: van Reuler and Prins (The role of for sustainable food crop in decomposition of fumigated organisms in soil. Soil Biol. Biochem. 8:203-208.

Janzen D. 1998.Gardenification of wildland nature and the human footprint. Science 279: 1312–1313.

Jobbaggy, E. G., and Jackson, R. B. (2000). The Vertical Distribution of Soil Organic Carbon and its Relation to Climate and Vegetation. Ecological Applications 10: 423–436. doi:10.1890/ 1051-0761(2000)010[0423:TVDOSO]2.0.CO;2.

Johnson, D.L., Amrose, S.H., Bassett, T.J., Baver, M.L., Grumnaey, D.E., Issacson, A.E., Johnson, D.N., Lamb, P., Saul, M. and Winter-Nelson, A.E. (1997) Meaning of Environmental Terms. Journal of Environmental. Quality 26:581-589.

Jones, D., and R. V. O'Neill. 1992. "Endogenous Environmental Degradation and Land Conservation: Agricultural Land Use in a Large Region." Ecological Economics 6:79-101.

Jordan, C. F.1982. Amazon Rain Forests. American Scientist 70: 394 - 401.

Jordan, C.P. (1985). Nutrient cycling in tropical forest ecosystems. John Wiley & Sons, New York, USA. 190p.

Jyotshi, Amalendu, 2000. Swidden cultivation: A review of concepts and Issues. Institute for social and Economic Change, Bangalore, India.

Kahn, J.R and McDonald, J. A. (1995). Third-world debt and tropical deforestation. Ecological Economics 12: 107-123. Elsevier Science B.V.

Kang, B.T., Grimme, H., Lawson, T.L. 1985. Alley cropping sequentially cropped maize and cowpea with Leucaena on a sandy soil in southern Nigeria. Plant and Soil, 85,267-276.

Kant P, Kleine M, Acharya KP, Hossain MK and Lee DK (2007). Rehabilitating forests and extending tree cover in South Asia - A Synthesis, (In) Keep Asia Green, Volume III "South Asia" Lee DK (edn.), IUFRO World Series Volume 20-III. Vienna, p. 221.

Karamesouti, M., Detsis, V., Kounalaki, A., Vasiliou, P., Salvati, L., and Kosmas, C.: (2015) Land-use and land degradation processes affecting soil resources: evidence from a traditional Mediterranean crop land (Greece). Catena, 132: 45-55.

Karim, S.M.R., A.A. Mamun and N. Islam, 1999. Agro-ecology of major crops and their weeds in Bangladesh. Pak. J. Sci. Ind. Res., 42: 295-300.

Karim RSM and Mansor M, 2011. Impact of Jhum Cultivation on the Agro-ecology of Mountains and Socio-economy of Tribal peoples. Asian Journal of Agricultural Research, 5: 109-114, DOI: 10.3923/ajar.2011.109.114

Karim, S.M.R. and A.A. Mamun, 1988. Crop-weed competition: Analysis of some perspectives. Bangladesh J. Agric. Sci., 15: 65-73.

Karim, S.M.R. and S. Ahmed, 1997. Effect of degree of land preparation and spacing on weed growth and yield of transplanted amon rice. Bangladesh J. Agric. Sci., 24: 40-42.

Kerkhoff E and Sharma E, 2006. Debating Shifting Cultivation in the Eastern Himalayas Farmers' Innovations as Lessons for Policy, International Centre for Integrated Mountain Development (ICIMOD), Kathmandu, Nepal

Kendawang, J., J.; Tanaka, S.; Ishihara, J.; Shibata, K.; Sabang, J.; Ninomiya, I.; Ishizuka, S.; Sakurai, K., Adam Burma Ahmed Burma / International Journal of Research Publications (IJRP.ORG) www.ijrp.org (2004). Effects of Shifting Cultivation on Soil Ecosystems in Sarawak, Malaysia, Social Sci., Plant Nutr., 50(5), 677-687, 2004.

Ketterings, Q. M., van Noordwijk, M., & Bigham, J. M. (2002). Soil phosphorus availability after slash-and-burn fires of different intensities in rubber agroforests in Sumatra, Indonesia. Agriculture, ecosystems & environment, 92(1), 37-48.

Klanderud K, Mbolatiana HZH, Vololomboahangy MN, Radimbison MA, Roger E, Totland O and Rajeriarison C, 2010. Recovery of plant species richness and composition after slash-and-burn agriculture in a tropical rainforest in Madagascar. Biodivers Conserv. 19:187–204, DOI 10.1007/s10531-009-9714-3.

Kleinman, P. J. A., Pimentel D. & Bryant, R.B. (1995).The ecological sustainability of slashand-burn agriculture. Agriculture, Ecosystems and Environment, 52(2-3): 235-249.

Knudsen, J. L. and N. A. Khan (2002). An Exploration of the Problems and Prospects of Integrated Watershed Development in the CHT. In: N. A. Khan, M. K. Alam, S. K. Khisa and M. Millat-e-Mustafa (eds.) Farming Practices and Sustainable Development in the Chittagong Hill Tracts (Chittagong, Bangladesh: CHTDB and VFFP-IC), pp. 165-180.

Körner, C (2004) _Mountain biodiversity, its causes and function.' Ambio Supplement 13: 11–17

Kosmas, C., Detsis, V., Karamesouti, M., Kounalaki, K., Vassiliou, P., and Salvati, L.: Exploring long-term impact of grazing management on land degradation in the socio-ecological system of Asteroussia Mountains, Greece, Land, 4, 541–559, 2015.

Kotto-Same, J., Woomer, P.L., Moukam, A. and Zapfack, L. (1997). Carbon dynamics in slash andbum agriculture and land use alternative of the humid forest zone in Cameroon. Agriculture,

Kowal NE. 1966. Shifting cultivation, fire, and pine forest in the Cordillera Central, Luzon, Philippines. Ecological Monographs; 36: 389-419.

Kummer DM. Deforestation in the Postwar Philippines. Chicago: University of Chicago Press, 1992(a): 1-177.

Kummer DM. The political use of Philippine forestry statistics in the postwar period. Crime, Law and Social Change 1995; 22: 163-180.

Kummer DM. Upland agriculture, the land frontier and forest decline in the Philippines. Agroforestry Systems 1992(b); 18: 31-46.

Kummer, D.M., 1992. Upland agriculture, the land frontier and forest decline in the Philippines. Agroforestry Systems, 18, 31-46.

Lal, R., 1974. Soil erosion and shifting agriculture. In: FAO. (Eds.), Shifting Cultivation and Soil Conservation in Africa. FAO, Rome, pp. 49–71.

Lal, R., 1981. Soil erosion problems on Alfisols in western Nigeria, VI. Effects of erosion on experimental plots. Geoderma, 25:215-230.

Lal R (1989). Potential of agroforestry as a sustainable alternative to shifting cultivation: concluding remarks. Agrofor. Syst. 8: 239-242.

Lal, R. 1975. Role of mulching techniques in tropical soil and water management. Technical Bulletin No. I, International Institute for Tropical Agriculture, Ibadan, Nigeria.

Lal, R. (2001), 'Managing world soils for food security and environmental quality: review article', Advances in Agronomy 74: 155–192. 368.

Lal, R. 1999. Soil quality and food security concept for combating soil erosion. In: Soil Quality and Soil Erosion (eds: Lal, R.), pp. 309-317. Soil and Water Conservation Society, Ankeny, Iowa.

Lambin E F *et al.* 2001 The causes of land-use and land-cover change: moving beyond the myths Glob. Environ. Change 11 261–9

Lasco RD, Visco RG, Pulhin JM. (2001). Secondary forests in the Philippines: formation and transformation in the 20th century. J Tropical Forest Sci 2001; 13: 652-670.

Laurance, W. F., M. A. Cochrane, S. Bergen, P. M. Fearnside, P. Delamonica, C. Barber, S. D'Angelo, and T. Fernandes. 2001. Environment—The future of the Brazilian Amazon. Science 291:438–439.

Lawrence, Deborah 2005. Biomass Accumulation after 10-200 Years of Shifting Cultivation in Bornean Rain Forest. Ecology, 86(1):26–33

Lawrence, D.C., M. Leighton, and D.R. Peart. 1995. Availability and extraction of forest products in managed and primary forest around a Dayak village in West Kalimantan, Indonesia. Conservation Biology 9: 76-88

Lawrence, D., Radel, C., Tully, K., Schmook, B., & Schneider, L. (2010). Untangling a decline in tropical forest resilience: constraints on the sustainability of shifting cultivation across the globe. Biotropica, 42(1), 21-30.

Lawrence, D., Vester, H. F., Pérez-Salicrup, D., Eastman, J. R., Turner II, B. L., & Geoghegan, J. (2004). Integrated analysis of ecosystem interactions with land-use change: The southern Yucatán peninsular region. Ecosystems and land use change, 277-92.

Lawrence D, Coe M, Walker W, Verchot L and Vandecar K (2022). The Unseen Effects of Deforestation: Biophysical Effects on Climate. Front. For. Glob. Change, https://doi.org/10.3389/ffgc.2022.756115

Leanman, J.G and Sedjo, R.A. (1994). Global Forest: Issues for six billion people. Mc Graw-Hill, ING. USA.

Lele, S., Kurien, A., 2011. Interdisciplinary analysis of the environment: insights from tropical forest research. Environ. Conserv. 38, 211–233.

Lestrelin G and and Giordano M, (2007). Upland development policy, livelihood change and land degradation: Interactions from a Laotian village. Land Degradation and Development 18(1):55 – 76, DOI:10.1002/ldr.756

Li, Tania Murray. 1991. Culture, Ecology and Livelihood in the Tinombo Region of Central Sulawesi. Jakarta: Environmental Management Development in Indonesia Project. [EMDI Environmental Report 6.]

Liu, X; Chen, B (2000) _Climatic warming in the Tibetan Plateau during recent decades.' International Journal of Climatology 20: 1729–174

Lowry PP II, Schatz GE, Phillipson PB (1997) The classification of natural and anthropogenic vegetation in Madagascar. In: Goodman SM, Patterson BD (eds) Natural change and human impact in Madagascar. Smithsonian Institution Press, Washington, DC, pp 93–123

Madagascar Action Plan 2007-2012: A Bold and Exciting Vision for Rapid Development. FAO, Rome, Italy.

Majumdar B, Kumar K, Venkatesh M.S, Patiram, Bhatt BP (2004) Effect of different agroforestry systems on soil properties in acid Alfisols of Meghalaya. Journal Hill Research 17(1): 1.5.

Majumdar B, Venkatesh M S, Satapathy KK, Kumar K, Patiram (2002) Effect of alternative farming systems to shifting cultivation on soil fertility. Indian Journal of Agricultural Sciences 72(2): 122.124.

Malhi Y, Roberts JT, Betts RA and Killene TJ (2008). Climate Change, Deforestation, and the Fate of the Amazon. Science 319(5860):169-72, DOI:10.1126/science.1146961

Mantel, S., M. Mohiuddin, M.K. Alam, J.R. Olarieta, M. Alam and F.M.A. Khan, 2006. Improving the jhum system in Bangladesh. LEISA, 22: 20-21.

Mapa, R. and D. Kumaragamage. (1996). "Variability of soil properties in a tropical Alfisol used for shifting cultivation." Soil Technology. Vol. 9 (1996): 187-197.

Marcus, R. R. (2001). Seeing the forest for the trees: integrated conservation and development projects and local perceptions of conservation in Madagascar. Human Ecology, 29(4), 381-397.

Maxwell SL, Evans T, Morel A, Grantham H, *et al.* (2019). Degradation and forgone removals increase the carbon impact of intact forest loss by 626%. Science Advances, 5: DOI: 10.1126/sciadv.aax2546

Mazoyer, M. and Roudart, L. Eds. 2006. A history of world agriculture from the neolithic to the current crisis, pp. 101-102, Earthscan publishers, London.

McClanahan, T. R. and T. P. Young 1996. East African ecosystems and their conservation. Oxford University Press US

McKinley, T. and D. Alarc´on (1995), 'The prevalence of rural poverty in mexico,' World Development, 23(9), 1575–1585.

Megevand C, Mosnier A, Hourticq J, Sanders K, Doetinchem N, & Streck C (2013). Deforestation Trends in the Congo Basin: Reconciling Economic Growth and Forest Protection. Washington: World Bank. ISBN 978-0-8213-9742-8 DOI:10.1596/978-0-8213-9742-8.

Mertz, O., Leisz, O., Heinimann, A., Rerkasem, K., Thiha, Dressler, W., Pham, V.C., Vu, K.C., Schimdt-Vogt, D., Colfer, C.J.P., Epprecht, M., Padoch, C., Potter, L., 2009b. Who counts? Demography of Swidden cultivators in southeast Asia. Hum. Ecol. 37, 281–289.

Mertz, Ole, Reed L. Wadley, Uffe Nielsen, Thilde B. Bruun, Carol J.P. Colfer, Andreas de Neergaard, Martin R. Jepsen, Torben Martinussen, Qiang Zhao, Gabriel T. Noweg and Jakob Magid. 2008. 'A Fresh Look at Shifting Cultivation: Fallow Length an Uncertain Indicator of Productivity'. Agricultural Systems 96: 75–84.552

Mertz, O. (2009). Trends in shifting cultivation and the REDD mechanism. Current Opinion in Environmental Sustainability, 1, 156–160.

Mertz, O., Birch-Thomsen, T., Elberling, B., Rothausen, S., Bruun, T. B., Reenberg, A., Fog, B., Egsmose, R. M. and Breuning-Madsen, H. (2012) 'Changes in shifting cultivation systems on small Pacific islands', The Geographical Journal 178(2), pp175-187, DOI: 10.1111/j.1475-4959.2011.00447.x

Mertz, O., C. Padoch, J. Fox, R. A. Cramb, S. J. Leisz, N. T. Lam, and T. D. Vien. 2009. Swidden change in Southeast Asia: Understanding causes and consequences. Hum. Ecol. 37(3):259–264.

Michon G. (ed., 2005). Domesticating forests: How farmers manage forest resources. IRD/ Cifor/Icraf, Bogor.

Michon, G. 1991. The damar gardens: existing buffer zones at Pesisir area of Sumatra Selatan National Park, Lampung. Symposium on rain forest protection and national park buffer zones. Ministry of Forestry, Government of Indonesia, Manggala Wanabakti, Jakarta, Indonesia.

Miller PM, Kauffman JB (1998) Seedling and sprout response to slash-and-burn agriculture in a tropical deciduous forest. Biotropica 30:538–546

Miracle, M. (1967). Agriculture in the Congo Basin. Madison: University of Wisconsin Press, 1967.

Mishra, B. P., Tripathi, O. P., Tripathi, R. S., & Pandey, H. N. (2004). Effects of anthropogenic disturbance on plant diversity and community structure of a sacred grove in Meghalaya, northeast India. Biodiversity and Conservation, 13(2), 421–436.

Mittermeier, R. A., C. G. Mittermeier, T. M. Brooks, J. D. Pilgrim, W. R. Konstant, G. A. B. da Fonseca, and C. Kormos. 2003. Wilderness and biodiversity conservation. Proc. Natl. Acad. Sci. USA 100:10309–10313.

MoEF (2006): Ministry of Environment and Forests Report of the National Forest Commission. Government of India, Ministry of Environment and Forests. website: www.envfor.nic.in/divisions/nfr.html.

Mohsin A 1997. The Politics of Nationalism: The Case of Chittagong Hill Tracts, Bangladesh,The University Press Limited, Dhaka, Bangladesh

Moorman FR and Kang BT, 1978. Microvariability of soils in the tropics and its agronomic implications with special reference to West Africa: In diversity of soils in the tropics.

Moraes, J.F.L. de, B. Volkoff, C. Cerri, and M. Bernoux (1996), 'Soil properties under Amazon forest and changes due to pasture installation in Rondˆonia, Brazil', Geoderma 70: 63–81.

Moran, E. F. 1981. Developing the Amazon. Indiana University Press: Bloomington.

Moran EF, Brondizio E, Mausel P, Wu Y. 1994. Integrating Amazonian vegetation,land-use,and satellite data. BioScience 44:329–338.

Morton O, Borah JR and Edwards DP, 2020. Economically viable forest restoration in shifting cultivation landscapes. Environmental Research Letters, 15: 1-12, 064017 https://doi.org/10.1088/1748-9326/ab7f0dMorton et al. (2020)

Mukul S.A. and Herbohn J. 2016. The impacts of shifting cultivation on secondary forests dynamics in tropics: A synthesis of the key findings and spatio temporal distribution of research. Environmental Science & Policy 55: 167–177

Mulongoy, K., Kunda, K.N. and Chiang, C.N.K. (1992). Effect of aUey cropping and fallowir some soil fertility parameters in southern Nigeria. In: Mulongoy and Merck (eds.). Soil orl matter dynamics and sustainability of tropical agriculture. Proceedings of an internat symposium organised by K.U. Leuven and IITA, 4-6 November 1991, Leuven, Belgium.

Mulyoutami, E., van Noordwijk, M., Dewi, S. and Agus, F. (2010) Perubahan Pola Perladangan: Perubahan Persepsimengenai Para Peladang di Indonesia (Shifting Patterns Change: Shifting Perceptions about the Cultivators of Indonesia), World Agroforestry Centre, Bogor, Indonesia

Myers N. 1989. Deforestation rates in tropical forests and their climatic implications. London: Friends of the Earth.

Myers, N. Ed. 1991. Tropical forests and climate, pp. 3-21, Kluwer Publishers, London, UK.

Myers N. 1993. Tropical forests: The main deforestation fronts.Environmental Conservation 20:9–16.

Myers N, Mittermeier RA, Mittermeier CG, da Fonseca GAB, Kent J. Biodiversity hotspots for conservation priorities. Nature 2000; 403: 853-858.

Nadkarni. M Vand V Govindaru (1995). 'Nobody's Child: The Economic and Institutional Aspects of Soil Conservation in India', Journal of Environment and Development. Vol. 4, No. I.

Nietschmann, B. 1973. Between Land and Water. New York and London: Seminar Press.

Nilsson, S. (1996). Do we have enough forests? IUFRO Occasional Papers No5. Wien, Austria.

Noble IR, Dirzo R. Forests as human-dominated ecosystems. Science 1997; 277: 522-525.

Nogues–Bravo, D; Araujo, MB; Errea, MP; Martınez–Rica, JP (2007) _Exposure of global mountain systems to climate warming during the 21st Century.' Global Environmental Change 17: 420–428

Nounamo, L. and M. Yemefack, M. 2000. Shifting Cultivation in the Evergreen Forest of Southern Cameroon: Farming Systems and Soil Degradation, Final Report, Tropenbos-Cameroon Reports 00-2, The Tropenbos-Cameroon Programme, Kribi, Cameroon.

Nounamo, L. and Yemefack, M. (1997). Shifting cultivation in the evergreen forest of Sou Cameroon: Farming systems results. Lu2 interim report. The Tropenbos-Cameroon PrograJ Kribi, Cameroon. 27pp.

Nye, P.H.; Greenland, D.J. 1960. The Soil Under Shifting Cultivation. Technical Communication no. 51. Commonwealth Bureau of Soils, Harpenden, Commonwealth Agricultural Bureaux, England

Ofori, C.S., Higgins, G.M. and Purnell, M.F. (1986). Criteria for choice of land suitable for clearing for agricultural production. In: Lal, R., Sanchez, P.A. and Cummings, R.W. Jr (eds). Land Clearing and Development in the Tropics. A.A. Balkema, Rotterdam: 19-28

Okigbo, B. N. 1982. Impact of agricultural systems and rural development on Nigerian forests. In MacDonald, L. H. in the African Humid Tropics. Tokyo: United Nations University. pp. 41 - 45.

Om Parkash, 1980. Studies on the ecological impact of shifting agriculture (jhum) on forested ecosystem. Thesis submitted to North-Eastern Hill University, Shillong, Meghalaya India.

OTA 1984. Farnworth, E. and Golley, F. (eds.). 1974. Fragile Ecosystems: Evaluation of Research and Applications in the Neotropics. New York: Springer-Verlag.

Otsuka, K. and Place, F. (2001) Land Tenure and Natural Resource Management: A Comparative Study of Agrarian Communities in Asia and Africa, Johns Hopkins University Press Baltimore, MD

Ohmagari, K. & Berkes, F. (1997). Transmission of indigenous knowledge and bush skills among the Western James Bay Cree women of subarctic Canada. Human Ecology 25: 197-222

Okigbo, B. N. 1982. Impact of agricultural systems and rural development on Nigerian forests. In MacDonald, L. H. in the African Humid Tropics. Tokyo: United Nations University. pp. 41 - 45.

Olson, D., and E. Dinerstein. 2002. "The Global 200: Priority Ecoregions for Global Conservation." Annals of Missouri Botanical Garden 89 (2): 199–224.

Oxford Advanced Learners Dictionary. (2005). Oxford University Press, Great Claredon Street, Oxford, UK.

Pacheco, P. 2009. Smallholder livelihoods, wealth and deforestation in the Eastern Amazon. Human Ecology 37: 27-41.

Padoch. C (1980). 'The Environmental and Demographic effects of Alternative Cash- Producing Activities Among the Shifting Cultivators of Sarawak' in I L Furtado (ed.) Tropical Ecoloy and Development. Proceeding of the Fifth International Symposium of Tropical Ecology. 16-2 I April 1979. Kuala Lumpur. Malaysia, International Socieo/ of Tropical Ecology. Kuala lumpur.

Padoch C and Peters C 1993. "Managed Forest Gardens in West Kalimantan." In C, Potter, J. Cohen, D. Janezewski, (eds). Perspectives on Biodiversity: Case Studies of Genetic Resource Conservation and Development. AAAS Press, Washington DC pp. 167-176.

Padoch, C., Coffey, K., Mertz, O., Leisz, S., Fox, J., Wadley, R.L., 2007. The demise of swidden in Southeast Asia? Local realities and regional ambiguities. Geografisk Tidsskrift-Danish Journal of Geography 107, 29–41.

Palm CA, Swift MJ,Woomer PL. 1996. Soil biological dynamics in slash- and -burn agriculture. Agriculture, Ecosystems and Environment 58: 61–74.

Palm, C. A.; Houghton, R. A.; Melillo, J. M.; Skole, D. L. (1986). Atmospheric Carbon Dioxide from Deforestation in Southeast Asia. Biotropica, v. 18, n. 3, p. 177-188,.

Palm, C. A., Vosti, S. A., Sanchez, P. A., and Ericksen, P. J (Eds). 2005. Slash-and-burn agriculture: The search for alternatives. New York, Columbia University Press.

Parry, L., B. Day, S. Amaral, and C. A. Peres. 2010. Drivers of rural exodus from Amazonian headwaters. Population Environ. 32:137–176.

Pascual U. and Barbier E, 2001. A Model of Optimal Labour and Soil Use with Shifting Cultivation. Fondazione Eni Enrico Mattei Corso Magenta, 63, 20123 Milano, Italy.

Paul SP, Hossain ATME (2001). Agroforestry research and development in the hill region of Bangladesh: experiences, problems and research needs. Proceedings of the National Workshop on 'Agroforestry Research and Development in Bangladesh. BARI, Gazipur.

Pedelahore, P. 2012. Stratégies d'accumulation des exploitants agricoles : L'exemple des cacaoculteurs du Centre Cameroon de 1910 à 2010. PhD dissertation. Toulouse: Université Toulouse II Le Mirail, Géographie-Aménagement, Mention Etudes Rurales.

Perrings, C. 1989. 'An optimal path to extinction? Poverty and resource degradation in the open agrarian economy,' Journal of Development Economics, 30, 1–24.

Persson, R. (1983). Forestry in Laos. Swedish International Development Co¬operation Agency (SIDA) 70p. Stockholm, Sweden.

Phongoudome, C., Park, P.S., Kim, H., Sawathvong, S., Park, Y.D., Combalicer, M.S., Ho, H.M., 2013. Changes in stand structure and environmental conditions of a mixed deciduous forest after logging and shifting cultivation in Lao PDR. Asia Life Sci. 22, 75–94.

Phuntsho, K; Aryal, KP; Kotru, R 2015. Shifting cultivation in Bangladesh,Bhutan and Nepal: Weighing government policies against customary tenure and institutions. ICIMOD Working Paper 2015/7. Kathmandu: ICIMOD

Pimentel, D., 2006. Soil erosion: a food and environmental threat. Environ. Dev. Syst. 8, 119–137.

Pimentel, D., Harvey, D., Resosudarmo, P., Sinclair, K., Kurz, D., McNair, M., Crist, S., Shpritz, L., Fitton, L., Saffouri, R., Blair, R., 1995. Environmental and economic costs of soil erosion and conservation benefits. Science 267, 1117–1123.

Piotto, D., Montagnini, F., Thomas, W., Ashton, M., Oliver, C., 2009. Forest recovery after swidden cultivation across a 40-year chronosequence in the Atlantic forest of southern Bahia, Brazil. Plant Ecology, 205, 261-272.

Poffenberger, M. 1990. Keepers of the Forest. Kumarian Press, West Hartford, Connecticut.

Pollini, Jacques. 2015. Shifting Cultivation, Gender and REDD+ in Cameroon and the Democratic Republic of Congo. USAID-supported Forest Carbon Markets and Communities (FCMC) Program. Washington, DC, USA.

Poulsen, G. 1978. Man and tree in tropical Africa: three essays on the role of trees in the African environment. Ottawa, Ont.: IDRC.

Prasad RN, Singh A, Verma A. 1986. Problem of Hill lands and their Management in North Eastern India. Journal of Soil Conservation; 14:66-72.

Prasad, V. K.; Kant, Y.; Gupta, P. K.; Sharma, C.; Mitra, A. P.; Badarinath, K. V. S. 2001. Biomass and combustion characteristics of secondary mixed deciduous forests in Eastern Ghats of India. Atmospheric Environment, v. 35, p. 3085–3095.

Prat, C. 1990. Relation entre erosion et systemes de production dans le bassin-versant sud dl Managua (Nicaragua). Cah. Orstom, ser. pedol., vol. XXV, no 1-2, 171-182.

Purba, C., Nanggara, S. G., Ratriyono, M., Apriani, I., Rosalina, L., Sari, N. A. and Meridian, A. H. 2014. Potret Keadaan Hutan Indonesia Tahun 2009-2013 (State of the Forest, Indonesia, Period of 2008-2013), Forest Watch Indonesia, Bogor, Indonesia

Pyne, SJ; 1982. Fire in America: A Cultural History of Wildland and Rural Fire. University of Washington Press, ISBN: 9780295975924

Raharimalala, O., Buttler, A., Ramohavelo, C.D., Razanaka, S., Sorge, J.P., Gobat, J.M., 2010. Soilvegetation patterns in secondary slash and burn successions in Central Menabe, Madagascar. Agriculture, Ecosystems & Environment, 139, 150–158.

Raharimalala, O., Buttler, A., Schlaepfer, R., Gobat, J.M., 2012. Quantifying biomass of secondary forest after slash-and-burn cultivation in central Menabe, Madagascar. J. Trop. For. Sci. 24, 474–489.

Ramakrishnan, P. S. 1992. Shifting agriculture and sustainable development: An interdisciplinary study from north-eastern India. Carnforth, Parthenon Publishing Group and UNESCO.

Raman, T. R. S., Rawat, G. S. and Johnsingh, A. J. T. 1998. Recovery of tropical rainforest avifauna in relation to vegetation succession following shifting cultivation in Mizoram, North-East India. Journal of Applied Ecology 35: 214-231.

Rambo AT.1996. The composite swiddening agroecosystem of the Tay ethnic minority of the northwestern mountains of Vietnam. Pages 69–89 in Rerkasem B, ed. Montane Mainland Southeast Asia in Transition. Chiang Mai (Thailand): Chiang Mai University Consortium.

Randrianarison, A., Schlaepfer, R., Mills, R., Hervé, D., Razanaka, S., Rakotoarimanana V., Buttler, A. 2016. Linking historical land use to present vegetation and soil characteristics under slash-and-burn cultivation in Madagascar. Applied Vegetation Science, 19(1), 40–52. https://doi.org/10.1111/ avsc.12202

Rao Y. 1989. Forest resources of tropical Asia. Pages 1–20 in Environment and Agriculture—Environmental Problems Affecting Agriculture in the Asia-Pacific Region. World Food Day Symposium; 11 October 1989; Bangkok (Thailand): Food and Agriculture Organization of the United Nations.

Read, L., Lawrence, D., 2003. Recovery of biomass following shifting cultivation in dry tropical forests of the Yucatan. Ecological Applications, 13, 85-97.

Reardon, T. and S. A. Vosti (1995), 'Links between rural poverty and the environment in developing countries: Asset categories and investment poverty,' World Development, 23(9), 1495–1506.

Renard, K.G., Foster, G.R., Weesies, G.A. and Porter, J.P., 1991. RUSLE: revised universal soil loss equation. J. Soil Water Conserv., 46(1): 30-33.

Repetto, R. 1988. The forest for the trees? Government policies and the misuse of forest resources. World Resources International, Washington, D.C.

Richards, P. W. 1973 b. Africa, the " Odd Man Out". In Meggers, B. J., Ayensu, E. S. and Duckworth, W. D. (eds.) Tropical Forest Ecosystems in Africa and South America: A comparative review. Washington, D. C.: Smithsonian Institution Press. pp. 21 - 26.

Richards, P. 1977. Ideas, environment and agricultural changes: a case study from western Nigeria. Ph. D. thesis, University of London.

Riswan, R., and L. Hartanti. 1995. Human impacts on tropical forest dynamics. Vegetatio 121: 41-52.

Roder, W., Phengchanh, S., Maniphone, S., 1997. Dynamics of soil and vegetation during crop and fallow period in slash-and-burn fields of northern Laos. Geoderma 76, 131–144.

Rudel, T., O. T. Coomes, E. F. Moran, F. Achard, A. Angelsen, J. Xu, and E. Lambin. 2005. Forest transitions: Towards a global understanding of land use change. Global Environ. Change 15:23–31.

Rudel, T.K., R.S. Defries, G.P. Asner, and W.F. Laurance. 2009. Changing drivers of deforestation and new opportunitiesfor conservation. Conservation Biology 23: 1396-1405.

Russell WMS. (1988). Population, swidden farming and the tropical environment. Population and Environment; 10: 77-94.

Ruthenberg, Hans.(1980). Farming Systems in the Tropics: Oxford: Clarendon Press, 1980

Saha R, Ghosh PK, Mishra VK, Majumdar B, Tomar JMS (2010) Can agroforestry be a resource conservation tool to maintain soil health in the fragile ecosystem of north-east India? Outlook on Agriculture 39(3): 191-196.

Saha R, Mishra VK (2007) Long term effect of various land use systems on physical properties of silty clay loam soil of NE hills. Journal of the Indian Society of Soil Science 55(2): 112-118.

Saha R, Tomar JMS, Ghosh PK 2007. Evaluation and selection of multipurpose tree for improving soil hydro-physical behaviour under hilly eco-system of north east India. Agroforestry Systems 69(3): 239.247.

Sajise PE, Orlido NM, Castillo LC, Lales JS and Atabay RC. (1976). Ecology of Philippine Grassland I. Floristic composition and community dynamics. Philippine Agriculturist 1976; 59:317-334.

Sajise PE, Tapay N, Pacardo EP, Briones ND, Jimenez R, Zamora P, Fortes M, Gomez E, Zafaralla M, Zosa-Feranil I. 1992. Saving the Present for the Future: the State of the Philippine Environment. Quezon City: UP Center for Integrative and Development Studies, 1992: 1-64.

Sala EO, Chapin FS III, Armesto JJ, Berlow E, Bloomfield J, Dirzo R, Huber-Sandwald E, Huenneke LF, Jackson RB, Kinzig A, Leemans R, Lodge DM, Mooney HA, Oesterheld M, LeRoy Poff N, Sykes MT, Walker BH, Walker M, Wall DH (2000) Global biodiversity scenarios for the year 2100. Science 287:1770–1774

Salafsky, N. 1994. Forest gardens in the Gunung Palung region of West Kalimantan, Indonesia. Agroforestry Systems 28: 237-268.

Sanchez PA. 1994. Alternative to slash and burn: a pragmatic approach for mitigating tropical deforestation. In Agricultural Technology: Policy Issues for the International Community, Anddrson JR (ed.). CAB international: Wallingford; 451–479.

Sanchez PA 1976. Propertie and management of soils in the tropics. John Wiley and Sons, New York, USA.

Sanchez P.A. 1987. Management of acid soil in the humid tropics of Latin America. In: Sanchez P.A., Stoner E.R. and Pushparajah E. eds. Management of acid tropical soils for sustainable agriculture. IBSRAM Proceedings 2:63-107. Bangkok, Thailand: IBSRAM.

Sanchez P.A., Palm C.A. and Smyth T.J. 1990. Approaches to mitigate tropical deforestation by sustainable soil management practices. In: Schapenseel H.W., Schomarker M. and Ayoub A. eds. Soils on a warmer earth. Developments in Soil Science 20:211-220. Amsterdam: Elsevier.

Sanchez, P. (1996), 'Introduction,' Agriculture, Ecosystems and Environment, 58(1), 1–2.

Sanchez, P.A. (1977). Soil management under shifting cultivation. In: Sanchez P.A. (eds.). Revit soils research in tropical Latin America. Tech. Bull. no. 219. North Carolina State Univel Raleigh, USA.

Sánchez, P.A., C.A. Palm, S.A. Vosti, T.P. Tomich, and J. Kasyoki. 2005. Alternatives to slash and burn: Challenge and approaches of an international consortium. In Slash-and-burnagriculture: The search for alternatives, edited by C.A. Palm, S.A. Vosti, P.A. Sanchez, and P.J. Ericksen. New York: Columbia University Press, 3-37.

Sanchez, P.A., D.A. Bandy, J.H. Villachica, and J.J. Nicholaides (1982), 'Amazon Basin soils: management for continuous crop production', Science 216: 821–827.

Scarascia-Mugnozza, G., Oswald, H., Piussi, P., and Radoglou, K.: (2000). Forests of the Mediterranean region: gaps in knowledge and research needs, Forest Ecol. Manag., 132, 97–109.

Scherrer, D; Körner, Ch (2011) _Topographically controlled thermal–habitat differentiation buffers alpine plant diversity against climate warming.' Journal of Biogeography 38(2): 406–416

Schlesinger WH, "Carbon sequestration in soils," Science, vol. 284, no. 5423, p. 2095, 1999.

Schmidt-Vogt, D. 1999. Swidden farming and fallow vegetation in Northern Thailand. Wiesbaden, Germany, Franz Steiner Verlag.

Seavoy, R.E. 1973. The transition to continuous rice cultivation in Kalimanthan. Annual Asso. Amer. Geography, 63: 218-225.

Seiler, W. and Crutzen, P.J. (1980) Estimates of Gross and Net Fluxes of Carbon between the Biosphere and the Atmosphere from Biomass Burning. Climatic Change, 2, 207-247.

Serajuddin, A. M. (1971). The Origin of the Rajas of the Chittagong Hill Tracts and Their Relations with the Mughals and the East India Company in the Eighteenth Century. Journal of the Pakistan Historical Society, Vol. 19, pp. 52-60.

Serrao, E.A.S. and A.K.O. Homma (1993), 'Country profiles: Brazil', in Sustainable Agriculture and the Environment in the Tropics, Washington, DC: National Academy Press for the National Research Council.

Serrão, E. A.; Nepstad, D.; Walker, R. T. (1996). Upland agricultural and forestry development in the Amazon: sustainability, criticality and resilience. Ecological Economics, v. 18, p. 3-13.

Shah, T., Roy, A., Qureshi, A., andWang, J. (2003). Sustaining asias groundwater boom: An overview of issues and evidence. In Natural Resources Forum, 27, pages 130{141, Wiley Online Library.

Sharma, U. C. & Prasad, R. N. (1995) Socio-economic aspects of acid sooil management and alternate land use systems for north-eastern states of India. In: Plant-Soil Interactions at Low pH. (ed. by R. A., Date; N. J. Grundon.; G. E. Rayment, . & M. E. Probert), 689-696, Kluwer Academic Press, The Netherlands.

Sharma U. C. and Vikas Sharma (2005). Traditional wisdom of tribal people in integrated water resources management and environmental quality – A case study from NE India. In MPMD – 2005, Monitorin, Prediction and Mitigation of Water Related Disasters (ed. by K. Takara *et al.*). Kyoto, Japan, pp. 709-712.

Sharma TC (1976). The pre-historic background of shifting cultivation in North-East India. North-East Indian Council for Social Science Research.

Sharma UC and Bujarbaruah KM 2016. Issues and Challenges in Shifting Cultivation and its Relevence in the Present Context, National Academy of Agricultural Sciences, Policy Paper No. 83.

Sharma UC, Prasad RN (1994) Potential indigenous farming systems of North Eastern Hill region.ICAR Research Complex for NEH Region, Shillong. 48p.

Sharma UC, Prasad RN, Sonowal DK (1994) An indigenous technique of soil and water conservation in north eastern region- The Zabo system of farming. Soil and water conservation challenges and opportunities. In: L.S. Bhushan, I.P. Abrol and M.S. Rama Mohan Rao (eds.) Proceeding of 8th ISCO Conference. Oxford and IBH Publication Co. Pvt. Ltd., New Delhi (India). pp. 969-975.

Sharma, U.C. and Sharma, Vikas (2003). The 'Zabo' soil and water management and conservation system in northeast India : tribal beliefs in the development of water resources and their impact on society- an historical account of a success story. In: The Basis of Civilization-water Science (edn.J.C. Rodda and L/Ubertini). IAHS Publ.286,pp.184-191, Rome, Italy.

Sharon, M. F and Kenneth A. F. (1988). Deforestation in South-east Asia, The Asian Forum of environmental Journalist and ESC,4P, Bangkok, Thailand. Shifting Cultivation Practices By Hmong, Khamu and Lao ethnic categories In The Nam Nane Watershed, Nane District, Luang Prabang Province, Lao PDR 20040530 68

Shearman, P.L., J. Ash, B. Mackey, J.E. Bryan, and B. Lokes. 2009. Forest conversion and degradation in Papua New Guinea 1972-2002. Biotropica 41: 379-390.

Shoaib J.U., Mostafa G. and Rahman M. (1998): Soil erosion hazard in Chittagong Hill Tracts: a case study. Annual report. Soil Resources Development Institute, Dhaka.

Shrestha, AB; Wake, CP; Dibb, JE; Mayewski, PA (2000) _Precipitation fluctuations in the Nepal Himalaya and its vicinity and relationship with some large scale climatological parameters.‘ International Journal of Climatology 20: 317–327

Shrestha, AB; Wake, CP; Mayewski, PA; Dibb, JE (1999) _Maximum temperáture trends in The Himalaya and its vicinity: An analysis based on temperature records from Nepal for the period 1971–94.‘ Journal of Climate 12(9): 2775–2789

Shukla, U.C. and Agrawal, R.P., 1984. The effect of shifting cultivation on physico-chemical properties of soils in tropical Africa and methods for their improvement--a review. In: A.H. Bunting and E. Bunting (Editors), The Future of Shifting Cultivation in Africa and the Task of Universities, Proc. Int. Workshop on Shifting Cultivation: Teaching and Research at University Level, Ibadan, Nigeria. FAO, Rome, pp. 64-76.

SIDA (2009) Natural Resource Tenure, position paper for SIDA, Swedish International Development Cooperation Agency, Stockholm

Silva JMN, Carreiras JMB, Rosa I and Pereira JMC, 2011. Greenhouse gas emissions from shifting cultivation in the tropics, including uncertainty and sensitivity analysis. Journal of Geophysical Research, 116: D20304, doi:10.1029/2011JD016056, 21p.

Singh, K. D. and Marzoli, A., Paper presented at the World Wildlife Fund Conference on the Potential Impact of Climate on Tropical Forests, San Juan, Puerto Rico, pp. 2–3.

Singh SK and Parihar A (2015). Strategies for combating land degradation in India. Journal of Agroecology and Natural Resource Management, 2: 360-364, p-ISSN: 2394-0786, e-ISSN: 2394-0794.

Singh A and Singh MD, “Effect of various stages of shifting cultivation on soil erosion from steep hill slopes,” Indian Forester, vol. 106, no. 2, pp. 115–121, 1981.

Singh J.S. and Singh S.P. (1992): Forests of Himalaya, Gyanodaya Prakashan, Nainital, p. 294. Singh and Singh 1992).

Siren, A. (1997). Saving the tropical rainforests: From what and how? Department of Rural Development Studies, Working paper No1;1997. SLU. Sweden.

Smith J, van de Kop P, Reategui K, Lombardi I, Sabogal C, Diaz A (1999) Dynamics of secondary forests in slash-and-burn farming: interactions among land use types in the Peruvian Amazon. Agric Ecosyst Environ 76:85–98

Sodhi, N.S., Koh, L.P., Clements, R., Wanger, T.C., Hill, J.K., Hamer, K.C., Clough, Y., Tscharntke, T., Posa, M.R.C., Lee, T.M., 2010. Conserving Southeast Asian forest biodiversity in human-modified landscapes. Biological Conservation, 143, 2375–2384.

Stearman, A.M. (1983) Forest to pasture: frontier settlement in the Bolivian Lowlands. In: The Dilemma of Amazonian Development. Moran, E.F. (ed.). Boulder, Colorado: Westview Press.

Styger E, Rakotondramasy HM, Pfeffer MJ, Fernandes ECM, Bates DM (2007) Influence of slash-and-burn farming on fallow succession and land degradation in the rainforest region of Madagascar. Agric Ecosyst Environ 119:257–269

Suarez R.K. and Sajise P. E. 2010. Deforestation, Swidden Agriculture and Philippine Biodiversity. Philippine Science Letters, 3: 91-99

Swaminathan. (1987). The promise of agroforestry for ecological and nutritional security. In: Steppler, H. A and P.K.R. Nair (eds), Agroforestry: A decade of development. International Council for Research in Agroforestry (ICRAF). Nairobi, Kenya.

Swamy SL, Puri S, Singh AK (2003) Growth, biomass, carbon storage and nutrient distribution in Gmelina arborea Roxb. stands on red lateritic soils in central India. Bioresource technology 90(2): 109-126.

Szott L.T. and Palm C.A. 1986. Soils and vegetation dynamics in shifting cultivation fallows. In: I Simposio do Tropico Umido 1:360-379. Belem, Brazil: EMBRAPA.

Szott, L.T., C.A. Palm, and R.J. Buresh. "Ecosystem fertility and fallow function in the humid and subhumid tropics." Agroforestry Systems. Vol. 47 (1999): 163-196.

Tanaka, S., Kendawang, J.J., Ishihara, J., Shibata, K., Kou, A., Jee, A., Ninomiya, I., Sakurai, K., 2004. Effects of shifting cultivation on soil ecosystems in Sarawak, Malaysia II. Changes in soil chemical properties and runoff water at Balai Ringin and Sabal experimental sites. Soil Sci. Plant Nutr. 50, 689–699.

Tanaka, S., Kendawang, J.J., Yoshida, N., Shibata, K., Jee, A., Tanaka, K., Ninomiya, I., Sakurai, K., 2005. Effects of shifting cultivation on soil ecosystems in Sarawak, Malaysia—IV. Chemical properties of the soils and runoff water at Niah and Bakam experimental sites. Soil Sci. Plant Nutr. 51, 525–533.

Tchoundjeu Z., Avana M-L., Simons A.J., Dawson I., Were J. and Jeanicke H. 1999. Domestication of Prunus africana: the contribution of vegetative propagation. Poster American Society AgronomyASB Symposium, Salt Lake City, USA, November, 1999.

Teegalapalli K., Gopi, G.V. and Prasanna K. Samal. 2009. Forest recovery following shifting cultivation: an overview of existing research. Tropical Conservation Science Vol.2(4):374-387.

Thein, H. M., Kanzaki, M., Fukushima, M., Minn, Y., & others. (2007). Structure and composition of a teak-bearing forest under the Myanmar selection system. Southeast Asian Studies 45: 303–316.

Thiele, G. (1991) Amazonian frontier development: the barbecho crisis revisited. Santa Cruz, Bolivia: International Centre for Tropical Agriculture [unpublished report].

Thomas, W. L. (ed.) (1956) Man's Role in Changing the Face of the earth. Man's Role in Changing the Face of the earth, The University of Chicago Press, Chicago and London.

Thrupp L. A., Hecht S. B. and Browder J. O. (1997), The diversity and dynamics of shifting cultivation: Myths, realities, and policy implications. World Resources Institute, Washington, DC.

Tilman, D., Balzer, C., Hill, J., and Befort, B. (2011). Global food demand and the sustainable intensification of agriculture. Proceedings of the National Academy of Sciences.

Tinker, P.B.; Ingram, J.S.I.; Struwe, S. Effects of slash-andburn agriculture and deforestation on climate change. Agr. Ecosyst. Environ. 1996, 58, 13–22.

Tivy, J. 1987. Nutrient cycling in agro-ecosystems. Applied Geography 7: 93 - 113.

Toky O.P. and Ramakrishnan P.S. 1981. Cropping and yields in agricultural systems of the northeastern hill region of India. Agro Ecosystems 7:11-25.

Toky, O. P. and Ramakrishnan, P. S. 1983. Secondary succession following slash and burn agriculture in north-eastern India: I. Biomass, litterfall and productivity. Journal of Ecology 71: 735-745.

Tomich, T.P., D.E. Thomas, and M. van Noordwijk (2004), 'Environmental services and land use change in Southeast Asia: from recognition to regulation or reward?', Agriculture, Ecosystems and Environment 104: 229–244.

Troeh, F.R. and Thompson, L.M. (2005) Soils and Soil Fertility. Sixth Edition, Blackwell, Ames, Iowa, 489.

Tubiello, F. N., Salvatore, M., Rossi, S., Ferrara, A., Fitton, N., and Smith, P. (2013). The FAOSTAT database of greenhouse gas emissions from agriculture. Environmental Research Letters, 8(1), [015009]. https://doi.org/10.1088/1748-9326/8/1/015009

Tulaphitak, T.; Pairintra, C.; and Kyuma, K. 1985. Changes in Soil Fertility and Tilth under Shifting Cultivation II: Changes in Soil Nutrient Status. Soil Science and Plant Nutrient, 31: 239-249.

Turner BL, Moss RH and Skole DL.1993. Relating land use and global landcover change: A proposal for an IGBP-HDP core project. International Geosphere-Biosphere Programme: A study of global change.

Uddin, M.S., M.S. Kamal and M.H. Mollah, 2000. Hill farming system and resource utilization in the Chittagong hill tracts: A base line survey. Hill Agricultural Research Station, Khagrachari, Research Report, pp: 4-7, http://himaldoc.icimod.org/record/5660.

Uhl, C. 1983. You can keep a good forest down. Natural History 92(4): 69 - 79.

Uhl, C. 1987. Factors controlling succession following slash-and-burn agriculture in Amazonia. Journal of Ecology 75: 377-407.

Unruh JD (1990). Iterative increase of economic tree species in managed swidden fallows of the Amazon. Agrofor. Syst., 11(2): 175– 197.

Unruh, J. D. 1988. Ecological aspects of site recovery under swidden-fallow management in the Peruvian Amazon. Agroforestry Systems 7: 161 - 184.

Upadhyaya M 2004..Traditional Farming Practices and Farmers' Rights in the HKH Region-Policy brief No. 6, SAWTEE, Kathmandu, Nepal.

Van Noordwijk M., Mulyoutami, E., Sakuntaladewi, N., & Agus, F. (2008). Swiddens in transition: shifted perceptions on shifting cultivators in Indonesia. Occasional Paper no.9. Bogor, Indonesia: World Agroforestry Centre.

van Reuler, H. and Prins, W.H. (eds.): (1996) The role of plant nutrients for sustainable food crop production in Sub-Saharan Africa. Vereniging van Kunstmest Producenten, Leidschendam, the Netherlands.

van Vliet, N., O. Mertz, A. Heinimann, T. Langanke, U. Pascual, B. Schmook, C. Adams, D. Schmidt-Vogt, P. Messerli, S. Leisz, J. C. Castella, L. Jørgensen, T. Birch-Thomsen, C.

Hett, T. B. Bruun, A. Ickowitz, K. C. Vu, J. Fox, R. A. Cramb, C. Padoch, W. Dressler, and A. Ziegler. 2012. Trends, drivers and impacts of changes in swidden cultivation in tropical forest-agriculture frontiers: A global assessment. Global Environ. Change 22:418–429.

Veldman JW, Putz FE. (2011). Grass-dominated vegetation, not species-diverse natural savanna, replaces degraded tropical forests on the southern edge of the Amazon Basin. Biol. Conserv.; 144.5: 1419–1429.

Vieira, I.C.G., Proctor, J., 2007. Mechanisms of plant regeneration during succession after shifting cultivation in eastern Amazonia. Plant Ecol. 192, 303–315.

Viloria BHR, Indelible DP, van Rij W, Tabbada AU. Balancing biodiversity conservation and incomes of upland communities: approaches and experiences of the upland development programme in Southern Mindanao. In: 14th Annual Philippine Biodiversity Symposium of the Wildlife Conservation Society of the Philippines. Crown Lodge, Tuguegarao, Cagayan; 2005: 1-28.

Volkoff, B., Cerri, C.C. and Andreux, F. (1989). Matiere organique et conservation des sols en tropicales forestieres : voies actuelles de recherches. geme reunion du Comite ouest et Cl africain de correlation des sols. 14-23 novembre 1988. Cotonou, Benin.

Vosti, S. and J. Witcover (1996), 'Slash and burn agriculture; household perspectives,' Agricultural Economics and Environment, 58, 23–38.

Vosti, S.A., C.L. Carpentier, J. Witcover, and J.F. Valentim (2001b), 'Intensifying small-scale agriculture in the western Brazilian Amazon: issues, implications and implementation', in D.R. Lee and C.B. Barrett (eds), Tradeoffs or Synergies? Agricultural Intensification, Economic Development, and the Environment, Wallingford, UK: CABI Publishing.

Wairiu, M., Lal, R., 2003. Soil organic carbon in relation to cultivation and topsoil removal on sloping lands of Kolombangara, Solomon Islands. Soil Tillage Res. 70, 19–27.

Wallace BJ. (1996). Fuelwood consumption and deforestation in th Philippines: a rejoinder. Human Organization; 55: 500-501.

Warner, K. (1991). Shifting Cultivators, Local technical knowledge and natural resource management in humid tropics. Food and Agriculture Organization of the United Nations, Via delle Terme di Caracalla, 00100 Rome, Italy.

Warner, K. 1981. Swidden strategies for stability in a fluctuating environment: The Tagbanwa of Palawan. In Olafson, H. (eds.) Adaptive Strategies and Change in Philippine Swidden-based Societies. Laguna, Philippines: Forest Research Institute. pp. 13 - 28.

Watters, R.F. 1960. The nature of shifting cultivation. Pacific Viewpoint 1:59-99.

Watters, R.F., 1971. Shifting Cultivation in Latin America. FAO Forestry Development Paper 17. FAO, Rome, 305 PP.

Went, G. W. and Stark, N. 1968. The biological and mechanical role of soil fungi. Proceedings of the National Academy of Science 60: 497 - 504. Cited in Beckerman 1987.

Whitmore, T. C. 1984. Tropical rainforests of the Far East, pp. 276-278, Oxford Science publications, Clarendon Press, Oxford.

Wiens, J. and Donoghue, M. (2004). Historical biogeography, ecology and species richness. Trends in Ecology & Evolution, 19(12):639-644.

Williams M. 1990. Forests. Pages 179–201 in Turner BL II, Clark WC, Kates RW, Richards JF, Mathews JT, Meyer WB, eds. The Earth as Transformed by Human Action. Cambridge (UK): Cambridge University Press.

Wilson, A.M., Latimer, A.M., Silander Jr, J.A., 2015. Climate controls on ecosystem resilience: Postfire regeneration in the Cape Floristic Region of South Africa. Proceedings of the National Academy of Sciences, USA, 112: 9058-9063.

Witcover J., Vosti S.A., Carpentier C.L., Gomes T.C.D. A. 2006. Impacts of soil quality differences on deforestation, use of cleared land, and farm income. Environment and Development Economics 11: 343–370

Wood, Charles H. and Porro, Roberto (Ed.). (2002) Deforestation and Land Use in the Amazon. University Press of Florida, FL, USA.

Wood, S., K. Sebastian, and S.J. Scherr. 2000. Pilot analysis of global ecosystems: Agroecosystems. IFPRI and WRI, Washington, DC.

World Resource Institute, World Resource, Oxford University Press, Oxford, 1996.

Wulandari C, 2017. Policies that Transform Shifting Cultivation and encourage community-based forest management in Lampung province, Indonesia. In: Shifting Cultivation Policies: Balancing Environmental and Social Sustainability. CABI, The United Kingdom.

Wulandari, C. and Cahyaningsih, N. (2010) Ketika Masyarakat Adat Mengelola Hutan: REDD Menjadi Suatu Pilihan (When Indigenous People Manage Forests, REDD Becomes an Option), Watala and partnership, Bandar Lampung, Lampung, Sumatra.

Wulandari, C. (2007) 'Local ecological knowledge on forest conservation: An important factor to consider in forest regulation development', Journal of Rimba Kalimantan 12(2), accredited under No. 26/DIKTI/Kep/2005, Faculty of Forestry, Mulawarman University, Samarinda, Indonesia.

Wulandari, C. and Zakaria, W. A. (2010) 'Agroforestry pattern alternatives for transitional grasslands (Imperata cylindrica sp.) areas', paper presented at an International Seminar on 'Social Forestry for Plantation Wood Production: Connecting Industry, Communities and Government to Generate Wealth and Protect the Environment' 14-15 April 2010, Sriwijaya University, Palembang, Indonesia

Xu J, Fox J, Xing L, Podger N, Leisz S and Xihui A (1999). 'Effects of Swidden Cultivation, State Policies and Customary Institutions on Land Cover in a t-lani Village. Yunan. China'. Mountain Research and Development. Vol. 19. No.2.

Xu, J; Grumbine, RE; Shrestha, A; Eriksson, M; Yang, X; Wang, Y; Wilkes, A (2009) _The Melting Himalayas: Cascading Effects of Climate Change on Water, Biodiversity And Livelihoods.' Conservation Biology 24(3): 520–530

Xu, J; Shrestha, A; Vaidya, R; Eriksson, M; Hewitt, K (2007) The melting Himalayas: Regional challenges and local impacts of climate change on mountain ecosystems and livelihoods, ICIMOD Technical Paper. Kathmandu, Nepal: ICIMOD

Yesilonis, I., Szlavecz, K., Pouyat, R., Whigham, D., & Xia, L. (2016). Historical land use and stand age effects on forest soil properties in the Mid-Atlantic US. Forest Ecology and Management, 370, 83–92. https://doi.org/10.1016/ j.foreco.2016.03.046

Young, V and Hyde, Mi. (1988). Southern area development master plan. Sector Report. Forestry. Prep. By Lavalin International Inc. and MPW Rural Development Pty for State Planning Committee, Lao PDR.

Zaman, M.M., A.S.M.H. Masum, N.U. Ahmed, A. Salam and M.H. Rahman, 2002. Effect of tillage and mulch on the growth and yield of ginger in the hilly area. J. Biol. Sci., 2: 121-123.

Zhang, Q., C. Justice, M. Jiang, J. Brunner, and D. Wilkie. 2006. "A GIS-Based Assessment on the Vulnerability and Future Extent of the Tropical Forests of Congo Basin." Environmental Monitoring and Assessment 114 (1–3): 107–21.

Ziegler, A.D., Bruun, T.B., Guardiola-Claramonte, M., Giambelluca, T.W., Lawrence, D., Lam, N.T. (2009). Environmental consequences of the demise in Swidden cultivation in montane mainland Southeast Asia: hydrology and geomorphology, Humn. Ecol. 37:361-373.

Ziegler, A.D., Fox, J.M., Webb, E.L., Padoch, C., Leisz, S.J., Cramb, R.A., Mertz, O., Brunn, T.B., Vien, T.D., 2011. Recognizing contemporary roles of swidden agriculture in transforming landscapes of Southeast Asia. Conservation Biology, 25, 846-848.

Ziegler, A.D., Giambelluca, T.W., Sutherland, R.A., Nullet, M.A., Yarnasarn, S., Pinthong, J., Preechapanya, P., Jaiaree, S., 2004. Toward understanding the cumulative impacts of roads in upland agricultural watersheds of northern Thailand. Agric. Ecosyst. Environ. 104, 145–158.

7

Decriminalisation of Shifting Cultivation

7.1 GENERAL DESCRIPTION

Shifting cultivation is a type of land use that dates back to the dawn of agriculture. It's done in places where forest resources are still plentiful. It usually refers to the gradual transformation of primary forests into a mosaic of secondary forests at varying stages of growth, rather than the complete abolition of forests. However, primary forest removal occurs gradually in order to preserve the same fallow time despite population development. As no primary forests remain (or as the remaining primary forests are entirely protected by local customs or conservation legislation), secondary forests fade away, and new techniques arise or are adopted, the fallow period shortens. Natural capital is gradually replaced by man-made productive capital, or the system approaches a state of crisis, depending on whether the limitations to these changes can be levered. In theory, the new land uses allow for more forest protection because they require less territory to provide the same output. In practice, they are triggered or triggered by a series of more radical changes, such as infrastructure development, migrations, financial investments, and new consumer habits, all of which frequently end in large-scale deforestation.

Despite this, shifting farming is still considered as the primary cause of deforestation. Shortsighted policymakers ignore the context of tree-cutting and are swayed by modernizing myths that portray slash-and-burn agriculture as a primitive land use. Slash-and-burn agriculture is a primitive land use in that it predates others in many parts of the world. However, there should be no value judgment linked to this statement. Shifting cultivation could be considered as an opportunity to avoid falling into productivism, unsustainable paradigms at a time when doubts about mainstream growth pathways are growing. But this isn't to argue that slash-and-burn farmers should be stuck with this "primitive" method of land management. They, like all communities before them, have the right to adapt and change. They should be provided access to the same variety of economic and technical possibilities that other farmers have in our globalized world. Smallholder farmers who practise slash-and-burn agriculture must be given special attention in the fight for sustainability and attempts to overcome poverty, as they will continue to be significant targets of conservation and development projects. If right decisions are to be made, that is, if individuals are to be given the chance to determine their own destiny and boost output while providing public goods, they must be heard, free of prejudiced attitudes that criminalise or idealise their choices, depending on which political objective is pursued. The migrants move on to a new plot of shifting agriculture after two or three years of farming vegetables and grains on cleared ground. To clear land, slash-and-burn techniques are often utilised, in which trees, bushes, and woodlands are sliced and the remaining vegetation is burned. The ashes add potassium to the soil. Simple tools such as machetes, sickles, axes, and sticks are used by the farmers. Rather than buying seeds, they store seeds from previous harvests. The removal and burning of trees and undergrowth, as well as the sowing of seeds, both involve a lot of effort.

7.1.1 Change with growing population pressure

In a closed system, the possibility of changing technology in shifting cultivation may be imagined as the population pressure developed. To address their food needs, some communities have turned some land to settled farming. The amount of rainfall received and the quality of land are two major factors that influence the shifting cycle as well as the time of occupancy of a certain place. The dynamics of forest regeneration are determined by these factors. People in high-rainfall areas cannot cultivate land for more than a year due to the high rate of soil erosion. The field can be farmed for more than one year in low-rainfall areas, but no more than three years in high-rainfall areas. Even in the second year, soil fertility begins to deteriorate. If the economic system remains open and shifting cultivators have alternative options, some of them may begin working as agricultural laborers in nearby locations. For some of them, forestry is also a source of income. Though shifting farming remains a primary source of income, other supporting areas may expand as well. Shifting cultivation can be classified into three types: (i) areas where the shifting cycle is very long and traditional equilibrium is maintained, (ii) areas where the shifting cycle is not very long but the equilibrium has reached a limiting stage and may become contaminated, and (iii) areas where the shifting cycle has become hazardous. Cultivation shifts Rural poor and indigenous peoples in developing countries rely on agroecosystems. Shifting agriculture that is ecologically sound is sustainable since it does not on fossil-fuel-based fertilizers, pesticides, or irrigation. One way to demonstrate the soundness of slash-and-burn agroecosystems is to demonstrate the ecological compatibility of this crop production strategy empirically (Kleinman *et al.* 1995).

Shifting cultivation is sometimes referred to be unsustainable (Cramb, 1989; Tobing, 1991), and it is the leading cause of tropical deforestation (World Resources Institute (WRI), 1990). Shifting cultivation is undoubtedly linked to low crop yields and quick soil degradation (El Moursi, 1984; Christanty, 1986). As one might expect, development programs favor continuous cropping methods over slash-and-burn. In addition, several developing countries have launched institutional campaigns denouncing all forms of shifting agriculture (Dove, 1984; Cramb, 1989; Shengji *et al.*, 1990; Myers, 1992). A rising community of academics and development professionals, on the other hand, now claims that some shifting cultivation agroecosystems are viable under minimal land use pressure (Grandstaff, 1980; Dove, 1983; Okigbo, 1984; Pimentel and Heichel, 1991). Myers (1992) distinguished between the "shifted cultivator" and the "conventional shifting cultivator." Migrants who have only recently began practicing slash-and-burn agriculture are known as shifted cultivators, and they often practice a type of slash-and-burn that ignores environmental limits. The public's perception of slash-and-bum agriculture has evolved to shifted cultivators rather than conventional shifting cultivators as the global number of shifted farmers has expanded in the previous four decades (Myers, 1992). Traditional shifting farmers practice a more complex type of shifting cultivation that is suited to the local climate (Pimentel and Heichel, 199 l; Kidd and Pimentel, 1992). Moreover, shifted cultivators are more likely to dwell in areas with high population density and high land use pressure, whereas traditional shifting cultivators are more likely to live in areas with low population density and limited land availability (Myers, 1980; Okigbo, 1984). The variability of shifting cultivation can be found at several levels, in addition to discriminating between traditional and shifting cultivators (Conklin, 1961). Shifting cultivation is today most common in the tropics and subtropics, where it supports at least 300 million people (Andriesse and Schelhaas, 1987), but it was once common in temperate areas (Okigbo, 1984; Hillel, 1992). This practice can be found in a variety of settings, including woods, grasslands, and savannas (Denevan, 1981). From tiny tribal populations like the Dayaks of Borneo to vast civilizations like the early Mayans of Mesoamerica, shifting cultivation has been practised by a varied spectrum of societies (Dove, 1985; Hillel, 1992).

7.1.2 Shifting cultivation is integrated with indigenous culture

In many cases, shifting agriculture demands the use of profound indigenous technical expertise. Cultivators can use up to forty distinct crops, each with its own set of desirable characteristics like taste, nutrition, medicinal value, production resilience, and so on. To boost output and follow-up operations, each planter might employ a range of plot management approaches. For example, retaining soil by laying logs and branches across the slope, no-till sowing, and bio-charging the slash. On the other hand, shifting cultivation is more than a technical activity. It's part of a social system: farmers need to know that if they leave a plot fallow, they'll be able to return to it in the future. This demands a tenure structure based on local authority. The social order, which has authority and decision-making capabilities over resource availability and distribution, as well as festivals, rituals, and a sense of holiness in many ethnic communities, is closely related to the material cultural practise of shifting agriculture. Reciprocal collective labour activities like as collective clearing, fire management, and collective harvesting are frequent around the agriculture calendar. Thus, shifting agriculture is more than an individual activity; it is a community-supported activity that takes place within the framework of traditional tenures and authority. Customary tenures, on the other hand, have not been recognized or maintained by the Myanmar government.

7.1.3 Future of shifting cultivation

The integral shifting cultivator succeeds by recognising and working within the restrictions of natural processes linked with the year-round growing season and rapid ecological succession in the humid tropics, according to an analysis of multiple examples of traditional techniques. Shifting cultivation/fallow can work where other land use systems have failed because it makes use of natural processes and has a deep understanding of the micro-environments of forest and field, as well as the micro-site needs of certain crops. Despite their historical effectiveness, shifting cultivation-based agroecosystems cannot be used as a paradigm for the tropics' future. The tropical forest, which is so important for the shifting cultivation/fallow agroecosystem, is rapidly shrinking as landless settlers, logging concerns, and national financial requirements put growing pressure on it. Local technical expertise, on the other hand, can help with better natural resource management and the establishment of sustainable agroecological systems in holistic swidden societies. Shifting cultivators can take an active role in building new agroecosystems to address the difficulties of a shrinking resource base. In shifting cultivation communities, on-farm research is essential to aid in the development of novel cropping systems for shifting agricultural intensification. Such research could lead to advancements that smallholders in the tropics who don't employ shifting agriculture could benefit from. Agricultural and forestry extension agents should also be taught in the broad principles of shifting cultivation systems, such as the use of microenvironment differences, the integration of trees into smallholder agroecosystems, and the perception of agriculture as part of a larger agroecosystem.

7.1.4 Shifting cultivation and climate change

Shifting agriculture entails the use of fire to clear forest, releasing carbon into the atmosphere, and is thus considered to be harmful to climate change (Brady 1996, Rastogi *et al.* 2002) and is being argued on various platforms. It is often regarded as a system characterized by inadequate natural resource management and a major contributor to deforestation, biodiversity loss, and ecological instability. Because it differs from the rest of the plains' agriculture practices, it is commonly misunderstood and misrepresented by the bulk of the populace. It is recognized as a practice of primitive peoples and individuals outside of mainstream society who are responsible for resource waste (Warner 1991). However, current research concur that the impact of shifting cultivation has been grossly exaggerated, based on limited and incorrect data (Cairns and Garitty 1999, Bruun 2009). Shifting cultivation is also being acknowledged for its contribution in environmentally sustainable and economically viable agriculture (Ziegler 2009). Though the practice was once thought to be a major contributor to biodiversity loss and ecosystem imbalance, the fact remains that the majority of the world's mega biodiversity area coincides with the territory occupied by indigenous peoples who have been practicing a similar system of agriculture for centuries (Sobrevila 2008). For these indigenous peoples, shifting cultivation is not merely an agricultural land use system, but also a way of life, with all of their customs, rituals, festivals, and legacy inextricably interwoven with it. The people of the hilly regions adopted this age-old practice due to the tough topography, hostile terrain, constant rains, and terrible weather conditions. It's their warm response to the challenges of creating an agro-ecology in the challenging tropical forest habitat, which is marked by generally poor acidic soils and a diversified flora and fauna, resulting in a plethora of potential food crop competitors. In heavy rainfall areas of the hill tracts, the technique was proven to be economically and energy efficient when compared to other forms of agriculture (terrace or valley cultivation) (Ramakrishnan 1992). The underlying science in location selection, planting date, crop and variety selection, cultural management approaches, and so on requires a great deal of research and study. The most notable feature is that, despite the fact that slush-burning has been practiced in these hilly places for centuries, they still have the same evergreen diversified forest that meets all of their resource needs for subsistence, from house construction to medicines, in a very sustainable manner. The investigation is being conducted to learn more about shifting cultivation in general, as well as the science underpinning its long-term viability and the ecosystem's viability.

Advantages

- It aids in the recovery of all lost nutrients on used land, and as long as no damage occurs, this technique of agriculture is one of the most sustainable.
- The land may readily be recycled or regenerated in this manner, as it absorbs seeds and nutrients from nearby vegetation or the environment.

7.1.5 Characteristics of shifting cultivation (Gogoi and Datta 2020)

1. It is normally practiced in hills and hill slopes.
2. A group of 10-15 families cultivate in one Jhum site for several years.
3. Mixed cropping of 15-20 crops are followed in an area of 1-4 ha (Raman, 2001).
4. The shifting farmers cultivate traditionally with locally available seeds, tools and implements.
5. Major crops grown are paddy, maize, ginger, turmeric, cucumber, chilli, beans, pumpkin, squash, chow-chow, colocasia, pineapple, banana, orange and lemon.
6. It is a labour-intensive farming.

Impact of shifting cultivation

1. Agro-biodiversity conservation, particularly of local and native crop species and types.
2. As they burn their fields before moving to other locations, they control weeds, pests, and diseases in crops.
3. Fire also adds ashes to the soil, which helps to neutralize acidity.
4. Soil erosion occurs due to vegetation loss.
5. Deforestation occurs due to land clearance for jhum.
6. Provides food security to the Shifting cultivation and their families.
7. Generate more income per unit area as they utilize every inch of land for cultivation as compared to monocropping in plains.

7.1.6 Strategies for improving shifting cultivation

Verma *et al.* (2017) suggested some strategies to improve or transform shifting cultivation to settled cultivation, as follows:

Fallow management: The soil fertility and nutrient status of Jhum fallows can be improved by soil incorporation of green manure crops like *Crotolaria juncea, Sesbania bispinosa, Thysanolaena maxima* etc.

Integrated farming system: The principle behind IFS is to use the by-product of one component as the input of another component for resource cycling within the whole farming system. The farming system of various components like crop production, horticulture, livestock, fishery and apiary.

Multistorey agroforestry: Crops of various canopy orientation and root length are to be grown in a particular area in order to maximize the use of vertical and horizontal spaces. For example- Coconut + Black pepper + Pineapple and Arecanut + Patchouli + Betelvine.

Watershed management: Contour trenches or terrace farming conserves both water and soil in hills. This is practiced in Nagaland and Meghalaya. Crops like paddy, cabbage, potato, carrot etc. are grown in this system. Jalkunds can also be excavated to harvest the monsoon rain water and utilize it in the dry spells.

7.2 WHY SHIFTING CULTIVATION IS CRIMINALISED?

Shifting cultivation is the most important land use problem in the tropics globe, according to the FAO (1957). Shifting cultivation has been criticised as damaging, and it is typically portrayed as the action of illiterate people who slash and burn without thinking, wasting vital forest and soil resources and increasing the risk of runoff and erosion (Grandstaff 1981). In the twenty-five years since the alarm was sounded, shifting cultivation has increased rather than decreased. Over 250 million shifting cultivators are estimated to exist worldwide, with 100 million in Southeast Asia alone (Hatch and Tie 1979). Shifting cultivation is, in reality, an optimum option for agriculture in the humid tropics (Watters 1971), as long as human population density is kept low and fallow intervals are long enough to recover soil fertility. This agricultural system is environmentally friendly and efficiently fulfils a wide range of human requirements, notably in terms of labour and other agricultural inputs. However, many locations where shifting agriculture has been effectively practised for millennia are experiencing fast population growth, owing partly to migration from other areas. People are also cultivating new places that have never been cultivated before and where shifting farming may not be feasible. As a result, many of the negative consequences of changing cultivation are gradually becoming a reality, even though they are sometimes more imagined than real. Unfortunately, for many tropical places, no alternative food production strategies have yet proven to be ecologically and economically viable (Eckholm 1976).

Shifting cultivation is a type of agro-forestry in which annual agricultural crops are mixed with a period of fallowing to allow trees to develop before the plot is cultivated again. It is well-known and widely used all throughout the world. For years, millions of rural workers in Myanmar have relied on shifting agriculture as their primary source of income. In many parts of the country, it still provides food security for families and communities, and it is seen as an important part of their cultural history (Springate-Baginski 2018). However, some governments have come to regard shifting agriculture as inferior to their preferred kind of farming, stationary cultivation, since the 19th century. Shifting cultivation is portrayed as a 'bad' thing that needs to be eliminated from this standpoint. Shifting cultivation is typically characterised as a primary cause of deforestation, and shifting cultivators' fallow lands are commonly labelled as "degraded forest" or even "empty land," and taken by the state, sometimes subsequently passed on to others to utilise. Poverty, resentment, and conflict have resulted as a result of these acts in numerous shifting agriculture areas (Springate-Baginski 2018). Shifting farmers face a challenging political environment that fails to reflect the democratic ideals of small-scale food producers that employ shifting agriculture (Springate-Baginski 2018). Positive outcomes are not yet guaranteed, even if the policy environment improves. International scientific opinion has shifted dramatically in recent decades in favour of a balanced understanding of shifting agriculture (e.g., Dressler *et al.* 2016), and both the National Land Use Policy 2016 and the current Federal Agriculture Policy provide good, albeit broad acknowledgement. However, a multitude of issues continue to impact land and agriculture policy, as well as climate and environmental policy and legislative procedures. Beyond official recognition, the government must continue to promote shifting cultivators' interests and help them achieve a bright future. In order to demand that their rights be recognised and protected, shifting cultivators must be well-organized and well-informed.

A global perspective on changing agricultural practices

- It is a very rudimentary kind of agriculture that is thought to have begun approximately 7000 B.C. in the Neolithic period.
- In Africa, Central and South America, Southeast Asia, and Oceania, a variety of shifting agriculture methods are used in varied historical, environmental, and socio-economic contexts. The Food and Agriculture Organization of the United Nations (FAO) published a report in 1974.
- Upland rice is the most common crop in Southeast Asia, maize in South America and millet and sorghum in Africa.
- Between 1980 and 1990, more than 6% of tropical forest land was converted to shifting agriculture in all tropical countries (Ranjan and Upadhyay, 1999)

- Between 1980 and 1990, over 10% of forest area in Asia was converted to shifting agriculture. Singh and Marzoli (Singh and Marzoli, 1995)
- The method is stable, ecologically sound, and reasonably efficient in pioneer or frontier circumstances when ratios of fallow periods to cultivation phases are up to 10 or more (Okigbo, 1981).
- Fallow times are substantially decreased as a result of population pressure and different factors for intensification of production.
- Why Short fallow periods are no longer enough to restore the soil's productive capability. According to the FAO, shifting agriculture is responsible for the loss of 1.9–3.6x10^6 ha of primary close forests, 3.4–40 x10^6 ha of secondary close forests, and 6.9–21.9 x10^6 ha of secondary open forests each year. Detwiler and Hall (Detwiler and Hall, 1988)
- This technique is widely used in the northeastern part of India. Shifting farming accounts for 85 percent of overall cultivation in northeast India.
- This approach is used by roughly 5.0 million tribal families in India on 4.37 million hectares of land spread over 11 states (Sahu *et al.* 2005).

7.3 PEOPLE'S VIEW OF SHIFTING CULTIVATION

Since the colonial era, shifting cultivation, the most complicated and multifaceted kind of agriculture, has been a source of dispute and intervention. Many scientists, environmentalists, foresters, developmental practitioners, policymakers, and the general public have decried the practise as primitive, backward, harmful or wasteful, unproductive, and exploitative, as well as the cause of extensive environmental damage. The destruction of much of the world's tropical forest, land degradation, air pollution, and global climatic change are all blamed on shifting cultivators. As a result, shifting cultivation in many forms has been labelled as environmentally detrimental since forest areas are being converted to farmland through its techniques, reducing the area under primary forest cover. Shifting cultivation detractors today use derogatory phrases to spread misconceptions about shifting cultivators (Thruppetal, 1997). It has resulted in several modifications in the surrounding environment's natural parameters. The decrease of forest cover has an impact on the climate, which contributes to biodiversity loss. Furthermore, siltation, flooding, and soil degradation have an economic impact on forest-dependent people's livelihoods and cultural integrity. Deforestation and forest degradation have several causes. One of the major underlying causes has been recognised as demographic considerations. While shifting agriculture has long been recognised as one of the primary causes of tropical deforestation.

Intensive shifting farming, deforestation concerns, rapid urbanisation, the occurrence of forest fires, and a high prevalence of unemployment are some of the demographic and economic variables that are already putting strain on rich natural resources.

It is believed that the land resource for agricultural activities should be evaluated at the micro-level on the basis of slope, dissection index, soil and vegetation cover, and land usage (Wagh *et al.* 1984). Because there is no viable alternative way to substitute, the farmers are left with no other option. The following are some examples of people's perceptions:

1. Because of the Hill topography, only slash and burn cultivation is possible; terrace cultivation necessitates low-lying areas and costs a lot of money. It was discovered that the government promised to give terrace cultivation projects, but that the schemes never reached the growers.
2. The majority of individuals living in rural areas have no source of income, whereas those living in urban areas rely on forest resources and encroachment of more land under agriculture as their only source of income.

Farmers also practise soil conservation through the use of physical and living barriers. To prevent soil erosion in the Jhum fields, they built a contour bund out of poles, logs, and bamboos across the slopes. Cucumber, ginger, yam, and other cash crops are grown in the contour bunds. This ensures food security as well as a sustainable economic livelihood for rural residents. To discourage shifting cultivation, alternative sources of revenue should be found that are properly aligned with proper land use practises. Efforts should be made to keep such agriculture contained within the areas that have already been impacted by the spread of improved agricultural methods. Damage caused by such farming should be repaired using social forestry and energy plantings (National Forest Policy, 1988). Shifting cultivation is seen as a wasteful and inefficient method of agricultural production. It is true that under shifting agriculture, production is low from the standpoint of land utilisation (Barkakoti, 1990).

Deforestation is a result of human intervention in a variety of activities. Deforestation is also exacerbated by logging and wildfires, both of which are linked to the practise of shifting cultivation. Due to the mountainous topography, terrace agriculture is low in almost all of the sampled settlements. The situation of deforestation was also investigated in various villages in order to learn more about forest management and deforestation. Forest cover has been noted to have reduced to a substantial extent in all of the different ranges due to the pressure of Jhum cultivators and logging. Landslides, which are widespread in these mountains, are exacerbated by the degradation of virgin forests. Due to the degradation of evergreen and moist-deciduous forests, many natural abodes with diverse wildlife and plants are also disappearing. One of the most pressing worldwide environmental challenges today is the devastation of the world's tropical forests, which are disappearing at an alarming rate (Agrawal, 1998).

Impact on environment

i) Shifting farming is no longer viable.
ii) As the world's population grows, so does the demand for agricultural land, which has resulted in the shifting cultivation of vast swaths of forest.
iii) Deforestation and biodiversity loss in the environment result from widespread forest devastation.
iv) Shifting cultivation burning has one of the worst climate consequences because it produces massive amounts of smoke that are emitted into the atmosphere.
v) A temperature increase of a few degrees Celsius.
vi) In all ranges, a decline in soil fertility accompanied by a low yield is common.
vii) Mild climate and late monsoon.

7.3.1 Shifting cultivation cause deforestation

Shifting cultivation is one of the most important causes of deforestation and forest degradation (Kleinman *et al.* 1995, Myers 1992). Deforestation, on the other hand, entails a long-term (>10 years) or irreversible loss of forest cover (Giri 2007). According to the FAO's definition of deforestation (Erni 2009), the clearance of forest for shifting cultivation is "forest change," not "deforestation." Clear-felling and replanting, according to van Noordwijk *et al.* (2008), is a regular part of forest management, whether done naturally or artificially, as long as the regrowth is permitted to reach woody vegetation. By definition, though, converting slashed-burn land to intensive cultivation can be classified as deforestation.

The encroachment of forest lands for agricultural purposes is at the heart of the tropical deforestation controversy. Despite the fact that swidden (shifting cultivation) is one of the most rudimentary kinds of agriculture in the tropics' forest-farming continuum. Some researchers believe it is a forest degrading agent. In tropical countries, shifting cultivation still covers a considerable portion of the forestlands. According to some estimates, the global proportion of swidden in deforestation is between 45 and 60 percent (UNEP reported in Angelson, 1995). (Myers. 1995). Shifting cultivation, it is said, is a key cause to deforestation, resulting in two main environmental issues: loss of biodiversity and increased soil erosion (Amalendu 2000). Apart from that, climate change as a result of forest loss has had a significant impact. The study of shifting cultivation as an ecological issue is critical because natural balance changes have direct and indirect implications on human life. Most typically, the most important causes of deforestation are mentioned as poverty, population growth, and swiddener migration into the forest (Amalendu 2000).

7.3.2 Shifting cultivation and GHG emission

During shifting cultivation, the GHG emission (which is primarily carbon dioxide) is not at all a luxury emission. Also, when we look at shifting cultivation's entire carbon dioxide (CO_2) emission budget, the net emission is minimal (Fearnside 2000, Lehsten *et al.* 2009). Land use, chemical-free management, practically permanent land cover with alternative crops, non-flooded fields, forest recovery in left-out places, and other factors cancel out whatever is emitted during burning. Because fallow periods allow for forest regrowth, forest removal for shifting cultivation releases less carbon than permanent forest clearing (IPCC 2006). Previous research has demonstrated that shifting farming can be carbon neutral while maintaining soil fertility provided the fallow intervals are long enough (Ruthenberg 1971, Van Noordwijk 2008). Tillage can result in considerable carbon loss (in the form of CO_2 bursts) right after tillage (La Scala *et al.* 2006). Reduced tillage or zero-tillage strategies have been observed to reduce net carbon dioxide emissions from soil (Noel and Herby 2000, Wan and Lin 2004), while also keeping plant residue on the soil surface (Noel and Herby 2000, Wan and Lin 2004). Shifting cultivators' dibbling method of seeding is zero tillage, emitting almost little CO_2 and retaining soil organic matter. According to the IPCC (2007), nitrogen fertilisers are responsible for 70% of overall GHG emissions from agriculture, with N_2O having a GHG potential of so and so times that of carbon dioxide. During shifting cultivation, chemical-free growing and management procedures result in nearly no emissions of these gases. Only farm scraps, animal wastes, and other natural materials are used in their production system. The majority of their insect and pest management tactics are based on observing natural phenomena and indicators in the field. They have a thorough understanding of the dynamics of insect pests and diseases as a result of years of observation. They reduce the prevalence of insect and pest infestations by carefully selecting the planting date and other cultural management practices based on the lunar cycle. Sowing over the days closest to the Full Moon has been observed to reduce pest assault and promote germination. Shifting cultivators has a unique pest management approach that involves slashing medium girth plants to a specific height based on the expected height of the crop to be presented. The shifting cultivators place a high value on fallow periods, with 15 years being proven to be sufficient for vegetation regeneration and soil rejuvenation. In the tropical forest, a fallow time of at least 10 years is usually sufficient for vegetation regrowth, but this depends on the type of the soil and plants (Aryal and Kerkhoff 2008).

7.3.3 Shifting cultivation cause loss of biodiversity

The indigenous communities who use shifting farming have long prioritised biodiversity conservation. Shifting cultivation, on the other hand, has been extensively criticised as a threat to biodiversity, both natural and agricultural (Terborgh 1999). Rather than providing more emphasis to particular selected groupings of biological assets, indigenous shifting farmers maintain natural biological asset balance (assets whose initial form was defined by the environment of the location). During the site selection process for shifting cultivation, they make a point of avoiding areas dominated by uncommon or medicinally significant plants. They avoided cutting certain large trees (which they saw as sprit abodes) because they were aware of their importance to the immediate ecology and its long-term viability. In addition to providing shelter for numerous birds, animals, and insects, such large trees serve as guides in agricultural chores. When such practices are linked to rituals and customs, they become the society's and individual's moral and spiritual obligation. In Malaysia, site selection based on some indicator plants was also recorded (Tanaka *et al.* 2007). They regard the forest as a protector, as well as a source of food and shelter for humans (Nimachow 2003). In contrast to modern techniques, the concrete and spiritual coexist in the indigenous community, complementing and enhancing rather than competing and opposing each other (Nakashima and Roue 2002).

Regardless of climate change, indigenous communities' crop wild cousins and landraces are deemed critical to the future viability of world food supply (Lane and Jarvis 2007). Agricultural diversity is conserved by people and the actions connected with them, not by landscapes. Environmental protection, sustainable development, good natural resource management, and biodiversity conservation are some of the emerging topics of public concern around the world. The topic of human impact on the rapid decline of biological diversity is receiving more attention. The world is home to a wide variety of living organisms whose genetic diversity interacts with one another and with their physical environment, resulting in a high level of biodiversity. Furthermore, human civilizations' ethical, aesthetic, spiritual, cultural, and religious values are an intrinsic element of these complex resources. The district is a veritable treasure trove of diverse biodiversity that has been protected for generations in its natural habitat. Changes in the environment occur as a result of the birth of modern civilization, population growth, and the continued use of traditional farming methods. One of the main causes of biodiversity loss is the physical modification of forests and jungles. It has seen a significant change in forest cover. Permanent tree and forest felling has resulted in significant loss, and recovery will take many years.

Within a 50-year period, the district's biodiversity status has altered dramatically. With the continued practice of traditional cultivation, there are negative effects of human impact due to the encroachment of forest resources. The villages' private forest is being used in an unplanned and unscientific manner. All kinds of restrictions are still in place in the Reserved Forest, including sanctuaries, tree felling, hunting, and fishing. It has risen substantially, posing a serious threat to the basic foundations of long-term development and quality of life. The ability of ecological communities to reset or recover from shock and environmental change, including long-term climatic change, is influenced by the loss, fragmentation, and diversity of species and genes in biodiversity.

7.3.4 Forest fire

Large amounts of smoke are released into the atmosphere when vegetation in forest cover is burned, which has an impact on the natural environment by causing climatic fluctuations in an area. Human-induced pollution of the environment is widespread, and it continues to grow. In 2009-2010, the district was hit by a slew of fires, many of which happened during the Jhum field fire. During these years, 80 hectares of forest cover were entirely destroyed. Many natural biodiversity abodes were lost as a result of this. With the green house effect, gases warm the earth's surface to -18°C, making it too cold for life to exist. However, the higher the concentration of these gases in the atmosphere, the more heat is trapped. Human activities add little amounts of water vapour to the atmosphere, but they also add significant amounts of carbon dioxide and methane, which are increasing global temperatures (Kran Arms, 1990). Many steps have been done by village communities, as well as government policies and non-governmental organisations (NGOs), to cope with fire accidents, however the laws and restrictions that have been enacted are also weaning local communities away. Since 1996, the state has prohibited jungle burning, although it has not been enforced. Measures and safeguards should be taken, particularly for rural populations who primarily cultivate Jhum crops. Local communities, in collaboration with the fire department and other organisations, must come out with additional rules and procedures to help protect our community's forest cover.

7.3.5 Effect on hydrological cycle

Deforestation caused by shifting farming interferes with the rain cycle since vegetation is essential to the rain cycle. Deforestation results from the wholesale removal of forest cover with forest canopy gaps, which limits rain creation due to low evapo-transpiration. During cloud formation, it is used as one of the principal sources of water vapour. When forest cover is destroyed, water catchment regions dry off, resulting in droughts.

7.3.6 Land degradation

In locations where shifting agriculture is prevalent, land productivity deteriorates quickly. shifting cultivation was heavily practised on hill slopes in India's Eastern and Northeastern areas. The cycle of cultivation is impacted by increased land use. Jhum cycle has been observed to have decreased from 10-15 years to 8-10 years. It has been reduced to 5-7 years in several ranges in the district. The resilience of ecosystems has been broken down as a result of the reduction in cycle, and the land is deteriorating. The ground deteriorates as a result, making it more prone to soil erosion and loss of soil fertility.

7.3.7 Shifting cultivation and soil fertility

It has been reported that shifting cultivation has a negative impact on soil fertility, increased soil erosion, reduced annual crop yields, and watershed siltation (Mertz *et al.* 2002, Ziegler *et al.* 2009), land degradation by encouraging erosion of fertile topsoil with eventual exposure of rocks (Mertz *et al.* 2002, Ziegler *et al.* 2009), and land degradation by encouraging erosion of fertile topsoil with eventual exposure of rocks (Mertz (Kleinman *et al.* 1995). Runoff sediments washed away a substantial amount of nutrients from the upper 10 cm soil, and a significant drop in organic matter content was also documented (Gafur 2001). (Wiel 1982). Fallow cycle reduction has been identified as a danger to biodiversity and soil protection (Duguma *et al.* 2001). Annual crops grown by slash and burn on high slopes have been reported to have harmed the forest, land, and environment (Knudsen and Khan 2002). There is widespread fear that shifting agriculture will result in the depletion of soil carbon and, as a result, an increase in CO_2 levels in the atmosphere (Bruce *et al.* 1999). According to Detwiler (1986), shifting cropping resulted in a 40 percent loss of soil carbon over the course of five years. According to the Indian Council of Agricultural Research (1983), shifting agriculture has resulted in a loss of organic carbon in the soil (Zodinpuii *et al.* 2016).

7.4 DECRIMINALISATION

7.4.1 Shifting cultivation does not induce deforestation

The clearance of forest for shifting cultivation is not deforestation, but "forest change," according to the definition of deforestation given by (Erni 2009).

Clear-felling and replanting, according to van Noordwijk *et al.* (2008), is a regular part of forest management, whether done naturally or artificially, as long as the regrowth is permitted to reach woody vegetation. By definition, though, converting slashed-burn land to intensive cultivation can be classified as deforestation. During shifting cultivation, the GHG emission (which is primarily carbon dioxide) is not at all a luxury emission.

Soil acidity may be one of the primary causes of disease and pest infestations in these places, however burning not only reduces soil acidity but also helps to eliminate pest spores in the soil.

The Jhumias place a high value on fallow periods, with 15 years being proven to be sufficient for vegetation regeneration and soil rejuvenation. In the tropical forest, a fallow time of at least 10 years is usually sufficient for vegetation regrowth, but this depends on the type of the soil and plants (Aryal and Kerkhoff 2008). The widths of the recovering plants' stems suggest that the soil and vegetation have been rejuvenated. As a result, the question remains as to why Jhum or other indigenous community-related systems have always been targeted for environmental deterioration or disaster. Is there any logic to the allegations, or is it just a case of transferring blame? Why are some ancient indigenous methods so sustainable and maintain ecosystem diversity from antiquity, yet our current development policies and technologies, in some way or another, cause problems for the natural environment or climate now and then, here and there? What do we mean by development? Is development a standardised procedure? What are the basic flaws in our current policies and strategies?

7.4.2 Shifting cultivation is not a wasteful practice

Shifting cultivation appears to be the most successful approach for coping with the ecological reality of the tropical forest, as opposed to the position that "shifting cultivation is wasteful and promotes environmental deterioration" (Cox and Atkins 1979). Shifting cultivation hasn't always been exclusive to the tropics. When faced with woods, agricultural cultures all across the world have utilised it since the Neolithic. The forests were removed and farms appeared as early agriculturalists migrated through Asia, Europe, Africa, and the Americas. Swidden was still in use in northern Europe's spruce pine woodland until relatively recently (Cox and Atkins 1979; Russell 1968; Ruddle and Manshard 1981). It persists in the tropics due to the region's environmental constraints. The difficulty of establishing an agroecosystem in the tropical forest has led to shifting cultivation. The tropical forest ecosystem is characterised by relatively poor but variable soils and an exceedingly diversified flora and fauna, which provide limited nutrients but a large number of potential food crop competitors. By chopping the forest and burning the felled trees and debris, the shifting cultivators utilise a "artificial energy pulse" to eradicate rival species and concentrate nutrients and energy flow into food crops (Odum 1971; Bodley 1976). It refers to a cultivator's active management of a forest part and its conversion to a more open and advantageous succession (Rambo 1981, Olafson 1983).

7.4.3 Energy efficiency of traditional agricultural systems

Scholarly interest in traditional subsistence farming techniques was reignited by the finding that energy-intensive mechanised agriculture was not always a viable model for developing Asia's energy-limited countries (Rambo 1981). At first look, these systems seem fairly energy-efficient, delivering anywhere from 5 to 50 food calories for every calorie of effort expended, even though many of them need a significant amount of human labour (Leach 1976; Steinhart and Steinhart 1974). Moreover, a different study (Ward, Sutherland, and Sutherland 1980) discovered that automated modern farming methods might not be any more energy-efficient than traditional ones based on the employment of draught animals. Claims of yields of 10 to 15 food calories per calorie of work used to operate the system indicate that shifting agriculture, however, appears to be highly energy efficient in tropical forest areas of Asia and the Pacific (Rappaport 1971; Pimentel and Pimentel 1979).

7.4.4 Ecological advantages of shifting cultivation

In the tough climatic conditions of the Asian tropical rain forest, shifting cultivation has a variety of ecological advantages. It produces consistent yields with limited sensitivity to insect outbreaks and, when used at suitable population densities, causes little long-term deterioration of the environment's productive capability. Barney (1970), Conklin (1957), and Geertz (1963), among others, offer detailed descriptions of these and other benefits of shifting cultivation. Agricultural researchers have failed to establish alternative methods with equivalent social and ecological virtues since the practice of shifting cultivation is so well adapted to the humid tropics. In fact, much current research focuses on improving shifting cultivation rather than attempting to replace it with fixed field techniques (Grandstaff 1980; Greenland 1975). Shifting cultivation thus merits a lot of the respect that ecological anthropologists currently give it (Bennett 1973, Bodley 1976). It does not, however, deserve the accolades it has gotten for being an energy-efficient farming system.

7.4.5 Burning and shifting cultivation

Given our relative lack of awareness of the role that burning plays in the running of shifting cultivation systems, it's not surprising that the energy value of fire has been disregarded. Although ethnographers who have examined certain shifting cultivation systems in Southeast Asia agree that the quality of the burning has a significant impact on crop success (Barney 1970), there is no agreement on why this is so. However, at least six key beneficial impacts of burning have been proposed by anthropologists, agronomists, and soil scientists: (1) removing unwanted vegetation from the field; (2) changing the soil structure to make planting easier; (3) increasing soil fertility with plant ashes; (4) decreasing soil acidity; (5) increasing soil nutrient availability; and (6) sterilising the soil and reducing microbial, insect, and weed populations. The detail about each of these proposed purposes of burning in shifting agriculture is given hereunder.

7.4.6 Clearance of unwanted vegetation

Farmers removing enormous amounts of vegetation from their fields when clearing new plots on forest lands have a difficulty. Given that tropical rain forests can have hundreds of metric tonnes of aboveground biomass per hectare, clearing the fields using human physical power alone would be impossible in any reasonable amount of time. This significant obstacle that the primitive farmer faces can be immediately solved by burning (Conklin 1957; Nye and Greenland 1960).

7.4.7 Change in soil structure

The heat of the fire softens and friablises the surface soil, resulting in a good seed bed that may be readily planted with a wooden dibble stick (Conklin 1957). If fire couldn't do it, the farmer would have to do it himself using a hoe or other manually operated cultivation tool. Hoe cultivation, on the other hand, promotes far more rapid disintegration of soil structure, and hence higher sensitivity to erosion, than dibbling, which generates only minor surface disturbance (Nye and Greenland 1960).

7.4.8 Enhancement of soil fertility

It is generally known that many important nutrient components are stored in plant biomass rather than soil in tropical rain forest ecosystems. For example, the biomass stores nearly all of the available phosphorus (Sanchez 1976). As a result, when vegetation is burned, vast amounts of nutrient-rich ashes are deposited on the soil surface, giving freshly planted crops with the equivalent of hundreds of kg of chemical fertiliser per hectare (Sanchez 1976). Although much of the carbon, nitrogen, and sulphur stored in vegetation is lost to the atmosphere during field burning, topsoil stocks of these nutrients are not depleted, and incorporation of unburned charcoal and plant residues into the topsoil may slightly increase soil carbon and nitrogen levels (Nye and

Greenland 1960, Sanchez, 1976). In any event, the majority of nitrogen is kept in the ecosystem's soil component rather than in plant components, so losses from burning are minimal. Of course, substituting composting for burning would reduce nutrient losses, but it would also necessitate more human labour, would make nutrients available slowly and gradually rather than all at once during the critical early weeks of cereal plant germination and growth, and, perhaps most significantly, would not change the pH of the soil, given the acidic character of many tropical soils.

7.4.9 Sterilization of Soil

Burning of shifting cultivation plots in northern Thailand results in a temperature of at least 75°C in the top 2 metres of the soil, according to studies (Zinke *et al.* 1978). Most roots, weed seeds, insect eggs and larvae in the surface soil are killed by this heat, and the microbial population of the plot is severely reduced (Nye and Greenland 1960; Conklin 1957). The greatest advantageous impact of this partial sterilization of the soil is the elimination of weed seeds, as the crop plants are able to establish themselves before new seed from external sources can infiltrate the plot. There is certainly solid evidence that the increased demand for human effort in weeding, rather than a reduction in soil fertility per se, causes shifting cultivation to be abandoned after only a single grain crop is harvested. Experiments with meticulously weeded plots have often maintained high levels of crop output for several years, but the human effort required for weeding far outweighs the labour required to clear a fresh plot in the forest (Janzen 1973; Nye and Greenland 1960; Sanchez 1976).

7.4.10 Shifting cultivation and livelihood

Shifting cultivation, in various variants of the slash-and-burn system, was a customary method of cultivating tropical highland and mountain soils until the 1950s, primarily for providing a minimal subsistence to the peasantry (Jasbir Singh, 1997). Cultivation was a noteworthy invention for early cultures, as it was based on the revolutionary transition from forest-based food collecting to agriculture-based food production. The rotation cycle was likewise long, allowing adequate time for the soil to rejuvenate. This is how a labor-efficient nomadic farming system for self-sufficiency in the subsistence economy evolved. The cultivation of Jhum has a number of negative consequences over time. As the world's population grew, so did the need for land, which resulted in shorter shifting cultivation cycles. Short fallow periods are no longer sufficient to recover soil productive capacity, resulting in a decrease in crop output. The Jhum cycle has slowed as population strain on the land has increased. The compatibility of this style of cultivation with the physical environment is the major reason for its persistence.

In the Northeast region, where shifting cultivation is practised by the majority of hilly people, the sparse population, steep and undulating topography, community land tenure system, and other socio-economic and cultural variables are some of the factors that went in its favour. Shifting agriculture has a variety of repercussions, including large-scale deforestation, which causes harmful ecological imbalances. Because the hilltops, particularly the catchment areas, are a source of water, deforestation in the hills has resulted in the removal of water sources, increasing run-off due to the soil's inability to absorb the water. Deforestation occurs as a result of forest clearing, which accelerates soil erosion and amplifies rainfall unpredictability. Burning Jhum land can be regarded one of the greatest climate impacts since it allows for a high rate of soil erosion and the structural stability of soil erosion due to runoff and winds. Water availability in the soil for crop production reduces as well, making it subject to climatic change. Fertility of the soil decreases with the detrimental impact of crop production as farmers cultivate the same plot of land for two years in some hamlet. Some of the environmental repercussions of shifting agriculture include the fragmentation of natural habitat, the invasion of more exotic weeds, and the burning of sun-dried plants, which pollutes the air with carbon monoxide, nitrous oxide, and a variety of other toxic chemicals.

7.4.11 Positive effects

In contrast to the above observations, Sharma (1976) found that shifting cultivation causes the least disturbance to soil, builds natural fertility through remnants of mixed cropping on moderate to steep slopes with minimal tillage under rain-fed conditions reliant on local resources, and causes the least disturbance to soil. A lot of scientists have stated that shifting cultivation has good environmental effects (Padoch & Pinedo-Vasquez 2010). Contrary to popular belief, shifting cultivation is now regarded as a very environmentally and economically efficient agricultural practice when the fallow period is long enough (Ramakrishnan 1992, Crasswell *et al.* 1997). One of the most important markers of soil health and fertility is soil organic carbon. Soil plays a critical role in the global carbon cycle because it is the biosphere's greatest stock of terrestrial organic carbon, holding more carbon than plants and the atmosphere (Schlesinger 1977). For more than a century, organic carbon has been recognised as a critical control of soil fertility and agricultural production (Tiessen *et al.* 1994).

7.4.12 Soil erosion and shifting cultivation

Soil erosion is also usually attributed to shifting farming. However, shifting cultivation is one of the most rudimentary land use methods. It is a well-known fact that the shifting cultivation system's long-term success was dependent on maintaining soil fertility. If nutrients were lost or shifted during the cycle's cultivation phase. Their equilibrium is maintained by those which are replaced during the fallow period, and the system remains in operation indefinitely (Xu *et al.* 1999: Ramakrishnan. 1992: Hong. 1990). However, moist tropical forests are magnificent and robust, and growth quickly covers the ground and reduces erosion for the remainder of the monsoon. According to field survey findings, topsoil loss under grass or shrub cover was around 0.9 tonne ha^{-1}, compared to 1.78 tonne ha^{-1} in a well-terraced tea garden (Xu *et al.* 1999). Other investigations found that, when compared to other modern agricultural systems, swiddening is the most efficient system of farming in the forest and on hill slopes (Hatch. 1980; Padoch. 1980: Nye and Greenland. 1960). This could be due to the shifting cultivation method, which involves only a few millimetres of soil disturbance when planting and weeding.

7.4.13 Economics of production in shifting cultivation

Dove discovered that the average production of grains per home is around 4270 kgs in one of his case studies (Dove, 1989). In another study, Chin discovered that the average annual production per household is 2300 kg (Chin, 1984), greatly above the consumption needs. Ramakrishnan (1992) also looked into average grain and seed yields. In the northeastern portion of India, leaves and fruits, as well as tuber and rhizomes, are subjected to a variety of shifting cycles. However, unless the output-input ratio is exceptionally high (both in energy and monetary terms) when compared to alternative agricultural systems, the average yield alone does not show the shifting system's economic efficiency.

7.4.14 Food energy use efficiency

Shifting agricultural systems have been praised for their high productivity, with 5 to 50 units of food energy produced for every unit of energy depleted (Rapport. 1971: Steinhart and Steinhart. 1974. In truth, alternative agricultural methods may be effective in terms of human labour and time, but they are extremely wasteful in terms of the entire energy equation. Other systems demand a greater amount of energy to produce a single unit of food energy. Ramakrishnan concluded in a comparative study in northeastern India that shifting cultivation has survived all these years primarily due to the system's high energy efficiency associated with longer cycles, where the farmer's only energy input is manual labour, a large portion of which goes to the slash-and-burn operation (Ramakrishnan, 1992).

7.4.15 Output efficiency in shifting cultivation

When compared to terrace and valley cultivation, the output-input ratio in shifting cultivation is also better (Banerjee, 1995). Dove (1984) conducted a similar study in Kalimanthan, Indonesia, and found that shifting cultivation is efficient in terms of net return. Dove states that not only are net returns higher than commercial logging in a comparison of shifting cultivation and contemporary logging. However, shifting cultivation employs more than three times the number of workers. For the sake of a developing economy. A higher level of subsistence support would be a more relevant criterion for land use selection.

7.4.16 Institutional aspects of shifting cultivation

Shifting cultivation is an age-old activity that is mostly practised by indigenous tribes in the forested and hilly regions of the tropics. It has grown up in institutional settings that are not typical of mainstream society. There are a few key features that highlight shifting cultivation's institutional face. Shifting cultivation was directed by communal laws and norms until other types of cultivation, such as market forces, the state, or any other institution, intervened. When these forces become a part of the shifting cultivators' way of existence. The structure of property rights as a common holding has also changed. Similarly, inside the community, forms of relation have shifted from reciprocity and redistributive forms to market forms of transaction. The shifting cultivators were also affected by the state's intervention in the forest in numerous ways (Jyotshi 2000). The colonial government's policy regarding changing cultivators experienced adjustments as well.

7.5 SHIFTING CULTIVATORS VIEW

Shifting farming, according to shifting cultivators, has numerous advantages. It is an indigenous knowledge-based production method that aids in the preservation of indigenous peoples' traditional culture and history. Shifting farmers harvest a variety of seasonal crops from fields, ensuring that they have enough rice, vegetables, and other foods for at least a few months of the year. Shifting cultivation is a low-tillage, low-nutrient-loss agriculture method that uses no tillage. Jum is rain-fed, thus no irrigation or watering is required on the field. Shifting cultivation has a higher agrobiodiversity than lowland farming systems. For future cultivation, moving growers have their own seed preservation strategy. In the local market, seasonal and yearly cash crops from jum, such as ginger, turmeric, chilli, cucumber, and others, are in high demand. In hilly locations, shifting farming is a source of income for rural communities. When crops are cultivated, the nutrients in the soil are depleted after a period of time. Shifting cultivation entails reserving a new, more nutrient-dense area of land for vegetable production. Farmers have previously shifted between different plots of land, allowing the soil in some plots to recuperate nutrients. Shifting cultivation, according to some proponents, has the advantage of not using fertilisers or chemicals to encourage crops to thrive in nutrient-depleted soil. It aids in the retention of a varied range of nutrients in the soil. One crop may remove all of the nitrates from the soil, resulting in no nitrates in that location after a year or two, and the plant will not grow properly. This is the point at which you should either leave the region alone for a few years or plant a crop that returns nitrates to the soil. In large-scale, long-term operations, crop rotation is critical. Shifting cultivation crops have adapted to their surroundings and are usually pest and disease resistant. As a result, the use of chemicals to manage pests and diseases is not required. Without employing outside help, shifting cultivation crops can be gathered in a systematic manner. Traditionally, shifting cultivation fields are only farmed for one year, and the fallow shifting cultivation fields, known locally as ranya ('vegetable market,') are harvested for a variety of food crops, spices, medicinal and aromatic plants. shifting cultivation is a multi-cropping agricultural system that is culturally linked and hence contributes significantly to the food security of the shifting cultivators, who do not have to worry about rice and vegetables for at least 6-9 months of the year. The shifting cultivators can readily store shifting cultivation veggies without the use of preservatives. Diverse crops can be grown on the same plot of land without the need for additional inputs. Other lowland agriculture practices make it impossible to have a wide variety of crop species in a compact area. Staggered planting, the adoption of rice varieties with varied maturity periods, and multiple cropping with crops with differing susceptibility to unfavourable climatic circumstances are all ways to increase resilience to fluctuating weather conditions. Shifting cultivation is a type of land management that boosts biodiversity. When shifting cultivation is replaced by permanent land use practices, plant diversity suffers dramatically in most locations. The loss of agrobiodiversity is particularly concerning. Shifting cultivators have protected agrobiodiversity through local regulations, practices, and informal seed and knowledge exchange networks, assuring their communities' food security. With the replacement of shifting farming, these networks collapse, resulting in a significant loss of crop genetic resources. The availability of significant genetic variety in agricultural plants, on the other hand, has been identified as a key component in climate change adaptation techniques.

For years, millions of rural workers in Myanmar have relied on shifting agriculture as their primary source of income. In many parts of the country, it still provides food security for families and communities, and it is seen as an important part of their cultural history. The fundamental concepts of shifting cultivation are still widely misunderstood, particularly at the level of the Union administration. Shifting farmers face a challenging political environment that fails to reflect the democratic ideals of small-scale food producers that employ shifting agriculture. Positive outcomes are not yet guaranteed, even if the policy environment improves. In recent decades, international scientific opinion has evolved significantly in favour of a balanced appreciation of shifting cultivation (Dressler *et al.* 2016), and the National Land Use Policy 2016 and the current draught Agriculture Policy both provide good, albeit wide acknowledgement. However, land and agriculture policy, as well as climate and environmental policy and legislative procedures, continue to be shaped by a variety of factors. In addition to legal recognition, continual government support is required to advocate for shifting growers' interests and assist them in achieving a prosperous future. Historically, practically all farmers left fields fallow, and we see a range of fallowing intensity in almost every agricultural system, with some proportion of cultivated lands kept fallow to recover productivity. Fallow seasons were shorter as pressures to enhance output grew over time, and with little land to shift to, new sources of nutrients had to be discovered as a substitute.

7.6 UN DECLARATION ON THE RIGHTS OF INDIGENOUS PEOPLES

The following articles demonstrate the United Nations' concern for indigenous peoples, among other things:

Article 10: No indigenous peoples should be forced to leave their native lands or territories. There shall be no relocation without the indigenous peoples' free, prior, and informed consent, as well as an agreement on equitable and reasonable compensation and, when possible, the right to return.

Article 11: Indigenous peoples have the right to continue to practice and renew their cultural traditions and customs. This includes the right to maintain, conserve, and develop the past, present, and future manifestations of their cultures, such as archaeological and historical sites, artefacts, designs, ceremonies, technologies, visual and performing arts, and literature.

Article 25: Indigenous peoples have the right to maintain and strengthen their unique spiritual relationships with their traditionally owned or otherwise occupied and used lands, territories, waters, coastal seas, and other resources, as well as to uphold their responsibilities to future generations in this regard.

Article 26:

1. Indigenous peoples have the right to the lands, territories, and resources that they have historically held, occupied, or obtained in any other way.
2. Indigenous peoples have the right to own, use, develop and control the lands, territories and resources that they possess by reason of traditional

ownership or other traditional occupation or use, as well as those which they have otherwise acquired.

3. States shall give legal recognition and protection to these lands, territories and resources. Such recognition shall be conducted with due respect to the customs, traditions and land tenure systems of the indigenous peoples concerned.

7.7 GOVERNMENT POLICIES

In upland locations with low population density, Tran (2007) believed that shifting cultivation was a very suited farming system since fallow periods may be lengthy enough to allow the recovery of soil nutrients and vegetation. It also provides a higher standard of life for shifting cultivators than other popular land uses, such as wet rice production, because the returns per unit of land labour are substantially higher (Tran, 2007). Moreover, recent disturbances associated with climate change have indicated that shifting cultivation can have a role in maintaining and enhancing carbon stocks (Moeliono *et al.*, 2016). Growing population pressures, along with expanding government controls over forest resources in various tropical regions, have in many ways compelled shifting cultivators to shorten the period for which their land can be fallow. This has resulted in adverse impacts on the environment, including deforestation, accelerated soil erosion, soil-nutrient depletion and biodiversity loss (Gafur *et al.*, 2000). In addition, under a high population density, shifting cultivation is incapable of improving the economic condition of practitioners because returns per unit of land labour are substantially reduced (Sanchez, 1994). The negative impacts of shifting cultivation on the environment and the associated decline of shifting farming into an unproductive system have led most governments in Southeast Asia to consider it as a primitive and uneconomical method of cultivation. Consequently, governments throughout the region have tried to ban shifting cultivation. In Malaysia, for example, huge development programmes have been implemented to improve shifting cultivators' livelihoods when they abandon shifting cultivation (Majid, 1983). Castella *et al.*, (2013) reported that in Laos, the government has issued land-use and forestry policies aimed at increasing the country's forest cover while eradicating shifting cultivation in order to promote intensification of agriculture. Similarly, in Vietnam, the government has for many years adopted stringent policies and measures to eradicate shifting cultivation. These measures include a large-scale programme called "Fixed Cultivation, Fixed Residence," which aims to permanently settle "shifting fields" and "shifting people" (McElwee, 2004), as well as a forestland-allocation policy that ties people to specific plots of land, effectively making shifting cultivation impossible (Castella *et al.*, 2005).

Studies conducted throughout Southeast Asia in recent decades have revealed that shifting cultivation has been rapidly transformed in response to countless social, economic and political challenges (Rasul and Thapa, 2003; Padoch *et al.*, 2007; Fox *et al.*, 2009). Despite the complexity of this transformation process, scholars generally agree that shifting cultivation is declining (van Vliet *et al.*, 2012), though with this trend, the practice remains important in many areas and will continue to be important throughout at least the current century in areas where alternatives are not yet available (van Vliet *et al.*, 2013). In the mid-1990s in the Lao PDR, 218,900 hectares of land was under shifting cultivation, involving a total of 124 million people, or about one third of the total population (Pravongviengham, 1998). McElwee (2004) reported that in the early 1990s in Vietnam, about 50 ethnic groups, out of the country's total of 54 ethnic groups, were engaged in shifting cultivation, employing 2.8 million people to work 3.5 million hectares in the country's uplands. Villagers continue to illegally cut down trees for shifting cultivation in numerous regions, according to the Ministry of Agriculture and Rural Development (MARD 2007). The Ministry of Natural Resources and Environment (Monre 2011) estimated that many villagers in Vietnam's uplands are still engaged in shifting cultivation on an area of more than one million ha. Recent government data shows that shifting cultivation accounted for 54% of total forest loss in the country in 2012, and 71.6% of total forest loss in the period between 2008 and 2011 (Forest Protection Department, 2013).

7.7.1 Shifting cultivation and planned change

It is wrong to believe that all forms of shifting farming are unsatisfactory land use. Bush fallow systems, according to Okigbo (1981), can be stable, ecologically sound, and efficient, however Nye and Greenland (1960) contend that there has yet to be a superior way of stable food cultivation in the tropics. It is important to note that land scarcity, soil degradation, and declining yields are linked to short fallow cultivation rather than shifting cultivation in locations where shifting agriculture was historically practised but conditions have changed. As a result, extensive fallow periods are no longer conceivable, and fertiliser and other inputs required for more intensive land use are either too expensive or unavailable. Priority planning should therefore focus on those "crisis" areas of short fallow agriculture, rather than those places where sufficient fallow periods may still restore soil fertility and forest growth. The improvement of fallow cultivation is one sort of planning; the replacement of fallow agriculture with other forms of farming is the other. Trees play a critical role in both circumstances, as they assist manage erosion, suppress or remove herbaceous and woody weeds, and restore soil fertility (Grinnell, 1975). Alternatives to fallow cultivation, such as complete land clearance or switching to other forms of agriculture without trees, such as annual crops, have largely failed.

Deforestation, land degradation, and, more recently, global warming are all seen as key contributors to shifting cultivation. The majority of shifting agriculture systems are sustainable and have excellent labour productivity at low population concentrations, therefore this reputation is mostly unwarranted. Destabilization of the formerly stable system as a consequence of variables such as rapidly growing populations, shifting production incentives as a result of market integration, or other significant changes in the institutional and regulatory environments in which swidden farmers labour causes distortions (Voss 2007). Despite the changing perceptions, policies and management approaches to shift ing cultivation have globally largely centred around replacement with alternate land uses than on finding ways to make the practice economically viable and productive and environmentally sound (Thomas 2003). The sole option for most governments dealing with shifting cultivation has been to 'wean' growers away from the practise and toward sedentarized agriculture, primarily horticulture and cash crops. Policymakers have frequently missed the need for support throughout the transition period, particularly for the least fortunate, in their zeal for fueling such revolutions to replace shifting cultivation. Many homes have been marginalised as a result of this shortsightedness, which has increased rather than decreased their vulnerability. The policy need for replacing shift ing cultivation with settled agriculture has also been to address the environmental concern of forest degradation and deforestation. However, as outlined in the earlier sections, replacement of shift ing cultivation has been counterproductive in most cases. Examples from South-East Asia, particularly China, Laos, and Indonesia, show that replacing shift ing cultivation has actually resulted in cash crop plantations—particularly rubber and oil palms—rather than in the regeneration of forests (Fox 2000). Such examples strongly indicate the need for revisiting policy approaches towards management of shifting cultivation and the strong policy support for agricultural transformations if deforestation is to be avoided and forest conservation ensured.

7.7.2 How shifting cultivation can be addressed

1. Because getting a high yield is the goal of most agronomists and extension agents working in agricultural projects, the recommended alternative land uses are frequently built on the assumption that yield per acre matters. It's what they've been taught to do. The above-mentioned labour and investment barriers, as well as the risk and discount rate aspects, are frequently disregarded, resulting in non-adoption of the

proposed strategies, with the exception of local elites, who are frequently foreigners or recent migrants with some investment capacity.

2. The strategies being presented are framed within a sustainability paradigm that believes that resource stocks should be kept constant, which is a common misconception, or at the very least a counter-productive approach. For example, it is suggested that forest biomass and soil nutrient content should not be lowered. This contradicts the logic of most economic actors, including smallholder farmers, who are devoted to substituting resources once they have been depleted rather than maintaining a continuous stock. The crisis can be avoided if the transition to a new resource (B) can be completed before the previous resource (A) is totally depleted. When resource B is used, resource A is relieved of its burden and is able to recover.
3. Policymakers frequently regard fallow land in a slash-and-burn system as idle land, as evidenced by the misleading term shifting cultivation. In "shifting" (slash-and-burn) agricultural methods, fields do not actually shift. They are continuously maintained, but cropping and fallowing periods alternate, with indistinct boundaries between the two. Fallow ground is frequently actively managed, with the addition of beneficial trees and the maintenance of any remaining planted perennials. Unless the plot is left uncultivated for longer than the normal fallow period, in which case it can be reverted to community land, it is typically administered by the same individual households that grow the crops. By ignoring this truth, policymakers may be inclined to deny slash-and-burn farmers their right to use fallow land, resulting in foolish measures that further isolate them.

Despite this, shifting cultivation is still held responsible for the majority of deforestation. Shortsighted policymakers ignore the context of tree-cutting and are swayed by modernising myths that portray slash-and-burn agriculture as a primitive land use. Slash-and-burn agriculture is a primitive land use in that it predates others in many parts of the world. However, there should be no value judgement linked to this statement. Slash-and-burn agriculture could be considered as a chance to avoid falling into productivist, unsustainable patterns at a time when doubts about mainstream development pathways are growing. But this isn't to argue that slash-and-burn farmers should be stuck with this "primitive" method of land management. They, like all communities before them, have the right to adapt and change. They should be provided access to the same variety of economic and technical possibilities that other farmers have in our globalized world.

Shifting cultivation is typically portrayed in a negative light, as are individuals who practise it. Many stakeholders regard them as primitive, backwards, unproductive, wasteful, exploitative, and harmful to the environment. They are supposed to be underprivileged and live on a subsistence level, regardless of their location. They've been blamed for much of the world's tropical deforestation, land degradation, and climate change. As a result, many contemporary national laws and regulations that influence shifting farmers are antagonistic to them, attempting to replace shifting cultivation with more modern farming practices. As a result, the government has asserted authority over lands used by shifting farmers in a number of regions, as well as the forcible displacement of local populations. There are also negative feelings about transferring cultivators in agricultural research and development institutions. Many scholars and policymakers believe that modern agriculture implies settled, intense farming with monocultures and sophisticated technologies. They typically overlook opportunities to learn from, adapt, and improve some of the most productive features of changing cultivation (Ramakrishnan, 1992; Alcorn, 1991; Redford and Padoch, 1992; Brookfield and Padoch, 1994). These myths regarding shifting cultivation and cultivators are based on historical ignorance and oversimplifications, and have resulted in laws and regulations that are antagonistic to the technique and its practitioners. The following allegations or myths about shifting cultivation are presented in Fig. 7.1, as well as how shifting cultivation and its practitioners are absolved/ acquitted of all allegations.

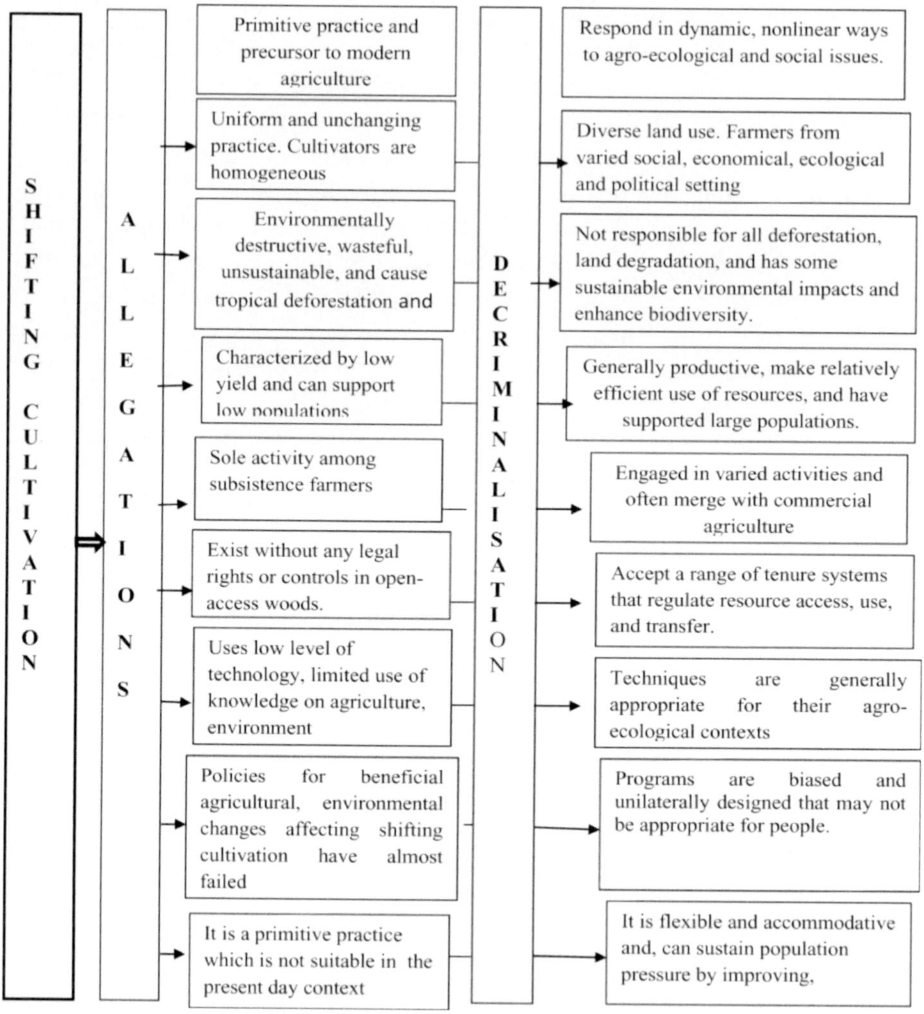

Fig. 7.1. Showing allegations against shifting cultivation and shifting cultivators and their exoneration. (Bassed on Thrupp *et al.* 1997)

References

Agrawal AA (1998). Induced responses to herbivory and increased plant performance. Science, 279: 1201-1202, DOI:10.1126/science.279.5354.1201

Alcorn. J.B. 1991. "Ethics, Economies and Conservation." In M.L. Oldfield and J.B. Alcorn, eds. Biodiversity, Conservation and Ecodevelopment. Boulder, Co: Westview Press.

Amalendu J. 2000). (2000). Swidden cultivation: A review of concepts and Issues. Institute for social and Economic Change, Bangalore, India.

Andriesse JP & Schelhaas RM (1987). A monitoring study on nutrient cycles in soils used for shifting cultivation under various climatic conditions in tropical Asia. III. The effects of land clearing through burning on fertility level. Agric Ecosys Environ 19(4), 311-332.

Angelson. Ariled (1995). 'Shifting cultivation and deforestation: A study from Indonesia', World Development. Vol. 23, No. 10, p. 1713 -29.

Aryal, K.P. & Kerkhoff, E. (2008). The right to practice shifting cultivation as a traditional occupation in Nepal. A case study to apply ILO Conventions No. 111 (Employment and Occupation) and 169 (Indigenous and Tribal peoples) Kathmandu, International Labour Office.

Banerjee. Ajit K (1995). 'Rehabilitation of degraded forests in Asia', World Bank Technical paper Number 270, World Bank, Washington, DC.

Barkakoti, S (1990).Alternative plan for jhum area development. In: Majumdar DN (ed.) Shifting Cultivation in Northeast India, Om Sons Publication, New Delhi, India.

Barney, GL. 1970 An Analysis of Swidden Cultures in Southeast Asia. Ph.D. dissertation, University of Minnesota.

Bennett, JW. 1973 Ecosystemic effects of extensive agriculture. Annual Review of Anthropology 2:36-45.

Bodley, JH. 1976 Anthropology and Contemporary Human Problems. Menlo Park, California: Cummings.

Brady, N.C. Alternatives to slash-and-burn: A global imperative. Agric. Ecosyst. Environ. 1996, 58, 3–11.

Brookfield, H. and C. Padoch. 1994. "Appreciating Agrodiversity: A Look at the Dynamism and Diversity of Indigenous Farming Practices." Environment 36(5):7-20.

Bruce JP, Frome M & Haite E (1999). Carbon sequestra-tion in soils. J Soil Water Cons, 54, 382-389.

Bruun, T.B., de Neergaard, A., Lawrence, D., Ziegler, A.D., 2009. Environmental consequences of the demise in swidden agriculture in Southeast Asia: soil nutrients and carbon stocks. Hum. Ecol. 37, 375–388.

Cairns, M., & Garrity, D. P. (1999). Improving shifting cultivation in Southeast Asia by building on indigenous fallow management strategies. Agroforestry Systems, 47(1), 37–48. Doi: 10.1023/A:1006248104991

Castella J.C., Lestrelin G., Hett C., *et al.* (2013) Effects of landscape segregation on livelihood vulnerability. Human Ecology, http://dx.doi.org/10.1007/s10745-012-9538-8

Castella, J. C., Boissau, S., Tran, N. T, and Dang, D. Q. (2005) 'Agrarian transition and lowland-upland interactions in mountain areas in northern Vietnam: Application of a multi-agent simulation model', Agricultural Systems 86(3), pp312-332

Chin, SC (1985) 'Do shifting cultivators deforest', in V. Shiva (ed.) Forest Resources: Crisis and Management (1992), Natraj Publisher: Dehradun.

Christanty L, 1986. Shifting Cultivation and Tropical Soils: Patterns, Problems, and Possible Improvements.. In Marten GG (edn) Traditional Agriculture in South East Asia: A Human Ecology Perspective. Westview Press, Boulder, Colorado.

Conklin, HC. 1957 Hanunoo Agriculture. FAO Forestry Development Paper no. 12. Rome: Food and Agriculture Organization.

Conklin. H C (1961). 'The Study of Shifting Cultivation'. Curren/Anthropology. Vol. 2. p. 27-61.

Cox, G. W. and Atkins, M. D. l979. Agricultural Ecology. San Francisco: W. H. Freeman and Co.

Cramb, R. A. (1989). The Use and Productivity of Labour in Shifting Cultivation: An East Malaysian Case Study. Agricultural Systems 29: 97–115. doi:10.1016/0308-521X(89)90057-7.

Craswell ET, Sajjapongse A, Howlett DJB & Dowling AJ (1997). Agroforestry in the management of sloping lands in Asia and the Pacific. Agrofor Syst 38,121–137.

Denevan, W.M., 1981. Swiddens and cattle versus forest: the imminent demise of the Amazon rain forest reexamined. In: V.J. Sultive, N. Altshuler and M.D. Zamora (Editors), Where Have All the Flowers Gone? Deforestation in the Third World, Studies in Third World Societies. College of William and Mary, Williamsburg, VA, pp. 25-44.

Detwiler RP (1986). Land use change and the global car-bon cycle: the role of tropical soils. Biogeochemistry, 2, 67-93.

Dove, M. R. (1984). Man, land, and game in Sumbawa: Some observations on agrarian ecology and development policy in eastern Indonesia. Dove, Michael R, 112-124.

Dove, M. R. (1992). Foresters' beliefs about farmers: a priority for social science research in social forestry. Agroforestry Systems, 17(1), 13-41.

Dove, MR 1983. 'Theories of Swidden Agriculture, and the political economy of ignorance' in Agroforestry Systems, June 1983, 1(2): 85-99.

Dove. Michael. R (1985). 'The Agroecological Mythology of the Javanese and the Political Economy of Indonesia'. Indonesia. Vol. 39.

Dove MR (1996). So far from power, so near to the forest: a structural analysis of gain and blame in tropical forest development, In: Padoch C, Peluso NL (eds.), Borneo in Transition: People, Forests, Conservation, and Development, Oxford University Press, Kuala Lumpur, pp. 41-58.

Dressler W. H., Wilson D., Clendenning J., Cramb R., Keenan R., Mahanty S., Bruun T. B., Mertz O. and Lasco R. D. 2016 The impact of swidden decline on livelihoods and ecosystem services in Southeast Asia: A review of the evidence from 1990 to 2015. Ambio 46(3): 291-310.

Duguma B, Gockowski J & Bakala J. (2001). Smallholder cacao cultivation in agroforestry systems of West and Central Africa: Challenges and opportunities. Agrofor Sys 51(3), 77–188.

Eckholm, E.P. 1976. Losing Ground. New York, N.Y.: Norton and Company.

El Moursi, A.W.A., 1984. The role of higher agricultural education in the improvement of shifting cultivation farming systems in Africa. In: A.H. Bunting and E. Bunting (Editors), The Future of Shifting Cultivation in Africa and the Task of Universities, Proc. Int. Workshop on Shifting Cultivation: Teaching and Research at the University Level, Ibadan, Nigeria. FAO, Rome, pp. 8-14.

Erni, C. (2009). Shifting the blame? Southeast Asia's indigenous peoples and shifting cultivation in the age of climate change. Paper presented at the seminar on "Adivasi/ ST Communities in India: Development and Change", Delhi, August 27-29, 2009, p. 6

FAO. 1957. Shifting cultivation. Unasylva 11: 1

FAO (2010) Global Forest Resources. Assessment, FAO forestry paper, 163, Main Report, Rome, Italy.

Fearnside, P. M. (2000), Global warming and tropical land□use change: Greenhouse gas emissions from biomass burning, decomposition and soils in forest conversion, shifting cultivation and secondary vegetation, Clim. Change, 46, 115–158, doi:10.1023/A:1005569915357.

Forest Protection Department (2013) Forest allocation and swidden management (Giao rừng và quản lý nương rẫy),www.kiemlam.org.vn, (Vietnamese language), accessed 29 October 2013

Fox, J. 2000, 'How Blaming "Slash and Burn" Farmers is Deforesting Mainland Southeast Asia', Asia Pacifi c Issues, 47: 1–8.

Fox, J., Fujita, Y., Ngidang, D., Peluso, N., Potter, L., Sakuntaladewi, N., Sturgeon, J. and Thomas, D. (2009) 'Policies, political-economy, and swidden in Southeast Asia', Human Ecology 37(3), pp305-322

Gafur A (2001). Effects of shifting cultivation on soil properties, erosion, nutrient depletion and hydrological responses in small watershed of the Chittagong Hill Tracts of Bangladesh.

Doctoral dissertation, The Royal Veterinary and Agricultural University, Copenhagen, Den-mark.

Gafur, A., Borggaard, O. K., Jensen, J. R. and Peterson, L. (2000) 'Changes in soil nutrient content under shifting cultivation in the Chittagong Hill Tracts of Bangladesh', Danish Journal of Geography 100, pp37-46

Geertz, C (1963) Agriculturallnvolution. Berkeley: University of California Press.

Giri, T. (2007). Strengthening Monitoring, Assessment and Reporting on Sustainable Forest Management in Asia (GCP/INT/988/JPN). Forest Department, Food and Agriculture Organization of the United Nation, Rome.

Grandstaff, T.B., 1980. Shifting Cultivation in Northern Thailand: Possibilities for Development. Resource Systems Theory and Methodology Series, 3. United Nations University, Tokyo, 44 pp.

GrandstafT. T.ll. 1981 Shifting Cultivation: a reassessment of stratcgics. In CERES. 14 (4).

Greenland DJ (1975). Bringing the green revolution to the shifting cultivator. Sci., 190: 841–844.

Grinnell, H.R. 1975. A study of agrisilviculture potential in West Africa. Ottawa, International Development Research Centre.

Hatch, T., and Y.L. Tie. 1979. Shifting cultivation in Sarawak and its effects on soil fertility. Paper presented at a Seminar on the Fertility and Management of Deforested Land, Kota Kinabalu, Malaysia.

Hatch. T (1980). 'Shifting Cultivation in Sarawak: Past. Present and Future' in J L Furtado (ed.) Tropical Ecolog and Development. Proceeding of the Fifth International Symposium of Tropical Ecology. 16-21 April 1979. Kuala Lumpur. Malaysia. International Sociery of Tropical Ecology.

Hillel, D., 1992. Out of the Earth: Civilization and the Life of the Soil. University of California Press, Berkeley, CA, 321 pp.

Hong. Eve~e (1990). 'Swidden as a Rational Agricultural System' in Return to the Good Earth, Damaging Effects of Modern Agriculture and the Case for Ecological Farming, Third World Network. Malaysia.

Indian Council of Agricultural Research (1983). Shifting cultivation North east India. Publication unit, ICAR, New Delhi. Pp - 68.

Intergovernmental Panel on Climate Change (IPCC) 2006. Land Use, Land-Use Change and Forestry. IPCC, accessed on 29 July 2009 at: http://www.grida.no/publications/other/ipcc_sr/?src=/climate/ipcc/land_use/008.htm#s6-1

IPCC. (2007). Synthesis report. An Assessment of the Intergovernmental Panel on Climate Change. Valencia.

Janzen, D.H. 1973. Tropical agro ecosystems. Science, 182, 1212-1219.

Jasbir Singh (1997). Agricultural development in South Asia: a comparative study in the Green Revolution experiences. National Book Organisation.

Jyotshi A (2000). Swidden cultivation: A review of concepts and Issues. Institute for social and Economic Change, Bangalore, India.

Kidd, C.V. and Pimentel, D. (Editors), 1992. Integrated Resource Management: Agroforestry for Development. Academic Press, San Diego, CA, 223 pp.

Kleinman PJ, Pimentel D & Bryant RB (1995). The eco-logical sustainability of slash-and-burn agriculture. Agric Ecosys Environ 52(2-3), 235-249.

Knudsen JL & Khan NA (2002). An exploration of the problems and prospects of integrated watershed develop-ment in the CHT. In: Khan, N.A., Alam, M.K., Khisa, S.K. and Millat-e- Mustafa M. (Eds.), Farming Practices and Sustainable Development in the Chittagong Hill Tracts, CHTDB and VFFP –IC, Chittagong, Bangladesh, pp. 165–180.

La Scala Jr. N, Bolonhezi, D. and Pereira GT (2006). Short-term soil CO2 emission after conventional and reduced tillage of a no-till sugar cane area in southern Brazil. Soil and Tillage Research, 91: (1–2), 244-248

Lane A and Jarvis A (2007). Changes in Climate will modify the Geography of Crop Suitability: Agricultural Biodiversity can help with Adaptation. Open Access Journal published by ICRISAT, 4(1): 1-12

Leach, G. 1976 Food production and consumption: the efficiency of industrial energy in human food chains. In Human Food Chains and Nutrient Cycles, edited by A. N. Duckham, J. G. W. Jones, and E. H. Roberts, pp. 371-382. Amsterdam: North Holland Publishing Co.

Lehsten, V., Tansey, K., Balzter, H., Thonicke, K., Spessa, A., Weber, U., Smith, B. & Arneth. A. (2009). Estimating carbon emissions from African wildfires, Biogeosciences, 6: 349–360, doi:10.5194/bg-6-349-2009.

Majid, Z. (1983) 'Malaysia', in Y. Atal and P. L. Bennagen (eds) Swidden Cultivation in Asia, Vol. 2, Country Profiles: India, Indonesia, Malaysia, Philippines, Thailand, Regional Office for Education in Asia and the Pacific,United Nations Educational, Scientific and Cultural Organization, Bangkok, pp157-238

Mard (2007) Proposal for Supporting Upland People in Sustainable Agroforestry Cultivation on Swidden Land in the period of 2008-2012, Ministry of Agriculture and Rural Development, Hanoi

McElwee, P. (2004) 'Becoming socialist or becoming Kinh? Government policies for ethnic minorities in the Socialist Republic of Vietnam', in C. Duncan (ed.) Civilizing the Margins, Cornell University, New York, pp182-213

Mertz O (2002). The relationship between length of fal-low and crop yields in shifting cultivation: a rethinking. Agrofor Sys 55(2), 149-159.

Moeliono, M., Thu, T. P., Ngoc, D. L., Brockhaus, M., Wong, G., Kallio, M. and Dinh, T. N. (2016) 'Local governance, social networks and REDD+: Lessons from swidden communities in Vietnam,' Human Ecology 44(4), pp435-448

Monre (2011) Decision 2282 dated 8 December 2011 on Approval and Announcement of Statistics on Land Areas as of 1 January 2011, Ministry of Natural Resources and Environment, Hanoi

Myers, N., 1980. Conversion of Tropical Moist Forests. National Academy of Sciences, Washington, DC, 205 pp.

Myers, N., 1992. The Primary Source: Tropical Forests and Our Future. W.W. Norton, New York, 416 pp.

Myers. Norman (1995). 'Tropical deforestation: rates and patterns' in Katrina Brown & David W Pearce (ed.) The Causes of Tropical Deforestation, The Economic and Statistical Anajysis of Factors giving Rise to the Loss of Tropical Forests. UCL Press London & Research Press New Delhi.

Nakashima D and Roue M (2002). Indigenous knowledge, peoples and sustainable practice. (ed.) Peter Timmerman, In: Encyclopedia of Global Environmental Change, John Wiley & Sons, Ltd, Chichester, 2002

Nimachow G, (2003). Akas and their forests: A study on traditional management of forest resources, Arunachal Univ Res J, 6 (2): 82.

Noel DU and Herby B (2000) Global climate change and the effect of conservation practices in US agriculture. Environmental Earth Sciences 40(3):197-209, DOI:10.1016/S0959-3780(00)00023-6

Nye. P H and D I Greenland (1960). The Soil under Shifting Cultivation. Technical communication no. 51. Farnham Royal. England: Commonwealth Agricultural -Bureaux.

Odum, H. T. 1971. Environment, Power, and Society. New York: Wiley.

Okigbo B. N. 1984. Improved permanent production systems as an alternative to shifting. International Institute of Tropical Aqriculture, Ibadan, Nigeria. In: Okigbo B. N. Improved permanent production systems as an alternative to shifting. FAO Soils Bulletin No. 53, Food and Agricultural Organization of the United Nations.

Okigbo, B.N. 1981. Alternatives to shifting cultivation. Ceres (FAO), 15 (6): 41-45. Rome.

Olofson, H. 1980. Swidden and kaingin among the southern Tagalog: a problem in Philippine upland ethno-agriculture. Philippine Quarterly of Culture and Society, 8: 168-180.

Padoch C & Pinedo-Vasquez M (2010). Saving slash-and-burn to save biodiversity. Biotropica 42(5), 550-552.

Padoch,C.,Coffey, K.,Mertz,O., Leisz, S.,Fox, J., andWadley, R. (2007). The Demise of Swidden in Southeast Asia? Local Realities and Regional Ambiguities. Danish Journal of Geography 107: (1)29–42.

Padoch. C (1980). 'The Environmental and Demographic effects of Alternative Cash- Producing Activities Among the Shifting Cultivators of Sarawak' in I L Furtado (ed.) Tropical Ecoloy and Development. Proceeding of the Fifth International Symposium of Tropical Ecology. 16-2 I April 1979. Kuala Lumpur. Malaysia, International Socieo/ of Tropical Ecology.

Pimentel, D, and Pimentel M (1979) Food, Energy and Society. New York: John Wiley

Pimentel, D. and Heichel, G.H., 1991. Energy efficiency and sustainability of farming systems. In: R. Lal and F.J. Pierce (Editors), Soil Management for Sustainability. Soil and Water Conservation Society, Ankeny, IA, pp. 113-123.

Pravongviengham, P. P. (1998) 'The role of animal husbandry and aquaculture in improvements of swidden-based livelihood systems in the Lao PDR', PhD thesis, Asian Institute of Technology, Bangkok.

Ramakrishnan PS (1992). Shifting Agriculture and Sus-tainable Development: an interdisciplinary study from north-eastern India. MAB Series, Volume 10, UNESCO, Paris.

Raman T.R.S. (2001): Effect of slash-and-burn shifting cultivation on rainforest birds in Mizoram, Northeast India. Conservation Biology, 15(3): 685-698.

Rambo T, (1981). Fire and the Energy Efficiency of Swidden Agriculture. East-West Environment and Policy Institute, Honolulu, Hawaii, 96848.

Ranjan R and Upadhyay, VP, 1999. Ecological problems due to shifting cultivation, Current Science, 77: 1246-1250, https://www.researchgate.net/publication/288180254

Rappaport, R.A. (1971) The flow of energy in an agricultural society. Scientific American ,225:116-132.

Rastogi, M., Singh, S. & Pathak, H. (2002). Emission of carbon dioxide from soil. Current Science, 82(5), 510-517. International Letters of Natural Sciences Vol. 46 27

Rasul G, Thapa GB. 2003. Shifting cultivation in the mountains of south and Southeast Asia: Regional patterns and factors influencing the change, Land Degradation and Development, 14: 495–508.

Redford, K. and C. Padoch, eds. 1992. Conservation of Neotropical Forests. New York: Columbia University Press.

Ruddle, K. and Manshard, W. 1981. Renewable natural resources and the environment. Dublin: Tycooly International (for UNU).

Russell, W. M. S. 1968. The slash-and-burn technique. Natural History 78(3): 58 - 65.

Ruthenberg, H. (1971). Farming Systems in the Tropics. Oxford: Clarendon Press.

Sahu, S.K., Nayak, R.K. and Sarangi, D. 2005. Sustainable Soil and Land Management Under Shifting Cultivation in Orissa. Orissa Review, pp. 49-52.

Sanches, Pedro A. 1976 Properties and Management 0/ Soils in the Tropics. New York: John Wiley.

Sanchez PA. 1994. Alternative to slash and burn: a pragmatic approach for mitigating tropical deforestation. In Agricultural Technology: Policy Issues for the International Community, Anddrson JR (ed.). CAB international: Wallingford; 451–479.

Schlesinger WH (1977). Carbon balance in terrestrial detritus. Ann. Rev. Ecol. Syst. 5: 51-81.

Sharma TC (1976). The pre-historic background of shift-ing cultivation in North-East India. North-East Indian Council for Social Science Research.

Sharma UC and Bujarbaruah KM (2016). Issues and Challenges in Shifting Cultivation and its Relevence in the Present Context, National Academy of Agricultural Sciences, Policy Paper No. 83.

Shengji, P., Huijon, G. and Zhiling, D., 1990. An introduction to the study of ethnobotany of Xishuangbanna. Proc. 4th Int. Conf. on Thai Studies, 11-13 May 1990, Kunming, Vol. 3. Institute of Southeast Asian Studies, Kunming, pp. 435-445.

Sobrevila, C. (2008). Role of Indigenous People in Biodiversity Conservation. The International Bank for Reconstruction and Development / THE WORLD BANK, 1818 H Street, N.W. Washington, D.C. 20433, U.S.A

Springate-Baginski O, 2018. Decriminalise agroforestry – A primer on shifting cultivation in Myanmar. Transnational Institute (www.tni.org www.tni.org), 36p.

Steinhart JS and Steinhart CE (1974). Energy use in the U.S. food system. S 184:307-316.

Tanaka, H. K., Nakano, T., Takahashi, S., Yoshida, J., Takeo, M., Oikawa, J. *et al.* (2007). High resolution imaging in the inhomogeneous crust with cosmic-ray muon radiography: The density structure below the volcanic crater floor of Mt. Asama, Japan. Earth and Planetary Science Letters, 263(1-2), 104-113.

Terborgh J (1999). Requiem for nature. Island Press, Washington, D.C.

Thomas, D.E. 2003, 'A Brief Review of Upland Agricultural Development in the Context of Livelihoods, Watersheds and Governance for Area-based Development Projects in the Lao PDR', ICRAF Report for the International Fund for Agricultural Development, Chiang Mai, Thailand.

Thrupp L. A., Hecht S. B. and Browder J. O. (1997), The diversity and dynamics of shifting cultivation: Myths, realities, and policy implications. World Resources Institute, Washington, DC.

Tiessen HE, Cuevas and Chacon P (1994). The role of soil organic matter in sustaining soil fertility. Nature 371, 783–785.

Tobing, P.L., 1991. Forests for the future. Garuda Indonesia, April: 23-30.

Tran, D. V. (2007) Solutions in Fallow Management: Swidden Agriculture Experience in Vietnam's Uplands, Center for Agricultural Research and Ecological Studies, Hanoi Agricultural University, Hanoi

Van Noordwijk M., Mulyoutami, E., Sakuntaladewi, N., & Agus, F. (2008a). Swiddens in transition: shifted perceptions on shifting cultivators in Indonesia. Occasional Paper no.9. Bogor, Indonesia: World Agroforestry Centre.

van Noordwijk, M., Suyamto, D. A., Lusiana, B., Ekadinata, A., & Hairiah, K. (2008b). Facilitating agroforestation of landscapes for sustainable benefits: tradeoffs between carbon stocks and local development benefits in Indonesia according to the FALLOW model. Agriculture, Ecosystems & Environment, 126(1-2), 98-112.

van Vliet N, Adams C, Vieira ICG, Mertz O. 2013. "Slash and Burn" and "Shifting" cultivation systems in forest agriculture frontiers from the Brazilian Amazon. Soc Nat Resour. 26:1454–1467.

van Vliet N, Mertz O, Andreas Heinimann A et al. 2012. Trends, drivers and impacts of changes in swidden cultivation in tropical forest-agriculture frontiers: A global assessment. Global Environmental Change, doi:10.1016/j.gloenvcha.2011.10.009.

Verma PK, Kumar V, Chandra A and Thounaojam B (2017). Alternatives of shifting cultivation in North-eastern region of India. Report and Opinion 9 (12): 1-8, doi:10.7537/marsroj091217.01.

Voss, Joachim. 2007, 'Foreword', in Malcolm Cairns (ed.), Voices from the Forest: Integrating Indigenous Knowledge into Sustainable Upland Farming, Washington, D.C.: Resources for the Future.

Wan, Y.F. & Lin, E.D. (2004). The influence of tillage on CH4 and CO2 emission flux in winter fallow cropland. Chinese Journal of Agrometeorology 3:8-10

Ward, G. M., Sutherland TM, and Sutherland JM (1980) Animals as an energy source in Third World agriculture. S 208:570-574.

Warner K, (1991). Shifting cultivators Local technical knowledge and natural resource management in the humid tropics. Food and Agriculture Organization of the United Nations, Rome.

Watters, R. F. (1971). Shifting Cultivation in Latin America. FAO, Rome.

Weil R (1982). Soils of the Upper Mahaweli reforestation and watershed management project. Project Report pub-lished by SECID, Department of Forest, Sri Lanka.

World Resources Institute (WRI), 1990. World Resources 1990-91. Oxford University Press, Oxford, 383 pp.

Xu. lainchu. lafferson Fox. Lu Xing, Nancy Podger. Stephen Leisz and Ai Xihui (1999). 'Effects of Swidden Cultivation, State Policies and Customary Institutions on Land Cover in a t-lani Village. Yunan. China'. Mountain Research and Development. Vol. 19. No.2.

Ziegler A. D., Fox J. M. and Jianchu Xu (2009), 'The rubber juggernaut', Science, vol 324, pp. 1024–1025.

Ziegler AD, Bruun TB, Guardiola-Claramonte M, Giam-belluca TW, Lawrence D & Lam NT (2009). Environ-mental consequences of the demise in Swidden cultivation in montane mainland Southeast Asia: hydrology and geo-morphology. Hum Ecol 37, 361–373.

Zinke, PJ., Sabhasri S, and Kunstadter P, (1978) Soil fertility aspects of the Lua' forest fallow system of shifting cultivation. In Farmers in the Forest, edited by P. Kunstadter, E. C. Chapman, and Sanga Sabhasri, pp. 134-159. Honolulu: University Press of Hawaii.

Zodinpuii, B., Lalnuntluanga and Lalthanzara. H. 2016. Impact of shifting cultivation on soil organic carbon in tropical hilly terrain of Mizoram, India. Science Vision 16 (3): 135-143

8

Dabbling, Decline and Demise

8.1 GENERAL DESCRIPTION

Based upon ethnic values and culture, both private and common property land tenure regime is found in existence in the shifting cultivation system. In Odisha state of India, shifting cultivation locally called as *podu*, there exists both private and common property regime (Dash 2006). In case of Meghalaya state, while the private land ownership pattern does exist among the Khasi and Karbi tribes, the *Garo* tribes, on the other hand, resort to community land ownership pattern (Deb *et al.,* 2013). Furthermore, an ethno-agrarian economy based on shifting cultivations is self-sufficient, and such shifting cultivators typically rely on outside sources for alleviation of their economic demands very infrequently, if at all (Gupta 1994). The use of human labour is the key input in the shifting cultivation system and the practice of mutual exchange of labour amongst these shifting cultivator communities is also not uncommon. In practices like harvesting, the groups of families join hands together to harvest the crop of each plot one by one. This practice of community labouring reduces drudgery and increases work productivity. During this harvesting time, it is said that farmers exchange their innovations and look at each other's innovations as they work. Further, it is customary to arrange food and drinks by the family of the plot which is being harvested. The community recognizes these harvest periods as festive events which makes their work enjoyable. The case of village named Nawalparasi, Nepal, shows that shifting cultivators work in an innovative way in their community in which the part of the village land is set aside and is cultivated by the members of whole community jointly. The benefit they harvest from this area is used to fund for teachers' salaries in community school (kerkhoff and Sharma 2006). Further, the close knit and kinship ties are not only necessary for labour sharing but also important for technology diffusion among the members (NAEB 2008). The division of labour exists between men and women. It varies from place to place. The Manipur shifting cultivation characteristics portray that the women predominate in seed selection and planting, weeding and other operations, the activities like cutting of the trees in jungle, clearing, burning are generally done by men. Both men and women participate in harvesting of produce and so harvested produces are carried to the villages by head loading.

8.1.1 Traditional institutions and its role in shifting cultivation society

Traditional institution plays an important role in the Jhum society. Institutions are defined as "common norms, values, traditions, beliefs, religion, rules, regulations, laws, civil society groups, and government agencies" in sociological terms (North 1990). The traditional institutions were nothing but the village councils which have been responsible for managing land resources at community level. Each and every clan was represented in this village council. However, the composition of the traditional institution varies between the tribes. The communities in Nagaland's Mokokchung district, Nagaland, India, practise collective decision making on land use for various purposes such as demarcation of burning areas, digging of fire lines, and controlled burning, all of which contribute to the state's continued sustainable and equitable use of land and resources (IWGIA 2014). Similarly in Kachin state of Myanmar, the community cooperation took place in fire management and fire control. In Chittagong hill tract of Bangladesh, the community decision making was involved in natural resources management. Social security was also major objective of local establishment of shifting cultivators. They make available the farmers with the relationship which they need for their living. The hangvu, a traditional village chief in Manipur, was similar to other shifting agriculture villages. The hangvu is comprised of chieftain and a council of clan elders (Kerkhoff and Sharma 2006).The indigenous knowledge occupies an important place among the communities of shifting cultivators. Land management, biodiversity conservation, water management, health care, and medicine are only a few of the activities covered by indigenous knowledge (Chin 2006). This indigenous knowledge has been acquired over centuries by community members through careful observation and experimentation, and it is passed down verbally or via experience learning.

8.2 DABBLING

Dabbling means to work at anything in an irregular or superficial manner. It is involving oneself superficially or intermittently especially in a secondary activity or interest. We are of the view that “the shifting cultivation may not be relevant today due to changed scenarios of population pressure, non-availability of land, intensification of agriculture, government policies etc. but, never-the-less it has been a remarkable land use system which has remained sustainable since Neolithic period – for such a long period that no other land use could match it. Now, so many improved land use systems come and go but the practice of shifting cultivation still exists in various forms”. Millions of farmers still utilise shifting agriculture, which was originally used all across the world. Slash and burn includes utilising a controlled burn to cultivate land for production. These fires function by burning fields, which gradually return to their natural state and can support new life such as native plants and trees. Shifting cultivation has various advantages, including the conversion of biomass or dead plant waste into ashes and the nutrient enrichment of soils. It can also help to keep pests and weeds at bay. However, shifting cultivation can be an unsustainable agricultural strategy, leaving forest regions unable to support plants, animals, and a variety of crops, when cropping period is dabbled and fallows are not managed appropriately

The history of shifting cultivation is as old as human civilization. Dabbling with or disinterest in shifting cultivation is on the increase these days. There are several reasons for this and the major factors are; socio-economic, very low yields not commensurate with the demand, race for alternate white collar jobs by the new generation, land tenure, small holdings, subsidiary sources of income like piggery, business, government service, awareness about the environmental consequences as a result of the practice. Moreover, the shifting cultivators, who practice piggery, use plant refuse and below quality grains from the shifting cultivation fields for feeding pigs. Dabbling with shifting cultivation is becoming common, accelerating the practice in its decline and finally demise. Out of all the activities or factors responsible for decline and demise of shifting cultivation, dabbling attitude holds the central position because if the people, particularly the youth of new generation are themselves not interested in the practice, no incentive can help in stopping decline. Other activities will assist in acceleration of the process of decline. It's worth noting that the land parcels designated for shifting cultivation aren't picked in any particular order or sequence, allowing for some flexibility. Jhum cultivators who relocate regularly do not follow cyclical cultivation methods on a single plot, according to research. The plots are often chosen depending on how easy it is to remove the vegetation and the soil fertility levels (Siddiqui & Chohan, 2015).

Though a simple method of utilizing land and forest resources, shifting agriculture illustrates a complicated link between environment, austerity, and fraternity (Kerkhoff & Sharma, 2006; Priyadarshni, 1995). In Northeast India, shifting farming has been defined as a way of life for the societies that practice it. In fact, shifting cultivation areas surrounded by woods and environment are discovered to give two alternate means of subsistence to the shifting cultivation community. Malevolent spirits, Gods, and Goddesses are worshipped and sacrifices are performed before planting to ensure a successful crop and family fortune. It is believed that if a cock's throat is half cut and left wandering in the field, and it dies resting to its right, the field will yield a plentiful crop and fortune to the family, and vice versa, in the inner sections of the Garo and Khasi hills of Meghalaya (Priyadarshni, 1995). However, sacrifices before sowing are no longer widespread in the region. To defend their shifting cultivation fields, many shifting farmers in the area follow the tradition of erecting a house. Many human activities, diminishing soil fertility, deforestation, degradation, agricultural intensification, government policies, availability of alternate farming systems and dabbling will eventually lead to decline and demise of shifting cultivation (Fig.8.1).

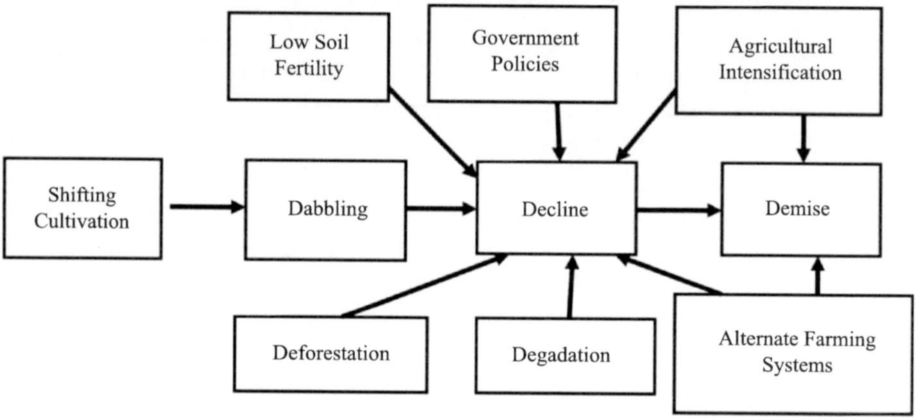

Fig. 8.1 Activities leading to decline and demise of shifting cultivation

Shifting agriculture is harmful for the environment. It is because in this type of cultivation, a piece of land is first cleared. The trees are felled and burnt down. Thus, the practice of shifting cultivation results in the degradation of soil and the forest land. By cultivating rice and maize in terraces, the emissions have been minimised. Rapid soil erosion due to deforestation of hill tops and slopes, as well as high runoff velocity and siltation of reservoirs, rivulets, and valleys, were all detrimental outcomes of shifting cultivation. As the soil loses its fertility, the land is abandoned, and the farmer relocates to a new plot. Shifting agriculture has a number of drawbacks, including deforestation, loss of land fertility, and soil erosion. For millennia, shifting agriculture has been an important part of upland Southeast Asia's agricultural systems, and the use of fire is crucial in these locations. External actors, including as governments, enterprises, and conservation groups, have been putting pressure on shifting cultivation systems for almost as long, with the goal of controlling or eliminating shifting farming (Scott 2009). In Southeast Asia, the shifting agricultural method is usually concentrated on the production of upland rice. The Dayak Tunjung community is one of the remaining aboriginal tribes in

East Borneo that still relies on shifting farming, fishing, hunting, and gathering with neighbours for their subsistence (Dove 1983). The activities of the Dayak people are centered on a shifting rice intercropping system with other staples as well as forest succession following abandonment.

8.2.1 Social dimension

Shifting cultivation holds a specific place in the ethos of concerned tribal societies, and it is intimately related to social relationships, cultural values, and mythical belief. The majority of Lao PDR's rural population is fully reliant on nature and what it can supply. Rural people are frequently forced to pick non-sustainable solutions in their search for a way to exist. To begin with, the shifting cultivator's social structure is based on households of various sizes with members who are generally closely related. Home heads are often between the ages of 20 and 45, and they rely on household labor for the majority of their income. As a result, over 300.000 families (or about 40% of the population) from all ethnic groups (Lao Sung, Lao Theung, Lao Lum) are involved in some sort of shifting farming. Aside from that, almost all rural people make use of what the forest has to offer. The freehold system of shifting cultivation has evolved dramatically as a result of recent population growth. The land tenure system is based on the freehold system, which allows family members to utilize but not sell their land. Because of the rising demand for land as a result of the growing population, farmers have cut the fallow period to the point where the soil does not have enough time to renew its fertility (Tommy 1984). For example, it is impossible to find a wide variety of forest items at the local market, such as flowers, leaves, herbs, Bamboo shoots, mushrooms, wild animals, and so on. Many of these items can only be made in the presence of steady forest conditions. For some, they provide a tiny income, while for others, they provide a means of survival.

Shifting cultivation is seen as an expensive and unscientific method of land use around the world. It is a general notion that the practice has a destructive and far-reaching negative influence on the ecosystem and ecology. This may hold good under higher population densities and scarcity of land. The increase in human population, particularly in the developing countries, has increased the demand for food and the practice is unable to fulfill the people's requirement for food and other services. The expansion of agriculture lands to increase food production has resulted in a direct decline in forest areas, resulting in forest degradation. By the early 1990s, about 40% of the earth's land surface had been turned to crop lands and permanent pastures, according to one estimate (WRI 1996).

Shifting cultivators throughout tropics are rapidly abandoning indigenous land use practices. While these changes have been quantified many a times but, not crystallized on global scale. There have been agreements and disagreements on several points with poor reliability from both sides. Coffey *et al.* (2007) identified three interconnected issues that account for at least some of the knowledge gap; (i) shifting cultivation is a diversified, complex, and dynamic land use that experts have trouble defining and quantifying; (ii) shifting cultivation is practiced by marginal farmers, and government officials find it hard to measure what is going on in dynamic and variable marginal lholdings; and (iii) national policies in almost all countries have attempted to reject shifting cultivation. They stated that a precise assessment of the extent and rate of changes in shifting cultivation on a regional scale is vital for identifying the processes that cause these shifts, as well as assessing their local and regional repercussions.

The shifting cultivators have an intimate relationship with the land and have a profound attachment and regard for it because it is a way of life, according to the report. Though there are certain patterns of change in shifting agriculture, the data is extremely variable. According to the research, several countries are transitioning from a subsistence-based system to a dual-economy one in diverse forms. The practice of shifting cultivation land use is controversial, as it is considered as an unscientific method of land use that results in land degradation and ecosystem fragility. As a result, erosion from steep slopes is causing massive losses of vegetation cover and top productive soils in the region. As a result, a comprehensive, multifaceted plan is needed to turn large-scale agriculturally degraded land from a financial and environmental liability into a profitable asset. However, in the tropics, a thorough and extensive study on the social dimension of shifting cultivation and methods for its improvement is required.

8.2.2 Why dabbling?

Due to societal compulsions, the new generation youths are not opting for shifting cultivation. The persons involved totally in shifting cultivation have the lowest social ranking as they are not financially sound. This has discouraged the youths of new generation for taking up the profession of their ancestors but somehow carrying on with it half-heartedly as either they have got a job in government service or started some skilled jobs for subsidiary source of income. For them shifting cultivation has become a side business which compensate some of their requirements of food grains, vegetables, timber, fuel wood and livestock fodder.

Low yields

Very low crop yields from shifting cultivation is another reason for disinterest of the people in continuing with this practice. With the fast increase in population, the demand for food grains and other services has increased tremendously. The shifting cultivation cannot cope with the increase in demand and the shifting cultivators have to depend on other sources of income to run the families. With this, people no more show interest in the practice and simply dabble with it.

Efforts for cash

The practice of shifting cultivation provides the produce, which is not even sufficient for the family for the whole year, not to think of selling the produce to get cash. Now-a-days, the requirement for cash has increased due to payments for school fee for education of the children in good schools, new clothing, purchase of fruits and vegetables and other requirements of the children and elder people. So, the people remain busy in doing such works which ensure remuneration in cash. Under such situations, shifting cultivation is relegated to the second place just to carry on with family profession, showing least interest for its improvement.

Race for government jobs

In rural areas, there is trend with the rural youths to go in search of government jobs which involve less physical labour than the shifting cultivation as well as ensure payments in cash. By and large, in all the countries where shifting cultivation is practsed, the literacy rate has increased and the youths of new generation have become eligible for white collar jobs. Again, with development, many jobs have been created by the respective governments and private agencies and youths always remain in the line to grab them. So, the intention of the shifting cultivators has been diverted from the shifting cultivation and just dabbling with it, which, I think, is doing more harm than good.

Land tenure

In many of the shifting cultivation areas, the shifting cultivators have usufructuary rights over land. The get the land from village chief or community only for the period of cultivation and if they keep the land fallow for 2 or 3 years, they loose the right over land. Further, they do not know when the allotted land will be taken back from them. For example in northeastern states of India, the main hub of shifting cultivation, the ownership of land falls under three categories, such as; (i) land belonging to village head or village chief, (ii) community owned land and, (iii) land belonging to individual farmers. Most of the land falls under first two categories, and the shifting cultivators have least interest in developing such land for higher production. The land ownership of most indigenous peoples in Bangladesh's Chittagong Hill Tract (CHT) is based on customary ownership rights. Shifting cultivators often do not own private land for shifting agriculture, but they see this land as 'de facto' local property. Unlike the shifting cultivators, who have very few individual land titles, the plough cultivators in the valleys have individual ownership titles (Roy 2002). Karbaries (village chiefs, elders), Headmen (mouza chiefs), and Rajas make up the indigenous peoples' traditional administration system in the CHT (circle chiefs). They are in charge of land and other resource management, revenue collection, law and order enforcement, and justice administration, particularly in indigenous peoples' customary concerns. When the shifting cultivators do not have rights over land, they only dabble with the practice and follow no conservation measures, do not apply adequate dose of compost, farmyard manure and/or fertilizers to exploit full potential of the land.

Small land holdings

Another reason for disinterest of the shifting cultivators in shifting cultivation is small land holdings. In fact, they are frequently termed as small land holders. These small holdings have been further fragmented with increase in population and number of households. The uneconomic returns from such holdings dampens the interest of the shifting cultivators and they do not fully engage themselves with the practice.

Search for subsidiary sources of income

For the last about five decades, and more vigorously during the last two decades, the rural youths have been busy in seeking jobs with different agencies. This is not their fault or out of way approach but, because of increase in population the shifting cultivation is not able to feed them. They have to get married and support a family, care for the nutrition and education of children, which is impossible without an avenue outside the family traditional profession, that is, shifting cultivation. This has led to a callous approach of the shifting cultivators towards the practice with which they only dabble.

Awareness about environmental consequences

Many shifting cultivators with education background, have realized the harmful environmental consequences of the shifting cultivation. This has led to their disinterest in shifting cultivation.

8.3 DECLINE

8.3.1 Economic rationality

Shifting agriculture is commonly performed in forest land with humid weather and low population density all over the world. In these circumstances, it is regarded to be the most sensible land use, and it can be maintained despite population expansion as long as forest land exists, through intensification (Pollini 2014). Different scholars have expressed different opinions and views about the continuance of shifting cultivation. Land is abundant when population density is low, but labour is usually a strong limiting factor in putting more land into production. Slash-and-burn agriculture producers, like other economic factors, want to maximise production for a given level of scarce resource input, such as labour (Pollini 2014). Clearing the forest is arduous work, but it is carried out by men who do not have substantial domestic responsibilities, and it is not subject to strict deadlines because it is carried out in the off-season, during the dry season (Pollini 2014). After the burning of vegetation, crops are sown by men and wome, with women in charge of the majority of operations until harvest, at least in African countries.

8.3.2 Decline in shifting cultivation

The evidence of decline

In most Asian countries, there is abundant evidence of processes commonly associated with shifting cultivation degradation. Fallow periods used to span 30 to 50 years in Sri Lanka's arid zone, but they have subsequently been replaced by 12-15 year fallows. Under the current dry conditions, shorter fallows may result in soil deterioration and a vegetation dominated by grasses such as Pteridium sp. and Imperata cylindrica (Perera 2001). In the Philippines, where the Mangyan's traditional fallow system is decreasing, similar activities are taking place. Previously, the Mangyan would leave their forest fallows for 10 to 19 years, but currently they only leave them for 1-3 years depending on farmer needs and seed availability (Rodel *et al.* 2001). Increased population density, as well as government limits on removing new forest areas, may have influenced those changes. Due to suitably long cycles, the traditional swidden agricultural technique, popularly known as shifting agriculture, in the hill region of northeastern India was formerly in peace with the environment. Fallow periods have been reduced to an average of 4-5 years (Ramakrishna & Kushwaha 2001). The area covered by swidden fallow secondary forests is shrinking as swidden cultivation declines.

8.3.3 Pressures on shifting cultivation fallow secondary forest

Aside from the loss in swidden agriculture, which is primarily due to shorter fallow times, there are additional direct risks to secondary forest cover. Clearing for housing and government development projects, for example, are serious threats to secondary forests in Sri Lanka. Pulp and oil palm plantations are the biggest obstacles to swidden fallow agriculture and alternative fallow management in Indonesia. According to de Jong (2001b), these projects have had an impact on the transformation of swidden fallow-agroforestry landscapes into mixed land use landscapes with permanent and semi-permanent agriculture and a mosaic of artificial forests.

Despite the low density of swidden farmers in India's Central Himalayan region, both fresh secondary forest development and land degradation occur, implying that timber harvesting and in-migration are the primary causes of deforestation. One of the major causes contributing to forest degradation, especially secondary forest degradation, is government policy. Furthermore, replacing traditional swidden agricultural production with pine plantations, which is heavily reliant on the production of keystone tree species, puts local communities' livelihoods in jeopardy and has a negative influence on biodiversity (Ramakrishnan and Kushwaha 2001). In Indonesia, mixed rice-tree crop production systems are replacing swidden agriculture, as evidenced by a number of cases from Kalimantan and Sumatra. The swidden farmers of Kalimantan have a long tradition of managing forests and planting trees (de Jong 1997, Padoch and Peters 1993). In Indonesia, the introduction of cash tree crops into swidden agricultural systems is a common occurrence. Rice cultivation shifts from swidden to mostly permanent wet rice agriculture, or it vanishes entirely, along this continuum (de Jong 2001b). With the help of external assistance, Thai farmers have transitioned from swidden agriculture to permanent agriculture and tree crop cultivation (Schmidt-Vogt 2001) People in Vietnam now have more opportunities to create industrial forest plantations, cash crops, and fruit trees thanks to the rise of the private sector and the reorganisation of the national economy.

8.3.4 The limits of the system

Although the sustainability of a shifting cultivation system is highly dependent on environmental conditions, it is generally recognised as a sustainable land use when the cropping phase is followed by a long fallow period, allowing the initial fertility to be recovered. With a population density of less than 35 people per square kilometer, this is usually doable (Mazoyer and Roudart, 2006). Weed invasion, pests, and diseases thrive in shorter fallow periods. When these factors are combined, it might result in a fertility and/or labour crisis. The system cannot be sustained beyond a certain point. It does not generate enough food to ensure the residents' food and livelihood security.

Agriculture is the sector of human activity in which the environment and the culture that has arisen in and from it interact the most. All cultures are intimately linked with the mode of subsistence. The environment, local conditions and the values are invariably connected with the agricultural practices that have evolved in the area. Shifting cultivation implies the whole nexus of people's religious belief, attitudes, self-image and tribal identity. All operations of shifting cultivation are associated with some religious rites or festivals. For example, in Garo Hills of India, the festivals such as galmaka, rongchugala and wangala are associated with different operations of shifting cultivation. Any abrupt deviation from shifting cultivation cut across such social festivities. The practice, which has evolved in the Neolithic age, is closely linked to socio-cultural life of the people and it is also responsible for their adherence to this system of cultivation. The tribal population in India remained almost constant for a very long period until inter-tribal war-fares, head-hunting, pestilence and epidemics got effectively controlled. The equilibrium between land and the people got upset when population started to increase paradoxically. The low yields, high population and the economic and social goals, are no longer compatible with shifting cultivation as a 'way of life'. The move from shifting to settled cultivation not only involves technical measures but also cultural and social changes. In shifting cultivation areas, it is mostly economic and social factors which delay change from subsistence economy to a commercial one. When cultural and social factors are involved, mere pouring of funds or fixing targets unrelated to people's attitude and capacities, do not succeed.

There are as many as 65 or more ethnic groups in India practicing shifting cultivation. Various tribes live in shifting cultivation zones, and their economies can be split into hunting, nomadism, pastoralism, shifting cultivation, and, to some extent, settled cultivation. For a long time, the social sanctions and belief systems in these areas maintained a balance between resource potential and usage, but an imbalance has been established due to increased demographic pressure and indiscriminate use of natural resources. The world's rapidly rising population has put strain on food production, and people have mishandled and exploited natural resources to meet their requirements. There has been widespread land and environmental degradation as a result of manmade and natural factors such as shifting agriculture, land tenure systems, free range grazing, deforestation, and severe rains. To sustain the fast growing population, there is strong need for integrated approach for natural resources management and livelihood security. The most essential issue is to promote natural resource conservation and sustainable use, which allows for long-term economic growth and increased productive capacity while remaining egalitarian and environmentally friendly. Socioeconomic restrictions, a scarcity of trustworthy data, and a lack of human and institutional capability are the key obstacles to the harmonious development and improvement of shifting cultivation in India.

Shifting cultivation, according to Schendel (1992), is not just an old agricultural practise, but is also inextricably linked to the culture and identity of individuals who practice it. As long as shifting farming is interwoven with their cultural diversity, it helps to preserve the rich culture and heritage of more than 200 tribal races who live in India's north-east region. As a result, the tribal people wisely utilised natural and forest resources, and they have the experience to deal with the instability and unpredictability that all ecosystems face (Kanagaraj 2014; Zaitinvawra & Kanagaraj 2008). The tribal people own and develop the land collectively. The number of persons supported by shifting agriculture can be used to estimate its productivity. The majority of shifting cultivation research in India has been done from an ecological standpoint, with only a few studies on the socioeconomic aspect of shifting cultivation. Recent research (Ramakrishnan 2003) and his group have looked into the significance of traditional ecological knowledge in the practice's variety.

Shifting cultivation supports 23 persons per square kilometre, which is more than double the amount of people supported by commercial cultivation in Kalimantan, Indonesia. However, the focus on modern ways labels indigenous people's shifting agriculture as a primitive approach (Das 2006), as well as traditional knowledge systems and their ability to improve existing methods (Lalengzama & Kanagaraj 2012). Because of the geographical and biological features of the area where shifting cultivation is conducted, a few studies have determined that it is one of the most scientific ways. According to the FAO, UNDP, and UNEP, agriculture intensification and large-scale direct conversion of forest for small-scale and large-scale industrial plantations, rather than shifting cultivation, are the main sources of deforestation and consequently carbon emissions in Asia. Shifting agriculture, according to the FAO's definition of forest, does not result in deforestation but rather forest alteration. Majid (1993) and Goswami (1980) mentioned that the shifting cultivation was sustainable as long as the shifting cycle was long enough due to low population pressure. Because the fallow time of shifting cultivation was extensive, shifting cultivation was able to last for decades (Cochet 2014; Luoga, 2000; Goswami 1980). However, the fallow period for shifting agriculture has been decreased from 20-25 years to 2-3 years, possibly as a result of growing population pressure leading in greater demand for grains and forest products (Mwampamba, 2009). Fallow land reduction has been advocated as a management strategy to mitigate the detrimental effects of shifting agriculture in the tropics (Nye and Stephens, 1962). Introducing cash crops is not the greatest strategy for stabilizing shifting farming since it necessitates careful consideration of the socioeconomic situation and appropriate research (Ducourtieux *et al.*, 2005).

There were very few researchers who looked upon the causes of shifting cultivation system. Rahman *et al.* (2011) elaborated the underlying causes of the shifting cultivation. They discovered that older and less educated farmers have a predisposition to use shifting cultivation as a result of their heritage. Variables including high family size and a lack of off-farm income opportunities considerably impacted the farmers' ability to choose shifting cropping. The likelihood of permanent agriculture increases as more investment and finance becomes available. Shifting agriculture is less likely to be practiced by younger farmers with small households. Shifting cultivation was used by the farmers because it provided a generally consistent net cash flow year after year. In addition, his logistic regression study demonstrated that tradition and custom play a critical role in the continued cultivation of shifting crops. The study conducted by Gupta (1994) on the shifting cultivators of Tripura revealed that 50.47 per cent of the Jhumia households continue jhuming as a way of life but attributed that the economic compulsions force them to continue. Despite keenness of the central and state governments in promoting various shifting cultivation programmes, 73.77 per cent of the respondents reported that the government assistance given to them for giving up the practice was inadequate. Furthermore, 60.46 percent of the shifting cultivator stated that if the practice of shifting cultivator was to be abandoned, they would need to find another source of income. It was also observed that as back as in 1982 itself, there were no core shifting cultivator i.e. shifting cultivator who entirely depend on shifting cultivation for livelihood. Since the shifting cultivator work as wage labourer and remain dependent on low income occupations to eke out their livelihood, their annual income were reported to be low.

The transition in shifting cultivation was also observed by few researchers in various countries. Fox *et al.* (2009) observed that there were transitions from rural to urban livelihoods in South East Asia. The reasons for transitions were shortage of land, employment opportunity for youth in urban centers or in the coast and inability of the farmers to depend only on shifting cultivation for livelihood. Similarly, Cramb *et al.* (2009) also pointed out that shifting cultivation was disappearing because of population decline. Moreover, he observed that many young people temporarily or permanently leave their villages to seek jobs in cities. This in turn reduces the labour force required for shifting cultivation. Also, the variable education contributed for the abandonment of shifting cultivation.

In the state of Nagaland of India, the current shifting cultivation is increasing and in case of Tripura it shows the decreasing trend. The per cent change is higher in the state of Manipur, Tripura compared to 2005-06. It also reveals that, the states like Manipur and Tripura may have contributed for diversification in terms of non-farm, off- farm and farm activities which may be responsible for reduction of current shifting cultivation in Manipur and Tripura. The shifting cultivators cultivating crops only for subsistence is rare and diversify to other crops and other off farm opportunities due to risk in cultivating upland rice population pressure, reduced fallow period (Mertz *et al.* 2005). In Bangladesh Cambodia, India, Indonesia, Laos, Nepal, and Thailand, the FAO conducted case studies to learn more about the changes in livelihood and food security among indigenous shifting cultivation communities in South and Southeast Asia as part of the 'Regional Support to Indigenous Peoples for Livelihood and Food Security' project. The findings of the study revealed that, shifting cultivation not only gave economic security but also a cultural identity with rituals and festivals revolving around it. Further, it was also reported that livelihood in indigenous communities became more diversified partly because of necessity and partly because of choice. The factors like scarcity of land, market integration, education, and mass media played an important role for the livelihood preferences especially among youth (Patel *et al.* 2013). or permanently leave their villages to seek jobs in cities. This in turn reduces the labour force required for shifting cultivation. Also, the variable education contributed for the abandonment of shifting cultivation.

There have been evolving institutional systems for land and forest management at the community level, as well as various degrees of engagement in or resistance to government schemes and programs. Swirling change has resulted in the loss of cultures and identities, as well as their reassertion, realignment and reinvention, all of which have crucial implications for resource access. These alterations have had a wide range of effects. Although cash crops have benefited lives in many cases, complete market specialization increases vulnerability (Cramb *et al.* 2009). As a result, shifting cultivation can still act as a valuable safety net in the event of market volatility. The wellbeing of previously marginalized populations has mostly improved as a result of better market access and social provision of education and health care. Growing gaps within and between communities, on the other hand, can leave certain groups excluded and worse off as a result of swidden transitions. Heavy-handed state interventions based on skewed stereotypes can amplify these differentiation processes. Nonetheless, communities have mobilized to safeguard their livelihood assets and techniques rather than passively accepting these challenges. As a result, swidden farmers are not opposed to development that is appropriate and beneficial (Cramb *et al.* 2009).

8.3.5 Change in population

Population changes affecting livelihood plamming is mainly due to a decrease in a specific area with increasing population, as well as in-migration and out-migration. As a result, swidden farmers adapt to demographic changes while also contributing to them (e.g., through intensification) (e.g., through migration). Understanding population-induced swidden agriculture intensification begins with Boserup's (1965) agricultural transformation model. This is possible whether the increasing demographic pressure is driven by natural growth in the original locality, as stated by Roder (1997) for northern Laos, or in-migration, both impulsive and government-supported, as appears to be the situation in northern Laos. Population growth and decline, as well as in-migration and out-migration, are all demographic patterns that affect livelihood options. Boserup's (1965) agricultural transformation model is the initial stage for knowing population-shifting cultivation expansion due to demographic pressure. In-migration decreases the area available for shifting cultivation as land is purchased by or captured by in-migrants, culminating in a succession of expansion to bush and grass fallow systems, and eventually to cultivation without break. Furthermore, the inmigrants may bring lowland farming technology with them, such as ploughing, which the local uplanders eventually adopt.

The shift of shifting cultivators to be closer to roads, markets, and public facilities, particularly health and education, can result in an increase in population in a specific place (Hansen and Mertz 2006). In this situation, uplanders migrate from more remote and isolated regions, increasing population pressure in more accessible areas, leading to increased farming in the destination site and the conversion of swidden lands to forest or other land uses in the origin site. In northeastern states of India. Migration from outside is almost zero because there were strict laws to curb migration from outside in most of the states. Anybody coming from other states has to take a permit of permission to enter. Although these rules have diluted now but still permanent residence in these states is banned. by the respective states.

As previously stated, shifting cultivation may be vanishing owing to localised population loss rather than population growth prompting its metamorphosis into more intensive agricultural systems. Outmigration, both temporary and permanent, may occur in shifting cultivation communities as a result of economic and social change, resulting in a decrease in the local resident population. Because there aren't enough young male employees to clear the forest, the amount of land cleared is limited, and site selection is limited to newly farmed plots with young secondary growth, which are simpler to clear by women and elder household members. These changes result in an increase in the age of fallow vegetation further out from the village by reducing the overall quantity of land cultivated in a village area while concentrating it closer to the village centre.

8.3.6 Political, social and cultural trends

The loss of shifting cultivation has been attributed to increased political and regulatory control. This stems from central government territorial policy, as discussed by Peluso and Vandergeest (2001) and Fox *et al.* (2009). Some jurisdictions have had regulations in place since colonial times that try to restrict or prohibit shifting cultivation as a land use. These rules come in a range of shapes and sizes. (i) outright prohibition—an outright prohibition on shifting cultivation at any time and in any place; (ii) prohibited without a permit-allowing government officials to determine whether and to what degree shifting cultivation should be authorised; (iii) prohibition in particular land classes-an attempt to limit shifting cultivation to regions outside of forest reserves and protected areas (sometimes coupled with relocation, as previously indicated); (iv) two examples of tenure-related disincentives are the safe acknowledgment of claims to land used for commercial tree crop cultivation, or "agroforestry," and the rejection of traditional land claims derived from shifting agriculture. For many ancient communities, shifting agriculture was both an economic and ceremonial undertaking (Condominas 2009).

8.3.7 Conversion

Young people's education away from the village in modern school systems not only reduces the availability of family labor for shifting cultivation, but also hinders them from learning the necessary technical and ritual skills for the practice. Simultaneously, it may expose them to development ideals that see shifting cultivation as backward and other forms of agriculture as contemporary and progressive. Changes in gender roles are a related socio-cultural issue. Swiddens are largely a women's realm of activity in various communities, despite the fact that both men and women labor on them (Colfer 2008). Their job commitments are to shifting cultivation, which they regard as their responsibility rather than those of males. This is demonstrated by their control over the rice types in the household and, as a result, over planting and harvesting rituals and patterns, as well as their commitment to the vital task of weeding (Geddes 1954; Cramb 2007). On the other hand, restricting shifting cropping without providing viable economic alternatives increases household pressure to exploit the main and secondary forest for timber and nontimber forest products.

8.3.8 Human and social capital

The time, skills, and work responsibilities of family members are referred to as human capital. For household members, shifting agriculture offers low-paying urban or plantation jobs (Nielsen *et al.* 2006) in addition to productive employment comparable to cash crops (Cramb 1989). There's also no need for group rituals to get good farming omens or appease soil and woodland spirits (Dove 1988; Sather 1994). The usage of labour exchange is also decreasing, especially if shifting cultivation is substituted with perennial cash crops. These factors contribute to a loss of social and cultural cohesion as well as a sense of belonging. Land rights may be weakened in the eyes of the state if shifting cultivation areas are underutilized (Scott 1998). The lack of state-recognized land rights could be a possible cause of shifting cultivation decline. The loss of a rich cultural repertoire related with beliefs and rituals centered on the shifting cultivation cycle, as well as the material culture that sustains activities of the practice, is likely linked to the extinction of shifting cultivation.

For a century or more, shifting farmers had been collecting, producing, maintaining, and selling forest items (such as rattan, damar, and bird's nests) as well as planting pepper, fruit, coconut, and rubber, thus their demise was not simply due to the entrance of "dollar crops." Shifting agriculture became less common after the advent of specialised crops and their production under specified farming regimes. Government officials and foreign planners in the so-called Golden Triangle (Thailand, Laos, and Myanmar) were (and still are) concerned about shifting cultivators producing opium on their shifting cultivation fields, despite the fact that some officials were players and partners in the opium traffic. As a result of these concerns, numerous opium replacement strategies and initiatives were implemented beginning in the 1960s and continuing till now. Enclosure or the creation of management territories by nation-states, or both, have been the predominant movements in the past. If the shifting cultivators who had cultivated those lands did not change them into commercial producers or communal foresters, these activities frequently culminated in their eviction. So, what happens to shifting cultivators who lose access to their property or are unable to support themselves solely or in conjunction with local off-farm opportunities? Rigg (2005, 2006) has released significant work demonstrating that agricultural livelihoods are no longer dominating for many rural people in most of Southeast Asia, even for those who live in rural regions as their primary residence. His argument is focused on Thailand, with evidence from Java and parts of the Philippines. Similar evidence has been presented by Vandergeest (2003) and Walker (2004), suggesting that upland populations, particularly in Laos and Thailand, are physically losing ground. Work opportunities in Taiwan, Hong Kong, or Saudi Arabia attract certain people. Rigg (2005) provides some helpful information on this subject.

In 2003, thanks to a new "Grain for Green" ad, the desire to produce rubber grew even stronger. Farmers were urged to plant trees on their sloped farms as part of the Grain for Green environmental protection scheme. Because rubber is designated a forest species in China, farmers in Xishuangbanna received grain and cash subsidies for growing rubber trees as part of the Grain for Green programme. In 2003, the global price of rubber began to skyrocket, motivating farmers to plant even more rubber. The Department of Forestry established additional conservation policies and legislation throughout the 1990s, restricting upland people's access to their swidden fields and fallows. Laos has seen an upsurge in investment from its neighbors, particularly China. Upland farmers, whose livelihoods were already in flux, saw new economic prospects as a result of these shifts.

8.3.9 The rise and fall of shifting cultivation

Though shifting cultivation started during Neolithic period, about 7000 BC, the area under the practice remained stagnant or increased marginally for a long period, that is, up to about 1000 AD. Then it stated increasing and peaked somewhere in 1950 ± 10 years (Sharma 2020) (Fig.8.2). The area under shifting cultivation is on the decline after 1950, and it is predicted that by 1990, this practice would become a land use of the past, going down deep in the history books.

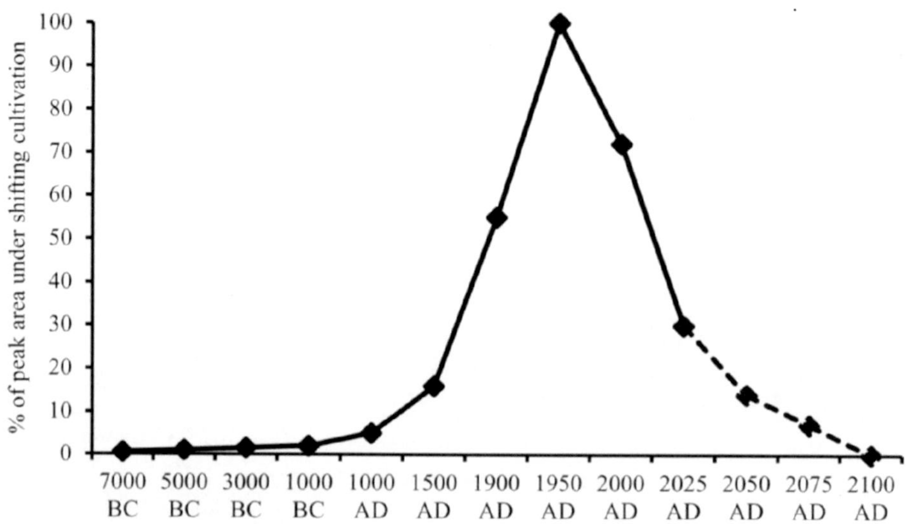

Fig. 8.2 Percent of peak area achieved in shifting cultivation so far during different periods.

8.4 ANTHROPOLOGICAL ASPECTS OF SHIFTING CULTIVATION

To a considerable part, a location's environment defines its habitat, manner of life, and civilization's advancement. For a long time, social penalties and belief systems kept a balance between resource potential and consumption, but due to increased demographic pressure and indiscriminate use of natural resources, the basin has become imbalanced (Sharma 1998; Sharma and Prasad, 1995). The region's rapidly growing population has put strain on the food production base, and people have mismanaged and exploited water resources to suit their needs (Sharma, 2003). 'Rather than physiography, culture and cultural history govern the wide environmental location of shifting agriculture as a cropping system,' Spencer (1966) observed. Many academics have argued and agreed that it represents a "unique stage in the transition from hunting and food collecting to sedentary fanning" (Geertz 1974), hence a "old" or "primitive system," and so a "remnant of the past" (Spencer 1966, Found 1987). 'There are evidences to imply that shifting cultivation extended progressively across practically the entire of southern and eastern Asia, Europe, and humid Africa in the early stages of agricultural community settlement of these countries,' Spencer adds. Despite the fact that shifting cultivation has a long history, nothing has been investigated or explored in an anthropological setting. Even now, nothing is known about the geographical range, characteristics, socio-cultural and ideological contexts, as well as the diversity and dynamics of dazzling cultivation. The reason for this, according to Dhakal (1990), is that studies on shifting cultivation have been limited, mostly describing its magnitude and ecological repercussions, with little effort made to compare analyze, and classify them.

8.4.1 On farm social constraints

The main problem of facing the harmonious development and management of acid soils in the northeastern region of India, apart from social and economic constraints, is the paucity of reliable data and lack of human and institutional capacity necessary for confronting the complex interactions of the social issues with acid soils management. The shifting cultivation prevailing in the region was a good land use system when the population was less and the shifting cycle used to be 25 to 30 years but, with increase in demographic pressure this cycle has come down to 2 to 10 years and the land does not get enough time for vegetation rejuvenation. The practice involves the whole nexus of tribal belief, self-image, attitude and identity. Since, the vegetation is the main component for in-situ burning to add nutrients for higher crop productivity, the production is now declining. This has caused environmental and soil degradation in the region. The farmers have no permanent proprietary ownership rights over the land they cultivate, but they do have usufructuary rights as recognized by tradition or clan (Sharma & Prasad 1995). The major practice of shifting cultivation is uneconomical and the farmers consider modification of soil acidity with correctives as additional burden. Lack of proper land use pattern restricts the options to replace shifting cultivation with settled one. However farmers in some areas of Nagaland state of the region, have developed excellent methods of soil and water management due to their ingenuity and skill ((Sharma 1998). The states of the region are predominantly agricultural areas and their current status is largely a product of socio-economic restrictions and under-development of their agriculture (Sharma 1988), (Sharma & Prasad 1995). It has been said that agriculture is that sector of human activity in which there is greatest interaction between the environment and the culture that has grown in and from it (Schlippe 1976).

8.4.2 Socioeconomic, policy and equity context

Gross disparities in land tenure are allowed by land-use rules. These causes have resulted in an increase in the number of landless rural people, who have the option of remaining in their current location, migrating to cities, or migrating to the rainforests that form the border of many developing countries. Similarly, land-tenure rules frequently do not provide individuals with secure title to land, which may deter producers from adopting long-term sustainable practices. Policies that promote strategies to discourage deforestation should continue, though these may not enough to aaleviate deforestation. Linking environmental and economic policies gives a viable and realistic strategy. have highlighted that fresh attempts in this approach are starting to appear, which has sparked heated arguments (Bouwman 1990, Gillis and Repetto 1988). In response to national and international environmentalist demands, a few deforestation-prone countries have created legislation to limit deforestation. Some of them are much too far-fetched to be practical. In certain circumstances, they backfire, causing severe economic consequences and encouraging deforestation in neighbouring countries. As vital as biophysical research is strong policy research on tropical deforestation. This should be done in conjunction with and in conjunction with study on technological and socioeconomic choices.

8.4.3 Landlessness and labour

In addition to working as cultivators on a part-time or full-time basis, many indigenous farmers work in seasonal or temporary on-farm or off-farm jobs to supplement their income. This is especially common in areas when cash cropping options are restricted. Indigenous farmers are the most economically vulnerable, as they do not have enough land to make a living as either subsistence or commercial farmers. Conditions deteriorated and food security was lost in regions where communities were resettled and were not supplied with enough land to make a good life. Regardless of the sort of land use they participate in, resettlement, dispossession, and privatisation of land, as well as the resulting accumulation of land in the hands of a few, have left other farmers landless or with land holdings too tiny to make a living. These (former) farmers have little choice but to work as farm labourers or on plantations, or to travel to cities or even abroad in search of jobs.

8.4.4 Economic drawbacks to shifting cultivation

Kamajou (1983) mentioned about some economic disadvantages of the practice of shifting cultivation, in addition to various agronomic drawbacks. Some are built into the system, while others are imposed by the rapid socioeconomic changes impacting areas under shifting agriculture. These drawbacks include; (i) poor remuneration for shifting cultivation in comparison to labor requirements. It's also low because shifting farmers can't receive a good price for their production because there are few markets for it, and (ii) shifting cultivation's huge and systematic destruction of forests and forest products, as well as the degradation of forest soils. This devastation results in a significant loss of precious resources, as well as the low investment capacities associated with shifting cultivation. This is due to poor pay, which renders all investments unattractive from an economic standpoint, resulting in low production.

The disadvantages imposed on shifting agriculture by numerous socio-economic and institutional changes are related to factors such as increasing population, changing lifestyles, a growing need for cash income, and changing wants of the younger generation. During the 1970s, the African population (particularly in Sub-Saharan Africa) increased by 2.7 percent, but food production decreased by roughly 10%. The rate of urbanization has accelerated dramatically, averaging around 6 to 8%. It is more difficult for the farmer to adhere to the appropriate fallow season, expand plot size, or relocate to new locations. Traditional farming methods, due to their poor productivity, are unable to produce enough to boost the cultivator's consumption above subsistence levels or to meet new needs that are reliant on cash flow. Shifting cultivators now require an increasing amount of cash to purchase new goods and services that are not generated by the family, making traditional shifting farming more challenging while producing the marketable surplus required to meet these new demands.

8.4.5 Shifting cultivation and climate change

In the context of climate change, Ziegler *et al.* (2012) argued that deforestation and forest degradation caused by shifting cultivation farming practices have had a significant influence on the global carbon sink, causing widespread alarm among worldwide populations. Intergovernmental organisations' policies have greatly benefited studies on the links between swidden agriculture, forest degradation, and global warming, from the United Nations Framework Convention on Climate Change in 1992 to the Kyoto Protocol in 1997. Furthermore, the UN-REDD Program, later called 'REDD+,' was suggested by the UN, beginning with the Bali Road Map in 2007 and ending with the Copenhagen Accord two years later. Following the Durban Climate Change Conference in 2011 and the Doha World Climate Summit in 2012, the UN-REDD+ project was critical in assisting humid nations in Africa, Asia-Pacific, Latin America, and the Caribbean in implementing national REDD+ programmes. According to Mertz (2009), REDD + will have a substantial future impact on swidden cultivation, either as a new difficulty or as a huge opportunity. There has been more original scientific effort regarding shifting cultivation in SEA and around the world after the establishment of REDD+.

Shiffting cultivation has been rapidly converting and replacing primary forest to monoculture plantations in SEA (Ziegler *et al.* 2012), such as oil palm (Stone 2007), rubber (Ziegler 2009), timber products (e.g., eucalyptus (Boulay *et al.* 2014) and teak (Newby *et al.* 2013), and other cash crops, as a traditional pan-tropical farming (Ziegler *et al.* (e.g., cassava and sugar). Shifting cultivation practices, on the other hand, are unquestionably employed to prepare a plot of land (shifting cultivation) for plantation cultivation. In the last two decades, considerable quantities of research have been devoted to quantifying the effects of plantation monoculture, in addition to the obvious environmental consequences (Hartemink 2005).

8.4.6 Policies

Here's a quick rundown of the several types of ignorance that characterize policy making.

Most agricultural experts and extension workers working on agricultural projects assume that yield per acre matters, thus the recommended alternative land uses are usually created with that premise in mind. It's what they've been taught to do. The above-mentioned labor and investment barriers, as well as the risk and discount rate aspects, are frequently disregarded, resulting in non-adoption of the proposed strategies, with the exception of local elites, who are generally recent migrants, capable of investment. "Support" for shifting cultivators can then favour social variations or who can entrap resources, resulting in shifting cultivators' marginalisation and continued reliance on natural resources, whereas "successful" intensive farmers, may acquire land by investing money and encourage more forest clearing.

The strategies being presented are framed under a sustainability paradigm that presupposes that resource supplies should be kept constant, which is a common mistake, or at least a counter-productive approach. The policymakers frequently regard fallow land in a slash-and-burn system as idle land, as evidenced by the misleading term shifting cultivation. In "shifting" (slash-and-burn) agricultural methods, fields do not actually shift. They are continuously maintained, but cropping and fallowing periods alternate, with indistinct boundaries between the two. Fallow ground is frequently actively managed, with the addition of beneficial trees and the maintenance of any remaining planted perennials. Unless it goes uncultivated beyond the customary fallow time, in which case the plot can be reintroduced into common land, it is typically administered by the same individual households that grow the crops. Overlooking this reality, policymakers may be tempted to restrict slash-and-burn farmers access to fallow land, resulting in erroneous policies.

So, how should policymakers deal with shifting cultivation? To begin with, it should not be classified as a land use that deforests more than others, nor should it be criminalised or penalised. The fundamental drivers of large-scale deforestation are commonly considered to be assets expansion and big agribusiness investments on forest frontiers, ranging from livestock rearing and growing of crops in the Amazon to palm oil extraction in Southeast Asia. Securing the rights of local farmers who use shifting agriculture, as well as hunting and gathering communities, could function as a deterrent to large-scale deforestation in this area. It could be easier for smallholder farmers to make the switch to more intensive land uses if they have access to a wider range of land use alternatives, such as new technology and inputs and "ecological intensification." It is possible that emphasizing ecological intensification too much will deny them the opportunity to follow a more conventional, straightforward route to "development," therefore proceed with caution.

The fall of the shifting cultivation system was marked by increasing tension between crofters (Swedish torpare, Finnish torppari), who are peasants in Finland and Sweden whose primary source of income is the cultivation of a small portion of an estate held in tenancy. The crofter paid the majority of his rent by working on the parent farm for so many days. Crofting is a type of land tenure found only in Scotland. A croft is a small agricultural land holding that is usually held in tenancy and may or may not have buildings or a dwelling attached to it.), as well as other claimants to the land. The crofters gradually lost out to the timber industry, which was able to make better use of the Nordic forests. Slash-and-burn was nearly extinct by the end of the nineteenth century. Landowners now backed the push for better forestry, and swidden agriculture and the use of fire in the farm cycle were quickly relegated to the past (Hamilton 1997). The last time a Swedish crofter used slash-and-burn methods was in 1937. Slash-and-burn is a dying art form these days.

8.4.7 Sustainable alternatives to shifting cultivation

Integral shifting agriculture has evolved through millennia in response to local environmental and societal constraints, demonstrating a deep and nuanced grasp of local flora, soils, and climate (Conklin, 1957; Kunstadter *et al.*, 1978). It has been proposed that, due to population pressures, the move from long-fallow shifting farming to short-fallow shifting cultivation, with associated land degradation, is inescapable. Finding an alternative land use approach that maintains the logic of conventional slope shifting farming is tough (Liang *et al.*, 2009). In mountainous places where sedentary agricultural adoption is severely constrained, the rotational agroforestry system may be pushed as a viable option for the growth of the shifting farming system (Liang *et al.*, 2009). For sustainable alternative to shifting cultivation, agroforestry has the requisite potential (Lal, 1989).

Because shortening the fallowing phase has negative repercussions, experts and government laws usually recommend stationary agriculture (continuous cultivation in the same fields without fallowing) as a feasible alternative to short shifting farming (Brady, 1996; Cairns and Garrity, 1999; Fox, 2000; Borggaard *et al.*, 2003). Nonetheless, in certain trials, artificial fallows did not exceed natural forest fallows in terms of nutrient immobilisation during the fallowing phase or crop yields during the cropping phase (Sanchez, 1976). Bringing in cash crops isn't a panacea for restoring agricultural stability. Their advancement demands a detailed analysis of the socioeconomic condition as well as proper on-farm research (Ducourtieux *et al.*, 2006). In Nepal, government policies remain indifferent to shifting cultivation and moving cultivators, despite the fact that thousands of indigenous peoples rely on it for subsistence. In the face of a changing climate, land use transitions such as altered fallow periods and agricultural intensification can worsen the vulnerability of resource-poor farmers while also reducing the mitigation (carbon sequestration) capability of soil in Nepal. Systematic research on shifting cultivation are urgently needed in Nepal, with the goal of providing result-based recommendations to the government for inclusion in national plans, policies, and priorities.

8.4.8 Demerits of shifting cultivation

Thousands of hectares of woodland are destroyed each year as a result of the shifting farming technique. Forest ecosystems are changing as a result of it (Henriques, 2007). Shifting agriculture has a number of severe environmental consequences, one of which is harm to the soil system. It has a significant impact on soil erosion. Shifting cultivation is responsible for the loss of soil nutrients, as well as useful soil fauna and microbes, in addition to air pollution caused by burning. Slash burning reduces soil acidity, organic matter, and total nitrogen levels. When compared to other farming systems/ methods, most shifting cultivation practises are a subsistence level agricultural system with a very low output/input ratio. Excessive shifting cultivation agriculture activity not only reduces forest area, but also transforms primary forest into secondary shrubland. It enhanced soil and gully erosion, as well as acidification, in the phase of soil property (Das *et al.*, 2012). The loss of forest cover causes climate variations such as unequal rainfall, precipitation, wind, and humidity, among other things. The loss of biodiversity has an impact on a region's ambient climatic circumstances, impacting the habitat of numerous faunas. Deforestation caused by shifting farming interferes with the rain cycle since vegetation is essential to the rain cycle. Deforestation results from the wholesale removal of forest cover with forest canopy gaps, which limits rain creation due to low evapo-transpiration.

The flaw of shifting cultivation is that, depending on the shifting cycle, the land can only yield crops once every few years (5 to 19 years). However, with settled fanning, a single area can be brought under multiple cropping with scientific management and irrigation, producing two or more crops. The expanding population has created a vicious cycle in which more land is jhumped, resulting in a shorter fallow period and deterioration of soil fertility, necessitating the shifting cultivation of more plots. Indigenous peoples across Asia have been subjected to a variety of human rights violations as a result of state policies and activities aimed at eliminating shifting cultivation.

8.4.9 Ecological implication

Shifting cultivation is one of the primitive forms of agriculture (Warner 1991), which has been considered obsolete and destructive by growing establishments (Brady, 1996; Thrupp *et al.*, 1997; O'Brien, 2002), devastating, extravagant, and a hazard to tropical forests regarded by biologists, foresters, and conservationists since the colonial era, destructive, wasteful, and a threat to tropical forests (FAO 1957; Raman, 2000). Because of the study of Thomas (1956) and Bartlett (1957), the notion of shifting cultivation's effect on ecological gained stronger in the 1950s (1956). For many years, scholars, agriculture scientists, anthropologists, and economists have been concerned about the ecological impact of shifting cultivation. They claim that shifting farming is essentially primitive, and that their attitudes toward shifting cultivators are typically political in nature (Gadgil and Berkes 1991; Lowenthal 1989). As a result, it should be approached with caution, as judgments based on scientific evidence and experiments conducted around the world have shown that shifting cultivation has extremely negative consequences (Rao & Hajra, 1986; Lal & Prajapathi, 1990; Gerber and Veuthey 2010; FAO, 1984; Tawnenga *et al.*, 1997). One of the biggest consequences of shifting agriculture is deforestation, which is often seen as one of Southeast Asia's most serious environmental issues (FAO, 1995; Seidenberg et.al. 2003; Temudo and Silva 2012; Navarrete and Jansen 2015). Between 1980 and 1990, more than 6% of tropical forest land was converted to shifting farming, while about 10% of forestland in Asia was transformed to shifting cultivation over the same time period (Van Vliet *et al.* 2013; Gerber and Veuthey 2010). Many researchers have studied the ecological impact of shifting cultivation, including Zaitinvawra and Kanagaraj (2008), Lalengzama (2011), and Lalengzama and Kanagaraj (2012). (2013). The shifting cultivation has resulted in low cereal crop productivity due to the loss of top soil (Prasad and Rinawma 2014). Meanwhile, several studies accept the relevance of shifting agriculture in the past, when land was plentiful and population was low. However, in the current scenario, they reject it (Webb 1966). According to Myers (1991), postcolonial development tactics have pushed traditional farmers into poverty, which has been pursued by national governments and international organisations. The World Commission on Environment and Development (1987) used economic and social development mechanisms to disrupt shifting cultivation practises.

8.4.10 Ecological problems

Shifting agriculture has disastrous and far-reaching consequences for the environment and ecology of these areas. Large-scale deforestation, soil and nutrient loss, and weed and other species invasion have all resulted as a result of this (Ranjan and Upadhyay 1999). The indigenous biodiversity has been severely impacted. The rise in human population, especially in emerging countries, has put enormous strain on land resources. The expansion of crop lands to increase food production has resulted in a direct decline in the area covered by woods and grasslands. By the early 1990s, about 40% of the earth's land surface had been turned to crop lands and permanent pastures, according to one estimate (World Resource Institute 1996). Between 1980 and 1990, more than 6% of tropical forest land was converted to shifting agriculture in all tropical countries. During the time period mentioned, almost 10% of forest land in Asia was converted to shifting cultivation (Singh and Marzoli 1988). According to FAO and other sources, shifting agriculture destroys around 1.9–3.6 x 10^6 ha of primary close forests, 3.4–40 x 10^6 ha of secondary close forests, and 6.9–21.9 x 10^6 ha of secondary open forests each year (Detwiler and Hall 1988). Tropical countries are the most common places to see shifting cropping. Shifting farming on hill slopes is practiced by people in India's eastern and north-eastern regions. Shifting farming accounts for 85 percent of overall cultivation in northeast India (Singh and Singh 1992). Due to the increased demand for land, the cycle of agriculture followed by fallow land has been decreased from 25–30 years to 2–3 years. Previously, the fallow period lasted 20–30 years, allowing the land to revert to its original state. The ecosystem's resilience has been broken down as a result of the decrease of the cycle to 2–3 years, and the land is rapidly deteriorating.

8.4.11 Eco-system development for shifting cultivation

These mountain eco-systems, which have shifting cultivation practises, must be made ecologically viable. In order to ensure the environmental sustainability, an eco-development plan could include totally replacing agricultural practises with farm forestry. In shifting-cultivation settings, the loss of forest energy per unit of agricultural production could be much higher. Farm forestry could be one of the ways to make up for this loss. The following are some of the benefits of farm forestry (Singh *et al.* 1986):

- The defensive value of trees much outweighs that of annual crops.
- Unlike annual crops, tree growing cause minimal soil erosion.
- Forests produce much more net primary production than agricultural crops and grassland.
- Farm forestry, by providing feed and firewood, would create a favourable condition for the regeneration of natural forests.
- A resurrected forest cover could provide a slew of other benefits.

Determining the population-supporting capacity (PSC) should be one of the most important components of preventing environmental deterioration and resource depletion. This research should take into account not only food production and land availability, but also other factors that affect carrying capacity, such as;

- Resource management should be the foundation of the population supporting capacity., in order to maintain and improve sustainable yields,
- The population's current living standards in relation to sustainable subsistence yields and advances in technology, innovation, and government policy
- Food trade flows take precedence over lumber, fuelwood, and fodder trade flows, with a focus on data on transportation infrastructure, administration, natural disasters, and regional conflicts in terms of import and export. The state of environmental equilibrium and the amount to which natural resources are used.

8.4.12 Effect of shifting cultivation decline on livelihoods

Transitions from long-fallow shifting cultivation (LFSC) systems to alternative land uses are being driven by global economic development and policy interventions in Southeast Asia's uplands. Dressler *et al.* (2017) conducted a comprehensive analysis of how these transformations affect livelihoods and ecological services in the region. Analysis of land-use transitions from swidden to intensified cropping systems revealed multiple outcomes: more households had improved overall income, but these advantages came at a large cost in terms of traditional practices, socioeconomic wellbeing, livelihood possibilities, and staple yields. Transitions have a negative influence on soil organic carbon, cation-exchange capacity, and aboveground carbon, according to research (Dressler *et al.* 2017). When the proximal and underlying causes of the transitions from LFS to alternative land uses, particularly enhanced perennial and annual cash crops, were considered together, they resulted in significant losses in pre-existing livelihood security and the ecological services that supported this security. Our findings suggest that policies requiring upland farmers to transition to different land uses in order to improve their livelihoods and the environment are misguided; in the context of multiple land uses, shifting cultivation can support livelihoods and ecosystem services, helping to mitigate the effects of climate change in Southeast Asia.

According to some researchers, the 'end of the peasantry' and 'de-agrarianisation' are distinctive characteristics of the current shift of rural lives (Rigg 2013). Furthermore, the agricultural change's complexity has been exacerbated by more explicitly coercive processes of 'territorialisation,' 'enclosure,' and 'exclusion' associated with state growth and conservation (Lund 2011). In upland Southeast Asia, agrarian transformation has had a significant impact on shifting farming, among other land uses. In the uplands, swidden (or shifting cultivation) and other forest uses continue to support 15 to 35 million people (Schmidt-Vogt *et al.* 2009). Mertz *et al.* (Mertz *et al.*, 2009b). For millennia, it has been a vital part of local life and the forest environment, integrating produce from cultivated fields and secondary woods and setting the foundation for upland land uses and customs (Conklin 1961; Cairns 2007). shifting cultivation is the process of cutting and burning small portions of forest for subsistence food crop cultivation, followed by long periods of fallow when forest regrowth restores the land's productivity (Cramb *et al.* 2009). Swidden is a land-use strategy that involves a natural or increased fallow period long enough for woody vegetation to dominate, followed by clearing and burning to allow for a shorter annual crop cultivation phase in this study (Mertz *et al.* 2009a). Long-fallow swidden (LFSC) systems have a fallowing phase of more than 10 years in the past, but this has been estimated to be closer to 5–10 years in recent decades. As a result, most research allocated fallow lengths within this range, with short fallow defined as shorter than 5 years. Long fallows are those that last at least 5 years, whereas short fallows are those that last less than

5 years. The evaluation was able to include more material with comparative data thanks to this fallow criteria.

8.4.13 Related drivers

Individualisation of tenure referred to situations in which traditional land rights were replaced by private land ownership (by an individual or a family), resulting in a shift away from traditional usage and toward activities that increased household income and advantages. Some research, on the other hand, showed situations in which this could be a result of cacao being planted on shared fallow lands by smallholders in order to remove it from the commons and personalise tenure (Belsky and Siebert 2003; Feintrenie *et al.* 2010). Shifting cultivation policies are likely more complicated now than in the past, as certain state and non-governmental programs support swidden farmers' livelihoods and land rights (Dressler and Pulhin 2010). Nonetheless, international agencies, governments, civil society organizations, and upland farmers themselves continue to engage in contentious policy disputes about the effects of shifting cultivation (Astuti 2015). While the above-mentioned policy interventions and other variables are definitely to blame for the loss of shifting cultivation fallow in Southeast Asia, our study (and that of the authors we cited) shows that LFSC continues to provide livelihood stability and ecological services for many impoverished uplanders today. In many ways, then, the discussion regarding the relative influence of LFSC in policy arenas and development initiatives remains unresolved, contentious, and ambiguous (Fox *et al.* 2014).

8.4.14 Integrating shifting cultivation knowledge into intensification

Until the second half of the 20th century, shifting cultivation was the dominant agricultural system in much of the tropics. However, since then Southeast Asia particularly China, Cambodia and Thailand – has experienced a decrease in swidden area at a pace faster than that of the rest of the world (Van Vliet *et al.*, 2012). This has been attributed to market development and population growth, as well as government policies that have encouraged or coerced the conversion of shifting cultivation lands into permanent and more intensive agricultural practices. Throughout the region, shifting cultivators have been marginalized by laws that criminalize their practices, or land laws that prohibit the use of land for anything other than permanent agriculture or forestry (Dressler and Pulhin, 2010). Much existing swidden cultivation is practised by ethnic minorities and indigenous groups using their own indigenous agricultural knowledge, developed over long histories of interaction with their natural surroundings (Xu *et al.* 2017).

Indigenous knowledge of this kind has inspired the emergent concept in agricultural ecology of farming systems that mimic nature (Xu, 2017). In this respect, the indigenous knowledge of swidden farmers – and the agro-ecological science it has helped to inspire – provides a welcome antithesis to modernist projects such as the Green Revolution and Chinese Maoism. Moreover, such an approach to indigenous knowledge opens up the potential for scientists – as well as policy-makers and other stakeholders – to work with and alongside indigenous peoples as co-theorists, not only of agriculture and ecology, but also of culture and society (Xu *et al.,* 2017).

8.4.15 Innovative solutions to improve livelihoods

Farming systems are changing rapidly around the world due to many factors, including commercial alternatives to swidden agriculture, large-scale land and infrastructure development, increasing regional market integration and agricultural intensification. Across Southeast Asia, swidden cultivation is being replaced by monoculture tree plantations and annual cash crops such as maize, rubber, bananas and cassava, as well as paddy rice (Van Vliet *et al.*, 2012). Increasingly, these crops are being grown for regional and global value chains and for a wide range of uses. Their production often involves trade-offs between subsistence-oriented and market-oriented activities and objectives. Although such commercialization has resulted in increased household incomes, what is less clear is its impact on positive livelihood outcomes, such as improved food and nutrition security. Moreover, equity – in terms of access to resources and markets and benefits received, is an issue that is often overlooked. Specialization for regional and global markets makes farmers vulnerable to both production and market risks. Monoculture cropping also leaves farmers vulnerable to extreme climatic events, plant diseases and food insecurity. Moreover, to achieve positive impacts for the poor and the vulnerable, research-for-development needs to be based on integrated agricultural systems research and natural resource management. Integration should also take place among different sectors both horizontally and vertically (Coe *et al.*, 2014).

8.5 DEMISE

Traditional land use methods are rapidly being abandoned by shifting cultivators across Southeast Asia. While these changes have been quantified in a number of local places, there is no credible regional data available. Three interconnected challenges that contribute to at least some of the knowledge gap are very obvious. To begin with, shifting cultivation is a diverse, complex, and dynamic land use that data collectors find difficult to perceive, characterise, and quantify, and as a result, it is frequently relegated to a "residual category" of land use. Second, shifting cultivation is a smallholder category, and it's tough for government officials to measure what's going on in so many diverse and dynamic smallholdings. Third, all Southeast Asian countries' national governments have attempted to prohibit shifting cultivation farming and encourage shifting cultivators to embrace permanent agriculture land use practices (Padoch *et al.* 2007).

Guo *et al.* (2002) recently asserted that, "from the mountains of Yunnan Province in southwestern China to the interior of Kalimantan," the ways of life of smallholder farmers throughout Southeast Asia are changing at rates and scales that "are unprecedented." They suggested that "hill farmers with their hundreds of rice landraces, their many intercropped vegetables and fruits, and their cyclic way of farming are disappearing throughout the region," and report on what they called a "landscape in transformation" using data from communities in the Xishuangbanna Dai Autonomous Prefecture of southern Yunnan Province in China. Over the last decade of the twentieth century, Padoch *et al.* (1998) reported the gradual, patchy extinction of shifting or swidden cultivation in the town of Tae in West Kalimantan, Indonesia. Shifting cropping is a vanishing land use in many communities across mainland Southeast Asia, according to other studies (Rerkasem & Rerkasem, 1994; Fox & Vogler, 2005). These insights, on the other hand, are neither novel nor startling. Shifting cultivation and shifting cultivators have always been dynamic and diverse, with households and groups migrating, adopting new crops, and altering their farming practices on a regular basis. Some farmers, as one might expect in Southeast Asia, have long abandoned shifting cultivation as a primary land use, while others are embracing it or returning to it. The scale and magnitude of the shift are now different. There have been previous attempts to estimate regional land use change in Southeast Asia, but the data is at many years old. Between 1880 and 1980, Richards & Flint (1994) offered a good review of land-use changes in Southeast Asia. However, their landuse/ land-cover analysis categories provide little insight into changes in swidden, which is seen in the overlap between these categories. "Coverage was frequently incomplete, especially for territories under shifting cultivation," they admitted. Uhlig *et al.* (1994) employed secondary sources to estimate the number of shifting cultivators and the area under shifting cultivation in Myanmar, Thailand, Sarawak, and Sabah, with the goal of quantifying the effects of shifting cultivation on the world carbon cycle. They calculated that shifting cultivation supported around 15-20 million people in the early 1980s, and that an area of 5.5-6 million hectares in the region was under some type of shifting cultivation management (though no area estimates were provided for Myanmar). Between 1993 and 1997, Mittelmann (2001) examined multiple data sets for changes in secondary forest cover in Cambodia, Laos, Thailand, and Vietnam. Many studies in the past predicted that when a certain threshold for reducing fallow was achieved by rising populations cultivating in constrained spaces, virtual environmental and social collapse would ensue (Mertz, 2002). Shifting cultivators around the world are progressively changing their habits, rather than collapsing, according to a large body of evidence.

Many shifting cultivators have established cultivation cycles that are more akin to crop rotation systems and agroforestry operations than what is commonly referred to as swidden, or they may have always done so but academics were distracted by the more dramatic "slash and burn" image (Wadley, 2007). According to a World Bank assessment, Indonesia has "few true examples of traditional farmers who practice rotational shifting cultivation" (World Bank, 2001), yet the challenges we've discussed illustrate how little we can realistically say about the magnitude of these alterations.

Commercial agriculture's hydrological consequences, such as soil erosion and run-off, may be even worse than those seen in secondary vegetation in traditional shifting cultivation systems (Zinke *et al.*, 1978; Forsyth, 1994). According to field surveys, the negative effects of shifting cultivation on water- and soil-holding capacities may be exaggerated, particularly when compared to more permanent alternatives such as commercial agriculture, plantation forestry (Zinke *et al.*, 1978; Alford, 1992; Forsyth, 1994, 1996;), or even highways (Ziegler *et al.*, 2004). As a result, shifting cultivation can be considered a significant source of environmental services when compared to alternative land use systems; however, quantifying these services is impossible without first understanding the scope and dynamism of shifting cultivation in the region. It may seem strange that such information isn't available in this age of GIS and remote sensing-based land use planning. However, there are at least three interconnected reasons why changing cultivation is difficult to measure.

1. Shifting cultivation is a diversified, complex, and dynamic land use that data collectors struggle to see and define, l*et al.*one measure. Shifting cultivation is generally represented in the landscape by a vast variety of unique landscape features, any of which may be "in transition" and readily confused with other land uses. Because of this complexity, as well as other aspects of the method and its practitioners, government officials frequently categorise shifting cultivation fallow lands as "wasteland/ abandoned/unused" or "residual/miscellaneous" in provincial, national, and regional reports. However, according to Fox (2000), one-third to half of "non-forested" land in many of these large-scale land-use

classifications for Southeast Asia may be categorized as "unclassified" or "other."

2. Shifting cultivation is a smallholder category, and government agents find it significantly more difficult to sum up what's going on in all those temporally and spatially divergent and dynamic smallholdings than it is to sum up a few hundred monocultural farms of a thousand hectares apiece.
3. Shifting cultivation's very existence is, and has long been, a politically sensitive matter (Thrupp *et al.*, 1997). Many swidden lands – particularly older fallows – are legally removed from the pool of lands accessible for future swidden farming due to regulations prohibiting shifting cultivation, forest removal, or encroachment on protected areas.

8.5.1 The changing shifting cultivation

As others have already noted, the extinction of shifting cultivation looks to be a reality in many forest agriculture frontiers, particularly in Southeast Asia (Padoch *et al.*, 2007). Southeast Asia appears to be changing its shifting cultivation faster than other regions, owing in part to government policies that have curtailed shifting cultivation by prohibiting it or rewarding its conversion to permanent agriculture (Fox *et al.*, 2009; Ziegler *et al.*, 2009b). Shifting cultivation has been marginalised throughout the region as a result of laws criminalising it. Swidden farmers have been labelled "poor quality individuals" in Southwest China, "isolated primitive groups" in Indonesia (Li, 1999), and "pyromaniacs" in Madagascar (Kull *et al.*, 2007). Furthermore, regulations such as agricultural revenue tax exemptions, rules governing land tenure security, progressive land taxes encouraging clearing and conversion to pasture, and credit schemes that finance corporate cattle ranches are all common in the Brazilian Amazon (Binswanger and Deininger, 1997). Similarly, Madagascar's government has a long history of striving to eliminate swidden agriculture, claiming that doing so will enhance biodiversity protection and increase food yields sufficiently to balance expanding populations (Hume, 2006). Farmers are occasionally enticed away from swidden to pursue other occupations as a result of integration with large regional markets. Due to land development for oil palm plantations, several swidden regions in Sarawak (Malaysia) have significantly declined (Fox *et al.*, 2009; Hansen and Mertz, 2006).

As a result of tight land allocation programmes and government-supported programmes emphasising on wet rice growing, shifting cultivation has been disappearing in many locations of Vietnam and Lao PDR (Thongmanivong and Fujita, 2006). Improvements in irrigation and infrastructure, as well as greater access to roads, markets, and services, have allowed Vietnam's agricultural output to support more people on roughly the same amount of land that shifting cultivators previously utilised (Muller and Zeller, 2002). As a result, a frequent tendency in Southeast Asia is the move from shifting cultivation to permanent agriculture, such as monoculture tree crop plantations, annual cash crops, and paddy rice, as well as the increase of protected areas. The majority of the case studies in East Africa focused on annual crop growth. In Zambia, agricultural input credits were introduced in the early 1990s, assisting in the progressive replacement of the traditional swidden system with semi-permanent hybrid maize varieties (Kakeya *et al.*, 2006). Meanwhile, Uganda's main land cover conversion has been from forests/woodlands to sugarcane fields, putting indigenous people's access to land in jeopardy (Mwavu and Witkowski, 2008). Swidden is still prevalent in Central Africa, but it faces competition from food crops (such as banana and plantain) (Sunderlin *et al.*, 2000). Although altered agriculture landscapes have been connected to higher earnings, they have frequently resulted in major environmental effects. These include, among other things, a permanent loss of forest cover on a landscape scale, significant losses of wild biodiversity and agro-biodiversity, increased weed pressure, decreased soil fertility, accelerated erosion, declining stream water quality, and likely reductions in carbon sequestration. Water extraction for irrigation of rice and annual crops, as well as increased use of fertilisers and pesticides in areas where commercial agriculture and monoculture tree crops have replaced shifting farming, are creating stream desiccation and jeopardising water quality (Dressler and Puhlin, 2010; Ziegler *et al.*, 2009b). Less intensive and more diverse land uses, such as managed fallows and mixed fruit trees, have, on the other hand, generally enhanced household income and food security while conserving forest cover, biodiversity, and soil fertility (Messerli, 2004; Porro, 2005; Sears *et al.*, 2007; Xu *et al.*, 2009). Short agricultural periods, extended fallow times, and the mosaic nature of traditional systems, according to Padoch and Pinedo-Vasquez, retain sufficient seed pools to permit the regeneration of various secondary forests (2010). (Rerkasem *et al.*, 2009a). However, in many locations, shifting agriculture with shorter fallow intervals has continued to have severe effects on forest cover and soil fertility (Lindell *et al.*, 2010; Ludewigs *et al.*, 2009). This study examines changes in shifting cultivation patterns around the world over the last 10–15 years, investigates the origins of observed changes, and considers how these patterns affect rural livelihoods and the environment. Household incomes have typically increased as a result of intensification in previously shifting cultivation environments, but this has not translated into land conservation, as it is often accompanied with agricultural growth and deforestation (van Vliet 2012). Recent Southeast Asian shifting cultivation study reveals a wide range of changes affecting current and previous shifting cultivation lands, as well as a lack of understanding of swidden locations and classification. These findings suggest that, while we know a lot about what's going on in discrete places, scaling out these observations with any precision remains a challenge. Indeed, our capacity to generalise to the regional level is hampered in part by the tremendous diversity of shifting cultivation on a local scale.

8.5.2 Shifting cultivation and its demise

The prevalent trend of events related to shifting cultivation show that the shifting cultivation or swidden agriculture has to ultimately embrace demise. Slash and burn practice has not been included in this since this is likely to continue and the operation can be done to enhance soil fertility and the practice has nothing to do with shifting cultivation or swidden. Development or so called 'improvement of shifting cultivation' means saying no to the age old practice to go for settled cultivation. The major causes of demise of shifting cultivation are decline in area and number of shifting cultivators in the tropics in due course, only dabbling with shifting cultivation and shifting cultivation development and intensification of agriculture (Sharma 2020) (Fig.8.3). The decline will be due to government policies, incentives from the government to leave the practice, awareness programmes and forest protection and conservation policies. The shifting cultivators are likely to loose interest in shifting cultivation due to generational gap, youth will try for subsidiary sources of income, white collar jobs, and do shifting cultivation in small areas to observe some rituals. Another good reason for the demise of shifting cultivation is that all agriculture developmental programme in most of the tropical countries, practicing shifting cultivation, are directed towards elimination of the practice. So much so that even FAO has admitted that shifting cultivation is a resources depletion practice. The intensification of agriculture, prevailing land tenure system and availability of alternate land uses would be responsible for the demise of shifting cultivation. So, ultimately demise of shifting cultivation appears apparent. In our opinion, 50 years from now, that is, up to 1972, the shifting cultivation will be a practice of the past in the world, citing only for a reference.

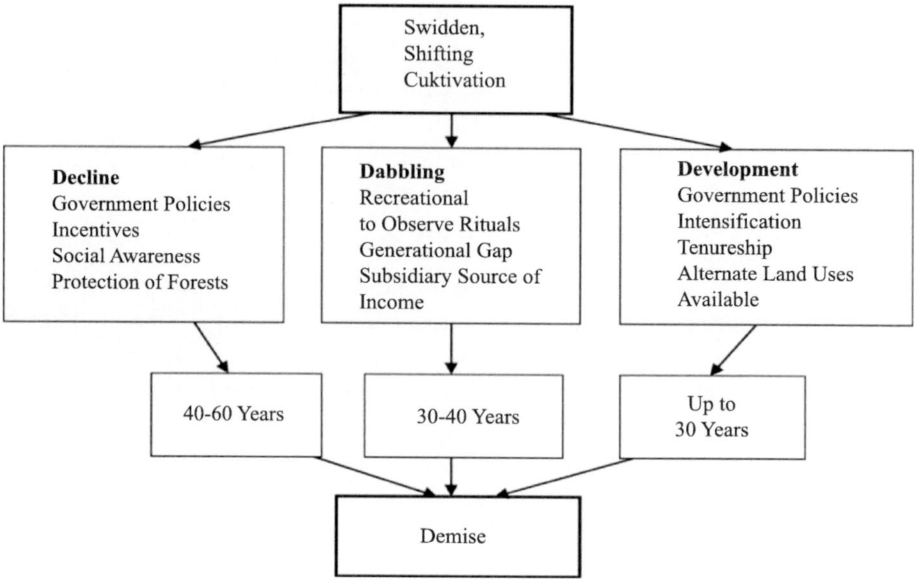

Fig. 8.3 Pathways to temporal demise of shifting cultivation.

8.5.3 Demise or continuance of shifting cultivation

The discussion and worry over the permanence or destruction of shifting cultivation among governments and scholars has never formally ended in the last century (Ellen *et al.* 2012). Governments' negative attitudes toward shifting agriculture in Southeast Asia have hastened the extinction of this traditional shifting system. In terms of the shifting cultivation debate, there is a prevailing dichotomy: on the one hand, the persistence of this farming practise due to biodiversity preservation (Fox 2000, Padoch 2010), livelihoods dependence (Cramb *et al.* 2009), and cultural identity (Ziegler *et al.* 2011); on the other hand, the demise of shifting cultivation due to large-scale plantations (Ziegler *et al.* 2012), economic development (van Vliet *et al.* 2013). Information on the dispute over whether shifting cultivation would change reveals not just the cognitive gaps between disciplines opposing it, but also implies that shifting cultivation is very controversial and warrants further investigation.

Scholars who believe shifting cultivation is condemned to extinction argue that this primitive and extensive agricultural practice is harmful and unsustainable, and that alternative agriculture should be developed. The three steps of the traditional swiddening process in the tropics are conversion (forest slashing/ felling, sun/air drying and burning during the dry season), cropping (settling familial farming for several years), and fallow (vegetation restoration and soil organic content rejuvenation) (Rossi *et al.* 2010). Human activities usually cause a variety of eco-environmental hazards during the conversion and cropping periods, such as forest degradation and deforestation (Tinker *et al.* 1996, Sodhi *et al.* 2010), total mineral nutrient loss (Roder *et al.* 1997, Alegre and Cassel 1996), soil biota communities decline (Alegre and Cassel 1996, Matsumoto *et al.* 2009), atmospheric pollutions (Brady 1996) and heavy metal contamination (Béliveau *et al.* 2009). The future of tropical forests was thought to be secondary vegetation/forest regeneration throughout varied fallow periods (Rerkasem *et al.,* 2009). The length of the fallow period, on the other hand, will be a critical component in vegetation restoration (Metzger 2002). Recent variables such as population increase, market forces, and forest conservation legislation have shortened the fallow period (Rossi 2010, Sanchez 2005) and had a significant impact on vegetation regrowth, carbon sequestration, and forest ecosystem resilience (Comte *et al.,* 2012). To counteract the negative consequences of swidden agriculture, governments and scientific organizations have long sought to eliminate the age-old farming practice and replace it with other farming techniques (Pascual 2005, Rasul and Thapa 2006). Padoch *et al.* (2007) examined the potential extinction of swidden agriculture in Southeast Asia, concluding that swidden cultivation is decreasing in several places of MSEA (Padoch 2007). A case study in a remote upland community in north central Vietnam found that swidden agricultural methods were significantly curtailed as a result of government restrictions Jacobsen *et al.* 2007). Furthermore, economic, cultural, ecological, and political variables such as market access, distance from main roads, off-farm job opportunities, ethnic minorities, and land policies will all contribute to the decline or disappearance of shifting cultivation around the world (Vliet *et al.* 2013). Access to markets, as well as conservation regulations and practices, have accelerated the collapse of swidden agriculture, according to a meta-analysis of global assessments on the patterns and drivers of changes in shifting cultivation (Vliet *et al.* 2012). Evidence from northeastern Vietnam [84] shows that ethnic groups' attitudes toward subsistence agriculture are influenced by their proximity to major roadways. Local populations who live near major roads are more likely to establish market-oriented farming practices, whereas those who live in remote mountainous locations are compelled to rely on shifting cultivation.

The shifting cultivation system is a dynamic and centuries-old land use type. In several mountainous areas of Southeast Asia, this practice has diminished or on the way to extinction. It is widely assumed that other intensive farming systems will eventually replace this approach. Scientists, on the other hand, argue that declaring the end of shifting cultivation is premature (Dressler and Pulhin 2010). In many sections, though, it still exists (Ellen 2012, Padoch 2007). Based on a summary of eight locality studies in montane MSEA,

Fox *et al.* (2000) determined that shifting cultivation has been a dominating land use mode in this region for the past half century (Fox and Vogler, 2005). Surprisingly, despite a clear drop in global tropical forest-agriculture boundaries, van Vliet *et al.* (2012) concluded, based on a global meta-analysis, that shifting cultivation will continue to exist and prosper in many parts of the world (van Vliet *et al.* 2013). Local farmers in Palawan Island, Philippines, say that shifting cultivation is still a de facto crucial and integral activity, despite the state's laws, policies, and practices prohibiting it (Dressler and Pulhin 2010). Other scholars, in contrast to the unfavorable attitude toward swidden farming, urged for changing those one-sided perceptions. They began to rethink and reconsider widely held beliefs about shifting cultivation's impacts. Several case studies from northern Thailand were presented to demonstrate the likelihood of biodiversity conservation under well-managed swidden farming (Schmidt-Vogt 1998). Shifting cultivation, according to Fox, is ecologically fit, culturally appropriate, and the most successful approach to preserving biological variety in Southeast Asia (Fox 2000). Because of their nutritional predilection for rice, smallholder subsistence, and inability to implement tillage technology on sloping soils, this is especially appropriate in MSEA's highland regions (Roder 1997).

Despite the state's efforts to prohibit it, shifting cropping is a common agricultural technique in several parts of North-East India.. Drawing from a survey of 500 respondents from 52 villages spread across six states in North-East India, Pandey *et al.* (2020) found that shifting cultivation is not just an alternative farming method, but also a form of landscape management that has evolved over centuries of experimentation and is inextricably linked to the culture and way of life of those who continue to practise shifting cultivation. Indigenous peoples undertake shifting agriculture since it is a part of their culture and customs, because shifting cultivation represents far more to them than a source of income, and because there are no viable alternatives. According to empirical research, there are three types of attachment or bonding: nature bonding (connection to the natural landscape), social bonding (attachment to the local society and customs), and economic bonding (attachment to the means of sustenance and location) (Pandey etal. 2020). The causes were social connection, a lack of other livelihoods, economic bonding, and natural bonding, in order of significance. Even more importantly, despite to widespread belief among state authorities and other organizations, shifting cultivation offers indigenous people with a sustainable source of income and food security.

Shifting cultivation, being the major agricultural system on sloping forested terrain in Southeast Asia, continues to provide local rural farmers with essential subsistence items. This man-land relationship has existed for a long time. In the meantime, swidden farming has played an important role in the preservation of ecological services and cultural identity (Ziegler *et al.* 2011). Shifting agriculture has evolved into various land use patterns in the last two decades as a result of developing infrastructure networks and accessible market opportunities, particularly since the establishment of the Greater Mekong Sub-regional (GMS) Economic Cooperation Program in 1992. Messerli and colleagues (Messerli *et al.*) (Messerli *et al.*, 2009). It's worth mentioning that more frequent and simpler engagement in a market economy may increase the risk.

8.5.4 Impacts of shifting cultivation changes

Increased human activity (e.g., road development) in SEA will undoubtedly have a significant impact on shifting cultivation systems in the future (van Vliet *et al.* 2012). Shifting cultivation has long been practised mostly by ethnic minority populations (Mertz *et al.* 2009). They typically settle in distant upland locations and are labelled as impoverished. Governments have long prioritised poverty reduction and enhancing local lives. Furthermore, SEA is one of the three primary tropical forest zones (Hansen. Hansen *et al.* 2008) and is home to a diverse range of species (Sodhi *et al.* 2004). How, in what ways, and to what extent will these extraordinary transformations have an influence on swidden farmers' livelihoods (Cramb *et al.* 2009) and cultural identity (Ziegler *et al.* 2011), biological diversity, and ecosystem services are among the pressing questions that need to be answered (Padoch and Pinedo-Vasquez 2010, Rerkasem *et al.* 2009).

The conversion of shifting cultivation to other land uses has a wide range of consequences for local livelihoods (Vongvisouk *et al.* 2014). Construction of traffic infrastructure facilitates the transition of the practice because it gives farmers additional opportunity to participate in market activities. Market engagement may tempt shifting cultivators to relocate to the lowlands, where their livelihood may be threatened and they may lose their cultural identity (Cramb *et al.* 2009). These consequences are deadly for farmers who are facing rapid shifting cultivation change. This may be true for shifting cultivators who live near a major highway, but not for those who live in isolated, inaccessible areas (Chi *et al.* 2013). According to Chi *et al.*, shifting cultivators in Northwest Vietnam faced various repercussions on their livelihoods as a result of traffic infrastructure and markets throughout the transition of shifting cultivation (Chi *et al.* 2013). More research should be done in the future to assess the effects of this land use change as a result of ASEAN's growing economic globalization and regional integration.

Development organisations have viewed shifting cultivation, sometimes known as "slash and burn agriculture," as damaging, inefficient, and wild since the colonial era (Dove, 1983) (Brady, 1996; Thrupp *et al.*, 1997; O'Brien, 2002). In the guise of forest protection and development, Asian colonial and post-colonial governments have implemented rules and regulations targeted at eradicating shifting cultivation for more than a century. According to the most typical explanations given by governments (AIPP/IWGIA/IKAP, 2009), shifting cultivation is technologically antiquated, inefficient, and wasteful, slows growth and thus keeps people in poverty, and is harmful to forests and soils. Despite the lack of comprehensive studies on the impact of existing policies on shifting cultivation practises in Nepal, numerous authors have highlighted the unsupportive policy and policy ignorance features (Regmi *et al.*, 2005; Kerkhoff and Sharma, 2006). The 'terminology or word "shifting cultivation' is not recognised by the Nepalese government in any land use legislation, and it has caused significant conflicts in resource management and profit sharing between indigenous shifting cultivators and government authorities.

8.5.5 Changes during transition

In recent years, shifting cultivation in India's northeastern states, which account for over 86 percent of the country's total shifting cultivation area, has posed a serious ecological danger to the ecosystem in mountainous places. Demographic density, trade network expansion, and economic and social change are all pushing "shifting cultivation" to the outskirts. Because shifting cultivations rarely provide a surplus, the majority of the locals are increasingly transitioning to wet or permanent agriculture. Instead of making the transition to settled agriculture, the entire agricultural yield is insufficient to meet local demand (Patel *et al.* 2013). The prevalence of shifting cultivation, mountainous terrain, unpredictable climate variations, low levels of modern input use, poor infrastructure, and other factors all contribute to low crop output (Karmakar 2008; Barah 2006). Shifting agriculture is still popular in Northeastern Region of India in some form or another, and it is practised by more than 100 tribal ethnic minorities (Singh *et al.* 1996; Ramakrishnan 1993). The region's most distinguishing features are its high rural population (89.86%), significant tribal population (Irshad Ali and Das 2003), and large forest area (63.9%) (Mishra and Sharma 2001). The geographical environment of this region is infinitely diverse.

The *Apatanis* and *Monpas* tribes relocated permanently for settled agriculture, preserving their rites and festivals but greatly improving their economic situation. Locals have become aware of the ecological imbalance caused by shifting farming in recent decades. Certainly, this is due to a tremendous improvement in literacy, which went from 20.4 percent in 1961 to 68.66 percent in 2001, which is greater than the national average of 65.50 percent (Gupta 2002). Shifting cultivation is a wasteful and unscientific method of land usage currently in use (Ranjan and Upadhyay 1999). In the instance of Northeastern Region of India, the net sown area under shifting agriculture ranges from 2.6 percent in Assam to 96.0 percent in Mizoram (Sharma and Prasad 1994). A substantial reduction of the shifting cultivation cycle is contributing to significant forest destruction (Table 8.1) (Mishra and Sharma 2001), resulting in climate change, biodiversity loss, and harmful effects on other dimensions of the broader ecosystem. According to Rahman *et al.* (2007), these negative downstream environmental effects contribute to low agricultural output, which disproportionately affects rural farmers. In the 8th and 9th five-year plans, the Ministry of Agriculture and Co-operations of the Government of India began socioeconomic upgradation of jhumia families in order to encourage sustainable agriculture and provide other livelihood options for the local population, but the results have been far from satisfactory.

Table 8.1: Ecologictransformations in Northeast India

State	**GA area**	**Forest area**	**Forest area as % of GA**	**Forest area annual decrease (%) (1991-95)**	**Soil loss**	**N+P+K loss**
	Km^2	Km^2			Million t yr-1	'000 t yr^{1}
Arunachal Pradesh	83600	68600	82.0	0.049	14.49	3.111
Assam	78400	24000	30.6	0.696	12.32	1.685
Manipur	22300	17500	78.6	0.179	20430	4.289
Meghalaya	22400	15700	70.0	0.250	14.15	2.999
Mizoram	21100	18600	88.1	0.367	13.04	2.738
Nagaland	16600	14400	86.2	0.052	7.96	1.432
Tripura	10500	5500	52.8	0.013	5.95	9.839
Total NER	262000	167400	63.9	0.211	88.34	17.092

Source: Mishra and Sharma, 2001

Northeast India's economy is predominantly agricultural, with agriculture accounting for around 30% of the gross domestic product. There are at least 100 indigenous tribes and 620,000 families who live here, each with their own language and culture. The majority of their food comes from shifting farming (Ramakrishnan 1992). Except in Assam and Tripura, India's northeastern region's population has quadrupled to 40 million in the last half-century (with annual growth rates ranging from 2.01 percent to 5.22 percent), causing the land-man ratio to become increasingly inverse (Barah 2006). Watershed Development Projects, Soil Conservation Projects, Shifting cultivation Management Projects, New Land Use Policy Scheme, and other rehabilitation plans have been undertaken by state and federal governments to control shifting cultivation (Tripathi and Barik 2003).

8.6 ENVIRONMENTAL CONSEQUENCES OF SHIFTING CULTIVATION DEMISE

Traditional shifting cultivation has negligible hydrological and geomorphological implications in Montane Mainland Southeast Asia, however the impacts of enhanced replacement agricultural systems are frequently much more significant. The negative perceptions of shifting cultivation in the region were largely based on cases of forest conversion and land degradation associated with (a) intensified shifting cultivation systems, which are characterised by shorter fallow and extended cropping periods, and/or (b) widespread opium cultivation for cash after WWII. Neither of these techniques should be called traditional subsistence farming techniques (Ziegler *et al.* 2009). Other intensive agriculture techniques in the region, such as semi-permanent and permanent cash cropping, monoculture plantations, and greenhouse complexes, are gradually replacing shifting cultivation. Changes in streamflow responsiveness, increased surface erosion, increased landslide danger, and a deterioration in stream water quality are all negative outcomes of these systems. Unlike traditional shifting cultivation, these effects are caused by a number of factors, such as; (1) large areas of upland catchments are cultivated at the same time; (2) accelerated water and soil erosion on plots that are continuously cultivated with short or no fallowing for recovery of soil properties, such as infiltration; (3) irrigation is typically done with surface and ground water; and (4) insecticides and herbicides are employed. Furthermore, the commercial viability of these systems is dependent on the presence of dense networks of roadways, which are known for disturbing hydrological and geomorphological systems. To mitigate the effects of these intensive upland farming practices, a new conservation focus is required (Ziegler *et al.* 2009).

The shifting cultivation system in Bangladesh is one example of swiddening causing considerable changes in stream flow variables. A hectare^{-1} sub-catchment cleared and cultivated resulted in a 600% increase in peak flow and a 160% increase in yearly runoff (Gafur *et al.* 2003). While certain changes in streamflow are expected as a result of converting such a large percentage of forest (i.e., the >33 percent requirement), the observed changes in Bangladesh could be due to the effects of a more intensive swidden system, with fallow period reduced to only 3–5 years. Despite this, after one year of cultivation, stream flow variables reverted to near normal in the first year of fallow, illustrating the evolving agriculture landscape's resilience to long-term hydrological changes. The rapid recovery of tropical flora, as well as the small decrease in infiltration, both in degree and space, produced by swiddening activities, contribute to this resilience. The move from shifting cultivation to more intensive forms of agriculture, on the other hand, may result in significant reductions in infiltration, resulting in enough surface runoff to enhance stormflow peaks and reduce dry season flows. Several elements of agricultural intensification make this scenario probable, despite the fact that it has rarely been proven in catchment studies (Valentin *et al.* 2008).

i) large parts of the catchment is changed to farming, which has induced soil erosion on large scale for long periods of time throughout the year (Janeau *et al.* 2003; Turkelboom *et al.* 2008);

ii) high turbulence into subsurface soils during the construction of terraces, such as; during rubber plantations in China and vegetables in the Cameron mountains of Malaysia (Midmore *et al.* 1996);

iii) accelerated erosion due to tillage which results in low-permeability (Turkelboom 1999; Ziegler *et al.* 2004b);

iv) due to very short fallow period, enough recovery of soil porosity and aggregate stability is not possible (Nye and Greenland 1965, Bronick and Lal 2005)

v) there is increase of processing areas, greenhouse complexes, and other low-infiltration surfaces and;

vi) there is spoilage of vegetative buffers which encourage infiltrating overland flow (Nikolic *et al.* 2008). (Ziegler *et al.* 2007c; Vigiak *et al.* 2008).

Despite evidence to the contrary, extreme floods are commonly considered to be caused by forest removal and/or agricultural conversion (van Dijk *et al.* 2009). Floods occur when a canal receives more water than it can store or move through in any basin downstream (van Dijk *et al.* 2009). Rainfall decreases have been linked to large-scale forest removal and agricultural conversion (van Dijk and Keenan 2007; Malhi *et al.* 2008).

8.6.1 Agricultural transformations: the early attempts and their impacts

Government programmes targeted towards transformation of shifting cultivation were initiated in the 1970s with a thrust on horticulture and cash crop promotion. Coffee, rubber, and cashew plantations were established under various government schemes as alternatives to shifting cultivation. The plantations, established by concerned departments under centrally sponsored programmes, were meant to have a demonstration eff ect and were largely planned and maintained by the departments on land made available by participating villages; the involvement of the villagers were limited to providing labour for establishment and maintenance of the plantations during the gestation period, aft er which the plantations were to be handed over to villagers to reap the benefi ts from the crops. In addition, horticultural crops—mainly pineapple—were also promoted as alternatives (Leduc and Choudhury (2012). Terrace farming schemes were also created with the twin goals of expanding wet rice cultiv ation (thus lowering the reliance on shifting rice agriculture) and boosting soil erosion control. The acceptance of such programmes by the villagers was not very encouraging until recently and few, if any, replicated these options.

The most critical factors that discouraged farmers from accepting these alternatives, however, were the long gestation period during which no returns could be expected and the total absence of market linkages once the crops started maturing. Further, restrictions on the sale of crops such as coffee in the open market during this period—coffee could be sold only to the Coffee Board till the late 1990s—frustrated the farmers to the extent that plantations set up with enthusiasm in many villages were abandoned and allowed to revert back into fallow forests. In many areas, plantations of coffee, cashew, and other crops were slashed down consequent to poor market linkages and farmers reluctantly reverted back to shift ing cultivation.

8.7 TRANSFORMING UPLANDS: THE DEMISE OF SHIFTING CULTIVATION?

Over the years, however, shifting cultivators have slowly revived the process of transformation, often with the active encouragement of different government programmes and the gradually increasing influence of markets (Leduc and Choudhury (2012). With the rapid penetration of market forces into the uplands of north-east India and the consequent transformation of the subsistence economy into a monetarized, cash-driven system, most upland households have little choice but to accept sedentarized options or rapidly transform conventional shifting cultivation into a 'commodified' form (Choudhury *et al.* 2003), often complimented with settled agricultural options that allow them a means of cash generation. Agricultural transformation seems to have taken roots in the uplands of north-east India, particularly the state of Meghalaya. Estimates reported by the Task Force on Shifting (Jhum) Cultivation, setup by the Government of Meghalaya in 2001, show a gradual decrease in the number of families practising shifting cultivation (Government of Meghalaya 2003). Citing estimates provided by the Agro-Economic Research Centre for North-East India of the Assam Agricultural University, Jorhat, the Task Force reports 40.57 per cent of the rural population to be dependent on shifting cultivation in 1971. Corresponding figures, based on census surveys conducted in 1981, puts 23.58 per cent of the rural population to be dependent on shifting cultivation. Based on a survey of 443 villages in thirty-two blocks of Meghalaya, conducted by the Task Force in 2001, the same figure was estimated at 12.57 per cent of the rural population. During 1971, the rural population dependent on shifting cultivation was 40.57 per cent.

8.7.1 Transition from shifting cultivation to settled agriculture

Are transformations really taking place?

While the Task Force estimates and agricultural statistics indicate a gradual transition from shifting cultivation to settled agricultural systems and suggest a resultant reduction on the dependency on shifting cultivation, the results of a field study conducted in six villages of West Garo Hills1 to get a 'reality check' seem to indicate otherwise and suggest the need for a critical re-examination of the statistics provided by the agencies (Choudhury *et al.* 2008). The study, covering eighty-one households in six villages, found that while transformations, particularly the induction of cash crops such as areca nut, were progressively increasing in the uplands, it did not necessarily reduce the dependency of households on shifting cultivation as nearly all the surveyed households continued with the practice, though exceptions did exist in a few villages. Contrary to the Task Force report and the agricultural statistics provided by the authorities, shifting cultivation seems to persist in the study villages. An interesting revelation of the survey is the fact that access to shifting cultivation has not changed despite transformations (and a reduction in area under shifting cultivation, as suggested by the agricultural statistics), and all households retain their access to shifting cultivation plots. Moreover, the study also revealed that a growing proportion of households had multiple farming systems managed simultaneously.

8.7.2 Transformations of the uplands

Despite the growing acceptance of alternate options, concerns still remain for the majority of shifting cultivators, in particular for the poor. The gradual transformation of upland agriculture has resulted in diversifi cation of farming systems and compelled many households to establish multiple farming systems. This has resulted in an increase in land use pressure, with a resultant reduction in fallow cycles, a corresponding decline in land recuperation, and consequently, a rapidly declining productivity, resulting in the marginalization of shifting cultivation (Choudhury and Sundriyal 2003). Transformations have also brought about increased demands on labour, adversely impacting management and productivity. The most severe implication of transformations, however, has been the drastic replacement of common property regimes with private property frameworks, posing serious threats to the continuity of equitable tenurial access to productive resources hitherto inherent in traditional access regimes. While implications of agricultural transition in the uplands has been elaborated by Choudhury *et al.* (2008) based on the study carried out in West Garo Hills.

8.7.3 Transformations: The demise of forests and ecosystem services

The conversion of shifting cultivation fallows into settled agricultural systems has an additional implication which is equally serious and having far more serious long-term ramifications, particularly in the context of climate change. As shifting cultivation fallows get converted to sedentarized systems, both the land use and the land cover undergo a permanent change. While shifting cultivation allows secondary forests to regenerate in the fallows, a temporary change in land use and land cover, settled systems permanently erase the chance of forests regenerating in such plots. The expansion of sedentarized systems to replace shifting cultivation thus replaces not only shift ing cultivation but also secondary forests. Thus, for every hectare of shifting cultivation transformed to settled agriculture, a proportionate area of forests is permanently lost. This reduction of secondary forests consequent to expansion of sedentary systems has serious consequences for hydrological regimes in the uplands (and subsequently downstream), gradually affecting discharge, stream flow, and water availability in the watersheds.

8.7.4 Transformations, risk management, and vulnerability

Although agricultural transformations hold out the promise of cash generation and improved incomes, shifting cultivators are apprehensive to embark on change. The bitter experiences of farmers who have tried to transform to cash crops, the uncertainty of markets, and unassured returns undermine the confidence on such options, making farmers hesitant in making the change.

Volatile markets, unattractive prices, and the exploitation at the hands of traders make shifting cultivators, particularly the poor, extremely wary of the high vulnerability that such transformations couldentail. During the initial transition years, the absence of returns from plantations and the inability to ensure food security makes farmers reluctant to change. For shifting cultivators, although the practice is subsistence, it gives them the confidence to avert risks. The diversity of crops within shifting cultivation is a strategy evolved to avoid several risks (Leduc and Choudhury (2012).

8.7.5 Unintended marginalization

Some policies in Southeast Asia have inadvertently accelerated the demise of shifting cultivation (Falvey 2018). These include policies that have classified them as ethnic minorities within nation-states, defined rural highlands as either forested or agricultural, promoted single-minded conservation ideals, assigned land tenure based on permanent land use and increased accessibility (Fox *et al.*, 2009). Yet it is generally agreed that these objectives provide the greatest benefit to the majority in a nation. However, on the basis of United Nations' and other ideals, today's arguments also include rights to protection from collateral policy damage.

Burning phase of shifting cultivation as a source of carbon emissions is reminiscent of forest-conservation policies from at least colonial times. Such scientism readily infuses development policy and has produced such aims as seeking to 'reduce emissions from deforestation and forest degradation' by outlawing shifting cultivation (Griffiths, 2008) even as intensification of industrial agriculture is shown to be the sector's major source of carbon emissions (FAO, UNDP, UNEP, 2008). In fact, routine shifting cultivation practices contribute to carbon sequestration by encouraging the regeneration of secondary forest by trimming trees to allow them to re-sprout; by maintaining long fallows with short cultivation periods and creating higher levels of soil organic matter than is normal in continuous cropping. 'Forest removal for shifting agriculture releases less carbon than permanent forest clearing since the fallow period allows some forest regrowth,' according to the Intergovernmental Panel on Climate Change (IPCC, 2006). But we must wonder what impact such a realization, buried in one paragraph of a long report, will have amid other ideas and detailed information in a politically charged agenda.

As an ICIMOD conference summarized: 'If we alter our name from slash-and-burn to rotational agroforestry, agroforestry with a burn cycle, or a type of forest gardening, we begin to use more positive phrases that emphasize the growing cycle rather than the cutting cycle.' Farmers spend significantly more time producing trees and crops than they do burning them, according to research, conserving soil, recovering nutrients, fallowing, and resting' (Kerkhoff and Sharma, 2006). The ICIMOD conference recommended that policies that discouraged shifting cultivation should be replaced by addressing land tenure, along with credit access, research and extension, market development and commercialization of new niche products; strengthening customary institutions; and coordination of government agencies. Such observations seem to be supported by a longer-term study in Laos, where changes in land cover, land use and livelihoods indicated a relationship between the vulnerability of highlanders and policies that aimed to increase forest cover, eradicate shifting cultivation and stimulate intensive and commercial agriculture. Rather than designating areas specific to agriculture or forests, the study suggested that the focus be shifted to means of maintaining diverse livelihoods and economic opportunities (Castella *et al.*, 2013). Similar experience in Kalimantan suggests that external guesses about suitable cash crops to grow in combination with food crops are seldom ecologically stable, although in that case locally known rattan production seemed to be an exception (Weinstock, 1996).

8.7.6 Shifting cultivation and recent trends

Africa is a humid tropical region with a wide variety of patterns. In West and Central Africa, shifting cultivation is mainly still practised, and in certain places it has even grown. This pattern contrasts with changes in other parts of Africa where declining and disappearing shifting agricultural areas have dominated during the previous two decades (van Vliet *et al.*, 2012). In many humid regions of West Africa, shifting cultivation is still prevalent; in Nigeria, however, it has virtually disappeared and is only found in isolated areas. Even more prevalent than in West Africa, shifting cultivation is also widespread in Central Africa, where it is still gaining traction in some regions like the Democratic Republic of the Congo (Molinario *et al.* 2015). In eastern and southern Africa, shifting cultivation is still used, although it is not very common; Tanzania and Kenya have the lowest percentages. According to Zaehringer *et al.* (2015), there have only been slight decreases in the area under shifting cultivation throughout the past 20 years, especially along Madagascar's eastern escarpment. Shifting cultivation is still widespread throughout Central America, and Fig. 6 and the survey data indicate that it has grown in certain regions well into the 2000s (Panama and Guatemala). Mexico has witnessed a push to decrease the area under shifting cultivation.

8.7.7 Shifting cultivation and its future trends

It is predicted that sustained rapid economic development and the ensuing changes in agricultural practises and, more importantly, in the economic structure will cause shifting cultivation to disappear more quickly in Asia than in Africa or the Americas. In contrast, it is anticipated that shifting cultivation will persist longer across Africa, especially in Central Africa. Shifting agriculture has a higher chance of surviving in areas with low population density and restrictions on alternative livelihoods or agricultural growth. Despite the inherent limits of these estimates, we can discern many more definite trends. According to the study results, shifting agriculture will continue to decline quickly as it currently is. By 2030, a sizable chunk of the land under shifting cultivation will likely have disappeared, and by 2060, the pockets that remain are likely to be almost completely gone. Additionally, trends for several Asian nations may be seen in the poll results:

- Vietnam and Laos: Shifting cultivation is likely to be greatly reduced by 2030 and completely gone by 2060.
- Myanmar: Shifting cultivation is estimated to mostly disappear sometime between 2060 and 2090 if conflicts between union government and ethnic armed groups are resolved.
- Borneo and Sulawesi: Shifting cultivation is expected to disappear sometime between 2030 and 2060.
- India and Bangladesh: Shifting cultivation is estimated to disappear by 2030.
- Papua New Guinea: Shifting cultivation may persist well into the second half of this century, may be even up to 2090.

Due to a lack of data, humid tropical Africa is most likely the location where future trends are most difficult to predict. According to the findings, shifting farming is most likely to survive in Africa. It is projected to increase in some locations, particularly in Central Africa, over the next decade before beginning to diminish.

Heinimann *et al.* (2017) used an expert survey to make a first attempt at forecasting recent and expected future trends in global shifting agriculture distribution until the end of the twenty-first century. Their prediction is shown through Table 8.2.

Table. 8.2 Estimated disappearing of shifting cultivation at global level

Domain	2030	2060	2090
Asia	30-50%	80-100%	100%
Central & South America	10-30%	60-80%	80-100%
Asia	30-50%	80-100%	100%

Source: Heinimann *et al.* (2017)

Recent international attempts to evaluate and synthesize various earth system models have placed a heavy emphasis on measuring land use's past, present, and future contributions to climate change (Silva *et al.* 2011, Hurttet a. 2011, Houghton *et al.* 2012). However, accurate forecasting of land-based emissions necessitates a better understanding of land-use system change megatrends (Ellis *et al.* 2013). There is a scarcity of worldwide data on specific land use regimes, such as shifting cultivation on a global scale.

References

AIPP/IWGIA/IKAP 2009. Shifting cultivation and climate change. Briefing paper for UNFCCC Intersessional Meeting, Bangkok. AIPP, IWGIA, IKAP.

Alegre, J.C.; Cassel, D.K. Dynamics of soil physical properties under alternative systems to slash-and-burn. Agric. Ecosyst. Environ. 1996, 58, 39–48.

Alford, D. 1992: Streamflow and sediment transport from mountain watersheds of the Chao Phraya Basin, northern Thailand: A reconnaissance study. Mountain Research and Development 12: 257-268.

Anonymous, 2009. Draft report of the inter-ministerial national task force on rehabilitation of shifting cultivation land. Report submitted to the Ministry of Environment and Forests, Government of India.

Anonymous, 2011a. Regional project implementation plan. North East rural livelihood project, Ministry of DONER, Govt. of India. http://sirdmizoram.org/. Accessed 2 March 2012.

Aryal KP, Kerkhoff EE (2008). The right to practice shifting cultivation as a traditional occupation in Nepal. A case study to apply ILO Conventions Nos. 111 (Employment and Occupation) and 169 (Indigenous and Tribal Peoples), Kathmandu, International Labour Office.

Astuti, R. 2015. REDD+ Governmentality: Governing Forest, Land and Forest Peoples in Indonesia. PhD Thesis, Department of Geography, Environment and Earth Sciences, Victoria University of Wellington.

Aumtong, S., Magid, J., Bruun, S., and de Neergaard, A. (2009). Relating Soil Carbon Fractions to Land Use in Sloping Uplands in Northern Thailand. Agriculture, Ecosystems & Environment 131: 229–239. doi:10.1016/j.agee.2009.01.013.

Baird, I. G., and Shoemaker, B. (2007). Unsettling Experiences: Internal Resettlement and International Aid Agencies in Laos. Development and Change 38: 5865–888. doi:10.1111/j.1467-7660.2007.00437.x.

Barah, B.C., 2006 Agricultural development in the north-east India: The challenges emerging opportunities. Presented at the 66th annual conference of the Indian Society of Agricultural Economics, November 8-10.

Bartlett, H. H. (1956). Fire, Primitive Agriculture and Grazing in the Tropics. In: W. L. Thomas (ed.) Man's Role in Changing the Face of the Earth (Chicago and London: The University of Chicago Press), pp. 692-720.

Bartlett, Harley H. (1957) Fire in Relation to Primitive Agriculture and Grazing in the Tropics: Annotated Bibliography. Vol. II. Ann Arbor: University of Michigan Press.

Béliveau, A.; Lucotte, M.; Davidson, R.; Do Canto Lopes, L.O.; Paquet, S. (2009) Early Hg mobility in cultivated tropical soils one year after slash-and-burn of the primary forest, in the Brazilian Amazon. Sci. Total Environ., 407, 4480–4489.

Belsky JM, Siebert SF. (2003). Cultivating cacao: Implications of sun-grown cacao on local food security and environmental sustainability. Agriculture and Human Values. 2003;20:277–285. doi: 10.1023/A:1026100714149.

Binswanger, H., Deininger, K., 1997. Explaining Agricultural and Agrarian Policies in Developing Countries. FAO, Rome.

Borggaard OK, Gafur A, Petersen L (2003). Sustainability appraisal of shifting cultivation in the Chittagong Hill Tracts of Bangladesh. Ambio. 32: 118–123.

Boserup, E. (1965). The Conditions of Agricultural Growth: The Economics of Agrarian Change under Population Pressure. Earthscan, London.

Boulay, J. N., Hellberg, M. E., Cortés, J., & Baums, I. B. (2014). Unrecognized coral species diversity masks differences in functional ecology. Proceedings of the Royal Society B: Biological Sciences, 281(1776), 20131580.

Bouwman A.F. ed. 1990. Soils and the greenhouse effect. New York: Wiley.

Brady NC (1996). Alternatives to slash-and-burn: a global imperative. Agric. Ecosyst. Environ., 58: 3–11.

Bronick, C. J., and Lal, R. (2005). Soil Structure and Management: A Review. Geoderma 124: 3–22. doi:10.1016/j.geoderma.2004.03.005.

Bruun TB, de Neergaard A, Lawrence D and Ziegler AD, 2009. Environmental consequences of the demise in swidden cultivation in southeast asia: carbon storage and soil quality. Human Ecology, 37:375–388, DOI 10.1007/s10745-009-9257-y

Cairn M (2007). Conceptualizing indigenous approaches to fallow management: a roadmap to this volume. in Cairn M editor. Voices from the forest. Resources for the Future, Washington, D.C. pp. 16– 36.

Cairns M, Garrity DP (1999). Improving shifting cultivation in Southeast Asia by building on indigenous fallowmanagement strategies. Agrofor. Syst., 47(1): 37–48.

Castella JC, Lestrelin G, Hett C, Bourgoin J, Fitriana YR, Heinimann A, Pfund JL. Effects of landscape segregation on livelihood vulnerability: Moving from extensive shifting cultivation to rotational agriculture and natural forests in Northern Laos. Human Ecology. 2013;41:63–76. doi: 10.1007/s10745-012-9538-8.

Chi, V.K.; van Rompaey, A.; Govers, G.; Vanacker, V.; Schmook, B.; Hieu, N. Land transitions in Northwest Vietnam: An integrated analysis of biophysical and socio-cultural factors. Hum. Ecol. 2013, 41, 37–50.

Chin Y W, Balunas M J, Chai H B and Kinghorn A D .2006. Drug discovery from natural sources. AAPS J 8: 239–53.

Chin, S. C. (1985). Agriculture and Resource Utilization in a Lowland Rainforest Kenyah Community. Sarawak Museum Journal Special Monograph No. 4. Sarawak Museum, Kuching.

Choudhury, D. and R. Sundriyal. 2003, 'Issues and Options for Improving Livelihoods of Marginal Farmers in Shift ing Cultivation Areas of Northeast India', Outlook in Agriculture, 32: 17–28.

Choudhury, D., Ingty, D. and Jamir, Sentikemba 2003 Managing marginalisation in shifting cultivation areas of Northeast India: Examples of community innovations and initiatives. In: Tang Ya and Pradeep M. Tulachan (eds.) Mountain Agriculture in the Hindu Kush Himalayan Region. ICIMOD, Kathmandu, pgs. 207-212.

Choudhury, D., N.S. Jodha, and E. Sharma. 2008, 'Policy Approaches in Management of Shift ing Cultivation: Compromising Equitable Access, Food Security and Common Property Institutions', Paper presented at IASC Biennial Conference 'Governing shared Resources: Connecting Local Experience to Global Challenges', Cheltenham, England, 14–18 July.

Cochran, D.M., 2008. Who will work the land? National integration, cash economies, and the future of shifting cultivation in the Honduran Mosquitia. Journal of Latin American Geography 7 (1).

Coe, R., Sinclair, F. and Barrios, E. (2014) 'Scaling up agroforestry requires research 'in' rather than 'for' development', Current Opinion in Environmental Sustainability, 6 February 2014, pp73-77

Coffey K, Mertz O, Leisz, SJ and Fox JM, 2007. The Demise of Swidden in Southeast Asia? Local Realities and Regional Ambiguities. Journal of Geography, 107: 29-41, DOI: 10.1080/00167223.2007.10801373

Colfer C, Peluso N, Chung C.S. 1997. Beyond Slash and Burn. Building on Indigenous Management of Borneo Tropical Rainforest. N.Y. Botanical Garden, New York.

Colfer, CJP. (2008). The Longhouse of the Tarsier, Changing Landscapes, Gender, and Wellbeing in Borneo. Borneo Research Council, Phillips, ME.

Comte, I.; Davidson, R.; Lucotte, M.; de Carvalho, C.J.R.; de Assis Oliveira, F.; Da Silva, B.P.; Rousseau, G.X. (2012). Physicochemical properties of soils in the Brazilian Amazon following fire-free land preparation and slash-and-burn practices. Agric. Ecosyst. Environ., 156, 108–115.

Condominas, G. (2009). Anthropological Reflections on Swidden Change in Southeast Asia. Human Ecology. doi:10.1007/s10745-009-9248-z.

Conklin HC (1957). Hanunoo Agriculture. a report on an integral system of shifting cultivation in the Philippines. FAO Forestry Development. FAO, Rome, p. 12.

Conklin, H.C. The study of shifting cultivation. Curr. Anthropol. 1961, 2, 27–61.

Cramb, R. A. (1989). The Use and Productivity of Labour in Shifting Cultivation: An East Malaysian Case Study. Agricultural Systems 29: 97–115. doi:10.1016/0308-521X(89)90057-7.

Cramb, R. A. (2007). Land and Longhouse: Agrarian Transformation in the Uplands of Sarawak. NIAS, Copenhagen.

Das A, Ram Krushna GI, Choudhury BU, Munda GC Patel DP, Ngachan SV, Ghosh PK, Tripathi AK, Das S, Kumar M (2012) Natural resource conservation through indigenous farming systems: Wisdom alive in North East India. Indian Journal of Traditional Knowledge11(3): 505-513.

Dash B. 2006. Shifting cultivation among the tribes of Orissa, 2006, odisha.gov.in/e-magazine/Orissareview/engpdf/76-84.pdf.

de Jong W, van Noordwijk M, Sirait M, Liswanti N, Suyanto, S. 2001. Farming secondary forest in Indonesia. Journal of Tropical Forest Science 13: 705–726

de Jong, W 1997. 'Developing Swidden Agriculture and the Threat of Biodiversity Loss'. Agriculture, Ecosystems and Environment 62: 187–97.

de Jong, W. (1995) Diversity, Variation and Change in Ribereno Agriculture and Agroforestry. Doctoral thesis, Wageningen University for Life Sciences, The Netherlands.

de Neergaard, A., Magid, J., and Mertz, O. (2008). Soil erosion from shifting cultivation and other smallholder land use in Sarawak, Malaysia. Agriculture, Ecosystems and Environment 125: 182–190.

Deb S, Lynrah M M, and Tiwari B K. 2013. Technological innovations in shifting agricultural practices by three tribal farming communities of Meghalaya, North East India. Trop. Ecol. 54(2): 133–48..

Detwiler, R. P., & Hall, C. A. (1988). Tropical forests and the global carbon cycle. Science 239(4835), 42-47.

Dhakal S, 1990. an anthropological perspective on shiiting cultivation: A Case Study of Khoriya Cultivation III the Al"Un Valley of Eastern Nepal 31. World Bank. (1990). World Developmcnt Report.

Dove M (1983). Theories of swidden agriculture and the political economy of ignorance. Agrofor Syst., 1: 85–99.

Dove MR (1996). So far from power, so near to the forest: a structural analysis of gain and blame in tropical forest development, In: Padoch C, Peluso NL (eds.), Borneo in Transition: People, Forests, Conservation, and Development, Oxford University Press, Kuala Lumpur, pp. 41-58.

Dove, MR (1988). The ecology of intoxication among the Kantu' of West Kalimantan. In Dove, M. (Ed.), The Real and Imagined Role of Culture in Development: Case Studies from Indonesia. University of Hawaii Press, Honolulu, pp. 139–192.

Dressler W, Pulhin J. (2010). The shifting ground of swidden agriculture on Palawan Island, The Philippines. Agriculture and Human Values.;27:445–459. doi: 10.1007/s10460-009-9239-0.

Dressler WH, Wilson D, Clendenning J, Cramb R, Rodney Keenan R, Mahanty S, Bruun TB, Ole Mertz O and Lasco RD, 2017. The impact of swidden decline on livelihoods and ecosystem services in Southeast Asia: A review of the evidence from 1990 to 2015. Ambio. 46: 291–310. doi: 10.1007/s13280-016-0836-z

Ducourtieux O (2005). Shifting cultivation: The Phunoy Traditional management System: In Improving Livelihoods in Uplands of the Lao PDR. NAFRI, pp. 71-77.

Ducourtieux O, Visonnavong P, Rossard J (2006). Introducing cash crops in shifting cultivation regions – the experience with cardamom in Laos. Agrofor Sys. 66: 65–76, DOI 10.1007/s10457-005-6645-1.

Dufumier M (1996). Ethnic minorities and slash and burn agriculture in Laos. Notebooks Humanities, 2(1): 195-208.

Eaton JM, Lawrence D (2009). Loss of carbon sequestration potential after several decades of shifting cultivation in the Southern Yucata. For. Ecol. Manage., 258: 949–958.

Ellen, R. (2012). Studies of swidden agriculture in Southeast Asia since 1960: An overview and commentary on recent research and syntheses. Asia Pac. World, 3, 18–38.

Ellen, R. F., & Soselisa, H. L. (2012). A comparative study of the socio-ecological concomitants of cassava (Manihot esculenta Crantz): Diversity, local knowledge and management in eastern Indonesia. Ethnobotany Research and Applications, 10, 15-35.

Ellis EC, Kaplan JO, Fuller DQ, Vavrus S, Klein Goldewijk K, Verburg PH. (2013). Used planet: A global history. Proc Natl Acad Sci;110: 7978–7985. pmid:23630271

Falvey JL, 2018. Policies impacting shifting cultivation - Getting them right. https://www.researchgate.net/publication/322298177.

FAO, UNDP, UNEP 2008. UN Collaborative Programme on Reducing Emissions from Deforestation and Forest Degradation in Developing Countries (UN-REDD). Framework Document 20 June 2008.

FAO. 1957. Shifting cultivation. Unasylva 11: 1.

FAO. 1984. Institutional aspects of shifting cultivation in Africa. FAO Human Resources, Institutions and Agrarian Reform Division. Based on work by Dr M. Tiffen. Rome.

FAO. 1995. Forest Resources Assessment 1990. Global synthesis. FAO Forestry Paper 124, Food and Agriculture Organization of the United Nations, Rome.

Feintrenie L, Schwarze S, Levang P. (2010). Are local people conservationists? Analysis of transition dynamics from agroforests to monoculture plantations in Indonesia. Ecology and Society. 2010;15:37. doi: 10.5751/ES-03870-150437.

Forsyth, T. (1996): Science, myth and knowledge: Testing Himalayan environmental degradation in northern Thailand. Geoforum 27: 375-392.

Forsyth, T.J. (1994): The use of Cesium-137 measurements of soil erosion and farmer's perceptions to indicate land degradation amongst shifting cultivators in northern Thailand. Mountain Research and Development 14: 229-244.

Found. WC 1987. Shifting cultivation in Jamaica – An invisible dilemma. Development Discussion Paper. No.238. Ilarvard Institute for International Development. Harvard Universitty.

Fox J, Castella JC, Ziegler AD. Swidden, rubber and carbon: Can REDD+ work for people and the environment in Montane Mainland Southeast Asia? Global Environmental Change. 2014;29:318–326. doi: 10.1016/j.gloenvcha.2013.05.011.

Fox J, Fujita Y, Ngidang D, Peluso N L, Potter L, Sakuntaladewi N, Sturgeon J and Thomas D. 2009. Policies, Political- economy, and swidden, in Southeast Asia. Hum. Ecol. doi: 10.1007/ s10745-009-9240-7.

Fox J, Truong DM, Rambo AT, Tuyen NP, Cuc LT and Leisz S. 2000. Shifting Cultivation: A New Old Pa r a d i g m for Managing Tropical Forests. BioScience, 50: 521-528

Fox, J. (2000). How Blaming 'Slash and Burn' Farmers is Deforesting Mainland Southeast Asia. Asia Pacific Issues 47: 1–8.

Fox, J. F., & Papanicolaou, A. N. (2008). An un-mixing model to study watershed erosion processes. Advances in Water Resources, 31(1), 96-108.

Fox, J., & Vogler, J. (2005): Land-use and land-cover change in montane mainland Southeast Asia. Environmental Management 36: 394-403.

Fujisaka S, Hurtado L, Uribe R (1996). A working classification of slashand-burn agricultural systems. Agrofor. Syst., 34: 151-169.

Gadgil, M., & Berkes, F. (1991). Traditional Resource. Adaptive Marine Resource Management Systems in the Pacific, 8(3-4), 127-141.

Gafur, A., Jenson, J. R., Borggard, O. K., and Peterson, L. (2003). Runoff and Losses and Nutrients from Small Watersheds under Shifting Cultivation (Jhum) in the Chittagong Hill Tracts of Bangladesh. Journal of Hydrology 279: 293–309. doi:10.1016/S0022-1694(03)00263-4.

Garrity, D. P. (1993). Sustainable land-use systems for sloping uplands in South East Asia. In Ragland, J., Lal, R. (eds.), ASA Special Publication No. 56, 41–66.

Geddes, W. R. (1954). The land Dayaks of Sarawak. HMSO, London.

Geertz. C. 1974 Agricultural involution: The process of ecological change in Indinesia. University of California Pres, Berkley.

Gerber JF and Veuthey S (2010). Plantations, Resistance and the Greening of the Agrarian Question in Coastal Ecuador. Journal of Agrarian Change 10(4):455 – 481, DOI:10.1111/ j.1471-0366.2010.00265.x

Gillis M. and Repetto R. 1988. Deforestation and wood, fodder and fibre, plus increased income and improved living standards. government policy. Cambridge: Cambridge University Press.

Griffiths, T. (2008). Seeing 'REDD'? Forests, climate change mitigation and the rights of indigenous peoples and local communities. In. Forest Peoples Programme, Wales, England.

Guo, H., Padoch, C., Coffey, K., Aiguo, C. & Yongneng, F. (2002): Economic development, land use and biodiversity change in the tropical mountains of Xishuangbanna, Yunnan, Southwest China. Environmental Science& Policy 5: 471-479.

Gupta M D. 1994. Continuity and change in tribal societies: A study of the economic aspects of tribal jhumia society in Tripura. (In) Sangma M B (Ed) Essays on North East India, Indus publishing company, New Delhi, pp 147–64.

Gupta, P.K., 2002. Education and socio-economic development of the north east. Dept. of Education, NEHU, Shillong, India.

Hamilton, H, 1997. Slash-and-Burn in the History of the Swedish Forests. Rural Development Forestry Network Paper 21f, ODI, Portland House, Stag Place, London SW1E 5DP, UK, pp 19-24.

Hanks, L. M. (1972). Rice and Man, Agricultural Ecology in Southeast Asia. University of Hawaii Press, Honolulu.

Hansen, M.C.; Stehman, S.V.; Potapov, P.V.; Loveland, T.R.; Townshend, J.R.; DeFries, R.S.; Pittman, K.W.; Arunarwati, B.; Stolle, F.; Steininger, M.K. Humid tropical forest clearing from 2000 to 2005 quantified by using multitemporal and multiresolution remotely sensed data. Proc. Natl. Acad. Sci. USA 2008, 105, 9439–9444.

Hansen, T. S., and Mertz, O. (2006). Extinction or Adaptation? Three Decades of Change in Shifting Cultivation in Sarawak. Land Degradation and Development 17: 135–148. doi:10.1002/ldr.720.

Hansen, T.S. (2005): Spatio-temporal aspects of land cover and land use changes in the Niah Catchment, Sarawak, Malaysia. Singapore Journal of Tropical Geography 26: 170-190.

Hartemink, A.E. (2005). Plantation agriculture in the tropics—Environmental issues. Outlook Agric., 34, 11–21.

Heinimann A, Mertz O, Frolking S, Egelund Christensen A, Hurni K, Sedano F, Chini LP, Sahajpal R, Hansen M and Hurtt G, (2017) A global view of shifting cultivation: Recent, current, and future extent. PLoS ONE 12(9): e0184479. https://doi.org/10.1371/journal.pone.0184479

Hill, R. D. (1998). Status and Change in Forty Years of Southeast Asian Agriculture. Singapore Journal of Tropical Geography 19: 1–25. doi:10.1111/j.1467-9493.1998.tb00247.x.

Houghton RA, House JI, Pongratz J, van der Werf GR, DeFries RS, Hansen MC, *et al.* Carbon emissions from land use and land-cover change. Biogeosciences. 2012;9: 5125–5142.

Hume D. W. 2006. Swidden Agriculture and Conservation in Eastern Madagascar: Stakeholder Perspectives and Cultural Belief Systems. Conservation and Society, 4: 287–303.

Hurtt GC, Chini LP, Frolking S, Betts RA, Feddema J, Fischer G, *et al.* Harmonization of land-use scenarios for the period 1500–2100: 600 years of global gridded annual land-use transitions, wood harvest, and resulting secondary lands. Clim Change. 2011;109: 117–161.

Intergovernmental Panel on Climate Change (IPCC). (2006). Land Use, Land-Use Change and Forestry. IPCC, http://www.grida.no/publications/other/ipcc_sr/?src=/climate/ ipcc/land_use/008.htm#s6-1

Irshad Ali, A.N.M., Das, I., 2003. Tribal Situation in North East India. Stud. Tribes Tribals 1(2): 141-148.

IWGIA. 2014. Shifting cultivation, livelihood and food security new and old challenges for indigenous peoples in Asia. International Work Group for Indigenous Affairs. AIPP Shifting cultivation livelihood security. www.iwgia.org./0694.

Jakobsen, J.; Rasmussen, K.; Leisz, S.; Folving, R.; Quang, N.V. The effects of land tenure policy on rural livelihoods and food sufficiency in the upland village of Que, North Central Vietnam. Agric. Syst. 2007, 94, 309–319.

Janeau, J. L., Bricquet, J. P., Panchon, O., and Valentin, C. (2003). Soil Crusting and Infiltration on Steep Slopes in Northern Thailand. European Journal of Soil Science 54: 543–553. doi:10.1046/j.1365-2389.2003.00494.x.

Kakeya, M., Sugiyama, Y., Oyama, S., 2006. The Citemene system, social leveling mechanism, and agrarian changes in the bemba villages of northen Zambia: an overview of 23 years of ''fixed point'' research. African Study Monographs 27, 27–38.

Kamajou, F. 1983. socio-economic and institutional considerations in improving shifting cultivation in tropical Africa. In: Okigbo B.N.(Edns) . Improved permanent production systems as an alternative to shifting intermittent cultivation. FAO, Soils Bulletin No. 53, International Institute of Tropical Aqriculture. Ibadan, Nigeria.

Kanagaraj, E. (2014) Human Development and Social Exclusion among Primitive Tribes in India, Deutschland / Germany : Lambert Academic Publishing.

Karmakar, K.G., 2008. Agriculture and rural development in north-eastern India: The role of NABARD. ASCI Journal of Management 37(2): 89-108.

Kerkhoff, E., and Sharma, E. (2006). Debating Shifting Cultivation in the Eastern Himalayas. Farmers' Innovations as Lessons for Policy. International Centre for Integrated Mountain Development (ICIMOD), Kathmandu.

Kull, C.A., Ibrahim, C.K., Meredith, T.C., 2007. Tropical forest transitions and glob-alization: neo-liberalism, migration, tourism, and international conservation agendas. Society & Natural Resources 20, 723–737.

Kunstadter P, Chapman E, Sanga S (eds) (1978). Farmers in the forest. University of Hawaii Press, Honolulu.

Lal J.B. and Prajapatu R.C. (1990): Space-borne monitoring of shifting cultivation in northern region of India. Van Vighan, 28: 125-126.

Lal R (1989). Potential of agroforestry as a sustainable alternative to shifting cultivation: concluding remarks. Agrofor. Syst. 8: 239-242.

Lalengzama C and Kanagaraj E (2014). Shifting Cultivation to Settled Agriculture : Agrarian Change and Tribal Development in Mizoram. Journal of Social Work and Social Development, 5: 14-43.

Leduc, B and Choudhury, D 2012 Agricultural transformations in shifting cultivation areas of Northeast India: Implications for land management, gender and institutions. In: Dev Nathan and Virginius Xaxa (Eds) Social Exclusion and Adverse Inclusion: Development and Deprivation of Adivasis in India OUP, New Delhi pgs 237-258

Li, T.M., 1999. Marginality, power and production: analyzing upland transforma-tions. In: Li, T.M. (Ed.), Transforming the Indonesian Uplands: Marginality, Power and Production. Harwood Academic, Amsterdam.

Liang L, Shen L, Yang W, Yang X, Zhang Y (2009). Building on traditional shifting cultivation for rotational agroforestry: Experiences from Yunnan. China For. Ecol. Manage., 257: 1989–1994.

Lindell, L., Astrom, M., Oberg, T., 2010. Land-use change versus natural controls on stream water chemistry in the Subandean Amazon, Peru. Applied Geochemistry 25, 485–495.

Lowenthal, D. (1989), The Timeless Past: Some Anglo-American Historical Preconceptions. The Journal of American History, 75 (4), 1263-80.

Ludewigs, T., D'antona, A.d.O., Brondizio, E., Hetrick, S., 2009. Agrarian structure and land-cover change along the lifespan of three colonization areas in the Brazilian Amazon. World Development 37, 1348–1359.

Lund C. (2011). Fragmented sovereignty: Land reform and dispossession in Laos. Journal of Peasant Studies. 38:885–905. doi: 10.1080/03066150.2011.607709.

Luoga E.J. (2000): Subsistence use of wood products and shifting cultivation within miombo woodland of eastern Tanzania, with special notes on commercial uses. South African Journal of Botany, 66: 72-85.

Ma, Q. (1999). Asia-Pacific Forestry Sector Outlook Study: Volume I— Socio-Economic, Resources and Non-Wood Products Statistics. FAO, Rome.

Majid, Z. (1983) 'Malaysia', in Y. Atal and P. L. Bennagen (eds) Swidden Cultivation in Asia, Vol. 2, Country Profiles: India, Indonesia, Malaysia, Philippines, Thailand, Regional Office for Education in Asia and the Pacific,United Nations Educational, Scientific and Cultural Organization, Bangkok, pp157-238

Malhi, Y., Roberts, J. T., Betts, R. A., Killeen, T. J., Li, W., and Nobre, C. A. (2008). Climate Change, Deforestation, and the Fate of the Amazon. Science 319: 169–172. doi:10.1126/science.1146961.

Matsumoto, T.; Itioka, T.; Yamane, S.; Momose, K. Traditional land use associated with swidden agriculture changes encounter rates of the top predator, the army ant, in Southeast Asian tropical rain forests. Biodivers. Conserv. 2009, 18, 3139–3151.

Mazoyer, M. and Roudart, L. Eds. 2006. A history of world agriculture from the neolithic to the current crisis, pp. 101-102, Earthscan publishers, London.

Mertz, O., (2002). The relationship between length of fallow and crop yields in shifting cultivation: a rethinking. Agrofor. Sys. 55: 149–159.

Mertz, O., Leisz S, Heinimann A, *et al.* Who counts? Demography of swidden cultivators in Southeast Asia. Human Ecology. 2009;37:281–289. doi: 10.1007/s10745-009-9249-y.

Mertz, O., Leisz S, Heinimann A, Rerkasem K, Thiha, Dressler W, Cu PV, Vu KC, Schmidt-Vogt D, Colfer CJP, Epprecht M, Padoch C, Potter L (2009b).Who Counts? The Demography of Swidden Cultivators. Hum. Ecol., 37: 281–289.

Mertz, O., Mueller D, Sikor T, *et al.* The Forgotten D: Challenges of addressing forest degradation in complex mosaic landscapes under REDD+ Geografisk Tidsskrift—Danish Journal of Geography. 2012;112:63–76. doi: 10.1080/00167223.2012.709678Rigg J. From rural to urban: A geography of boundary crossing in Southeast Asia. TRaNS. 2013;1:5–26.

Mertz, O, Padoch C, Fox J, Cramb RA, Leisz S, Nguyen TL, Tran TD (2009a). Swidden Change in Southeast Asia: Understanding Causes and Consequences. Hum. Ecol., 37: 259–264.

Mertz, O., Trends in shifting cultivation and the REDD mechanism. Curr. Opin. Environ. Sustain. 2009, 1, 156–160.

Mertz, O., Christensen, A.E., Højskov, P. & Birch-Thomsen, T. (1999): Subsistence or cash: strategies for change in shifting cultivation. Geografisk Tidsskrift, Danish Journal of Geography, Special Issue: 133-142.

Mertz, O., Wadley, R. L., & Christensen, A. E. (2005). Local land use strategies in a globalizing world: Subsistence farming, cash crops and income diversification. Agricultural systems, 85(3), 209-215.

Messerli, P., 2004. Alternatives a la culture sur brulis sur la falaise Est de Mada-gascar: strate′ gie en vue d'une gestion plus durable des terres. University of Bern, Bern.

Messerli, P.; Heinimann, A.; Epprecht, M. Finding Homogeneity in Heterogeneity-A New approach to quantifying Landscape mosaics developed for the Lao PDR. Hum. Ecol. 2009, 37, 291–304.

Metzger, J.P. Landscape dynamics and equilibrium in areas of slash-and-burn agriculture with short and long fallow period (Bragantina region, NE Brazilian Amazon). Landsc. Ecol. 2002, 17, 419–431.

Midmore, D. J., Jansen, H. G. P., and Dumsday, R. G. (1996). Soil Erosion and Environmental Impact of Vegetable Production in the Cameron Highlands, Malaysia. Agriculture Ecosystems and Environment 60: 29–46.

Mishra, A.K., Sharma, U.C., 2001. Traditional wisdom in range management for resource and environment conservation in northeastern region of India. ENVIS Bulletin: Himalayan Ecology & Development Volume 9, No. 1.

Mittelmann, A. (2001): Secondary forests in the lower Mekong subregion: an overview of their extent, roles and importance. Journal of Tropical Forest Science 13: 671-690.

Molinario G, Hansen M, Potapov P. Forest cover dynamics of shifting cultivation in the Democratic Republic of Congo: a remote sensing-based assessment for 2000–2010. Environ Res Lett. 2015;10: 94009.

Muller, D., Zeller, M., 2002. Land use dynamics in the central highlands of Vietnam: A spatial model combining village survey data with satellite imagery interpre-tation. Agricultural Economics 27, 333–354.

Mwampamba H.T. (2009): Forest Recovery and Carbon Sequestration under Shifting Cultivation in the Eastern Arc Mountains, Tanzania: Landscape and Land use Effects. Un published dissertation submitted in partial satisfaction of the requirements for the degree of Doctor of philosophy in Ecology in the Office of graduate studies of the University of California.183 p.

Mwavu, E.N., Witkowski, E.T.F., 2008. Land-use and cover changes (1988–2002) around Budongo Forest Reserve, Nw Uganda: implications for forest and woodland sustainability. Land Degradation & Development 19, 606–622.

Myers, N. 1991. Tropical forests: Present status and future outlook. Climatic Change 19:3–32.

NAEB (National Afforestation and Eco development Board). 2008. Report of the inter-ministerial national task force on rehabilitation of shifting cultivation areas, naeb.nic. in/.../ Shifting%20Cultivation%20Task%20Force%20Repo.

Navarrete AC and Jansen K (2015). Oil palm expansion without enclosure: smallholders and environmental narratives, The Journal of Peasant Studies, 42:3-4, 791-816, DOI: 10.1080/03066150.2015.1016920

Newby, J., Cramb, R., & Sakanphet, S. (2014). Forest transitions and rural livelihoods: multiple pathways of smallholder teak expansion in Northern Laos. Land, 3(2), 482-503.

Ngidang, D. (2002): Contradictions in land development schemes: the case of joint ventures in Sarawak, Malaysia. Asia Pacific Viewpoint 43: 157-180.

Nielsen, U., Mertz, O., and Noweg, G. T. (2006). The Rationality of Shifting Cultivation Systems: Labor Productivity Revisited. Human Ecology 34: 210–218. doi:10.1007/ s10745-006-9014-4.

Nikolic, N., Schultze-kraft, R., Nicolic, M., Böcker, R., and Holz, I. (2008). Land Degradation on Barren Hills: A Case Study in Northeast Vietnam. Environmental Management 42: 19–36. doi:10.1007/s00267-008-9099-1.

North D. 1990. Institutions, institutional change and economic performance. Cambridge University Press, New York.

Nye P, Greenland D (1960). The soil under shifting cultivation. Technical communication 51. Commonwealth Agricultural Bureau, Harpenden, United Kingdom.

Nye PH, Stephens D (1962). Soil fertility. In: Wills JB, ed, Agriculture and Land Use in Ghana, pp 127--143. Oxford University Press, London.

Nye, P. H., and Greenland, D. J. (1965). The Soil Under Shifting Cultivation. Jarrold and Sons, Norwich.

O'Brien WE (2002). The nature of shifting cultivation: stories of harmony, degradation, and redemption. Hum. Ecol. 30: 483–502.

Olofson, H. Swidden and kaingin among the southern Tagalog: A problem in Philippine upland ethno-agriculture. Philipp. Q. Cult. Soc. 1980, 8, 168–180.

Padoch C (2010), Saving slash and burn to save diversity. Biotropica, 42: 550-552.

Padoch C, Coffey K, Mertz O, Leisz SJ, Fox J and Wadley RL, 2007. The Demise of Swidden in Southeast Asia? Local Realities and Regional Ambiguities. Danish Journal of Geography 107: 29-41.

Padoch, C., & Peters, C.M. (1993): Managed forest gardens in West Kalimantan, Indonesia. Pages 167-176 In: C.S. Potter, J.I. Cohen, & D. Janczewski (eds.) Perspectives on Biodiversity: Case Studies of Genetic Resource Conservation and Development. American Association for the Advancement of Science Press, Washington DC.

Padoch, C., Harwell, E. & Susanto, A. (1998): Swidden, sawah, and in-between: agricultural transformation in Borneo. Human Ecology 26: 3-20.

Padoch, C.; Pinedo-Vasquez, M. Saving slash-and-burn to save biodiversity. Biotropica 2010, 42, 550–552.

Pandey DK, Det HK, Dubey SK, Kumar B, Dobhal S and Adhiguru P, 2020. Indigenous people's attachment to shifting cultivation in the Eastern Himalayas, India: A cross-sectional evidence. Forest policy and economics, 111: https://doi.org/10.1016/j.forpol.2019.102046

Pascual, U. Land use intensification potential in slash-and-burn farming through improvements in technical efficiency. Ecol. Econ. 2005, 52, 497–511.

Patel T, Karmakar S, Sanjog J, Kumar S and Cowdhury A, 2013. Socio-economic and environmental changes with transition from shifting to settled cultivation in North-eastern India: an ergonomics perspective. International Journal of Agricultural Science and Research, 3: 117-136.

Peluso, N. L., and Vandergeest, P. (2001). Genealogies of the Political Forest and Customary Rights in Indonesia, Malaysia, and Thailand. The Journal of Asian Studies 60: 3761–812. doi:10.2307/2700109.

Perera, G.A.D. 2001. Secondary forest situation in Sri Lanka: A review. Journal of Tropical Forest Science, 13: 768-785.

Pollini J, 2014. Slash-and-Burn Agriculture. In: Springer Reference - Encyclopedia of Food and Agricultural Ethics. (edn) Thompson PB and Kaplan DM, DOI: 10.1007/978-94-007-0929-4_87

Porro, R., 2005. Palms, pastures, and swidden fields: the grounded political ecology of agro-extractive/shifting-cultivator peasants in Maranhao, Brazil. Human Ecology 33, 17–56.

Potter, L. (2003): Supplement to Chapter 2: Forest degradation, deforestation, and reforestation in Kalimantan: Towards a sustainable land use. Pages 74-76 in: Padoch, C. & Peluso N.L. (eds.) Borneo in Transition: People, Forests, Conservation, and Development. Oxford University Press, Kuala Lumpur.

Priyadarshni. (1995). Shifting cultivation: cropping patterns, jhum cycle and problems. Accessed on January 6, 2019 at http://www.yourarticlelibrary. com/cultivation/shifting-cultivation-cropping-patterns-jhum-cycle-and-problems/44650.

Rahman S A, Rahman F and Sunderland T. 2011. Causes and consequences of shifting cultivation and its alternatives in the hill tracts of eastern Bangladesh. Agrofor. Sys. doi 10.1007/10457-011-9422-3.

Rahman, S.A., Rahman, M.F., Codilan, A.L., Farhana, K.M., 2007. Analysis of the economic benefits from systematic improvements to shifting cultivation and its evolution towards stable continuous agroforestry in the upland of eastern Bangladesh. Int for Rev 9(1): 536-547.SS

Ramakrishnan P S. 2003. Linking natural resource management with sustainable development of traditional mountain societies. Trop. Ecol. 44: 54.

Ramakrishnan PS (1992). Shifting Agriculture and Sustainable Development: An Interdisciplinary Study from Northeastern India, Man and the Biosphere Book Series 10. Paris: UNESCO, and Caernforth (UK): Parthenon Publishing (Republished by Oxford University Press, New Delhi, 1993).

Ramakrishnan, P.S. and Kushwaha, S.P.S. 2001. Secondary forests of Himalaya with emphasis on north-eastern hill region of India. Journal of Tropical Forest Science, 13: 727-747

Ramakrishnan, P.S., 1993. Shifting agriculture and sustainable development: an interdisciplinary study from Northeastern India. Unesco and Oxford University Press, New Delhi.

Raman TRS (2001). Effect of slash and burn shifting cultivation on rainforest birds in Mizoram northeast India. Conservation Biology, 15: 685-698.

Ranjan R, Upadhyay VP, 1999. Ecological problems due to shifting cultivation. Current Science 77(10): 1246-1250.

Rao RR, Hajra PK. 1986. Floristic diversity of eastern Himalaya e in a conservation perspective. In: Proceedings of Indian Academic of Sciences (Animal/Plant Science), Supplementary. pp. 103e125.

Rasul, G.; Thapa, G.B. Financial and economic suitability of agroforestry as an alternative to shifting cultivation: The case of the Chittagong Hill tracts, Bangladesh. Agric. Syst. 2006, 91, 29–50.

Regmi BR, Subedi A, Aryal KP, Tamang BB (2005). Shifting cultivation systems and innovations in Nepal. Unpublished report. LIBIRD, Pokhara.

Rerkasem K, Yimyam N, Rerkasem B (2009). Land use transformation in the mountainous mainland Southeast Asia region and the role of indigenous knowledge and skills in forest management. For. Ecol. Manage., 257: 2035–2043.

Rerkasem, K., & Rerkasem, B. (1994): Shifting Cultivation in Thailand: Its Current Situation and Dynamics in the Context of Highland Development. IIED Forestry and Land Use Series No. 4. IIED, London.

Rerkasem, K., Lawrence, D., Padoch, C., Schmidt-Vogt, D., Ziegler, A. D., and Bruun, T. B. (2009a). Consequences of swidden transitions for crop and fallow biodiversity in Southeast Asia. Human Ecology, this issue.

Richards, J.F., & Flint, E. (1994): A century of land-use change in South and Southeast Asia. Pages 15-57 in: Lange, O.L., Mooney, H.A. & Remmert, H. (eds.) Effects of Land-use Change on Atmospheric CO2 Concentrations. Ecological Studies 101. Springer-Verlag, New York.

Rigg, J. (2005). Poverty and Livelihoods after Full-Time Farming: A South-East Asian View. Asia Pacific Viewpoint, 46: 2173–184. doi:10.1111/j.1467-8373.2005.00266.x.

Rigg, J. (2006). Land, Farming, Livelihoods, and Poverty: Rethinking the Links in the Rural South. World Development, 34: 1180–202. doi:10.1016/j.worlddev.2005.07.015.

Rigg, J. (2013). From rural to urban: A geography of boundary crossing in Southeast Asia. TRaNS: Trans-Regional and-National Studies of Southeast Asia, 1(1), 5-26.

Rodel, D. L., Visco R.G., Pulhin J. M. 2001. Formation and transformation of secondary forests in the Philippines. Journal of Tropical Forest Science.

Roder, W. (1997). Slash-and-Burn Rice Systems in Transition: Challenges for Agricultural Development in the Hills of Northern Laos. Mountain Research and Development 17: 1–10. doi:10.2307/3673908.

Roder, W., Phengchanh, S., and Maniphone, S. (1997). Dynamics of Soil and Vegetation During Crop and Fallow Period in Slash-and-Burn Fields of Northern Laos. Geoderma 76: 131–144. doi:10.1016/S0016-7061(96)00100-0.

Rossi, J.P.; Celini, L.; Mora, P.; Mathieu, J.; Lapied, E.; Nahmani, J.; Ponge, J.F.; Lavelle, P. Decreasing fallow duration in tropical slash-and-burn agriculture alters soil macroinvertebrate diversity: A case study in southern French Guiana. Agric. Ecosyst. Environ. 2010, 135, 148–154.

Rossi, J.P.; Celini, L.; Mora, P.; Mathieu, J.; Lapied, E.; Nahmani, J.; Ponge, J.F.; Lavelle, P. (2010). Decreasing fallow duration in tropical slash-and-burn agriculture alters soil macroinvertebrate diversity: A case study in southern French Guiana. Agric. Ecosyst. Environ. 2010, 135, 148–154.

Roy RD, 2002. Land and Forest Rights in the Chittagong Hill Tracts. International Centre for Integrated Mountain Development (ICIMOD), Kathmandu, Nepal, ICIMOD Talking Points-4/02, 44 pp.

Rudel TK, Schneider L, Uriarte M, Turner BL, DeFries R, Lawrence D, *et al.* Agricultural intensification and changes in cultivated areas, 1970–2005. Proc Natl Acad Sci USA. 2009;106: 20675–20680. pmid:19955435

Ruthenberg, H. (1980). Farming Systems in the Tropics. Clarendon, Oxford.

Sanchez PA (1976). Properties and Management of Soils in the Tropics. John Wiley & Sons, New York, pp. 346–412.

Sanchez, P.A.; Palm, C.A.; Vosti, S.A.; Romich, T.; Kasyoki, J. Alternatives to Slash and Burn: Challenge and Approaches of an International Consortium; Columbia University Press: New York, NY, USA, 2005.

Sather, C. A. (1990). Trees and Tree Tenure in Paku Iban Society: The Management of Secondary Forest Resources in a Long-Established Iban Community. Borneo Review 1: 16–40.

Sather, C. A. (1994). The one-sided one: Iban rice myths, agricultural ritual and notions of ancestry. In Walker, A. (Ed.), Rice in Southeast Asian Myth and Ritual. Contributions to Southeast Asian Ethnography vol. 10. Ohio State University, Columbus, OH, pp. 119–150.

Sati VP and Rinawma P (2014). Practices of Shifting Cultivation and its Implications in Mizoram, North-East India: A Review of Existing Research. Nature and Environment, 19: 179-187.

Schendel van W (1992). The Invention of the 'Jummas': State Formation and Ethnicity in Southeastern Bangladesh. Modern Asian Studies, 26: 95-128., DOI:10.1017/S0026749X00015961.

Schlippe, P. de. 1956. Shifting cultivation in Africa : The zande system of agriculture. London, Routledge.

Schmidt-Vogt D (1999). Swidden Farming and Fallow Vegetation in Northern Thailand. Geoecological Research Franz Steiner, Stuttgart.

Schmidt-Vogt D, 2001. Secondary forests in swidden agriculture in the highlands of Thailand. Journal of Tropical Forest Science, 13: 748-767.

Schmidt-Vogt, D. Defining degradation: The impacts of swidden on forests in northern Thailand. Mt. Res. Dev. 1998, 18, 135–149.

Schmidt-Vogt, D.; Leisz, S.J.; Mertz, O.; Heinimann, A.; Thiha, T.; Messerli, P.; Epprecht, M.; Cu, P.V.; Chi, V.K.; Hardiono, M.; *et al.* An assessment of trends in the extent of swidden in southeast Asia. Hum. Ecol. 2009, 37, 269–280.

Scott JC. 2009. The art of not being governed: an anarchist history of upland Southeast Asia. New Haven (CT): Yale University Press.

Scott, J. C. (1998). Seeing Like a State: How Certain Schemes to Improve the Human Condition Have Failed. Yale University Press, New Haven.

Sears, R.R., Padoch, C., Pinedo-Vasquez, M., 2007. Amazon forestry tranformed: integrating knowledge for smallholder timber managemet in eastern Brazil. Human Ecology 35, 697–707. Sturgeon, J.C., 2005. Border Landscapes: The Politics of Akha Land Use in China and Thailand. University Washington Press, Seattle.

Seidenberg, C., Mertz, O., and Kias, M. B. (2003). Fallow, labour and livelihood in shifting cultivation: implications for deforestation in northern Lao PDR. Geografisk Tidsskrift-Danish Journal of Geography 103: (2)71-80.

Sharma, U. C. & Prasad, R. N. (1995) Socio-economic aspects of acid sooil management and alternate land use systems for north-eastern states of India. In: Plant-Soil Interactions at Low pH. (ed. by R. A., Date; N. J. Grundon.; G. E. Rayment, . & M. E. Probert), 689-696, Kluwer Academic Press, The Netherlands.

Sharma, U.C. (1988). Land use trends and agricultural production in north-eastern hill states. Environment and Development, Gangtok, Sikkim, India, pp 160-167.

Sharma, U.C. (2003) Impact of population growth and climate change on the quantity and quality of water resources in the northeast of India. In: Water Resources Systems - Hydrological Risks, Management and Development(eds.), IAHS Publ. No. 281, 240-247, IAHS Press, Wallingford, U.K.

Sharma, U.C., Prasad, R.N., 1994. Potential indigenous farming systems of North-Eastern Hill region. ICAR Research Complex for NEH Region, Shillong, India.

Siddiqui, S & Chohan, S. (2015). Legal response to institutionalising participatory land use planning in nagaland: developing strategies for mainstreaming sustainable jhum practices into existing policy and legal framework of nagaland. UNDP Report, The Printing Wala.

Sidle, R. C., and Ochiai, H. (2006). Landslides: Processes, Prediction, and Land-use. American Geophysical Union, Washington DC.

Sidle, R. C., Pearce, A. J., and O'Loughlin, C. L. (1985). Hillslope Stability and Land Use, Water Resources Monograph, vol. 11. American Geophysical Union, Washington, DC, p. 140 pp.

Sidle, R. C., Ziegler, A. D., Negishi, J. M., Abdul Rahim, N., and Siew, R. (2006). Erosion Processes in Steep Terrain—Truths, Myths, and Uncertainties Related to Forest Management in Southeast Asia. Forest Ecology and Management 224: 199– 225. doi:10.1016/j.foreco.2005.12.019.

Silva JMN, Carreiras JMB, Rosa I, Pereira JMC. Greenhouse gas emissions from shifting cultivation in the tropics, including uncertainty and sensitivity analysis. J Geophys Res. 2011;116.

Singh, J. S. and Singh, S. P., Forests of Himalaya, Gyanodaya Prakashan, Nainital, 1992, p. 294.

Singh, J. S. *et al.*, (1986) in Eco-Development Guidelines and Model of Development of the Central Himalaya, Department of Botany, Kumaun University, Nainital, p. 48.

Singh, J., Borah, I.P., Barua, A., Baruah, K.N., 1996. Shifting cultivation in north east India - An Overview. Newsletters, Rain Forest Research Institute 'Deovan', Jorhat, Assam, India.

Singh, K. D. and Marzoli, A., (1988). Paper presented at the World Wildlife Fund Conference on the Potential Impact of Climate on Tropical Forests, San Juan, Puerto Rico, pp. 2–3.

Sodhi, N.S.; Koh, L.P.; Brook, B.W.; Ng, P.K. Southeast Asian biodiversity: An impending disaster. Trends Ecol. Evol. 2004, 19, 654–660.

Sodhi, N.S.; Posa, M.; Lee, T.M.; Bickford, D.; Koh, L.P.; Brook, B.W. The state and conservation of Southeast Asian biodiversity. Biodivers. Conserv. 2010, 19, 317–328.

Spencer, J.E.(1966). Shifting Cultivation in Southeastern Asia; University of California Press: Berkeley, CA, USA, 1966; p. 247.

Stone, R. (2007) Can palm oil plantations come clean? Science, 317, doi:10.1126/science.317.5844.1491.

Sturgeon, J. (2005): Border Landscapes: The politics of Akha land use in China and Thailand. University of Washington Press, Seattle Szott LT, Palm CA, Buresh RJ (1999). Ecosystem fertility and fallow 154 Int. J. Biodvers. Conserv. function in the humid and subhumid tropics. Agrofor. Syst., 47: 163-196.

Sunderlin, W.D., Ndoye, O., Bikie, H., Laporte, N., Mertens, B., Pokam, J., 2000. Economic crisis, small-scale agriculture, and forest cover change in southern Cameroon. Environmental Conservation 27, 284–290.

Szott, L. T., Palm, C. A., and Buresh, R. J. (1999). Ecosystem Fertility and Fallow Function in the Humid and Subhumid Tropics. Agroforestry Systems 47: 163–196. doi:10.1023/A:1006215430432Sturgeon, 2005).

Tawnenga, Shankar U. and Tripathi R.S. (1997): Evaluating second year cropping on jhum Fallows in Mizoram, Northeastern India: Energy and economic efficiencies. Journal of Biosciences, 22: 605-613.

Temudo, M. P., & Silva, J. M. (2012). Agriculture and forest cover changes in post-war Mozambique. Journal of Land Use Science, 7(4), 425-442.

Thomas, W. L. (ed.) (1956) Man's Role in Changing the Face of the earth. Man's Role in Changing the Face of the earth, The University of Chicago Press, Chicago and London.

Thongmanivong, S., Fujita, Y., 2006. Recent land use and livelihood transitions in northern Laos. Mountain Research and Development 26, 237–244.

Thrupp LA, Hecht SB, Browder JO (1997). The Diversity and Dynamics of Shifting Cultivation: Myths, Realities, and Policy Implications. World Resources Institute, Washington DC, p. 48.

Tinker, P.B.; Ingram, J.S.I.; Struwe, S. Effects of slash-and-burn agriculture and deforestation on climate change. Agric. Ecosyst. Environ. 1996, 58, 13–22.

Tommy, J. L. (1984). Shifting cultivation and possibilities for improving it in Sierra Leone. In: Improved Production systems as an alternative to shifting cultivation. Soil Bulletin No 53, Rome, FAO.

Tripathi, R.S., Barik, S.K., 2003. Shifting Cultivation in North East India. Proc. Approaches for increasing agricultural productivity in hill und mountain ecosystem. ICAR research complex for NEH region, Umiam, Meghalaya, India.

Turkelboom, F. (1999). On-Farm Diagnosis of Steepland Erosion in Northern Thailand. Ph.D. Thesis, Faculty of Agricultural and Applied Biological Sciences, K.U. Leuven.

Turkelboom, R., Poesen, J., and Trebuil, G. (2008). The Multiple Land Degradation Effects by Land-Use Intensification in Tropical Steeplands: A Catchment Study from Northern Thailand. Catena 75: 102–116. doi:10.1016/j.catena.2008.04.012.

Uhlig, J., Hall, C.A.S., & Nyo, T. (1994): Changing patterns of shifting cultivation in selected countries in Southeast Asia and their effect on the global carbon cycle. Pages 145-200 in: Lange, O.L. Mooney, H.A. & Remmert, H. (eds.) Effects of Land-Use Change on Atmospheric CO2 Concentrations. Ecological Studies. Springer-Verlag, New York.

Valentin, C., Agus, F., Alamban, R., Boosaner, A., Bricquet, J. P., Chaplot, V., de Guzman, T., de Rouw, A., Januea, J. L., Orange, D., Phachomphonh, K., Do Duy, Phai, Podwojewski, P., Ribolzi, O., Silvera, N., Subagyono, K., Thiebaus, J. P., Tran Duc, Toan, and Vadari, T. (2008). Runoff and Sediment Losses from 27 Catchments in Southeast Asia: Impact of Rapid Land Use Changes and Conservation Practices. Agriculture Ecosystems and Environment 128: 225–238. doi:10.1016/j.agee.2008.06.004.

van Dijk, A. I. J. M., and Keenan, R. J. (2007). Planted Forests and Water in Perspective. Forest Ecology and Management 251: 1–9. doi:10.1016/j.foreco.2007.06.010.

van Dijk, A. I. J. M., van Noordwijk, M., Calder, I. R., Bruijnzeel, L. A., Schellekens, J., and Chappell, N. (2009). Forest–Flood Relation still Tenuous—Comment on "Global Evidence that Deforestation Amplifies Flood Risk and Severity in the Developing World" by C.J.A. Bradshaw, N. S. Sodi, K. S.-H.Peh, and B.W. Brook. Global Change Biology 15: 110–115. doi:10.1111/j.1365-2486.2008.01708.x.

Van Vliet N, Mertz O, Heinimann A, *et al.* (2012) Trends, drivers and impacts of changes in swidden cultivation in tropical forest-agriculture frontiers: A global assessment. Global Environmental Change. 22:418–429. doi: 10.1016/j.gloenvcha.2011.10.009

van Vliet, M. J., Harmsen, H. J., de Bont, E. S., & Tissing, W. J. (2010). The role of intestinal microbiota in the development and severity of chemotherapy-induced mucositis. PLoS pathogens, 6(5), e1000879.

van Vliet, N., 2010. Participatory vulnerability assessment in the context of conservation and development projects: a case study of local communities in South-west Cameroon. Ecology and Society 1, 5.

Van Vliet, N.; Mertz, O.; Birch-Thomsen, T.; Schmook, B. Is there a continuing rationale for swidden cultivation in the 21st century? Hum. Ecol. 2013, 41, 1–5.

Vandergeest, P. (2003). Land to Some Tillers: Development-Induced Displacement in Laos. International Social Science Journal 175: 47–56. doi:10.1111/1468-2451.5501005.

Vigiak, O., Ribolzi, O., Pierret, A., Sengtaheuanghoung, O., and Valentin, C. (2008). Trapping Efficiencies of Cultivated and Natural Riparian Vegetation in Northern Laos. Journal of Environmental Quality 37: 889–897. doi:10.2134/jeq2007.0251.

Vliet NV,. Mertz O, Birch-Thomsen T and Schmook B, 2013. Is There a Continuing Rationale for Swidden Cultivation in the 21st Century? Human Ecolology, 41: 1–5, DOI 10.1007/s10745-013-9562-3.

Vongvisouk, T.; Mertz, O.; Thongmanivong, S.; Heinimann, A.; Phanvilay, K. Shifting cultivation stability and change: Contrasting pathways of land use and livelihood change in Laos. Appl. Geogr. 2014, 46, 1–10.

Wadley R, Mertz O. (2009). Pepper in a time of crisis: Smallholder buffering strategies in Sarawak, Malaysia and West Kalimantan, Indonesia. Agricultural Systems.;85:289–305. doi: 10.1016/j.agsy.2005.06.012.

Wadley, R. L., and Colfer, C. J. P. (2004). Sacred Forest, Hunting, and Conservation in West Kalimantan, Indonesia. Human Ecology 32: 313–338. doi:10.1023/B:HUEC.0000028084.30742.d0

Wadley, R.L. (1997): Circular Labor Migration and Subsistence Agriculture: A Case of the Iban in West Kalimantan, Indonesia. Arizona State University, Tempe, Arizona, USA.

Wadley, R.L. (2002): Coping with crisis – smoke, drought, flood and currency: Iban households in West Kalimantan, Indonesia. Culture and Agriculture 24: 26-33.

Wadley, R.L. (2007): Slashed and burned: War, environment and resource insecurity in West Borneo during the late nineteenth and early twentieth centuries. The Journal of the Royal Anthropological Institute 13 (1): 109-128.

Walker, A. (2004). Seeing Farmers for the Trees: Community Forestry and the Arborealisation of Agriculture in Northern Thailand. Asia Pacific Viewpoint, 45: 3311–324. doi:10.1111/j.1467-8373.2004.00250.x.

Warner, K., 1991. Shifting Cultivators: Local Technical Knowledge and Natural Resource Management in the Humid Tropics. FAO Community Forestry Note, 8, Rome, Italy.

Watters RF (1971). Shifting Cultivation in Latin America. FAO Forestry Report Development Paper No. 17, Rome.

Webb EJ (1966). Unobtrusive Measures; Nonreactive Research in the Social Sciences. Rand McNally, the University of Wisconsin - Madison

Weinstock, J. (1996) 'Rattan: A complement to swidden agriculture in Borneo', Journal of Southeast Asian Social Science and Humanities 48, pp51–61, http://ejournals.ukm.my/akademika/article/view/3114, accessed 18 September 2015

Whitmore, T. C. (1998). An Introduction to Tropical Rain Forests. Oxford University Press, Oxford.

Willis K, Gillison L, Brncic T (2004). How "virgin" is virgin rainforest? Sci., 304: 402–403.

World Bank. (2001): Indonesia: Environment and Natural Resource Management in a Time of Transition. World Bank, Washington DC.

World Bank. (2007). www.worldbank.org; accessed October 2020.

World Commission on Environment and Development (1987)

WRI (World Resource Institute), (1996) World Resource, Oxford University Press, Oxford

Xu J, Mclelian T and Hiwasaki L, 2017. Integrating swidden agricultural knowledge systems into sustainable intensification in the Central Mekong region. World Agroforestry Centre, East and Central Asia

Xu, J; Grumbine, RE; Shrestha, A; Eriksson, M; Yang, X; Wang, Y; Wilkes, A (2009) _The Melting Himalayas: Cascading Effects of Climate Change on Water, Biodiversity And Livelihoods.' Conservation Biology 24(3): 520–530

Zaehringer J, Eckert S, Messerli P. Revealing Regional Deforestation Dynamics in North-Eastern Madagascar—Insights from Multi-Temporal Land Cover Change Analysis. Land. 2015;4: 454–474.

Zaitinvawra, D. and Kanagaraj, E. (2008), Shifting Cultivation to Settled Agriculture: Rural Livelihood Patterns in Mizoram, The Eastern Anthropologist, 61 (2), 201-226.

Ziegler A. D., Fox J. M. and Jianchu Xu (2009), 'The rubber juggernaut', Science, vol 324, pp. 1024–1025.

Ziegler A. D., Fox J. M., Webb E. L., Padoch C., Leisz S. J. and others (2011), 'Recognizing contemporary roles of swidden agriculture in transforming landscapes of southeast Asia', Conservation Biology, vol 25, no 4, pp. 846–848.

Ziegler AD, Phelps J, Yuen JQ, *et al.* Carbon outcomes of major land-cover transitions in SE Asia: Great uncertainties and REDD plus policy implications. Global Change Biology. 2012;18:3087–3099. doi: 10.1111/j.1365-2486.2012.02747.x.

Ziegler, A. D., Bruun, T. B., Guardiola-Claramonte, M., Giambelluca, T. W., Lawrence, D., and Nguyen, T. L. (2009). Environmental consequences of the demise in swidden agriculture in Montane Mainland SE Asia: hydrology and geomorphology. Human Ecology 37: 361–373

Ziegler, A. D., Giambelluca, T. W., Plondke, D., Leisz, S., Tran, L. T., Fox, J., Nullet, M. A., Vogler, J. B., Dao Minh, Truong, and Tran Duc, Vien (2007c). Hydrological Consequences of Landscape Fragmentation in Mountainous Northern Vietnam: Buffering of Accelerated Overland Flow. Journal of Hydrology 337: 52–67. doi:10.1016/j.jhydrol.2007.01.031.

Ziegler, A. D., Giambelluca, T.W., Tran, L. T., Vana, T. T., Nullet,M. A., Fox, J. M., Vien, T. D., Pinthong, J., Maxwell, J. F., and Evett, S. (2004b). Hydrological Consequences of Landscape Fragmentation in Mountainous Northern Vietnam: Evidence of Accelerated Overland Flow Generation. Journal of Hydrology 287: 124–146. doi:10.1016/j.jhydrol.2003.09.027.

Ziegler, A. D., Sutherland, R. A., and Giambelluca, T. W. (2000). Runoff Generation and Sediment Transport on Unpaved Roads, Paths, and Agricultural Land Surfaces in Northern Thailand. Earth Surface Processes and Landforms 25: 519–534. doi:10.1002/(SICI)1096-9837(200005)25:5<519::AID-ESP80>3.0. CO;2-T.

Ziegler, A.D., Giambelluca, T.W., Sutherland, R.A., Nullet, M.A., Yarnasarn, S., Pinthong, J., Preechapanya, P., Jaiaree, S., 2004. Toward understanding the cumulative impacts of roads in upland agricultural watersheds of northern Thailand. Agric. Ecosyst. Environ. 104, 145–158.

Zinke, P. J.; S. Sabhasri; and P. Kunstadter (1978). Soil Fertility Aspects of the Lua' Forest Fallow System of Shifting Cultivation. In: P. E. Kunstadter, E. C. Chapman and S. Sabhasri (eds.) Farmers in the Forest: Economic Development and Marginal Agriculture in Northern Thailand (Honolulu, Hawaii: University Press of Hawaii).

9

Development and Improvement

9.1 GENERAL DESCRIPTION

Education can help farmers in impoverished nations move away from environmentally damaging slash-and-burn agriculture (Nganje *et al.* 2001). The goal of their study was to see how much extension education can help people embrace agricultural strategies other than slash and burn. In Cameroon, farmers' cropping patterns are modelled as a function of farm size, farmer educational level, and extension professional visits, whether slash and burn, varied crops, or monocropping. According to the data, higher extension visiting rates reduce not just the risk of farmers choosing slash and burn agriculture, but also encourage them to switch to monocropping. Because monocropping is a trend toward export-oriented agriculture in Cameroon, it may help to promote greater economic development in Western Africa (Nganje *et al.* 2001). Continued extension education activities are therefore vital in decreasing the environmental damage caused by slash and burn agriculture while also encouraging the adoption of more profitable cropping strategies.

The negative impacts of shifting cultivation on the environment as farmers strive to provide for higher populations and the associated decline of shifting cultivation into an unproductive system have led most governments in Southeast Asia to perceive it as a backward method of cultivation (Thanh and Vien 2017). Consequently, governments throughout the region have tried to ban shifting cultivation. In Malaysia, for example, huge development programmes have been implemented to improve shifting cultivators livelihoods when they abandon shifting cultivation (Majid, 1983). The government of Laos has enacted land-use and forestry policies aiming at expanding forest cover, eliminating shifting farming, and supporting intensive commercial agriculture (Castella *et al.*, 2013). For many years, the Vietnamese government has implemented strict rules and procedures to combat shifting cultivation. These include a large-scale programme called 'Fixed Cultivation, Fixed Residence' (officially known as Resolution No. 38/CP), which aims to permanently settle 'shifting fields' and 'shifting people' (McElwee, 2004), as well as a forestland-allocation policy that ties people to specific plots of land, effectively making shifting cultivation impossible (Castella *et al.*, 2005).

It has been found that shifting cultivation has been rapidly transformed in response to social, economic and political challenges (Rasul and Thapa, 2003; Padoch *et al.*, 2007; Fox *et al.*, 2009). Despite the complexity of this transformation process, researchers generally agree that shifting cultivation is declining (van Vliet *et al.*, 2012). Yet even with this trend, shifting cultivation remains important in many forest-frontier areas and will continue to be important throughout at least the current century in areas where alternatives are not yet available (van Vliet *et al.*, 2013). To put figures to this assertion, in the mid-1990s in the Lao PDR, 218,900 hectares of land was under shifting cultivation, involving a total of 1.24 million people, or about one third of the total population (Pravongviengham, 1998). In the early 1990s in Vietnam, about 50 ethnic groups (out of the country's total of 54 ethnic groups) were practising swidden cultivation, employing 2.8 million people to work 3.5 million hectares in the country's uplands (McElwee, 2004). The government said that about 50% of the 110,000ha of forest was lost every year during the 1980s and 1990s because of shifting cultivation (Do, 1994). 'In many regions, locals continue to unlawfully cut down the forest for shifting cultivation,' according to the Ministry of Agriculture and Rural Development in 2007. (MARD, 2007). The Ministry of Natural Resources and Environment estimated in 2011 that hundreds of thousands of villagers in Vietnam's uplands still engaged in shifting cultivation on an area of more than one million hectares (MONRE, 2011). Recent government data shows that shifting cultivation accounted for 54% of total forest loss in the country in 2012, and 71.6% of total forest loss in the period between 2008 and 2011 (Forest Protection Department, 2013).

The persistence and prevalence of shifting cultivation in the uplands of Southeast Asia clearly indicates the failure of government measures to halt the practice. In Lao PDR, the overall area under slash and burn farming is estimated to range from 1.4 to 2.5 million hectares (including fallow areas). The population affected is projected to be between 1 and 1.9 million individuals (Kenney-Lazar 2018).

Shifting cultivation was identified by the Food and Agriculture Organization more than 50 years ago as "the greatest hurdle not only to current increases in agricultural productivity, but also to the conservation of future production potential in the form of soils and forests" (FAO 1957). In Mizoram, India, around half of all households engage in shifting farming (Anonymous 2009c), particularly in distant communities that are underdeveloped (Singh *et al.* 2010). However, the temperature and, as a result, the vegetation in these areas differ significantly from that of Mizoram. As a result, while surveys of the literature provide a plethora of conclusions and a variety of possibilities for enhancing or replacing shifting farming practises, each must be carefully evaluated in the context of Mizoram's topography and climate. It goes without saying that if such basic data is lacking, planners and policymakers would be unable to make informed judgments on land use and development in their countries' poorest areas. Because agriculture is enhanced and confined in a small area while broader regions are allowed to regenerate, it is widely assumed that replacing shifting farming increases environmental services. Although data suggests that this is taking place in Vietnam, the transition has had a negative impact on local livelihoods (Jakobsen *et al.* 2007). While some research imply that swidden has a positive influence when compared to other land use systems, such as forests, the importance of swidden in soil carbon conservation is yet unknown (Lawrence *et al.* 2005; Sommer *et al.* 2000).

The genesis of agriculture occurred in the Near East, in a place where wild relatives of cultivated plants grew in a more or less open setting. Agriculture's growth to the northwest into Europe covered forested areas, necessitating clearing in order to practise agriculture (IFFN 2004). According to some older theories, Neolithic agriculture was rudimentary and lacked sophisticated manuring and tillage techniques. As a result, to avoid low agricultural output, people had to relocate their fields on a frequent basis (Childe 1929, 1960). Quantitative models of Neolithic agriculture and nutrition are based on these yields. Permanent fields and communities were established as a result of these factors. Neolithic agriculture was distinct from recent shifting cultivation in the tropics, as well as slash-and-burn agriculture on poor soils in the mountains in the early modern period.

To deliver new and different arguments, agricultural experiments were started in 1994 to reconstruct Late Neolithic land use. Since 1998, these trials have been carried out in Forchtenberg, northeast of Stuttgart, where the Baden-Württemberg Forest Department has provided us with a 3.5-hectare mixed deciduous forest (Rösch *et al.* 2002a, b). According to WRB (FAO 1998), the area's soil type is Luvisol and Stagno-Luvisol; the climate is somewhat suboceanic, with an annual average temperature of around 9°C and an annual precipitation of about 850mm. The land has been wooded for at least 200 years and is completely encircled by forest on all sides. Celtic fields reflect former agriculture in the eastern section, where the richest soils are found. The experiment's key findings were; (i) the temperature of fire in the soil varies greatly, but it is usually above 80°C at a depth of 1 cm and above 60°C at a depth of 2 cm. So nearly all weeds are suppressed, (ii) average yields of autumn-sown cereals were low on the poorer and higher on the better soils with burning, (iii) yields were very low without burning, and (iv) the main reasons for the high yields after burning are the mobilisation of nutrients and the rise in pH, which improved nutrient availability. Clearing the forest takes time, but the most significant drawback is a lower total yield in a fully utilised environment, due to the extensive use of fallow land for the production of new wood for fuel during long-term fallow periods. As a result, an increasing human population will require more advanced but even more costly technologies to replace this form of agriculture.

The Lao government's policy on shifting cultivation has mostly stayed unchanged over the last three decades, however particular policy documents show some shifts in approach. The Lao government has been concerned about slash-and-burn agriculture since 1975. The Lao government has maintained a tight policy toward shifting cultivation, aiming to eliminate the practise entirely and replace it with modern, sedentary agricultural production methods. Low productivity, poverty, deforestation, and environmental deterioration were all thought to be related with shifting cultivation. During this time, the government's major policy for dealing with the problem was to relocate villages from upland to lowland areas. The government intended to decrease shifting cultivation and manage villagers' agricultural practises in the uplands through forestland allocation as a result of the 1989 National Forestry Conference. The Land and Forest Allocation Programme was established in the 1990s with the goal of stabilising and eventually eliminating shifting cropping. Resettlement, village consolidation, development village clusters, and focus sites were all employed to replace shifting cultivation with various types of agriculture and livelihood possibilities (Kenney-Lazar 2018).

Although the government's strategy on shifting cultivation has remained centred on stabilisation and eventual eradication, the techniques and mechanisms through which the policy is executed have become more moderate. This has happened in four significant ways:

1. The government has refocused its upland development priorities on providing food security and economic growth, which are now regarded the two most essential upland development goals.
2. Some forms of shifting cultivation are permissible, at least in the short term. Rotational upland agriculture is a recognised form of shifting cultivation in which farmers rotate in a cycle within established agricultural land use boundaries. Although this is not considered the best method of cultivation, it is accepted when other options are unavailable, yet it is still only a temporary solution.
3. The government has placed a greater emphasis on attaining the goal of stabilising shifting cultivation before completely eliminating it. While the exact meaning of the term'stabilization' is unclear, it can be seen as a stage of transition between shifting cultivation and eradication.

4. The government is increasingly aware that policies relating to shifting cultivation must be implemented in a pluralistic manner to account for regional differences in agro-ecological and socio-economic issues.

9.1.1 The early attempts and their impacts

Government programmes targeted towards transformation of shifting cultivation were initiated in the 1970s with a thrust on horticulture and cash crop promotion. Coffee, rubber, and cashew plantations were established under various government schemes as alternatives to shifting cultivation. In addition, horticultural crops, mainly pineapple, were also promoted as alternatives (Leduc and Choudhury 2012). Terrace farming schemes were also created with the twin goals of expanding wet rice cultivation (thus lowering the reliance on shifting rice production) and increasing soil erosion control. The acceptance of such programmes by the shifting cultivators was not very encouraging until recently and few, if any, replicated these options.

9.1.2 Fate of shifting cultivation

Over the years, however, shifting cultivators have slowly revived the process of transformation, often with the active encouragement of different government programmes and the gradually increasing influence of markets. With the rapid penetration of market forces into the uplands of north-east India and the consequent transformation of the subsistence economy into a monetarized, cash-driven system, most upland households have little choice but to accept sedentarized options or rapidly transform conventional shifting cultivation into a 'commodified' form (Choudhury *et al.* 2003), often complimented with settled agricultural options that allow them a means of cash generation. Estimates reported by the Task Force on Shifting Cultivation, set up by the Government of Meghalaya in 2001, show a gradual decrease in the number of families practising shifting cultivation (Government of Meghalaya 2003). Citing estimates provided by the Agro-Economic Research Centre for North-East India of the Assam Agricultural University, Jorhat, the Task Force reported 40.5% of the rural population to be dependent on shifting cultivation in 1971, which reduced to 23.6% in 1981, based on census surveys.

9.1.3 Transformation of slash-and-burn systems

Reduced fallow periods, along with this diversification, can cause, or contribute to, a fundamental transition of shifting cultivation systems. The fallow is too brief to allow the establishment of woody perennials below a specific threshold. Large tree stumps deteriorate over time, and the root mat becomes less dense. As a result, paring and burning can be done, as well as ploughing the field. As intensification continues, new measures for preserving soil fertility will be required. Fallow vegetation brings nutrients to the plot in shifting cultivation systems, which pumps them deep into the soil. Through burning, they are made available to the crops in the form of ash. This nitrogen supply is lost when there is a brief herbaceous fallow or no fallow at all. To maintain a good output, more efficient nutrient management (crop rotation, ploughing, paring and burning, zero burning) may suffice at initially. However, in the long run, a new supply of nutrients is required. Typically, farmers expand animal husbandry and begin to employ animal dung, taking use of the additional grazing resources of short fallow dominated by grasses. They see that crops planted in corrals yield better, so they alternate corrals and fields at first. They begin to actively manage dung by collecting it in the corral and transporting it to fields as land pressure and fertiliser requirements increase. Grazing animals in fields increase nutrient management but do not result in a significant net gain because nutrients are withdrawn from grass growing on the same ground as the crops. If they graze on uncultivated pastures, this is a net benefit since nutrients are transferred horizontally from pastures to agricultural land in the form of manure, replacing vertical transfer in the form of ash by fallow trees (Mazoyer and Roudart 2006).

Farmers are being encouraged to invest in permanent agriculture through the continuing land-forest allocation. It has become a crucial instrument for the government to confront evolving agricultural practises (Sandewall *et al.*, 1999). Although the government owns all land in Laos, farmers can be awarded private land-use rights, which can be sold and inherited. Although official ownership certificates have not been produced in the pilot region, paddy field ownership is widely recognised and rarely a source of contention. In practice, both local communities and government institutions have recognised customary land rights over changing agriculture areas. People used to acquire land by clearing forests or claiming unclaimed property. The family who cleared the land retains ownership when it is left fallow.

The policy principle of effectively controlling shifting cultivation has been endorsed by governments around the world because of the claimed negative impacts of shifting cultivation on the environment and the belief that it is an unproductive system of agriculture. Despite this official attitude, shifting cultivation is still widespread in the uplands of Southeast Asia, with dynamic expansion or contraction (van Vliet *et al.*, 2013). Inadequate understanding of the dynamic and complex nature of shifting cultivation systems, a lack of collaboration and shared vision among stakeholders and uncertain information are the key reasons why government initiatives to crack down on shifting cultivation have failed. The relationships between the key drivers and inhibitors of shifting cultivation are not linear, but are causal feedback links that attribute the existence and continuation of shifting cultivation to a variety of interdependent factors, categorized here as environmental, economic, arising from government policy and socio-cultural factors. The model may serve as a platform for dialogue, shared visions and collaboration among stakeholders. This makes it a potentially powerful tool for policy-makers and planners who are seeking to share responsibility for the management of shifting cultivation.

In India, the government's shifting cultivation management effort, which began in the mid-1950s, gained traction in the 1970s with the adoption of a range of programmes by state and central governments. Although the hill tribes are progressively becoming aware of the negative effects of this primitive kind of agriculture, the impact of these programmes in terms of weaning the people away from shifting cultivation and improving their socio-economic situation has yet to be assessed.

9.1.4 Shifting cultivation control programmes

The government's shifting cultivation control effort, gained momentum with the deployment of a range of programmes by various states and the central government. According to the suggestion of the task force on shifting cultivation (1983), a scheme for shifting cultivation control was in place from 1987-88 to 1990-91 in the seventh five-year plan. The Government of India has started the WDPSCA (Watershed Development Project for Shifting Cultivation Areas) in the Northeast with a 100% funding to the state plan. These programmes have failed to make a difference in the larger problem, and their influence on weaning people off shifting cultivation and improving their socio-economic situation has yet to be determined. The shifting cultivation settlement has not always been successful, and in many cases, they have abandoned new settlement to return to shifting cultivation. Below are some case studies of various programmes that have had varying degrees of effectiveness.

9.1.5 Policy Context

Making the "correct" policies is a laudable objective, but only if we acknowledge those goals, not destinations, are the goal. With this insight, we can accept the unavoidable compromises that come with policymaking. When single-issue counsel influences policy, however, unexpected and undesirable effects might occur (Falvey 2018). In the first case some development agencies resisted funding dairy development in tropical nations because of the misguided belief that milk production in the tropics could not be efficient and that milk products were best purchased from temperate dairy nations.

9.1.6 The highlands

Assuming that the majority of shifting cultivators are in mountains Southeast Asia (Mertz *et al.*, 2009a), even though their numbers are apparently in decline (Padoch *et al.*, 2007), policy discussions will necessarily focus on highland regions. While highlands or mountains receive some specialized attention, they are a poor cousin to lowland agriculture and cultures. The distribution of highlands in the world are shown in Fig. 9.1.

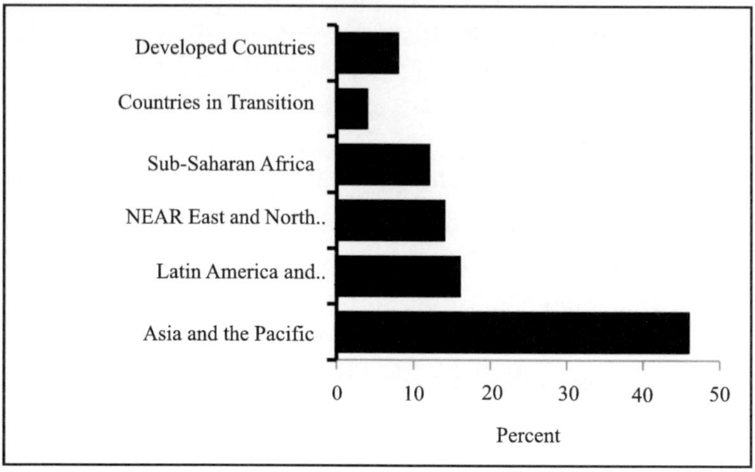

Fig. 9.1. Location of the World's highlands (Huddleston *et al.*, 2003)

9.1.7 Rationale for continuing shifting cultivation

Shifting cultivation continues to occupy large sections of tropical forest landscapes, either partially or entirely, although it is obvious that the scope and intensity of shifting cultivation is changing in many regions (Vliet *et al.* 2013). This is not a new concept; it has always been a dynamic system that is used when it is appropriate and then adjusted or even abandoned when better options for land and labour utilisation, income, and so on become available. These changes appear to be happening faster in some areas and slower or not at all in others (van Vliet *et al.* 2012). The main drivers of such shifts, notably towards more intensive farming methods or other sorts of land usage, have been political and economic forces (urbanization, large scale plantations, protected areas, extractive concessions, etc.). Because shifting cultivation is often not a desirable type of farmer, there is little political interest in understanding the logic behind it or estimating the number of people who rely on it (Mertz *et al.* 2009a).

Traditional shifting cultivation systems, according to Grogan *et al.* (2013), remain a significant component of subsistence strategies in Zambia and, to a lesser extent, Tanzania, but in a stage of advanced decline in more populous areas. Based on data collected in the 1990s and 2010 from two villages in Zambia's Northern Province and two villages in Tanzania's Rukwa Region, some authors suggest that a variety of factors contribute to the slow transition from shifting cultivation crops to more permanent land use practises (demographic, ecological, economic, and political). Traditional shifting agriculture practises are being phased out, and general living conditions are improving. This has been made possible by adaptation and diversification in agricultural practices and livelihood activities. Due to rapidly changing external circumstances, swidden cultivation communities, on the other hand, continually shift their attention between alternate livelihood possibilities (market conditions and policies).

The key drivers of the transition process are market access and government policies, according to these scholars. Conservation policies continue to limit production, while increased cash crop income, aided by greater market access and agricultural policies, has driven farmers to convert shifting plots into permanent agricultural fields. Agricultural diversification and off-farm income prospects, on the other hand, have reduced the burden on land resources, and households are less reliant on subsistence agriculture. As a result, swidden agriculture is sometimes only retained as a supplement to food security or cultural activities. Small portions are preserved in some circumstances to keep the shifting cultivation system and crops 'alive' in the event that other activities fail and shifting cultivation must resume (Vliet *et al.* 2013). Traditional shifting agricultural practices have declined in recent years, however this has been accompanied by an improvement in people's living standards and income.

9.1.8 Crop production

Under particular socioeconomic conditions of the ethnic people that practised it in a customary manner, the shifting cultivation agriculture system was well fitted to the environment. Traditional rituals and people's understanding of how to raise crops aided in the preservation of ecological equilibrium. Crop production is mostly focused on shifting cultivation, which involves only a few years of fallow time followed by one or two years of cultivation. New high forest areas are being cleared, and fallow seasons are increasing shorter. Forest lands are gradually being converted to shrub or grassland, resulting in a reduction in production potential. Crop production is poor, and alternate sources of income are limited, with farm labour in rice being the most common. The main causes are that the output was low due to excessive weed infestations (mainly *Imperata cylindrica*). The relationship between shifting cultivation and the fast expansion of Imperata grasslands in Southeast Asia is complicated (Powell, 1990). Weeds are one of the most important issues in any upland farming, whether shifting cultivation or continuing agriculture. Land degradation is the most serious environmental issue affecting a large region of land. Traditional labour exchange is eroding, forcing shifting cultivators to rely on hired labour. When the demand for weeding labour rises, poor shifting cultivators are likely to abandon or lease their plots in favour of a hired income (Uraivanh. 1998).

9.1.9 Animal husbandry

Livestock production for internal consumption and export has emerged as a significant current and future economic opportunity (Chapman, 1998). Crop-livestock integration is uncommon, yet it is considered as inextricably linked. Almost every farmer values livestock productivity. For most moving growers, animal husbandry is a crucial occupation. The demand for cattle production is high in both the lowlands and the highlands. Animal husbandry is an essential source of income for Lao PDR's moving cultivators (Sodarak *et al.*, 1998).

9.1.10 Transformations of the uplands: Implications

Despite the growing acceptance of alternate options, concerns still remain for the majority of shifting cultivators, in particular for the poor. The gradual transformation of upland agriculture has resulted in diversification of farming systems and compelled many households to establish multiple farming systems. This has resulted in an increase in land use pressure, with a resultant reduction in fallow cycles, a corresponding decline in land recuperation, and consequently, a rapidly declining productivity, resulting in the marginalization of shifting cultivation (Choudhury and Sundriyal 2003). Transformations have also brought about increased demands on labour, adversely impacting management and productivity. The most severe implication of transformations, however, has been the drastic replacement of common property regimes with private property frameworks, posing serious threats to the continuity of equitable tenurial access to productive resources hitherto inherent in traditional access regimes. While implications of agricultural transition in the uplands has been elaborated by Choudhury *et al.* (2008) based on the study carried outin West Garo Hills, a rapid survey was conducted in early 2009, covering seven villages in three districts of Meghalaya to compliment the earlier study and assess emerging concerns in regard to agricultural transformations. Distribution of work among men and women in different farming systems is given in Table 9.1.

Table 9.1. Distribution of work among men and women in different farming systems

Agricultural activities	Farming system		
	Shifting cultivation	Wet rice cultivation	Orchard plantation
Plot clearing	Men	Men	Men
Burning	Men	-	-
Seed-bed prep.	-	Women	Women
Sowing/transplanting	Women	Women	Women
Weeding	Women	Women	Women
Harvesting	Both	Both	Both
Transportation	Women	Both	Both

Source: Rural appraisal (2009).

Household labour availability for agricultural activities has progressively decreased with family members moving out of the village for education or wage-earning opportunities. Often, households are left with the very young members, aging members, and the women, or only the aging parents. Scarcity of able-bodied labour is becoming an issue in most villages as the young migrate to urban areas for education or to seek employment. With diversification of farming systems, particularly during transition from shifting cultivation into settled systems, demand on labour has become a serious issue affecting management of the agricultural systems (Choudhury *et al.* 2008). For households that are in transition, diversification of the farming system has increased the burden of agricultural workload. While this is true for both men and women, the increase in workload for women is spread throughout the year, while that of the men seems more concentrated in one season (Table 9.2). Women's workload has worsened with the increasing decline in productivity from shifting cultivation and inadequate returns from other systems. Declining productivity compounds the situation forcing the menfolk to migrate from the village for seasonal work in mines and other wage-earning opportunities. Of late, such migration, though seasonal, has left the burden of agricultural management on women whose workload has increased to levels where the poor and small families are forced to altogether give up agriculture and look for wage earnings as the sol livelihood option.

Table 9.2. Time schedual of different operations in three land use systems.

Agricultural calendar	**Farming system**		
	Shifting cultivation	**Rice cultivation**	**Orchard plantation**
Jan-Feb	Burning (men) Plot perp. (women)	-	Clearing (women)
Mar-Apr	Sowing (women)	-	-
May-Jun	Weeding (women)	Transplanting (w)	Weeding (women)
Jul-Aug	Weeding (women)	Weeding (women)	Harvesting (women)
Sep-Oct	Harvesting (both)	Harvesting (both)	Harvesting (both)
Nov-Dec	Clearing (men)	Harvesting (both)	Harvesting (both)

Source: Rural appraisal (2009).

Choudhury (nd) presented the following schematic diagram for the provisioning, regulating and cultural functions of the mountain ecosystems (Fig. 9.2). Hunzai *et al.* (2011) reported that the poverty percent among people was more as national average than prevalent in the HKH countries. Interestingly, the mountain people were relatively more poor than average poverty in India; while in rest of the countries of the region the case was reverse.

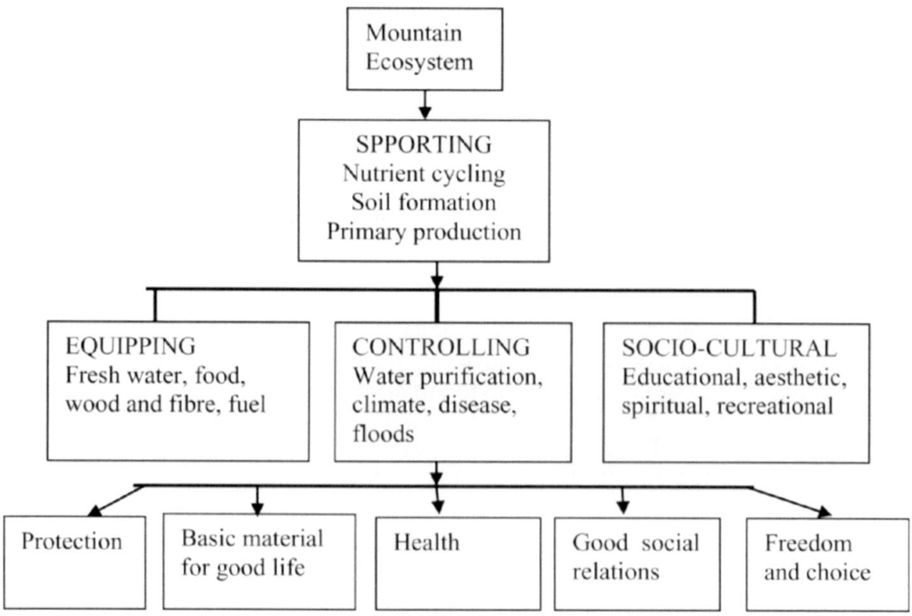

Fig. 9.2. Mountain ecosystem services and people's well-being
Source: Choudhury, ICIMOD

For ages, shifting cultivation has been a common farming practise throughout Southeast Asia. These systems, on the other hand, are evolving and sometimes disappearing at a rate never seen before. Shifting cultivation is considered subsistence, economically unviable, and environmentally destructive by the majority of development planners and policymakers, and hence a substantial hindrance to agricultural expansion in states where it is practised. As a result, governments have continuously attempted to replace the practise with established agriculture, investing significant financial resources to do so. Despite the community's desire for change and the government's efforts to bring about change, shifting cultivation remains a mystery and is still practised in significant portions of the region today. The NITI Aayog (Planning Commission), Government of India, formed a theme Working Group on 'Shifting Cultivation: Towards a Transformation Approach' in order to enhance people's livelihoods, alleviate poverty, and stop land degradation. Measures needs to e undertaken to; (i) consolidate knowledge of the problem's magnitude, (ii) identify viable best practices with upscaling potential, (iii) assess institutions (formal and traditional) and the need for transformation, (iv) determine to what extent and which "co-benefits" could be delivered (to jhumias and State agencies), and (v) propose an action agenda (short, medium and long term). The majority of plantation programmes lasted three to five years. After the initiative ended, there was no way to track the results, and farmers were sometimes uninterested in continuing to plant trees. After the programmes ended, a number of the afforested fields reverted to shifting cultivation (Pant *et al.* 2018).

The land tenure system is based on the freehold system, which allows family members to utilise but not sell their land. A small charge is normally paid to the chief or tribal leader for the use of the land in the event of a communal bush (Tommy 1983). Grazing pastures for cattle have been difficult to come by for traditional livestock producers. Another issue that has harmed the shifting agriculture system is a lack of home labour in rural areas. As a result of the lack of effective infant care, there is a higher rate of infant death. Because small farmers have limited access to operating capital, they are unable to use necessary inputs efficiently, and their farms have become less productive. Furthermore, the inefficiency of traditional hand tools is hampered by their poor quality. About half of the world's woods have been lost to agriculture and other purposes, and 78% of what remains has been drastically transformed, resembling nothing like the original forests (Bryant *et al.* 1997). According to more recent studies, deforestation rates in the 1990s decreased by about 10%. (Durst 2000). Previous techniques, according to Achard *et al.* (2002), overstated tropical deforestation rates by up to 25%.

9.2 ALTERNATIVES TO SHIFTING CULTIVATION

The present status of research, technology, and policy indicates the principle of 'controlling deforestation from shifting cultivation in situ by eliminating the need to clear more land and by reclamation degraded land and resources,' which can be attained by (i) offering sustainable alternatives to shifting cultivation, (ii) rehabilitating and managing left-over and degraded lands with meager production, such as regenerated forest fallows and low producing lands,' and, (iii) taking care of policy factors that affect shifting cultivation and the adoption of alternatives. Options for land-use management that enhance the economic situation of subsistence farmers, sustain agricultural output on deforested fields, and recover the productivity of degraded lands are urgently needed. Such choices will allow for the long-term development of forest areas while both meeting human requirements and preserving the ecology. These alternatives must be compatible with the various socioeconomic needs of individual locations in order to be generally adopted.

Marginalisation of Indigenous Peoples

(Karkhoff and Sharma 2006) were of the view that integrated shifting cultivation was mainly practised by indigenous peoples, who were condemned as "inappropriate and primitive in current times" by mainstream society. Their profound understanding of their surroundings was rarely valued, and it was frequently ignored or dismissed. When policies are designed for mountain areas by people from the plains, they are frequently based on extrapolated perceptions of environmental and local people from the plains experience, and are meant to give what plains people consider as benefits. Several national governments in the eastern Himalayan region have recently begun to take aggressive actions to ensure sustainable development for their people, the majority of whom come from a variety of ethnic groups and minorities. The governments are implementing a set of policies and practices aimed at ensuring the region's bio-resources are managed sustainably (Karkhoff and Sharma 2006). The present advancements in shifting cultivation show that both farmers and policymakers are involved. Shifting cultivators, like farmers all throughout the world, are continually modifying their cultivation to meet the current needs of growing communities. This is accomplished by an inventive method based on guiding principles gained from prior experiences, as well as prevalent ideas about what is required and suitable.

9.2.1 Ecosystem-Related Constraints to Agricultural Intensification

Shifting cultivation methods are environmentally feasible as long as there is enough land for a long-term restorative fallow (10–20 years) and crop yield goals and living standards are not unnecessarily high. These systems are tailored to the harsh environment and delicate ecosystems of the tropics. As a result, efforts to find effective crop shifting alternatives have had mixed results. There are a number of soil and climate constraints that must be addressed. As a result of these limits, soil degradation becomes a serious issue when changing from shifting cultivation to mechanised intensive agriculture. As a result, many large-scale mechanical farming plans that were implemented in the humid tropics have failed. There are a number of soil and climate constraints that must be addressed. As a result of these limits, soil degradation becomes a serious issue when changing from shifting cultivation to mechanised intensive agriculture. Low inherent soil fertility, vulnerability to soil erosion, soil physical degradation, and nutrient imbalance are the soil-related restrictions in developing alternatives to shifting agriculture. Low levels of radiation throughout the growth season, large storage losses, high relative humidity, insect and disease incidence, and high rainfall are all examples of climatic restrictions.

There can be many mixed land use systems where forestry can find appropriate place for improvement of shifting cultivation (Fig. 9.3).

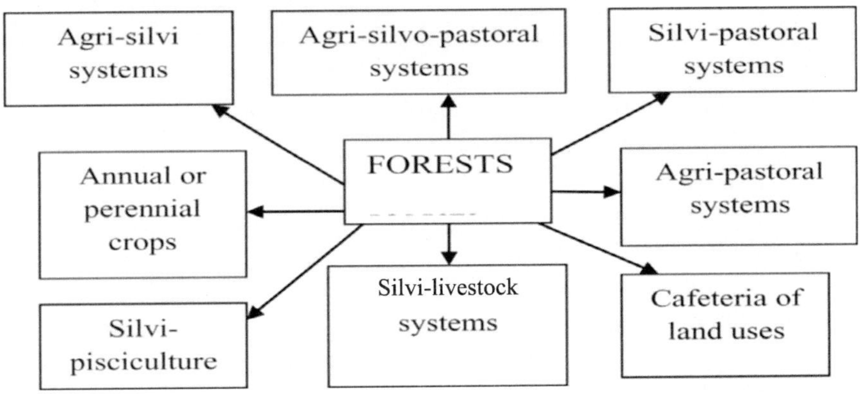

Fig. 9.3. Mixed farming systems (modified and redrawn from Serrao and Homma (1993)

9.2.2 Alternatives to shifting cultivation

The taungya is an agroforestry system in which short-term crops are produced in the early years of a woody perennials species' planting to use the land, manage weeds, reduce establishment costs, provide early income, and stimulate the woody perennials species' development. The British are thought to have invented taungya in Burma (now Myanmar) during the nineteenth century. Ethnic minorities have been using successional forest management strategies that match the taungya pattern for at least three centuries, according to historical studies. The Burmese word taungya (taung = hill, ya = cultivation) was coined in Burma in the 1850s. Brandis introduced the taungya system to India in 1890, and the first taungya plantations were established in North Bengal in 1896. The system's benefits include: artificially lowering the cost of forest regeneration, addressing the issue of unemployment to some extent, better land utilisation, and remunerative to the forest department. However, the approach has certain drawbacks, including the creation of legal concerns, the risk of epidemic, and the loss of soil fertility.

9.2.3 Farming systems research to develop alternatives to shifting cultivation

The challenge of establishing farming practises that are viable alternatives to shifting cultivation has either remained in its infancy or has been carried out in the field with insufficient zeal. There are several explanations for this condition, according to Okigbo (1981):

i) To encourage production of commercial crops in developed countries of Europe and North America, better marketing and other services were developed in support of industries, sometimes to the detriment of food crops;

ii) Agricultural productivity is site-specific, and transferring results and practises from temperate countries to the tropics is generally unsuccessful. Even if such results could be modified, until recently, adaptive research was impossible due to a scarcity of skilled scientists and engineers.

iii) Agricultural researchers were frequently cut off from the farmers whose farming practises and wellbeing they sought to enhance. As a result, just a few relevant technologies have been created to help farmers in the tropics solve their food crop and animal production concerns. Even then, farmers were unable to readily adopt the potentially promising new technologies developed due to fragmentary findings, the lack of location-specific technology, innovations unrelated to the farmer's needs and not socially and culturally acceptable; and innovations that were not economically viable (Wharton, 1969).

In the uplands, managing transitions in shifting cultivated zones is critical to agricultural development. As a result, shifting cultivation transformation is critical to agricultural transformation (Pant *et al.* 2018).

9.2.4 Options and research priorities for replacing shifting cultivation techniques

The purpose of research should be to develop practical, achievable, and financially viable technologies or alternatives to shifting cultivation. Only by incorporating and integrating the economic, ecological, and social aspects of this challenge, including land ownership constraints, will we be able to achieve success (Ramakrishnan 1992, Cairns and Garrity 1999; Singh *et al.* 2010). They must also be acceptable to farmers and conveniently accessible to them (Craswell *et al.* 1997). "The message that comes through clearly is that, while many technologies may be viable otherwise, unless the technology is appropriately integrated in the social and cultural context, it may not find acceptance and may not be able to come established," Ramakrishnan (1992) concluded after reviewing at least 25 years of fundamental and applied research on shifting cultivation.

Financial incentives must be included in initiatives to promote specific land-use management solutions (Anonymous 2009c) that are aimed at both promoting and maintaining change. Despite the fact that they may not be politically palatable, policies that guarantee annual payments over many years and maintain a significant portion as a bonus for farmers who meet the scheme's conditions at the conclusion of the 5-year programme are more likely to succeed. In the 1960s and 1970s, the 'green revolution,' which relied on substantially increased use of commercial fertilisers and pesticides, as well as novel crop strains developed using genetics and biotechnology, addressed the issue of feeding a growing population from a finite land supply (Mooney *et al.* 2005). The shorter fallow difficulty in shifting farming is how to sustain fertiliser availability on steep slopes throughout multiple year cropping cycles. Organic fertiliser additives (such as bovine manure, pig dung, composted litter, and agricultural waste) are considerably more environmentally benign than commercial fertiliser alone (Eastmond and Faust 2006; Kato *et al.* 1999).

9.2.5 Initiatives and strategies for controlling shifting cultivation

Various actions for the rehabilitation of shifting agriculture regions have been proposed by the National Mission on Greening India (MoEF, 2010). The main rehabilitation operations are:

i) providing marginal farmers involved in shifting agriculture with job and income creation options. These opportunities can be supplied by making good use of land resources in order to maintain control over shifting cultivation. Various government projects under the tribal plan will need to pump in enough resources to properly reclaim and rehabilitate the wasteland. They can benefit from agroforestry approaches. It helps to improve food and fodder production while also protecting the current forest, which is used by unemployed and disadvantaged people to make a living (Elevitch and Wilkinson, 2000).

ii) Efforts should be made as soon as possible to incorporate community members in forest-based activities. It should be made financially viable by putting in place suitable marketing infrastructure. This exercise will help to alleviate the load on shifting cultivation practices. It will also give community members more opportunities to participate in non-timber forest products.

iii) Through the development of village forest committees, degraded forestland should be safeguarded and developed. Farmers can be diverted from shifting agriculture by these committees. At the time of harvesting, appropriate incentives should be supplied to them.

iv) The mountain eco-systems should be rendered ecologically sound through changing agriculture practices. Through the formulation of an eco-development strategy for ecological sustainability, shifting cultivation to farm forestry can be partially substituted.

v) Land resource management will ensure the long-term viability of shifting farming. Furthermore, improved sustainable yields, technological advancements, innovative legislation, and improved transportation infrastructure will all contribute to the long-term viability of shifting cultivation.

vi) The woodlands that surround a hill village are referred to as the 'support area.' Farmers rely on the forest for firewood, feed, lumber, water, and animal bedding. Forest resources are mostly used by marginal farmers. They do not want to leave their natural environment.

vii) Specific policies should be formulated and executed, taking into account the unique characteristics of the mountain niche, to avoid any inconsistencies in shifting farming practices.

viii) Terracing fields, encouraging the production of vegetables, fruits, and high-value crops, and making proper use of timber and non-timber forest products can all help to make shifting farming more sustainable.

9.2.6 Resettlement

Another geographical trend affecting the practice of shifting cultivation is resettlement. This can take several forms. Shifting cultivators may be relocated from the uplands, woodlands, or international boundaries. Alternatively, lowland peoples may be relocated to the uplands, into the territory of shifting cultivators (Vargas 1985; Hardy 2003; Peluso and Vandergeest 2009). The causes for the latter rules and practices were numerous and intertwined with other current events. Many governments have wished for shifting cultivators to settle down and engage in sedentary agriculture for a long time (Wadley and Eilenberg 2005; Potter *et al.* 1995). Farmers resettled from the lowlands, according to government agents, might act as role models for shifting cultivators, demonstrating them "how to do settled agriculture."

Laws like this gave Indonesian bureaucrats the power to evict people and relocate them at will. Li (2007) cited a study that found "land allocated to large-scale plantations was 1.2 million hectares; land allocated to timber concessions and industrial timber plantations was 2.2 million hectares; land under protected forests and wildlife reserves was 2.5 million hectares; and land granted as mining concessions was 1.1 million hectares" in her study of Central Sulawesi, Indonesia. As a result, the total 'state' land legally assigned was 7 million hectares, despite the fact that the province's entire land area is just 6.8 million hectares (Li 2007).

Land is a non-expandable capital asset which renders innumerable services to the human and animal population as well as all other living bodies. There is need to manage our land resources optimally in order use them efficiently. The Government of India set up National Land Use Conservation Board (NLCB) in May 1985. The Ministry of Environment and Forestry has established a comparable National Wastelands Development Board (NWDB). Both bodies are overseen by the National Land Use and Wasteland Development Council (NLWC), which is chaired by Prime Minister David Cameron (Subramaniyan, 1990). Land use policy must be cooperatively developed by land users and the government, and it must be enforced through legislation and the encouragement of optimal usage (Subramanyan 1990).

9.2.7 Basic objectives of land use policy

The basic objectives of land use policy explains the various steps proposed in this direction. Some programmes which emphasize the various details of land use policies and includes; (i) Completion of surveys for a reliable data base and setting up of necessary organizational infrastructure, (ii) Launching a national campaign for educating farmers and officials concerned with the programme, (iii) The legal and administrative structure under any land use policy to adequately protect community rights on common land, particularly when it concerns the rights of tribals and poorer sections, (iv) Restructuring urban planning so as to ensure that productive land is not taken away, (v) Suitable cropping patterns for agriculture, (vi) Emphasis for rehabilitation of degraded lands, (vii) emphasis on national dry land project, and (viii) planting of appropriate species of fodder crops and trees on lands supported by suitable technologies.

9.2.8 State Land Use Board (SLUB)

To ensure continuous watch on land management and adequate attention at all levels, SLUBs were set up in different states of India. The man task of SLUBs is to make and enhance the stock of natural resources of land, water and vegetation for meeting the demands of people. SLUBs are confronted with the identification of the problems relating to, (a) managsement and care of resources, (b) management of land to meet competing demands, (c) increasing crop yields, (d) preserving and enhancing the quality land resources, and (e) rehabilitation and restoration of degraded lands.

Since shifting cultivation is the main-stay of life of the people, all the sectoral programmes such as; soil conservation, agriculture, horticulture, animal husbandry, water supply, housing, roads and communication, industry etc. should be oriented towards control of the practice. Control of the practice should be considered as major thrust area of all development programmes. Creation of an appropriate Authority, namely, Shifting Cultivation Control Authority Board may be considered for effective implementation of shifting cultivation control programmes. The shifting cultivation control organization at regional and district level in each state should be multidisciplinary team comprising officers and staff of planning, land based development and other research supporting services, departments. To ensure proper technical input there is need for:

- In each block, one micro-watershed for demonstrations.
- Ensuring better and timely supply of inputs such as seeds, fertilizers etc.
- Review of transferring rights of common property, resources etc.
- Amendment of existing laws, which should be acceptable to the community.
- Involvement of voluntary agencies for motivation of people for the control of SC.
- Curriculam on SC control needs to be included in training programmes.
- Upgradation of skilled staff at field and district level.

The central and state government in India have started many schemes to wean away the shifting cultivators from the practice. A few field surveys have been conducted for the appraisal of the schemes under operation. These schemes have met with little success and the results have been far below the expectations. In some of the instances it was found that those farmers who abandoned shifting cultivation for settled cultivation, kept small area to do shifting cultivation in order to celebrate festival and observe associated rituals for their well-being, as believed by them. The primary causes for the reluctance of shifting cultivators to do away with the practice were (Sarkar 1990); (i) the new settlements cuts their socio-cultural life abruptly, (ii) the farmers were not used to cultivation in terraces and using bullocks and new implements, (iii) the farmers found the production from newly constructed terraces too low due to removal of top soil during the process of development, (iv) the production technology for cultivation on terraces is also not well developed and, (v) dearth of well trained extension workers.

Shifting agriculture is not recognised as a lawful land use in Nepal, according to government policies. The Forest Act of 1993, which considers shifting cultivation land to be part of the national forest because it is not registered as private land, is the main policy that oversees shifting cultivation. This and other methods render customary tenure over shifting cultivation land and shifting cultivation institutions 'illegal.' Shifting cultivators, on the other hand, continue to use conventionally owned shifting cultivation land, but they are unable to invest in boosting the land's productivity due to tenure insecurity. Alternatives to shifting cultivation are specified by government policies in Bangladesh and Bhutan. Government support for alternative agricultural extension and research services, marketing, credit schemes, and financing, on the other hand, has to be improved. Because the government rules that affect shifting farming in Nepal do not recognise it as a land use, shifting growers have established alternate solutions. Annual agricultural income crops, rapid growing fruit, mixed fruit gardens with perennials, perennial cash crops, and forest tree plantations are some of the alternative possibilities in Bangladesh's research locations. Dryland farming, horticulture, and private forestry are all viable alternatives in Bhutan. Cardamom-alder agroforestry, chiraito (Swertia, a medicinal herb) plantations, and horticulture-based integrated agroforestry are some of the alternative solutions used by farmers in Nepal (Phuntsho *et al.* 2015).

As was the case with the rest of the Indian subcontinent, land tenure and property rights in the Eastern Himalayas remained mainly outside of land survey, titling, and construction of settlements based on revenue collection to the state. As a result, individual land ownership became the exception, and communal ownership became the norm. Assam, Meghalaya, Arunachal Pradesh, Tripura, Nagaland, Manipur, and Mizoram were all brought under the British India umbrella by the Chin Hill Regulation of 1896. The Bengal Eastern Frontier Regulation of 1874 (also known as the Inner Line Regulation) prohibited land acquisition in mountainous areas outside of Bengal's borders even before the Chin Hill Regulation. Both of these Regulations acknowledged the rights of the local population to their own properties. As a result, while transferring land administration to the people, the British respected communal ownership and customary rights. The Chittagong Hill Tracts Regulation of 1900 formally recognised Bangladesh's Chittagong Hill Districts as part of British India. The British authorities recognised the tribes' self-management of land over the areas they claimed as tribal domains under this Regulation. For years, shifting farming known as Khoriya and Tseri has been conducted in the hills and mountains of Eastern Nepal and Bhutan on unsurveyed probably state territory and forests. Because of the inaccessibility and remoteness of such property, the state had little or no influence or control over such actions. Much of the history of the Eastern Himalayas mimics that of a frontier colony. Bangladesh's CHT Due to a sparse population and sufficient vegetative cover due to frequent rainfall and a warm, humid environment, North Western Myanmar remains mostly under forest cover, except for small pockets of settlements. The land is more fertile here than in the Western and Trans Himalayas, the hills are smaller and less stony, the rainfall is higher, and the climate is milder. These areas are also remarkable in that the region is still largely occupied by Tibeto-Burman natives. There were no Indo-Aryan incursions, unlike in the Western Himalayas, and battles and colonisation were limited among the region's dominant ethnic groups.

Human influences on the hills include agriculture in the Apatani plateau of Arunachal Pradesh, and the construction of terraces for wheat cultivation by the Monpas tribes of Arunachal Pradesh, 'Zabo' system of soil and water conservation by Chakhesang tribe of Phek district of Nagaland and 'Agriculture with alder' by Angamis of Kohima district of Nagaland (Sharma and Sharma,2003 ;Prasad and Sharma,1994 ; Sonowal *et al.*,1989) and terraced agriculture in Sikkim hills may be the result of positive human endeavour. On the other hand, the prevalence of shifting cultivation has resulted in severely eroded hills without good vegetation. In Northeast India, there is a high percentage (12.2 to 56.7 %) of area under wasteland due to shifting cultivation in Nagaland, Assam hills, Manipur, Meghalaya, Mizoram and Tripura. Crop production development, crop diversity, and basic income augmentation to meet individual requirements are the farmers' immediate concerns.

9.2.9 Farming systems research

In a nutshell, farming systems research seeks to (Okigbo 1984):

i) Perform research to better understand farmers' physical, biological, and socioeconomic surroundings in a given ecological region or country;
ii) Conduct research on the farmer's resource base, farm enterprise structure and operation;
iii) Identify cultivators' limits in crop production and new technology adoption;

Prehistoric hunting and gathering, subsistence level crop and animal husbandry, early consumer stage, and primary mechanised stage were the stages of development Harwood (1976) postulated for a taxonomy of tropical Asian farming systems.

Following the statements made by Flinn, *et al.,* (1976), Ruthenberg (1980), and Okigbo (1981), these are the main physical, biotic, and social restrictions that farming systems in West Africa face:

A. *Physical Restraints*

a) Meteorological conditions, such as;

  i) Rains are unreliable in terms of start, period, and intensity;

  ii) Uncertainty of drought and flood times;

  iii) In light soils and sharp slopes, lower effective rainfall;

  iv) High soil temperature, which is detrimental to various crops and biological activities (such as nitrogen fixation);

  v) Fast decomposition rates and low OM levels;

  vi) Lower photosynthetic efficiency due to cloudiness;

  vii) Severe moisture shortages during several months of the year, with dry spells lasting more than three months;

b) Soil: In the humid and sub-humid tropics, most soils are

  i) Weathered, sandy, and poor in clay;

  ii) Have a low CEC, making colloidal complex less active;

  iii) Of low inherent fertility; except on hydromorphic and young volcanic soils,

  iv) High acidity of the soil;

  v) Subjected to a variety of nutrient shortages and toxins as a result of constant cultivation;

  vi) High P-fixation rate;

  vii) Prone to heavy leaching and high erosion risk during heavy rainstorms.

B. Biological Constraints

i) Unimproved crops and livestock;

ii) Poor yields imply a limited potential;

iii) Illness and pest susceptibility;

iv) A high prevalence of illnesses, pests, and weeds as a result of the favourable environment;

v) Significant environmental changes as a result of human activity, which have a negative impact on natural ecological equilibrium or balance.

C. Economic and Social Constraints

i) Reduction of farm size due to population pressure;
ii) Land tenure not favouring farmer, fragmentation of holding;
iii) Labour scarcity;
iv) Lack of loans, poverty;
v) Lack of market; pricing system;
vi) Unavailability and high cost of inputs;
vii) Poor extension services;
viii) Illiteracy, ignorance, social taboos;
ix) Lack of communications
x) Until recently, there was a lack of a comprehensive approach to manufacturing systems, technology, design, assessment, and application;
xi) There is a lack of strong farmer groups and political 'voice.'

9.2.10 Soil development and management

Expanding the quantity of land under cultivation or using a number of other technologies that produce higher productivity per unit area can help increase food and agricultural output. The former is the most important method of increasing food production in tropical Africa. The amount of land and labour available each year to clear new area will determine the strategy's effectiveness. Simultaneously, land demand for diverse development activities is increasing. As a result, if farm sizes are to be increased and the current trend of shrinking farm sizes is to be avoided, approaches for preserving soil fertility and removing or drastically reducing the period spent in bush fallows as part of permanent cropping systems must be developed.

9.2.11 Why continuing with shifting cultivation

Crop diversity and food availability

By providing financial aid for the construction of terraces and the growing of plantation crops, the government's initiatives to regulate shifting cultivation have promoted permanent agriculture. The programmes have mostly focused on grain and plantation crops, resulting in a reduction in the number of crops that farmers may produce and access.

Tenurial security

Despite the fact that land has traditionally been administered as a common property under shifting cultivation, the transition to private ownership has weakened tenurial access to land, leaving farmers, particularly women farmers, with less or no land. For affected households, this has ramifications for both tenurial access and livelihood security. Such developments alter institutional regulatory frameworks (sometimes introducing competing functions), needing a comprehensive understanding of how changes in the institutions that oversee these resources should be managed. Furthermore, a lack of tenure security deters farmers from investing in land upkeep, hastening soil deterioration. In locations where communal institutions have lost ownership of land, landlessness has become a serious socioeconomic challenge. Many farmers continue to jhum to avoid losing their land rights and becoming landless. Landlessness is becoming a major worry among indigenous populations in many northeast Indian states. Any jhum change that leads to a loss of land cannot be sustained.

Ecosystem services

Promotion of cash crops and general changes in land use patterns are often driven by community aspirations for better integration with expanding market forces, rather than by government programmes and plans. However, the emphasis on the economics and the resulting changes in land usage have ecological consequences, particularly for vegetative cover. Regenerative fallow cycles are commonly changed into non-forest vegetation as a result of established farming, resulting in a loss of essential ecosystem services and land degradation. Under the conditions of agricultural intensification, drying up of water supplies, depletion of soil fertility (and its consequences), and reduced availability of fuel wood, fodder, and wild delicacies are all important concerns. Farmers do not view the shifting cultivation solely as a food producing system. They also benefit from the system in a variety of different ways. Any proposed land use modification for shifting cultivation transformation must therefore assess the influence on ecosystem services in general and hydrology in particular.

Access to various schemes

Successful shifting cultivation transformation necessitates a variety of enabling programmes and policies that can create the optimal atmosphere for communities to overcome transformational problems. Most programmes and methods for tackling shifting agriculture transitions are developed for specific watersheds or locales and do not necessarily span vast areas, limiting their application to a subset of the population. Jhum is still practised in areas and communities that are not covered by such programmes. As a result, access to and benefits from such programmes are limited to residents of these specific places, and they are not generically applicable to the entire community. This raises the issue of equity, and the issue is only addressed in a limited way. Gender is also a key concern when it comes to programme and scheme access. As a result, even if shifting cultivators want to make revolutionary changes, they don't have access to the support resources that would enable them to do so.

9.2.12 An innovative alternative to shifting cultivation

Shifting cultivation is still an important livelihood for millions of people in the tropics. External constraints on this practice in Indonesia's Kalimantan Island include significant development of oil palm plantations, concessions for industrial plantations, and a lack of financial and political backing from the government. External forces such as those mentioned above, which disregard local expertise in forest and land management, have limited the amount of land accessible for swidden cultivation (Imang *et al.* 2018). As a result of the external influences, the shifting cultivation cycle has been shortened, and the yield of this practice has declined. The research was carried out in East Kalimantan, Indonesia, in an indigenous Bahau Dayak village.

9.2.13 Agroforestry, an alternative to shifting cultivation

Land use techniques based on age-old practices of actively blending or keeping trees in crop/animal producing fields are known as agroforestry (Nair 1983). Agriculture and forestry are combined in sustainable production systems on the same piece of land, either simultaneously or sequentially, in sustainable production systems. Because a modern idea of agroforestry is still being created, there isn't yet a commonly accepted definition, however numerous have been proposed. - One of the more recent definitions is "Agroforestry is a collective name for land use systems and practices where woody perennials (trees, shrubs, palms, bamboos, etc.) are deliberately used on the same land management unit with agricultural crops and/or animals, either in some form of spatial arrangement or temporal sequence. In agroforestry systems there are both ecological and economical interactions between different components."

9.2.14 Shifting to settled cultivation

The absence of financial resources for the initial investment and regular purchase of chemical weedicides, as well as the communities' reliance on a variety of crops tailored to grow in shifting cultivation sites, were major impediments to permanent agriculture. The size of the area under settled cultivation possessed by a household was determined by the family's economic level: households with higher economic standing had greater areas under settled cultivation. Five houses in Bomdo village were fully reliant on shifting farming and had a low economic position. This suggests that the inability to make the first investment is another stumbling block to the transition to settled agriculture. In settled cultivation fields, chemical weedicides like Roundup were also increasingly widespread, adding to the cost of the operation. The terrain may also play a role in determining whether a household engages in established agriculture; two households from the village that was carefully surveyed indicated that they engaged in agriculture in the neighbouring village area by purchasing property suited for settled farming.

9.2.15 Alternatives and land use change

Land Use Changes

Shifting cultivation is a land use in which annual cultivars are planted on plots to provide vegetative cover and secondary forest in various stages of regrowth. Increased population, market integration, and environmental regulations have all contributed to the intensification of traditional agriculture and, as a result, changes in soil use and cover (Walker and Homma 1996). Long-term intensive agricultural or agroforestry systems could be the result of this strategy. Shifting cultivation looks to be dying out in most parts of Southeast Asia, with only a few places remaining stable (Padoch *et al.* 2007). Another consequence of intensification under the shifting cultivation method is a decline in agricultural diversification (Almeida and Uhl 1995). This technique has harmed the system's production and long-term survival by disrupting the ecosystem and reducing crop genetic diversity. The replacement of locally eaten species with market-valued species is another factor leading to the loss of biodiversity (Altieri *et al.* 1987). Myers (1993) makes a distinction between "shifting cultivators" and "shifted cultivators," the latter being migrants who have recently begun practising slash-and-burn agriculture without considering the region's environmental potential and constraints. The public's attention has changed away from traditional shifting cultivators as the number of shifted farmers has expanded considerably over the previous four decades.

9.2.16 Alternatives to shifting cultivation

Agroforestry systems encompass a wide range of techniques whose success is decided by their economic viability (Hardter *et al.* 1997), which is defined not only in terms of production stability but also in terms of agroecosystemic efficiency in recycling mineral nutrients (Hardter *et al.* 1997). (Hardter *et al.* 1997). Juo and Manu (1996; Juo and Manu 1997; Juo and Manu 1998; Juo and Agroforestry systems can produce a variety of goods at different times of the year while also retaining portion of the land under secondary forest cover of varying ages through a rotational system, in which different sections are in different stages of succession at any given moment (Perz and Walker 2002). Sanchez (2000) lists two studies that showed agroforestry alternatives to shifting agriculture practises, both of which centred on policies that provided smallholders with appealing soil use options in exchange for carbon sequestration services. Among these alternatives were the use of fertiliser inputs to increase agroforestry sustainability, such as:

i) leguminous tree species planting (replacing approximately 35 percent of the cleared forest's original carbon stock),
ii) phosphorous application from local phosphoric rock, and
iii) biomass transfer from nearby nutrient-rich ruderal species that mineralize quickly to nourish the soil.

The local administration made a concerted effort to explore alternatives to the shifting cultivation practice after realising that fire is one of the main causes of environmental degradation and biodiversity loss in Mexico. To prevent more deforestation and soil degradation, forest rangers in a community in northern Vietnam limited the area on which shiting cultivation may be conducted.

9.2.17 Alternate options to shifting cultivation in Northeast India

To limit the area under shifting agriculture, the Indian government and state governments in the region devised policies and passed legislation.The intents of the governments' policies have been to transform shifting cultivation to sedentary agriculture and cash crop plantations, and horticulture with an object of improving economic efficiency and reducing environmental degradation (Tiwari 2014). While governments have been trying to 'wean away' shifting cultivators from this supposedly ecologically and economically inefficient and degradative system of cultivation, the shifting cultivation farmers also engaged themselves in innovating their crop production system to adapt it to the changing socio-economic conditions particularly for producing more food for increasing population and in order to meet the needs of cash for modern living (Tiwari 2014).

According to literature reviews, the net area of land under shifting agriculture was 87220 ha in 1970-71 and 84002 ha in 1995-96. During the same era, however, area under permanent agriculture expanded from 28006 ha to 81281 ha (Patel *et al.*, 2013). The area under shifting cultivation declined at an annual compoung growth rate (ACGR) of -0.15%, whereas; land under permanent cultivation and total area under cultivation increased at an ACGR of 4.35% and 1.45 %, respectively. Due to demographic pressure, the land under cultivation increased at a very faster rate. Since the land under shifting cultivation decreased at very slow rate, more and more forest area was encroached to feed the growing population. Further, it is also noted that there was a drop in area of land per family in shifting cultivation

(1.10 to 0.7 ha) and increase in permanent agriculture (0.35 to 0.74 ha). The Planning Commission, Government of India supported Watershed Development Project in Shifting Cultivation Areas (WDPSCA) during 8th plan period in the year 1994-95 with the objective to rehabilitate the lands affected by shifting cultivation and support the shifting cultivation farmers to move to settled cultivation. According to a WDPSCA evaluation study conducted in Nagaland and Tripura, shifting cultivation area has decreased by 30%, and 27% of shifting cultivation farmers have quit practising shifting agriculture. shifting cultivation land per family has likewise shrunk from 0.84 to 0.56 hectares. This has been largely attributed to the move from shifting to established agriculture. Furthermore, the *Apatani* and *Mongpa* tribes in Arunachal Pradesh have abandoned shifting cultivation in favour of established agriculture, posing a threat to the continuation of their rites and celebrations. As a result, their financial situation has significantly improved. In recent decades, the tribal people of northeast in general have become aware about adverse ecological impact of short cycle shifting cultivation. Certainly, this has been possible due to social development as reflected in the phenomenal increase in the levels of literacy in the region, which was 20.4% in 1961 and increased to 68.66% in 2001, higher than the national average of 64.80%. As of 2013 except Assam and Arunachal Pradesh all states of northeast has literacy rate higher than national average (77.3%) and Mizoram and Tripura are among the most literate states of the country. There are no authentic data available on the number of families practicing shifting cultivation. According to a 2007 publication of Ministry of Environment and Forests, Government of India, 600,000 families are practicing shifting cultivation on 3800 thousand hectare of land (MoEF, 2007). Strengthening rather than replacement of shifting cultivation has been accepted by governments of Nagaland and Tripura who have made substantial stride towards improving the economic condition of shifting cultivators. However, policymakers and experts have recently concluded that, in addition to encouraging farmers to settle down, it is critical to use technology to increase the productivity of land under shifting cultivation by implementing site-specific innovations and interventions. Several national and international organisations have created programmes and provided grants for increasing productivity in shifting agriculture areas and mitigating land degradation through state governments and non-governmental organisations. However, nothing has been done to collect and compile the policies and alternative solutions to shifting agriculture that have been supported and are currently used in northeast India. Chane in area under shifting cultivation and permanent agriculture between 1970-71 and 1995-96 is shown in Fig. 9.4.

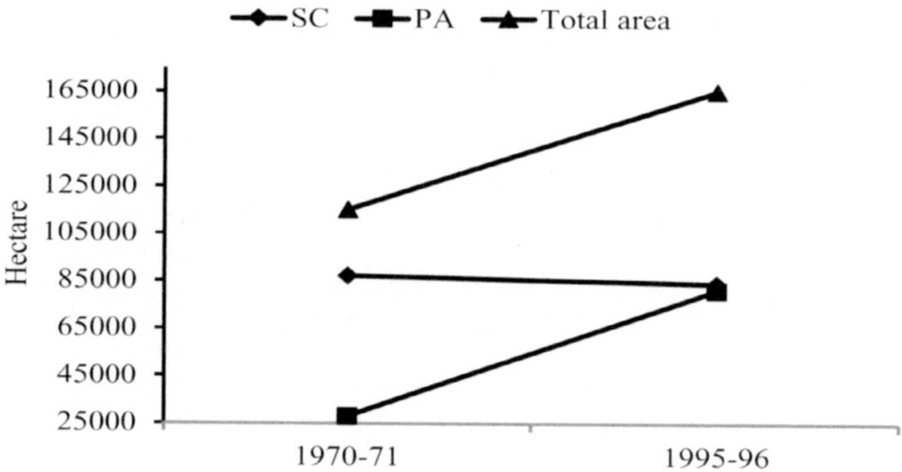

Fig. 9.4. Change of area under shifting cultivation (SC) and permanent agriculture(PA).

9.2.18 Rubber plantations for rehabilitation

Rubber plantations are one of the most prominent measures used in Tripura state of India as an alternative for shifting cultivation and shifting cultivators's rehabilitation. The rubber plantation project was designed to provide a profitable alternative. The rubber option appeared appealing since it had the potential to provide shifting cultivators with a significant income, as well as cash crops like pineapple and pepper, which could be planted alongside it. By tapping the major industrial and export markets, the Tripura government would be able to improve its revenue potential by using tilla or wastelands for cash crops. Rubber plantations cover a total of 29,120 hectares out of which 15660 hectares are mature plantation, while 13,460 hectares are immature plantation, defined as plantation that is less than 7 years old and cannot be tapped for latex on a regular basis.

9.2.19 Shifting cultivators rehabilitation through horticulture, tea and coffee

The horticultural programme is another important tool for rehabilitating shifting growers. Tripura's edapho-climatic conditions are ideal for growing horticulture crops such as bananas and pineapples, with pineapple being one of the most profitable and viable crops to encourage among shifting cultivators. The Horticulture as well as tea and coffee plantation has not been picked up by the jhum farmers on any large scale as it needs support in terms of finance and technology which is not available in similar measures as that for rubber (Prasad 2004).

9.2.20 Integrated shifting cultivators rehabilitation programme

In Tripura, "Integrated shifting cultivators Rehabilitation Program" is supported by Government of India through Tribal Welfare Department to improve the livelihood of tribal shifting cultivation farmers. The object of the program is to take up both rehabilitative as well as supportive measures for the shifting cultivators of the state. This is a multi-pronged programme that includes land allotment and development, as well as the formation of micro-enterprises for economic progress (i.e. pig rearing, poultry farming, apiculture, fishery, carpentry, tailoring, handlooms and handicrafts). By offering productive employment, the initiative aids in the relocation of shifting cultivation farmers. It improves the state's poor tribal groups' overall economic and social status. The economic development of tribal communities practicing shifting cultivation through this scheme has reduced the dependency of such people on shifting cultivation in parts of the state where this program has been implemented (IFAD *et al.*, 2001).

9.2.21 Programmes of North Eastern Council, Government of India

North Eastern Council, Government of India during 5th five year plan initiated a project entitled Regional River Basin Scheme for Controlling Shifting Cultivation in North-East India. Christened as Regional River Basin Scheme for Control of Shifting Cultivation in northeast India, this project also consisted of eight schemes for the treatment of catchment of the rivers. North Eastern Council had implemented 8 pilot projects for implementation during the sixth and seventh five year plan period (Table 9.3).

Table 9.3. Shifting cultivation control project of North Eastern Council

State / River basin	No. of families	Area (ha)
Arunachal Pradesh (Siang)	1613	1893
Assam (Ranga)	NIL	1880
Manipur (Iril)	1369	1379
Meghalaya (Didram)	679	1315
Mizoram (Palak)	350	700
Mizoram (Dheleshwari)	800	1818
Nagaland (Magalew)	132	295
Tripura (Haora)	594	1118
Total	5187	9598

Source: Shifting cultivation in North-East India: Policy Issue and options by B. P. Maithani, 2005.

9.2.22 Alternative options promoted by the central & state governments

Farming System Models of Indian Council of Agricultural Research (ICAR): The Indian Council of Agricultural Research had developed 8 viable alternatives farming systems models that could be introduced in areas under shifting cultivation (ICAR, 1980). Field trial done by ICAR in their research complex at Barapani near Shillong (Meghalaya state of India) had shown that watershed based farming system coupled with mechanical soil conservation measures such as contour trenches, - contour bunds, bench terraces, half-moon terraces, grass waterways, etc. at appropriate locations can retain maximum rainfall within the slopes, safely disposing off the excess runoff from the slopes to the foothills with non-erosive velocity were found suitable for replacing jhum in several locations. These models however, did not become popular with the jhum farmers mainly because of poor extension and requirement of external inputs. ICAR has also evolved several agroforestry models for shifting cultivation areas which also have not been accepted by the farmers in any significant scale (Bhatt *et.al.*, 2001).

9.3 TRANSITION AND LIVELIHOOD

People in the miombo forests of southern, central, and eastern Africa have traditionally relied on shifting cultivation as a source of income. However, as deforestation and forest degradation worsen across the region, traditional shifting agricultural systems are under increasing strain. Indeed, agricultural intensification and expansion are thought to be the principal source of miombo degradation, which is influenced by a variety of endogenous and exogenous variables working at different scales. The study will chronicle the transition from shifting agriculture to more intensive land use practices using data collected between the 1990s and 2010 from two communities in Zambia's Northern Province and two in Tanzania's Rukwa Region. Growing population, government policies, and increased commercialization/market integration are said to be the key causes promoting miombo deterioration, and hence the transition process of traditional shifting agriculture practices (Grogan *et al.* 2013).

The change of shifting agriculture systems is a long-term process that dates back to pre-colonial times. Land allotment and development, as well as the construction of micro-enterprises for economic improvement, are all part of this multi-pronged approach (i.e. pig rearing, poultry farming, apiculture, fishery, carpentry, tailoring, handlooms and handicrafts). The initiative supports in the resettlement of jhum farmers by providing useful jobs. It raises the overall economic and social status of the state's disadvantaged tribal groups.

9.3.1 Agricultural change and its main drivers

The development of ideas of how and why agricultural transformation and land use intensification occur has been aided by two conceptual frameworks that focus on the impact of population dynamics. The first major contribution to the subject was Thomas Malthus' (1798) essay on the inherent imbalance between population growth rates and food production. This early study later inspired the formation of a neo-Malthusian paradigm, which contends that resources are limited within specific geographies, and thus have an intrinsic population carrying capacity, which, if surpassed, leads to inevitable calamity (Tiffen and Mortimore 1994). The response to endogenous and external agents of change has been the subject of theoretical and empirical research. Although agricultural development is rarely assigned to a single cause, population pressure is frequently cited as one of the key drivers of change, typically inextricably linked to variations in natural resource availability and/or the introduction of new technologies (Turner *et al.* 1977; Bennett 1976; Adams and Mortimore 1997). In addition to these 'physical' driving forces, Stone (2001) claims that social/cultural and economic systems can have a significant impact on agricultural development, and that they can often contribute to, and even override, population density as a driving force. Farmers are frequently enticed to intensify their agricultural by market forces, even when there is no lack of land (Turner and Brush 1987; Lambin *et al.* 2001). Farmers rarely cultivate exclusively for subsistence, and even fewer grow solely for sale, with the exception of extremely remote and sparsely populated locations. Markets, according to Netting (1993), may not be able to override Boserupian processes, but they do have the ability to affect smallholder decisions. Brookfield (2001) goes a step further, arguing that market pressures (such as cash cropping incentives) can be a more effective trigger of change than population growth in some circumstances. Land tenure in Zambia is divided into two categories: State Land and Customary Land. Adams (2003) provides an outline of Zambia's land tenure policy and practice. Almost all of the land in the study region is classified as customary land and is managed by traditional authority. The village headman distributes land to residents in a typical community. Land conflicts are growing more common as the population grows, but they are frequently resolved by local governments. During the late 1980s and 1990s, structural adjustment policies (SAP) in both nations had a significant impact on small-scale farmers' market participation.

Shifting cultivation, one of the oldest practices in agriculture is intrinsically related with social institutions of human societies particularly the tribal societies. Understanding tribal culture without shifting cultivation is incomplete. shifting cultivation never fails to produce. In the patterning of crop distribution, sometime some crop may fail, but some surely bring certain amount of yield. Due to climate change, towards jhum. Besides this, Scientific and technological achievements in turns also creates new needs, new adjustment. Yet the practice of shifting cultivation still continues giving security to the farmer in the upland. But some changes have taken place in the cropping patterns. The traditional multi-cropping pattern has been replaced by mono-cropping of various cash crops resulting in manifestation of indigenous food system (Chongloi 2016).

9.3.2 Rituals for good harvest

The *Thadou-Kukis* of Manipur grew numerous crops in the same plot using their labour as the primary input, using instruments of their unique cultural traditions. Shifting cultivation provides them with a source of income, food security, and social security. This system/practice is linked to their socio-cultural life, rituals, annual events, celebrations, and day-to-day living. It also has a strong connection to their cultural identity. That is a way of life for them because it is the only way they can support themselves. Man and woman, young and elderly, including the chief and the priest, each have a specific role to play in this process based on their talents. shifting cultivation has left an indelible mark on the cultural landscape. They have also maintained a specific fascination with this method of gardening due to their historic habit and belief. As a result, shifting cultivation promotes communal bonding and develops village-level institutions, increasing social capital. The Forest Survey of India (FSI) has also stated repeatedly that the loss of forest cover is primarily due to shifting cultivation.

9.3.3 Policies to control shifting cultivation in Odisha, India

Given the size of the territory and population affected by shifting cultivation in Odisha, India, the state government has attempted to address the issue by regulating or nationalising the practice. During the 1960s, the colonisation programme was implemented in troubled areas to help primitive tribes transition to settled agriculture by giving cultivable land, essential inputs, and housing (Biswal and Kumar 2013). A number of colonies were founded in tribal areas during the first four programmes. However, the strategy has failed to produce the expected benefits in the majority of areas. Aside from the colonisation project, the State Government's Soil Conservation Department has taken up a programme of rational land use based on watersheds. The following were among the programmes: (a) providing land to tribals prepared to give up farming on high slopes. (b) Planting of economic species that are beneficial to the tribal group. (c) The introduction of conservation farming to enable tribal people to get more yield from their cropland. (d) The use of steep slopes for timber production. A appropriate watershed in the shifting cultivation region is chosen, and a sensible land use programme is implemented on the existing catchment, according to the aforesaid scheme.

9.3.4 Task forces

The Ministry of Agriculture was created in 1983, the Ministry of Environment and Forests was established in 2002, and the Inter-ministerial Task Force of the Ministry of Environment and Forests was established in 2008. Each of them suggested steps for their respective ministries. Each work group could only achieve limited aims since shifting cultivation is a problem that impacts food security, livelihood, culture, and land tenure, as well as climate and land topography, and the area of shifting cultivation remained essentially unaltered even after their recommendations were followed. These task teams' limited outcomes have showed us that any transformation approach must be broad, address numerous problems, and meet the majority, if not all, of shitting cultivators' objectives and ambitions.

The important takeaway from this study is that shifting cultivation management requires social, economic, and infrastructure development. Poor connectivity, in example, restricts people's access to markets and technologies, perpetuating shifting cultivation. Government plans to regulate shifting cultivation have promoted permanent agriculture by giving financial assistance for the construction of terraces and the growth of plantation crops. The programmes have primarily focused grain and plantation crops, resulting in a decrease in the variety of crops that farmers could previously produce and access. This has implications for both tenurial access and livelihood security for affected households. Such changes modify institutional regulatory frameworks (often introducing competing functions), necessitating a thorough understanding of how changes in the institutions that oversee these resources should be managed. Furthermore, a lack of tenure security deters farmers from investing in land upkeep, hastening soil deterioration. In locations where communal institutions have lost ownership of land, landlessness has become a serious socioeconomic challenge. Many farmers continue to shifting cultivation to avoid losing their land rights and becoming landless. Landlessness is becoming a major worry among indigenous populations in many northeast Indian states. Any shifting change that leads to a loss of land cannot be sustained. Successful shifting cultivation transformation necessitates a variety of enabling programmes and policies that can create the optimal atmosphere for communities to overcome transformational problems.

Shifting cultivation diminishes in landscapes with access to local, national, and international markets, which stimulate cattle production and cash cropping including biofuels, according to this meta-analysis of land-cover transitions in tropical forest-agricultural borderlands around the world. By reducing forest destruction and fostering commercial agriculture, conservation policies and practices hasten changes in shifting cultivation. Shifting cultivation is significant in some locations simply because intensification is not an option such as where population density and/or food market demands are low (van Vliet *et al.* 2012).

9.3.5 Multifaceted land use?

Multifaceted land use is based on management strategies that aim to provide many products or services (Simelton *et al.* 2020). Grain, fodder, timber firewood, biofuel, fruits, and flowers are examples of products, while services include water infiltration, wind breaks, microclimate regulation, carbon storage, erosion control, groundwater recharge, and soil conservation, to name a few. Landscapes are described as multifunctional by Mander *et al.* (2007) because they simultaneously serve habitat, production, regulatory, social, and economic activities. They pointed out that heterogeneity (diversity) is a fundamental feature of landscapes, and that heterogeneity indicates the landscape's ability to support many, sometimes contradictory uses at the same time.

9.3.6 Traditions

Many traditional land uses arose via interactions between environmental functions and cultural benefits over extended periods of time. Shifting cultivation is an example of this, which has existed in nearly all agroecological zones throughout history, mostly for subsistence farming. A cyclic slash-and-burn approach is commonly used, with fallow times to restore soil fertility and a crop sequence that responds to falling soil fertility during the cultivation phase. Fallow seasons become shorter as land scarcity increases, soil fertility falls, and altering farming practices can no longer maintain production. Before scientific agricultural research and extension methods gained traction, these traditional farming practices were essentially the result of accumulated indigenous knowledge, culture, and adaptations passed down from generation to generation. The word 'agroforestry' was coined in the 1970s to describe farming practices that involved farmers intentionally planting or maintaining trees on their land (Nair 1993). However, the overall practices dates back thousands of years, as farmers discovered early on that simulating natural-forest systems with numerous canopy levels, keeping animals close to trees, and producing homegardens provided multiple benefits, products, and services.

9.4 AGRICULTURAL INTENSIFICATION

9.4.1 Agricultural intensification and improved fallow management

Terracing

In the evolution of shifting cultivation, the formation of terraces along the slopes has been a regular occurrence. In Andhra Pradesh and northeastern states of India, it is quite common. However, due to the high cost of physical development, which necessitates a significant amount of labour capital, and unequal access to both natural water flow to such terraces and government and non-governmental watershed development initiatives, the elite in the villages have been able to intensify while the non-elite poor have been left on the periphery of shifting cultivation evolution. In non-terraced locations, this will also mean that the poor will bear a greater share of the cost of enhanced fallow management. Despite the fact that intensified terraces may generate cash-labor demand and lessen the poor's reliance on shifting cultivation, allowing the fallows to rest and regenerate more quickly, this may be the case.

Vegetable cultivation

Many shifting cultivation groups have adopted vegetable farming on the uplands and in home gardens due to surplus capital and market connectivity to address the issue of demand and perishability. Vegetable growing has reduced shifting cultivation in an unexpected way, in addition to increasing income and decreasing reliance on shifting agriculture for fundamental living needs. It's tough for a family to conduct both shifting agriculture and vegetable growing since the labour markets are so competitive. Families are more likely to choose vegetable farming if there is easy access to markets.

Horticulture

Most government and other formal entities have taken the approach of promoting horticulture crop plantations in shifting agricultural regions. Horticulture has given the much-needed funding with which land intensification has been performed in many cases, in addition to increasing the value of the fallow or shifting to settled cultivation. Villagers, for example, have purchased wet paddy lands using horticultural surplus money, so addressing not just their food security problem but also lowering their vulnerability to the vagaries of the horticultural produce markets.

9.5 STRATEGY FOR SETTLED CULTIVATION

Sharma and Bujarbaruah (2016) reported the following strategies for a change from shifting to settled cultivation.

i) Rehabilitating degraded jhum regions using a watershed strategy.

ii) The integration of sectoral initiatives for the long-term development and management of watersheds

iii) Provision of means for permanent agriculture,

iv) A multidisciplinary approach at the district level, incorporating several line departments and evolving programme measures.

v) Adoption of better technology and suitable land use based on land capability. Pathways to intensification

9.5.1 Pathways to intensification

In general, farmer responses to intensification pressures can be characterised as innovations aimed at:

i) More 'effective' fallows, in which the biological efficiency of fallow functions is improved, resulting in greater production benefits in a shorter time frame.

ii) More 'productive' fallows, in which the length of the fallow remains constant or even increases as the farmer adds value to it by adding perennial economic species.

iii) A hybrid of the two, in which productive perennials are alternated occasionally with annual crop farming to get a degree of both biophysical and economic benefits.

Increased output per unit land area and cropping frequency are possible with more effective or quicker fallows. They are frequently used as a stopgap between shifting cultivation and continuous production of annual crops. To ensure the continual and efficient replacement of shifting cultivation and related bush fallow systems, significant importance should be given to technology packages that contain the following:

i) Managing land and soil on sound ecological principles;

ii) Conservation tillage;

iii) Managing crop left-overs suitably;

iv) Affordable energy technology;

v) Enhancing fertiliser use efficiency with appropriate technologies;

vi) Encouraging biological N fixation and nutrient recycling;

vii) Suitable crop rotations to keep the soil covered for most of the year;

viii) Integrated management of pests and diseases;

ix) Proper care of farm produce during storage and transportation; and

x) Encouraging horticulture crop production systems.

9.5.2 Changing shifting cultivation systems

In contrast to growing evidence of this agricultural system's extinction in other regions of the tropics, there is minimal information on change in shifting cultivation systems of small Pacific islands. The farming system's productivity remains high, despite signs of soil degradation in the island's central region. However, there are no signs that Bellona's traditional shifting farming system will become unsustainable in the foreseeable future, and there are no drastic land use changes like those witnessed on other Pacific islands (Mertz *et al.* 2012). In recent decades, shifting agricultural methods have undergone tremendous modifications. Other land use regimes, like as plantations, protected forests, or pasture, are rapidly replacing them in many regions of the world. This trend has been studied in Southeast Asia (Mertz *et al.* 2009b; Padoch *et al.* 2007), where shifting cultivation appears to be on its way out. However, there is a lack of understanding of these processes at the subnational level, as well as the ramifications of changes for livelihoods and the environment (Bruun *et al.* 2009; Mertz *et al.* 2009a; Schmidt-Vogt *et al.* 2009). On a worldwide basis, shifting cultivation does not exhibit similar patterns of change in other locations (van Vliet *et al.* forthcoming), and this study adds to the general paucity of understanding regarding the national and local-scale repercussions of shifting farming to other land uses.

9.6 IMPROVEMENT OF SHIFTING CULTIVATION

9.6.1 The acceptance of shifting cultivation

In 2006, the government of Meghalaya, India, promised to stop fighting shifting cultivation and instead look into ways to incorporate soil and water conservation measures into it. Shifting cultivation was covered extensively in the Shillong Declaration on shifting agriculture in 2004, and various governments in the participating nations have put it on their agenda. Many initiatives have been undertaken in India, both at the central and state government levels. The proposals have been promoted in key policy meetings and in reports, such as the Farmers' Commission Report on North-East India and a participatory forestry programme. The Ministry of Environment and Forests has established a task group on 'Rehabilitation of Shifting Cultivation Fallows.' Nagaland's state government has educated government extension workers in participatory mapping, while Tripura's government is looking for ways to start shifting cultivation development initiatives.

9.6.2 What can be done about it?

The National Forest Commission has included ICIMOD's recommendations in its advise to the Government of India, emphasising that the security of land tenure for shifting cultivators should be increased during both the crop and fallow phases. This can be accomplished by reclassifying shifting cultivation sites as agricultural land with adaptive forest management during the fallow period, rather than as shifting cultivation areas. The panel also requested that customary institutions be strengthened and capacitated in order to improve local level governance, tribe and community-based natural resource management, and tenurial access and control. Existing lending policies must be reoriented to be more sensitive and proactive in instances where common property regimes apply, and coordination among government entities with responsibility for shifting cultivation must be promoted. It also agreed to ICIMOD's recommendation to propagate medicinal plants and bamboo the North-most East's adaptable crop. The necessity for stronger regulation appears to be critical. When it comes time for cropping, farmers should be able to remove the fallows, and species selection should be geared toward soil conservation.The timber trade should be regulated so that farmers can sell their timber once it has been cleared. Existing common property regulations should be reinforced to prevent elites from within communities from capturing land and improving fallows. Overall, tenure security, as well as access to finance in instances where common property regimes apply, should be improved. Policies and programmes should be implemented that better recognise the importance of customary institutions in community resource management, fire control and forest resource protection against the timber mafia and elite capture.

A viable solution to the problem of shifting cultivation has become essential, not only for biodiversity conservation but also for the region's profitable agriculture. Rather than imposing a foreign technology from outside the region, the solution to the problem of the shortening fallow cycle is to enhance this already weakened agro-forestry system through fallow management employing appropriate tree species. Local people have access to scientifically analysed TEK (traditional ecological knowledge), which can be used to reconstruct shifting agriculture as part of a short-term strategy or as part of a larger landscape management plan. To put it another way, the transition from slash-and-burn agriculture to other methods isn't just a case of using less land to produce the same amount of product. Other changes, such as a rise in farmer demands and total output, may occur. The landscape of slash-and-burn agriculture becomes suddenly appealing. The forest frontier is pushed forward as outmigration declines and immigration rises, often with the help of government incentives. The poorest farmers, who lack the resources to pursue this road, are frequently outcompeted by the wealthiest portions of the population.

9.6.3 Improvements to shifting cultivation

As the world's population grows, so does the need for food crops. As a result, the fallow period is gradually decreased, and the acacia's soil-enriching contribution is lost. The project's long-term goal of switching from fallow fields to planting acacia in cycling rows with food crops may be hampered by a lack of outlets for extra produce. Making greater use of removed vegetation is a third way to improve shifting farming. A lot of initiatives to replace fire with other clearing procedures are based on the premise - which, as previously said, is partly inaccurate - that burning plants is wasteful. The "Subri conversion approach," named after the Subri Forest Reserve in Ghana where it was established, is one such effort. The majority of the cut vegetation is used to make sawn lumber, charcoal, fuelwood, and minor local industries like basket making and carving. The little branches and leaves that are left are used as green mulch for the agricultural produce.

There have been evolving institutional systems for land and forest management at the community level, as well as various degrees of engagement in or resistance to government schemes and programmes. Swirling change has resulted in the loss of cultures and identities, as well as their reassertion, realignment, and reinvention, all of which have crucial implications for resource access.

The principal roadblocks to enhancing shifting agriculture in the African tropics are, for the most part, the same roadblocks that stymie agricultural progress in those regions in general. The constraints listed below deal with the problem's socioeconomic components (Kamajou 1983).

Economic constraints

i) Technology that isn't up to par. Appropriate technology is now widely recognised as a requirement for long-term agricultural success.

ii) Farmer's financial resources are limited. Due to the peasant's historically low productivity, it is practically impossible for him or her to save or make any productivity-enhancing investments with personal funds.

iii) Inadequate marketing infrastructure, particularly for food producers. Marketable products must be accessible in order to create and coordinate such markets. If the peasant is to produce more for family consumption as well as for the growing urban population, both supply and demand must be adjusted at the same time. On the supply side, peasant production must rise, which may necessitate a significant amount of product specialization. On the demand side, potential urban customers must be able to get marketable surpluses at prices that are advantageous to the producers. Storage and transportation facilities for inputs and goods should be given high attention when it comes to marketing infrastructures.

Social constraints

A number of socioeconomic developments have rendered traditional shifting farming outdated.

i) Political progress and government strengthening have resulted in a transfer of power or authority from traditional leaders to government or political officials, who form a new elite.

ii) The emergence of an open economy has exacerbated the disintegration of the traditional extended family. This has an impact on the availability of family labour and the organization of family production.

iii) Rural areas have seen a significant decline in active manpower due to education and training in the form of new off-farm employment. Because the average age of peasants is growing, this migration of the young to cities affects not just the amount but also the quality of labour available to farmers.

iv) The need for hired labour is inextricably linked to this rural-urban mobility. Many peasants, however, do not wish to labour for money, regardless of how much they can earn, for socio-cultural and historical reasons (Kamajou 1983).

9.6.4. Institutional constraints

The performance of shifting cultivation may be influenced by five different types of institutional restrictions. They are as follows (Kamajou 1983):

i) Constraints relating to land tenure systems, for example. Land ownership or tenure, as the most important aspect in agricultural production systems, impacts the performance and improvement of any farming system. In tropical and sub-tropical Africa, tribal, communal, or collective land tenure systems (in all of its variations) are not favourable to individual producers making long-term investments in land, such as soil conservation or fertility-supplementing or-balancing techniques. Nonetheless, these are widely used alternatives to traditional shifting cropping.

ii) Training, research, and extension institutions. These organizations are either non-existent, understaffed (both quantitatively and qualitatively), under-equipped, or uncoordinated. It is critical to provide support and coordination for these institutions.

iii) Organizations and structures at the local or village level. For the enhancement of farm systems, local organisation at the village level is equally critical. Depending on whether villagers can organise themselves freely and easily and whether there are permanent village level production structures, shifting farming could be developed more or less quickly. These structures must emerge naturally in the hamlet and cannot be imposed from without.

iv) Institutions that manufacture, sell, and service novel agricultural inputs and farm goods. Producers will be rewarded for upgrading their farming methods and cultural practises by these entities. Because of increased market opportunities, their greater output will have a higher economic worth. The peasant will see no motive to develop farming and cropping techniques unless these institutions are established and functioning.

v) Government measures affecting the agricultural-non-agricultural income balance (such as subsidies, taxes, and agricultural input prices) have a significant impact on peasants' efforts to boost productivity. These strategies are still insufficient in most areas of Africa with shifting farming.

Improving shifting cultivation in Southeast Asia

In many Southeast Asian countries, shifting cultivation remains the economic backbone of highland populations. However, the circumstances that once supported the long-fallow rotation's long-term viability have essentially evaporated. Rapid population increase, the gazettement of remnant wildlands into protected areas, and state measures to sedentarize agriculture and prohibit the use of fallows and fire have all intensified the need to establish more permanent forms of land use. There are numerous compelling examples of shifting cultivators managing local resources to solve local concerns. Technical initiatives to stabilise and increasing the yield of shifting agriculture systems have largely failed (Cairns and Garrity 1999). Farmers' rejection of researcher-driven solutions has resulted in a clearer understanding of their limitations. This experience highlighted the importance of using participatory, on-farm research to find solutions. The task is to document and assess indigenous techniques for shifting agriculture intensification through a research and development process. Identification of promising indigenous practices, characterization of the practices, validation of the practice's utility for other communities, extrapolation to other areas, verification with key farmers, and wide-scale extension are all part of this process.

9.6.5 Indigenous strategies in a regional context

Mountainous terrain dominates much of Southeast Asia, which is home to a broad range of cultural minority communities. Because of the vast forests and limited people, these mountain villages were able to adopt a variety of shifting agriculture techniques, allowing them to survive in relative harmony with their surroundings (Cairns and Garrity 1999). However, the annual cycle of slashing and burning that characterises land preparation in shifting agriculture systems has long been criticised as wasteful and a major contributor to tropical deforestation. The relationship between foresters and shifting cultivators has been marked by mutual animosity, deep suspicion, and open conflict. Finding strategies to prevent shifting cultivation and encourage the development of more permanent forms of agriculture remains a top concern for Southeast Asian governments. Detailed anthropological investigations, beginning with Conklin's work in the Philippines in 1954 and 1957, led to a far more favourable view of shifting agriculture. They gave convincing evidence that it is a logical agricultural system in the context of the limits and opportunities that come with remote upland terrain, and they cited its lengthy history as proof of its long-term viability. Shifting cultivation, they said, is a land-use technique that reflects; (i) indigenous knowledge gained through centuries of trial and error, (ii) a nuanced balance between product yield and ecological resilience, and (iii) an outstanding level of agrodiversity. The custodial role often played by moving farming communities in sustaining forest ecosystems and wild species, as well as the intimate links between ecological and cultural variety, has been highlighted in recent work (for example, De Foresta and Michon, 1997). This shows that attempts to conserve biodiversity will be ineffective unless they are broadened to include cultural conservation. It seems unlikely that these opposing viewpoints will be harmonised anytime soon. To go forward in identifying research and development interventions that can stabilise declining upland agroecosystems and enhance the living conditions of the most marginalised upland communities, we need to reframe the argument. Regardless of one's feelings about the virtues or drawbacks of shifting cultivation as a land-use technique, there is widespread agreement that a human and environmental tragedy is unfolding in Southeast Asia's uplands (Cairns and Garrity 1999). Farmers can no longer afford the luxury of protracted fallow seasons that allow secondary forest recovery and rejuvenation of tired soils in the face of rising land-use constraints. Shortening fallows, longer cropping periods, and rapidly deteriorating habitats are all too common in Southeast Asia's highland regions. Crop yields are decreasing, weed control work is growing, and household food security is in jeopardy.

The overall goal of the project is to (i) document and evaluate indigenous strategies for shifting cultivation intensification in Southeast Asia, and (ii) strengthen regional institutions' and researchers' capacity to examine and illustrate the role of indigenous knowledge and innovations in improved agricultural land management. This indigenous fallow management (IFM) study is part of ICRAF's larger Alternatives to Slash-and-Burn (ASB) initiative, which links it to global efforts to mitigate the consequences of degraded shifting agriculture systems.

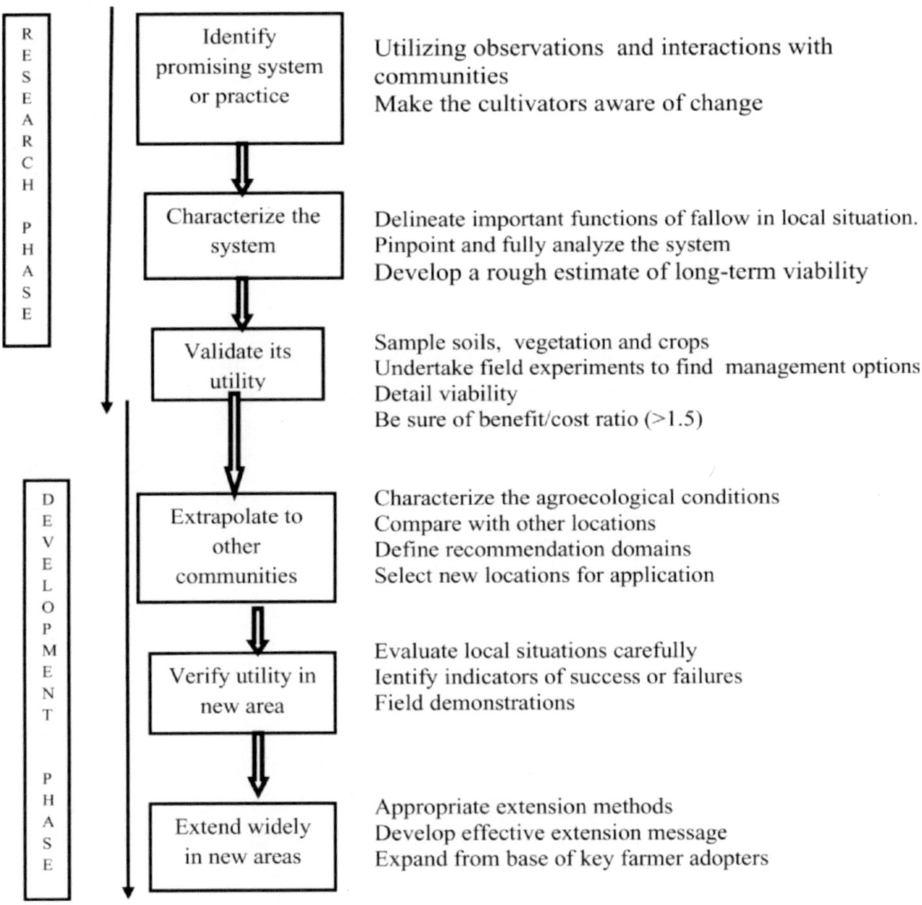

Fig. 9.5. Process for research and development on indigenous fallow management Systems (Improved over Cairns and Garrity, 1999).

9.6.6 Constraints in improving shifting cultivation

Land intensification is a historical fact, and land use in shifting cultivated zones is altering as well in response to local conditions (typically in the direction from long-cycle fallows to continuous, annual farming). As a result, each level in this historical land use intensification's evolutionary phase is only brief. Because shifting cultivation cycles take such a long time to complete, there may be several transitory periods, and different populations and locales may be at different stages of evolution (e.g. composite shifting agriculture). It's also worth noting that a single fallow management system innovation may be relevant/beneficial at one stage of this evolutionary process but not at others. As a result, the adoption and replication of fallow management improvements is challenging.

In many Southeast Asian countries, shifting cultivation remains the economic backbone of highland populations. However, the circumstances that once supported the long-fallow rotation's long-term viability have essentially evaporated. Rapid population increase, the gazettement of remaining wildlands into protected areas, and state measures to sedentarize agriculture and restrict the use of fallows and fire have all intensified the need to establish more permanent forms of land use (Cairns and Garrity 1999). There are numerous compelling examples of shifting cultivators managing local resources to solve local concerns. Technical methods to stabilising and boosting shifting agriculture systems' productivity have been largely unsuccessful. Farmers' rejection of researcher-driven solutions has resulted in a clearer understanding of their limitations. This experience highlighted the importance of using participatory, on-farm research to find solutions. The task is to document and assess indigenous techniques for shifting agriculture intensification through a research and development process. Identification of promising indigenous practices, characterization of the practices, validation of the practice's utility for other communities, extrapolation to other areas, verification with key farmers, and wide-scale extension are all part of this process (Cairns and Garrity 1999).

9.6.7 An alternative approach

In addition to the improvement of shifting agriculture, an alternative approach may be thought upon to bring about cropping on a sustainable basis in the hilly terrain of the NE region and the approach should be based on the participatory mode in order to avoid any sort of contradictions and resistance from the participating farmers.

Introduction of Plantation and Horticultural Crops

A shift to horticulture/plantation crops along with forestry would ensure plant cover which might, perhaps, be the ideal alternative to jhum in the hilly terrain. The north eastern region of India offers much scope for a variety of tropical and temperate fruits. Plantation and horticultural crops such as rubber, coffee, tea, banana, citrus, black pepper, cashew, spice trees, pineapple, and others can be planted in jhum fields in sloping terrains. In this regards, first step should, therefore, be of determining with degrees of firmness the kind of plantation that would be taken up in the location specific area. During the establishment of plantation crops, immediate plan for resettlement of jhumias on permanent terraced cultivation can be achieved through providing them sufficient employment for their subsistence. Action should be taken simultaneously to discourage whatever possible to prohibit the practice of the shifting agriculture in the area. The success of the strategy would depend on whether the plantation could provide sustainablity, sufficient income to the families to buy their own food and sufficient surplus for better economic conditions. Locals can achieve this without breaking their traditions by receiving a fair portion of revenues after processing and marketing. Tree and bush crops would represent a relatively high replacement of original forest land put under jhum and can sustain a large number of people for whom food may be purchased without loss of land resources.

Fig. 9.6. Terrace cultivation with a water harvesting source (See colour version on page 654)

Encouragement of Livestock Farming

Dairy based farming is an alternative to jhum cultivator, living nearby road heads to improve the deteriorating fragile hill ecosystem and is, therefore, one of the best utilization of wastelands. The land can be utilized for silvi-pasture and pasture to meet the feed and fodder requirement of reared animals. Now many people are engaged in dairy farming particularly those residing nearby road heads with the help of Governments. This would not only overcome the problems of land degradation and overgrazing but would meet the need of fodder, fuelwood, and timber. The characteristics of the silvi-pasture should be with the objective of fast growth in poor soil, good under growth and over ground biomass production, palatable to animals, good nutritive value, and can survive in water deficient winter months. Fodder trees around the field all along the stream banks and drainage channels, on field bunds and stony patches can provide fodder during winter season by partial lopping of the trees. Soil fertility gets improved through recycling of nutrients via animal manures. The stall feeding of animal can check (Ives and Messerli, 1989) the environmental deterioration through various ways. They are: 1) rotational grazing is easy to manage 2) young tree seedlings are not killed 3) All excrement is dumped on arable land 4). (Livestock parasite burdens are lower 5). There is less damage to standing crops and terrace risers. The usage of trees, which supply green leaf fodder during the dry season and have the extra benefit of generating lumber and fuelwood as a by-product, should be part of the rehabilitation programme in the interest of effective livestock feeding. Farmers can keep animals for milk and meat to create financial income, as well as excreta to recycle nutrients lost from the ground, by using contour grass strips on mildly sloping farms with fodder trees.

Proper land use planning on watershed basis

The best way to plan the development of an area is to use natural drainage units called 'watersheds,' with the goal of developing resources in such a way that they provide maximum benefits to people while maintaining ecological balance through long-term efforts and commitments, such as infrastructure maintenance, land, water, and forest resource protection, and so on. The following goals should have been set in order to implement land use planning in catchments for hilly terrains:

- Optimization of production from agriculture, forests, plantations, mixed farming systems, and other sources for self-sufficiency in basic necessities on a long-term yield basis.
- Land degradation must be controlled in order to maximise primary production potential.
- Wasteland development for economic biomass production
- Exploitation of key mineral resources with suitable planning for mined areas rehabilitation.
- Reduced run-off and sedimentation allows for more efficient use of perennial water supplies.
- To provide the security of food, fodder, fibre, fuel, and timber, among other things.
- Preservation of the region's scenic splendour, natural vegetation, wildlife and birds.The modification of indigenous technological knowledge (ITK) for the benefit of local inhabitants.

Many watershed initiatives also include the protection of natural ecosystems, scenic regions, and wildlife habitat. Human interference should be limited to the selection of a crop, livestock, or forest type based on the climate, soil, landform, and hydrology of an area, according to accepted practise. Soil survey and geological information system (GIS) data can be used to determine the appropriateness of soil and land for various uses based on household and community demands for a sustainable hill environment.

9.6.8 Cooperative programme for improving shifting cultivation systems

Before crops to be successfully grown again in some areas of West Africa's humid tropics, at least five years of fallow are required (Nye and Greenland, 1959). Of course, this changes depending on the soil type and ecological zone. Fallow periods have even been cut to fewer than two years in some locations. Soil erosion, loss of fertility, reduced yields, rampant weed growth pest infestation, and other issues beset farmers in such conditions. As a result many sections of the world are in desperate need of improved or more efficient systems for long-term yields that are scientifically solid, economically scale farmers, who grow the majority of the food in developing countries, is extremely crucial. Large-scale mechanised farms, on the other hand, should not be overlooked, despite receiving significant support in the past and producing poor outcomes in places with shifting crops. These issues can't be solved in pieces. Recognizing this, delegates at the FAO's informal meeting on shifting cultivation in February 1983 determined that an international initiative including a variety of organizations would be the most promising solution. They came up with this Cooperative Program as a result of their discussions.

Programme Goals

A initiative to develop more efficient, continuous production technologies that could replace shifting farming has three overall goals. Continuous production systems, on the other hand, are superior to more traditional methods only provided they receive the greater levels of inputs that they require to function.

1. To accommodate population increase and modernization, systems should strengthen their capability for shifting agriculture and related intermittent production systems.
2. Farming systems in emerging nations, particularly small-scale farmers should be suitable. This is made possible by a diverse research strategy.
3. Systems should be environmentally friendly while still meeting the needs of farmers for food and income, which they need to buy other goods and services.

9.6.9 Horticulture for shifting cultivation improvement

The age-old methods of Assam's Bari (Backyard farming) system, Nagaland's Zabo system, and Arunachal Pradesh's Apatani system are some of the most visible instances of effective bioresource exploitation in farming systems mode. Based on the underlying ecology, the vast agro-climatic variance from subtropical to alpine provides considerable scope for cultivating various horticulture crops with little modification in abandoned jhum lands and current jhum fields. One of the potential solutions is to develop a horticultural-based agricultural system in various topography, soil, and environmental circumstances (Jha *et al.* 2017). As a result, horticulture can be a very realistic choice for shifting cultivation development and rehabilitation that is both environmentally and economically sustainable. Horticulture can offer the much-needed finance with which land intensification will be done in many circumstances, in addition to increasing the value of the fallow or shifting to settled cultivation. Despite the fact that sufficient efforts have been made to control soil/land deterioration, the intended results have yet to be achieved. The region's altitudinal, climatic, and socio-ecological diversity allowed for the growth of a wide range of forest species, cereals, and horticultural crops such as fruits, vegetables, spices, flowers, and a wide range of foliage ornamentals in various agro-climatic zones. The NE Region covers 1.37 million hectares and produces 11.50 million tonnes of horticultural crops, accounting for 5.66 percent and 4.15 percent of national area and production, respectively (NHB, 2015).

Although the region is abundant in natural resources and biodiversity, its exploitation and extraction, combined with insufficient replenishment, has become a source of concern for their long-term viability. The degradation process has been accelerated by large-scale interference caused by shifting farming and growing population pressure. The need of the hour is to develop a horticultural-based agricultural system that can operate in a variety of topographic, soil, and environmental circumstances. The social, economic, and cultural traditions of the shifting cultivators must be considered while developing such programmes (Jha *et al.* 2017). Furthermore, approaches and practices that can increase production and productivity while providing consistent revenue at the farm level should be prioritised. Horticulture has supplied the much-needed finance for further intensification of lands, in addition to increasing the value of the fallow or shifting to settled cultivation. Horticulture has the potential to play a critical role in jhum development and rehabilitation for long-term ecological and economic viability.

9.6.10 Policies of government of LAO on shifting cultivation

Although the government's policy on shifting cultivation has remained centered on stabilization and eventual eradication, the techniques and processes by which the policy is executed have become more moderate over the previous two decades, as evidenced by numerous policy documents. Kenney-Lazar (2013) identifies four significant ways in which this has occurred:

1. The government's top priority in upland development have shifted away from shifting cultivation reduction and toward guaranteeing food security and economic growth. These are now widely regarded as the two most critical upland development objectives.
2. Some forms of shifting cultivation are permissible, at least in the short term. As a result of low population density and available secondary forest, 'rotational upland agriculture' is an acceptable kind of swiddening, in which farmers rotate in a cycle within set bounds of agricultural land, providing adequate time for appropriate fallow regrowth. Although this is not considered the best method of cultivation, it is accepted when other options are unavailable, yet it is still only a temporary solution. Due to a misunderstanding of English and Lao words, GOL officials frequently interpret the term "shifting cultivation" to refer to the pioneering form of shifting agriculture.
3. The government has placed a greater emphasis on attaining the goal of stabilising shifting cultivation before completely eliminating it. While the exact meaning of the term 'stabilization' is unclear, it can be thought of as a stage that falls between between unfettered swiddening and total eradication, and is most likely considered as a step toward the latter. It's most likely some kind of restricted, sedentarized, and limited shifting cultivation within agreed-upon zones.
4. The government is increasingly aware that policies relating to shifting cultivation must be implemented in a pluralistic manner to account for regional differences in agro-ecological and socio-economic issues. Shifting cultivation has become widely recognised, partially as a result of several research and conference conclusions over the years, that it is more appropriate in some regions than others. Local biophysical and socioeconomic elements, such as population density, ecological circumstances, cultural practices, and potential alternative livelihoods, should influence policies.

9.6.11 The Forestry Law of 2007

The Forestry Law has been updated since its first publication in 1996. The types of forest classifications it establishes and how they relate to shifting cultivation are the most interesting aspects of the law. The forest classifications of regeneration forest and degraded forestland, as defined in Article 3, Explanation of Terms, are particularly applicable. It is defined as a forest that can deteriorate into an old fallow forest before being transformed into a rich natural forest. Degraded forestland, on the other hand, is made up of regions that have been significantly and repeatedly harmed to the point where they are unable to regenerate naturally into a lush forest (Kenney-Lazar).

9.6.12 Fallows improvement

A site specific diagnosis is thus needed to evaluate the nature of decline in crop productivity and thereafter to find out the type of fallow. Crop/fallow rotation management should establish a balance between crop exploitation and soil fertility restoration during the fallow time. Determinants of biophysical limits to improved fallow depending on the main constraints to crop production under the given soil and climatic conditions are given in Table (Van Noordwijk 1999). Functions of fallow would augment the soil fertility during the cropping period (N and P supply, cation supply, soil acidity, and soil organic matter), improve soil physical structure (which in turn interacts with soil fertility), reduce the incidence of soil borne pests and restrict the weed growth ((Buresh and Cooper, 1999). The effectiveness of fallow in overcoming these constraints depends on environmental factors, such as climate, soils, and landscape position, as well as fallow species and its duration. Important species selection criteria based on fallow function may include: rooting patterns, mycorrhizae nitrogen-fixing as well as nutrient scavenging ability, allelopathic potential, resistance/antagonism to and ability to harbour pests, rapidity to establishment of leaf area, the quality and quantity of litter production, and water relations. However, high economically productive fallow may be compromised in their ability to restore fertility if nutrient removal in harvested product is more. Indigenous knowledge for improving the soil fertility and eagerness of the farmers to find a substitute for nutrient inputs can contribute greatly to their enthusiasm for improved fallow. Table 9.4. shows determinants of biophysical limits to improved fallow.

Table 9.4. Determinants of biophysical limits to improved fallow

Constraints to crop yield	Functions of the fallow	Determinants of biophysical limits to fallow 'functions'	Technical alternatives
N supply	N2 fixation Recover N leached into subsoil during cropping period Capture on N from subsurface lateral flow	Occurrence of effective Rhizobium strains Soil depth, rainfall, hydrology, nitrification, apparent adsorption constant As 'b' and slope	a-c) N fertilizer
P supply	Increase soil organic P-based on P capture by fallow species from occluded soil P sources Decreased P sorption during crop phase- due to organic acids from fallow residues Release of organically bound P during fire used for clearing of fallow vegetation Facilitate mycorrhiza development	P capture strategy (root exudates, mycorrhiza) of fallow vegetation, presence of occluded soil P fractions, soil mineralogy Soil P sorption characteristics Changes in organic P pools, fire temperatures Occurrence of effective mycorrhizal infection in crop	a-c) P fertilizer D) vesicular-arbuscular mycorrhiza inoculation
Cation supply and acidity	Recycle cations from soil depth Reduce acidity-based on interactions of organic acid from fallow residues on soil Al	Recharge of subsoil cation content, leaching, lateral flow Chemical composition of fallow residues interacting with Al	Lime, dolomite K fertilizer
Nematodes and other soil-borne pests	'Fool' propagules to reduce their infection potential Stimulation of antagonists	Specific biological properties of soil and fallow species	Nematicides, crop rotations
Weeds including parasites	'Fool' propagules into germination (e.g., Striga seeds) Reduce vigor of propagules (e.g., Imperate rhizomes)	Effectiveness of root exudates from fallow vegetation Shade duration and intensity	Herbicides
Soil structure affecting infiltration	Create old tree root channels Facilitate soil biota- through inputs of organic materials	Rooting pattern and fine root turnover of fallow vegetation Specific effects of fallow litterfall on soil fauna	Use of crop residues, soil tillage

Source: Van Noordwijk, 1999.

Effective resource management

Any farming system requires effective resource management. Crop mechanisation, for example, is expensive in tropical Africa, but with the right labor-intensive equipment, it can considerably enhance yields. Crop rotation may partially replace shifting cultivation throughout the automation phase. Land is used more intensely here, and management is improved. Improved water control and land development, for example, are necessary for effective mechanisation.

Extension and training

Adult literacy programmes, which can improve farmers' labour performance, can lessen the risks of shifting crops. Farmers need more information about shifting agriculture, which will require more research and extension activity. The sequential operations involved in shifting cultivation, the challenges at each stage, and remedial methods to alleviate existing or emerging problems will all need to be documented here. Simultaneously, Sierra Leone's land tenure structure requires practical and realistic reforms to accommodate shifting cultivation, as well as alternative techniques to replace it in the future.

Shifting cultivators and transmigrants are two types of Indonesians who reside in less developed areas and are unable to improve their life without support from the outside world. However, there are numerous distinctions between the two groups (Hardjosoediro 1983). A comprehensive overview of each of them will help to identify the type of institution that would be most beneficial to enhancing their farming operations. Furthermore, the efforts made in Indonesia to date to build institutions capable of addressing this important challenge are briefly summarised below.

Despite constraints of remoteness and land tenure, the following strategies may be useful to improve shifting cultivation cultivation.

9.6.13 Institutional transformations

The following institutional transformation is required for improvement of shifting cultivation:

- The scholars, state governments and other stakeholders involved in shifting cultivation should strongly create awareness on "shifting cultivation improvement" from the present understanding of "shifting cultivation as destructive" form of shifting cultivation.
- The shifting cultivated land may be transformed from liabilities to assets, because shifting cultivators lives resolve around the shifting cultivation.
- Frequent interface meeting of the line departments-research institutions-NGOs-Stakeholders to take stock of the situation and discuss various options/development programmes related to shifting cultivation.
- The schemes which have existed for shifting cultivation has an agenda of shifting cultivation eradication than livelihood improvement.

Hence, a suitable scheme in a mission mode should be implemented for the livelihood improvement of shifting cultivators and their shifting cultivation based on the location specific needs and willingness of the community members.

- The customary laws of the community govern the ownership of land and individual rights with constitutional sanction. Reforms with regard to allocation of ownership of land are needed through introduction of new government policy. While doing so, sentiments of tribal community should be given priority.
- There is an urgent need for evolution of proper strategic plans and policies based on reliable and sound data based upon land records to find compatible land uses for shifting cultivation land management.
- Participatory land use planning and fallow management is considered essential for the long term ecological sustainability of shifting cultivation lands, especially in areas where soil erosion has led to decline in crop yield threatening local food security.
- Shifting cultivation continues to be addressed by adhoc schemes and programmes with serious limitation because of both financial coverage and technical capacity. Schemes/ programmes should focus integrated vision about the ecological viability of the options and alternatives and their co-relations with the food security and cultural ethos.
- Access to better market opportunities and accompanied infrastructure such as cold storage facility should be planned to promote entrepreneurship as an alternative to subsistence farming.

9.6.14 Constraints in weaning away cultivators from shifting cultivation

The negative environmental effects of shifting farming are well-known scientific findings. These results are based on scientific facts and study carried out all around the world, especially in north-east India (FAO, 1984; Tawnenga *et al.* 1997). To address the issues of shifting cultivation, the Indian government established two task groups, one in 1983 and the other in 2001 both in the Ministry of Environment and Forests. In addition to calculating the area under jhum or shifting agriculture in the country, the task forces proposed a long-term strategy for managing shifting cultivation that was comprehensive and integrated.

A - Socio-economic

i) The environment, beliefs and other socio-economic conditions of the people are not totally similar and, so, it is difficult to generalize the same for all the areas. Yet, there are many things are common, to enable some degree of common issues, challenges, constraints and threats. All operations of shifting cultivation are associated with some religious rites, festivals and festivities. For example, in Garo hills of Meghalaya state of India, various festivals namely; galmaka, miamla, rongchugala, hia and wangala, are associated with various operations of shifting cultivation (Majumdar 1976). Any abrupt deviation from shifting cultivation will prevent the cultivators from these celebrations. However, this does not stand as a major disincentive for a change over to other forms of cultivation.

ii) Sense of ownership As the land mostly belongs to the community of headman of the village, the farmer allotees of the developed land do not have a sense of ownership or proprietary over the land they are cultivating. Due to the deprivation of sense of ownership, the cultivators do not feel a sense of reposibility towars land, thereby leading to soil erosion and decline in soil fertility.

iii) Favouritism in land distribution In many cases, the farmers complain of allotment of degraded land to them while influential farmers and relatives get highly fetile and productive plots. These people get a preferential treatment.

iv) Cropping patterns Shifting cultivation is generally a mixed cropping system of agriculture. A mixture of different crops seed is sown in the field. The proportion of different types of seeds is according to the requirement of the family. From the field, a farmer can get maize, paddy, millets, chillies, melon, tapioca, pumpkin, cotton etc., sown in different compatible mixture which satisfy simple need of the family. On the other hand, in settled ccultivation, the pattern is generally a monoculture where one or at the most two crops are sown at one point of time. The farmer may get more of a particular crop but kinds are limited.

v) Socio-cultural and religious functions Certain social, cultural and even religious functions are associated with the practice of shifting cultivation. Many socio-cultural-semi religious functions are held according to the season of agricultural operations. In Garo hills, (a) Agalmaka, is held before burning of vegetation, (ii) Rongchugala, before harvesting, (iii) Ahaia, after harvesting and, (iv) Wangala, after completion of all works in shifting cultivation. Again, in weeding also, certain local names have been closely associated with the time of weeding, such as, first weeding – Ajakra, second weeding as, Bamil, third weeding as, Diri, fourth weeding as, Rata and fifth weeding as Murat.

vi) Shifting cultivation as part of life The practice of shifting cultivation is being followed since time immemorial. Many facets of its operatios have become part and parcel of the life of the shifting cultivators. It is deeply and very strongly inter-twined with various socio-cultural rituals of the people. The system has travelled through many generations and is deeply rooted and has imbibed in life structure of the people. To wean away the shifting cultivators from this deep bond is simply unthinkable. It may take time and constant persuation and proper and adequate guidance to convince the people of the new ideas and gradually wean them away from the practice.

B-Technological

i) Low technological absorption capacity The shifting cultivators being poor, their level of absorption of new systems of agriculture or progressive scientific and technological innovations or changes is still in poor stage. Inspite of many successful demonstrations it takes a very long time for them to be fully convinced and to adopt to new changes.

ii) Soil preparation In shifting cultivation, practically no soil preparation is done. On the other hand, in terraced cultivation, lot of soil working is required for optimum crop production. To traditional and chronic shifting cultivators, this would mean additional inputs of money, labour and energy. The farmer is not ready to change so easily.

iii) Infrastructural constraints The primary factors governing the shifting cultivation control programmeis the availability of suitable land for permanent cultivation, roads as well as suitable marketing facilities.

iv) Lack of diversification enterprises All areas are not suitable for all crops. Rice being the staple food of the people of Northeast India, where shifting cultivation is the mainstay, the farmers are more inclind towards growing of rice crop for food security. The cultivators are bound to grow rice even in those areas where other kind of crops are more suitable and highly paying like, some horticultural and cash crops. The feeling of security from being able to produce their own food, even in a very subsistence way, has in majority of cases made farmers reluctant to switch over to other forms or types of cultivation, even more profitable.

9.6.15 The transmigrants

The majority of transmigrants are landless peasants from Indonesia's most densely populated islands, Java, Bali, and Lombok (Hardjosoedirо 1983). They have some experience with soil tilling and crop cultivation. They live in communities that are already market-oriented. They have limited land and are untrained in fields other than subsistence agriculture. With the rising rate of population increase, there is even more competition for land, resulting in a large percentage of people living on a bare subsistence level. They do, however, have a better agricultural potential. If only the landless peasants of Java, Bali, and Lombok had adequate land, they could expand their productivity. Currently, the only way to achieve this is to transmigrate to the less heavily inhabited islands, where there is still land suitable for farming. Institutions tasked with implementing and supporting transmigration schemes have well defined responsibilities. They must encourage individuals to migrate by preparing them to leave their birthplaces, assisting them in adjusting to new circumstances, and raising their living standards. Land allocation and clearing in new areas must be completed, as well as the construction of houses, road systems, and transportation infrastructure. Furthermore, during the first year, these institutions must give top attention to food supply; in the following four years, other programmes must be introduced before they are given to local government authorities, who will be responsible for their ongoing development (Hardjosoediro 1983). In Indonesia, the transmigration program's efforts are given top attention. As a result, the programme has widespread public support as well as support from all government departments, and it has been a huge success.

9.6.16 Shifting cultivators

Shifting cultivators, unlike transmigrants, come in a variety of forms. They can be divided into stages based on where they are in their growth. Unfortunately, most persons dealing with the matter do not understand the problem or the many classes, as evidenced by the usual phrase for them, "wild shifting cultivators." The most common interpretation of shifting cultivation focuses on its environmentally harmful and primitive technique of food production, its remote, inaccessible location, the growers' low living standards, their "unwillingness" to follow regulations, and their traditional way of life. Unfortunately, many people, including policymakers, assume that these difficulties are mostly owing to the growers' lack of education. Furthermore, some anthropologists advocate for the cultivators' way of life to be preserved. The modern world, according to anthropologists, is responsible for upsetting the balance between man and nature through environmental exploitation. The cost of those projects, as well as the achievements obtained thus far, have recently been heavily criticised. Any institution tasked with improving a shifting cultivation system should be modelled after the relevant shifting cultivation's primary qualities. The approach for research and development on improved fallow systems involves (1) characterization of promising IFM (indigenous fallow management) systems, (2) definition of their extrapolation domain and generation of preliminary extension materials, (3) design and setup of agronomic trials to validate farmers' perceptions of these systems' superiority and test technological refinement, (4) extrapolation through adaptive trials in other locations, and (5) generation and dissemination of latest extension materials (ICRAF,1997; Cairns and Garrity, 1999). These characteristics must be evaluated, and special attention paid to them.

9.6.17 Improved Fallows

The soil nutrient depletion process which continues through cropping phase under shifting cultivation also continues through the fallow phase up to about 5-10 years. So, 10-years shifting cycle is critical from the point of view of sustainable cropping under this system. Under short term shifting cycle (5 years or less), the forest completely loses its ability to regenerate , thus creating a desert like situation (Ramakrishnan, 1992). The first 4-5 years of secondary succession is dominated by weeds like Eupatorium odoratum. *Imperata cylindrical, Succharum spontaneum* and *Borreria hispida* (Ramakrishnan and Toky, 1981). These are soon replaced by seedlings and sprouts from underground rhizomes of bamboo (*D. hamiltonii*) reaching its dominance between 10-20 years. Only shade tolerant species are present along with bamboos but none is able to achieve dominance in the area. Indeed, traditional shifting cultivation with longer cycles of 10 years or more is closer to natural ecosystems where nutrient cycling and soil fertility are maintained properly. Short-term weedy species fallow periods (3-5 years) not only result in poor soil fertility build-up, but also pose a severe threat to the cropping phase due to heavy weed infestation. Improved fallow with woody and herbaceous legumes can help to restore and maintain soil fertility quickly. *Sesbania, Tephrosia, Leucaena, Mucuna, Centrosema, Pueraria, Crotolaria, Cajanus, Indigofera,* and *Mimosa* are the principal legumes that can be successfully employed as a short fallow to replenish soil fertility lost during cropping by shifting agriculture (Sanchez, 1999). Furthermore, fallow systems may have the added benefit of reducing weed damage to crops. Herbaceous cultivated fallow species need less time to establish and recover fertility than woody species, but they do not produce the same secondary products (such as wood for building and fuel). Planting an upgraded fallow, on the other hand, differs from planting a conventional jhum and may not be culturally acceptable. Litter fall from nitrogen-fixing tree species can offer carbon to soil microbes as an energy source, improving nutrient recycling and soil physical characteristics.

A site specific diagnosis is thus needed to evaluate the nature of decline in crop productivity and thereafter to find out the type of fallow. Crop/fallow rotation management should establish a balance between crop exploitation and soil fertility restoration during the fallow time. Functions of fallow would augment the soil fertility during the cropping period (N and P supply, cation supply, soil acidity, and soil organic matter), improve soil physical structure (which in turn interacts with soil fertility), reduce the incidence of soil borne pests and restrict the weed growth ((Buresh and Cooper, 1999).

9.6.18 Economic drawbacks to shifting cultivation

Shifting cultivation has a number of economic drawbacks in addition to other agronomic drawbacks. Some are built into the system, while others are imposed by the rapid socioeconomic changes impacting areas under shifting agriculture. The following are some of the inherent disadvantages (Kamajou 1983):

i) Shifting cultivation's poor compensation in relation to its labour requirements and the shifting cultivator's labour supply. It's also low since there are no markets for shifting growers' produce, so they can't get a good price for it.

ii) Shifting cultivation's large and systematic destruction of forests and forest products, as well as the degradation of forest soils. This devastation represents a massive waste of important resources.

iii) Shifting cultivation is characterized by low investment capacities. This is due to inadequate pay, which makes all investments undesirable economically, resulting in low production (thus completing a vicious circle).

The drawbacks that diverse socio-economic and institutional changes have imposed on shifting agriculture are related to two phenomena: expanding population and a growing desire for cash revenue. During the 1970s, Africa's population (particularly in Sub-Saharan Africa) increased by 2.7 percent, while per capita food output fell by roughly 10%. The rate of urbanisation has been around 6% to 8% on average. With such rapid population growth, it is becoming increasingly difficult for peasants to adhere to the appropriate fallow season, expand plot size, or relocate to new locations. Traditional farming methods, due to their low yield, are unable to expand peasant consumption above subsistence levels or meet new cash-based needs. Today's shifting cultivators require an increasing amount of money to purchase new commodities and services that are not generated by the family, such as transistor radios, gas lighting, sugar, schools, medical bills, security, and so on. Traditional shifting farming is becoming more difficult for peasants to practise while producing the marketable surplus required to meet these new needs.

9.6.19 Institutional constraints

The performance of shifting cultivation may be influenced by five different types of institutional restrictions. They are as follows (Kamajou 1983):

i) Constraints relating to land tenure systems, for example. Land ownership or tenure, as the most important aspect in agricultural production systems, impacts the performance and improvement of any farming system. In tropical and sub-tropical Africa, tribal, communal, or collective land tenure systems (in all of its variations) are not favourable to individual producers making long-term investments in land, such as soil conservation or fertility-supplementing or-balancing techniques. Nonetheless, these are widely used alternatives to traditional shifting cropping.

ii) Training, research, and extension institutions. These organizations are either non-existent, understaffed (both quantitatively and qualitatively), under-equipped, or uncoordinated. It is critical to provide support and coordination for these entities.

iii) Organizations and structures at the local or village level. For the enhancement of farm systems, local organization at the village level is equally critical. Depending on whether villagers can organize themselves freely and easily and whether there are permanent village level production structures, shifting farming could be developed more or less quickly. These structures must emerge naturally in the hamlet and cannot be imposed from without.

iv) Institutions that manufacture, sell, and service novel agricultural inputs and farm goods. Producers will be rewarded for upgrading their farming methods and cultural practices by these entities. Because of increased market opportunities, their greater output will have a higher economic worth. The peasant will see no motive to develop farming and cropping techniques unless these institutions are established and functioning.

v) Government policies affecting the agricultural-non-agricultural income balance (such as subsidies, taxes, and agricultural input prices) have a significant impact on peasants' efforts to boost productivity. These strategies are still insufficient in most areas of Africa with shifting farming.

9.7 PROGRAMMES TAKEN UP FOR DEVELOPMENT OF SHIFTING CULTIVATION

9.7.1 Northwest Brazil integrated development programme

In Amazonia, the Brazilian government has worked to encourage perennial crop-based agricultural systems that are sustainable. In 1981, it established the Northwest Brazil Integrated Development Program (Polonoroeste), which covered the entire state of Rondonia as well as the western section of Mato Grosso. The goal was to promote tree crop farming while reducing forest removal on areas with limited long-term promise. These efforts have failed to decrease deforestation or change colonial farming practices. The early 1980s saw a drop in cash available for credit and subsidised inputs as a result of macroeconomic austerity measures. Farmers were hesitant to engage in perennial crops when finance was available because the subsidies offered were insufficient to cover the risks connected with the cultivation of new crops in which the new.

Colonist had little or no experience (Mahar, 1991). Where inflation is high and tenure is precarious, the long-term investment required for sustainable land use is unappealing, and short-term profit seeking has become the prevailing option (Anderson, 1990). The ultimate result has been the abandoning of land after years of farming or the selling of such property to cattle ranchers, as well as a migration of colonists to the frontier.

9.7.2 Research priorities

Further research is needed to develop a good design for sustainable agriculture in Latin America's tropical rainforests based on the variables listed above. Below is a list of the major research areas.

1. As interaction with foreigners develops, there is a risk that the indigenous inhabitants' agricultural systems will vanish in the not-too-distant future. Although market-oriented systems derived from Amerindian swidden management techniques are not subject to the same challenges as traditional practices and are likely to persist, they are modifications of traditional practices. It is also critical that more research on indigenous management methods be commissioned while this is still viable.
2. Microeconomic study is required to determine resource requirements and productivity, particularly for land and labour, in both Amerindian and derivative production systems. Improved performance and replication require research that identifies the socio-economic and biological limits.
3. More research is needed to determine the market potential for traditional agroforestry products and to examine the impacts of market development; for example, will market development result in large-scale forest clearance? There is little reason to believe that any Latin American farmer would refuse to take advantage of market prospects.
4. There is a need for more formal and participatory research in the subject of sustainable agroforestry. The Overseas Development Administration

is already addressing this issue with its 'Flood Plain Forest Ecology and Management Project, Belem' and the 'Tocantins Forest Management and Rural Development Project,' though the latter does not explicitly intend to use indigenous technical knowledge as part of its goal to encourage @ sustainable farming systems. The demand is so great, though, that many more projects like this may be established across the Amazon region and Latin America as a whole.

9.7.3 Impact of government policies on landscape changes

During the 1980s, the entire western region was seeing considerable changes in farming methods (Castella *et al.* 2013). The landscape was dominated by the mosaics 'Swidden 5' and 'Forest fallows,' both of which had been fallow for more than ten years, while some pioneering shifting cultivators had begun to open up the dense forest of the Phou Loey mountain group to the east. Agricultural systems and landscape mosaics largely remained intact during the 1990s, with agriculture steadily expanding into woodland areas to the east. Long fallow systems had covered the whole area suitable for shifting cultivation by the end of the 1990s (only isolated and sharp relief areas remained densely forested) (Castella *et al.* 2013).

9.7.4 Factors responsible for the failure of shifting cultivation rehabilitation scheme

Although the Jhumia rehabilitation plan has partially fulfilled its purpose of resettling shifting cultivators in alternative livelihoods, indigenous people continue to engage in shitting cultivation in the hilly areas of Tripura state. The following are the facts that are accountable for this:

- The Jhumias were not involved in the conceptualization or formulation of this strategy because the government made this decision without knowing their interests.
- The system has not been tested as pilot projects, which is required prior to implementation.
- The target population's successful engagement in order to develop a sense of belongingness has not been done correctly.
- Poor research back-up.
- Less integrated aapproach
- Lack of harmonization.
- Lack of popularization of alternative livelihood approach for eradication of poverty.

References

Achard, F., Eva, H. D., Stibig, H.-J., Mayaux, P., Gallego, J., Richards, T., and Malingreau, J. P. (2002). Determination of Deforestation Rates of the World's Humid Tropical Forests. Science 297: 999–1002. doi:10.1126/science.1070656.

Adams, M. (2003). Land Tenure Policy and Practice in Zambia: Issues Relating to the Development of the Agriculture Sector, Draft 2003. Mokoro Ltd, Oxford.

Adams, W. M., and Mortimore, M. J. (1997). Agricultural Intensification and Flexibility in the Nigerian Sahel. The Geographical Journal 163(2): 150–160.

Almeida, O. T.; Uhl, C. (1995). Developing a Quantitative Framework for Sustainable Resource-Use Planning in the Brazilian Amazon. World Development, v. 23, n. 10, p. 1745-1764.

Altieri, M. A.; Anderson, M. K.; Merrick, L. C. (1987) Peasant agriculture and the conservation of crop and wild plant resources. Conservation Biology, v. 1, n. 1, p. 49-58.

Anderson, A.B. 1990. Alternatives to deforestation: steps toward sustainable use of the Amazon rain forest. Columbia University Press, New York.

Anderson, B. (1991). Imagined Communities: Reflections on the Origin and Spread of Nationalism. Verso, London.

Anonymous (2009c) New land use policy: executive summary. Government of Mizoram, Aizawl

Ansari M A, Saraswat P K, Sharma S K, Prakash N, *et al.* (2017). Changing from subsistence to remunerative cropping: A success story of diversification through legume in Jhum land. (In) Book of Abstract of National Seminar on Smart Farming for Enhancing Input Use efficiency, Income and Environmental Security (SFEIES) during 19-21st September 2017 at ICAR RC for NEH Region, Umiam, Meghalaya. Pp: 160.

Baird, I.G. & B. Shoemaker. 2005. Aiding or Abetting?: Internal Resettlement and International Aid Agencies in the Lao PDR. Probe International. Toronto.

Baishya L K, Ansari M A, Sarkar D, Ghosh M, Kumar S and Prakash N. 2016. Productivity enhancement in shifting cultivated lands through biofertilizers and micro-dosing of NPK in Eastern Himalayan Region. Research on crops 17(2): 268–75.

Ball, J.B. and Umeh, L.I. 1981. Development trends in taungya systems in the moist lowland forest zone of Nigeria between 1975 and 1980. Paper presented at the Workshop on Agroforestry in the African Humid Tropics, University of Ibadan. Ibadan, Nigeria, Federal Department of Forestry.

Barber, C. V., and Churchill, G. (1987). Land Policy in Irian Jaya: Issues and Strategies. Jakarta: GOI-UNDP/IBRD. Government of Indonesia, United Nations Development Program, and International Bank for Reconstruction and Development, Jakarta.

Bennett, J. W. (1976). The Ecological Transition. Pergamon, Oxford.

Bhadana V P, Sharma P K, Ansari M A, Baishya L K, Punitha P, Shiv Dutt, Prakash N and Rana K S. 2013. Food legumes for livelihood and nutritional security in North Eastern Himalayan Region: Prospects and constraints. Indian Journal of Agricultural Sciences 83(9): 899–906.

Bhatt B. P., Singh R., Mishra L. K., Tomar J. M. S., Singh M., Chauhan D. S., Dhyani S. K., Singh K. A., Dhiman K. R. and Datta M. 2001. Agroforestry practices and research: an overview. In: Verma ND, Bhatt BP (eds) Steps towards modernization of agriculture in NEH region. ICAR Research Complex for NEH region, Umiam, pp 365–392

Biswal D.K. and Kumar, S. 2013. Shifting cultivation and policies of sustainable development: a meaningless obsession. Journal of Economic & Social Development, 9: 132-138.

Brookfield, H. C. (2001). Intensification, and Alternative Approaches to Agricultural Change. Asia Pacific Viewpoint 42(2/3): 181–192.

Bruun T B, de Neergaard A, Lawrence D and Ziegler A (2009) Environmental consequences of the demise in swidden agriculture in Southeast Asia: carbon storage and soil quality Human Ecology 37 375–88

Bryant, D., D. Nielsen, and L. Tangley. 1997. The last frontier forests: Ecosystems & economies on the edge. World Resources Inst., Washington, DC.

Buresh, R.J. and Cooper, P.J.M. (1999) The science and practice of short-term improved fallows: symposium synthesis and recommendations. Agroforestry Systems. 47:345-356.

Butterfield, R. (1997). Land-Use Allocations and Community Livelihoods: Policy Issues. Associates in Rural Development, Forest Policy Reform Process Project, Phnom Penh

Cairns M and Garrity DP, 1999. Improving shifting cultivation in Southeast Asia by building on indigenous fallow management strategies. Agroforestry Systems 47: 37–48.

Castella, J. C., Boissau, S., Tran, N. T, and Dang, D. Q. (2005) 'Agrarian transition and lowland-upland interactions in mountain areas in northern Vietnam: Application of a multi-agent simulation model', Agricultural Systems 86(3), pp312-332

Castella, J. C., Lestrelin, G., Hett, C., Bourgoin, J., Fitriana, Y. R., Heinimann, A. and Pfund, J. L. (2013) 'Effects of landscape segregation on livelihood vulnerability', Human Ecology 41, pp63-76

Catacutan D, Finlayson R, Gassner A, Perdana A, Lusiana B, Leimona B, Simelton E, Oborn I, Galudra G, Roshetko J *et al.* 2018. ASEAN Guidelines for Agroforestry Development. ASEAN Secretariat. Jakarta, Indonesia. www.world agroforestry. org/region/sea/ publications/detail?pubID=4392.

Chamberlain, J. 2001. Participatory Poverty Assessment, the Lao PDR. ADB, SPC, National Statistics Center. Vientiane. Department of Environment (DOE),

Chapman, E.C. (1998). Upland fanning system in Lao PDR-Problem and opportunities for livestock. ACIAR Proceeding No 87, Canberra, Australia.

Childe, V. G. (1929). The Danube in prehistory. Oxford: Clarendon Press.

Childe, V.C. 1960. Vorgeschichte der europäischen Kultur. Hamburg.

Chongloi, S. 2016 Shifting cultivation and the issue of food security: a case study of Thadou-Kukis of Manipur. South -Asian Journal of Multidisciplinary Studies (SAJMS), 3: 75-86, ISSN:2349-7858:SJIF:2.246.

Choudhury, D. and R. Sundriyal. 2003, 'Issues and Options for Improving Livelihoods of Marginal Farmers in Shift ing Cultivation Areas of Northeast India', Outlook in Agriculture, 32: 17–28.

Choudhury, D., Ingty, D. and Jamir, S. 2003 Managing marginalisation in shifting cultivation areas of Northeast India: Examples of community innovations and initiatives. In: Tang Ya and Pradeep M. Tulachan (eds.) Mountain Agriculture in the Hindu Kush Himalayan Region. ICIMOD, Kathmandu, pgs. 207-212.

Choudhury, D., N.S. Jodha, and E. Sharma. 2008, 'Policy Approaches in Management of Shift ing Cultivation: Compromising Equitable Access, Food Security and Common Property Institutions', Paper presented at IASC Biennial Conference 'Governing shared Resources: Connecting Local Experience to Global Challenges', Cheltenham, England, 14–18 July.

Conklin HC (1957) Hanunoo Agriculture: A Report on an Integral System of Shifting Cultivation in the Philippines. FAO Forestry Development Paper No 12, Food and Agricultural Organization of the United Nations, Rome, Italy

Conklin, H. C. (1963). The Study of Shifting Cultivation. Routledge & Kegan Paul, London.

Conway, R.G. (1985). Agro-ecosystem Analysis. Agricultural Administration 20:31-35

Craswell ET, Sajjapongse A, Howlett DJB, Dowling AJ (1997) Agroforestry in the management of sloping lands in Asia and the Pacific. Agrofor Syst 38:121–137

De Foresta H and Michon G (1997) The agroforest alternative to Imperata grasslands: when smallholder agriculture and forestry reach sustainability. In: Garrity DP (ed) Agroforestry Innovations for Imperata Grassland Rehabilitation, pp 105–120. Kluwer Academic Publishers, Dordrecht, the Netherlands and the International Centre for Research in Agroforestry, Nairobi, Kenya

Dennis, R. A., and Colfer, C. P. (2006). Impacts of Land Use and Fire on the Loss and Degradation of Lowland Forest in 1983–2000 in East Kutai District, East Kalimantan, Indonesia. Singapore Journal of Tropical Geography 27: 130–48. doi:10.1111/j.1467-9493.2006.00238.x.

Do, D. S. (1994) Shifting Cultivation in Viet Nam: Its Social, Economic and Environmental Values Relative to Alternative Land Use, International Institute for Environment and Development, London

DoF (Department of Forestry) 1997. Plan to the year 2000 for stabilising shifting cultivation by providing permanent occupation. Paper presented at the Stakeholder Workshop on Shifting Cultivation Stabilisation, Vientiane, Laos, 6-7 February 1997. Ministry of Agriculture and Forestry and Asian Development Bank.

Durst, P. 2000. Forest news. Tigerpaper 27(3 July–Sept. 2000).

Eastmond A, Faust B (2006) Farmers, fires, and forests: a green alternative to shifting cultivation for conservation of the Maya forest? Landsc Urban Plan 74:267–284

Eder, J. F. (2003). Land use and economic change in the post-frontier upland Philippines. In Mertz, O., Wadley, R. L., and Christensen, A. E. (eds.), Local Land Use Strategies in a Globalizing World: Shaping Sustainable Social and Natural Environments. Proceedings of the International Conference, August 21–23, 2003. Volume 1.. Institute of Geography, University of Copenhagen, Copenhagen, pp. 79–101.

Elevitch C.R., Wilkinson K.M. (2000): Economics of farm forestry: financial evaluation for landowners. In: Elevitch CR, Wilkinson KM (eds) Agroforestry guides for Pacific Islands. Permanent Agricultural Resources (PAR), Hawaii

Flinn JC, Jellema BM, and Robinson KL (1976), Barriers to increasing food production in 1976 Che lowland humid tropics of Africa. (Mimeo). Ibadan, IITA.

Foley, J. A., DeFries, R., Asner, G. P., Barford, C., Bonan, G., Carpenter, S. R., Chapin, F. S., Coe, M. T., Daily, G. C., Gibbs, H. K., Helkowski, J. H., Holloway, T., Howard, E. A., Kucharik, C. J., Monfreda, C., Patz, J. A., Prentice, I. C., Ramankutty, N., and Snyder, P. K. (2005). Global Consequences of Land Use. Science 309: 570–574. doi:10.1126/science.1111772.

Forest Protection Department (2013) Forest allocation and swidden management (Giao rừng và quản lý nương rẫy),www.kiemlam.org.vn, (Vietnamese language), accessed 29 October 2013

Fox J, Fujita Y, Ngidang D, Peluso N, Potter L, Sakuntaladewi N, Sturgeon J, Thomas D, *et al.* Policies, political-economy, and swidden in Southeast Asia. Human Ecology. 2009;37:305–322. doi: 10.1007/s10745-009-9240-7.

Fox, J. (2000). How blaming 'slash and burn' farmers is deforesting mainland Southeast Asia. Asia Pacific Issues 47: 1–8.

FSI (Forest Survey of India) (2015) State of Forest Report, 2015. Forest Survey of India, Dehradun, Uttarakhand, India.

Geist, H.J., and E.F. Lambin. 2002. Proximate causes and underlying driving forces of tropical deforestation. BioScience 52(2):143–149.

Gleave, M. B. (1996). The length of the fallow period in tropical fallow farming systems: a discussion with evidence from Sierra Leone. The Geographical Journal 162: 14–24.

Government of Meghalaya. 2003, Draft Report of the Task Force on Shift ing (Jhum) Cultivation. Ministry of Environment and Forests. 2003, Report of the Task Force on Shift ing Cultivation, New Delhi: Ministry of Environment and Forests, Government of India.

Grinnell, H.R. 1975. A study of agrisilviculture potential in West Africa. Ottawa, International Development Research Centre.

Grogan K, Birch-Thomsen T and Lyimo J. 2013. Transition of Shifting Cultivation and its Impact on People's Livelihoods in the Miombo Woodlands of Northern Zambia and South-Western Tanzania. Human Ecology, 41:77–92, DOI 10.1007/s10745-012-9537-9.

Grogan P, Lalnunmawia F and Tripathi SK, 2012. Shifting cultivation in steeply sloped regions: a review of management options and research priorities for Mizoram state, Northeast India. Agroforestry Systems, 84:163–177, DOI 10.1007/s10457-011-9469-1.

Guo, H., Padoch, C., Coffey, K., Aiguo, C., and Yongneng, F. (2002). Economic development, land use and biodiversity change in the tropical mountains of Xishuangbanna, Yunnan, Southwest China. Environmental Science & Policy 5: 471–479. doi:10.1016/ S1462-9011(02)00093-X.

Hansen PK (1998). Shifting Cultivation Development in Northern Laos. Canberra, Australia: Australian Centre for International Agricultural Research.

Hansen, T. S., and Mertz, O. (2006). Extinction or adaptation? Three decades of change in shifting cultivation in Sarawak, Malaysia. Land Degradation and Development 17: 135–148. doi:10.1002/ldr.720.

Hardjosoediro, S. 1983. Institutional aspects of improvements in shifting cultivation with special reference to transmigration projects. In: Okigbo B.N.(Edns). Improved permanent production systems as an alternative to shifting intermittent cultivation. FAO, Soils Bulletin No. 53, International Institute of Tropical Aqriculture. Ibadan, Nigeria.

Hardter, R.; Chow, W. Y.; Hock, O. S. (1997). Intensive plantation cropping, a source of sustainable food and energy production in the tropical rain forest areas in southeast Asia. Forest Ecology and Management, v. 93, p. 93-102, 1997.

Hardy, A. (2003). Red Hills: Migrants and the State in Vietnam. University of Hawaii Press, Honolulu.

Harwood, R.R. and Price, Multiple cropping in tropical Asia. _In: 1976 Papendick, R.I. _et _al. eds. Multiple Cropping. ASA special publication 27, Madison. ASA/CSSA/SSA.

Hauck F.W. 1974. Shifting cultivation and soil conservation in Africa. FAO Soils Bull. 24:1–4.

Heinimann A, Hett C, Hurni K, Messerli P, Epprecht M, Jørgensen L, *et al.* Socio-Economic Perspectives on Shifting Cultivation Landscapes in Northern Laos. Hum Ecol. 2013; 41: 51-62.

Higashi S. 2015. An alternative approach to land and forest management in northern Lao PDR. In: Shifting Cultivation, Livelihood and Food Security: New and Old Challenges for Indigenous Peoples in Asia. (edn.) Erni, C., FAO, IWGIA and AIPP publication, 434p.

Huddleston, B., Ataman, E. and d'Ostiani, L. (2003) Towards a GIS-based Analysis of Mountain Environments and Populations, Environment and Natural Resources Working Paper No. 10, Food and Agricultural Organization of the United Nations, Rome.

Hunzai K, Gerlitz JY and Hoermann B, (2011). Understanding mountain poverty in Hindukush Himalayas – Regional report for Afghanistan, Bangladesh, Bhutan, China, India, Myanmar, Nepal and Pakistan, Kathmandu, ICIMOD.

ICAR 1980. Model land use- An alternative to Jhum (Agricultural extension programme) Extn. Folder No. 3 1980

Ickowitz A. Shifting cultivation and deforestation in tropical Africa: Critical reflections. Development and Change. 2006;37:599–626. doi: 10.1111/j.0012-155X.2006.00492.x.

ICRAF (1997) Indigenous strategies for intensification of shifting cultivation in southeast Asia. Programme and Abstract for a workshop, 23-27 June 1997. International Centre for Research in Agroforestry, Bogor, Indonesia, p. 125.

IFAD, IDRC, CIIFAD, ACRAF and IIRR 2001. Strategies for developing areas practicing shifting cultivation in Northeast India. In Shifting cultivation: Towards sustainability and resource conservation in Asia. pp 365-369

IFAD, IDRC, CIIFAD, ICRAF and IIRR (eds) (2001 a, b): Shifting Cultivation: Towards Sustainability and Resource Conservation in Asia. IIRR Publication, Manila.

IFFN (International Forest Fire News) (2004). ASEAN Agreement on Transboundary Haze Pollution, UN, Geneva 2005.

Imang N, Rujehan and Duakaju N. N. 2018. Assessment of daleh swidden agriculture as an innovative alternative to conventional swidden under conditions of external pressure on local forest management in Kalimantan, Indonesia. Biodiverstas, 19: 840-848. DOI: 10.13057/biodiv/d190312.

Ives, J.D. and Messerli, B. (1989) The Himalayan Dielemma, Reconciling development and conservation. Routledge, 11 New Fetter Lane, London.

Jakobsen, J., Rasmussen, K., Leisz, S., Folving, R., and Quang, N. V. (2007). The effects of land tenure policy on rural livelihoods and food sufficiency in the upland village of Que, North Central Vietnam. Agricultural Systems 94: 309–319. doi:10.1016/j.agsy.2006.09.007.

Jamir A. (2015). Shifting options: A case study of shifting cultivation in Mokokchung district in Nagaland, India. In: Erni C, editor. Shifting cultivation, livelihood and food security: New and old challenges for indigenous people in Asia. Bangkok: Food and Agriculture Organization of the United Nations and International Work Group for Indigenous Affairs and Asia Indigenous Peoples Pact;

Jha AK, Rymbai H, Verma VK, Deshmukh NA, Talang H, Assumi SR and Devi MB, 2017. Role of horticulture in jhum improvement and rehabilitation. In: Prakash N, Roy S S, Ansari M A, Sharma S K, Punitha P, Sailo B and Singh IM (edn). Jhum Improvement for Sustaining Farm Livelihood and Natural Resource Conservation in North Eastern Hill Region : Vistas And Frontiers, ICAR Research Complex for NEH Region, Meghalaya, India, 110p.

Jha LK and Lalnunmawia F (2003) Agroforestry with bamboo and ginger to rehabilitate degraded areas in North East India. J Bamboo Rattan 2:103–109

Juo, A. S. R.; Manu, A. (1996). Chemical dynamics in slash-and-burn agriculture. Agriculture, Ecosystems & Environment, v. 58, p. 49-60,

Kamajou, F. 1983. socio-economic and institutional considerations in improving shifting cultivation in tropical Africa. In: Okigbo B.N.(Edns) . Improved permanent production systems as an alternative to shifting intermittent cultivation. FAO, Soils Bulletin No. 53, International Institute of Tropical Aqriculture. Ibadan, Nigeria.

Kato MSA, Kato OR, Denich M, Vlek PLG (1999) Fire-free alternatives to slash-and-burn for shifting cultivation in the eastern Amazon region: the role of fertilizers. Field Crops Research 62:225–237

Kenney-Lazar, M, 2013. Shifting Cultivation in Laos: Transitions in Policy and Perspective. Report commissioned by the Secretariat of the Sector Working Group For Agriculture and Rural Development (SWG-ARD), Government of Lao PDR.

Kenney-Lazar, M, 2018. Shifting Cultivation in Laos: Transitions in Policy and Perspective. Report commissioned by the Secretariat of the Sector Working Group For Agriculture and Rural Development (SWG-ARD), Government of Lao PDR.

Kerkhoff E and Sharma E, 2006. Debating Shifting Cultivation in the Eastern Himalayas Farmers' Innovations as Lessons for Policy, International Centre for Integrated Mountain Development (ICIMOD), Kathmandu, Nepal

Kitamura, 2004. "Forest Policy Development in Laos", Inoue, M., and Isozaki, H. eds. People and Forest – Policy and Local Reality in Southeast Asia, the Russian Far East and Japan, pp. 113-116. Kluwer Academic Publishers. Dordrecht. Lao-Swedish Forestry Programmeme. 2001. Land Use Planning and Land Allocation. Vientiane. 288 Lao PDR

Kitamura, N. 2003. "Sinrin no Riyou to Hozen" ("Utilization and Protection of Forests"). Nishizawa, N., Furukawa, H., and Kiuchi, Y. eds. Raosu no Kaihatsu to Kokusai-kyouryoku (Development and International Cooperation in Laos). pp. 209-235. Mekong Publishers, Tokyo.

Lal R., P.A. Sanchez and R.W Cummings Jr. eds. 1986. Land clearing and development in the tropics. Rotterdam: Balkema.

Lambin, E.F., *et al.*, 2001. The causes of land-use and land-cover change: moving beyond the myths. Global Environmental Change 11, 261–269.

Lawrence D, Schlesinger WH (2001) Changes in soil phosphorus during 200 years of shifting cultivation in Indonesia. Ecology 82:2769–2780

Lawrence, A., 1997. Kaingin in the Philippines: is it the end of the forest? Rural Development Forestry Network Paper 21, London, UK

Lawrence, D., Suma, V., and Mogea, J. P. (2005). Change in species composition with repeated shifting cultivation: limited role of soil nutrients. Ecological applications 15: 1952–1967. doi:10.1890/04-0841.

Lawrence, D.C. 1996. Trade-offs between rubber production and maintenance of diversity: the structure of rubber gardens in West Kalimantan Indonesia. Agroforestry Systems 34(1): 83-100.

Lawrence, D.C. and J. Mogea. 1996. A preliminary analysis of tree diversity under shifting cultivation north of Gunung Palung National Park. Tropical Biodiversity 3 (3): 297-319.

Laxminarayana K, Bhatt B P, and Tulsi R. 2005. "Soil fertility buildup through hedgerow intercropping in integrated farming system: a case study," in Agroforestry in North East India: Opportunities and Challenges, pp 495–506. B P Bhatt and K M Bujarbaruah (Eds.) ICAR Research Complex for NEH Region, Umiam, Meghalaya.

Leduc, B and Choudhury, D 2012 Agricultural transformations in shifting cultivation areas of Northeast India: Implications for land management, gender and institutions. In: Dev Nathan and Virginius Xaxa (Eds) Social Exclusion and Adverse Inclusion: Development and Deprivation of Adivasis in India OUP, New Delhi pgs 237-258

Li, T. M. (2007). The Will to Improve: Governmentality, Development, and the Practice of Politics. Duke University Press, Durham.

MAF. 1999. The Government's Strategic Vision for the Agricultural Sector. Lao PDR: Ministry of Agriculture and Forestry.

Maithani BP (2005) Shifting cultivation in north-east India: Policy, issues and options. Mittal publications, New Delhi

Majid, Z. (1983) 'Malaysia', in Y. Atal and P. L. Bennagen (eds) Swidden Cultivation in Asia, Vol. 2, Country Profiles: India, Indonesia, Malaysia, Philippines, Thailand, Regional Office for Education in Asia and the Pacific,United Nations Educational, Scientific and Cultural Organization, Bangkok, pp157-238

Majumdar DN, 1976. Shifting cultivation Is it a way of life? An analysis of Garo data: shifting cultivation in Northeast India. Indian Coucil of Social Science Research, Shillong. 14-18p.

Malthus T (1798) . An essay on the principle of opulation. London, Printed for J. Johnson, in St. Paul's Church-Yard, 1798..

Mander Ü, Helming K, Wiggering H. 2007. Multifunctional land use: meeting future demands for landscape goods and services. In: Mander Ü, Wiggering H, Helming K (eds). Multifunctional Land Use. Springer. Berlin, Heidelberg. P 1–13.

Mann, H.S. and Saxena, S. K. eds. Khejri (Prosopis cineraria) 1981 in the Indian Desert. CAZRI Monograph No. 11. Jodhpur Central Arid Zone Res. Inst., 77 p.

Manning, C. (1971). The Timber Boom in East Kalimantan. Bulletin of Indonesian Economic Studies 7: 30–61. doi:10.1080/00074917112331331902.

MARD (2007) Proposal for Supporting Upland People in Sustainable Agroforestry Cultivation on Swidden Land in the period of 2008-2012, Ministry of Agriculture and Rural Development, Hanoi

Mazoyer, M., Roudart, L., 2006. A history of world agriculture : From the neolithic age to the current crisis. New York, Monthly Review Press.

McElwee, P. (2004) 'Becoming socialist or becoming Kinh? Government policies for ethnic minorities in the Socialist Republic of Vietnam', in C. Duncan (ed.) Civilizing the Margins, Cornell University, New York, pp182-213

McNeely, J.A., and S.J. Scherr. 2003. Ecoagriculture: Strategies to feed the world and save biodiversity. Island Press, Washington, DC.

Mertz O (2002) The relationship between length of fallow and crop yields in shifting cultivation: a rethinking. Agrofor Syst 55:149–159

Mertz O, Birch-Thomsen T, Elberling BO, Rothausen S, Bruun TB, Reenberg A, Fog B, Egsmose RMR and Breuning-Madsen H, 2012. Changes in shifting cultivation systems on small Pacific islands. The Geographical Journal, Vol. 178:.175–187, doi: 10.1111/ j.1475-4959.2011.00447.xg

Mertz O, Leisz S, Heinimann A, Rerkasem K, Thiha T, Dressler W, Cu P V, Vu K C, Schmidt-Vogt D, Colfer C J P, Epprecht M, Padoch C and Potter L 2009a Who counts? Demography of swidden cultivators in Southeast Asia Human Ecology 37 281–9

Mertz O, Padoch C, Fox J, Cramb R A, Leisz S J, Nguyen T L and Vien T D 2009b Swidden change in Southeast Asia: understanding causes and consequences Human Ecology 37 259–64

Mertz O, Padoch C, Fox J, Cramb R, Leisz S, Lam N, Vien T (2009) Swidden change in Southeast Asia: understanding causes and consequences. Hum Ecol 37:259–264 Michon G (ed) (2005) Domesticating forests: how farmers manage forest resources. IRD, CIFOR and ICRAF, Nairobi

Mettrick, H. (1993). Development oriented research in Agriculture: An ICRA TextBook.

MoEF 2007. India's Forests. (Kishwan J., Pandey D., Goyal A. K. And Gupta A. K. (eds.). Xtreme Office Aids (Pvt.) Ltd., New Delhi.

MoEF (2010): Ministry of Environment and Forests National Mission for a Green India. (Under the National Action Plan on Climate Change), draft submitted to Prime Ministers Council on Climate Change. Government of India, Ministry of Environment and Forests.

Moniaga, S. (1993). Toward community-based forestry and recognition of adat property rights in the Outer Islands of Indonesia. In Fox, J. (Ed). Legal Frameworks for Forest Management in Asia: Case Studies of Community State Relations. Program on Environment Occasional Paper No. 16. East-West Center, Honolulu, HI.

Monre 2011 Decision 2282 dated 8 December 2011 on Approval and Announcement of Statistics on Land Areas as of 1 January 2011, Ministry of Natural Resources and Environment, Hanoi

Mooney HA, Cropper A, ReidW(2005) Confronting the human dilemma: how can ecosystems provide sustainable services to society? Nature 434:561–562

Myers N. 1994. Tropical deforestation: Rates and patterns. In K. Brown and D.W. Pearce (eds.). The causes of tropical deforestation. UCL Press, London.

Myers, N. 1993. Tropical forests: The main deforestation fronts. Environmental Conservation, v. 20, p. 9-16.

Nair P. 1993. An introduction to agroforestry. Kluwer Academic Publisher. Dordrecht. www.worldagroforestry.org/units/library/books/PDFs/32_An_introduction_

Nair, P.K.R. 1983. Agroforestry as an alternative to shifting cultivation. In: Okigbo B.N.(Edns). Improved permanent production systems as an alternative to shifting intermittent cultivation. FAO, Soils Bulletin No. 53, International Institute of Tropical Aqriculture. Ibadan, Nigeria.

Netting, R. M. (1993). Smallholders, Householders: Farm Families and the Ecology of Intensive, Sustainable Agriculture. Stanford University Press, Palo Alto.

Nganje W, Schuck EC, Yantio D, and Aquach E, 2001. Farmer Education and Adoption of Slash and Burn Agriculture. Agribusiness & Applied Economics Miscellaneous Report No. 190. Department of Agribusiness and Applied Economics, Agricultural Experiment Station, North Dakota State University.

NHB (National Horticultural Board) (2015). Ministry of Agriculture and Farmers Welfare, Government of India, New Delhi, India.

Nielsen, U., Mertz, O., and Noweg, G. T. (2006). The rationality of shifting cultivation systems: labor productivity revisited. Human Ecology 34: 210–218. doi:10.1007/s10745-006-9014-4.

Nye, P.H. and Greenland DJ. 1959 Increases in the carbon and nitrogen contents of tropical soils under natural fallows. European Journal of Soil Science, 10: 284-299.

Nye, P.H. and Greenland DJ. 1960. The Soil Under Shifting Cultivation. Tech. Comm. 51, Commonwealth Agricultural Bureau, Harpenden, England.

Okigbo B.N. (eds) (1984): Improved Permanent Production Systems as an Alternative to Shifting Intermittent Cultivation. FAO Soils Bulletin, No. 53, FAO, Rome.

Okigbo, B.N. 1981. Alternatives to shifting cultivation. Ceres (FAO), 15 (6): 41-45. Rome.

Olawoye, O.O. 1975. The agri-silvicultural system in Nigeria. Commonw. For. Rev., 54 (3/4): 229-236.

Olofson, H. 1982. An anthropological approach to social forestry: the study of indigenous agroforestry systems. Paper presented to the Monthly Forum in Social Forestry of the Bureau of Forest Development, Asian Institute of Tourism, Quezon City, Philippines.

Oram, P. A. Production potentials in Africa: issues and strategies. 1981 _In. IFPRI. Food Policy Issues and Concerns in Sub-Saharan Africa. Washington, D.C., IFPRI. pp. 45-80.

Padoch C, Coffey K, Mertz O, Leisz S, Fox J and Wadley R L 2007 The demise of wwidden in Southeast Asia? Local realities and regional ambiguities Geografisk Tidsskrift-Danish Journal of Geography 107 29–41.

Pant RM, Tiwari BK and Chaudhury D, 2018. Shifting Cultivation: Towards a Transformational Approach Contributing to Sustainable Development in Indian Himalayan Region. Report of Working Group III, Niti Aayog, New Delhi, India.

Patel T., Karmakar S., Sanjog J., Kumar S. and Chowdhury A. 2013. Socio-economic and environmental changes with transition from shifting to settled cultivation in northeastern India: an ergonomics perspective. International Journal of Agricultural Science and Research 3 (2): 117-136.

Peluso, N. L., and Vandergeest, P. (2009). Taking the Forest out of the Jungle: Counter-insurgency and the Political Forest in Southeast Asia. In Hecht, S., Padoch, C. (Eds). The Social Lives of Forests. University of Chicago Press, Chicago.

Perz, SG and Walker, RT. (2002). Household Life Cycles and Secondary Forest Cover Among Small Farm Colonists in the Amazon. World Devel, v. 30, n. 6, p. 1009-1027.

Phuntsho, K; Aryal, KP; Kotru, R (2015) Shifting cultivation in Bangladesh,Bhutan and Nepal: Weighing government policies against customary tenure and institutions. ICIMOD Working Paper 2015/7. Kathmandu: ICIMOD

Potter, L., Brookfield, H., and Byron, Y. (1995). In Place of the Forest: Environmental and Socioeconomic Transformation in Borneo and the Eastern Malay Peninsula. United Nations University Press, Tokyo.

Powell, N. (1990). The Encroachment of Fire Disclimax Grasslands in the Humid Tropics: associated Problems and Research Recommendations, Department of Forest Soil Science (Unpublished draft). SLU, Sweden.

Prasad A, 2004. Forestry and Tribal development –a background note for Tripura state human development report.Government of Tripura. Annual Tribal Sub Plan, 2003-04. 10th Five year plan. 10p.

Prasad, R. N. and Sharma, U. C. (1994) Potential Indigenous Farming Systems of North Eastern Hill Region, ICAR Research Complex for NEH Region, barapani, Meghalaya, pp. 48.

Pravongviengham, P. P. (1998) 'The role of animal husbandry and aquaculture in improvements of swidden-based livelihood systems in the Lao PDR', PhD thesis, Asian Institute of Technology, Bangkok

Rahman, S. A., Jacobsen, J. B., Healey, J. R., Roshetko, J. M., & Sunderland, T. (2017). Finding alternatives to swidden agriculture: does agroforestry improve livelihood options and reduce pressure on existing forest? Agroforestry Systems, 91(1), 185–199. https://doi.org/10.1007/s10457-016-9912-4

Raintree, J.B. 1980. Conservation farming with soil improving legumes: a farming systems approach to tropical agroforestry. IITA Agric. Econ. Discussion Paper 9/80.

Raintree, J.B. 1981. Conservation farming with multipurpose tree legumes: an underdeveloped branch of tropical agroforestry research. In L. Buck, ed. Proceedings of the Kenya National Seminar on Agroforestry (12-22 Nov. 1980). Nairobi, ICRAF.

Ram M and Singh BP, (1993) "Soil fertility management in farming systems," Lectures notes, off campus training on farming system, Aizawl, India.

Ramakrishnan P.S. (1992): Shifting cultivation and sustainable development: an interdisciplinary study from north-eastern India. Man and Biosphere Series. Vol. 10. United Nations Educational, Scientific and Cultural Organization, Paris.

Ramakrishnan PS, Toky OP (1981) Soil nutrient status of hill agro-ecosystems and recovery after slash and burn agriculture (jhum) in northeastern India. Plant Soil 60:41 64

Rasul, G (2005) State policies and land use in the Chittagong Hill Tracts of Bangladesh, Gatekeeper Series-119. London, UK: International Institute for Environment and Development (IIED)

Rasul, G. and Thapa, G. B. (2003) 'Shifting cultivation in the mountains of South and Southeast Asia: Regional patterns and factors influencing the change', Land Degradation and Development 14, pp495-508

Rasul, G.; Thapa, G.B. (2006) Financial and economic suitability of agroforestry as an alternative to shifting cultivation: The case of the Chittagong Hill tracts, Bangladesh. Agric. Syst. 2006, 91, 29–50.

Roder, W. 2001. Slash-and-Burn Rice System in the Hills of Northern Lao PDR: Description,

Rösch, M, O. Ehrmann, L. Herrmann, E. Schulz, A. Bogenrieder, J.G. Goldammer, M. Hall, H. Page, and W. Schier. 2002. An experimental approach to Neolithic shifting cultivation. Vegetation History and Archaeobotany 11, 448-450.

Roy NC. Agricultural productivity in Arunachal Pradesh. In: Behera MC, Roy NC, editors. Trends in agrarian structure in the hills of north-east India. New Delhi: Commonwealth Publishers; 1997. pp. 55–59

Roy, RCK (2000) Land rights of the indigenous peoples of the Chittagong Hill Tracts, Bangladesh. Copenhagen, Denmark: International Work Group for Indigenous Affairs (IWGIA)

Ruthenberg, H. (1980). Farming Systems in the Tropics. Clarendon, Oxford.

Saha, R., Chaudhary, R.S. and Somasundaram, J. 2012. Soil Health Management under Hill Agroecosystem of North East India. Applied and Environmental Soil Science. pp 1-9, doi:10.1155/2012/696174.

Sahu, S.K., Nayak, R.K. and Sarangi, D. 2005. Sustainable Soil and Land Management Under Shifting Cultivation in Orissa. Orissa Review, pp. 49-52.

Sanchez, P. A. (1999). Improved fallows come of age in the tropics. Agroforestry systems, 47(1), 3-12.

Sanchez, P. A. (2000). Linking climate change research with food security and poverty reduction in the tropics. Agriculture, Ecosystems & Environment, v. 82, p. 371-383, .

Sandewall, NI., Ohlsson, B., and Sawathvong, S., (1999). Assessment of historical land use change for purposes of strategic planning and research - a case study in Laos. (Unpublished articles Draft). Department of Forest Resource Management and Geomatics, SLU, Sweden.

Sarkar AN, 1990. Role and programme implementation by North East Council (NEC) for shifting cultivation contrl in North Eastern Region. Paper presented at a seminar on 'Land Use Planning and Watershed Management for Control of Shifting Cultivation' held on 8-9 March, 1990 at Dimapur, Nagaland, India.

Schmidt-Vogt D, Leisz S, Mertz O, Heinimann A, Thiha T, Messerli P, Epprecht M, Cu P V, Vu K C, Hardiono M and Truong D M 2009 An assessment of trends in the extent of swidden in Southeast Asia Human Ecology 37 269–80

Scott, J. C. (1976). The Moral Economy of the Peasant: Rebellion and Subsistence in Southeast Asia. Yale University Press, New Haven.

Serrao EAS and Homma AKA, 1993. Combination of enterprises in mixed arming systems. In: National Research Council (edn) Sustainable Agriculture and the environment in the Humid Tropics, pp265-351, Washington, DC.

Sharma UC and Bujarbaruah KM 2016. Issues and Challenges in Shifting Cultivation and its Relevence in the Present Context, National Academy of Agricultural Sciences, Policy Paper No. 83.

Sharma, U.C. and Sharma, Vikas (2003). The 'Zabo' soil and water management and conservation system in northeast India : tribal beliefs in the development of water resources and their impact on society- an historical account of a success story. In: The Basis of Civilization-water Science (edn.J.C. Rodda and L/Ubertini. IAHS Publ.286,pp.184-191, Rome, Italy.

Simelton E, Ostwald M and Osiru M, 2020. Multifunctional land-use systems – a solution for food security in Africa? In: Multifunctional Land Uses in Africa – Sustainale Food Security Solutions, Simelton E and Ostwald (eds.), Routledge, 2 Park Square, Milton Park, Abingdon, Oxon OX14 4RN

Singh KD, Sinha B, Ashutosh S (2010) Techniques of survey and planning for conservation and sustainable use of biodiversity in Mizoram. Ministry of Environment and Forests, Delhi

Sodarak, H., Vaya., Soulignavongsy, S., Ditsaphone, Ch. and Hansen, P.K. (1998). Livestock development by the Shiftine cultivation research project in Luang Prabang province, Lao PDR. In: Upland farming system in the Lao PDR¬Problem and opportunities for livestock. ACIAR Proceedings No 87, Canberra, Australia.

Sommer, R., Denich, M., and Vlek, P. L. G. (2000). Carbon storage and root penetration in deep soils under small-farmer land-use systems in the Eastern Amazon Region, Brazil. Plant and Soil 219: 231–241.

Sonowal, D.K., Tripathi, A.K.& V.Kirekha (1989). Zabo, an indigenous framing system of Nagaland. Indian J.Hill Res, 2:1-8.

Souvanthong, P. 1995. Shifting Cultivation in Lao PDR; An Overview of Land Use and Policy Initiatives. International Institute for Environment and Development (IIED). London.

Srikham, W. 2010. The Effects of Commercial Agriculture and Swidden-field Privatization in Southern Laos. Paper presented at the RCSD International Conference, May 2010, Chiang Mai, Thailand.

Stone, G. D. (2001). Theory of the Square Chicken: Advances in Agricultural Intensification Theory. Asia Pacific Viewpoint 42 (2/3): 163–180.

Stromgaard P (1992) Immmediate and long-term effects of fire and ash fertilization on a Zambian miombo woodland soil. Agric Ecosyst Environ 41:19–37

Subramaniyan S, 1990. National policy and programme on land use. Paper presented at a seminar

Sun H, Tang Y, Xie JS (2008) Contour hedgerow intercropping in the mountains of China: a review. Agrofor Syst 73: 65–76

Tacio HD (1993) Sloping agricultural land technology (SALT): a sustainable agroforestry scheme for the uplands. Agrofor Syst 22:145–152

Task force on shifting cultivation (1983). Government of India, New Dwelhi.

Tawnenga SU, Shankar U. and Tripathi R.S. (1997): Evaluating second year cropping on jhum fallows in Mizoram, north-east India: Soil fertility, Journal of Bioscience, 22: 615-625.

Thanh MV and Vien TD, 2017. Assessing shifting cultivation policies through the lens of systems thinking. In: Shifting cultivation policies: balancing environmental and social sustainability, Edition: 1st, Chapter: A1, CABI, pp.1-15, ISBN 9781786391797

Thomas, D. 2004. Review of Policies and Practices in Upland Areas of the Lao PDR. Paper presented at the 2004 NAFRI Workshop on Shifting Cultivation and Poverty Eradication in the Uplands of the Lao PDR. Luang Prabang, Laos.

Thomaz EL (2009) The influence of traditional steep land agricultural practices on runoff and soil loss. Agric Ecosyst Environ 130:23–30

Thongchai, W. (1994). Siam Mapped. University of Hawaii Press, Honolulu

Tiffen, M., and Mortimore, M. (1994). Malthus Controverted: The Role of Capital and Technology in Growth and Environment Recovery in Kenya. World Development 22(7): 997–1010.

Tiwari BK, 2014. Policies and alternative options to shifting cultivation in northeast india - a review 2004-2014, Department of Environmental Studies, North–Eastern Hill University, Shillong -793022, India.

Toky OP, Ramakrishnan PS (1981a) Cropping and yields in agricultural systems of the northeastern hill region of India. Agro-Ecosystems 7:11–25

Tommy J.L. 1983. Shifting cultivation and possibilities for improving it in sierra leone. In: Okigbo B.N.(Edns). Improved permanent production systems as an alternative to shifting intermittent cultivation. FAO, Soils Bulletin No. 53, International Institute of Tropical Aqriculture. Ibadan, Nigeria.

Turner, B. L., and Brush, S. B. (1987). Comparative Farming Systems. The Guilford Press, New York.

Turner, B. L., Hanham, R., and Portararo, A. (1977). Population Pressure and Agricultural Intensity. Annals of the Association of American Geographers 67: 384–396.

Uraivanh, T. (1998). Socio-ecom_ lie Research Development and Linkages. Shifting Cultivation Research Sub-program, Luang Prabang, Lao PDR.

Van Noordwijk, M. (1999). Productivity of intensified crop—fallow rotations in the Trenbath model. Agroforestry Systems, 47(1), 223-237.

Van Vliet N, Mertz O, Heinimann A, Langanke T, Pascual U, Schmook B, and Castella JC (2012) Trends, drivers and impacts of changes in swidden cultivation in tropical forest-agriculture frontiers: a global assessment. Global Environmental Change doi. org/10.1016/j.gloenvcha.2011.10.009.

van Vliet, N., Mertz, O., Birch-Thomsen, T., Schmook, B., 2013. Is there a continuing rationale for swidden cultivation in the 21st century? Hum. Ecol. 41, 1–5.

Vargas, D. (1985). The Interface of Customary and National Land Law in East Kalimantan, Indonesia. Ph.D. dissertation, Yale University.

Vergara, N.T. 1981. Integral agroforestry: a potential strategy for stabilizing shifting cultivation and sustaining productivity of the natural environment. Honolulu, Hawaii, East-West Center, Environment and Policy Institute.

Wadley, R. L., and Eilenberg, M. (2005). Autonomy, Identity and 'Illegal' Logging in the Borderlands of West Kalimantan, Indonesia. The Asia Pacific Journal of Anthropology, 6: 119–34. doi:10.1080/14442210500074853.

Walker, R. T.; Homma, A. K. (1996). Land use and land cover dynamics in the Brazilian Amazon: an overview. Ecological Economics, v. 18, p. 67-80.

Wharton, C.R. Jr., (ed) (1996).. Subsistence aqriculture and economic 1969 development. Chicaqo, Aldine

Wilson, G.G. and Kang, B. T. Developing a stable and productive 1981 biological cropping system for the humid tropics. In: Stonehouse, B. ed- Biological husbandry: A scientific approach to organic farming, pp. 193-203. London, Butterworth.

World Bank, 1990. 'Environmental Action Plan for the Republic of Madagascar." Evaluation Report. Washington, D.C.: World Bank.

Ziegler, A. D., Bruun, T. B., Lawrence, D., Nguyen, T. L. (2009). Environmental consequences of the demise in swidden agriculture in Montane Mainland SE Asia: hydrology and geomorphology. Human Ecology, 37: 361-373.

10

Directives and Policies

10.1 GENERAL DESCRIPTION

The soil, water, and biodiversity of the land make up a complex composition. Together, these three components produce goods and services that lay the groundwork for people to live sustainably and in peace with one another. Yet an estimated 3.2 billion people's security, livelihoods, and health are at stake due to land degradation. Any decrease or loss in the biological or economic productive capability of the land resource base is referred to as "land degradation." Natural processes contribute, but humans are primarily to blame for the harm. Degradation is frequently intricately connected to biodiversity loss and the effects of climate change.

Restored lands provide access to clean water, reduce erosion, preserve livelihoods and biodiversity, create forest products, and supply biomass fuel. Trees in agricultural environments can enhance soil fertility, hold onto moisture, and boost food output. Trees and forests slow climate change by storing carbon; intensive restoration efforts may lead to a reduction in the atmospheric concentration of carbon dioxide. Restoration can assist communities in adjusting to the effects of climate change by ensuring water supplies or reducing the effects of severe storms.

Plant cultivation was only a modest activity of primarily nomadic populations in the second millennium BC, and it had little impact on the vegetation. After 0 BC/AD, sedentary civilizations created enormous settlement mounds, and on the dunes, a shifting agricultural system with fields and fallows was constructed. Park savannas, an agroforestry system with protected wild fruit trees scattered throughout the fields, were used for cultivation (Höhn and Neumann 2012). Extensive farming with protracted fallow periods was conducted during the first millennium AD. The composition of the woody vegetation changed as fallow seasons shrunk and the impact of cattle herding on the vegetation increased towards the start of the second millennium AD. The substantially lower species richness of the contemporary woody forest was eventually brought about by declining precipitation and anthropogenic pressure (Höhn and Neumann 2012).

Even while shifting cultivation is ecologically sustainable overall, there has frequently been significant ecosystem degradation as a result of attempts to keep up with rural growth and prospects for livelihood in other habitats. High importance has been placed on enhancing productivity while also maintaining environmental sustainability, and the Rio+20 call for green development supported these efforts (Srivastva 2017). Strong enabling policies are required to address the problems of supporting shifting cultivation and its dependent people as well as concerns for secure ecosystems. Geographically, countries with slopes and highlands that once had a healthy forest cover primarily use shifting cultivation as a farming method (FSI, 2013). Shifting agriculture, often known as swidden farming, was an integral part of indigenous communities' customs. Slash-and-burn agriculture, which is both economically and environmentally sound, can be accurately described as shifting cropping (Srivastva 2017).

According to ecologists (Fujisaka, 1991), shifting cultivation in a traditional setting must be viewed as a possibility under low population, availability of land, smallholding forest-based land use system, satisfactory regeneration of forest species and long fallow periods. However, shifting agriculture is now a significant factor in the deterioration of natural resources, and a sustainable development agenda still needs effective development strategies and a decision-making process (Daly, 1991). The large horizontal range of variation in tribal populations, their farming methods, their environments, and the opportunities available to them must be taken into account by such policies (NAEB, 2007). Given the current circumstances, a thorough policy paper on shifting cultivation and related challenges is required. Any plan to alter shifting agriculture must be properly thought out, take the long view into account, and include logical progressions of modification, change, and diversification. However, in order to limit shifting cultivation and the ensuing soil degradation, research and technology development are predominantly focused on the advancement of settled agriculture (Fonseka, 2010).

10.2 COMMUNITIES AND ECOSYSTEMS POLICY IN SHIFTING CULTIVATION

Calls for a fusion of traditional knowledge and modern technology have arisen in response to the need to increase system production and sustainability (National Commission on Farmers, 2005). Shifting cultivation is a method of tribal farming and community life that needs to be better understood from the perspectives of ecology, degradation and erosion, and trans-regional, on site-off site livelihoods. Policy actions should guarantee both a sufficient and guaranteed periodic cash income and food security at the home level. Additionally, they must to offer chances for gradual intensification of livelihoods. Greater focus should be placed on managed forest restoration after the end of swidden cultivation, as opposed to the simple abandonment of a degraded shifting cultivation area, in terms of the sustainable use of forests and other natural resources. A suitable compensation regime should exist to promote biodiversity and resource preservation. Establishing structures and recognizing lead communities and elders is important. Sustainable shifting cultivation and environmentally friendly diversification should not only receive the proper encouragement, acknowledgement, and financial rewards, but also social and national acknowledgment.

As in many other parts of the world, population growth, agricultural expansion, rising demand for wood products, and rapid economic growth all contribute to deforestation and forest degradation (Wulandari 2017). In fact, the UN's Food and Agriculture Organization recognized in 2005 that there were numerous factors contributing to deforestation (FAO, 2005). However, it claimed that there was growing understanding that diversifying tenure arrangements by giving local stakeholders secure tenure rights was a key mechanism for enhancing local accountability and control of forestry operations, thereby improving the conditions for sustainable forest management. A common explanation for intercommunal strife and even separatist violence is problems with land tenure. Land and resource disputes are typically seen as the primary causes of conflict. According to Herrera and Da Passano (2006), the conflicts over land tenure fell into three categories: (i) political influence, (ii) legal issues, and (iii) economic factors. The first one was a contributing factor in almost all land-tenure disputes; the second category appeared when parties involved were not aware of their legal rights or the various legal frameworks that governed access to certain areas and the use of natural resources in conflicting or divergent ways; and the third category was the primary cause of land-tenure disputes.

Act No. 41 of 1999 (Government of Indonesia, 1999), which governs the management of all types of forests, including production, protected, and conservation forests, serves as the theoretical foundation for sustainable forest management in Indonesia. Protected forests are those that manage water, prevent flooding, control erosion, stop sea water intrusion, and preserve soil fertility with the primary goal of protecting life-support systems. The term "conservation forest" refers to a certain type of forest. Preserving the variety of flora and wildlife and their ecosystems is their main objective. Act No. 5 of 1990 states that hunting parks, nature reserves, and environment preservation zones all fall under the category of conservation forests (Government of Indonesia, 1990). National park areas fall under the category of nature preservation areas; they are characterised as woods with particular features whose major purpose is the preservation of biodiversity of flora and fauna as well as the protection of ecosystems. Forests that have been set aside to promote hunting tourism are known as hunting parks. Act 41/99's Section 5 addresses the preservation of nature and the protection of forests. According to its Article 46, the implementation of protection and conservation measures aims at the best possible and long-term preservation of forests, forest regions, their surroundings, and their productive activities. Article 48 mentions that:

- Both inside and outside designated forest regions, the government will enact regulations to protect forests.
- The government will oversee the protection of state forests.
- Parties with authority to manage forests and those with permits to utilise forest areas are both expected to preserve the forests in their respective work zones.
- The owner of the rights is responsible for protecting private forests.
- Communities will be included in forest-protection initiatives to guarantee the thorough implementation of forest protection measures.
- Additional clauses in all of these cases will be governed by government regulation.

Agriculture and livestock rearing have had a significant cultural impact on the current West African savanna. Large wild mammals are all but extinct in West Africa as a result of severe population pressure and a shifting cultivation-based agricultural system, in contrast to the East African savannas with their diverse wildlife. This system must have fallows, and although the West African savannas appear to be natural, they are essentially an accumulation of fallows in various phases of development and with various purposes. Human activities such as farming, the gathering of firewood and other natural resources, extensive pastoralism with cattle, sheep, and goats, and man-made bushfires are all irreversibly affecting the vegetation.

Shifting cultivation has had a considerable impact on policy views thanks to an FAO publication from the late 1950s that called for its replacement since it was deemed "the most serious land use problem in the tropical world" (FAO 1957). This appeal has an impact on perceptions that characterize the practice as archaic, economically unviable, and a contributor to tropical deforestation. Policies based on this misunderstanding continue to work towards replacing the practice with established agriculture even in the face of mounting scientific evidence to the opposite and a reversal of this position by a number of nations and international organizations (FAO, IGWIA, and AIPP 2015). In shifting cultivation, with farmers slowly taking up modified farming (Heinimann *et al.*

2017), such policies have helped the rapid expansion of sedentary agriculture across large areas in South and Southeast Asia over the past half a century (Mertz *et al.* 2009, Fox *et al.* 2009). Terrace cultivation and plantations have become noticeable elements of village landscapes throughout the region, despite the fact that the practice continues (Singh *et al.* 2015). The change has not come without a price, however, as evidenced by upland landscapes, which show that regenerating fallows are also being converted to terraces and plantations. This disrupts the natural process of land recuperation and erodes the land's ability to regenerate into forests, which results in a sharp decline in forest cover. Therefore, the transition from fallows to settled agriculture has long-term effects on the availability of land and forest resources, as well as on ecosystem services like carbon and hydrological storage (Bruun *et al.* 2009; Cramb *et al.* 2009). South and Southeast Asia are already noticing the effects of the transition from fallows to settlement systems, with an alarming rise in the loss of forest cover.

10.2.1 Policy Context

What would be the "right" policy in regards to shifting cultivation? By outlining what has been discovered and providing a context for discussion with policy-makers, this essay explores the topic. Making the "right" decisions in terms of policy is a noble goal, but it is only noble inasmuch as we understand that ideals are targets rather than final destinations. We can accept the compromises that are a necessary part of creating policies if we are aware of them (Falvey 2018). It is possible for unintended and undesirable results to occur when single-issue counsel shapes policy, according to Falvey (2013). Because of the mistaken notion that milk production in the tropics could not be effective and that it was preferable to buy milk products from temperate dairy nations, some development agencies refused to sponsor dairy development in tropical countries. Informed experts were requested to contribute to a book that was eventually published by the International Livestock Research Institute, a member centre of the Consultative Group for International Agricultural Research (Falvey and Chantalakhana, 1999), noting that India was the world's largest producer and consumer of dairy products. Together with other factors, this book assisted in broadening the development agenda and moving it away from outdated Western ideas of efficiency and market growth. In the second illustration, policies that support large-scale agriculture often framed in modernizing terms that resemble Western development have marginalized smallholder farmers (Falvey, 2010). This bias persists despite the fact that small farmers in developing countries generate greater total economic benefits than would be feasible if they were simply replaced by large-scale farming.

Such policies drive small farmers into metropolitan surroundings where they face increased welfare and food needs while sustaining themselves and their families on less than two hectares of marginal land. There are around two billion small farmers in the globe, and an unknown proportion of them engage in shifting cultivation. The policies that impact these families are, however, enacted in the general interests of the countries in which they live despite the fact that they are dispersed over several, frequently isolated border regions and hilly terrain. Additionally, with new technologies creating a background of constant change, the pace of economic transformation has quickened in the majority of countries. Because of these factors, the first and most obvious point to be made is that policies affecting moving cultivators have not been designed with their welfare or environment in mind.

It was recommended at the ICIMOD conference that shifting cultivation-discouragement policies be replaced with ones that addressed rights over land, loan availability from banks, R and D, market facility and, strengthening of traditional institutions, and synergy among government agencies. These findings appear to be corroborated by a longer-term study conducted in Laos, where changes in land cover, land use, and livelihoods revealed a connection between highlanders' vulnerability and policies intended to boost forest cover, end shifting cultivation, and promote intensive and commercial agriculture. The study recommended that the emphasis be switched to ways to sustain a variety of livelihoods and economic opportunities rather than designating areas specifically for agriculture or forests (Castella *et al.*, 2013). Similar observations made in Kalimantan imply that external assumptions about appropriate cash crops to grow alongside food crops are rarely environmentally stable, albeit in that instance locally recognized rattan production appeared to be an exception (Weinstock, 1996).

We can gain a lot from research and development experience to help us better comprehend the laws governing shifting cultivation. A single "right" approach on a topic as diverse and marginalized as shifting cultivation, however, is impossible since policy must be as dynamic as the environment to which it applies. Perhaps it can be said that ineffective policy has decreased the amount of land available for shifting cultivation, which has led to an increase in the practice through shorter fallow periods, a breakdown in communal duty, a reliance on outside markets, and the threat to sensitive environments. According to numerous arguments, shifting cultivation handles agricultural fallows as medium-term forests, making it dynamically beneficial for both the environment and human livelihoods. Additionally, according to the argument, it should be backed up by pertinent research relating to farmer innovation and awards for contributions to biodiversity protection, fertility management, and pest control through controlled burning, as well as assistance for the growth of niche products. However, as this article has shown, policy is largely national in scope, and issues affecting modest populations or limited geographic areas are likely to be absorbed into higher-level objectives.

Falvey (2018) emphasized fifteen policy-related aspects

i) Shifting cultivators' wellbeing or the environment have not been the focus of policies that affect them;

ii) Adopting the current state of affairs as the context for policy requires taking a historical view of change;

iii) Policy is an ongoing process that is conducted in accordance with basic principles that have been decided upon and is complicated by a variety of forces;

iv) Long forest fallows are cleared and burned to make way for brief cropping phases in shifting agriculture, a traditional agricultural practice that is poorly understood but is essential to cultural identity; national policies influenced by global forces usually take precedence over specific policy proposals oriented to shifting cultivators and their environment;

v) The "highlands" policy grouping is favoured by relocating cultivators and is a policy sector that is especially impacted by international agendas;

vi) We shouldn't carry on the same studies and projects from earlier decad

vii) Some national programmes' effectiveness likely depends on the strength, independence, and riches of the nation;

viii) There are numerous sources of pertinent information and expertise available worldwide;

ix) Plans must be sufficiently flexible to allow for changes during implementation in order to be realistic

x) It would be wise to steer clear of policies that specifically address shifting agriculture regions and swidden cultivators;

xi) Integrated methods might be necessary;

xii) Research may advantageously be directed towards practical policy rather than a single ideal;

xiii) Given that institutional memories in assistance environments might be unproductively brief, short-term foreign interventions may be best centred on research to promote national aims;

xiv) In order to construct its technical trials, development research has to have a thorough understanding of the socio-economic values and limits and;

xv) Development depends on real information from research.

The state government of Odisha in India has made an effort to address the issue by regulating or nationalizing shifting agriculture in light of the scope of the region and population affected. In order to transition the nomadic tribes to settled agriculture, the colonization plan was initiated in the 1960s in problem areas by offering cultivable land, essential inputs, and housing. Several colonies have been developed in the tribal territories over the course of the first four programmes. The plan, however, has not produced the intended effects in the majority of the areas. The State Government has launched a policy of rational land use based on watersheds in addition to the colonization plan through the Department of Soil Conservation (Biswal and Kumar 2012). The projects included (a) giving land to tribe members who agreed to give up farming on sloping terrain. (b) Planting of commercially valuable species for the tribal community. (c) The use of conservation farming to give tribal members access to crop land with higher yields. (d) Making use of steep slopes to grow timber. The aforementioned plan chooses a suitable watershed in the shifting cultivation area and implements a programme for rational land use on the catchment already in place. For watershed areas, it was advised to rationalize shifting agriculture through a schematic land classification of watershed management units and land use programmes (Biswal and Kumar 2012). The aforementioned plan was initially launched on a pilot basis in a few catchments in the Koraput district, and it was later expanded to other trouble spots in the Phulbani, Kalahandi, and Keonjhar districts. However, given the size of the population and the region impacted by shifting cultivation, these programmes were insufficient. The Indian government's ministry of agriculture proposed a pilot programme in June 1987 to regulate shifting cultivation. Its dual goals are to improve the socioeconomic circumstances of tribal households who cultivate podu by weaning them off of it with 100% Central Assistance and to restore ecological balance in the hill areas. Additionally, the World Bank and the ICAR have suggested a number of methods to lessen shifting cropping. These are: (a) to encourage the establishment of silvi-pasture in the upper reaches of forestry. B) dividing the middle slope length into sections for annual or perennial fruit trees and intercrop, and C) putting agricultural crops on the lower slopes.

In Vietnam, shifting agriculture has long been considered incompatible with government initiatives for modernization, development, and environmental preservation. Targets, equipment, and tactics changed over time as a result of local officials' latitude and target communities' resistance, but these changes also lengthened overall anti-swidden campaigns by allowing for constant reinvention. Over time, evolving reasons have made room for fresh funds and methods while continuing to be based on misunderstandings and cultural chauvinism (McElwee 2022).

Despite a rise in interest in and understanding of indigenous populations around the world, there have been no changes in how swidden is approached. In contrast, due to their adaptable native practices, some shifting cultivators have been hailed as "true environmentalists" in other regions. The culturally specific indigenous relationships with nature that swidden frequently encompasses have never been acknowledged by any anti-swidden campaigns in Vietnam, and anti-swidden policies have had many unfavourable effects related to the issues of indigenous peoples, including the loss of indigenous knowledge, social systems, and community norms, as well as intergenerational learning. Targets of interventions have also changed their own ideas of subjectivity as a result of anti-swidden politics, with changing farmers frequently accepting the discourses and narratives about their social practices and lives as "backwards" and unsuited to modernity (Dressler *et al.* 2016).

According to estimates, 10–20% of upland lands in Vietnam are still used for shifting cultivation, albeit with shorter fallow cycles and different agricultural crops than in the past (Nguyen *et al.* 2004). This is despite decades of anti-shifting cultivation policy, and swidden persists strongly across Southeast Asia as a region and in many pockets of upland Vietnam. Without a doubt, shifting cultivation still exists today, but it is substantially different from what it was a century ago in terms of shrinking land area, decreasing crop variety, and decreasing population. Reviewing the main trends that have contributed to the loss of swidden practices in Southeast Asia, Fox *et al.* (2009) identified the following as the main offenders: resettlement and sedentarization projects; discriminatory actions against ethnic minorities; enforcing distinctions between zoning lands for forests and for agriculture; the rise of conservation programmes; privatization of land; and the development of markets. All of these processes have taken place in Vietnam and have either been the consequence of intentional measures targeted at moving cultivators or have assisted them. Given the absence of alternatives to swidden agriculture, poor investments, a lack of participatory planning, poor recipient targeting, and competition for land use in upland areas over the long period of time that anti-swidden policies have been in place, there have been few successes and many failures. Vietnam is not the only country with this issue; many other nations have discovered that it is exceedingly challenging to sedentarize and resettle populations, as well as change their traditions and agricultural practices (Cramb *et al.* 2009). Therefore, we must inquire as to why a practice like swidden was so despised and despised, and why anti-swidden legislation have endured despite being manifestly ineffective for such a long time. Cultural chauvinism is a significant motivator of these policies, as evidenced by the shifting justifications for them, which reveal a variety of motives. However, these anti-swidden policies have mostly remained the same over time due to misconceptions and the difficulty to transform roving cultivators into the precise subjects the state sought to achieve settled and 'legible' peoples (Scott 2009). The understanding that access to resources is necessary for all nations, especially developing ones, to achieve the sustained social and economic growth. Developing countries will need to utilise more energy while simultaneously looking into ways to improve energy efficiency and lower greenhouse gas emissions if they want to go any closer to meeting this objective.

10.3 POLICY OPTIONS AND RECOMMENDATIONS

10.3.1 Policies

Governments must implement policies and legal frameworks that recognize, protect, and reward the following farmer rights (Upadhayay 2004):

1. Farmers' right to prevent the misappropriation of their traditional knowledge related with plant genetic resources.
2. Farmers' rights to plant varieties and local knowledge, as opposed to corporate breeders' rights.
3. Rights of farmers enshrined in indigenous set-up to preserve, use, barter, sell, and proper care of crop seeds of any plant type.
4. Farmers' right to equitable benefit sharing in the production of new commercial varieties using plant genetic resources established and conserved by them, taking into consideration the economic value of plant and seed types developed by them over generations of evolution.
5. Farmers' right to be educated about market opportunities so that they can better assess the marketing choices for the products they produce using traditional knowledge.
6. Farmers' right to be protected from biopiracy and the theft of traditional knowledge.

Farmers' right to know about national and international agreements and legal systems that directly affect their livelihoods

- The following 9 general recommendations were given by Phuntsho *et al.* (2015) for all countries in the Eastern Himalayan region:

- Individual ownership or communal property can be used to secure the land tenure of shifting cultivators,
- Shifting cultivation must be recognized as a legal land-use category by policymakers in the region, and such farmers' land tenure must be secured, whether through individual or group/community ownership,
- Appreciate the importance of small farmers as founders in general, and shifting cultivators in particular,
- Create agricultural alternatives that are compatible with the local environment,
- Assist organizations and agencies tasked with the development of farming communities, specially those tasked with forestry and agriculture,
- Ascertain if policies are in accordance with international treaty provisions,
- The terms of such conventions must be considered in the changing agricultural plans of these countries,
- Create policy frameworks that encourage community-based, long-term land use plans that include watershed management and agroforestry,
- Shifting cultivation must be classified as a distinct type of agroforestry with a brief agriculture phase and a long forestry phase.

The process of policy development is equally as significant as the policies themselves. Traditional institutions develop policies that include all members of the community in conversations and decision-making. This increases community buy-in and develops a sense of shared responsibility, ensuring that policies are adhered to.

10.3.2 Recent Developments

Several national governments in the eastern Himalayan region have recently begun to adopt active measures to safeguard the long-term growth of their people, the bulk of whom are ethnic minorities and ethnic groups. Most importantly, the governments are establishing a set of policies and practices aimed at guaranteeing the sustainable management of the region's bio-resources (Karkhoff and Sharma 2006). Farmers and policymakers are both important stakeholders in the present transition in farming. Shifting cultivators, like farmers all throughout the world, are continually modifying their cultivation to meet the current needs of growing communities. This is accomplished by an inventive method based on guiding principles gained from prior experiences, as well as prevalent ideas about what is required and suitable. Their actions are constrained by changes and policies emerging from the efforts of policymakers responsible for the long-term development of shifting cultivation areas. However, in the eastern Himalayas, this innovation process does not appear to be functioning well for either communities or policymakers. Many policymakers believe that farmers are reluctant to respond to their instructions, but new studies reveal that existing policy and government practice misunderstand and undercut many well-intentioned farmers' efforts. The greatest danger in the current situation is that shifting cultivators are forced to abandon their traditional ways without being offered viable alternatives, which would take a long time to implement in practice.

10.3.3 The ICIMOD initiative

Despite extensive and long-term efforts by the government to eliminate shifting cultivation in the eastern Himalayan region, the practice has remained entrenched in significant regions. Recognizing this, the International Centre for Integrated Mountain Development (ICIMOD) launched a new initiative with support from the International Fund for Agricultural Development (IFAD) and participation from five eastern Himalayan countries (Bangladesh, Bhutan, India, Myanmar, and Nepal) on the premise that shifting cultivation would be meaningful if millions of farmers proceeded to practice it in-spite-of all motivations and stimulants. The purpose of the study was to take a fresh, unbiased look at the practice, with an emphasis on farmer-led improvements in response to present challenges and limits. The goal was to identify new ideas to assist solve the current scenario, in which outside interventions are failing to work and shifting cultivation is declining due to the restrictions placed on it. The goal was to increase awareness about shifting cultivation challenges, create a venue for idea exchange, and develop specific policy suggestions to assist governments in their work. Fallow forest area looks to be unmanaged 'open access' property that is available for usage, notably for commercial exploitation, from the outside. There is a pervasive lack of understanding of how forest and crops are intertwined. Similarly, shifting cultivation is frequently misinterpreted as removing forests to make space for agriculture, when in reality, farmers are cultivating trees and forests on their farmland.

10.4 FOREST PROTECTION AND LAND-USE POLICY

For at least two millennia, kings and governments have managed and maintained the region's forests through policies and legislation. These attempts were motivated not only by religious piety for the preservation of animals or to assure the provision of goods and services to society, but also by the desire to gain ownership of assets not claimed by particular rural communities. Since the dawn of industrialization, which was accompanied by population growth and rapidly increasing natural resource consumption, policies and regulations for forest conservation and sustainable use have been formulated and continuously updated and revised in response to changing societal requirements, particularly in the second half of the twentieth century. Today, each South Asian country has its own set of rules and regulations governing forest and tree management, administration, organization, social, and economic elements. In India, policies have been developed and provided to address not only public forestry conception, forest-life conservation, and wood harvesting, but also tenurial rights and institutional management such as the formation of forest state corporations, ministry reorganization, and the creation of new research entities. Bhutan and Nepal have established new forest policies that prioritise mountain ecosystem conservation and forest protection from natural threats like soil erosion. These new policies include decentralization of forest management as well as significant improvements in access rights and management duties for local populations. Forest-related land-use strategies in Bangladesh focus on increasing tree cover within agricultural land and urban areas due to the country's high population density. Similarly, Sri Lanka's forestry strategy aims to conserve surviving natural forests, empower people to safeguard and manage multiple-use forests, and develop and improve the productivity of traditional home gardens, agroforestry, and forest plantations in order to meet people's fundamental requirements. Forest policy directions in all countries prioritise the preservation and extension of existing forest cover. Forest tenure, benefit sharing between government and people, inter-sectoral cooperation, and other managerial and budgetary challenges have all been addressed through comprehensive exercises to develop forest sector master plans. Each country in the region's legislative framework and law has evolved over time to favour distributed forest management, permitting rural communities to invest in forest rehabilitation for long-term benefits. This marked a significant transition in forest governance from the forest department's conventional caretaker function to a participatory management paradigm engaging various stakeholders.

10.5 DEFINITIONS

1. The entire interaction between the geosphere, biosphere, hydrosphere, and atmosphere is referred to as the climate system.
2. "Emissions" means the emission of greenhouse gases and/or their forerunners into the atmosphere over a spatial and temporal scales.
3. The natural and artificial gaseous parts of the atmosphere that absorb and emit infrared radiation are referred to as "greenhouse gases".
4. "Regional economic integration organisation" refers to a group of sovereign States from a particular region that has the authority to

sign, ratify, accept, approve, or accede to the relevant instruments and is competent to deal with matters covered by this Convention or its protocols.

5. "Reservoir" refers to a part or parts of the climate system that store greenhouse gases or their precursors.
6. A process, activity, or mechanism is referred to as a "sink" if it eliminates greenhouse gases, aerosols, or precursors of greenhouse gases from the atmosphere.
7. "Source" refers to any procedure or activity that causes the atmosphere to be exposed to greenhouse gases, aerosols, or their precursors.

10.6 UNFCCC

10.6.1 Shifting cultivation and climate change

Resource use and management methods that rely on fire and hence produce carbon are under more strain in the era of global climate change. Particularly with shifting crops, this is true. Shifting cultivation is one of the least known land use methods since it differs so greatly from the agricultural practices used in the lowlands and by the bulk of the population. Therefore, colonial and post-colonial governments in Asia have developed policies and regulations trying to eliminate shifting agriculture for more than a century in the name of forest protection and development. Shifting agriculture is typically blamed for poverty-inducing state policies because it uses outdated technology, is inefficient, wasteful, and hinders development. Both woods and soils are destroyed by it. A thorough condemnation of shifting cultivation by government officials, legislators, or professionals is based on scant and incorrect information, or just a myth, according to decades of research on almost every facet of the practice. Despite all evidence, however, state policies and decision-makers' views have hardly changed (IWGIA/ AIPP 2009).

The climate of the Earth has changed and developed over time. Others can be ascribed to human activities like deforestation and atmospheric emissions from things like industry and transportation, which have resulted in gases and aerosols being stored in the atmosphere. Some of these changes have been caused by natural factors. They are referred to as greenhouse gases (GHGs) because they function as a greenhouse on the surface of the globe, trapping heat and raising air temperatures close to the ground. Sea level, snow cover, ice sheets, and rainfall are already being impacted by global warming. Watersheds and ecosystems are already being impacted by changes in regional climate patterns marked by rising air temperatures in various parts of the world (UNFCCC Handbook 2006).

The framework for coordinated international action to slow down climate change and prepare for its effects is provided by the United Nations Framework Convention on Climate Change. Its provisions are forward-thinking, creative, and deeply rooted in the idea of sustainable development. The Convention's membership is almost universal with 189 Parties.

10.6.2 Objective and principles of the convention

The UNFCCC describes how laws and regulations may affect these developments in climate change. In order to offer a reliable source of current, interdisciplinary knowledge on climate change, the IPCC was established in 1988. "Policy-relevant, but not policy prescriptive" is how it describes its conclusions. In the end, science cannot answer the question of what amounts of GHG are tolerable. Additionally, it must be the focus of political choices. But natural science provides proof of the existence, underlying mechanisms, and magnitude of human influence on the climate system, which has grown dramatically since the Convention's adoption. The findings of the IPCC's earlier Assessment Reports (1990 and 1995) were verified in its Third Assessment Report (TAR) (2001). The TAR concludes that it is highly likely that ecosystems, human life, and the economy are already suffering because of human-induced climate change. The TAR also offers a number of scenarios that estimate the GHG emission cuts required to stabilise atmospheric GHG concentrations at specific levels.

10.6.3 Reducing global warming

Mitigation of climate change and its effects can be done in one of two ways. First, sources of anthropogenic GHG emissions must be restricted or appropriately reduced. Second, GHG sinks and reservoirs must be preserved or appropriately improved. The entire atmosphere, geosphere, hydrosphere, and biosphere along with their interactions are referred to as the climate system. This includes not just the atmosphere but also the earth, water, and living things.

Taking mitigation measures and adapting to climate change is crucial to maintaining inclusivity and equality, as well as the fact that climate impacts do not discriminate and that climate action is necessary for reaching all SDGs. The Talanoa Dialogue, the UNFCCC Momentum for Change initiative, the Marrakech Partnership for Climate Action, and relevant activities on adaptation, mitigation, means of implementation, gender, empowering youth, the Loc-Action Initiative have all taken place or are currently taking place to enable countries to advance climate action, capture and share relevant successful experiences and lessons learned, and ensure inclusivity and equality in the context of climate action under the UNFCCC process. The Talanoa

Dialogue, which was part of the UNFCCC process, gave nations a chance to assess their progress and aim to raise global ambition in order to achieve the goals of the Paris Agreement (UN 2019). Fiji, which was in charge of the COP 23 UN Climate Change Conference, suggested the idea of "Talanoa," a Pacific island. Three key questions centred the Talanoa Dialogue's contributions from nations and other stakeholders, including business, investors, cities, regions, and civil society. Where are we trying to go? Just how do we get there? The Talanoa Dialogue increased participation, increased openness, and ensured that no one was left behind in the effort to boost global ambition to meet the goals of the Paris Climate Change Agreement. Nearly 100 actions from different sectors, levels, and regions were identified. Some of these actions have already been successfully carried out with visible results, inspiring others to use similar practices. The practices identified are success stories and sources of inspiration, showing what is effective and what might be applied elsewhere. These initiatives are carried out in the context of the 2030 Agenda for Sustainable Development and are directed by the long-term objectives of the Paris Agreement. To help the NDCs and National Adaptation Plans succeed and exceed their goals, the Partnership focuses on immediate climate efforts between now and 2020. The UN Climate Change secretariat has launched an effort called Momentum for Change to highlight the massive groundswell of initiatives taking place around the world that are guiding the world towards a highly resilient, low-carbon future. Innovative and ground-breaking responses to broader economic, social, and environmental issues as well as climate change are honoured by Momentum for Change. Some of the most realistic, scalable, and replicable examples of what people, organizations, governments, and corporations are doing to combat climate change may be found in these "Lighthouse Activities." Regional Climate Weeks encourage people and organizations to join the movement started by the Paris climate accord. It is a distinctive venue for collaboration where representatives of the government and non-Party stakeholders come together to discuss the whole spectrum of pertinent climate issues under one roof. The major objective of the Regional Climate Weeks is to unite various public and commercial sector players behind a shared objective of combating climate change. The Weeks are a collection of activities that offer a forum for a regional grassroots exchange of information and best practices on the implementation of the NDC, the SDGs, and global climate action. The following regions host the Regional Climate Weeks annually: Africa, Latin America and the Caribbean, and Asia-Pacific. It is a distinctive programme that seeks to increase regional climate neutrality by the middle of the century by educating and assisting people of our society in developing nations. The Regional Climate Weeks in 2019 act as vital preludes to the UN Secretary-General's Climate Summit in September. All nations are required to provide information regarding greenhouse gas (GHG) emissions, steps taken to reduce them, adaptation, and means of implementation, including funding, the transfer of technology, and capacity building. The data assembled and submitted by nations serve as a crucial starting point for understanding current emission levels, the scope of ongoing activities, and the development of national and international support for climate action. Additionally, it can give governments the chance to show that they are accountable to the general public and civil society by disclosing and accounting for their efforts to lessen the effects of climate change and GHG emissions as well as any assistance that was given, mobilised, or required. The Least Developed Countries Expert Group (LEG) makes ensuring that developing nations receive support for their climate change initiatives. The main goal of the LEG's work is to guarantee that the least developed nations receive appropriate assistance in order to adapt to the effects of climate change, which is one of the factors contributing to inequality. The LEG, which was established in 2001 as a component of the UNFCCC's support system for LDCs, has the responsibility of giving the LDCs technical advice and assistance regarding the creation and execution of national adaptation plans (NAPs), the creation and execution of national adaptation programmes of action (NAPAs), and the execution of the LDC work programme. In partnership with the GCF secretariat, the LEG also offers technical direction and assistance on how to obtain funds from the Green Climate Fund (GCF) for the procedure to develop and implement NAPs. NAPs continue to be the principal tool for planning and implementing adaptation measures under the UNFCCC process and serve as a springboard for coordinated, strategic planning, encouraging additional adaptation action across sectors, levels, and regions. The UN climate regime supports efforts to create and carry out training and education initiatives on climate change. The Local Communities and Indigenous Peoples Platform was founded at COP21 in accordance with the Paris Agreement, which acknowledges the crucial role that indigenous peoples and local communities play in creating a world that is climate resilient. The Platform's three functions related to knowledge, engagement capacity, and climate change policies and actions were agreed upon by all parties, with ongoing input from representatives of indigenous peoples. These principles call for the full and effective involvement of indigenous peoples as well as their equal standing with other Parties, including in positions of leadership. The secretariat and other international organizations may make sure that countries are assisted in achieving the SDGs in a coordinated and collaborative approach by working more closely with other United Nations bodies.

10.7 THE GOALS FOR 2050

By 2050, the draught framework aims to realise the 196 member parties of the CBD's vision of "humanity living in harmony with nature," which is accomplished through four goals.

Goal 1: To increase the connection and integrity of natural ecosystems, with a rise of at least 15% in the area, supporting robust and healthy populations of all species.

Goal 2: For the benefit of everyone, nature's contributions to humanity have been acknowledged, preserved, or improved via conservation and sustainable usage;

Goal 3: The exploitation of genetic resources should result in a fair and equitable distribution of financial and non-financial gains, particularly those related to biodiversity protection and sustainable usage.

Goal 4: The gap between the financial and other implementation methods that are currently available and those that are required to realise the 2050 Vision is closed.

10.7.1 Achieving certain benchmarks by 2030

Each of the four goals has two to three broad milestones that must be attained by 2030 (10 total milestones):

Goal 1

Milestones

1a: A net gain of at least 5% in the area, connectedness, and system integrity of natural systems.

1b: Reducing or reversing the risk of extinction of threatened species by at least 10%.

1c: Protection of genetic variety of wild and domesticated species.

Goal 2

Milestones

2a: All pertinent public and private decisions take into account nature's benefits to humans in their deliberations.

2b: Each applicable Sustainable Development Goal is furthered by ensuring the long-term sustainability of all categories of nature's contributions to humans, including the restoration of those that are currently in decline.

Goal 3

Milestones

3a: The proportion of financial rewards that providers including those who possess conventional knowledge receive has increased.

3b: Non-financial gains have risen, such as the involvement of suppliers including those who possess traditional knowledge in research and development.

Goal 4

Milestone

4a: Enough financial resources are available and deployed to implement the framework, gradually bridging the financing deficit to at least US $700 billion year by 2030.

4b: Sufficient alternative methods, such as technology transfer, technical and scientific cooperation, and capacity-building and development, are available and in use to carry out the 2030 framework.

4c: By 2030, sufficient financial and other resources have been committed or planned for the years 2030 to 2040.

10.7.1 What is the UNFCCC ?

In particular, economic development is crucial for the world's poorest countries. Development of any kind is challenging to achieve even in the absence of the extra challenges posed by climate change. The Convention acknowledges that in the years to come, emerging nations will contribute a growing portion of the world's greenhouse gas emissions. However, it attempts to assist these nations in reducing emissions in a manner that won't impede their economic development in order to achieve its ultimate goal. Later, as the Kyoto Protocol to the Convention was being created, one such win-win solution was to be found.

The Convention acknowledges that all countries are vulnerable to the consequences of climate change and calls for further action to mitigate such effects, particularly in less developed countries that are unable to do so on their own. Mitigation received greater attention than adaptation early in the Convention when Parties attempted to increase their understanding of the consequences of and vulnerability to climate change. The problem of adaptation gained traction following the release of the IPCC Third Assessment Report, and the Parties agreed on a framework for handling unfavourable effects and developing adaptation funding strategies. Several Convention entities are in charge of overseeing the adaption process.

The UNFCCC secretariat is in favour of the Paris Agreement, Kyoto Protocol, and United Nations Framework Convention on Climate Change. The UNFCCC secretariat carries out numerous mandates for diverse initiatives to combat climate change that have been formed as a result of more than 20 years of development under the UNFCCC process. According to the United Nations (UN 2019), the UNFCCC is the primary venue for giving leadership to the UN system and aiding Parties and other stakeholders in the transition to a world that is climate-resilient.

10.7.2 Work of the secretariat

With the acceptance of the Katowice Climate Package at COP 24, the world ushered in a new era of climate ambition:

- Urging all nations to put the Paris Agreement into effect and maximise its potential;
- For countries to considerably update their nationally determined commitments by 2020;
- To assist the least fortunate and developing nations in receiving the assistance they require;
- To achieve the 2030 Agenda's Sustainable Development Goals.

The Katowice package outlines the crucial processes and frameworks needed to make the Paris Agreement workable. Achieving the transition to a low-emissions, climate-resilient society promises to increase trust and deepen international collaboration. This task is one of the biggest of our time. The Katowice agreement is a comprehensive document that was reached after extensive technical deliberation and political compromise. It contains practical guidelines on (UN 2019):

- The data that governments will include in their Nationally Determined Contributions (NDCs) regarding domestic mitigation and other climate goals and initiatives;
- The guidelines for the operation of the enhanced transparency framework, which will reveal to the world what governments are doing about climate change; how to communicate about adaptation measures to climate effects;
- The creation of a committee to aid in the Paris Agreement's implementation and encourage adherence to the commitments made therein;
- How to carry out the worldwide assessment of overall progress towards the Paris Agreement's objectives;
- How to measure technological development and transmit advancement;
- The method for creating new financial targets starting in 2025 and how to provide advance information on financial assistance to developing nations.

10.8 THE UNCCD FRAMEWORK (2018-2030)

1. Drought, land degradation, and desertification (DLDD) are problems with a worldwide scope. They contribute to and exacerbate economic, social, and environmental issues like forced migration, poor health, a lack of food security, biodiversity loss, and water scarcity. They continue to pose significant obstacles to the sustainable development of all nations, but especially those who are impacted (UNCCD 2017).
2. Long-term integrated policies that prioritise increased land productivity as well as the management of land and water resources in a sustainable manner, including restoration and conservation will be necessary to address DLDD.
3. The strategy will help with:
   i) accomplishing the goals of the Convention and the 2030 Agenda for Sustainable Development, in particular with regard to Sustainable Development Goal (SDG) 15 and target 15.3: "by 2030, combat desertification, restore degraded land and soil, including land affected by desertification, drought, and floods, and strive to achieve a land degradation-neutral world" and other connected SDGs;
   ii) improving the living conditions of affected populations.
   iii) improving services of the eosystem.
4. A future that, within the purview of the Convention, eradicate the land degradation from the planet.
5. From 2018 to 2030, all UNCCD partners and stakeholders shall operate in accordance with the following "strategic objectives." By taking into account specific regional and national circumstances, achieving these long-term goals will help achieve the aforementioned vision within the parameters of the Convention.

10.8.1 Strategic objective 1

To protect threatened ecosystems, stop desertification and land degradation, encourage sustainable land use, and help achieve land degradation neutrality.

Impact anticipated

1.1 The productivity of the land and associated ecosystem services are preserved or improved.

1.2 Ecosystems are more resilient and have less vulnerability than unaffected ecosystems. anticipated effect.

1.3 Countries that desire to do so develop and accept national voluntary land degradation neutrality targets, identify and put into action relevant actions, and set up essential monitoring systems. anticipated effect.

1.4 Strategies for mitigating desertification and land degradation are exchanged, promoted, and put into practice.

10.8 2 Objecive 2 To enhance the quality of life of the people

Expected impact

2.1 The affected populations have better food security and appropriate access to water.

2.2 The quality and variety of the local populations' means of subsistence have improved.

2.3 Local residents, particularly women and young people, are empowered and take part in decision-making in the fight against DLDD.

2.4 Desertification and land degradation significantly diminish forced migration.

10.8.3 Strategic objective 3

To lessen the effects of drought, adapt to it, and manage it in order to increase the resilience of vulnerable individuals and ecosystems:

Impact anticipated

3.1 Ecosystems' susceptibility to drought is decreased, in part thanks to sustainable methods of water and land management.

3.2 Communities become more drought-resistant.

10.8.4 Objective 4

To provide advantages for the environment on a worldwide scale by effectively implementing the UNCCD

Impact anticipated

4.1 The conservation and sustainable use of biodiversity and addressing climate change are facilitated by sustainable land management and the fight against desertification and land degradation.

4.2 Enhancement of synergies with other multilateral environmental procedures and agreements.

10.8.5 Objective 5

To create powerful relationships at the international and national levels in order to organize substantial additional financial and non-financial assets to help the Convention's implementation.

Expected impact

5.1 Additional financial resources, including domestic resources, are promptly and adequately mobilized and made available to the affected country Parties.

5.2 Effective and targeted capacity-building initiatives as well as "on-the-ground interventions" are supported by international assistance, particularly through triangular, North-South, and South-South cooperation, in impacted country Parties.

5.3 Considerable work is done to encourage transfer of technology, notably on advantageous conditions, including concessionary and preferential terms, as mutually agreed, and to mobilize extra non-financial resources.

10.8.6 Framework execution

Initiatives at the national or subregional level will be the main means by which the strategy is carried out, with support from UNCCD institutions, partners, and relevant stakeholders. This section describes the tasks and responsibilities of the various Parties, UNCCD institutions, partners, and stakeholders in order to accomplish the previously specified strategic objectives. In particular, Parties bear primary responsibility for implementing The Strategy through their national action plans. Parties are required to oversee implementation with a mindset of international cooperation and solidarity as well as in accordance with their own national objectives. The Strategy will be a helpful tool for directing national, subnational, and local implementation because Parties will be able to evaluate implementation at the level of the Committee for the Review of the Implementation of the Convention (CRIC). Each Party may implement the plan through joint ventures, public-private partnerships, and innovative agreements. In general, the parties will look for ways to mobilise both financial and non-financial resources; develop and implement safety-net programmes, monitoring systems, and early warning systems for drought; support alternative revenue streams; establish mechanisms for knowledge and information sharing; and promote networking regarding best practices and strategies for managing the drought.

10.8.7 Global mechanism

The Global Mechanism's primary duties include helping to raise money, promoting technology transfer, and looking into ways to raise money for the Convention and The Strategy's implementation. It accomplishes this by assisting affected country Parties in exploring resource mobilization options that meet the needs outlined in their UNCCD action programmes;

10.8.8 Paris agreement

The Paris Agreement was approved by the Conference of the Parties at its 21st meeting on December 12, 2015. The Paris Agreement will enter into force 30 days after being ratified, accepted, approved, or acceded to by a minimum of 55 Parties to the UNFCCC (United Nations Framework Convention on

Climate Change), which collectively account for at least an estimated 55% of total worldwide greenhouse gas emissions. The Secretary-General of the United Nations is the Depositary. The agreement was made available for signature in New York on April 22, 2016. The Paris Agreement came into force on November 4, 2016, following the occurrence of the threshold for entrance into force on October 5, 2016.

The GHG emissions due to human activities are the major cause for global warming because of increase in temperature over the 20th century. In theory, the process by which some atmospheric gases radiate heat back to the Earth's surface is a natural occurrence; otherwise, our planet would be much colder and probably uninhabitable (de Chazournes 2008). By the middle of the 1980s, scientists had issued a warning that human activity and the rise in anthropogenic greenhouse gas (GHG) emissions were largely to blame for the global warming that had gone beyond natural fluctuation. In a conference held in Villach, Austria in 1985 titled "International Conference on Assessment of the Role of Carbon Dioxide and of Other Greenhouse Gases in Climate Variations and Associated Impacts," scientists urged politicians to work with them in examining possible policies to counteract human-caused climate change. An further sense of enhanced urgency for action was produced in 1988 by the revelation of the ozone hole and a heat wave.

According to numerous decades' worth of scientific research, human activities contribute to global climate change through producing greenhouse gases (GHGs) through the burning of fuel, specific industries, deforestation, and other processes. Before government leaders worldwide agreed to work together to discuss how to address the hazards that GHG-induced climate change poses to people and ecosystems, scientists researched and evaluated the science for more than 150 years. Following a number of international scientific conferences in the years 1985–1987, under the auspices of the WMO and the UNEP, governments made the decision to establish the IPCC to provide them with assessments of the science underlying climate change, forecasts of its social and economic effects, and potential response strategies.

10.8.9 Kyoto Protocol (KP)

The 1997 KP, which became operative in 2005, served as the first UNFCCC subsidiary agreement. United States is not a Party to KP because it was signed but not ratified by this country. To cut their GHG emissions by an average of 5% below 1990 levels between 2009 and 2012, 37 high-income nations and the European Union (EU) were given legally enforceable targets under the Kyoto Protocol. For developing countries, it eliminated GHG mitigation commitments. For the period of 2013 to 2020, the majority of the high-income

Parties mostly the EU members and other European countries took on new GHG reduction targets. Individual Parties have made varying degrees of progress towards reduction in GHG emissions. The KP was not joined by the United States, and Canada withdrew before the first commitment term was up. Australia, Japan, and other Parties made the decision to seek a deal that contained commitments on the same conditions from all Parties during the negotiations for the second KP commitment period. As a result, it was decided to establish a protocol, another legal instrument, or an agreed-upon resolution with legal effect under the UNFCCC that would apply to all Parties no later than 2015 during the 2011 COP17 in Durban, South Africa.

10.8.10 Nationally Determined Contributions (NDC) Guidelines

New rules for reporting NDCs were accepted by the PA's Parties. Every five years, NDCs are to be updated, and they "will" show an increase in ambition to reduce GHG emissions over the previous NDC. The Parties concurred that they "should" utilize a predetermined framework for sharing NDCs, although the specifics have yet to be determined. On "common timeframes" for NDCs, such as whether to look five or ten years into the future, there is still no consensus. The United States is one of the Parties that submitted NDCs with a deadline of 2025, and these Parties must communicate "new" NDCs by 2020. NDCs with time spans up to 2030 shall be communicated or updated by the Parties by 2020. If the US remained in the PA, a new NDC would have to be submitted in 2020. NDCs continue to have a national determination and are not legally binding, but they should reflect a Party's goals. The recommendations are applicable to NDCs filed in 2025, but Parties are encouraged to update their NDCs in 2020 using a format that has been agreed upon. For Parties who want to incorporate adaptation measures in their NDCs, the guidelines also provide information on how to report them.

10.8.11 Enhanced Technology Framework (ETF)

Since the UNFCCC was being negotiated, the United States has placed a high focus on establishing strict guidelines for each Party's attempts to be as transparent as possible. The details that Parties must include in their NDC reports are laid out in the ETF rules. Along with biennial transparency reports, this data is anticipated to enable "facilitative multilateral consideration of progress." The IPCC's technical recommendations will continue to be used in methods for GHG emission estimation and other technical matters. As applicable to their NDCs, all Parties are required to give information on the following, according to the U.N. Climate Change Secretariat:

- Time frame for implementation;
- Scope and coverage of the NDC

- Quantitative data on the benchmark for GHG mitigation efforts or goals.
- National planning procedures for creating the NDC and, if available, plans for its implementation that take into consideration local conditions;
- All presumptions and theoretical frameworks;
- How the Party decides whether its NDC is just and aspirational;
- What role the NDC plays in reaching the UNFCCC's goal.

The guidelines are designed to make it easier for the committee to examine and consult in order to promote adherence to promises. A technical team of specialists will examine whether a Party submits a timely NDC and reports its NDC in accordance with the rules. Under the ETF, the sufficiency and suitability of the Parties' NDCs are not up for assessment. Only those provisions in the modalities, processes, and guidelines that are specifically stated to allow flexibility are given flexibility in reporting under the PA. The use of flexibilities is at the discretion of the Parties. Having said that, there are safeguards in the review processes when exploiting the flexibilities. A developing country that chooses to apply a flexibility because it believes it lacks the capacity to comply with the rules must make it clear in its Biennial Transparency Report that it has done so. It must describe the capacity constraint, how it plans to resolve the constraint, and the projected time period for addressing the constraint or constraints. These flexibilities might not be examined by the technical review teams.

10.9 LAND DEGRADATION NEUTRALITY (LDN)

The UN Conference on Sustainable Development (Rio+20) in 2012 gave rise to the LDN idea. LDN addresses a pressing issue: how to increase food, fuel, and fibre production to satisfy future demand without further depleting our limited land resource base. In other words, LDN imagines a scenario in which human activity has a neutral or even beneficial effect on the environment. In human terms, soil is a limited resource, and every year, agricultural land loses around 24 billion tonnes of fertile soil. Because it supplies the ecosystem services that people rely on, soil is a crucial component of human development for numerous reasons (Sheals, 1969). More than 90% of our food comes from the soil, making it crucial for water filtering and groundwater recharge. Land degradation frequently happens in regions where population growth and land scarcity are already prevalent. Due to the lack of incentives to invest in sustainable land management and the increased danger that responsible farmers won't be able to profit from their investment, insecure land rights contribute to the degradation of the environment (Davies *et al.* 2016, p. 11). If we are to use soil sustainably, stopping and correcting land degradation must be a top concern. Though the idea of LDN was first presented in 2012 (United Nations 2012), the UNEP Global Environmental Outlook (UN Environment 2019) emphasised that it would be difficult to meet the land degradation neutrality target set by the United Nations Conference on Sustainable Development (Rio+20) by 2019. Satellite-based analyses estimate that 29% of the world's land surface is made up of hotspots for land degradation. Keesstra *et al.* (2018) also state that "Land is under pressure and is rapidly degrading as a result of poor use, which will put even more pressure on the remaining land."

A solid legal framework that can support national LDN aims, support from international and national organizations, and the LDN target setting programme that UNCCD implemented in 2015 are all necessary for achieving LDN. The understanding that better land management is essential to the UN Sustainable Development Agenda 2030 is an important part of the link between soil governance and LDN. "Highlighted the role of well functioning soils that may liberally deliver their ecosystem services for the accomplishment of the SDGs," Keesstra *et al.* (2018) wrote.

10.9.1 Global initiatives

Numerous worldwide initiatives have produced regulations that, when applied at the national level, can be used to stop and prevent land degradation (Boer and Hannam, 2019). As the main international tool for LDN, UNCCD encourages nations to take a wide range of actions to prevent, slow down, and stop land degradation. Planning, legislation, and sustainable land management techniques are some of these approaches, "combined with localized action to reverse past degradation, through land restoration and rehabilitation, to achieve a state of no net loss of healthy and productive land" (UNCCD 2019). However, it has been claimed that the UNCCD would need to be modified to include an LDN mechanism in the Convention in order to adequately address the special legal requirements of LDN (Boer and Hannam 2021).

Under the UNCCD, nations are urged to take a wide range of steps to prevent, slow down, and stop land degradation. The quantity and quality of land resources required to support ecosystem activities and services and improve food security have been listed as the main components of LDN. Within a set of predetermined temporal and spatial scales and ecosystems, these must remain stable or grow. This chapter makes the case that these elements would also serve as a helpful foundation for organizing the process of developing national legal instruments for LDN implementation.

The Global Environment Facility (GEF) is aiding in the implementation of Land Degradation Neutrality as a financing vehicle for UNCCD. A voluntary organization of more than 60 nations with various socio-ecological situations received technical help from the Land Degradation Neutrality Target Setting Project at first. It assisted them in integrating LDN into their national action plans in accordance with the strategic plan of UNCCD. The project provides assistance to nations in planning how to restore their landscapes through the GEF.

The United Nations Convention to Combat Desertification (UNCCD)'s 12th Conference of the Parties (COP12) introduced the idea of land degradation neutrality (LDN) in October 2015. LDN was added as one of Sustainable Development Goal 15's aims by the UN in the same year. According to Feng *et al.* 2022, LDN is essential to study on sustainable development and worldwide environmental change. The available LDN evaluations primarily concentrate on the LDN idea and provide little detail regarding the development of LDN research. Trade-offs and synergies exist between LDN and other SDGs, and LDN is a dynamic balance between land degradation and restoration. According to Feng *et al.* (2022), it is necessary to: identify the LDN indicators in various locations; create a technique to qualitatively and quantitatively monitor LDN changes; quantify the LDN drivers and describe the elements that affect LDN; clarify future LDN trends; and encourage the implementation of SDG 15.3.

LDN is "a state in which, within specified temporal and spatial scales and ecosystems, the quantity and quality of land resources required to support ecosystem functions and services to enhance food security are stable or growing."

The UNCCD's LDN goals consist of

i) Ensuring that ecosystem services are delivered sustainably.
ii) Maintaining or increasing land production to strengthen food security around the world.
iii) Making the environment and the people who depend on it more resilient.
iv) Looking for ways to combine these goals with other social, economic, and environmental ones.
v) Strengthening and advancing inclusive, accountable land governance.

10.9.2 LDN Principles (UNCCD)

A set of 19 ideas known as the Land Degradation Neutrality (LDN) principles directs the LDN implementation procedure. According to Orr *et al.* (2017), they are created to prevent or minimize unwanted or bad events while ensuring that LDN achieves its intended good outcomes. The guidelines are:

1. Maintain or improve land-based natural capital: Despite the effects of climate change, LDN is achieved when the amount and quality of land-based natural capital are steady or growing.

2. LDN-related acts shall not violate the rights of land users, particularly small-scale farmers and indigenous populations, in order to protect human rights and improve human welfare. These activities should support livelihoods and should not have an adverse impact on the cultural or agricultural value of the land.
3. Adherence to national sovereignty: Governments define national goals while taking into consideration local conditions, led by the global level of ambition. The aspiration level and the manner in which LDN targets are implemented into national planning procedures, programmes, and strategies are determined by the governments.
4. The baseline is equivalent to the LDN target: To preserve objectivity, the baseline (when the LDN conceptual framework was applied, the land-based natural capital as determined by a set of globally recognized LDN indicators) is the goal to attain.
5. Neutrality is typically the minimum goal; however, nations may choose to establish a higher bar. This frequently occurs when nations make efforts to raise the bar for their land-based natural capital and increase the amount of wholesome and productive land. In exceptional cases, a nation may choose (and defend) its LDN aim while accepting that advantages may occasionally outweigh drawbacks. This is permissible if they predict that some future land degradation brought on by prior choices or circumstances cannot currently be offset.
6. A framework for classifying and accounting for land use decisions, as well as the effects of land use and management, with respect to a "no net loss" aim should serve as the foundation for the neutrality mechanism. An integrated land use planning principle is this.
7. To achieve neutrality, balance planned increases elsewhere within the same land type with anticipated losses in natural capital based on the land. In order to achieve LDN, it could be necessary to balance increases elsewhere within the same land type during the same time period with predicted losses in land-based natural capital.
8. In order to ensure effective implementation, counterbalancing should be managed within national or subnational borders at the same scale as the biophysical or administrative areas where land use choices are made.
9. Balance "like for like": Gains and losses should be balanced "like for like" whenever possible. As a result, exchanges between various ecosystem-based land types typically do not take place, unless there is a net gain in land-based natural capital as a result of the exchange. Clear criteria should be established based on forecasts for determining what

types of "net gains" enable crossing land type boundaries in order to ensure that there is no unanticipated change in the overall ecosystem composition of a country and no harm to endangered ecosystems.

10. Maintain or improve the quality of all ecosystem services while balancing economic, social, and environmental sustainability. LDN does this by maximizing the trade-offs between environmental, economic, and social results. By fusing societal, economic, and environmental sustainability, LDN implementation supports sustainable development. Additionally, it prevents the load from being transferred to other areas or future generations while managing the land for ecological services.
11. Base land use decisions on multi-variable assessments: When making decisions about land use, it is important to take into account gender as well as other social, cultural, and economic variables as well as the potential and condition of the land. Before beginning interventions, these should be locally vetted to ensure that decisions are supported by facts and lower the chance of annexation of land.
12. Use the response hierarchy: When developing interventions and planning for LDN, the response hierarchy of Avoid > Reduce > Reverse land degradation should be used, where "avoid" and "reduce" have priority over "reverse" in order to identify and pursue the best possible combination of actions with the goal of achieving no net loss across the landscape.
13. Planning and implementing LDN requires thoughtful participatory processes that involve stakeholders, particularly land users, in creating, putting into practice, and overseeing interventions to achieve LDN. To ensure that local, traditional, and scientific knowledge is taken into account in the decision-making process, processes should employ multi-stakeholder platforms. The procedure needs to guarantee equal access to information and be gender sensitive.
14. Implement strong governance: Since LDN is based on good governance, planning and execution should include:

- Eliminating and reversing policy drivers that contribute to subpar land management.
- Ensuring tenure rights and security in the pursuit of LDN by adopting the guidelines and standards of the Voluntary Guidelines on the Responsible Governance of Tenures
- Taking into consideration the availability of resources (both human and financial) for putting into practice effective strategies to address land degradation and desertification

- Establishing procedures for LDN implementation monitoring and reporting
- Creating a system to coordinate integrated land use and management planning across scales and sectors to enable stakeholder input in national and global decision-making and reporting.
- Creating a system for the timely evaluation of implementation results and suggestions for improvement.
- Guaranteeing transparency and accountability at all levels.

15. For reporting and understanding the level of degradation, use the minimum set of internationally recognized indicators/metrics, which were adopted by the UNCCD.
16. A loss is deemed to have occurred if any of the three indicators/metrics (land cover change, land productivity, and carbon stocks) shows a significant decline. The "one-out, all-out" technique is the name of this tactic. A gain, on the other hand, is considered to have happened if at least one indicator/metric shows a significant positive change and none a significant negative change.
17. Use additional national and sub-national indicators, along with qualitative and quantitative data, to aid in interpretation and close any gaps for ecosystem services that the minimal global set does not adequately address.
18. Utilize regional expertise acquired through regional multi-stakeholder forums to interpret monitoring data in accordance with regional context and goals, within predetermined parameters.
19. Monitoring should be seen as a means of education. Opportunities for capacity building, the basis for testing the theories that guide counterbalancing choices and actual interventions, the LDN idea, and this conceptual framework are all provided by monitoring, along with knowledge that can guide adaptive management.

To guide the application of the conceptual framework and aid in preventing unexpected consequences when implementing Land Degradation Neutrality (LDN), LDN Principles are prese Another crucial LDN value is the sharing of knowledge and learning. Future LDN projects will benefit from the lessons learned from what works and what doesn't by monitoring and verifying the results.nted. Gains in natural capital should help to improve and sustain livelihoods in order to accomplish the broader development goals of the UNCCD and the Sustainable Development Goals (SDGs) through LDN interventions. Making sure there is multi-stakeholder participation backed by national coordination, which should include existing local and regional governance systems, is an essential component of these principles. This will guarantee that marginalised groups are not uprooted when a piece of land is chosen for restoration.

10.9.3 LDN assessment methods

Indicators and questionnaires are the two primary components of land deterioration neutrality evaluation methodologies. There are many social and environmental aspects to land degradation, and no one measure can accurately reflect the health or state of the land. Different ways to assessing land degradation have emerged as a result of different definitions of land degradation. The UNCCD has established three indicators for assessing LDN: vegetation productivity, land cover, and soil organic carbon, in an effort to analyse LDN consistently and improve the process of achieving the sustainable development goals (SDGs). According to Orr *et al.* (2017), these metrics show how well the land can deliver ecological services. The questionnaire approach is a crucial addition to the indicator method as well. According to several studies (Nzuza *et al.*, 2021) it is easier to evaluate land degradation when the questionnaire approach and the indicator method are combined.

Indicator method

The three assessment indicators recommended by the UNCCD have shown preliminary improvements thanks to a number of research. Degraded and undamaged land are separated by the UNCCD's LDN indicator. This distinction, though, is oversimplified for describing land deterioration. To supplement the dichotomous classification details, several research have integrated LDN evaluation with land degradation factors as wildfire, landslides, and drought to define 12 types of land degradation (Al Sayah *et al.*, 2021). The UNCCD promotes the creation of regional indicators that might improve LDN evaluation. New indicators, such as land contamination, gully erosion, soil loss rate, total soil loss (Tsymbarovich *et al.*, 2020), and atmospheric humidity (Kuderina *et al.*, 2020), have been added to local LDN assessments at the regional scale in accordance with the various land degradation characteristics. LDN assessment is facilitated by an expanded and enhanced LDN indicator, which is essential for determining the various forms of local land degradation.

Important LDN indicators include productivity, vegetation, and soil OC. The three indicators, however, have a wide range of data items. While LDN is a term that includes both deterioration and restoration, the majority of current research methodologies attempt to evaluate land degradation. To track LDN progress, it is necessary to create methodologies that take land degradation and restoration into account. To determine the status of LDN, deterioration and repair must be graded simultaneously.

Questionnaire method

To increase stakeholders' involvement in LDN evaluations, a number of studies have combined participatory mapping, farmer interviews, and field surveys with the LDN results (Crossland *et al.*, 2018). Utilizing a mapping questionnaire (QM) based on the World Overview of Conservation Approaches and Technologies (WOCAT), the location and degree of degradation in South Africa were ascertained. The results showed that the areas of land degradation identified by this method differed from those found by taking vegetation productivity, land cover, and soil organic carbon into account (von Maltitz *et al.*, 2019). Therefore, more efficient methods of assessing LDN need to be created. Furthermore, other studies have shown that integrating information from questionnaire surveys and remote sensing might improve the evaluation of land degradation (Nzuza *et al.*, 2021).

10.9.4 Elements of the conceptual framework

LDN aspires to preserve or improve the natural capital of the land and related ecosystem services provided by the land. Therefore, pursuing LDN necessitates making an effort to prevent additional net loss of the land-based natural capital in comparison to a reference condition, or baseline. The LDN promotes a dual-pronged approach of steps to avoid or mitigate land degradation, paired with efforts to correct past degradation, thereby, in contrast to previous approaches, creating a target for land degradation management. The goal is to produce a situation where there is no net loss of healthy and productive land by balancing losses and gains. The counterbalancing mechanism and integrated land use planning In order to achieve neutrality at the national level, it will be necessary to implement an optimal mix of interventions designed to prevent, reduce, or reverse land degradation, as well as track land use changes where degradation is anticipated. As a result, the conceptual framework presents a novel strategy in which soil degradation control and land use planning are combined. It is urged and advised for decision-makers to take into account the cumulative consequences of their individual actions that affect the management of specific land parcels on the health and productivity of a nation's land resources. Thus, LDN encourages integrated land use planning that takes into account potential long-term effects of climate change. The counterbalancing mechanism necessitates the adoption of actions that will result in gains in land-based natural capital that are at least as big as the predicted losses owing to environmental degradation elsewhere. Establishing neutrality In order to achieve LDN, sustainable land management strategies that prevent or slow down deterioration must be combined with initiatives to stop degradation by restoring or rehabilitating degraded land. The Avoid > Reduce > hierarchy of responses Reverse land degradation describes the priorities in designing LDN interventions: the majority of work should go into preventing land degradation on the principle that "prevention is better than cure" because repairing damaged land is time- and money-consuming. The landscape scale is used to manage LDN implementation. Within each land type, actions to generate similar gains must be implemented to offset predicted losses. Land potential, which is a reflection of innate qualities including soil type, topography, hydrological, biological, and climatic factors, defines different types of land.

1. Avoid – By addressing the causes of land degradation and taking proactive efforts to prevent negative changes in the quality of undamaged land and to confer resilience, through appropriate regulation, planning, and management practices, land degradation can be avoided.
2. Reduce – Applying sustainable land, water, and forest management practices can decrease or alleviate land degradation on agricultural and forestry land.
3. Reverse - When possible, degraded land's productive capacity and ecological services can be partially (rarely completely) restored or rehabilitated by actively promoting the restoration of ecosystem processes.

10.9.5 Monitoring LDN

Over 20% of the world's vegetated land surface and 1.5 billion people have been impacted by land degradation in the past 20 years, which is commonly characterised as a loss in biological production and an ecosystem's complexity (UNCCD, 2017). Due to unsustainable farming practices and improper land and water management on agricultural land, the majority of soil degradation is anthropogenic. Political and socioeconomic considerations are also underlying causes (Vorovencii, 2016). Globally speaking, agricultural landscape degradation threatens food security, lowers carbon sequestration in soils and biomass, and results in significant economic losses, particularly in already underdeveloped regions (Muchena *et al.*, 2005; UNCCD, 2017). The main objective of LDN monitoring is to evaluate changes from baseline values in the metrics designated as indicators of land-based natural capital. Planning for neutrality and monitoring neutrality are independent but complementary processes that use the LDN conceptual framework.

10.9.6 Indicators of LDN

The ability to provide services based on land-based ecosystems is reflected by essential factors and driving variables, which are monitored through indicators of LDN. Metrics that are generally applicable, comprehensible, and, ideally, quantifiable using the data sets at hand should be used to evaluate indicators. To track advancements in LDN implementation and socioeconomic effects of LDN, additional metrics are needed.

10.9.7 The three global indicators

Due to the enormous variation in land degradation, as well as highly practical factors about capability in data collection, analysis, interpretation, and reporting of governments and stakeholders, globally applicable indicators for land degradation have inherent limitations. The foundation for identifying and choosing the most appropriate indicators, as well as the related data and information required for their development, is a goal-driven conceptual framework.

10.9.8 Comparative measures of ecological services

The three global indicators focus on the most important elements of land-based natural capital, but it's possible that other indicators are needed to thoroughly analyse developments in land-based ecosystem services. In order to cover the ecosystem services provided by the land that are significant in each setting, national (or sub-national) level indicators should be added to the global indicators. Countries are urged to choose supplementary indicators that take into account their unique national and subnational characteristics that will improve how the global LDN indicators are interpreted in relation to the accomplishment of LDN's objectives. These could include metrics or indicators for extra SDGs (ECOSOC, 2016) or other national metrics or indicators, such as measures that address locally relevant issues such as heavy metal contamination and biodiversity evaluation through the Red List Index (Bubb *et al.*, 2009).

10.9.9 Indices of social and economic progress

The indicators that assess the short-, medium-, and long-term social and economic effects of LDN policies and initiatives are the final group to be taken into account. They might include measures of human well-being including the protection of land rights and impacts on local communities.

10.9.10 Principles related to monitoring

1. Report using the minimal set of globally recognized indicators/metrics that the UNCCD adopted in order to understand the state of deterioration (UNCCD, 2013).

2. National data sources, including aggregated sub-national data, should constitute the main foundation for monitoring and reporting.
3. Use additional national and sub-national indicators, along with quantitative and qualitative data and information, to fill in any gaps in the minimal global set's coverage of ecosystem services and to aid in interpretation.
4. All nations need to have access to surveillance techniques. Monitoring can be done at various levels of intensity and with the participation of many stakeholders and does not require complex equipment or expensive cost. There are numerous monitoring methods available, and each nation should choose the one(s) that best suit its priorities and resources and implement it consistently throughout time.
5. It is important to pursue harmonization in monitoring methods across countries, with the potential for standardization where appropriate and practical, while taking into account variation in the causes and effects of land degradation among countries, as well as in their capacity to measure and monitor change.
6. The three global indicators' results should be combined using a "one-out, all-out" approach, where any significant decline in any of the metrics is considered a loss; however, if at least one metric shows a significant rise and none shows a significant fall, the result is considered a gain.
7. Within the bounds of established criteria, use in-situ validation and local expertise acquired through regional multi-stakeholder forums to evaluate monitoring data.
8. Given that a minimal set of internationally recognized indicators may occasionally result in "false positives" (for instance, shrub encroachment may result in higher NPP and SOC), the monitoring system must allow users to report false positives that are backed up by national data and contextual information.
9. Monitoring should be viewed as an educational tool. The LDN idea and this conceptual framework are all provided through monitoring, together with insights that may direct adaptive management, opportunities for capacity building, the basis for testing the theories that lead counterbalancing options and practical interventions.

Land Degradation Neutrality (LDN, SDG Target 15.3) by 2030 may be challenging to achieve given present trends in land degradation. The major issue is land governance, which is preventing solutions to prevent, minimise, and reverse land degradation from being implemented at sufficiently broad scales. In this essay, we examine the dynamics of agricultural land governance and how they could contribute to or prevent land degradation (Debonne *et al.*, 2021). According to the literature, agency structures are changing, giving value chain actors more influence over land governance. Two related trends that reflect these agency shifts are as follows: First, actors in the agricultural value chain increasingly have an impact on decisions about land management through cooperation in the value chain, such as contract farming. Second, considerable amounts of land are being replaced by domestic and foreign large-scale farms, both of which represent enhanced direct participation of the value chain with land management (Debonne *et al.*, 2021). These new agreements are associated with both increased agricultural production and unsustainable land management because of absent landowners, short-term goals, and high-intensity agriculture. Debonne *et al.* (2021) claim that value chain participants have the resources and the business case to spur the development of LDN solutions. For LDN to be implemented successfully, efforts must be refocused on empowering and, where necessary, constraining all entities with influence over land management, including value chain actors.

A variety of LDN brokers, actors working to advance LDN, assist nations in setting goals and putting them into action. These actors include international NGOs, knowledge centres, and funding sources. By making up for any land degradation with land restoration, within predetermined time- and space-scales, LDN conceptually aims to, on balance, retain or improve the quantity and quality of land resources (Cowie *et al.*, 2018). At the national level, LDN interventions are technically implemented by compensating any ongoing land degradation within a land type by restoring and rehabilitating an equivalent amount of this type of land elsewhere. In order to achieve this, target setting programmes and national action programmes are created to visualise pathways towards LDN (Global Mechanism, 2019a). LDN is including a substantial role for soil governance into this objective by taking part in the global conversation under UN Sustainable Development objective 15 and the UN Convention to Combat Desertification (UNCCD) in 2015 to support a more effective policy response to land degradation. Governments, landowners and managers, and the community should be able to share responsibility for managing land degradation thanks to legislative provisions that enable stakeholder participation in the creation of policy, guidelines, and ecological standards for LDN (Hannam 2022).

10.10 THE SHILLONG DECLARATION

Recognizing that shifting cultivation evolved as a traditional practice and is now an institutionalized resources management technique that ensures ecological and food security while also providing a social safety net at the species, ecosystem, and landscape levels, Being aware of the various Traditional Institutions and Tenurial Systems associated with Shifting Cultivation in the Eastern Himalayan region, which includes Bangladesh, Bhutan, China, India, Myanmar, and Nepal.

Recognizing that the Institutional Mechanisms embedded in Traditional Shifting Cultivation Systems may ensure that everyone in the community, including the landless and the most marginalized, has access to productive resources, Recognizing that for many indigenous, tribal, and other poor and marginalized upland people, shifting agriculture is a way of life, Recognizing that external and internal pressures have put pressure on traditional shifting cultivation systems, Knowing about existing rules on shifting farming in the Eastern Himalayan countries.

10.10.1 Declaration

a) Shifting cultivation should be recognized as a scientifically sound land use system.

b) It's vital to create an enabling environment in order to solve the urgent livelihood and environmental concerns that have arisen as a result of rapid transformations brought on by development and other externalities like market pressures.

c) Shifting cultivators as rotating agroforestry practitioners must be allowed to actively engage in policy processes that directly affect them.

d) It is critical to raise awareness of the demands and challenges of shifting cultivation and shifting cultivators among existing research and extension services.

e) At the local, national, regional, and global levels, it is vital to develop interactive platforms and environments for varied stakeholders to access and share information.

f) Recognize that women have the most essential role in shifting cultivation.

The second year of cultivation has been discovered to be more dangerous than the first. From the first year of cultivation onwards, it is critical to implement soil and water conservation measures, as well as the introduction of income-generating crops. The following actions must be implemented to improve the existing slash and burn agriculture area.

- Before seeding different crops, contour bunds must be created at appropriate intervals.
- To boost the moisture regime and stabilise the land slope, intensive soil and water conservation techniques, as well as water collection, must be implemented at appropriate sites.

- Planting appropriate grasses and pineapple, among other things, on the contour bunds is required.
- In the areas between two contour bunds, certain fruit trees with various gestation times should be planted in half moon terraces in clocks. Assam lemon, guava, and papaya will be flowering and fruiting in two years. Farmers would be encouraged to stay in the watershed as a result of this.
- Grass on the contour bund that is suitable for livestock will supply fodder, and granular fertiliser dilution will increase crop output.
- It is necessary to introduce high-yielding seeds of various crops.
- To accommodate the farmers' needs, suitable feed, lumber, and fuel trees must be planted along the boundaries.
- Farmers' traditional knowledge-based techniques could be transformed into appropriate technology for the establishment of changing agriculture zones.

References

Adeel Z, Safriel U, Niemeijer D and White R (2005). Ecosystems and human well-being: desertification synthesis. The Millennium Ecosystem Assessment, World Resources Institute, Washington,

Akinyemi FO, Ghazaryan G and Dubovyk O (2020). Assessing UN indicators of land degradation neutrality and proportion of degraded land for Botswana using remote sensing based national level metrics. DOI: 10.1002/ldr.3695, at: https://www.researchgate.net/publication/342185517.

Al Sayah MJ, Abdallah C, Sarkissian RD, Abboud M (2021). A framework for investigating the land degradation neutrality-Disaster risk reduction nexus at the sub-national scales. J Arid Environ, 195.

Aynekulu, E., Lohbeck, M., Nijbroek, R., Ordóñez, J. C., Turner, K. G., Vågen, T., and Winowiecki, L. (2017). Review of methodologies for land degradation neutrality baselines: Sub-national case studies from Costa Rica and Namibia. International Center for Tropical Agriculture publication no. 441. Retrieved from http://hdl.handle.net/10568/80563.

Biswal DK and Kumar S (2012). Shifting cultivation and policies of sustainable development: a meaningless obsession. Journal of Economic and Social Development, Vol - IX, No. 1, 132-138. at: https://www.researchgate.net/publication/236855087.

Boer, B. and Hannam, I. (2021). Restoration of ecosystems and land degradation neutrality. In Fulfilling the Sustainable Development Goals (pp. 392-404). Routledge Abingdon.

Bruun, T. B., de Neergaard, A., Lawrence, A. and Zeigler, A. D. (2009). Environmental consequences of the demise of swidden cultivation in Southeast Asia: Carbon storage and soil quality. Human Ecology, 37, 375-388.

Bryan BA, Gao L, Ye YQ, Sun XF *et al.* (2018). China's response to a national land-system sustainability emergency. Nature, 559 (7713): 193-204

Bubb, P.J, Butchart, S., Collen, B., Dublin, H., Kapos, V., Pollock, C., Stuart, S.N., and Vie, J.C. (2009). IUCN Red List Index-Guidance for National and Regional Use. Gland, Switzerland: IUCN.

Burrell AL, Evans JP, Y. Liu Y (2017). Detecting dryland degradation using Time Series Segmentation and Residual Trend analysis (TSS-RESTREND). Remote Sens Environ, 197: 43-57.

Castella, J-C., Lestrelin, G., Hett, C., Bourgoin, J., Fitriana, Y., Heinimann, A. and Pfund, J-L. (2013) 'Effects of landscape segregation on livelihood vulnerability: Moving from extensive shifting cultivation to rotational agriculture and natural forests in northern Laos', Human Ecology 41, pp63- 76.

Chen C, Park T, Wang XH, Piao SL *et al.* (2019). China and India lead in greening of the world through land-use management. Nat Sustain, 2:122-129

Chen GZ, Li X, Liu XP, Chen YM *et al.* (2020). Global projections of future urban land expansion under shared socioeconomic pathways. Nat Commun,11: 1-12.

Clewella, A., Aronson, J., Winterhalder, K., 2004. The SER international primer on ecological restoration. https://www.ctahr.hawaii.edu/littonc/PDFs/682_SERPrimer.pdf

Cowie AL, Orr BJ, Castillo Sanchez VM, Chasek P (2018). Land in balance: the scientific conceptual framework for Land Degradation Neutrality.Environ. Sci. Policy, 79: 25-35, 10.1016/j.envsci.2017.10.011.

Cramb, R. A., Pierce Colfer, C. J., Wolfram, D., Laungaramsri, P., Trang, Q., Muloutami, E., Peluso, N. L. and Wadley, R. L. (2009). Swidden transformations and rural livelihoods in Southeast Asia. Human Ecology, 37, 323-346. DOI: 10.1007/s10745-009-9241-6.

Crossland M, Winowiecki LA, Pagella T, Hadgu K and Sinclair F (2018). Implications of variation in local perception of degradation and restoration processes for implementing land degradation neutrality. Environ Dev, 28: 42-54

Daly, H. E. (1991) Steady-state Economics (second edition), Island Press, Washington, DC

Davies, J., Herrera, P., Ruiz-Mirazo, J., Mohamed-Katerere, J., Hannam, I., Nuesri, E., and Batello, C. (2016). Improving governance of pastoral lands. Governance of Tenure Technical Guide (FAO) eng no. 6.

de Chazournes LB (2008). United nations framework convention on climate change. Copyright © United Nations, 2008. United Nations Audiovisual Library of International Law.

Debonne N, van Vliet J, Metternicht G and Verburg P (2021). Agency shifts in agricultural land governance and their implications for land degradation neutrality. Global Environmental Change, Vol. 66, https://doi.org/10.1016/j.gloenvcha.2020.102221.

Desa, U. 2016. Transforming our world: The 2030 agenda for sustainable development. https://stg-wedocs.unep.org/bitstream/handle/20.500.11822/11125/unepswiosm1 inf7sdg.pdf?sequence=1

Ding D, Peng J, Qiu S and Zhao Y (2020). Nearly Half of Global Vegetated Area Experienced Inconsistent Vegetation Growth in Terms of Greenness, Cover, and Productivity. Earths Future, 8 (10) e2020EF001618

Donohue RJ, M.L. Roderick, T.R. McVicar, G.D. Farquhar (2013). Impact of CO2 fertilisation on maximum foliage cover across the globe's warm, arid environments. Geophys Res Lett, 40: 3031-3035

Dressler W, Wilson D, Clendenning J *et al.* (2016). The impact of swidden decline on livelihoods and ecosystem services in Southeast Asia: A review of the evidence from 1990 to 2015. 10.1007/s13280-016-0836-z

ECOSOC. (2016). Report of the inter-agency and expert group on sustainable development goal indicators. Note by the SecretaryGeneral. E/CN.3/2016/2/Rev.1, Statistical Commission 47th Session, 8-11 March 2016. New York: United Nations Economic and Social Council (ECOSOC).

Falvey JL (2018). Policies impacting shifting cultivation: getting them right. DOI: 0.1079/9781786391797.0043. https://www.researchgate.net/publication/322298177

Falvey, J. L. (2010) Small Farmers Secure Food: Survival Food Security, the World's Kitchen and the Critical Role of Small Farmers, Thaksin University Press, Songkhla, Thailand, https://www. researchgate.net/publication/236684402

Falvey, J. L. (2013) Beliefs that Bias Food and Agriculture: Questions I'm Often Asked, Institute for International Development, Adelaide, Australia, https://www.researchgate.net/ publication/261357446.

Falvey, L. and Chantalakhana, C. (1999) Smallholder Dairying in the Tropics, International Livestock Research Institute, Nairobi, https://www.researchgate.net/publication/236737261

FAO (1957). Shifting cultivation. Unasylva, 11(1), 9-11

FAO (2005) State of the World's Forests, 2005, Food and Agriculture Organization of the United Nations, Rome.

FAO, IGWIA and AIPP (2015). Shifting Cultivation, Livelihood and Food Security: New and Old Challenges for Indigenous Peoples in Asia. Bangkok: FAO, IGWIA and AIPP.

Feng S and Fu Q (2013). Expansion of global drylands under a warming climate. Atmos Chem Phys, 13: 10081-10094

Feng S, Zhao W, Zhan T, Yan Y and Pereira P (2022a). Land degradation neutrality: A review of progress and perspectives. Ecological Indicators, 144: 109530, https://doi.org/10.1016/j.ecolind.2022.109530.

Feng S, Liu X, Zhao W, Yao Y, Zhou A, Liu X and Pereira P (2022b). Key Areas of Ecological Restoration in Inner Mongolia Based on Ecosystem Vulnerability and Ecosystem Service.Remote Sensing, 14: p. 2729.

Fonseka, H. (ed.) (2010) Hill Agriculture in SAARC Countries: Constraints and Opportunities, proceedings of a regional workshop, 3-5 November 2009, Chittagong, Bangladesh,

Fox, J., Fujita, Y., Ngidang, D., Peluso, N., Potter, L., Sakuntaladevi, N., Sturgeon, J. and Thomas, D. (2009). Policies, political-economy and swidden in Southeast Asia. Human Ecology, 37, 305-322. DOI 10.1007/s10745- 009-9240-7

FSI (2013) India State of Forest Report 2013, Forest Survey of India, Dehradun, Ministry of Environment, Forests and Climate Change, India.

Fujisaka, S. (1991) 'A diagnostic survey of shifting cultivation in northern Laos: Targeting research to improve sustainability and productivity', Agroforestry Systems 13, pp95-109

Gassner A, D. Harris, K. Mausch, A. Terheggen, C. Lopes, R.F. Finlayson, P. Dobie (2019). Poverty eradication and food security through agriculture in Africa: Rethinking objectives and entry points. Outlook Agr, 48: 309-315

Gichenje H and Godinho S (2018). Establishing a land degradation neutrality national baseline through trend analysis of GIMMS NDVI Time-series Land Degrad Dev, 29: 2985-2997

Gichenje H, Pinto-Correia T and Godinho S (2019). An analysis of the drivers that affect greening and browning trends in the context of pursuing land degradation-neutrality. Remote Sens Appl, 15, 100251

Giger G, Liniger H, Sauter C and Schwilch G (2018). Economic Benefits and Costs of Sustainable Land Management Technologies: An Analysis of Wocat's Global Data. Land Degrad Dev, 29: 962-974.

Global Mechanism (2019a). Land Degradation Neutrality Target Setting: Initial Findings and Lessons Learned. Bonn.

Gomes E, M. Inácio, K. Bogdzevič, M. Kalinauskas, D. Karnauskaitė, P. Pereira (2021). Future land-use changes and its impacts on terrestrial ecosystem services: A review. Sci Total Environ, 781, Article 146716

Government of Indonesia (1990) Act Number 5/1990 on Conservation of Biodiversity and Its Ecosystems, Ministry Secretariat, Jakarta.

Government of Indonesia (1999) Act Number 41/1999 on Forestry, Ministry Secretariat, Jakarta.

Hannam I (2022). Soil governance and land degradation neutrality. Soil Security, 6, 100030.

Hannam, I. D., and Boer, B. W. and Hannam I (2019). Land Degradation. In : E Lees, J. Viñuales (Eds.), The Oxford Handbook of Comparative Environmental Law, Oxford University Press (2019), pp. 438-459

Heinimann, A., Mertz, O., Frolking, S., Egelund, Christensen A., Hurni, K., Sedano, F., Chini, L. P., Sahajpal, R, Hansen, M. and Hurtt, G. (2017). A global view of shifting cultivation: Recent, current and future extent. Plos ONE, 12(9), e0184479. https://doi. org/10.1371/ journal.pone 0184479.

Herrera, A. and Da Passano, M. G. (2006) Land Tenure Alternative Conflict Management, Food and Agriculture Organization of the United Nations, Rome.

Höhn A and Neumann K (2012). Shifting cultivation and the development of a cultural landscape during the Iron Age (0–1500 AD) in the northern Sahel of Burkina Faso, West Africa: Insights from archaeological charcoal, Quaternary International,249: 72-83. https://doi. org/10.1016/j.quaint.2011.04.012.

Huang JP, H.P. Yu, X.D. Guan, G.Y. Wang, R.X. Guo (2016).Accelerated dryland expansion under climate change. Nat Clim Change, 6: 166-171

IWGIA/AIPP (2009). Shifting cultivation and climate change. UNFCCC Intersessional Meeting, Bangkok 2009 Briefing Paper

Jiang LL, A.M. Bao, G. Jiapaer, R. Liu, Y. Yuan, T. Yu (2022). Monitoring land degradation and assessing its drivers to support sustainable development goal 15.3 in Central Asia. Sci Total Environ, 807, Article 150868

Jiang LL, G. Jiapaer, A. Bao, Y.M. Li, H. Guo, G.X. Zheng, T. Chen, P. De Maeyer (2019). Assessing land degradation and quantifying its drivers in the Amudarya River delta. Ecol Indic, 107 , Article 105595

JiangC, H.Y. Zhang, L.L. Zhao, Z.Y. Yang, X.C. Wang, L. Yang, M.L. Wen, S.B. Geng, Q. Zeng, J. Wang (2020). Unfolding the effectiveness of ecological restoration programs in combating land degradation: Achievements, causes, and implications. Sci Total Environ, 748 (2020), Article 141552

Kerkhoff E and Sharma E, 2006. Debating Shifting Cultivation in the Eastern Himalayas Farmers' Innovations as Lessons for Policy, International Centre for Integrated Mountain Development (ICIMOD), Kathmandu, Nepal

Keesstra, S., Mol, G., De Leeuw, J., Okx, J., Molenaar, C., De Cleen, M., and Visser, S. (2018). Soil-related sustainable development goals: Four concepts to make land degradation neutrality and restoration work. Land, 7(4), 133.

Kuderina TM, S.B. Suslova, V.N. Lunin, A.V. Kudikov (2020).Atmospheric Moisture as a Factor of Land Degradation Neutrality in Forest-Steppe Landscapes. Arid Ecosyst, 10: 156-160Kust G, O. Andreeva, V. Lobkovskiy, N. Telnova (2018). Uncertainties and policy challenges in implementing Land Degradation Neutrality in Russia. Environ Sci Policy, 89: 348-356.

Kust GS, O.V. Andreeva, V.A. Lobkovskiy (2020). Land Degradation Neutrality: the Modern Approach to Research on Arid Regions at the National Level. Arid Ecosyst, 10: 87-92

Lal R, Bouma J, Brevik E, Dawson L *et al.* (2021). Soils and sustainable development goals of the United Nations: An International Union of Soil Sciences perspective. Geoderma Reg, 25, Article e00398.

Law EPA 1994). Elaboration of an international convention to combat desertification in countries experiencing serious drought and/or desertification, particularly in Africa. Environmental Policy and Law, 24, p. 1

Li CJ, B.J. Fu, S. Wang, L.C. Stringer, Y.P. Wang *et al.* (2021). Drivers and impacts of changes in China's drylands. Nature Reviews Earth and Environment, 2: 858-873

Li XS, Lu Q and Jia XX (2021). Harnessing Big Earth Data to Facilitate Land Degradation Neutrality Goals Practices and ProspectsBulletin of the Chinese . Academy of Sciences, 36: 896-903.

Liao WL, Liu XP , Xu XY, Chen GZ, Liang X, Zhang HH and Li X (2020). Projections of land use changes under the plant functional type classification in different SSP-RCP scenarios in China.Sci Bull, 65: 1935-1947.

Liniger H, Harari N, van Lynden G, Fleiner R *et al.* (2019). Achieving land degradation neutrality: The role of SLM knowledge in evidence-based decision-making. Environ Sci Policy, 94: 123-134.

McElwee P (2022). Shifting policies for shifting cultivation. A history of anti-swidden interventions in Vietnam. Journal of Southeast Asian Studies, 53: 153-182.

Mertz, O., Padoch, C., Fox, J., Cramb, R. A., Leisz, S. J., Thanh, N. and Duc Vien, T. (2009). Swidden change in Southeast Asia: Understanding causes and consequences. Human Ecology, 37, 259-264. DOI 10.1007/s10745-009-9245-2.

Mitri G, Nasrallah G, Gebrael K, Nassar MB *et al.* (2019). Assessing land degradation and identifying potential sustainable land management practices at the subnational level in Lebanon. Environ Monit Assess, 191: 1-22

Muchena FN, Onduru DD, Gachini GN, de Jager A (2005). Turning the tides of soil degradation in Africa: capturing the reality and exploring opportunities. Land Use Policy, 22, pp. 23-31, 10.1016/j.landusepol.2003.07.001

NAEB (2007) Report of the Inter-Ministerial Task Force on Rehabilitation of Shifting Cultivation Areas, National Afforestation and Eco-Development Board, Ministry of Environment, Forests and Climate Change, Government of India, New Delhi

National Commission on Farmers (2005) Serving Farmers and Saving Farming: Crisis to Confidence (second report), Ministry of Agriculture, Government of India, New Delhi .

Nguyen, T.L., Patanothai, A., and Rambo, A.T. (2004). Recent Changes in the Composite Swidden Farming System of a Da Bac Tay Ethnic Minority Community in Vietnam's Northern Mountain Region. Southeast Asian Studies, 42, 273-293.

Nkonya, E., Johnson, T., Kwon, H. Y., and Kato, E. (2016). Economics of land degradation in sub-Saharan Africa. In E. Nkonya, A. Mirzabaev, and J. von Braun (Eds.), Economics of land degradation and improvement – A global assessment for sustainable development (pp. 215–259). Cham: Springer. https://doi.org/10.1007/978-3-319-19168-3_9.

Nzuza P, Ramoelo A, Odindi J, Kahinda JMM and Lindeque L (2021).A triangulation approach for assessing and mapping land degradation in the Lepellane catchment of the greater Sekhukhune District. South Africa. S Afr Geogr J, pp. 1-25

Orr BJ, A.L. Cowie AL, Castillo Sanchez VM, Chasek P *et al.* (2017). Scientific Conceptual Framework for Land Degradation Neutrality. A Report of the Science-Policy Interface. United Nations Convention to Combat Desertification (UNCCD), Bonn, Germany.

Pereira P (2020).Ecosystem services in a changing environment. Sci Total Environ, 702, Article 135008.

Pereira P, Barcelo D and Panagos P (2020). Soil and water threats in a changing environment. Environ Res, 186, Article 109501

Phuntsho, K; Aryal, KP; Kotru, R (2015) Shifting cultivation in Bangladesh,Bhutan and Nepal: Weighing government policies against customary tenure and institutions. ICIMOD Working Paper 2015/7. Kathmandu: ICIMOD

Rabehi, W., Bentekhici, N., Bouhlala, M.A., Benharrats, F. *et al.*, (2020). Monitoring and Estimation of the Sustainable Development Goal -Fifteen- by Remote Sensing Tools, Assessment of Change in Land Cover (Sub-Indicator 15.3.1), Case of Algeria. 2020

Mediterranean and Middle-East Geoscience and Remote Sensing Symposium (M2garss), 239-241.

Schulze K, Malek Z and Verburg PH (2021).How will land degradation neutrality change future land system patterns? A scenario simulation study. Environ Sci Policy, 124: 254-266

Scott J (2017). The Art of Not Being Governed: An Anarchist History of Upland Southeast Asia, DOI - 10.12987/9780300156522.

Sheals, J. G. (1969). The Soil Ecosystem: Systematic Aspects of the Environment, Organisms and Communities. The Systematics Association, Publication No. 8, Staples Printers Ltd, London.

Sietz D, Fleskens L and Stringer LC (2017). Learning from NonLinear Ecosystem Dynamics Is Vital for Achieving Land Degradation Neutrality. Land Degrad Dev, 28: 2308-2314Song XP, Hansen MC, Stehman SV, Potapov PV *et al.* (2018). Global land change from 1982 to 2016. Nature, 560 (7720), pp. 639-643

Sims, N. C., England, J. R., Newnham, G. J., Alexander, S., Green, C., Minelli, S., and Held, A. (2019). Developing good practice guidance for estimating land degradation in the context of the United Nations Sustainable Development Goals. Environmental Science and Policy, 92, 349–355. https://doi.org/10.1016/j.envsci.2018.10.014.

Singh, L. Jitendro and Choudhury, D. (2015). Fallow management practices among the Tangkhuls of Manipur: Safeguarding provisioning and regulatory services from shifting-cultivation fallows. In Malcolm F. Cairns (Ed.), Shifting Cultivation and Environmental Change: Indigenous People, Agriculture and Forest Conservation (pp 449-467). London and New York: Routledge.

Speranza CI, Adenle A and Boillat S (2019). Land Degradation Neutrality-Potentials for its operationalisation at multi-levels in Nigeria. Environ Sci Policy, 94: 63-71

Srivastva AK (2017). Policy issues for shifting cultivation communities and their ecosystems The case of northeast India. In: Cairns, M. (ed.) Shifting Cultivation Policies: Balancing Environmenta and Social Sustainability. CABI (2017), 1115p.

Tsymbarovich P, Kust G, Kumani M, Golosov V and Andreeva O (2020). Soil erosion: An important indicator for the assessment of land degradation neutrality in Russia. Int Soil Water Conse, 8 (4): 418-429.

UN (2016). UNFCCC – The Paris Agreement

UN (2019). United Nations Framework Convention on Climate Change (UNFCCC). Department of Economic and Social Affairs Sustainable Development, United Nations.

UN Environment (2019). Global Environment Outlook – GEO-6: Healthy Planet Healthy People, Nairobi, p. 745, 10.1017/9781108627146.

UNCCD (2017). 2030 Agenda for Sustainable Development: implications for the United Nations Convention to Combat Desertification.

UNCCD (2017). Global Land Outlook. The scientific conceptual framework for LDN, pp312-318.

UNCCD (2019). Land Degradation Neutrality for Biodiversity Conservation: How Healthy Land Safeguards Nature, Global Mechanism of the UNCCD, Technical Report. Bonn, Germany, p. 48.

UNCCD. (2013). Report of the Conference of the Parties on its eleventh session, held in Windhoek from 16 to 27 September 2013. 124 124, Part two: Action taken by the Conference of the Parties at its eleventh session. ICCD/ COP(11)/23/Add.1. Bonn: United Nations Convention to Combat Desertification (UNCCD).

UNFCCC (1992). United nations framework convention on climate change. FCCC/ INFORMAL/84 GE.05-62220 (E) 200705.

UNFCCC (2020). The United Nations Framework Convention on Climate Change, the Kyoto Protocol, and the Paris Agreement: A Summary. Congressional Research Service https://crsreports.congress.gov R46204.

UNFCCC Handbook (2006). Climate Change Secretariat (UNFCCC) Martin-Luther-King-Strasse 8 53175 Bonn, Germany.

United Nations (2012). G.A. Res. A/RES/66/288, the Future We Want.

Upadhyaya M 2004..Traditional Farming Practices and Farmers' Rights in the HKH Region-Policy brief No. 6, SAWTEE, Kathmandu, Nepal.

Von Maltitz P, Gambiza J, Kellner K, Rambau T *et al.* (2019). Experiences the South African land degradation neutrality target setting process. Environ Sci Policy, 101: 54-62

Vorovencii I (2016). Assessing and monitoring the risk of land degradation in Baragan Plain, Romania, using spectral mixture analysis and Landsat imagery. Environ. Monit. Assess., 188, 10.1007/s10661-016-5446-5.

Webb, N. P., Marshall, N. A., Stringer, L. C., Reed, M. S., Chappell, A., and Herrick, J. E. (2017). Land degradation and climate change: Building climate resilience in agriculture. Frontiers in Ecology and the Environment, 15(8), 450–459. https://doi.org/10.1002/fee.1530.

Weinstock, J. (1996) 'Rattan: A complement to swidden agriculture in Borneo', Journal of Southeast Asian Social Science and Humanities 48, pp51–61, http://ejournals.ukm.my/akademika/article/ view/3114, accessed 18 September 2015.

Wuepper D, Borrelli P, Panagos P, Lauber T *et al.* (2021). A 'debt' based approach to land degradation as an indicator of global change. Global Change Biol, 27: 5407-5410

Wulandari C (2017). Policies that transform shifting cultivation - And encourage community-based forest management in Lampung province, Indonesia. Department of Forestry, Faculty of Agriculture, Lampung University, Bandar Lampung, Lampung, Indonesia.

Yailymov BY, Lavreniuk MS, Shelestov AY, Kolotii AV, Yailymova HO and Fedorov OP (2018). Methods of Essential Variables Determination for the Earth's Surface State Assessing. Space Sci Technol, 24: 24-37

Yuan S, Chen JH and Lu Q (2020). Conceptual Framework of Land Degradation Neutrality and Its Implications for China's Practice. World Forestry Research, 33: 7-12

Yuan S, Cheng LL, Xu J and Lu Q (2022). Evaluation of Land Degradation Neutrality in Inner Mongolia Combined with Ecosystem Services. Land, 11: p. 971.

Colour Plates

Chapter 1: Introduction

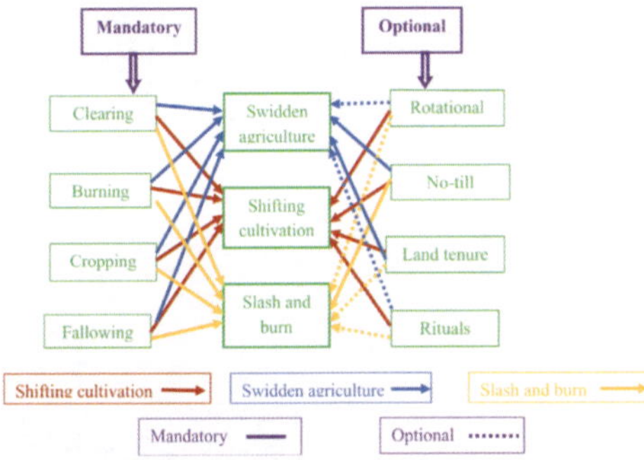

Fig. 1.2. Major facets of shifting cultivation, swidden agriculture and, slash and burn agriculture

Fig. 1.4. Burning of forest vegetation to add ash to soil for improving soil fertility

Fig. 1.5. A typical shifting cultivation field in Northeast India.

Chapter 3: Dimensions of Shifting Cultivation

Fig. 3.9. A view of shifting cultivation in hills of Tripura. (*Source*: Datta *et al.*, 2004)

Fig. 3.23. A view of the well laid Bamboo Drip Irrigation system
Source: ICAR Research Complex for NEH Region.

Chapter 4: Diversity in Shifting Cultivation

Fig. 4.4. Forest vegetation

Fig. 4.5. Burning of vegetation

Fig. 4.6. Readying to prepare land

Fig. 4.7. Land preparation

Chapter 6: Deforestation and Degradation

Fig. 6.10. Forest and soil degradation due to shifting cultivation in Northeast India
Source: Datta et al. 2004

Fig. 6.12. Cropping phase

Fig. 6.13. Cropping phase

Fig. 6.14. Land degradation during fallow phase

Fig. 6.15. Degradation of shifting cultivation areas if fallow phase is not managed properly.

Chapter 9: Development

Fig. 9.6. Terrace cultivation with a water harvesting source

Index

E

L

M

N

O

P

R

S